International Conference on Multidisciplinary Aspects of Materials in Engineering (IC-MAME 2021)

IOP Conference Series: Materials Science and Engineering
Volume 1225

Chandigarh, India
8 - 9 October 2021

Editors:

Sunil Agarwal
Rajesh Kumar
Gaurav Sapra

ISBN: 978-1-7138-4683-3
ISSN: 1757-8981

Printed from e-media with permission by:

Curran Associates, Inc.
57 Morehouse Lane
Red Hook, NY 12571

Some format issues inherent in the e-media version may also appear in this print version.

Institute of Physics
Dirac House, Temple Back
Bristol BS1 6BE UK

Phone: 44 1 17 929 7481
Fax: 44 1 17 920 0979

techtracking@iop.org

Additional copies of this publication are available from:

Curran Associates, Inc.
57 Morehouse Lane
Red Hook, NY 12571 USA
Phone: 845-758-0400
Fax: 845-758-2633
Email: curran@proceedings.com
Web: www.proceedings.com

TABLE OF CONTENTS

Preface

It is indeed a matter of immense pleasure and distinguished honor to bring to your kind attention that UIET, Panjab University Chandigarh has successfully organized the International Conference on Multidisciplinary Aspects of Materials in Engineering (IC-MAME 2021) from 8th – 9th October, 2021 in hybrid mode.

This conference aimed to bring together the experts from the varied disciplines involved in engineering and ancillary aspects of materials engineering thus making ways for youngsters to learn, interact and share their experiences, research methods & results in the multidisciplinary aspect of material science and their contemporary applications.

Prof. Gautam Biswas (J.C Bose Fellow and Professor, IIT Kanpur) and Prof. Rajeev Ahuja (Director, IIT, Ropar) graced the event as Chief Guest during Inaugural Session and valedictory Session of the conference respectively.

The scientific program included keynote lectures, invited talks by academic and industry professionals and 12 different thematic sessions consisting of peer-reviewed full length paper presentations & abstract presentations in parallel technical sessions.

This conference provided an opportunity for young researchers and scientists to discuss scientific topics and research results, to share their knowledge and experiences, to collaborate in innovative designs and proposals.

Editors

Prof. Sunil Agarwal

Professor, Electronics and Communication Engineering, UIET, Panjab University, Chandigarh, India

Dr. Rajesh Kumar

Associate Professor, Mechanical Engineering, UIET, Panjab University, Chandigarh, India

Dr. Gaurav Sapra

Assistant Professor, Electrical and Electronics Engineering, UIET, Panjab University, Chandigarh, India

IOP Conf. Series: Materials Science and Engineering 1225 (2022) 011001 doi:10.1088/1757-899X/1225/1/011001

ABOUT UIET

The University Institute of Engineering and Technology was established in 2002 and is constituent institute of Panjab University, Chandigarh. The institute has maintained high quality in technical education. The well qualified faculty is the backbone of the institute. Well equipped laboratories provide the exposure to the students towards practical aspects of engineering. State-of-art computing facilities and internet connectivity through 1Gbps National Knowledge Network (NKN) provide ample scope for students to learn round the clock. The whole campus is Wi-Fi enabled and students having laptops avail this facility in the institute as well as in the hostels. With the regular revision of syllabi, more industry oriented programme and open Electives have been introduced in the curriculum. UIET is spread over 10.5 acres of land with lush greenery. It has two academic blocks which include labs, office, library, and lecture halls. There are blocks for mechanical labs and workshops. There is a branch of State Bank of India and an ATM of State Bank of India on the campus. Students are from all over the nation, few are day scholars and many live in hostels which provide accommodation, food, washrooms, and sporting facilities. University Institute of Engineering & Technology (UIET) is a institute of Panjab University which has prospered by leaps & bounds over the years achieving great academic heights besides foraying into the top 30 finest engineering colleges of the country. UIET offers Undergraduate (B.E.), Postgraduate (M.E./M.Tech. degree) as well as Doctoral courses.

IOP Conf. Series: Materials Science and Engineering 1225 (2022) 011001 doi:10.1088/1757-899X/1225/1/011001

ABOUT PANJAB UNIVERSITY

Panjab University (PU) established in1882 at Lahore (Now in Pakistan) was shifted to independent India in 1947. It is one of the four universities established in India before partition by the then Federal Government. In 1956, PU shifted to Chandigarh and in 1966, after re-organizationof Punjab and was made as an Inter State Body Corporate under the Act of Parliament. PU is imparting education to over 2.5 lakh students through its 78 teaching and research departments located on the twin campus at Chandigarh and through 195 affiliated/constituent colleges spreadover Punjab and Chandigarh besides 4 Regional centers at Mukatsar, Ludhiana, Hoshiarpur and Kauni.

PU has a long tradition of pursuing excellence in teaching and research in Science & Technology, humanities, social sciences, performing arts and sports. By virtue of its history, experience, achievements, and philosophy, PU has a national character and it enjoys an international staturedrawing both the faculty and students from all over the country and different parts of the globe. PU faculty includes some of the most distinguished scientists and academicians. Over the years, the reputation of the PU has grown to emerge as an institution at the pinnacle in innovative teaching, research and community outreach. Its alumni include two former Prime Ministers and several pastand present cabinet ministers of the country, two Noble Laureates, a large number of Fellows ofvarious Academies, both in Science and Social Science, Theatre Artists, Actors. There is hardly anysphere where the university has not made its mark. The university is ranked amongst the top Universities in India and Internationally by various agencies/bodies.

IOP Conf. Series: Materials Science and Engineering 1225 (2022) 011001 doi:10.1088/1757-899X/1225/1/011001

ABOUT IC-MAME 2021

International conference on multi-disciplinary aspects of Materials in Engineering (IC-MAME 2021) is a flagship event of the University Institute of Engineering and Technology, Panjab University, Chandigarh.

This conference is a multidisciplinary event encompassing themes related to innovative research related to Materials in Engineering.

The objective of the conference is to bring professional engineers, academicians and research scholars of matching interests on a common platform to share new ideas, experiences, and knowledge in various fields of electrical and mechanical engineering.

The scientific program will consist of peer-reviewed paper presentations in parallel technical sessions. In addition, keynote lectures, presentations by academic and industry professionals will be conducted during the conference. Such interactions will facilitate better understanding about technological developments all across the globe amongst the peers. This conference will certainly ignite the minds of the researchers for undertaking more interdisciplinary collaborative research for up gradation of the technology.

IOP Conf. Series: Materials Science and Engineering 1225 (2022) 011001 doi:10.1088/1757-899X/1225/1/011001

Conference Theme

International conference on multi-disciplinary aspects of Materials in Engineering (IC-MAME 2021) is a flagship event of the University Institute of Engineering and Technology, Panjab University, Chandigarh. This conference is a multidisciplinary event encompassing themes related to innovative research related to Materials in Engineering.

The objective of the conference is to bring professional engineers, academicians and research scholars of matching interests on a common platform to share new ideas, experiences, and knowledge in various fields of electrical and mechanical engineering.

The scientific program will consist of peer-reviewed paper presentations in parallel technical sessions. In addition, keynote lectures, presentations by academic and industry professionals will be conducted during the conference. Such interactions will facilitate better understanding about technological developments all across the globe amongst the peers. This conference will certainly ignite the minds of the researchers for undertaking more interdisciplinary collaborative research for up gradation of the technology.

IOP Conf. Series: Materials Science and Engineering 1225 (2022) 011001 doi:10.1088/1757-899X/1225/1/011001

Call for Papers

Original and unpublished research or review papers are invited related to following sub-themes of the conference, but not limited to:

- Composite Materials
- Design, modeling and simulation, Multiscale Materials Modelling
- Industrial, Production, Energy and environmental applications
- Novel materials, manufacturing, joining and characterization
- Smart materials, sensors, instrumentation, AI and IoT
- Thermal and mechatronics
- Biomaterials and Bioengineering.
- Nanomaterials and Nanoengineering
- Functional Materials and Devices
- Advanced Structural Materials

Conference Dates: **October 08-09, 2021**

Note: There will be choice for the participants to present papers **online or offline.**

Kindly refer to the conference website *https://www.icmame2021.com* for more details. The information about the conference is also available at *https://easychair.org/cfp/icmame2021*

IOP Conf. Series: Materials Science and Engineering 1225 (2022) 011001 doi:10.1088/1757-899X/1225/1/011001

COMMITTEES

ADVISORY COMMITTEE

- Prof Mohsen Rahmani, NTU, UK
- Prof. Naveen Kumar, IIT Ropar
- Prof. Parveen Kalra, PEC Chandigarh
- Prof. P. C. Tewari, NIT Kurukshetra
- Prof. A. Manna, PEC Chandigarh
- Prof. H.C. Garg, GJUS &T, Hisar
- Prof Neeraj Dilbagi, GJUS&T, Hisar
- Prof. Sanjeev Kumar, PEC Chandigarh
- Prof. Jatinder Madan, CCET Chandigarh
- Prof. R. S. Walia, PEC Chandigarh
- Dr. Vijayender Bhalla, CSIR-IMTech, Chandigarh
- Dr. Vinay Sharma, IIT Jammu
- Dr. Harlal Singh Mali, MNIT Jaipur
- Dr. Sandeep Kumar, GJUS &T, Hisar
- Dr. Parveen Kumar, CSIO, Chandigarh
- Dr. Ayan Karmakar, SCL, Mohali
- Dr. Parveen Kumar, Head, R & D, Exigo Recycling
- Dr. Sumit Jindal, VIT, Vellore

Registration and Certificate Committee

- Dr. Shuchi Gupta (Convener), Associate Professor, Applied Science, UIET
- Dr. Preetika Sharma, Assistant Professor, EEE, UIET
- Dr. Sarpreet Kaur, Assistant Professor, EEE, UIET
- Er. Parul Gaur, Assistant Professor, EEE, UIET
- Dr. Vivek Pahwa, Assistant Professor, EEE, UIET
- Dr. Sarvjit Singh, Assistant Professor, ECE, UIET
- Dr. Garima Joshi, Assistant Professor, ECE, UIET

Stage Management Committee

- Dr. Preeti Singh (Convener), Assistant Professor, ECE, UIET
- Dr. Deepti Gupta, Assistant Professor, CSE, UIET
- Er. Sukhvir Singh , Assistant Professor, IT, UIET
- Er. Kuldeep Singh, Assistant Professor, EEE, UIET
- Ms. Kamal, Technical Officer, EEE, UIET

Hospitality Committee

- Er. Aman Wadhwa (Convener), Assistant Professor, MECH, UIET
- Er. Sumit Budhiraja, Assistant Professor, ECE, UIET
- Er. Jaswinder Singh Mehta, Assistant Professor, MECH, UIET
- Er. Neeraj Sharma, Assistant Professor, ECE, UIET
- Er. Anjali Gupta, Assistant Professor, MECH, UIET
- Er. Rajkumari, Assistant Professor, IT, UIET
- Mr. Lokesh, Programmer, CSE, UIET

Technical Management Committee

- Dr. Shankar Sehgal (Convener), Associate Professor, MECH, UIET
- Dr. Veenu Mangat, Associate Professor, IT, UIET
- Dr. Parveen Goyal, Assistant Professor, MECH, UIET
- Dr. Makhan Singh, Assistant Professor, CSE, UIET
- Dr. Sukesha Sharma, Assistant Professor, EEE, UIET
- Dr. Neelam Goel, Assistant Professor, IT, UIET
- Dr. Amit Thakur, Assistant Professor, MECH, UIET
- Mr. Arun Dhawan, Programming Assistant, IT, UIET
- Mr. Satish Verma, Programming Assistant, CSE, UIET
- Ms.Sunaina, Programmer, CSE, UIET

Sponsorship Committee

- Prof. Manu Sharma (Convener), Professor, MECH,UIET
- Dr. Rajesh Kumar , Associate Professor, MECH, UIET
- Dr. Gaurav Sapra, Assistant Professor, ECE, UIET
- Dr. Hitesh Kapoor, Assistant Professor, Applied Science, UIET
- Dr. Garima Joshi, Assistant Professor, ECE, UIET
- Dr. Amit Thakur, Assistant Professor, MECH, UIET

IOP Conf. Series: Materials Science and Engineering 1225 (2022) 011001 doi:10.1088/1757-899X/1225/1/011001

Publication and Award Committee

- Prof. Vinay Kanwar (Convener), Professor, Applied Science,UIET
- Dr Shankar Sehgal, Associate Professor, MECH,UIET
- Dr Deepak Kumar, Assistant Professor, EEE,UIET
- Dr. Preeti Singh , Assistant Professor, ECE, UIET
- Dr. Suresh Kumar, Assistant Professor, Applied Science,UIET
- Dr Madhu Khatri, Assistant Professor, Biotech,UIET
- Dr Nishima Wangoo, Assistant Professor, Applied Science,UIET

Purchase Committee

- Prof. Sunil Agrawal (Convener), Professor, ECE, UIET
- Prof. Vinay Kanwar , Professor, Applied Science, UIET
- Prof.Manu Sharma, Professor, MECH, UIET
- Dr. Parveen Goyal, Assistant Professor, MECH, UIET

IOP Conf. Series: Materials Science and Engineering 1225 (2022) 011001 doi:10.1088/1757-899X/1225/1/011001

Professor Raj Kumar
Vice Chancellor

PANJAB UNIVERSITY
CHANDIGARH, *India* 160 014

MESSAGE

It is a matter of pride that the University Institute of Engineering and Technology (UIET), Panjab University, Chandigarh is organizing an International Conference on Multidisciplinary Aspects of Materials in Engineering (IC-MAME 2021) on October 08-09, 2021. This International conference is a flagship event of the UIET of Panjab University, Chandigarh. The conference is designed as a multidisciplinary event covering themes related to innovative research on materials in engineering.

This international conference aims to bring together the research interests of professional engineers, academicians and scholars in the field of material science and engineering; in order to facilitate better understanding of technological developments globally.

I hope that there would be a constructive dialogue and contribution by the participants during the conference.

(Raj Kumar)

IOP Conf. Series: Materials Science and Engineering 1225 (2022) 011001 doi:10.1088/1757-899X/1225/1/011001

Professor V.R Sinha
Dean University Instructions
Panjab University

MESSAGE

Material science and engineering contribute towards our everyday life. Therefore, it is important that the current and future resources are required and explored for conducting research through wide deliberations by the scientific institutions, as ours is.

It is a matter of great pleasure and pride that the UIET Panjab University, Chandigarh, a prestigious institution, has taken the initiative of organizing two day's international level conference on **Multidisciplinary Aspects of Materials in Engineering (IC-MAME 2021)**. This is a multidisciplinary event and the key focus would be on themes related to research work in the fields of electrical and mechanical engineering. It is believed that the deliberations made during the conference would reach out to researchers working in this field, in particular.

I congratulate the organizing committee of the UIET for organizing this conference. Such like events are very prestigious and bring huge laurels to the institution to which they belong.

My sincere appreciation and best wishes to the organizing team of the UIET for its successful accomplishment.

Sd/-
Dean University Instructions

Professor Rajesh Gill
Dean Research
Panjab University

Dean Research

PANJAB UNIVERSITY
CHANDIGARH, India 160014

178/Dean. Res.
29/09/2021

It is a matter of great pride for the Panjab University that University Institute of Engineering & Technology is holding the prestigious International Conference on "Multi-disciplinary aspects of Materials in Engineering", under the dynamic leadership of Prof. J.K. Goswamy, Director, UIET, P.U. The kind of meticulous planning which has gone into organising this event, is evident from the minute details including the choice of currently relevant sub-themes, the noted key speakers and finally, the publication of papers presented in the Conference. I am confident that participation in this Conference will be enormously engaging for all the delegates not only in the theoretical sense but in application as well, given the multi-disciplinary essence of the deliberations and the quality of speakers.

I wish the organisers and my colleagues an extremely successful and fruitful Conference.

Dean Research

IOP Conf. Series: Materials Science and Engineering 1225 (2022) 011001 doi:10.1088/1757-899X/1225/1/011001

Professor J. K. Goswamy
Director UIET
Panjab University

UNIVERSITY INSTITUTE OF ENGINEERING & TECHNOLOGY
(UIET)

PANJAB UNIVERSITY, CHANDIGARH
Tel : 0172-2541242, 2534995, Fax : 0172-2547986
Web : www.uiet.puchd.ac.in

MESSAGE

No. 12382\UIET Date 4\10\2021

It is my pleasure to welcome the delegates to participate in the IC-MAME, 2021 being organized by the University Institute of Engineering & Technology (UIET), Panjab Univesity, Chandigarh on 8-9 October, 2021. The field of materials is mother to all Sciences as it encompasses the synthesis, characterization and application of materials wide and diversification application. The present day economy of any nation is strongly built on know how and judicious usage of materials.

As I have been informed that there is overwhelming response from within and outside India. It is expected that planned 12 sessions over a period of two days shall be very interesting and mind opening. This Conference is expected to be a useful platform to exchange ideas and also forge collaborations in the frontiers of material sciences research. The paper presented in this Conference shall be published in reputed Journals on basis of their merit.

On this occasion, I extend my best wishes to the organisrs and delegates for stupendous success of IC-MAME, 2021.

(Prof. J.K. Goswamy)
Director

Director,
University Institute of
Engineering & Technology,
Panjab University, Chandigarh.

IOP Conf. Series: Materials Science and Engineering 1225 (2022) 011001 doi:10.1088/1757-899X/1225/1/011001

Professor Renu Vig
Co-Patron
IC-MAME 2021

Message

UIET, Panjab University, Chandigarh was established nearly two decades ago, and has emerged as a leading technical institute in the region in teaching and research. Holding an international level conference is a matter of great pride for the students as well as faculty.

Now a day's material science and engineering has become a very important area across globe as new materials result in products with better functionality. It is an interesting and attractive area of study. Manipulation of materials, their properties and processes are the need of the hour to make the products which are cheaper, stronger and lighter. There are many other reasons for an in-depth and advance research in this emerging field. The study of material science and engineering has the potential to generate a large number of jobs, to provide an excellent skill for jobs and give platform to those students who wish to opt for an international career. Above all, research in this area requires a strong link with the industrial sector.

Faculty of UIET and the organizing committee has taken this initiative in the right direction where a platform is being provided to researchers to share their findings. I convey my best wishes to the faculty, staff, students and the organizing team, in particular, for its big success.

Prof. Renu Vig
Electronics and Communication Engineering, UIET,
Panjab University, Chandigarh

IOP Conf. Series: Materials Science and Engineering 1225 (2022) 011001 doi:10.1088/1757-899X/1225/1/011001

Professor Sanjeev Puri
Co-Patron
IC-MAME 2021

Message

Multidisciplinary approach equips a learner with cache of skills which are much needed in the current scenario when technology is fast shaping our society. This aspect in research and training empowers the youngsters with much required virtues of critical thinking, analytic and problemsolving mindset, research design & process and above all the team work. Bringing together the thinkers and doers under the one umbrella are the hallmark of achieving these virtues in the youngsters.

I am indeed happy to learn that the young faculty members at University Institute of Engineering & Technology brain stormed to bring forth these virtues through organizing an International Conference on Multidisciplinary Aspects of Materials in Engineering on 8th and 9th October 2021 at Panjab University, Chandigarh, India.

I rather feel honored to be associated with this event and extend a very hearty welcome to all the participants to this conference.

This Conference aims to bring together the experts from the varied disciplines involved in engineering and ancillary aspects of materials engineering thus making ways for youngsters to learn, interact and share their experiences, research methods & results in the multidisciplinary aspect of material science and their contemporary applications. I wish all very fruitful deliberations during this conference.

Bringing out an abstract book, prelude to conference proceedings to be published in IOP Conference Series: Materials Science and Engineering, will encourage the youngsters and faculty to showcase their dexterity in the field of material sciences and also opening newer vistas for research.

I wish the organizers all the success and commend them for their ever preparedness towards this conference. I am sure their efforts would make this conference academically enriching and hope that all the participants enjoy this conference and go back richer in their knowledge and collaborations for a fruitful tomorrow.

Prof Sanjeev Puri
FISN, FPAS
Biotechnology, UIET, Panjab University, Chandigarh

IOP Conf. Series: Materials Science and Engineering 1225 (2022) 011001 doi:10.1088/1757-899X/1225/1/011001

Professor Sunil Agrawal
Chair
IC-MAME 2021

Chair Message

It is my deep pleasure to announce that UIET, Panjab University, Chandigarh is organising an "International Conference on Multidisciplinary Aspects of Materials in Engineering" on 8th - 9th October 2021 in hybrid mode.

The objective of the conference is to bring professional engineers, academicians and research scholars of matching interests on a common platform to share new ideas, experiences, and knowledge in various fields of electrical and mechanical engineering.

The scientific program will consist of peer-reviewed paper presentations in parallel technical sessions. In addition, keynote lectures, presentations by academic and industry professionals will be conducted during the conference. Such interactions will facilitate better understanding about technological developments all across the globe amongst the peers. This conference will certainly ignite the minds of the researchers for undertaking more interdisciplinary collaborative research for up gradation of the technology.

I welcome all the participants of IC-MAME 2021 and hope that this conference will provide stimulating deliberations to enrich their scientific experience.

I wish all the participants a thriving and vivacious moment at the conference.

Best Wishes,

(Prof. Sunil Agrawal)
Conference Chair

IOP Conf. Series: Materials Science and Engineering 1225 (2022) 011001 doi:10.1088/1757-899X/1225/1/011001

Dr Rajesh Kumar
Co-Chair
IC-MAME 2021

It is a great honor that two days multidisciplinary conference is being organized at UIET, Panjab University on 8th-9th October, 2021. The objective of the conference is to provide opportunity for every individual to explore new windows for innovation and advancements, promoting the contribution of Materials and Engineering for a sustainable future. The conference aims to bridge the researchers working in academia and other professionals through research presentations, keynote addresses and discussions.

This conference is a platform where scholars can share their views and ideas on the related themes and exchange ideas for further progress in research and development for the betterment of society and mankind. I thank the conference committee, all the authors, reviewers, and other contributors for their constant efforts and their belief in the excellence of IC-MAME2021. Lots of appreciations for reviewers as without their cooperation and full support, this conference would not have been possible.

Rajesh Kumar

IOP Conf. Series: Materials Science and Engineering 1225 (2022) 011001 doi:10.1088/1757-899X/1225/1/011001

Dr. Gaurav Sapra
Co-Chair
IC-MAME 2021

Message

It is a matter of great pleasure for all of us that an International Level Conference on Multidisciplinary Aspects of Materials in Engineering (IC-MAME 2021) in hybrid mode is being organized at UIET, Panjab University, Chandigarh, India. This event will provide a platform to professional engineers, academicians and research scholars in this field to discuss and share their knowledge, new ideas, experiences, innovations made in different fields of electrical and mechanical engineering. New materials technologies developed through engineering and science shall help in making innovative and advance changes in future which would have great impact on our life. Apparently, the research in materials science and engineering will be the key factor in bringing out these changes.

It is in this context that the UIET, Panjab University, Chandigarh is organizing this two days conference on Materials Sciences and Engineering. A large number of research papers have been received from within the country and abroad for which the faculty of UIET expresses its sincere thanks and gratitude to all of them. The proposed plan is to get published the accepted, registered and presented papers of IC-MAME 2021 in the IOP: Materials Science and Engineering Journal.

We do hope that the deliberations made in this conference would be very beneficial to the scholars working in this field.

Gaurav Sapra

IOP Conf. Series: Materials Science and Engineering 1225 (2022) 011001 doi:10.1088/1757-899X/1225/1/011001

Chief Guest (Inaugural Session)

Prof. Gautam Biswas
J C Bose National Fellow and Professor
Department of Mechanical Engineering
Indian Institute of Technology Kanpur

About Chief Guest

Prof. Gautam Biswas is presently a J C Bose National Fellow and Professor, Department of Mechanical Engineering, Indian Institute of Technology Kanpur. Earlier, he has been the director of Indian Institute of Technology Guwahatiand director of the CSIR - Central Mechanical Engineering Research Institute at Durgapur. Prof. Biswas was the occupant of the position of GD and VM Mehta Endowed Chair Professor of Mechanical Engineering at the Indian Institute of Technology Kanpur. He was a Humboldt Fellow in Germany in 1987-88 and JSPS invited fellow in Japan 1994. He is a Fellow of the American Society of Mechanical Engineers (ASME). He has served a full term as the Associate Editor of the Journal of Heat Transfer (Trans ASME). He was a Guest Professor at the University of Erlangen-Nuremberg in 2002. He was the Dean of Academic Affairs at IIT Kanpur for three years since January 2003.

Prof Gautam Biswas is a Fellow of the all three major science academies of India, such as, the Indian National Science Academy (INSA), New Delhi, the Indian Academy of

Sciences (IAS, Bangalore) and the National Academy of Sciences India (NASI, Allahabad). He is a Fellow of the Indian National Academy of Engineering (INAE) and Institution of Engineers (India). He delivered prestigious Prof. CNR Rao Lecture in 2010. He has been awarded the esteemed J.C. Bose National Fellowship by the Department of Science and Technology, New Delhi in 2011.

Prof. Biswas completed his Baccalaureate in Mechanical Engineering from Calcutta University (IIEST, Shibpur) in 1979. In the year 2013, he was bestowed with the Distinguish Alumni Award of BESU (currently IIEST, Shibpur) by the Hon'ble President of India. He completed his MTech and PhD from the Indian Institute of Technology Kharagpur in 1981 and 1985 respectively.

IOP Conf. Series: Materials Science and Engineering 1225 (2022) 011001 doi:10.1088/1757-899X/1225/1/011001

Chief Guest (Valedictory Session)

Prof. Rajeev Ahuja
Director
Indian Institute of Technology Ropar

About Chief Guest

Prof. Rajeev Ahuja has joined Indian Institute of Technology (IIT) Ropar director from April 1st, 2021 Before joining as director at IIT Ropar, Rajeev Ahuja was professor of computational Materials science at Uppsala University, Sweden. He is one of the most highly cited researchers in Sweden. He has done his Ph.D. from I.I.T. Roorkee in India in 1992. Same year he has joined Uppsala University, Sweden as postdoctoral fellow. He became Assistant Professor in 1996, Associate Professor in 2002 and become Professor in 2007 at Uppsala University, Sweden. His main area of interest is computational materials science with focus on energy such Batteries, Hydrogen Storage & production, sensors as well high pressure physics. He has published 977 scientific papers in peer reviewed journals H-Index of 86 (Google Scholar), i-10 index (Google Scholar) 585 & no. of citations more than 34100, of which more than100 are in high profile journals (like Science , Nature , Nature Materials, PNAS etc.).

He was recently elected FRSC (fellow of Royal Society Chemistry), London, UK, APS-Fellow by American Physical Society (APS), USA and Appointed in the advisory Board of Journal of Materials Chemistry A (IF=10.8) & Materials Advances from Royal Society of Chemistry (England). He is Associate editor of Nano Energy (IF=16.6). He was also awarded Beller Lectureship for the APS March Meeting, 13-17 March 2017, in New Orleans, Louisiana. He been awarded the Wallmark prize for 2011 from KVA (Royal Swedish Academy of Sciences), and has previously received the Eder Lilly & Sven Thureus prize and the Benzelius prize from KVS. Ahuja is an elected member of the Swedish Royal Society of Sciences (KVS), of the board of the European High Pressure Research Group as well as of the executive board of the International Association for the Advancement of High Pressure Science and Technology.

IOP Conf. Series: Materials Science and Engineering 1225 (2022) 011001 doi:10.1088/1757-899X/1225/1/011001

Keynote Speaker

Prof. S. Anantha Ramakrishna
Director, CSIR-CSIO
Email: director@csio.res.in

About Speaker

Prof. S. Anantha Ramakrishna received the M.Sc. (integrated 5 years) degree in Physics from IIT Kanpur, in 1995 and the Ph.D. degree from the Raman Research Institute, Bengaluru, in 2001. He worked on theoretical aspects of light propagation in random media including temporal signatures of Anderson localization during his PhD and was supervised by the Late Prof. N. Kumar. After working as a Post-Doctoral Researcher with Sir John Pendry at Imperial College London, he joined IIT Kanpur as an Assistant Professor in 2003 and became a Professor in 2012. He joined CSIR-CSIO as Director on 27 July 2020 on lien from IIT Kanpur.

Keynote Speaker

Prof. Mohsen Rahmani
Professor
School of Science & Technology
Nottingham Trent University (NTU), UK
Email: mohsen.rahmani@ntu.ac.uk

About Speaker

Prof. Rahmani obtained his PhD from the National University of Singapore in 2013, followed by postdoc fellowship at Imperial College London, and the Australian Research Council Early Career Fellowship at the Australian National University (ANU). In 2020, he moved to the Nottingham Trent University as a Royal Society Wolfson Fellow, and has recently been awarded the UK Research and Innovation Fellowship. Prof. Rahmani has been invited to present his research findings at >30 international conferences, and he has published more than 60 peer-reviewed journal papers, with over 3k citations (H-index=34). He is the recipient of several prestigious awards and prizes, including the Eureka Prize for Outstanding Early Career Academics (Australian Oscar of Science), Early Career Medal from the International Union of Pure and Applied Physics, and the Australian Optical Society Geoff Opat Award. He is an editorial board member of Opto-Electronic Advances. Currently, he serves as the chair of the IEEE Nanotechnology Chapter in the UK and Ireland Session.

Abstract of Talk

Light-matter interactions can be highly controlled via nanostructured thin films on the surface of objects. Indeed, a single-layer of designed and engineered subwavelength nanostructures, so-called metasurfaces, can resonantly couple to the incident light and manipulate the light's behaviour on demand. Metasurfaces can reproduce the functions of bulk optics, and on occasions, can offer new functionalities that are not possible with conventional diffractive optics. In this presentation, Prof. Mohsen Rahmani from Nottingham Trent University will review his journey in employing metallic to dielectric and semiconductor metasurfaces to control the light intensity, frequency and propagation direction. In addition, Rahmani will discuss how metasurfaces can lead to several exciting applications, including night vision, flat optics and ultra-sensitive biochemical sensing.

IC-MAME 2021: Session-Wise Schedule

Session I: Composite Materials

1. ***Paper ID: 001_Nanocomposites.Sensing***
   Title of Paper: Microwave Assisted Synthesis of Molybdenum Disulphide/Tungsten Trioxide/Reduced Graphene Oxide (MoS2/WO3/RGO) Nanocomposites for Organic Vapor Sensing
   Authors: Jyoti Gupta, Prachi Singhal, Sunita Rattan
   First Author Affiliations: Amity Institute of Applied Sciences, Amity University Uttar Pradesh, Noida, UP, India

2. ***Paper ID: 002_E.Glass.Composites***
   Title of Paper: Fabrication and characterization of e-glass composites cylindrical tubes using three different methods
   Authors: Krishnan Ganesh Kumar, Lau Zheng Loong, Megat Ahmad Megat Mohamad Hamdan and Yaakob Mohd Yuhazri.
   First Author Affiliations: Faculty of Engineering & Technology (FET), Multimedia University, Jalan Ayer Keroh Lama, 75450, Melaka, Malaysia

3. ***Paper ID: 003_Failure.Behavior.Epoxy***
   Title of Paper: Effect of ply orientation, staking sequence and notch geometry on the failure behavior of glass fiber-reinforced epoxy composite laminates: A numerical investigation
   Authors: Krishna Prasad G.R and Faizan Mohammad Rashid
   First Author Affiliations: Department of Mechanical Engineering, Birla Institute of Technology and Science (BITS-Pilani) Pilani Campus, Jhunjhunu, Rajasthan 333031, India

4. ***Paper ID: 004_Metal.Oxides.Nanocomposites***
   Title of paper: A review on transition metal oxides based nanocomposites, their synthesis techniques, different morphologies and potential applications
   Authors: Sapna Yadav, Nutan Rani and Kalawati Saini.
   First Author Affiliations: Department of Chemistry, Miranda House, University of Delhi, Patel Chest Marg, New Delhi-110007, India

5. ***Paper ID: 005_Polymer.Composite.Carbon.Fiber***
   Title of paper: Flexural and shear strength properties of unidirectional carbon fiber reinforced polymer composite interleaved with recycled carbon fiber and short virgin aramid fiber non-woven mats
   Authors: Cephas Yaw Attahu, Jian Yang, Kok-Hoong Wong and Chungket Thein.
   First Author Affiliations: Faculty of Science and Engineering, University of Nottingham Ningbo China, Ningbo 315100, China

6. ***Paper ID: 006_Tool.Flank.Wear***
   Title of paper: The effect on the tool flank wear width VBmax during turning of Hybrid Mg/(8wt% SiC(p) +2wt% Al2O3(p) +1wt% Gr(p) MMC using PVD coated cermet tool
   Authors: Jaspreet Hira, Alakesh Manna, Deepak Bhutani, Anup Malik, Binit Kumar Jha and Rohit Sharma.
   First Author Affiliations: Assistant Professor, Department of Mechanical Engineering, Amity University Noida, India

7. ***Paper ID: 007_Red.Mud.Composite***
   Title of paper: Performance optimization of operating parameters: Al 6061-Red Mud composite under sliding wear
   Authors: Narender Panwar, Saroj Bala and Amit Chauhan
   First Author Affiliations Skill faculty of engineering and technology, Shri Vishwakarma Skill University, Dudhola, Palwal, Haryana-121102 (India)

8. ***Paper ID: 008_Tribological.Behavior.Polymer.Composite***
   Title of paper:A fuzzy logic approach for investigating the tribological behavior of polymer composite
   Authors: Gogineni Venkata Jagadeesh, T Teja Rakesh Kumar, T Sai Aneesh, N Dheeraj Srinivas, V Harish and A Ravi Kumar.
   First Author Affiliations Department of Mechanical Engineering, Seshadri Rao Gudlavalleru Engineering College, Gudlavalleru, Andhra Pradesh, India-521356

Session II: Design, modeling and simulation, Multiscale Materials Modelling

1. ***Paper ID: 009 PVA.Hydrogel***
   Title of paper: Print fidelity evaluation of PVA hydrogel using computational fluid dynamics for extrusion dependent 3D printing
   Authors: Ratnesh Raj, Siriki Vamsi Venkata Krishna, Akshat Desai, Chintapalli Sachin and Amit Rai Dixit.
   First Author Affiliations: Department of Mechanical Engineering, Indian Institute of Technology (ISM), Dhanbad, Jharkhand 826004, India.

2. ***Paper ID: 010_ Nano.Photo.Diode.Arrays***
   Title of paper: Modelling, Simulation and Analysis of Nano Photo Diode Arrays using CNT and Graphene Nano Materials for Sub-retinal implant
   Authors: M Vijai Meyyappan and R Joseph Daniel
   First Author Affiliations: Assistant Professor, M.I.E.T. Engineering College, Dept. of Electronics and Communication Engg., Gundur, Trichy, 620 007

3. ***Paper ID: 011_Nanofluid.Darcy.Brinkman***
Title of paper: Thermal Convection in a Casson Nanofluid Layer Saturating Porous Medium: Darcy-Brinkman Model
Authors: Mamta Devi, Jyoti Sharma and Urvashi Gupta.
First Author Affiliations: Energy Research Centre, Panjab University, Chandigarh-160014, India.

4. ***Paper ID: 012_Triboelectric.Nanogenerator.Simulation***
Title of paper Simulation and Modelling of Triboelectric Nanogenerator for Self-powered Electronic Devices
Authors Jashan Kumar Mainra, Akshpreet Kaur, Gaurav Sapra and Parul Gaur
First Author Affiliations UIET, Panjab University, Chandigarh, India

5. ***Paper ID: 013_ FDSOI.MOSFETS***
Title of paper Investigation of Self-heating Effects in Ultrathin FDSOI MOSFETS
Authors Vinod Kumar and Ashwani K. Rana
First Author Affiliations NIT Hamirpur, Hamirpur, India

6. ***Paper ID: 014_Single.Photon.Source***
Title of paper: Eight Wave Mixing Process is suitable to make Single-Photon Source for Quantum Cryptography
Authors Priyanka and Savita Gill
First Author Affiliations: Department of Applied Science, University Institute of Engineering and Technology, Kurukshetra, 136119, India

Session III: Industrial, Production, Energy and environmental applications

1. ***Paper ID: 015_Electric.Vehicles***
Title of paper: Revolution in electric vehicles
Authors: Amandeep Sagwal, Mangladeep Bhullar, Paras Agrawal and Randeep
First Author Affiliations: Physics Department, DAV College, Sector 10, Chandigarh 160011, India.

2. ***Paper ID: 016_Halide.Perovskite.Solar.Cell***
Title of paper: Numerical Assessment of Highly Efficient Tin-Halide Perovskite Solar Cell
Authors:Ajay Kumar, Uddeshya Gupta, Neha Gupta and Amit Kumar Goyal
First Author Affiliations: Electronics and Communication Department, Jaypee Institute of Information Technology, Noida, India.

3. ***Paper ID: 017_Bulk.Heterojunction.Solar.Cell***
Title of paper: Active layer thickness optimization for Maximum efficiency in bulk heterojunction solar cell

Authors: Susan G Daniel, B Devu and C O Sreekala.
First Author Affiliations: Department of Physics, Amrita School of Arts and Sciences, Amritapuri, Amrita Vishwa Vidyapeetham, Kollam, India

4. ***Paper ID: 018_MXene.Heterostructures.Electrode***
Title of paper: MXene based Heterostructures for electrode materials of Batteries: A Review
Authors: Md Abid Hasan Shanto, Muzaddidul Islam Chowdhury, Arijit Biswas Antu, Md Naimur Rahman Niloy, Nazmul Alam, Md Atiq Ullah and Md. Sakib Anowar.
First Author Affiliations: Ahsanullah University of Science and Technology Dhaka, Bangladesh.

5. ***Paper ID: 019_Simulation.Concentric.Tube.Heat.Exchanger***
Title of paper: Exergy modeling and simulation of a horizontal concentric tube heat exchanger
Authors: Ankit Upadhyay, Sankalp Gour, Ankul Kumar, Deepak Kumar and Aseem Chandra Tiwari.
First Author Affiliations: Department of Mechanical Engineering, UIT-Rajiv Gandhi Proudyogiki Vishwavidyalaya Bhopal, Madhya Pradesh, India-462033

6. ***Paper ID: 020_DFT.Investigations***
Title of paper: DFT Investigations of BeSnN2 Chalcopyrite Compound for Optoelectronic Applications
Authors Sanjay Lathwal, Aditi Gaur, Karina Khan, Sunil Kumar Goyal, Amit Soni and Jagrati Sahariya
First Author Affiliations: Birla Institute of Technology & Science, Pilani-333031, Rajasthan, India

7. ***Paper ID: 021_CFD.Analysis.Nanofluid***
Title of paper: Numerical and CFD analysis of a heat transfer enhancement in turbulent flow through a circular pipe using nanofluid
Authors S Gupta, J Singh and B K Gill
First Author Affiliations: Dr. S. S. B. U. I. C. E. T, Panjab University, Chandigarh, Punjab, India

8. ***Paper ID: 022_Bilayer.Organic.Solar.Cell***
Title of paper: Effect of Alkyl Substitution to the Active Layer Material for Improved Efficiency in Bilayer Organic Solar Cell
Authors S Rugma, B Devu and C O Sreekala
First Author Affiliations: Department of Physics, School of arts and sciences, Amritapuri, Amrita Vishwa Vidyapeetham, Kollam, Kerala.

Session IV: Novel materials, manufacturing, joining and characterization

1. ***Paper ID: 023_Raman.Photoluminescence***
   Title of paper: Raman Photoluminescence - An Efficient way to distinguish Natural Diamonds and Lab Grown Diamonds
   Authors: Akesh Kumar G, Prabavathy N and Laxmidhar Biswal
   First Author Affiliations: Centre for Innovation and Technology Excellence , Titan Company limited (Jewellery Division, Brand TANISHQ), Hosur, INDIA

2. ***Paper ID: 024_End.Milling.Al4032SiC.MMC71***
   Title of paper: A study on bi-objective optimization for end milling of Aluminium based composite
   Authors: Pardeep Saini, Deepak Kumar, Anuj Maurya and Pradeep Singh.
   First Author Affiliations: Department of Mechanical Engineering, Sant Longowal Institute of Engineering & Technology, Longowal, 148106, India

3. ***Paper ID: 025_Graphene.Quantum.Dots***
   Title of paper: Up-conversion and Tunable Up-conversion Emission from Nitrogen Functionalized Graphene Quantum Dots
   Authors: Abu Bakar Siddique, K L V Prasad, Sneha M Joseph, Dannina Kishore, Bhaskar Das, Suhail Mubarak, KingshukMukhuti and Mallar Ray
   First Author Affiliations: Aditya College of Engineering and Technology, Surampalem -533437. Andhra Pradesh, India

4. ***Paper ID: 026_Electrochemical.Micromachining***
   Title of paper: Electrochemical Micromachining of Linear Micropattern on Stainless Steel Surface
   Authors: Sandip Kunar, V. Pragna, A.K. Singh and M.S. Reddy
   First Author Affiliations: Department of Mechanical Engineering, Aditya Engineering College, Surampalem, India

5. ***Paper ID: 027_Focusing.Radially.Polarized.Light***
   Title of paper: Focusing of Radially polarized Light Using Gaussian laser beam Near its focal length
   Authors: Getu Endale, Devendra Mohan and Sandeep Yadav.
   First Author Affiliations: Department of Physics, Wolkite University, Wolkite, Ethiopia

6. ***Paper ID: 028_Silicon.Armchair.Graphene.Nanoribbon***
   Title of paper: Integrating silicon in armchair graphene nanoribbon for solar cell applications
   Authors: Preetika Sharma and Gaurav Sapra.
   First Author Affiliations: Assistant Professor, UIET, Panjab University, Sector 25, Chandigarh, India-160014.

7. ***Paper ID: 029_Mo.Nanopartcles.Leadfree.Solder.Joint***
   Title of paper: Impact of 3% Molybdenum(Mo) nanoparticles on the interfacial and shear properties of lead-free Sn58Bi/Cu solder joint.
   Authors: Amares Singh, Rajkumar Durairaj, K Ganesh Kumar and Seng How Kuan1
   First Author Affiliations: Faculty of Engineering Technology and Built Environment, UCSI Heights, 1, Jalan Puncak Menara Gading, Taman Connaught, 56000 Cheras, Wilayah Persekutuan Kuala Lumpur, Malaysia.

8. ***Paper ID: 030_Lower.Limb.Prosthetics***
   Title of paper: FEA simulations of Lower Limb Prosthetics
   Authors: Jashan Jyot Singh, Jaswinder Singh Mehta, Rajesh Kumar and Gaurav Sapra
   First Author Affiliations: UIET, Panjab University, Sector 25, Chandigarh, India-160014.

Session V: Functional Materials and Devices

1. ***Paper ID: 031_Wearable.Human.Motion.Monitoring***
   Title of paper: Wearable Human Motion Monitoring using Vertical Contact Separation mode Triboelectric Nanogenerator
   Authors: Akshpreet Kaur, Ankur Gupta, Cuifeng Ying, Mohsen Rahmani and Gaurav Sapra
   First Author Affiliations: UIET, Panjab University, Chandigarh, India
2. ***Paper ID: 032_Antenna.DCSRR***
   Title of paper: Dual band compact antenna design using DCSRR
   Authors: Ambika Chandru, M. Mohamed Humaid, Kowshic S., S. Mohamed Afrath and Abdul Rahman.
   First Author Affiliations: Electronics And Communication Engineering, BSA Crescent Institute Of Science And Technology, Chennai, Tamil Nadu, India

3. ***Paper ID: 033_Plastic.Polymers***
   Title of paper: Characterization of Plastics And Polymers: A Comprehensive Study
   Authors: Tarik Hassan, Ambikesh Kumar Srivastwa, Subhasis Sarkar and Gautam Majumdar
   First Author Affiliations: Department of Mechanical Engineering, Jadavpur University, Kolkata- 700032, India

4. ***Paper ID: 034_Proton.Irradiation.Effects***
   Title of paper: Study of the proton irradiation effects in n-Fz DSSD detectors for the Si tracker at R3B experiment
   Authors: Puspita Chatterjee, Nitu Saini, Ajay Srivastava, Shilpa Patyal, Balwinder Kaur, Thresia Michael and Sonia Raheja

   First Author Affiliations: Department of Physics, University Institute of Sciences, Chandigarh University, 140413, Punjab, India.

5. ***Paper ID: 035_Miniaturized.DGS.Antenna***
   Title of paper: Metamaterial-Based Miniaturized DGS Antenna for wireless Applications
   Authors: Jaget Singh and Fateh Lohar
   First Author Affiliations University Institute of Engineering and Technology, Panjab University, Chandigarh 160014, India

Session VII: Thermal and mechatronics

1. ***Paper ID: 036_Change.Gear***
   Title of paper: Analysis of Effects of Change of Gear Parameter Module on Transmission Error in Spur Gear using Interference Volume Method
   Authors: Vijay Karma, Govind Maheshwari and Sunil Somani.
   First Author Affiliations: Mechanical Engineering Department, Institute of Engineering & Technology, Devi Ahilya University, Indore, Madhya Pradesh, 452001, India

2. ***Paper ID: 037_Fem.Sandwich.Spherical.Shells***
   Title of paper: Finite element modelling for bending and vibration analysis of composite and sandwich spherical shells.
   Authors: Vikram Parmar, Najeeb Ur Rahman, NausadAlam, Varun Sharma and Aftab Alam
   First Author Affiliations Department of Mechanical Engineering, ZHCET, AMU, Aligarh, UP-202002, India

3. ***Paper ID: 038_Weld.Bead.Shape***
   Title of paper: Experimental investigation and optimization of weld bead shape factor and form factor using recycled steel slag as a flux in submerged arc welding process
   Authors: Sumit Saini, Kulwant Singh and Jatinder Garg.
   First Author Affiliations: Department of Mechanical Engineering, SLIET Longowal, Punjab, India-148106

4. ***Paper ID: 039_Fiber.Metal.Laminates***
   Title of paper: Numerical studies on mode I delamination and its effect on the vibrational characteristics in fiber metal laminates
   Authors: Naresh Kali and Srikanth Korla.
   First Author Affiliations: National institute of Technology Warangal, Warangal, India
5. ***Paper ID: 040_Washing.Machine***
   Title of paper: Investigation of jump resonance of a horizontal axis washing machine for nonlinear vibration using the harmonic balance method
   Authors: Mohit Dhanda and Arpan Gupta
   First Author Affiliations: School of Engineering, Indian Institute of Technology, Mandi, Himachal Pradesh, India

6. ***Paper ID: 041_Mechanical.PDMS***
   Title of paper: Static and dynamic mechanical characterization of polydimethylsiloxane (PDMS) under uniaxial tensile loading
   Authors: David Kumar and Sarthak S. Singh
   First Author Affiliations: Department of Aeronautical Engineering and Center for NDT, Chaoyang University of Technology, Taichung 413310, Taiwan

7. ***Paper ID: 042_Particle.Swarm.Optimization***
   Title of paper: Optimization of reinforced concrete cantilever retaining wall using particle swarm optimization
   Authors: Shubham Srivastava , Saurabh Pandey , Rajesh Kumar
   First Author Affiliations: Department of Civil Engineering, IIT BHU Varanasi, India

Session VIII: Biomaterials and Bioengineering.

1. ***Paper ID: 043_Anticancer.Agent***
   Title of paper: Biogenic Metal and Metal Oxides Nanoparticles as Anticancer Agent: A Review
   Authors: Nutan Rani and Kalawati Saini.
   First Author Affiliations: Department of Chemistry, Miranda House, University of Delhi, Patel Chest Marg, New Delhi-110007

2. ***Paper ID: 044_Yttrium.Oxide***
   Title of paper: Probe into DNA interaction, cell toxicity and antifungal activities of phyto-synthesized yttrium oxide (Y2O3) nanoparticles
   Authors: Rugmani Meenambal, Hema S K, Vedika Tomar and Anita Puyam
   First Author Affiliations: Department of Clinical Psychopharmacology and Neurotoxicology, National Institute of Mental Health and Neuro Sciences, Bangalore, India

3. ***Paper ID: 045_Maize.Seed***
   Title of paper: Comparative Study of Chemo- Bio Synthesized MgO Nanoparticle on Maize Seed Germination
   Authors: Karanpreet Kaur, NaradalaJayarambabu and KalagaddaVenkateswara Rao.
   First Author Affiliations: Department of Nanotechnology, Sri Guru Granth Sahib World University, Fatehgarh Sahib, Punjab, India.

4. ***Paper ID: 046_Milk.Adulteration***
   Title of paper: A Study on Milk Adulteration and methods of detection of various Chemical Adulterants qualitatively
   Authors: Riya Chugh and Gurmeet Kaur
   First Author Affiliations: Chemistry Department (UIS), Chandigarh University, Gharuan, Mohali – 140413,India

5. ***Paper ID: 047_Boron.Neutron***
   Title of paper: A Mini Review on Application of Boron Neutron Capture therapy in Cancer Treatment
   Authors: Shruti Namdev and Gurmeet Kaur
   First Author Affiliations: Chemistry Department (UIS), Chandigarh University, Gharuan, Mohali – 140413,India

6. ***Paper ID: 048_Rod.Shaped.Zno***
   Title of paper: Green Synthesis of Rod Shaped ZnO using Extract of Origanum majorana Leaf and Investigation for Antibacterial Applications
   Authors: Mona Saini, Asifa Mushtaq, Sapna Yadav, Seema Rawat, Kalpna Gupta and Kalawati Saini
   First Author Affiliations: Department of Chemistry, Miranda House, University of Delhi, Patel Chest Marg, New Delhi-110007, India

Session IX: Nanomaterials and Nanoengineering

1. ***Paper ID: 049_Antifreezing.Agent***
   Title of paper Role of nanofluids as an antifreeze agent
   AuthorsShivamSoni, Jodh Singh and Baljinder Kaur Gill.
   First Author Affiliations Dr S.S.B.UICET, Punjab University Chandigarh-160014.

2. ***Paper ID: 050_Graphene.Reducing.Agent***
   Title of paper Synthesis and characterization of Graphene Oxide and its reduction with different reducing agents
   Authors Karan Bansal, Jagdeep Singh and A S Dhaliwal.
   First Author Affiliations Department of Physics, Sant Longowal Institute of Engineering and Technology, Longowal-148106, Punjab, India

3. ***Paper ID: 051_Nio-Graphene***
   Title of paper Electro-Magnetic switching in NiO-Graphene film
   Authors Suchandra Goswami, Dr. Manashi Chakraborty and Dr. Debajyoti De
   First Author Affiliations Material Science Research Lab, The Neotia University, Sarisa, D.H. Road, 24 Pgs (South) West Bengal 743368, India

4. ***Paper ID: 052_Lubricating.Engine.Oil***
   Title of paper Experimental study of Carbon Nanotubes to enhance Tribological Characteristics of Lubricating Engine Oil SAE10W40
   Authors Nitesh Kumar and Parveen Goyal.
   First Author Affiliations UIET, Panjab University, Chandigarh-160014, INDIA

5. ***Paper ID: 053_PEG***
   Title of paper Effect of Surface Modification of Silica Nanoparticles with Thiol group on the Shear Thickening Behaviors of the Suspensions of Silica Nanoparticles in polyethylene glycol (PEG)
   Authors Shuchi Sharma, Yogesh Kumar Walia, Gunjan Grover and Sanjeev Kumar Verma.
   First Author Affiliations Career Point University, Hamirpur, 176041,India

6. ***Paper ID: 054_Fuel.Cell***
   Title of paper Synthesis and Characterization of $La_{0.70}Ca_{0.30}MnO_3$ System for Solid Oxide Fuel Cell Applications
   Authors Surinder Paul, Manokamna, Arun Kumar and Arvind Kumar
   First Author Affiliations Department of Physics, Arni University, Kathagarh Indora-176401, (H.P) India

Session X: Smart materials, sensors, instrumentation, AI and IoT

1. ***Paper ID: 055_Electrochemical.Sensing***
   Title of paper Electrochemical sensing of malathion using doped MOFs
   Authors Shibyendu Nikhar and Pawan Kumar
   First Author Affiliations Materials Application Research Lab. 13 Department of Nano Sciences and Materials, Central University of Jammu, 15 Jammu - 181143 (India)

2. ***Paper ID: 056_Piezoelectric.Sensor***
   Title of paper Piezoelectric sensor voltage variation with different adhesives at elevated temperature
   AuthorsSukesha Sharma
   First Author AffiliationsUIET, Panjab University, Chandigarh, India

3. ***Paper ID: 057_Ifttt.Implementation***
   Title of paper IoT based Resuscitation and Hypoxemia Diagnosis System using Embedded Microcontroller Architecture and IFTTT Implementation
   AuthorsNikita Mohanty, Auhona Ghosh and Sumit Kumar Jindal

   First Author Affiliations School of Electronics Engineering, Vellore Institute of Technology, Vellore, Tamil Nadu, India

4. ***Paper ID: 058_Iot.Control.System***
   Title of paper Design of low cost IoT enabled embedded control system for covid free smart home
   Authors Khushi Singh and Sumit Kumar Jindal.
   First Author Affiliations School of Electronics Engineering, Vellore Institute of Technology, Vellore, Tamil Nadu, India

5. ***Paper ID: 059_Ammonia.Sensing***
   Title of paperMoS_2/CeO_2 Based Composite for Ammonia Sensing
   Authors Nitesh Dogra, Kajal Dadwal, Arun Kumar and Sandeep Sharma
   First Author Affiliations Department of Physics, Guru Nanak Dev University Amritsar, Punjab-143005, India

6. ***Paper ID: 060_Semiconductor.Wafer***
   Title of paper Semiconductor wafer defect classification using convolution neural network: a binary case
   Authors Krishan Kumar Chauhan, Garima Joshi, Manjeet Kaur and Renu Vig.
   First Author Affiliations UIET, Panjab University, Chandigarh, India

7. ***Paper ID: 061_Xylene.Sensing***
   Title of paper Xylene sensing using Dy-doped NiO nanoparticles
   Authors Shailja, Kanwarjit Singh and Sandeep Sharma
   First Author Affiliations Department of Physics, Guru Nanak Dev University, Amritsar 143005, India

8. ***Paper ID: 062_Thin.Film.Sensor***
   Title of paper Green synthesis of vanadium oxide nanoparticles for thin film based sensor application
   Authors T.R. Kishan Chand, H.M. Kalpana and H.S. Lalithamba
   First Author Affiliations Department of Electronics and Instrumentation, Siddaganga Institute of Technology, Tumakuru 572103, Karnataka, India.

Session XI: Advanced Structural Materials

1. ***Paper ID: 063_Orthopaedics.Implants***
   Title of paper Electrochemical corrosion behaviour analysis of Mg-alloys used for orthopaedics and vascular implants
   Authors Rohit Kumar, Puneet Katyal, Munish Gupta and Vijender Singh
   First Author Affiliations Department of Mechanical Engineering Guru Jambheswar University of Science and Technology, Hisar Haryana, INDIA – 125001

2. ***Paper ID: 064_Vacuum.Annaeling***
   Title of paper Evolution of magnetic and structural properties of MgO thin film with vacuum annealing
   Authors Harsh Vardhan, Gagan Sharma, Kavita Sharma, R. J. Choudhary, D. M. Phase, V. Raghavendra Reddy and Ajay Gupta
   First Author Affiliations Amity Center for Spintronic Materials, Amity University UP, Sector-125, Noida-201313, India.

IOP Conf. Series: Materials Science and Engineering 1225 (2022) 011001 doi:10.1088/1757-899X/1225/1/011001

3. ***Paper ID: 065_Ni.Cr.Alloy***
   Title of paper Study on Defect Reduction in Casting of Ni-Cr alloy (UNS N06003)
   Authors Kaustubh Samvatsar and Harsh Dave
   First Author Affiliations Faculty of Technology and Engineering, The Maharaja Sayajirao University of Baroda, Vadodara, Gujarat 390001, India

4. ***Paper ID: 066_GaN.Hetrostructure***
   Title of paper Study of electronic structure and transport coefficients of GaN-MoS_2 heterostructure.
   AuthorsGurpal Singh Khosa, Ranjan Kumar and Shuchi Gupta
   First Author Affiliations Department of Physics, Panjab University, Chandigarh 160014, India

5. ***Paper ID: 067_Electrical.Discharge***
   Title of paper Wear and Corrosion Behaviour of Wire Electrical Discharge Machined Ti-6Al-4V alloy
   Authors Sanyukt Kumar Manderna, Puneet Katyal, Munish Gupta and Vijender Singh
   First Author Affiliations Department of Mechanical Engineering, Guru Jambheswar University of Science and Technology, Hisar, Haryana, INDIA – 125001

IOP Conf. Series: Materials Science and Engineering 1225 (2022) 011001 doi:10.1088/1757-899X/1225/1/011001

INVITED SPEAKER

Dr Navin Kumar
Associate Professor
Department of Mechanical Engineering
IIT Ropar, Punjab, India
Email: nkumar@iitrpr.ac.in

About Speaker

Dr. Navin Kumar is an Associate Professor in the Department of Mechanical Engineering at IIT Ropar. Prior to joining IIT Ropar, he was working as a Research Scientist at Stevens Institute of Technology, New Jersy, USA. He has completed masters in Mechanical Engineering from IIT Kharagpur and Ph.D. (Mechanical Engineering) from IIT Delhi. Dr. Navin Kumar's research interests are related to both theoretical and experimental aspects of mechanics and dynamics of nano, biological and bio materials and structures, Bone Properties Characterization, Biomedical Engineering, Biomechanics, Biomedical Instrumentation, smart structures and materials, fault diagnosis and condition monitoring. In particular he is interested in the areas, Biomaterial Characterization. Experimental and simulation of nano and bio materials and structures, active vibration control and fault diagnosis.

IOP Conf. Series: Materials Science and Engineering 1225 (2022) 011001 doi:10.1088/1757-899X/1225/1/011001

INVITED SPEAKER

Dr. Lalit M Bhardwaj
Chief Executive President, RML Consultancy & Fabrication, Global Director General, Rejoice Health Foundation.
Former
Vice President, Amity Science, Technology & Innovation Foundation, Deputy Director General, Amity Directorate of Science & Innovation
Prof. & Director Nanoscience & Nanotechnology, Amity
Dean, Research, Lovely Professional University, Jalandhar
Addl. Director, CSIO/CSIR, Chandigarh

About Speaker

Former Vice President, Amity Science, Technology & Innovation Foundation (ASTIF), Deputy Director General, Amity Directorate of Science & Innovation (ADSI), Professor of Nanotechnology, Noida and former, Additional Director and Head, Biomolecular Electronics and Nanotechnology, Central Scientific Instruments Organisation (CSIR), Chandigarh, India, has excellent academic record and worked in leading national and international institutions of repute and has acquired leadership qualities for multi-disciplinary research.

INVITED SPEAKER

Dr Parveen Kumar
Head R&D
Exigo Recycling Pvt. Ltd.
Noida, UP - 201309
Email: parveenkaushik7@gmail.com

About Speaker

Dr Parveen Kumar did his Ph.D. from Punjab University and CSIR-CSIO in year 2014. He worked on various analytical instruments to complete his Ph.D. in the field of biosensing. Another area of interest is urban mining to recover valuable high end products such as, nanoparticles. He achieved INSPIRE Faculty award from Department of science and technology, New Delhi, Dr. D. S. Kothari Post doc fellowship from UGC, New Delhi and MHRD Post doc fellowship. He has contributed in various CSIR, ICMR, DBT and DST funded projects as PI, Co-PI and team member. He is keen to work in collaboration with industries and provided consultancies and accomplished industry sponsored project in strict deadlines. He is currently supervising 4 Ph.D. students and guided many Graduates and post graduate students for their thesis/ dissertation. He has around 35 international publications in various reputed journals and published 2 book chapters, one patent and transferred 3 technologies. He has delivered multiple invited talks at prestigious forums.

Abstract

Traditionally mining is supplying all materials, especially metals for various industrial and domestic applications. These generate high carbon footprints even reducing forest area, disturbing all ecological aspects, and finally responsible for the ill health of mother earth. One initiative to reduce the mining of natural mines is to bring the mined material back into work. All the metal/nonmetal waste accumulated in urban areas can prove to be a better resource in the material supply chain. The government has introduced Extended Producer Responsibility to bring back the material produced by Original equipment manufacturers. These are recycled by professionals to use with pristine/virgin material to bring value back to recycled materials. Electronic waste management is one of such industries which is recycling/ recovering and refurbish to reuse the material back. The electronic devices after their end of life are collected and dismantled by the recycler to segregate metallic and nonmetallic components. Ferrous metals can be directly sent to metal refineries whereas nonmetals like plastic, glass, etc. are recycled by other industries. Some materials like Lithium-Ion batteries require special attention to recycle as the electric vehicle era is about to begin. These batteries can supply fine battery-grade material for upcycling as novel electrode materials. Exigo has scaled up technologies for graphite recovery, which is a precursor to making graphene oxide, reduced graphene oxide, quantum dots, etc.

INVITED SPEAKER

Dr. Vijayender Bhalla
Principal Scientist
CSIR - Institute of Microbial Technology
Ministry of Science & Technology, Govt. of India
E-mail: vkbhalla@imtech.res.in

About Speaker

His research focuses on detection technologies using nanomaterials for a range of analytes i.e. blood biomarkers, pathogens, pesticides and small molecules. The currently undergoing projects include Point of care detection of cardiac biomarker peptides using natural and synthetic receptors (aptamers) on nanostructured electrochemical based biochips. We also work on Purification and characterization of surface biomarkers to enable selective detection of water borne pathogens. We develop new kinds of optical and electrochemical immunoassays for pathogen detection in food and water samples. We synthesize different kinds of nanoparticles, nanoclusters, carriers, haptens, bioconjugates and work on biointerface development to enable low-cost diagnostics.

Abstract

Nanomaterials have opened new avenues for researchers working in the area of biosensors. By taking examples from my own research, I will explain how the different nanomaterials such as Carbon nanotubes, Graphene, Molybdenum disulfide can be used to realize high sensitivity of detection. The high surface area and presence of functional groups on nanomaterials allow facile immobilization of bioreceptors onto the sensing layer. The nanomaterials are easy to synthesize in lab and they have an indispensable role to play in sensing. Not only electrochemical and FET based sensors but also fluorescence and optical sensing have been demonstrated by employing nanomaterials.

IOP Conf. Series: Materials Science and Engineering 1225 (2022) 011001 doi:10.1088/1757-899X/1225/1/011001

INVITED SPEAKER

Dr Vinay Sharma
Ramanujan Faculty
Department of Materials Engineering
IIT Jammu, Jammu, India
Email: vinay.sharma@iitjammu.ac.in

About Speaker

Dr. Vinay Sharma currently works at the Indian Institute of Technology Jammu (IIT Jammu), India. He was previously associated with University of California, Riverside as a postdoctoral researcher and Indian Institute of Technology Indore as a research scholar. He obtained Ph.D. in the area of nanomaterial-based biosensor and bioimaging probes from IIT Indore. He was awarded with the prestigious Ramanujan fellowship by DST-SERB, Govt. of India in the year 2020. He has also received ESONN-CEFIPRA Fellowship award in the year 2016 by CEFIPRA, Govt of India. Dr. Sharma's research interests are in Nanopores, Single molecule techniques and nanomaterials for healthcare. He has extensive experience in fluorescent quantum dots, bioimaging and nano-bio-hybrids for bio-medical applications. He has published more than 30 research articles, review articles and book chapters in high impact journals of international repute. Dr. Sharma is also associated with various international journals as editor and reviewer. He works with a vision of making healthcare accessible and affordable with the intervention of nanotechnology.

Abstract

The unprecedented chemical and physical properties of nanomaterials have the potential to revolutionize the field of biology and medicine. Nanomaterials have found applications in the disease biosensing, bioimaging and even therapeutics. The nano material-based biosensors have the promise to provide portable alternatives to lab diagnostics. The high sensitivity, selectivity, high throughput and the potential to bring Lab-on-a-Chip have brought significant attention to the field of nanosensors. Similarly, the photodynamic and photothermal nanoparticles have shown immense potential in therapeutics. The field of DNA sequencing, protein folding and biomolecule interaction is hugely benefitted by the advent of nanopore technology which can make modern diagnostics fast, accurate and inexpensive. The present talk discusses various application of nanoparticles in sensors, bioimaging probes, therapeutics and envision the potential of nanopore technology to lead nanodiagnostics.

INVITED SPEAKER

Dr Sumit Jindal
Assistant Professor (Sr.)
School of Electronics Engineering
Vellore Institute of Technology, Vellore, Tamil Nadu, India
Email: sumitjindal08@gmail.com

About Speaker

Dr. Sumit Kumar Jindal received his Bachelor's degree in Electronic and Instrumentation Engineering from B.P.U.T, Orissa, India and Ph.D. from Indian Institute of Technology(ISM), Dhanbad, India in 2009 and 2017 respectively. He was University topper during his undergraduate studies. He specializes in the area of MEMS Sensors. In particular, his research interest is MEMS Sensor Technology, Electronics and Instrumentation & Embedded Systems. He has a total of twelve years of experience in Teaching and Research. Currently, he is an Assistant Professor (Sr.) with School of Electronics Engineering (SENSE) in Embedded Technology Division, Vellore Institute of Technology,Vellore. He is an active reviewer for renowned journals in his field of interest such as IEEE Transactions on Instrumentation and Measurement, IEEE Sensors Journal, Mechatronics (Elsevier), Microsystem Technologies (Springer) and Journal of Micro/Nanopatterning, Materials and Metrology (SPIE). He has published three patents with Intellectual Property, Govt. of India. He is faculty co-coordinator for IEEE Robotics and Automation Society Chapter at VIT. This chapter organizes a lot of technical workshops and hands on training sessions throughout the year. He is also an active member of Placement and Training cell at VIT.

Abstract

Modelling and simulation are changing the way modern materials science work. Over the years it has become one of the most vital tools for discovering new materials and material phenomena. It is increasingly being used for obtaining information about the processes that determine the behaviour of materials. It is also being used for quantitative forecasting which can then be utilized for design in association with synthesis and experimental set-up. Modelling and simulation acts as the essential bridge between science and engineering. It ranges from a basic understanding of the behaviour of materials to the cognizant plan to design new material technologies using newer properties as well as procedures. This work provides a wide outline of the huge impact of simulation and modelling on materials science over the past years and provides centred viewpoints on the way forward. This field continues to make rapid advancements and evolves to address the difficulties of materials for the upcoming decades. This work gives valuable insights into the advancements made in disciplines which ranges from phase field methods to model behaviour at the mesoscale; to molecular dynamics method in order to derive the basic dynamic processes at the atomic level that control the reaction of material, to the challenges of interdisciplinary research that targets complicated material problems that require multi-scale approaches. The shift from understanding the basic behaviour of the material to developing quantitative approaches in order to explain and foresee experimental observations require advancements in these methods and practice of simulation in terms of reproducibility and integrity. It also includes interaction with different computational ecosystems which allow to develop new theories, unique applications along with an integrated computing and software facility using efficient computational methods and highly powerful computational hardware resources.

IOP Conf. Series: Materials Science and Engineering 1225 (2022) 011001 doi:10.1088/1757-899X/1225/1/011001

INVITED SPEAKER

Mr. Ayan Karmakar
Scientist
Semi-Conductor Laboratory (SCL), Chandigarh
Indian Space Research Organization (ISRO), India
Email: ayan_ns@scl.gov.in

About Speaker

Ayan Karmakar received the B.Tech degree (ECE) from WBUT, Kolkata in 2005. Later he did his masters from NIT, Durgapur. He joined Indian Space Research Organization (ISRO) as a 'Scientist' and subsequently posted to Semi-Conductor Laboratory (SCL), Chandigarh. His research interests include design and development of L, S, C, X and K-band passive microwave integrated circuits and antennas using silicon based MIC and RF-MEMS technology. His field of expertise are expanded in fabrication technology of various MEMS based sensors and evolving the test strategies for the same. Currently, he engages himself in high frequency device modelling and development of electronic interfaces for various MEMS based sensors.

He has more than 40 publications in reputed journals and conferences at various national and international levels. Recently, he has authored one technical book named 'Si-RF Technology' from Springer Publication, Singapore. He received multiples prizes/awards from Department of Space, Govt. of India for his active involvement and technical contributions in Rajbhasha (Hindi). He has shown keen interest to build a strong communication link between academics and industry. In this context, he has delivered numerous lectures in the invited talk series of various institutes of national importance and Webinars. He serves as a reviewer in reputed referred journals, like-IEEE Access, RF-MiCAE, Springer, EJAET, etc. He is a Fellow member of IETE, India and life member of Indian Science Congress Association, ISSE and Bangiya Bigyan Parishad.

Abstract

Micro-electro Mechanical System (MEMS) is an agglomeration of sub-micron mechanical moving parts and electronic components on a single module. MEMS devices operated at Radio Frequency (RF) are emerging as a propitious technology. In near future

5G, 6G wireless network (WSN), IoT, and even in the potential millimeter and sub-millimeter wave applications, this technology will play a pivotal role. Lots of passive communication building blocks of the system can be realized utilizing this upcoming technology. Starting from the basic building blocks to system level approach can be adopted with MEMS. In terms of antenna engineering or any other signal processing blocks like-filter, oscillator, switching matrix or phased array antenna system, this technology can be implemented in multifaceted applications. Bottom to top approach is adopted in this field. Using the standard CMOS foundry and few additional post-CMOS fabrication techniques the unique MEMS devices are realized. Device modeling plays an important role. Vertical integration approach can also be materialized with MEMS, which is inevitable for higher frequency applications. MEMS using the silicon substrate or silicon micromachining can even be used to develop Terahertz (THz) devices. In this talk, the speaker will cover the basic understanding of this technology in context with RF/microwave engineering and then its applied fields.

INVITED SPEAKER

Dr. Harlal Singh Mali
Associate Professor,
Department of Mechanical Engineering
Malaviya National Institute of Technology,
Jaipur, India
Email: harlal.singh@mnit.ac.in

About Speaker

Alumni of PEC and PU Chandigarh, Dr Harlal Singh Mali has experience of 15 years in academics and 10 years in the aviation industry. He holds two patents and has applied for another seven. He held multiple research grants from Indian and international funding agencies, including BIG of BIRAC (DBT), ECR, EMR of DST (SERB), DRDO (ARMREB), NSF (USA) and various industries. Has supervised 08 PhD and 32 M. Tech. theses and has more than 100 peer-reviewed publications. He is also the founder-director of startup; Jaipur Club Foot Pvt. Ltd., which is an outcome of NISP 2019 to solve a complex socio-medico-technological problem called Club Foot treatment. He did set up an Advanced Manufacturing & Mechatronics Lab and founded CAM Society. MNIT CAM Society is a student-driven club dedicated to teaching and learning manufacturing technologies.

Abstract

State-of-the-art and high-performance textile composites are used worldwide due to their enhanced applicability in defense, automotive, aviation, sports, and marine sectors where high performance of a material is required. Textile composites have an architecture in the form of unidirectional, woven, braided, knitted, and stitched fabrics. High-performance textile reinforcements such as Aramid, Dyneema, Spectra, Zylon, Vectran, Boron, Carbon, and Glass are impregnated with a polymer matrix to fabricate the advanced polymer textile composites with high-end specifications. Glass, Carbon, and Kevlar fiber reinforcements are popular due to their high strength to weight ratio, modulus of rigidity, heat and chemical resistance, toughness, dimensional stability, creep resistance, and moisture insensitive properties.

Hybrid textile fabrics are developed by weaving more than one type of synthetic fiber to obtain the optimum properties of individual fiber material. Hybridization at interply and intraply levels can obtain different engineering properties on a specific characterization and be further optimized. The polymer textile composites can be processed using hand layup, spray layup, filament winding, autoclave, resin transfer molding, vacuum-assisted resin infusion, thermoforming, and compression molding techniques. Alike all other engineering materials, developed composites are further used for several experimental characterizations: tension, compression, flexural, fracture toughness, dynamic mechanical analyzer, high and low-velocity impact test, hardness, creep, machinability, fatigue, and wear.

Numerical characterization of textile composites has the advantage over the costly experimental setup. The numerical model can be able to predict the various engineering properties. Numerical simulation of textile composites entirely depends on the geometrical model of textile composites, which is a challenging task due to the complicated architecture of textile reinforcements. Few geometrical modeling tools such as Texgen, Wisetex, and TechText CAD can model the textile reinforcements.

Once modelled, the textile composites are numerically characterized using finite element simulation tools such as ABAQUS®, ANSYS®, HyperWorks®, and Nastran® to predict the many engineering properties. The finite element simulation tool, ABAQUS® is very popular among researchers due to the accuracy of results. Numerically simulated results can further be used for experimental validation, hence making the developed material usable for the intended application.

IOP Conf. Series: Materials Science and Engineering 1225 (2022) 011001 doi:10.1088/1757-899X/1225/1/011001

INVITED SPEAKER

Dr Balwinder Singh
Joint Director & Head, ACSD
Centre for Development of Advanced Computing (C-DAC)
Mohali, Punjab, India
E-mail: balwinder@cdac.in

About Speaker

Dr.Balwinder Singh has completed his PhD in 2014 from GNDU Amritsar & also has obtained his Bachelor of Technology degree from National Institute of Technology, Jalandhar and Master of Technology degree from University Centre for Inst. & Microelectronics (UCIM), Panjab University, Chandigarh in 2002 and 2004 respectively. He is currently serving as Joint Director and Head (ACSD) in Center for Development of Advanced Computing (CDAC), Mohali. He has successfully executed many Govt funded projects. He has 15+ years of teaching experience to both undergraduate and postgraduate students.Singh has published three books and about 120+ papers in the International & National Journal and Conferences.

He was also awarded with Best Teacher Award by ISTE,New Delhi in 2015 and COSMIC – Young Scientist Award for Contribution in the field of “Engineering and Technology 2015 atAETM 2015 – Bangkok,Thailand. His current interest includes IoT based smart Agri-systems, AI based agriculture and health systems, Low Power Techniques and VLSI Design & Testing

Abstract

Human beings have always been fascinated by nature and especially by advanced technology and their evolutionary process. This has resulted into Inspirations drawn from natural or technological systems, and innovations in agriculture, for problem solving and has seen an emergence of a new paradigm for revolutionary technology using Internet of Things (IoT) known as Smart Agriculture with Precision farming as its subset. Just as any of the smart technology are poised to change the way we get around similarly smart technology in the field of agriculture is set to revolutionize the way we cultivate the food we eat.

The main challenge towards the food and agricultural industry ponder is that, how to double food production by 2050 with less land available with the increase of population in the country every year. Moreover, according to the United Nations, the planet's global population is expected to reach 9.6 billion by 2050. There is an extensive demand of food products, along With require growth in the production itself, which requires the smart technologies for precision farming and further reach the food products in the market. However, in the food supply chain, it is believed that loT will help to revolutionize the technology towards food desirability for suppliers and supermarkets alike.

The broadly used approaches are artificial neural network, Precision farming, Remote Sensing, Robotics. IoT enabled Agri products etc. counts to be few technological developments in the field of agriculture. It also allows for much more reliable production ands, which in turn enables greater management of demand.

IOP Conf. Series: Materials Science and Engineering 1225 (2022) 011001 doi:10.1088/1757-899X/1225/1/011001

INVITED SPEAKER

Dr Akash Deep
Principal Scientist
Materials Science and Sensor Applications (MSSA) Division,
CSIR-CSIO
E-mail: dr.akashdeep@csio.res.in

About Speaker

Dr. Akash Deep received his Ph.D (Chemistry) degree in 2004 from the Indian Institute of Technology, Roorkee, India after which he worked as a postdoctoral research fellow till 2008. He joined CSIR-CSIO, Chandigarh in 2008, where he is presently working as a Principal Scientist. Dr. Akash Deep has worked in the fields of analytical and material chemistry, biochemistry, and nanosensors for environmental and health applications. He has expertise in the synthesis and applications of advanced materials, such as 2-dimensional layered materials, fluorescent nanoparticles, metal-organic frameworks, graphene composites, and conducting polymers. Dr. Akash Deep has also worked as a Visiting Professor at the Hanyang University, Seoul, South Korea and CSIR-Raman Research Fellow at RWTH Aachen University, Germany. He has published more than 180 research articles in reputed international journals. He has also transferred technologies to the industries and provided consultancy in the areas of nanomaterial synthesis, biosensors and waste treatment.

Abstract

The increasing demand for clean and green energy necessitates the development of efficient energy storage devices such as batteries, fuel cells, and supercapacitors. Batteries deliver high energy density values but are not necessarily efficient in terms of their cycle life and power density. Supercapacitors are the emerging energy storage devices which can exhibit high values of power density. Nonetheless, a moderate energy density limits the applications of supercapacitors in many fields. Therefore, the researchers are exploring the applications of new materials and composites to design superactions with combined features of high energy and power densities along with sufficient cycle life. Graphene and metal-organic frameworks (MOFs) are the class of advanced functional materials that are characterized with many attractive features such as high surface to volume ratio, accessible porosity, electrocatalytic activity, etc. The individual properties of graphene and MOFs can be further augmented by forming their composites. As our research studies have revealed, graphene, MOFs, and their derived structures/composites pool the benefits of high electrical conductivity, high specific surface area, and hierarchal pore size distribution. These features are useful in enhancing the specific capacitance of the supercapacitors. The graphene and MOF based supercapacitor electrodes can be operated in different electrolytes (e.g., redox additive and polymer gel electrolytes) to attain wide potential windows, high specific capacitance, and several other desirable supercapacitor performance parameters.

IOP Conf. Series: Materials Science and Engineering 1225 (2022) 011002 doi:10.1088/1757-899X/1225/1/011002

Peer review declaration for IC-MAME 2021

All papers published in this volume of *IOP Conference Series: Materials Science and Engineering* have been peer reviewed through processes administered by the Editors. Reviews were conducted by expert referees to the professional and scientific standards expected of a proceedings journal published by IOP Publishing.

- **Type of peer review** : Single-blind
- **Conference submission management system** : Easy Chair (https://easychair.org/cfp/icmame2021)

- **Number of submissions received** : 150 (One hundred fifty)
- **Number of submissions sent for review** : 140 (One hundred fourty)
- **Number of submissions accepted** : 67 (Sixtyseven)
- **Number of submissions accepted for IOP** : 67 (Sixty seven)
- **Acceptance Rate** : 44.66%
- **Average number of reviews per paper** : 02 (Two)
- **Total number of reviewers involved** : 197 (One ninety seven)
- **Any additional info on review process:**

Manuscripts submitted to the IC-MAME 2021 conference were initially screened by the editorial team. After initial screening, manuscripts were sent to at least two experts for peer review. All manuscripts submitted to the conference were single-blind peer-reviewed and were carried out to meet the scientific criteria of novelty and academic excellence. The review process was conducted according to the review policy of IOP conference series. Peer reviewers independently made a recommendation to the conference editors as to whether the manuscript should be rejected or accepted (with or without revisions). Manuscripts with revision were submitted again, accompanied by a file describing the changes wherever needed, and finally decision were made by the editors. The editors informed decision to accept or reject the manuscript.

- **Contact person for queries:**

Dr. Rajesh Kumar
(+91 9501019371; rajeshmadan@gmail.com.)

Dr. Gaurav Sapra
(+919815858051; gaurav.sapra@pu.ac.in)

Microwave Assisted Synthesis of Molybdenum Disulphide/Tungsten Trioxide/Reduced Graphene Oxide (MoS2/WO3/RGO) Nanocomposites for Organic Vapor Sensing

Jyoti Gupta, Prachi Singhal, Sunita Rattan S*

Amity Institute of Applied Sciences, Amity University, Uttar Pradesh, Noida, UP, India

*Corresponding Author mailing ID: srattan@amity.edu

Abstract. Hybrid nanostructures based on two-dimensional (2-D) materials such as transition metal dichalcogenides (2-D TMD's), transition metal oxides (2-D TMO's), and graphene are gaining significant attention for their potential application as semiconductor sensing devices owing to their excellent structural, electronic, and optical characteristics. The present work reported 2- D ternary nanocomposites (MoS2/WO3/RGO) as organic vapor sensing material, synthesized using microwave assisted method. Formation of ternary composite system proves to be an efficient strategy to develop novel high performance vapour sensors, providing multiple degrees of freedom and opportunity to exploit the synergistic properties of the resulting material to its full potential for a particular application. The synthesized MoS2/WO3/RGO nanocomposites has been characterized for morphological, and structural analysis through SEM, XRD and FTIR spectroscopic techniques. The electrical and sensing studies of synthesized material is evaluated through IV characteristics using Keithley electrometer. The sensing characteristics has revealed that the synthesized sensor has high sensitivity, good stability at room temperature and outperforms the result obtained from single or binary counterparts. The formation of the heterojunction and electronic modulations at the interface of MoS2, WO3 and RGO results in improved surface charge transfer mechanism and enhancement of sensing performance.

Keywords: 2-D materials, TMDs, TMOs, Graphene, Ternary Nanocomposite, organic vapor sensing

1. Introduction

Selective, fast, sensitive, and reliable alcohol monitoring is significantly needed in many industries including pulp and paper industries, food industries, pharmaceutical industries, etc., as long-term exposure of even low concentrated alcohol may result into acute illness related to eyes, nasal function, respiratory tract, nerve functioning etc.

In recent years, two-dimensional (2-D) materials emerges as the finest gas sensing materials owing to their substantial characteristics including room temperature operation, high surface to volume ratio, outstanding electron mobility, tunable band-gap and good electrical properties, as compared to conventional-metal-oxide vapor sensors. The conventional-metal-oxide vapor sensors require high temperature operation (200 to 300 °C) to trigger the chemical absorption process of organic vapor to produce sensor response [1] so it is not feasible according to power consumption point of view. One decade ago, many researchers have reported graphene and their derivatives-based sensing materials for gas sensing application. In recent years, some other 2-D materials besides graphene derivative such as 2-D TMDs and 2-D TMO's are gaining attention because of their extremely good properties such as tunable bandgap and excellent electrical properties [2-6]. As compared to individual 2-D materials, the hybrid nanostructures of 2-D materials improve the sensing response because of the formation of heterojunction between two different 2-D material which enhance the surface charge transfer mechanism [3,4]. Hence, the many researchers are trying to develop novel hybrid nanostructures to achieve the collaborative properties of 2-D materials and improve their sensing performance. Generally, binary hybrid nanostructures of graphene derivatives and 2-D TMDs (MoS2, WS2, MoSe2) or 2-D TMOs (WO3, MoO3) were reported for gas sensing applications [7-9]. As compared to binary hybrid nanostructure, ternary nanostructures develop multiple nanoheterojuctions that cause the modulation in interfacial electronics such as modulation in charge carriers, charge carrier densities, fermi-level and their charge carrier separation for the transportation of charges across the heterojunctions [10-12]. All these modulations are expected to enhance the sensing performance of the ternary nanocomposite.

In this work, attempts have been made to prepare microwave assisted ternary (MoS2/WO3/RGO) hybrid nanomaterial. The microwave synthesized MoS2 is hybridized with reduced graphene oxide and tungsten trioxide. RGO is prepared through Modified Hummers method. The synthesized nanostructure is evaluated for its application as organic vapor sensor.

2. Experimental details

2.1 Materials used:

The analytical grade reagents such as sulfuric acid, hydrogen peroxide, HCl (3.6 wt.%), Nitric acid (55%), hydrazine hydrate, sodium borohydride, Hydrogen peroxide Ammonium tetra thiomolybdate, Potassium permanganate (KMnO4), isopropanol and tungsten trioxide powder were procured from Sigma Aldrich for performing the experiment. Acidified graphite flakes (Analytical Grade 1721) were obtained through Asbury.

2.2 Preparation of Reduced graphene Oxide (RGO)

Initially, GO was prepared by modified hummers method from Acidified carbon graphite flakes as described in our previous research publications [13]. RGO was synthesized by dissolving GO (100mg) in water. (100 ml). After that, 10 micro liter of hydrazine hydrate was added dropwise into this prepared GO solution and refluxed it for 1 hour at 80 oC. After that, 1 mg of NaBH4 (sodium borohydride) was added into this solution and again refluxed it for 36 hours at 100 oC and cool down to 25 oC. The synthesized material was filtered through Millipore filter paper and washed it with deionized water and dried it.

2.3 Synthesis of MoS2/WO3/RGO nanocomposites

The pure MoS2 is synthesized by direct microwave assisted reaction. Ammonium tetra thiomolybdate is taken for synthesis of MoS2. 10mg of Ammonium tetra thiomolybdate (NH4)2MOS4 crystals are mixed in 20ml hydrazine solution and 10 ml distilled water and mixed well. The solution is kept in microwave at 800W for 1 minute after that it is filtered off and washed three times with isopropanol. The as obtained pure MoS2 is kept in oven for air drying. 10 mg of each MoS2, WO3 and RGO is added into 20 ml of hydrazine and 10ml of distilled water solution. The solution is subjected to microwave radiation treatment at 800W for 1 minute. The solution obtained is washed three times with isopropanol. The detailed synthesis procedure along with schematic representation is described in our previous publication [14]. The dilute solution of RGO/MoS2/WO3 (10 mg each) nanocomposite was prepared in dimethyl formamide (50 ml in vial) with and spin coated on glass slide. The same method is used to prepare RGO/MoS2 nanocomposite.

2.4. Characterizations

The surface morphology, structural behaviour and functional group of synthesized MoS2/WO3/RGO nanocomposite was analysed for scanning electron microscopy through SEM, ZEISS EVO, X-ray diffraction through XRD, Rigaku Smart lab using Cu Kα as source radiation (2θ range: 10° to 55°) and FTIR through Perkin Almer.

2.5 Gas sensing and device fabrication

Sensing device was fabricated by drop-casting the solution on a properly cleaned and dried glass substrate. The synthesized nanocomposite solution was spin coated on glass slide for IV characteristics. After drying the nanocomposite film, silver paste was deposited onto glass slide to make contacts using copper wires. A vapour delivery and measurement system assembled in the laboratory, was used for sensing evaluation in which electrical connection, exhaust flow of gas, inlet of gas is provided. The synthesized device was mounted inside the gas chamber and their silver contacts were connected to Keithley Source meter (6517 A/B)

3. Results and Discussion

3.1 Morphological structure of MoS2/ WO3/RGO

Figure 1 (a & b) represents the SEM image of microwave synthesized MoS2 and MoS2/WO3/RGO ternary nanocomposites respectively depicting their morphological characteristics. SEM image of nanocomposite film shows uniform distribution of MoS2 and WO3 on RGO sheets. The uniform distribution of MoS2, WO3 and RGO in nanocomposite has been confirmed through EDS mapping as shown in Figure 2.

Figure 1. SEM image of (a) Microwave synthesized MoS2 and (b) MoS2/WO3/RGO Ternary nanocomposites

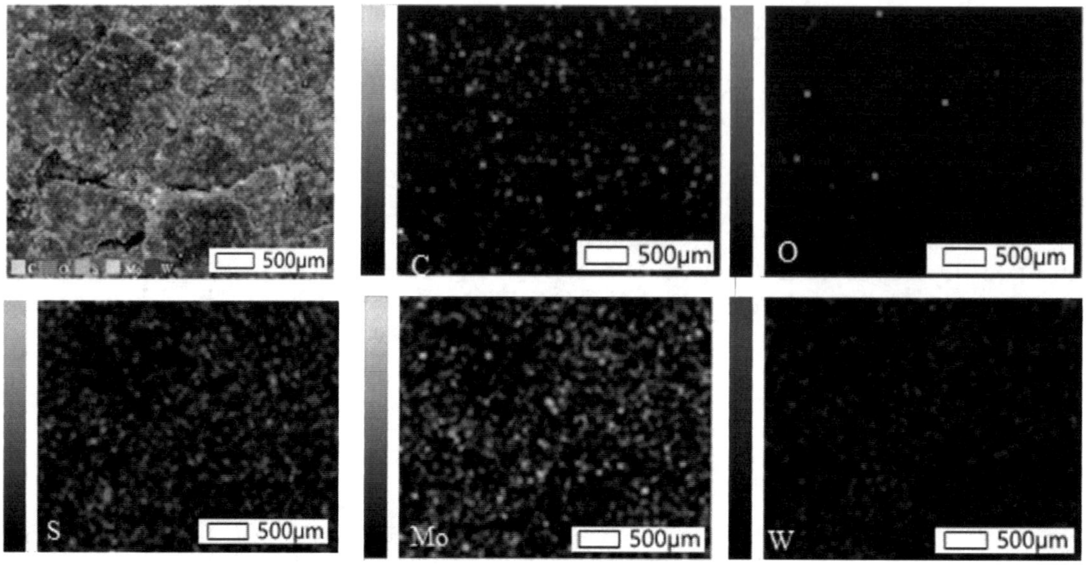

Figure 2. EDS Mapping Image of RGO/MoS2/WO3 ternary Nanocomposite

3.2 Crystalline Structural and Functional Group Evaluation

The FTIR spectra for RGO/MoS2/WO3 film shown in Figure. 3 (a). The various absorption band of RGO were revealed by the distinctive peaks at 3437.51 cm−1, 3046.36 cm−1 and 1719.12 cm−1 that were attributed to dissociation of hydroxide ions and C–H stretching vibrations as well as C=O, Carboxyl group, bond respectively. RGO peak at 1569.59 cm−1 was allied with the C–H bending. The peak at 1402.02 cm−1 represented Mo-S stretching mode in synthesized RGO/MoS2/WO3 nanocomposite film. Characteristic peak at 789.02 cm−1 is ascribed to O-W- O stretching band of WO3 [15]. The XRD pattern of synthesized RGO/MoS2/WO3 film shown in Figure. 4(b). The peak of synthesized RGO at 26.65o was assigned to (002) plane. The diffraction characteristic 2θ peaks at 23.1°, 23.57°, 24.53°, 29.04°, 34.19°, 43.56° and 49.79° were indexed to WO3 basal planes 002, 020, 200, 112, 202,222 and 440 respectively [16-17]. The characteristic peaks at 2θ = 14.5 o, 32.4 o were assigned to 002, 100 planes of MoS2 [18].

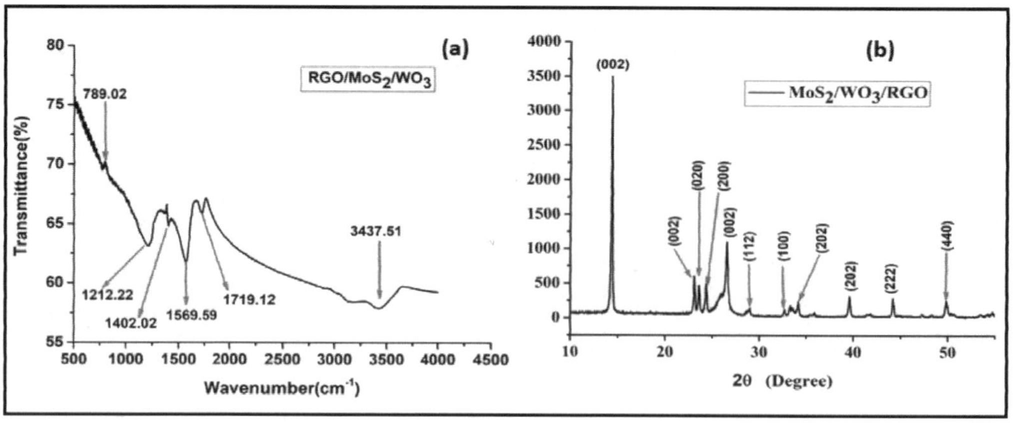

Figure 3. (a) FTIR spectrum of synthesized RGO/ MoS2 and RGO/MoS2/WO3 nanocomposite film.

IOP Conf. Series: Materials Science and Engineering 1225 (2022) 012001 doi:10.1088/1757-899X/1225/1/012001

3.3 Characteristics

The I-V characteristics of synthesized device was performed on Keithley 6517 A/B. The constant voltage of 1V was applied in the range from -1V to 1V and the sensing current was measured from -1V to 1 V which was shown in Figure 5. The current flow of synthesized nanocomposite film of MoS2/WO3/RGO in the range of -0.02 to 0.16 μA was plotted from −1 V to 1 V. The nanocomposite film of MoS2/WO3/RGO shows maximum conductivity of 0.16 × 10−6 S cm−1.

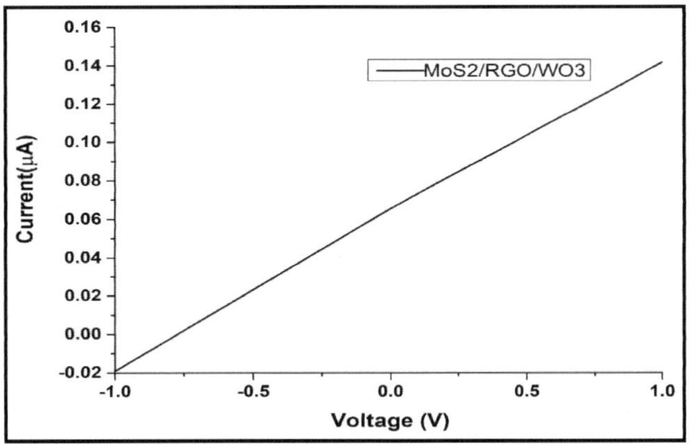

Figure 4. The I-V characteristics of synthesized RGO/MoS_2 and RGO/MoS_2/WO_3 nanocomposite films.

3.4 Organic vapour sensing Studies

The RGO, RGO/MoS2 and RGO/MoS2/WO3 nanocomposite films are evaluated for sensor response recovery characteristics towards various reducing organic vapours such as ethanol, methanol, acetone, propanol and oxidizing vapours such as nitrobenzene, formaldehyde, and benzaldehyde vapours. RGO and RGO/MoS2 does not respond to any applied organic vapours while Figure 6, which depicts the typical sensor response-recovery characteristics of RGO/MoS2/WO3 ternary nanocomposites, clearly indicates that the sensor is highly responsive to alcohols and only slightly to acetones, but not to other vapours. The sensor response (%) of 3.15%, 2.36%, 1.36% and 0.43% towards ethanol, methanol, propanol and acetone respectively are depicted in Figure 6.

While developing organic vapour sensor devices, especially hand-held devices for different applications, response time and recovery time being two significant parameters to characterize a sensor. The RGO/MoS2/WO3 nanocomposite film shows fast response and recovery time of ~1 sec and ~10 sec respectively at room temperature for ethanol. The mechanism responsible for sensing can be illustrated based upon the surface adsorption of vapour and the resulting charge transfer in the nanocomposite structure. The alcohol (ethanol, methanol and propanol) vapours are reducing in nature, so it is assumed that RGO/MoS2/WO3 exhibits p-type semiconducting nature, whereas WO3 are responsible for providing active sites where alcohol replace adsorbed oxygen and release free electrons that results in the subsequent decrement of charge carrier concentration, and hence decrease in electrical conductivity of nanocomposite sensor film [19].

The nanosheets of RGO provides faster conducting network and its mechanical strength enhances the durability of the sensor material. The interfacial electronic modulation at number of heterojunctions developed in RGO/MoS2/WO3 hybrid nanostructure decides the overall electronic structures that results in fast recovery and response time to the organic vapours.

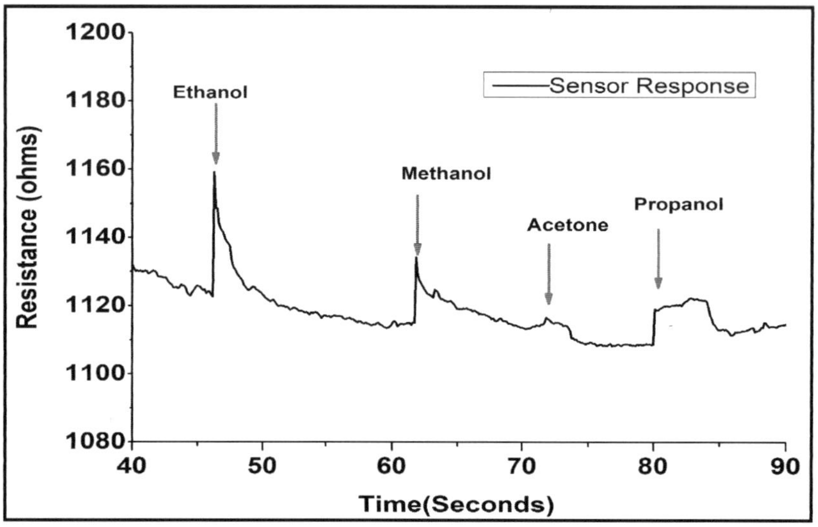

Figure 5. Resistance changes in RGO/MoS2/WO3 nanocomposite film on exposure towards various organic vapour at room temperature.

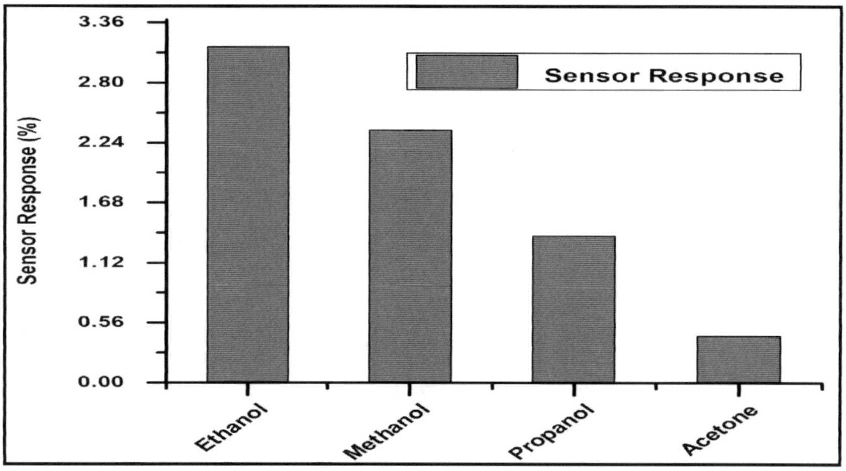

Figure 6. Sensor Response of RGO/MoS2/WO3 sensor towards ethanol, methanol and acetone

4. Conclusion

In the present work, RGO/MoS2/WO3 ternary nanocomposites are developed through microwave assisted method as potential candidate for chemiresistive alcohol vapour sensing. The vapour sensing response has been evaluated towards a series of reducing and oxidizing organic vapours. The ternary nanocomposite film shows selective response towards alcohol vapours with very fast response (1 sec for ethanol) and recovery time (10sec for ethanol) at room temperature. The selective sensing response of RGO/MoS2/WO3 nanostructures film towards alcohol with quick response time probably attributed to the synergistic exploitation of the advantages offered by all three components, RGO, MOS2 and WO3. The resultant electronic structure of the nanocomposite due to electronic modulations at multiple number of heterojunctions formed at interface is responsible for the specific sensing response shown by RGO/MOS2/WO3 ternary nanocomposites.

References

[1] Zhang J, Liu X, Pinna N 2016 Nanostructured materials for room-temperature gas sensors *Adv. Mater.* **28** 795–831.
[2] Yang W, Gan L, Li H, Zhai T 2016 Two-dimensional layered nanomaterials for gas- sensing applications *Inorg. Chem. Front. 3 433.*
[3] Liu X, Ma T, Pinna N, Zhang J 2017 Two-dimensional nanostructured materials for gas sensing *Adv. Funct. Mater.* **27** 1–30.
[4] Ghosh R, Midya A, Santra S, Ray S K, Guha PK 2013 Chemically reduced graphene oxide for ammonia detection at room temperature *ACS Appl. Mater. Interfaces* **5** 7599– 603
[5] Burman D, Ghosh R, Santra S, Ray S K, Guha P K 2017 Role of vacancy sites and UV- ozone treatment on few layered MoS2 nanoflakes for toxic gas detection *Nanotechnology* **28** 435502
[6] Ghosh R, Gardner JW, Guha PK 2019 Air pollution monitoring using near room temperature resistive gas sensors: a review IEEE Trans. *Electron Devices* **66** 3254–64.
[7] Singhal P, Mazumdar P, Rattan, S 2017 One pot synthesis of free standing highly conductive polymer nanocomposite films: Towards rapid BTX vapor sensor. Polym Eng. Sci. **58** 2018, 1074-1081.
[8] Singhal P, Rattan S 2016 Swift Heavy Ion Irradiation as a Tool for Homogeneous Dispersion of Nanographite Platelets within the Polymer Matrices: Toward Tailoring the Properties of PEDOT:PSS/Nanographite Nanocomposites *J. Phys. Chem. B* **12** 3403–3413.
[9] Mazumdar P, Singhal P, Diwan, R K, Rattan S 2016 Poly(4-vinylpyridine)/nanographite Nanocomposites as Organic Vapour Sensors *AIP Conf. Proc.* **1724** 020068.
[10] Lee SW, Lee W, Hong Y, Yoon DS 2018 Recent advances in carbon material-basedNO2 gas sensors *Sensors Actuators, B Chem.* **255** 1788–804.
[11] Mukherjee A, Jaidev L R, Chatterjee K, Misra A 2019 Modified fermi level in strontium nanoparticles decorated reduced graphene oxide for wide concentration detection of nitrogen dioxide at room temperature *Mater. Res. Express* **6** 065611.
[12] Tjoa V, Jun W, Dravid V, Mhaisalkar S Mathews N 2011 Hybrid graphene-metal nanoparticle systems: electronic properties and gas interaction *J. Mater. Chem.* **21** 15593–9
[13] Gupta J, Singhal P, Rattan S 2020 Chitosan/Nanographiteplatlets (NGP)/Tungsten Trioxide (WO3) Nanocomposites for Visible Light Driven Photocatalytic Applications. In: Jain V.K., Rattan S., Verma A. (eds) Recent Trends in Materials and Devices. *Springer Proceedings in Physics* 256 Springer, Singapore.
[14] Saifi A, Singhal P, Rattan S 2020 MoS2/NGP hierarchical hybrid composites synthesized via in–situ microwave method, *AIP Conference Proceedings* 020147
[15] Shukla S, Chaudhary S, Umar A 2014 Tungsten oxide (WO3) nanoparticles as scaffold for the fabrication of hydrazine chemical sensor. *Sens Actuators B* **196** 231–237. https://doi.org/10.1016/j. snb.2014.02.016
[16] Tehrani F S, Hydrothermal Synthesis and characterization of WO3 nanostructures: Effect of reaction time 2020 *Mater. Res. Express* **7** 015911.
[17] Boruah P J, Khanikar R R, Bailung H 2020 Synthesis and Characterization of Oxygen Vacancy Induced Narrow Bandgap Tungsten Oxide (WO3−x) Nanoparticles by Plasma Discharge in Liquid and Its Photocatalytic Activity, *Plasma Chemistry and Plasma Processing*.
[18] Ramalingam V, Indrajit M P, Alagarsamy P, Bhalchandra A K 2016 Synergistically Enhanced Electrocatalytic Performance of an N-Doped Graphene Quantum Dot-Decorated 3D MoS2−Graphene Nanohybrid for Oxygen Reduction Reaction, *ACS Omega* **1** 971−980.
[19] Zhao P X, Tang Y, Mao J, Chen Y X, Song H, Wang J W, Song Y, Liang Y Q, Zhang X M, 2016 One-Dimensional MoS2-Decorated TiO2 nanotube gas sensors for efficient alcohol sensing, *Journal of Alloys and Compounds* **674** 252-258.

Fabrication and characterization of e-glass composites cylindrical tubes using three different methods.

K. Ganesh Kumar[1,2], Z.L.Lau[1], M.M.H. Megat Ahmad[2], M Y Yuhazri[3]

[1] Faculty of Engineering & Technology (FET), Multimedia University, Jalan Ayer Keroh Lama, 75450, Melaka, Malaysia.

[2]Department of Mechanical Engineering, Faculty of Engineering, National Defence University of Malaysia, Sungai Besi Camp, 57000, Kuala Lumpur, Malaysia.

[3]Faculty of Mechanical and Manufacturing Engineering Technology, Universiti Teknikal Malaysia Melaka, Hang Tuah Jaya, Durian Tunggal, 76100, Melaka, Malaysia.

ganesh.krishnan@mmu.edu.my

Abstract. The mechanical properties of the composites might have variations due to their manufacturing methods. In this paper, the glass fiber reinforced epoxy composite tubes are fabricated by using three different methods which are hands lay-up, vacuum bagging, and compression molding. The tensile properties such as the tensile strength, Young modulus, and the tensile response of thermal spiking were determined by carrying out the tensile testing. The test was conducted according to ASTM D3039 standard. The specimens were undergone heat treatment with 50°C, 100°C, and 150°C before undergoing the test. All the results were compared with the untreated specimen at room temperature. The untreated compression molding fabricated composite has obtained the highest tensile strength with a value of 273.518 MPa. The tensile strength of all types of specimens starts to reduce when the treatment temperature reaches to 150°C. The performance in terms of the specimen's tensile properties shows that the compression molding method is the best approach.

1. Introduction

Nowadays, the implication of fibrous reinforced polymeric (FRP) composites tubes has been applied for an extensive variety of applications due to their exceptional combination properties. The composites tubes are widely used in the field of sporting goods, aerospace, structural, construction, and automobiles [1,2,3]. However, the effects of thermal spiking temperatures are encountered in real conditions which have highly affected the mechanical properties of the components. Aerospace components are always subjected to the thermal spike by exhaust fumes and different ranges of temperatures in the earth's atmosphere. The changing temperature in the environment alters the mechanical behavior of FRP composites.

In many previous types of research, most of the researchers carry out the investigation on the failure mode of fiber composites polymer based on the interfacial adhesion effects as it will contribute to the stiffness and mechanical properties of the composites [4,5]. Kumar et al [6] have described that most of the aerospace composites failed under tensile loads. The effects of tensile behavior of composites will be affected due to the thermal spike of the changing temperature environment and this mechanical behavior is due to the interfacial adhesion effects of the composites [7]. Different composite fabrication methods will lead to the different mechanical properties of the composites which are mainly based on the interfacial adhesion effects of the composites as it controls the overall mechanical behavior of fiber-reinforced composites. Thus, an investigation about the tensile behavior of FRP based on the different manufacturing processes will be carried out which all of the specimens will go through the thermal spike to investigate the effects of interfacial adhesion on the components on the mechanical behavior of the composites tubes.

2. Materials and Experimental Procedure

The materials that had been selected to fabricate cylindrical composite tubes are epoxy and E-glass fiber supplied by Chemi-Bond from Selangor, Malaysia. The resin used in this project was Auto-Fix 1345 B hardener and Auto-Fix 1710 A epoxy. The epoxy and hardener are mixed by the ratio of 1:1 by weight as recommended by the supplier. Electric stirring was used in the mixing to reduced bubble formation. For E-glass fiber, it was cut in 200 mm in width and the length of the perimeter for the 50 mm inner diameter tube depending on the number of layers used. For the current study, composite with 1, 2, and 3 plies were investigated.

In the hand lay-up process, a 50 mm diameter steel mandrel was used. The mandrel was wrapped with PTFE plastic film and a release agent was applied. On the other hand, epoxy was applied to the E-glass fiber surface. The wetting of the glass fiber was rolled on the mandrel. At the same time, a steel roller was used to compress the fiber and it also helps to release bubbles form in between layers. After the fiber rolled on the mandrel, PTFE film again rolled on the mandrel against the fiber to produce a small compression force to hold the fiber in place. The composite was left to cure for 24 hours at room temperature. After curing, the composite was separated from the mandrel and cut to 100 mm in length as testing specimens

The bagging process starts with a 50mm outer diameter hollow cylindrical tube that was prepared as a mold as same as the hand layup fabrication process. A layer of wrapping paper was wrapped around the mold. After that, the coated laminate fabric was wrapped around the shape of the mold. Another layer of wrapping paper acts as the release fabric was wrapped around the mold to separate the bleeder layer laminate fabric. After that, the thick type of wrapping paper with its length longer than the mold was acting as the vacuum bag. The vacuum bag was then covered over the mold which acts as a membrane to protect the laminate and facilitate the air sucking process. The sealing tape was used to seal the vacuum bag to prevent the air from flowing in. The mold pipe was then connected from the vacuum bag to the resin trap to prevent the extra resins from flowing into the vacuum pump. Another flexible tube was connected to the vacuum pump to allow the air sucking process. The connecting point was also sealed with industrial clay and double side tape to prevent void which causes the air to flow in. After the set-up of the vacuum bagging process is done, the one-sixth horsepower vacuum pump was allowed to operate to suck the air out from the setup for about 45 minutes. The vacuum pump was then turned off and the laminate is cured for at least six hours.

Unlike hand layup and vacuum bagging fabrication processes, the fabrication process does not require the cylindrical tube as a mold. The mold for the compression molding was a true close-set molding with an inflatable flexible bladder. The mold was first designed and fabricated with the dimension of 320mm x 100mm x 100mm rectangular solid box with a hollow cylindrical shape in the middle. The mold was made with aluminum 6061. An inflatable bladder was used to allow the inflation to the desired consolidation pressure so that the prepreg can be fully inflated in the hollow cylindrical shape. The overall setup was shown in figure 1.

a) b) c)

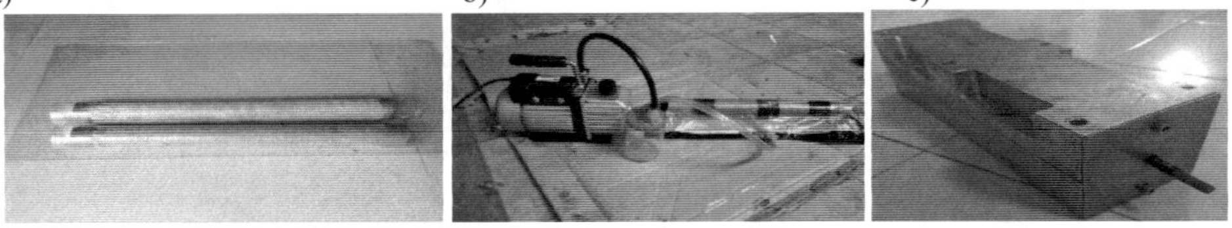

Figure 1. a) Hands layup process b) Vacuum bagging arrangement c) bladder assisted compression molding

3. Results and Discussion

3.1 Tensile properties

In this section, the comparison will be made based on different fabrication methods and the analysis will be carried out to determine the causes. The microscopic view of the void content will be shown in the later section. The strength of the mechanical properties of the composites will be highly dependent on the internal structure especially the void content between the matrix and fiber layer [8]. Through a simple observation on the diagram, it can be observed that the hand lay-up specimen had a very high void content area which was distributed along with the internal surface. The void content of the vacuum bagging specimen has a better void content distribution than the hand lay-up specimen while the compression molding specimen had a very low void content distribution. More details of the microscopic view with the use of an optical microscope will be shown in section 3.2.

The experiment data shows that different fabrication methods led to different mechanical properties of the specimen. The curing or treated temperature also brings some affection to the mechanical properties. The data comparison between the Hands lay-up (H1), Vacuum bagging (V1), and Compression molding (C1) specimen with its corresponding temperature (RT= Room Temperature), (50=50°C), (100=100°C), and (150=150°C) was shown in table 1 to table 3. Figure 2 shows the comparison of tensile stress versus strain curve with room temperature, treated at 50°C, 100°C, and 150°C temperature respectively.

Table 1. Mechanical properties of H1 composites

Specimens	Max Load (N)	Max Tensile Stress (MPa)	Max Strain	Young Modulus (GPa)
H1_RT	1510.300	180.874	0.0312	5.789
H1_50	1566.250	187.575	0.0325	5.732
H1_100	1707.525	204.494	0.0407	5.028
H1_150	1494.790	179.017	0.0362	4.944

Table 2. Mechanical properties of V1 composites

Specimens	Max Load (N)	Max Tensile Stress (MPa)	Max Strain	Young Modulus (GPa)
V1_RT	2004.920	266.789	0.0331	8.060
V1_50	2123.390	282.554	0.0358	7.898
V1_100	2397.577	319.039	0.0401	7.956
V1_150	2145.840	285.541	0.0326	8.759

Table 3. Mechanical properties of C1 composites

Specimens	Max Load (N)	Max Tensile Stress (MPa)	Max Strain	Young Modulus (GPa)
C1_RT	2055.490	273.518	0.0368	7.433
C1_50	2235.550	297.537	0.0369	8.063
C1_100	2523.730	335.826	0.0401	8.191
C1_150	2041.220	271.619	0.0399	6.811

From figure 2, it can be observed that the C1 specimen had the highest tensile stress, followed by the V1 and H1 specimens with the value of 273.518 MPa, 266.789MPa, and 180.874 MPa. The variation of the tensile stress between the 3 fabrication methods might be due to the void content [9] . The C1 specimen had the lowest void content percentage while comparing to the H1 and V1 specimens. The individual void that is present in the specimen will cause a high-stress concentration limited to a small neighbourhood of the void which will decrease the interlaminar stress between the matrix and fiber phase [10]. However, the void with a smaller area will not have much effect on the laminate. The laminate strength will depend on the size of the area over where it exists. From the observation, the H1 specimen has a very large void content thus the laminate strength between the matrix and fiber phase decrease, leading to the low tensile stress. Besides, the was only a small variation of the tensile stress between the V1 and C1 specimens. The void content area of V1 was slightly better than the H1 specimen but still higher than the C1 specimen.

From figure 2, it can be easily observed that the tensile stress of all types of specimens increases gradually with the 50°C and 100°C treated temperature. The improving effect of the heat treatment depends on the improvement in the fiber-matrix interfacial shear strength. When the specimens had undergone the heat treatment, the crystallinity increased significantly which cause the new arrangement

IOP Conf. Series: Materials Science and Engineering 1225 (2022) 012002 doi:10.1088/1757-899X/1225/1/012002

of the molecules that adjacent to the fiber to a more compact way, thus reduce the void content in the interface. The decreasing of void content in the interface further leads to the increasing of interfacial shear stress between the fiber and matrix phase. Thus, the H1, V1, and C1 specimens were able to withstand more force and thus increase the tensile strength.

However, the tensile stress started to drop when the treated temperature comes to 150°C as shown in figure 2. From the research of Jeng-Shyong Lin [11], he experimented on the effect of heat treatment on the glass fiber reinforced polypropylene. The results show that the polypropylene started to melt when the temperature comes to 120°C. Due to the similar results with the researcher, it might be concluded that the epoxy molecules start to melt between the 100°C and 150°C which the interfacial bonding between the interface was weakened. By observation, the tensile stress of the specimen with 150°C treated temperature was even lower than the specimen with untreated temperature.

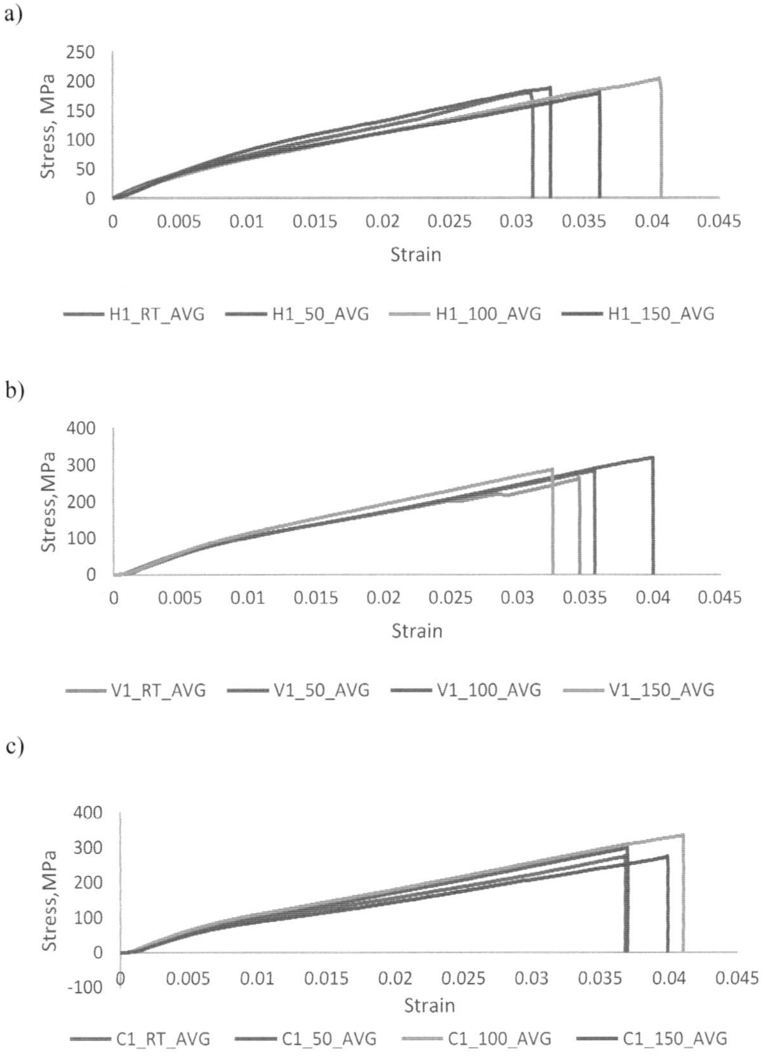

Figure 2. Average (AVG) comparison of tensile stress versus strain curve with room temperature, treated at 50°C, 100°C and 150°C temperature for a) H1 b)V1 and c) C1 respectively.

From figure 3, just focusing on 100°C it can be observed that the C1 specimen had the highest tensile strength, followed by V1 and H1. C1 is a manufacturing method that reflects that integrated highest pressure throughout the process compares to the other two processes. Pressure implied has led to a lower up amount of void in the material. The pressure inside the mold is to clear out and air gap and between the fiberglass and epoxy during the curing process. By doing this, the material is well compressed within the mold surface area which imparts greater strength. It is also evident in the structure of the C1 specimen which has less void percentage compares to the other two processes.

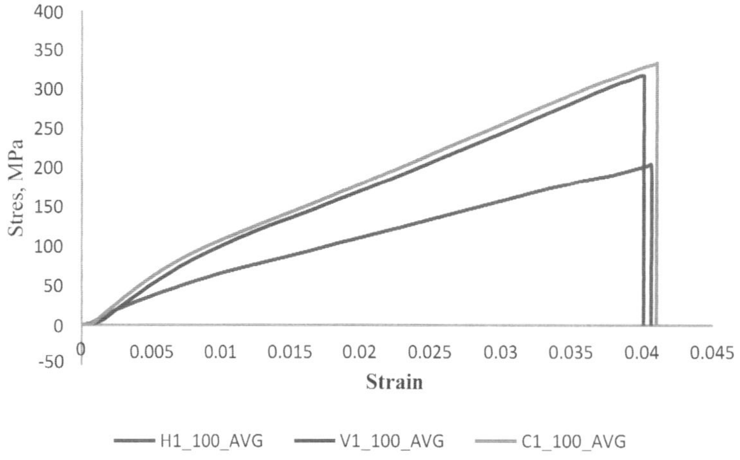

Figure 3. Average (AVG) comparison of tensile stress versus strain curve for H1, V1, and C1 at100°C

3.2 Microstructural

Using an optical microscope, the void content of each sample was analyzed under 40x magnification. The location of each measurement was chosen without discrimination and in approximately the same locations for each of the samples. The void content of the four pipes was measured using image analysis[12]. From figure 4, it can be observed that the H1 specimen had a large area of void content percentage on the interface compare to the other 2 specimens. The V1 specimen had a better void content percentage while the C1 specimen had only a few void contents on the interface. As mention, the void content had a significant effect on the interlaminar shear stress between the matrix and fiber interface[13]. The tensile strength of the composite was then affected by the interlaminar shear stress between the laminar. The results explained that why the C1 specimen has the highest tensile strength, followed by the V1 and H1.

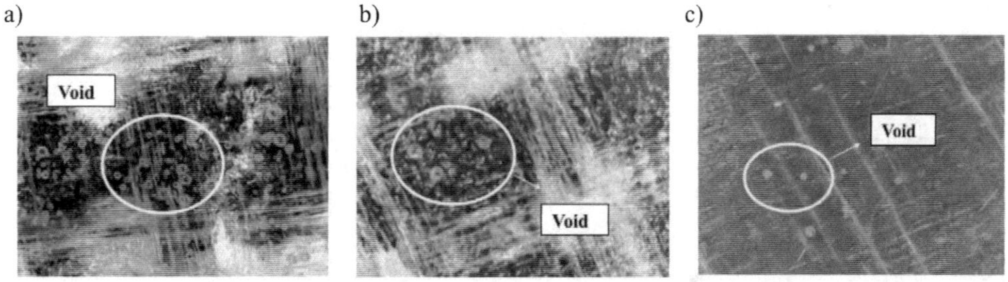

Figure 4. Void content in a) H1 b)V1 and c)C1 specimens

Figure 6 shows the surface failure of the V1 specimen. In figure 6 a) it can be observed that the fiber pull out might not only be due to the pulling force, it might also be due to misalignment of the fiber. The misalignment might be due to the error in the composite preparation process or the curing process. All the V1 specimens were failed under the fiber pull-out as it was the main reason for the tensile failure. From observation, the semitransparent area in figure 6 d) might be due to the molecules of matrix melting and cracking. It had the same condition as Figure 5 d) for the H1 specimen.

The microstructure of surface failure of the C1 specimen was shown in figure 7. The failure was similar to the H1 and V1 specimens. All the C1 specimen meet failure due to the fiber pull out. By observation, figure 7 c) shows a semitransparent area. By comparing Figure 5 d) and Figure 6 d), it might be the matrix melting area. It was observed that the matrix phase of the C1 specimen had started to melt at 100 degrees Celsius. From all the figures and data above, it can be agreed that the reduction of tensile strength of the composite under the treatment temperature was due to the melting of matrix molecules.

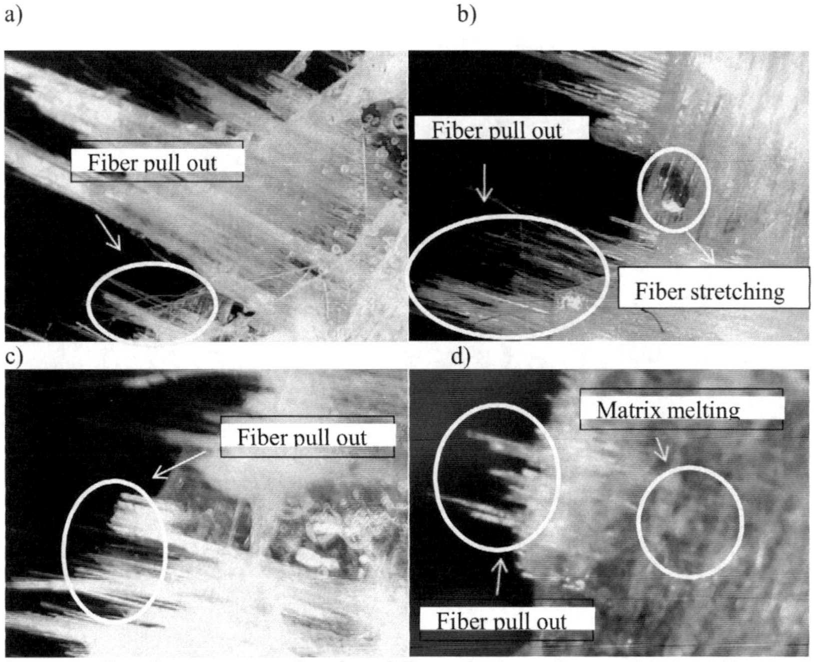

Figure 5. The microstructure of surface failure of H1 specimens; a) H1_RTb) H1_50, c) H1_100 and d) H1_150

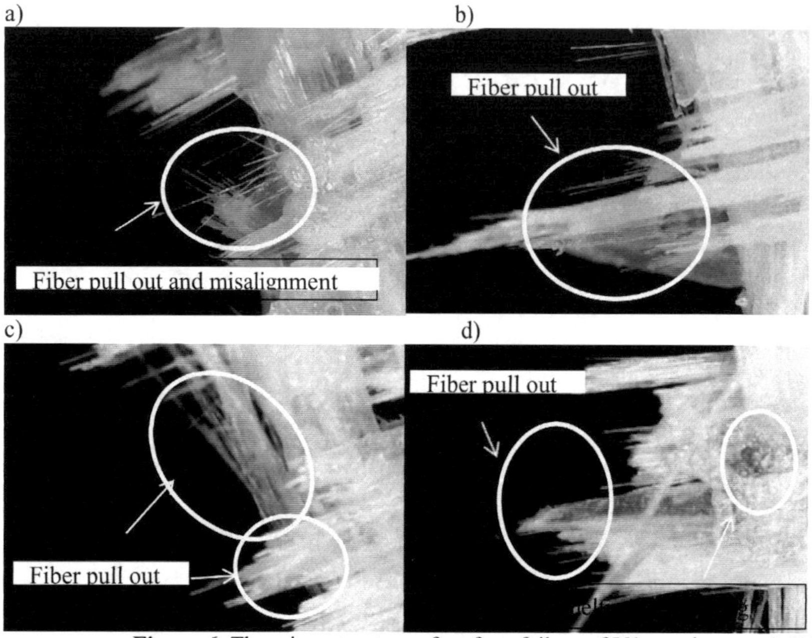

Figure 6. The microstructure of surface failure of V1 specimens; a) V1_RT, b) V1_50, c) V1_100 and d) V1_150

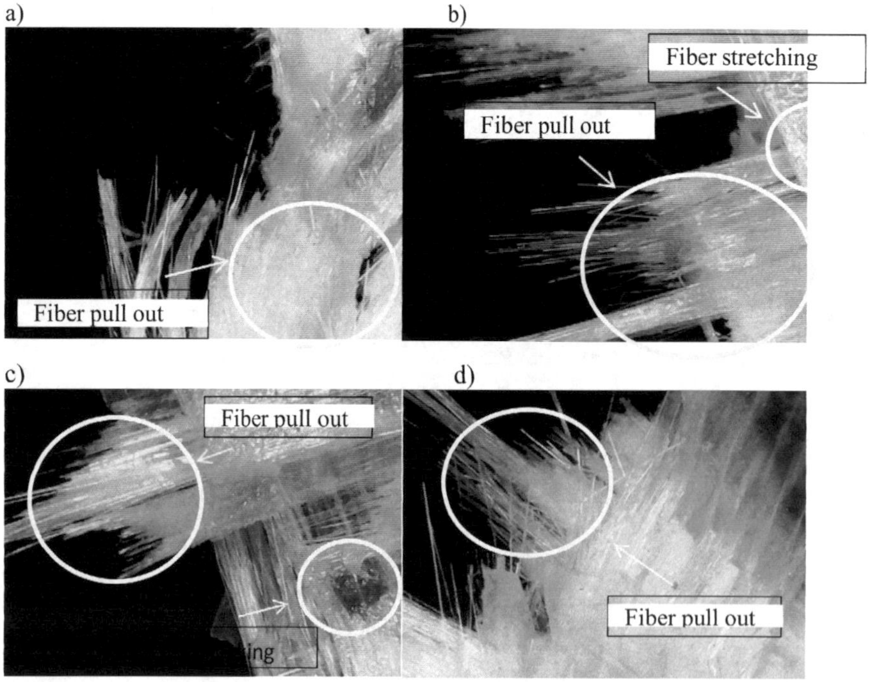

Figure 7. The microstructure of surface failure of C1 specimens; a) C1_RT,b) C1_50, c) C1_100 and d) C1_150

4. Conclusion

The glass fiber reinforced epoxy composites tubes were fabricated through three different manufacturing processes as Hands lay-up, Vacuum bagging, and Compression molding methods. The comparison of the composites showed that the C1 specimen had the highest tensile strength and young modulus among all other specimens while the H1 specimen had the lowest tensile properties. The observation showed that the C1 composites had the lowest void content percentage, followed by the V1 and H1 composites. The void content had a significant effect on the tensile strength of the composites as the void will reduce the interlaminar shear stress between matrix and fiber interface thus reducing the tensile properties of composites. The experimental data showed that the tensile strength of the composites increases generally and gradually with the increased temperature up to 100°C as the interlaminar shear stress or interface bonding force between the matrix and fiber layer improved. However, the tensile strength of the composites started to decrease when the treatment temperature come to 150°C.In conclusion, the overall performance of C1 composites was better than the V1 and H1 composites in terms of their mechanical properties and the tensile response towards the treatment temperature. The results show that the glass fiber reinforced composites had a strong tensile strength which was able to withstand a larger tensile load. Thus, the glass fiber reinforced composite was suitable for applications such as large as bridge hangers and smaller like a fishing rod in terms of their strong tensile strength and response towards the thermal spiking.

Acknowledgments
The authors acknowledge the assistance provided by Multimedia University (MMU) for financial and equipment aid.

References

[1] Babazadeh J , Rahmani K , Hashemi S J , and Sadooghi A , 2021 Effect of glass, carbon, andkevlar fibers on mechanical properties for polymeric composite tubes produced by a unidirectional winding method *Mater. Res. Express* **8** 45301
[2] Saffar A , Darvizeh A , Ansari R , Kazemi A , and Alitavoli M 2020 Damage analysis of fiber–metal laminate patches as a repair system for surface defects of steel pipelines: *Proc. Inst. Mech. Eng. Pt. L J. Mater. Des. Appl.* **235** 4 868–879
[3] Koli D K , Agnihotri G , and Purohit R , 2014 A Review on Properties, Behaviour and Processing Methods for Al- Nano Al2O3 Composites *Procedia Mater. Sci.* **6** 567–589
[4] Rahmani K , Wheatley G , Sadooghi A , Hashemi S J , and Babazadeh J 2021 The experimental analysis of creep and corrosion properties of polymeric tube reinforced by glass, carbon and Kevlar fibers *Mater. Res. Express* **8** 65307
[5] Wu W , Xie L , Jiang B , and Ziegmann G 2013 Simultaneous binding and toughening concept for textile reinforced pCBT composites: Manufacturing and flexural properties *Compos. Struct.* **105** 279-287
[6] Mahato K K , Dutta K , and Ray B C , 2018 Effect of Thermal Spike Conditioning on theTensile Behavior of Glass/Epoxy Composites *Mater. Today Proc.* **5** 12109–12114
[7] Wang W *et al* , 2020 Surface Modification of Flax Fibers with Isocyanate and Its Effects onFiber/Epoxy Interfacial Properties *Fibers Polym.* **21** 12 2888–2895
[8] Lekube B M , Hermann W , and Burgstaller C , 2020 Partially compacted polypropylene glassfiber non-woven composite: Influence of processing, porosity and fiber length on mechanical properties and modeling *Compos. Part A Appl. Sci. Manuf.* **135** 105939
[9] Mehdikhani M , Gorbatikh L , Verpoest I , and Lomov S V , 2019 Voids in fiber-reinforced polymer composites: A review on their formation, characteristics, and effects on mechanical performance *J. Compos. Mater.* **53** 12 1579–1669
[10] Zhang A Y , Li D H , Zhang D X , Lu H B , Xiao H Y , and Jia J , 2011 Qualitative separation of the effect of voids on the static mechanical properties of hygrothermally

conditioned carbon/epoxy composites *Express Polym. Lett.* **5** 8 708–716

[11] Lin J S , 2003 Effect of Heat Treatment on the Tensile Strength of Glass Fibre Reinforced Polypropylene: **11** 369–381

[12] Harris W , Soutis C , Gresil M , and Atkin C , 2020 Pressure response and life assessment of filament-wound composite pipes after impact *Int. J. Light. Mater. Manuf.* **3** 365–375

[13] Van De Werken N , Tekinalp H , Khanbolouki P , Ozcan S , Williams A , and Tehrani M ,2019 Additively manufactured carbon fiber-reinforced composites: State of the art and perspective *Additive Manufacturing* 31:100962

[14] Viel Q , Esposito A , Saiter J M , Santulli C and Turner J A 2018 Interfacial characterization by pull-out test of bamboo fibers embedded in Poly(lactic acid) *Fibers* **6** 1

Effect of ply orientation, staking sequence and notch geometry on the failure behaviour of glass fibre-reinforced epoxy composite laminates: A numerical investigation

Krishna Prasad G. R. and Faizan Mohammad Rashid

Department of Mechanical Engineering, Birla Institute of Technology and Science (BITS-Pilani), Pilani Campus, Jhunjhunu,Rajasthan, India. Pin Code: 333031

E-mail: faizan.rashid@pilani.bits-pilani.ac.in, faizan.iitm@gmail.com

Abstract. A detailed numerical failure analysis was conducted for unidirectional glass fibre-reinforced epoxy composite laminates having an elliptical cut-out at the centre using Hashin's damage model. The major axis, *a*, of the elliptical cut-out was fixed while the minor axis, *b*, was varied in such a way, that its upper extreme represented a situation of the plate with a circular hole, while the lower extreme of it represented the presence of a sharp crack in the structure. For angle ply cases, plies oriented along the material direction predicted an ever-increasing load replicating an elastic analysis without any damage accumulation. Such exceptionally higher stiffness and resistance to deformation were much anticipated, as the uni-axial tensile load was applied along fibre direction which shares a large portion of the load in comparison to the matrix material. While the plies oriented perpendicular to the direction of the material behaved as the weakest stack-up sequence in resisting uni-axial tensile load. Multi-angle hybrid composites predicted uneven failure behaviour. For the different combinations of multi-angle layups and minor axis, b, different trends for peak loads were observed. Interestingly, extension-shear coupling was observed for a few configurations which were marginally negligible for others. The variation of opening stress along the transverse direction predicts some very interesting observations. For the higher value of *b*, the opening stress decreases as we moved away from the irregularity. However, the situation of a plate with a sharp crack ($b \rightarrow 0$) showed the effect of crack tip blunting. Near the vicinity of the sharp crack, the opening stress first dropped, then picked up, and again kept on decreasing as we move away. The drop in the stress near the vicinity of the sharp crack is attributed to the blunting of the crack tip.

1. Introduction

Owing to their excellent weight to strength ratio, physical, chemical and mechanical properties fibre reinforced hybrid composite materials are used extensively in engineering components such as aircraft, automobiles, and wind turbines. Irregularities such as notches, voids, and cracks are inevitable in the structures. Industrial application of composite materials requires drilling of holes or grooves on it for fixation and connectivity ([1], [2], [3]) which act as favourable sites for crack initiation and propagation. A century ago, people started studies addressing the issues of stress localization in the vicinity of the hole. Forchet [4] being the front runner in this field studied stress concentration factors for holes, grooves, and fillets under different loading (pure tension, compression, and bending) using the photo-elasticity method and determined the maximum value of stress experienced by the body under elastic conditions. The impact of shape and size of notches and loading conditions on the failure of composite materials has been the theme of various research activities conducted in the recent past ([5], [6]). Composite laminates with an elliptical hole at the centre were also investigated ([7], [8]).

The objective of the article: Effect of ply orientation and stacking sequence on the failure of glass fibre-reinforced epoxy composite plats having elliptical cut-out of varying minor axis is the objective of the article in hand. Such studies have been carried by various researchers ([9], [10]).

2. Hashing damage criteria

Due to involved anisotropy, the mechanical behavior and failure of composite materials are difficult to understand as compared to homogenous and isotropic materials ([11], [12]). Various failure criterions have been developed to address failure behavior in composite materials. They are broadly classified into two types: failure criterion non-associated with failure modes and failure criterion associated with failure modes. The non-associated failure criterion is not concerned with the actual failure mechanism, instead, they focus on the gross macroscopic failure events. They are phenomenological models based on curve fitting. Tsai-Wu and Tsai-Hill are commonly used first-generation failure criteria for predicting failure of orthotropic materials ([13], [14], [15]). Though these criteria could predict failure of material point they don't account for the influence of strength and stiffness degradation. Hence, they are non-associated with failure modes. The associated failure criterion known as the second generation failure criteria can capture progressive material damage by the degradation of stiffness and strength after material failure. Hashin's damage criteria are the associated failure criterion ([16], [17]). In recent times extended finite element method has also been used for modelling damage imitation and propagation in composite structures.

3. Geometry, loading and boundary condition

A thin composite plate of uniform thickness of height, 'h = 25 mm', width 'w = 25 mm' and thickness, 't = 1 mm' has been modelled in Abaqus 6.20 [18] as a 2-dimensional plane stress continuum shell element. Different elliptical cut-out of major axis fixed as a = 5 mm and varying minor axis, b = 0, 1.25, 2.5, 5 mm has been considered at the centre of the plate, as shown in Figure 1(a). With a fixed major axis, a = 5 mm, the upper limit of the minor axis, b = 5 mm replicate plate with a circular cut-out, while the lower limit b = 0 mm introduces a sharp crack in the plate. The geometry has meshed in such a way that, enough elements are present near the cut-put to effectively capture the damage as shown in Figure 1(b). One end of the composite plate is completely restrained against all the rotational and translational degrees of freedom while a uni-axial displacement controlled monotonic load is applied on the other edge.

4. Material properties

Declaration for unavailability of in-house test data and future test plan: Respecting the time and resource constraint in the current affiliate organization, the orthotropic property of the unidirectional glass fibre-reinforced epoxy composite material has been adapted from the literature [19] as listed in Table. 1. Meanwhile we have started building a sophisticated composite test facility. In the subsequent manuscript, we would be able to show in-house test results. All the orthotropic elastic properties of the composite plate were assumed in the principal material directions and defined as '*Lamina Properties*' in Abaqus 6.20 [18].

5. Stacking sequences for composite plate

The composite plate has been modelled with eight plies of glass uni-directional fibre-reinforced epoxy layers. Each ply of ply thickness t = 0.125 mm was considered making the total thickness of the composite plate 1 mm. The fibre's orientations angle, θ, measured in anticlockwise direction is the angle between the longitudinal direction, L, and the principal material directions, F, of the composite plate as shown in Figure 2(a)-(b).

IOP Conf. Series: Materials Science and Engineering 1225 (2022) 012003 doi:10.1088/1757-899X/1225/1/012003

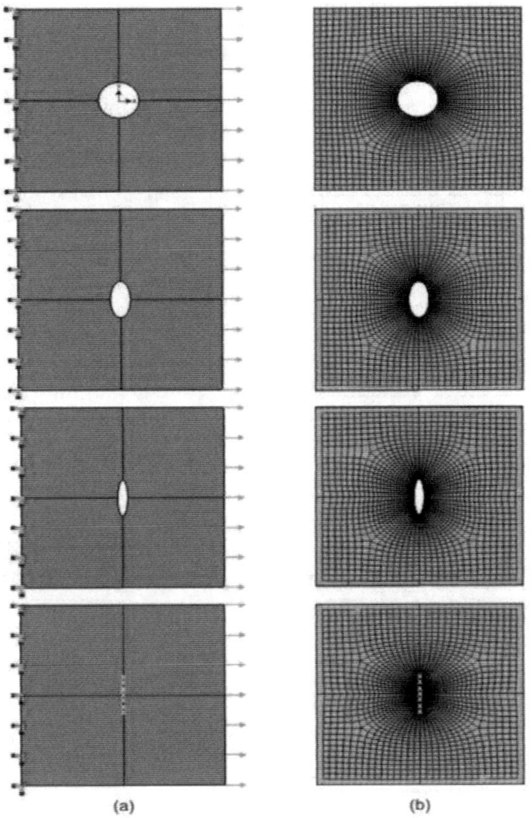

Figure 1. Composite plate with a various elliptical cut-out (a) Load and boundary condition (b) Finite element mesh.

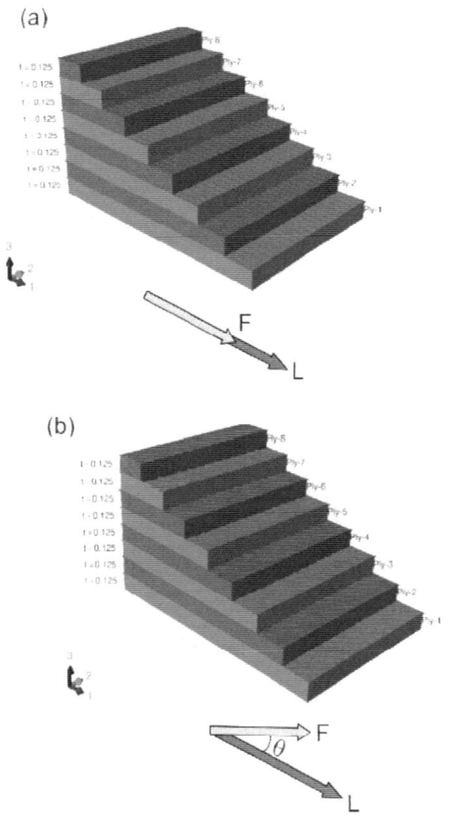

Figure 2. Stacking sequences for angle ply laminates.

Table 1. Material properties of glass fibre-reinforced epoxy laminate.

Mechanical Property	Value	Unit
E_1	55000	MPA
E_2	9500	MPA
$G_{12} = G_{13}$	5500	MPA
v_{12}	0.33	MPA
X^T	2500	MPA
X^C	2000	MPA
Y^T	50	MPA
Y^C	150	MPA
$S12$	50	MPA
$G_{ft,c}$	12.5	N/mm
G	12.5	N/mm
G	1.0	N/mm
G	1.0	N/mm

6. Damage stabilization

The composite plate has been modelled with eight plies of glass uni-directional fibre-reinforced epoxy layers. Each ply of ply thickness t = 0.125 mm was considered making the total thickness of the composite plate 1 mm. The fibre's orientations angle, θ, measured in anticlockwise direction is the angle between the longitudinal direction, L, and the principal material directions, F, of the composite plate as shown in Figure 2(a)-(b).

To avoid the convergence issue which is inevitable in highly non-linear and non-smooth numerical failure simulation, viscous damage stabilization was invoked. For a = 5 mm and b = 5 mm, replicating the situation of the plate with a circular hole, a parametric study was carried out for θ = 45° angle ply laminate to look into the impact of damage stabilization factor on the macroscopic force-deflection response as shown in Figure 3. A smaller value of the damagestabilization factor predicted higher peak load at the cost of computational time. '*Viscosity* = 0.001' sec. was considered as the damage stabilization factor for all the cases as reported by Lapczyk and Hurtado [19].

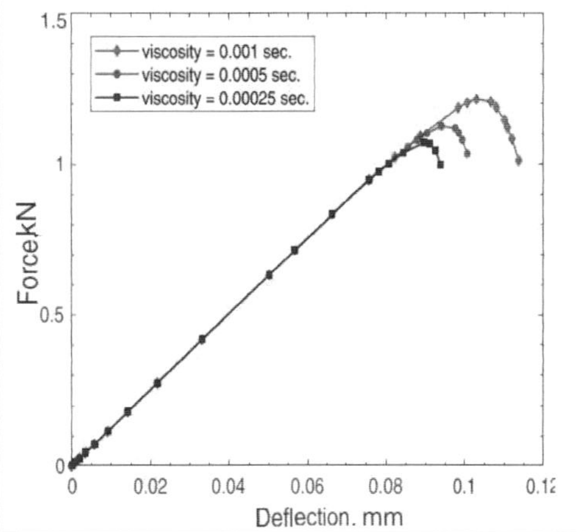

Figure 3. Effect of viscosity on the macroscopic force-deflection response of θ = 45° angle plylaminate.

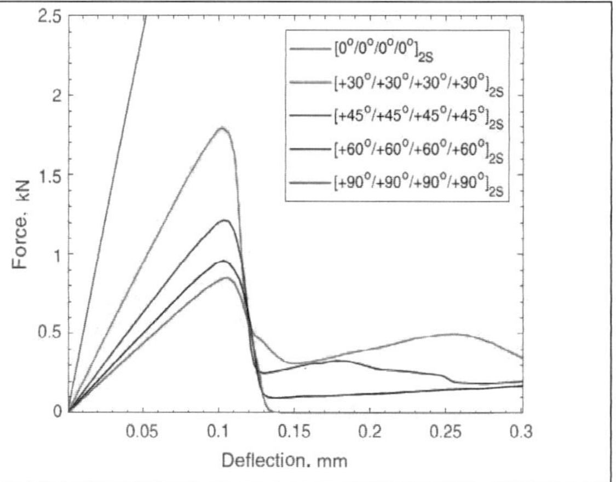

Figure 4. Macroscopic force-deflection for different angle ply orientations.

7. Results and discussion

Angle ply laminate analysis

For a = 5 mm and b = 5 mm, replicating the situation of the plate with a circular hole, different angle ply situations such as θ = 0°, 30°, 45°, 60°, 90° to the principal material the direction was considered as shown in Table. 2

Table 2. Stacking sequences for angle ply analysis.

Case	Stacking Sequences
1	$[0°/0°/0°/0°]_{2S}$
2	$[+30°/ +30°/ +30°/ +30°]_{2S}$
3	$[+45°/ +45°/ +45°/ +45°]_{2S}$
4	$[+60°/ +60°/ +60°/ +60°]_{2S}$
5	$[+90°/ +90°/ +90°/ +90°]_{2S}$

For all the angle ply cases, the macroscopic force-deflection response was linear until the damage was initiated. Plies oriented as $[0°/0°/0°/0°]_{2S}$, predicted an ever-increasing load replicating an elastic analysis without any damage accumulation as shown in Figure 4. Such an exceptionally higher stiffness and resistance to deformation were much anticipated, as the uni-axial tensile load was applied along fibre direction which shares a large portion of the load in comparison to the matrix material. On the other hand, plies oriented as $[+90^o/+90^o/+90^o/+90^o]_{2S}$, behaved as the weakest stack-up sequence in resisting uni-axial tensile load. Intermediate ply orientations, $[+30^o/+30^o/+30^o/+30^o]_{2S}$, $[+45^o/+45^o/+45^o/+45^o]_{2S}$ and $[+60^o/+60^o/+60^o/+60^o]_{2S}$ showed damage accumulation lying in between the two extreme cases of $[0^o/0^o/0^o/0^o]_{2S}$ and $[+90^o/+90^o/+90^o/+90^o]_{2S}$. The preliminary study demonstrated that structures having more plies aligned along the fibre direction will have a better resistance of the structure against deformation.

Multi-angle ply laminate analysis

In actual situations, engineering structures are acted by complex loading conditions. Hence, the plies are arranged in different orientations in the actual structure. In the second phase of our study, a numerical failure study was conducted on the multi-angle, $[0^o/+\theta/-\theta/+90^o]_{2S}$ composite laminate layups as listed in Table. 3.

Table 3. Stacking sequences for multi-angle ply analysis.

Case	Stacking Sequences
1	$[0^o/+30^o/-30^o/+90^o]_{2S}$
2	$[0^o/+45^o/-45^o/+90^o]_{2S}$
5	$[0^o/+60^o/-60^o/+90^o]_{2S}$

Case 1: For, b = 0 mm and b = 1.25 mm, the lay-up $[0^o/+30^o/-30^o/+90^o]_{2S}$ predicted almost the same peak load. However, there was a stiff fall from the peak load for b = 2.5 mm and b = 5 mm as shown in Figure 5(a). Though the fibre damage in tension was identical for all the values of b the matrix damage in tension was maximum observed for b = 0 mm as shown in Figure 7(a). No extension-shearcoupling was observed for the lay-up $[0^o/+30^o/-30^o/+90^o]_{2S}$ and any combination of minor axis, b.

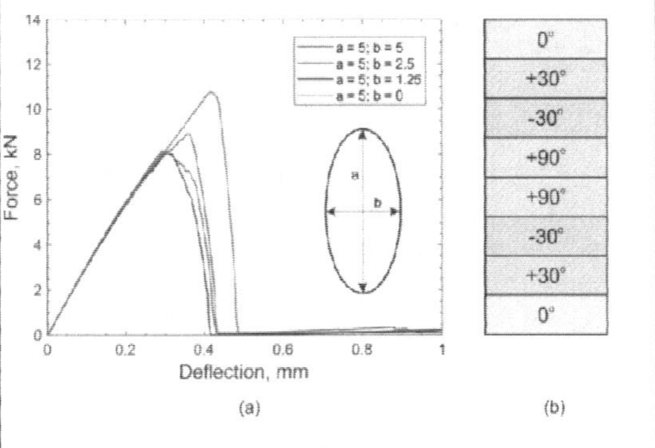

Figure 5. Macroscopic force-deflection response for $[0^o/+30^o/-30^o/+90^o]_{2S}$ layup analysis.

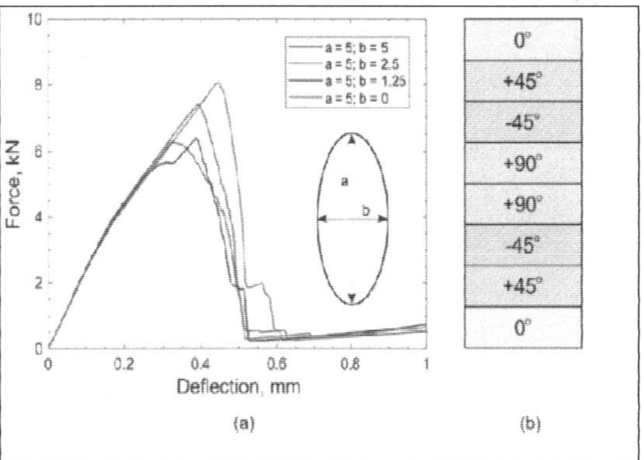

Figure 6. Macroscopic force-deflection response for $[0^o/+45^o/-45^o/+90^o]_{2S}$ layup analysis.

Case 2: As compared to $[0^o/+30^o/-30^o/+90^o]_{2S}$, the next lay-up $[0^o/+45^o/-45^o/+90^o]_{2S}$ predicted entirely different trend. The peak load was much higher for b = 0 mm than for b = 1.25 mm and b = 5 mm as shown in Figure 6(a). Fibre damage in tension was unevenly distributed for all the values of b. However, the matrix damage in tension was widespread for b = 5 mm, but it was much localized for b = 0, 1.25, 2.5 mm as shown in Figure 8. There were very minimal extension-shear coupling phenomenain this configuration.

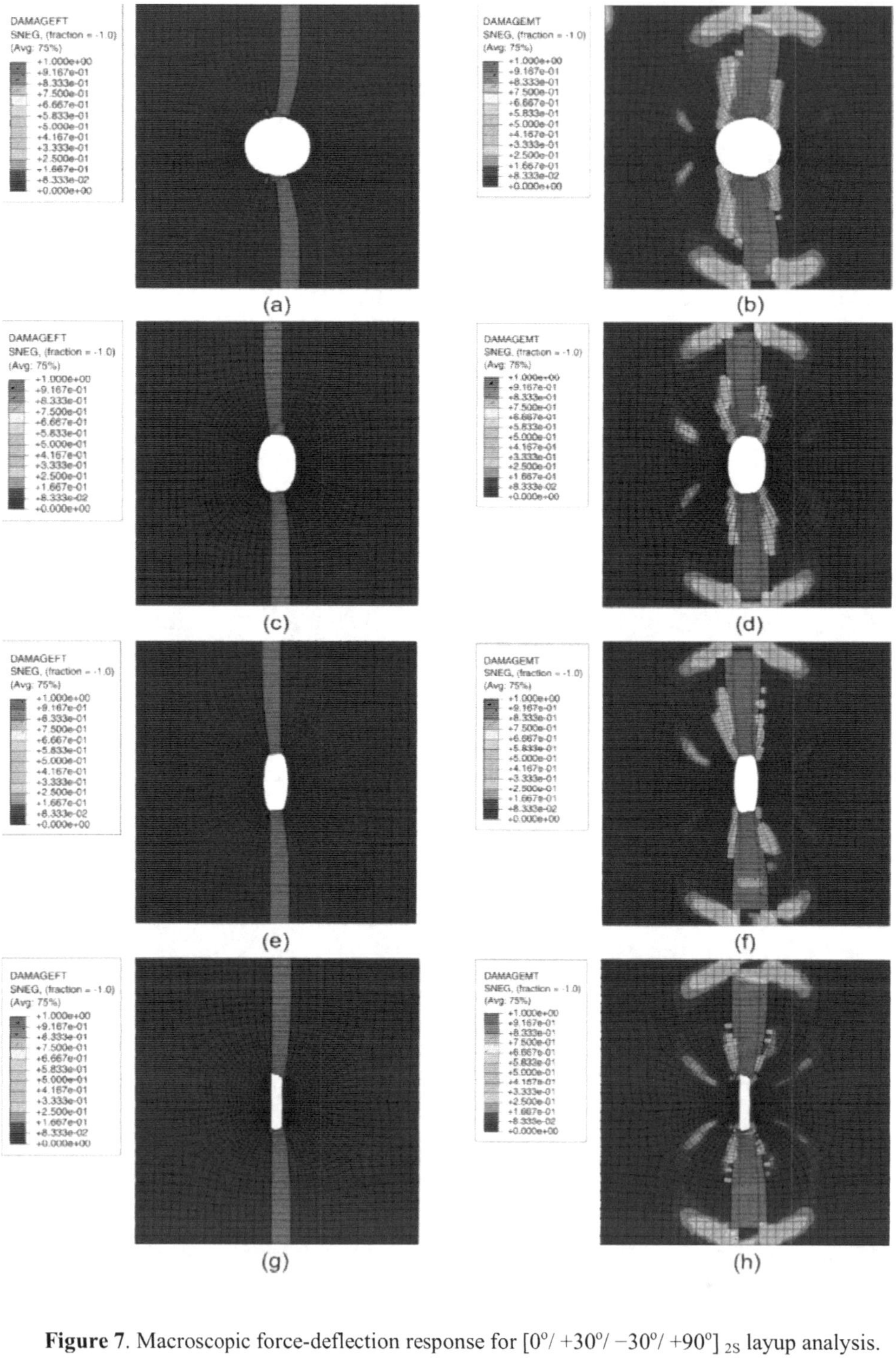

Figure 7. Macroscopic force-deflection response for [0°/ +30°/ −30°/ +90°] $_{2S}$ layup analysis.

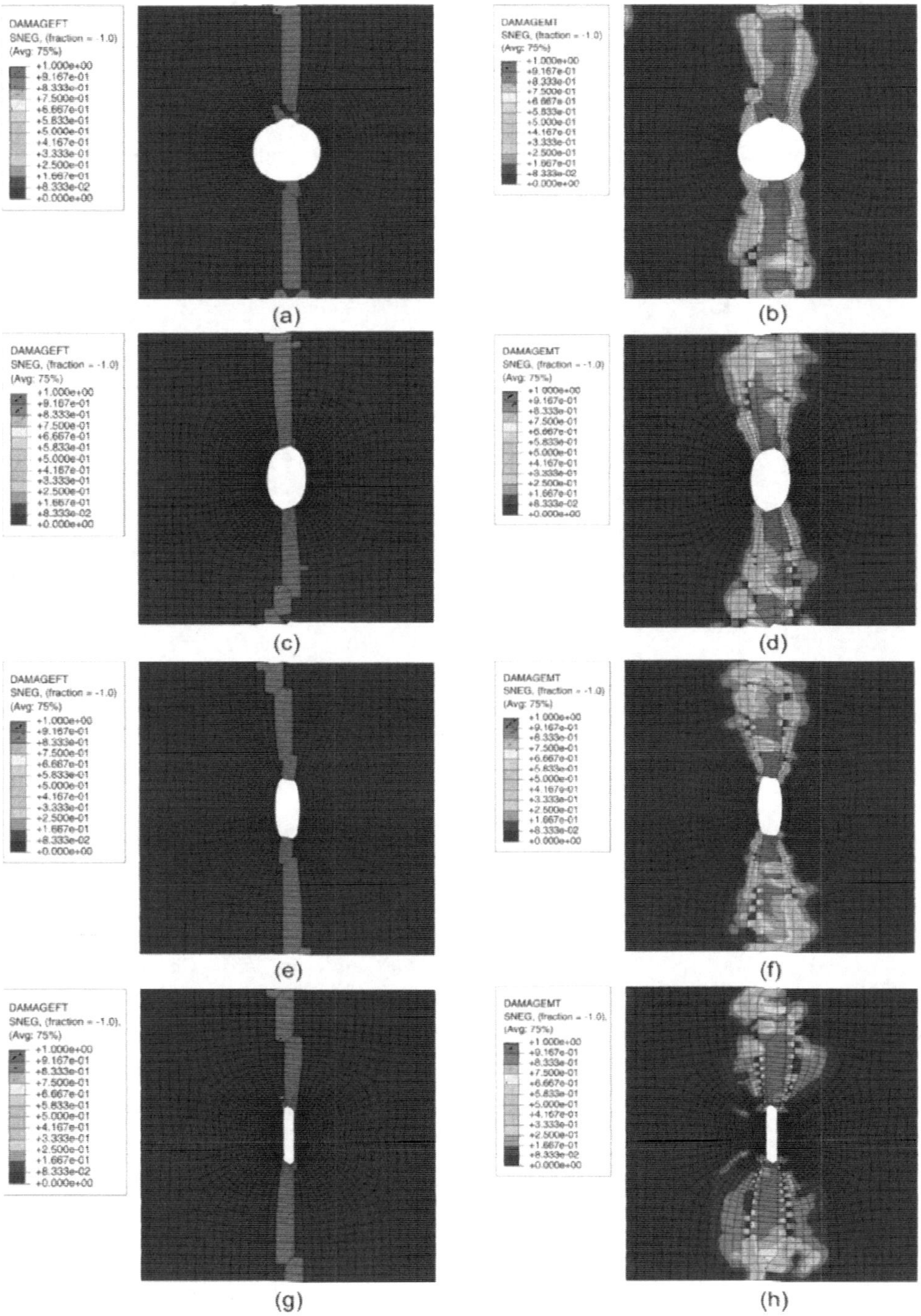

Figure 8. Contour of fibre and matrix damage for $[0^{o}/+45^{o}/-45^{o}/+90^{o}]_{2S}$ layup analysis.

Case 3: As compared to [0°/ +30°/ −30°/ +90°] $_{2S}$, the next lay-up [0°/ +45°/ −45°/ +90°] $_{2S}$ predicted entirely different trend. The peak load was much higher for b = 0 mm than for b = 1.25 mm and b = 5 mm as shown in Figure 6(a). Fibre damage in tension was unevenly distributed for all the values of b. However, the matrix damage in tension was widespread for b = 5 mm, but it was much localized for b = 0, 1.25, 2.5 mm as shown in Figure 8. There were very minimal extension-shear coupling phenomenain this configuration.

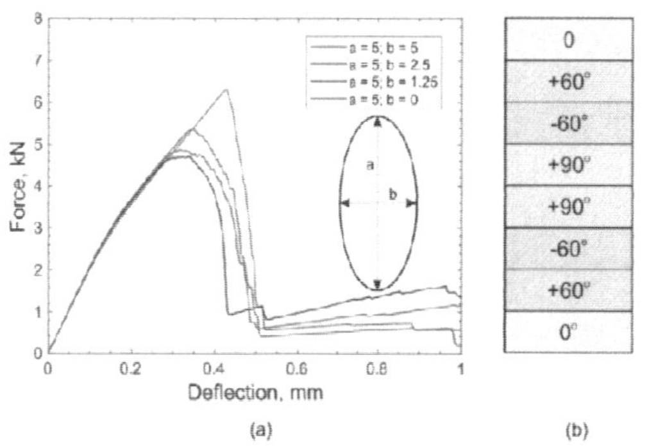

Figure 9. Macroscopic force-deflection response for [0°/ +60°/ −60°/ +90°] $_{2S}$ layup analysis.

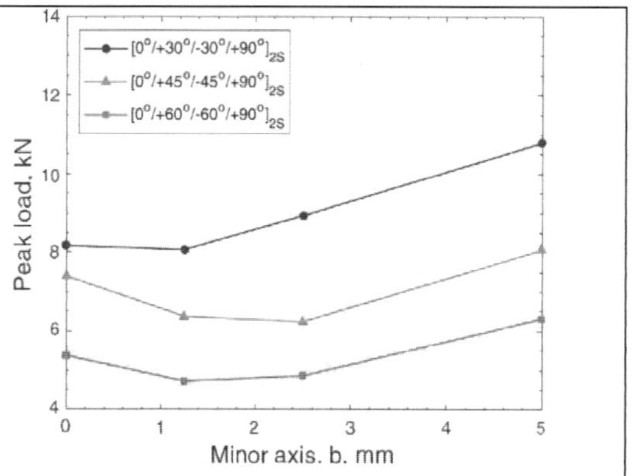

Figure 10. Peak load vs minor axis, *b*.

The numerical failure study of the composite notched plate predicted increasing peak load with increasing b for the stacking sequence, [0°/ +30°/ −30°/ +90°] $_{2S}$, as shown in Figure 10. However, the stacking sequences [0°/ +45°/ −45°/ +90°] $_{2S}$ and [0°/ +60°/ −60°/ +90°] $_{2S}$ showed a distinct trend. With increasing b first the peak load decreased and then increased. This trend was more prominent in the stacking sequences [0°/ +45°/ −45°/ +90°] $_{2S}$ than [0°/ +60°/ −60°/ +90°] $_{2S}$.

The variation of opening stress, σ_{11}, along the transverse direction predicts very interesting behaviour. For a = 5 mm and b = 5 mm replicates the situation of the plate with a circular hole. Opening stress, σ11 increases as the stacking sequence changes from [0°/ +30°/ −30°/ +90°] $_{2S}$ to [0°/ +60°/ −60°/ +90°] $_{2S}$ as shown in Figure 12. For a = 5 mm and b = 2.5 mm, a similar trend was observed. However, there was a stiff fall in the values of opening stress, σ_{11}, as we moved away from point, A, asshown in Figure 13. The situation a = 5 mm and b = 1.25 mm predicted almost same trend for [0°/ +30°/−30°/ +90°] $_{2S}$ and [0°/ +45°/ −45°/ +90°] $_{2S}$ as shown in Figure 14. Finally, the combination a = 5 mm and$b$ = 0 mm replicating the situation of a plate with a sharp crack predicted a different behaviour. Near the vicinity of the sharp crack, the opening stress, σ_{11}, first dropped, then picked up and again kept on decreasing as we move away from point, A as shown in Figure 15. The drop in the stress near the vicinityof the sharp crack is attributed to the blunting of the crack.

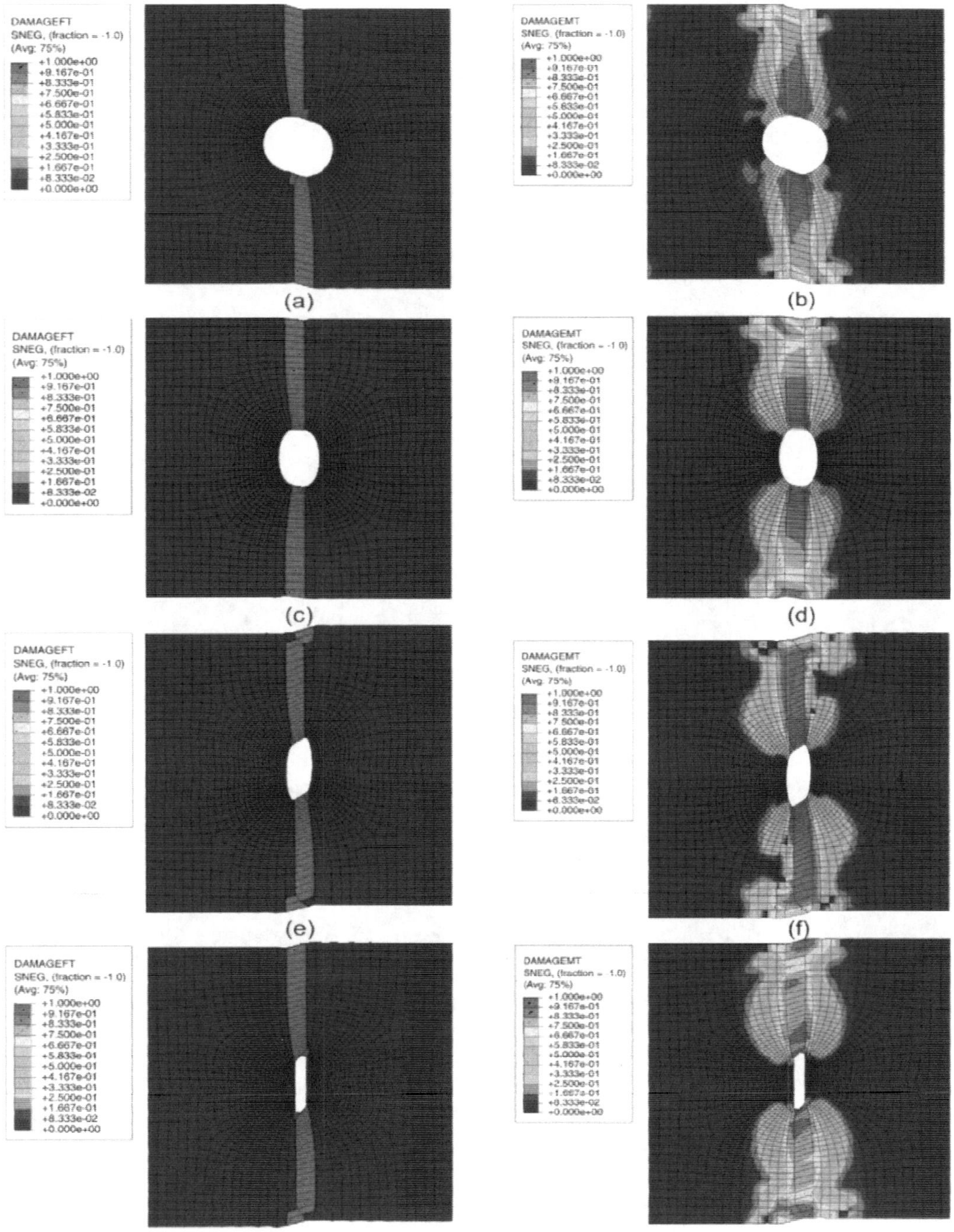

Figure 11. Contour of fibre and matrix damage for $[0^{\circ}/ +60^{\circ}/ -60^{\circ}/ +90^{\circ}]_{2S}$ layup analysis.

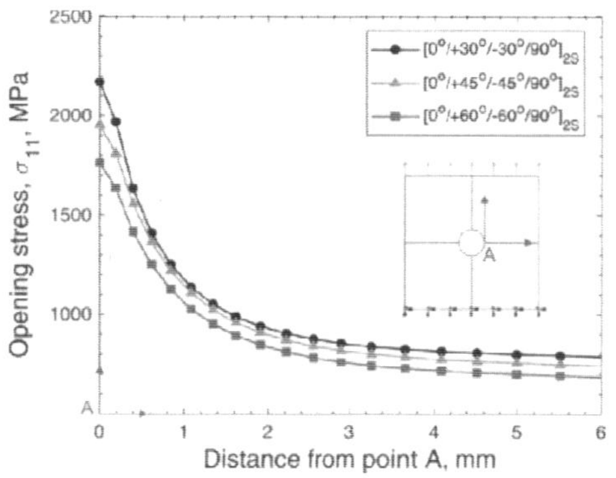

Figure 12. For a = 5 mm, b = 5 mm: opening stress, σ_{11}, at the instance of peak load.

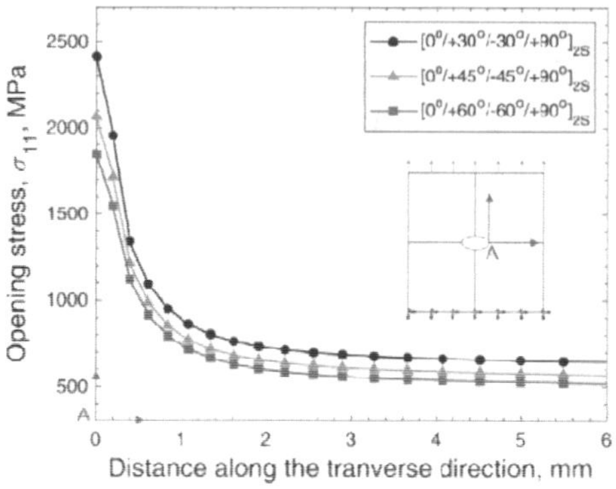

Figure 13. For a = 5 mm, b = 2.5 mm: opening stress, σ11, at the instance of peak load

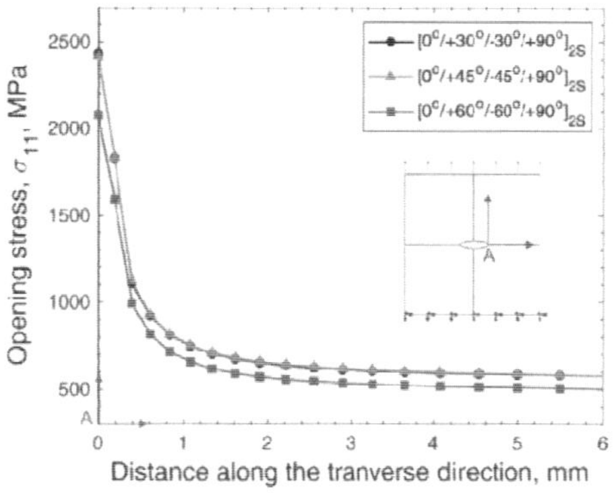

Figure 14. For a = 5 mm, b = 1.25 mm: opening stress, σ_{11}, at the instance of peak load.

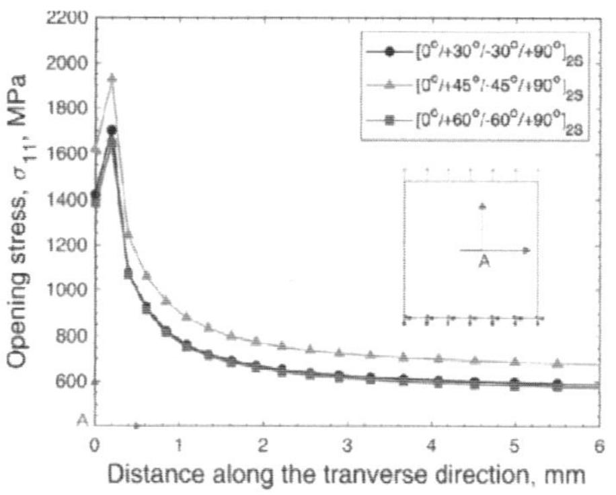

Figure 15. For a = 5 mm, b = 0 mm: opening stress, σ_{11}, at the instance of peak load.

8. Conclusions

In the present work, Hashin's damage model has been used in modelling the failure behaviour of unidirectional glass fibre-reinforced epoxy composite laminates having an elliptical cut-out at the centre. Following are the inferences that can be drawn from the study:

- For angle ply cases, plies oriented along the material direction predicted an ever-increasing load replicating an elastic analysis without any damage accumulation. Such exceptionally higher stiffness and resistance to deformation were much anticipated, as the uni-axial tensile load was applied along fibre direction which shares a large portion of the load in comparison to the matrix material.

- For angle ply cases, plies oriented perpendicular to the direction of the material behaved as the weakest stack-up sequence in resisting uni-axial tensile load.

- For multi-angle composites, different fibre orientation and stacking sequences predicted very uneven failure behaviour.

- For different combinations of multi-angle layups and minor axis, *b*, different trends for peak loads were observed.

- Interestingly, extension-shear coupling was observed for a few configurations which were marginally negligible for others. The variation of opening stress along the transverse direction predicts some very interesting observations.

- For the higher value of *b*, the opening stress along the transverse direction decreases as we moved away from the notch.

- Composite plate with a sharp crack predicted and showed the effect of crack tip blunting. Near the vicinity of the sharp crack, the opening stress first dropped, then picked up, and again kept on decreasing as we move away. The drop in the stress near the vicinity of the sharp crack is attributed to the blunting of the crack tip.

9. Future Work

Though, the engineering components are subjected to multi-axial loading and experience a triaxial state of stress. However, for ease of understanding, the materials are mainly characterized experimentally in idealized uni-axial tests. A detailed experimental and numerical study has been planned for the future, where complex loading conditions and the effect of an aggressive the environment will be considered.

Data availability

Dr Faizan Mohammad Rashid, Assistant Professor, Dept. of Mechanical Engineering, Birla Institute of Technology and Science (BITS-Pilani) holds the propriety right of the numerical data generated in the above findings. It may be shared based on request.

CRediT author statement

Conceptualization, Investigation, Original draft preparation, Reviewing and **Editing**: Faizan Mohammad Rashid; **Methodology, Validation, and Data analysis**: Krishna P.G.R and Faizan Mohammad Rashid.

Conflicts of interest

The authors declare no conflict of interest.

Funding

Present research activity was carried out as one of the research activities under the ongoing broad project titled 'Numerical failure analysis of structures at different stress-state and subjected to different strain-rate using the finite element methods' supervised by Dr Faizan Mohammad Rashid as Principal Investigator. The project is funded by the Birla Institute of Technology and Science, Pilani (BITS Pilani), Pilani Campus, Rajasthan, India via the Researcher Initiation Grant scheme (Serial No: 146).

Acknowledgements

The corresponding author, Dr Faizan Mohammad Rashid, would like to thank his wife Mrs. Tahsin Rakhshan, daughter Zaina Faizan and son Ammar Mohammad Rashid for their support and encouragement while preparing the manuscript

References

[1] Zecheru G, Dumitrescu A, Draghici G, Yukhymets P 2016 Reinforcement effects obtained by applying composite material sleeves to repair transmission pipelines *Materiale Plastice* **53** 252-258.

[2] Majumder P, Pandey K M, Deshpande N V 2021 Numerical investigation to study effect of number of blade in propeller loading with composite material *Journal of* The Institution of Engineers (India) Series C **102** 741-751.

[3] Budhe S, Banea M D, de Barros S 2020 Analysis of failure pressure of defective pipes repaired with compositesystems considering the plastic deformation of pipe *Journal of* The Institution of Engineers (India) Series C **101** 929-936.

[4] Forchet MM 1936 Photoelastic studies in stress concentration *Mechanical Engineering* 58 485.

[5] Khechai A, Tati A, Beleadri M Q, Guettala A 2019 Numerical analysis of stress concentration in isotropic and laminated plates with inclined elliptical Holes *Journal of The Institution of Engineers (India) Series C*, **100** 511-522.

[6] Voicu R, S Vinila, Vilag V, Mihalache R 2016 CAD Modelling Light weight composite centrifugal rotor manufacturing for energy efficiency *Materiale Plastice* **53** 623-625.

[7] Xiwu X, Liangxin S, Xuqi F 1995 Stress concentration of finite composite laminates with elliptical hole *ComputersStructures* **57** 29-34.

[8] Srivastava V K 2002 Notched strength prediction of laminated composite under tensile loading *Materials Science and Engineering A* **328** 302–309.

[9] Mandal A, Ray C, Haldar S 2020 Experimental and numerical free vibration analysis of laminated composite plates with arbitrary cut-outs *Journal of The Institution of Engineers (India) Series C* **101** 281-289.

[10] Sutar M, Nayak A 2020 Composites and effects of discontinuities on its performance A Review *Journalof The Institution of Engineers (India) Series C* **101** 203-217.

[11] Stanescu M M, Bolcu D, Manea I, Ciuca I, Bayer M, Semenescu A 2009 Experimental researches concerning the properties of composite materials with random distribution of reinforcement *Materiale Plastice* **46** 179-183.

[12] Iliesc M, Spanu P, Nutu E, Mihon L 2009 Experimental and theoretical studies on mechanical characteristicsof an important composite *Materiale Plastice* **46** 62-66.

[13] Tsai SW 1965 Strength Characteristics of Composite Materials, First Edition. *Philco Corporation Newport Beach* CA USA.

[14] Tsai SW, Wu E M 1971 A general theory of strength for anisotropic materials *Journal of Composite Materials* **5** 58-80.

[15] Jagannatha Guptha V L J, Sharma R S 2019 Experimental studies on strength behaviour of notched glass/epoxylaminated composites under uni-axial and bi-axial loading *Journal of The Institution of Engineers (India) Series C* **100** 75-81.

[16] Hashin Z, Ritem A 1973 A fatigue failure criterion for fiber reinforced materials *Journal of Composite Materials* **7** 7448-7464.

[17] Hashin Z 1980 Failure criteria for unidirectional fiber composites. *Journal of Applied Mechanics* **47** 329-334

[18] ABAQUS v 6.20 user's manual and Dassault Systems Inc. 2020.

[19] Lapczyk I, Hurtado J A 2007 Progressive damage modeling in fiber-reinforcedmaterials *Composites Part A* **38** 2333-2341.

A review on transition metal oxides based nanocomposites, their synthesis techniques, different morphologies and potential applications

S Yadav[1], N Rani [2] and K Saini[3*]

[1, 2, 3] Department of Chemistry, Miranda House, University of Delhi, New Delhi-110007, India

*Corresponding Author mailing ID: kalawati.saini@mirandahouse.ac.in

Abstract. In the field of nanotechnology and nanoscience, transition metal oxides based nanocomposites (TMONCs) are promising for various application uses such as Supercapacitors, Sensors, Bactericidal properties, Photocatalytic Degradation, Solar Cells etc. Modification of transition metal oxide nanoparticles (TMONPs) to TMONCs by doping/mixing of another transition metal and metal oxide, carbon based nanoparticles, conducting polymers etc. to achieve enhanced surface area, increasing surface activities or number of active surface sites, reducing electron- hole recombination, increasing charge transfer processes etc. have been reported in literature. These improved properties are the possible reason for the enhancement in its practical applications efficiencies. This review summarizes recent development on transition metal oxides based nanocomposites for different potential applications. Also synthesis methods of transition metal oxide based nanocomposites have obtained an increasing attractions to achieve cost effectiveness and environment friendly routes of synthesis with high rate of production, high yield of product and also less toxic waste production. Transition metal oxides nanocomposites have been fabricated by various methods such as Microwave assisted synthesis technique, Sol-Gel method, Biosynthesis method, Co-precipitation process, Simple Chemical method etc. Different morphologies of transition metal oxides based nanocomposites have been summarized in this review article.

Herein, this paper discuss about several reported synthesis techniques, various characterization techniques used for structural and surface properties identifications, different morphologies and various potential applications of transition metal oxide based nanocomposites.

Keywords: Nanocomposites; Bioynthesis; Co-precipitation; Sol-Gel method; Supercapacitors; Sensors; Photocatalytic Degradation

1. Introduction

A field of applied science which outlines fabrication, characterization and application of materials within at the minimum one dimension less than 100nm is referred to nanotechnology. Herein, a particle size within 1-100nm range displays drastically changes in chemical properties and physical properties contrast to corresponding bulk materials that bring about many opportunities as well as challenges. Nanoscience and Nanotechnology is a multidisciplinary field which integrate chemistry, physics, material science and biology etc. The use of nanoscience and nanotechnology for biomedical, sensing, energy storage, energy generation, water treatment and other applications have potential changed the landscape of many issues of environment, energy, health etc [1-6].

Nanocomposites are the composites with at the minimum one of the dimensions is less than 100nm. Nanocomposites fascinated the attention of scientists worldwide by their unique design possibilities and multifunctional properties. Nanocomposites are the materials of the era due to their multifunctional capabilities. Researchers all over the world are working on maturing of new approach of synthesis of nanocomposites to achieve nanocomposites with unique properties. Research is focused on improvement in effectiveness, efficiency, durability of practical application of nanocomposites by mixing suitable constituents for the synthesis of particular nanocomposites [7-9].

Transition metal oxide nanoparticles turned out to be widely explored for many applications such as supercapacitors, sensors, solar cell and Photocatalytic applications etc [10-14]. To improve their efficiency further research have been reported modification of these transition metal oxides mixed/doping with other nanoparticles such as carbon based nanoparticles, metal oxide nanoparticles, conducting polymers etc, it might be useful for new applications in different field of sciences such as chemistry, physics, biology, material science etc. Modification of transition metal oxide based nanoparticles achieve new sets of properties such as enhance surface area, enhanced surface activities, enhanced porosity, decrease in electron hole recombination, increase in conductivities etc [13,24,25,140,141]. In recent studies, transition metal oxide based nanocomposites are attracted considerable attention as versatile and favorable material in almost all fields of sciences for their wide range of practical applications. Many transition metal oxides based nanocomposites are reported in literature for their useful vital practical applications such as Carbon nanotube - ZnO nanocomposite reported as electrode material for supercapacitor escorted by maximum calculated specific capacitance of 323.9 Fg^{-1} [15], Synthesized CuO-ZnO nanocomposite were reported as antibacterial agent [16], ZnO/CuO nanocomposites was reported with humidity sensor property [17]. And many other such as GO/ZnO nanocomposite for phtocatalytic degradation of basic fuchsin dye [18], Polyaniline/ZnO nanocomposite as sensor [19], Graphene – ZnO nanocomposite as Photovoltaic cells [20] etc were reported in last few years.

In this Review we are going to study brief about synthesis techniques of TMONCs, such as co-precipitation technique, hydrothermal technique, sol-gel technique and wet chemical technique. Also we are going to study different characterization techniques with have been used in literature for structural and surface properties identifications of TMONCs. Morphologies have been proven as an important parameter for nanomaterials study; we are going to reviewed some of the reported morphologies of TMONCs. This article also reviewed various practical applications of various transition metal oxide based nanocomposites in the field of health, sensors, energy storage/generation and nature remediation.

2. Several synthesis techniques of TMONCs

The most common nanocomposites currently implemented are transition metal based nanocomposites, Transition Metal Oxide based nanocomposites (TMONCs), Carbon based nanocomposites and Polymer based nanocomposites. Among these, Transition metal oxide based nanocomposites are most interested due to their unique properties. As nanomaterials shows various sized based properties, the synthesis of nanomaterials has been considered as the preference field in the nano-science. Method of synthesis governs morphologies, physical properties and chemical properties of synthesized nanomaterials which further govern various potential applications of these nanomaterials. Researcher are focused to introduce new method of synthesis which are cost efficient, fast, environment friendly and products show good practical application with high efficiency. This article summarizes the potentially used synthesis method of transition metal oxide based nanocomposites. Currently, the developed method are like Hydrothermal method, Sol – Gel method and Chemical vapor deposition. Now, we discuss these methods one by one in brief.

2.1. Co-precipitation technique

This method is a very important technique to synthesize the nanomaterials. It have been explored that fabrication of nanomaterials in solution chemistry is preferred over solid and vapor phase for its properties such as short diffusion paths, more purities in products, less agglomeration, it simplicity, uniform and controllable particle size and cost effective [21-25]. Hadi Eslami et al. has reported aeration co-precipitation technique for mesoporous Iron-Manganese bi-metal oxide nanocomposite fabrication. Where, molar ratio of 2:1 of Manganese sulfate and ferric chloride were separately prepared and mixed at 70°C with stirring at 50rpm. 10% NH_4OH was added to get pH of 9 and kept at 50rpm for 4 hours to get precipitated nanocomposites [26]. In a similar manner $NiFeO_4$ nanocomposite was fabricated by chemical precipitation by using 1:1 molar ratio of $FeSO_4.7H_2O$ and $NiSO_4.7H_2O$ at pH= 12. Herein, hydrazine hydrate was acting as a reducing agent [27].

Many practically applicable transition metal oxide based nanocomposites were reported in literature with co-precipitation route of fabrication such as CO_2 hydrogenation to methanol was reported by Thongthai witoon on Cu/ZnO [29], Zno-CuO nanocomposite had show antibacterial properties [25], Graphene oxide / Fe_3O_4 nanocomposite used for waste water treatment was synthesized by shengyan Pu with In situ co-precipitation method [29] etc. Also several type of transition metal oxide combination were also reported in several years such as $CuWO_4$, $CuWO_4$ / NiO [21], NiO-ZnO [24], CeO_2.CuO.ZnO [30], CeO_2/ZnO/$ZnAl_2O_4$ ternary nanocomposite [23], ZnO/Ag [22] and many more. This shows that co-precipitation method is one of the effective, easy and vital techniques for fabrication of transition metal based nanocomposites.

2.2. Sol–Gel method

Physical method like (ball milling, laser ablation, sputtering) and chemical methods like (sol-gel method, co-precipitation method, chemical vapor deposition) are the general method of synthesis for transition metal oxide based nanocomposites. Where, sol-gel method is another fascinating type of wet chemical methods. Sol-Gel method is consist of a route where, formation of a Sol followed by formation of a Gel and then drying of the solvent. Various applications such as catalysis, sensing, antibacterial etc. are widely depends on size, morphology, dimensions, composition and structure of nanomaterial. Sol-Gel technique of synthesis is vital technique, because it is easy to operation, low temperature reaction conditions, defined size, high purity and homogeneity of product [31-34].

Ardiansyah Taufik et al. was able to fabricate ternary CuO/TiO_2/ZnO nanocomposites by simple sol-gel method where, that synthesized transition metal oxide based nanocomposite enhanced Photocatalytic degradation of organic dye (methylene blue (MB)) [35]. Degradation of MB organic dye was also reported by NiO/TiO_2 nanocomposite which also prepared by sol-gel method [36]. Alexis Lavin et al. had prepaired ZnO/CuO nanomaterial by sol-gel technique dissolved and dispersed of calculated amount of zinc nitrate and copper oxide in the solvent, afterword the PVA was add on to that previous solution. And followed by stirring with in ultrasound bath at 353K to get homogeneous gel. That gel was dried at 773K for 8 hours to get required nanocomposite [37]. TiO_4-Ag nanocomposite was reported with antibacterial activity against E. coli and was also fabricated by sol-gel route of synthesis [38]. Literature shows sol-gel synthesis technique is a flexible approach to obtaining a diverse range of transition metal oxide based nanocomposites such as CNTs/TiO_2 [39], WO_3/TiO_2 [40], ZnO/Reduced grapheme oxide [41], ZnO-SiO_2 [42], CdO-ZnO [43], $CoTiO_3$/$CoFe_2O_4$ [44] and ZnO-SnO_2 [45].

2.3. Chemical vapor deposition

Nowadays, the researchers have shown a big attraction in refining and developing efficient technique for fabrication of TMONCs. One such method is chemical vapor deposition. This method has matured significantly over last few years. Chemical vapor deposition is deposition of different precursors onto a substrate [46-49].

Literature is full of transition metal oxide synthesized by chemical vapor deposition. Jian–Shan Ye et al. prepared a supercapacitor electrode material of Aligned carbon nanotube-Ruthenium Oxide by chemical vapor deposition [50]. Davide Barreca et al reported ZnO-TiO_2 synthesized by same technique and shows application as gas sensor. Where, TiO_2 nanoparticles were dispersed on ZnO nanoplatelets [51]. Aerosol Assisted chemical vapor deposition of Slippery Liquid Infused porous TiO_2/SnO_2 nanocomposite was reported by Frances L. Heale et al [52]. Chemical vapor deposition is a vital route of synthesis of TMONCs. As Lei Ma et al. fabricated Fe-Ni doped TiO_2 and used this nanocomposite to synthesized CNTs/Fe-Ni/ TiO_2 by fluidized bed chemical vapor deposition [53]. One-step synthesis of nanocomposite MWCNT/ZnO also reported in literature by taking co-chemical vapor deposition route [54].

2.4. Hydrothermal technique

Several strategies were tested for synthesizing transition metal oxide based nanocomposites. Hydrothermal method was extensively utilized in the fabrication of transition metal oxide based nanocomposites. Hydrothermal method is easy, uncomplicated, fast and cost-effective in contrast with other sophisticated synthesis routes. Also products are formed directly from the solution with control size, shape, and composition. Even calcinations and milling is not required in many cases of hydrothermal synthesis. Hydrothermal technique of synthesis takes place at above ambient temperature and pressure. Where, a sealed solution is heated [55-59].

Yupeng Gao et al. synthesized TiO_2/Ti_3C_2 nanocomposite by hydrothermal method. Where, he prepared titanium sulfate solution in deionized water and afterward added 500mg Ti_3C_2 to that solution. That mixture was stirred to get homogeneous solution and kept in Teflon autoclave. That autoclave was kept in oven at 180°C for 18h. TiO_2/Ti_3C_2 then separated, washed and dried in vacuum at 80°C for 10 hours. The fabricated TiO_2/Ti_3C_2 nanocomposite was able to degrade organic dye (Methyl orange (MO)) photocatalytically [60]. Similar manner using Teflon-lined autoclave Zhang et al. reported synthesis of $ZnFe_2O_4$ nanocomposites. Certain amount of $ZnSO_4.7H_2O$, $FeCl_3.6H_2O$, and sodium dodecyl sulfate were added and dissolved in deionized water. After that, 10ml of NaOH was also added, after stirring solution was poured to Teflon lined autoclave followed by heating for 15h at certain temperature. Transition metal oxide based nanocomposite then washed and dried [61]. Herein, we reviewed hydrothermal synthesis of some transition metal oxide based nanocomposites. Aijian Wang et al prepared TiO_2-reduced grapheme oxide nanocomposites by facile hydrothermal method [62]. Hydrothermal synthesis of $CoFe_2O_4$/graphene nanocomposites by using graphene oxide and aqueous precursors solution of $Fe(NO_3)_3.9H_2O$ and $Co(NO_3)_2.6H_2O$ [63]. From the literature, there have been reported synthesis of number of TMONCs by simple hydrothermal method such as Carbon nanotube/Cubic Fe_3O_4 [64], α-Fe_2O_3-ZnO [65], MnO_2-CNT [66], TiO_2/reduced graphene oxide [67] and CuO-ZnO [68].

2.5. Wet Chemical synthesis technique

Wet chemical synthesis technique of fabrication of TMONCs is very simple, cost effective and reproducible. It also has some disadvantages of producing chances of having defects within the crystal because of low temperature reaction conditions than other methods [69].

Different researchers have been prepared transition metal oxide based nanocomposites by simple wet chemical method using different precursors. L. Zgura et al prepared ZnO-CdS nanocomposites by two step procedure where in first step he was synthesized CdS nanoparticle using Na_2S and $Cd(NO_3)_2$.$4H_2O$ in 1:1 molar ratio at room temperature with 30min stirring. After that, in second step this synthesized CdS nanoparticle were dispersed in water and $Zn(NO_3)_2$ and NaOH were added drop wised, then this whole solution has been stirred for 30min at room temperature. Then that obtained powder form product was dried at 100ºC for 2 hours to get ZnO-CdS nanocomposite [70]. Alejandra Mazabuel-Collazos et al synthesized ZnO-TiO_2 nanocomposite by using Zinc acetate dehydrate and titanium terabutoxide as precursors [71]. Wet chemical method is very useful and easy. Many transition metal oxide based nanocomposites are reported in literature. As, CdO-MgO-Fe_2O_3 multi metal based nanocomposite was synthesized by simple wet chemical method which was showing photo-catalytic properties [72]. Sharif Hussein Sharif Zein et al reported wet chemical synthesis of manganese oxide/carbon nanotube nanocomposite [73]. SnO_2/Fe_2O_3 nanocomposites also synthesized by wet chemical method [74]. Wet chemical method synthesized nanocomposites with different practical applications such as, High quality MgO:CuO nanocomposite were obtained by wet chemical method was reported with complete catalytic degradation of Rhodamine B within 15 minutes and 91% removal of 4-chlorophenol in 300 minutes [76]. Tongtong Jiang et al prepared S doped Co_3O_4 nanosheets/reduced graphene oxide by wet chemical method to achieve a dye sensitized solar cell [77].

2.6. Microwave assisted Method

Process of absorption and transformation of electromagnetic energy into heat energy is possesses heating from microwave radiation. Microwave radiations have been use in many organic and inorganic reactions. Fabrication of TMONCs by microwave assisted method is a greener and environment friendly route of synthesis. Microwave assisted heating for the fabrication of TMONCs have several advantages like faster the reaction, less energy required, controllable and uniform particle size compared to conventional heating [78-83].

K. Karthik et al synthesized CdO-NiO-ZnO using simple Microwave assisted synthesis for removal of MB dye photocatalytically and for antibacterial activity. They were mixed Cadmium acetate dihydrate solution and Nickel acetate tetrahydrate solution and kept for stirring for 30 minutes at 303K. After that $Zn(CH_3COO)_2.2H_2O$ and NaOH solutions were added to aforementioned solution and kept for stirring for 1hour at 303K and irradiated for 30 minutes in a microwave oven in the convection mode. After washing with ethanol and double distilled water, they once more kept that powder for irradiation for 15 minutes. Synthesized powder was calcinated at 673K for 2 hours [84]. K. Karthik et al have also synthesized CdO-ZnO TMONCs by simple microwave assisted synthesis technique for bactericidal activity. He took certain ratio of $ZnCl_2$ and $CdCl_2.2H_2O$ and dissolved them in water. After stirring for 15min he added ammonia solution drop wise to maintain pH 8. Then that solution was irradiated in microwave oven for 20minutes. After washing precipitate again irradiated with microwave for another 10 minutes and then that powder was kept in 773K for 4 hours and was to get CdO-ZnO nanocomposite [85].

Many TMONCs have been synthesized by microwave assisted synthesis technique for their different potential uses, for example Pingtao Dou et al. were able to fabricated Ag/ZnO/graphene for Photocatalytic activity [86], Li et al prepared Ag/ZnO-TiO_2 for Photocatalytic degradation of Rhodamine B [87], Rajesh Kumar et al reported Mn_3O_4-Fe_2O_3/Fe_3O_4/rGO nanocomposites for supercapacitor application [88].

2.7. Biological Method

The well recognized classical chemical methods of nanomaterials synthesis have their own disadvantages as, using hazardous reagents, mostly high temperature reaction conditions, chemical waste production and other. Wherein, biological method of nanomaterials synthesis plants and microorganisms are using a capping agents. Therefore this method is environmental friendly, non-toxic and simple method of synthesis of nanomaterials. Nowadays, the biological method of synthesis of transition oxide based nanocomposites is an interesting issue of nanoscience. Among both organisms and plants, the plants are easy available, easy to handle and required no special conditions that why they seem to be best candidates for biological synthesis of transition metal oxide based nanocomposites [89-92]. Different parts of plants such as roots, bark, leaves, fruits etc can be used for the biological method of transition metal oxide based nanocomposites synthesis, some of the examples are given below.

Qiujie Liu et al synthesized Fe,Cu oxide nanocomposite by using *Loquat* leaf extract through biological synthesis method [93]. Ramin Mohammadi-Alouchech et al have been prepared ZnO/CuO nanocomposite by dissolving certain amounts of zinc nitrate and copper acetate in deionized water, after that some amount of *Mentha longifolia* leaf extract was added followed by NaOH to achieve pH of 10. That solution was then irradiated for 10 minutes in the microwave oven to get ZnO/CuO nanocomposite with antibacterial applications [94]. Elias E. Elemike et al also fabricated Cu_2O/CuO-ZnO nanocomposites with Anticancer application by simply using *Alchornea cordifolia* leaf extract [95]. Yoki Yulizar also reported synthesis of TiO_2 nanoparticle using *Averrhoa bilimbi* fruits extract, Au nanoparticle using *Pandanus amaryllifolius* leaves extract and by using these already synthesized TiO_2 nanoparticles he was able to synthesized Au/TiO_2 nanocomposite with help of $HAuCl_4$ and *Pandanus amaryllifolius* leaves extract [96].

Many plant extract mediated synthesized nanocomposites have been reported in literature such as Ag-ZnO nanocomposite synthesized by using leaf extract of *Trigonella foenum-graecum* [97], Fe_2O_3-Ag TMONCs by using leaf extract of *Psidium guajava* [98], ZnO-NiO nanocomposite by using *Azadirachta indica* (Neem) leaves extract [92], $ZnFe_2O_4$ nanocomposite by using *Moringa Oleifera* extract [91], ZnO/CuO nanocomposite prepared by using seed bark extract of *Theohroma cacao* [89] and more. Figure1. Shows the general mechanisms involved in synthesis of nanomaterial by plant extract.

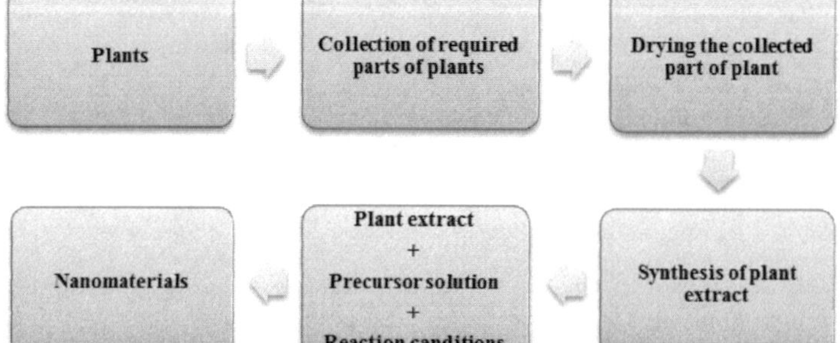

Figure 1. Simplified block diagram of general mechanisms involved in synthesis of nanomaterials by plant extract using above literature survey [89-96].

2.8. Ball-milling method

Ball-milling is a physical or mechanical method of transition metal oxide based nanocomposite synthesis. In this process reduction of particle size to get more fine particles by applying impact force and frication by blending or crushing them. Also size of product and time of synthesis highly depend on milling energy applied during milling. Ball-milling synthesis route is an important type of top-down approach of nanomaterials fabrications. Ball-milling technique of synthesis of transition metal oxide based nanocomposite has its own advantages such as environmental friendly, cost effective, fine particle size etc [99-102].

Henry Kahimbi et al. synthesized NiO/reduced graphene oxide by ball-milling method. For that, Graphene oxide and nickel powder in weight ratio 1:1 were kept in ball-mill and milled for total 1h under dried condition at 1200rpm to get NiO/Reduced graphene oxide nanocomposite [103]. Funda Aksoy Akgul et al also prepared Ti doped ZnO nanocomposite using ball milling method. He has taken certain amounts of ZnO and TiO_2 in a ball-mill and milled for 20h with 500rpm to get desired product [104]. Rahul Mundiyaniyil Thankachan et al. fabricated $ZnFe_2O_4$-C nanocomposite by ball milling method of synthesis. Already synthesized $ZnFe_2O_4$ nanoparticle and carbon black powder were taken in a milling container. And milling was done for 24h at 75 rpm under argon atmosphere of 100 kPa to get $ZnFe_2O_4$-C nanocomposite [102]. Other researcher also reported synthesis by ball milling method few of them are, Xingxing Gu et al. synthesized ZnO/sulphur/carbon nanotubes nanocomposite [105], Sunil P. Lonkar et al. prepared ZnO-graphene nanocomposite [100] and many more.

It is clear from above study that TMONCs have been synthesized by numerous synthesis methods with many application properties. Previous researchers have reported various synthesized transition metal oxide based nanocomposite by various routes of synthesis methods as shown in Table1.

Table1. Synthesized transition metal oxide based nanocomposites by various synthesis routes.

S.N.	Transition metal oxide based nanocomposites	Precursor used	Synthesis Routes	Ref.
1	ZnO-CuO	$ZnCl_2$ and $CuCl_2$	Co-precipitation method	25
2	$NiFe_2O_4$	$FeSO_4.7H_2O$ and $NiSO_4.7H_2O$	Co-precipitation method	27
3	$CeO_2/ZnO/ZnAl_2O_4$	$Ce(NO_3)_3.6H_2O$, $Al(NO_3)_2.9H_2O$ and $Zn(NO_3)_2.4H_2O$	Co-precipitation method	23
4	$ZnO-SnO_2$	$Zn(CH_3OO)_2.2H_2O$ and $SnCl_4.5H_2O$	Sol-Gel technique	45
5	CdO-ZnO	$Cd(NO_3)_2.4H_2O$ and $Zn(NO_3)_2.6H_2O$	Sol-Gel technique	43
6	TiO_2-Ag	Titanium tetraisopropoxide and Silver nitrate	Sol-Gel technique	38
7	$ZnO-TiO_2$	$Ti(OiPr)_2(dpm)_2$ and $Zn(hfa)_2$.TMEDA (OiPr: iso-propoxy; dpm: 2,2,6,6-tetramethyl3,5-heptanedionate; hfa: 1,1,1,5,5,5-hexafluoro-2,4-pentanedionate; and TMEDA: N,N,N′,N′-tetramethylethylenediamine	Chemical vapor deposition	51
8	TiO_2/SnO_2	Titanium isopropoxide and Butyltin trichloride	Chemical vapor deposition	52

9	CNT/Fe-Ni/TiO_2	Tetrabutylorthotitanate, $Ni(NO_3)_2.6H_2O$, $Fe(NO_3)_3.9H_2O$ and acetylene	Chemical vapor deposition	53
10	$CoFe_2O_4$/graphene	Graphite powder, $Co(NO_3)_2.6H_2O$ and $Fe(NO_3)_3.9H_2O$	Hydrothermal technique	63
11	α-Fe_2O_4-ZnO-Au	$FeCl_3.6H_2O$, $ZnSO_4.7H_2O$ and $HAuCl_4.3H_2O$	Hydrothermal technique	65
12	MnO_2-CNT	Multi-wall carbon nanotubes (CNTs) and $MnSO_4.H_2O$	Hydrothermal technique	66
13	ZnO/Au	$HAuCl_4.3H_2O$ and Zinc Acetate	Wet chemical method	76
14	MgO:CuO	$Mg(NO_3)_2$ and $Cu(NO_3)_3$	Wet chemical method	75
15	SnO_2/Fe_2O_3	$SnCl_4.5H_2O$ and $Fe(NO_3)_3$	Wet chemical method	74
16	Ag/ZnO-TiO_2	Zinc acetate dehydrate, silver nitrate and titanium isopropoxide	Microwave assisted method	87
17	CdO-ZnO	$ZnCl_2$ and $CdCl_2.2H_2O$	Microwave assisted method	85
18	CdO-NiO-ZnO	$Cd(CH_3COO)_2.2H_2O$, $Zn(CH_3COO)_2.2H_2O$ and $Ni(CH_3COO)_2.4H_2O$	Microwave assisted method	84
19	ZnO/CuO	$Zn(NO_3)_2.6H_2O$ and $Cu(NO_3)_2.3H_2O$	Biological method	89
20	$ZnFe_2O_4$	$Fe(NO_3)_3.9H_2O$ and $Zn(NO_3)_2.6H_2O$	Biological method	91
21	Fe_2O_3-Ag	$AgNO_3$ and $Fe(NO_3)_3$	Biological method	98
22	Ti doped ZnO	ZnO Powder and TiO_2 powder	Ball-milling technique	104
23	NiO/reduced graphene oxide	GO powder and Nickel powder	Ball-milling technique	103
24	ZnO-graphene oxide	ZnO powder and Graphene oxide powder	Ball-milling technique	100

3. Some basic characterization techniques and morphology of TMONCs

Characterization techniques do have an important role in the study of TMONCs. TMONCs we have reported in this review were synthesized by different method and even different reaction conditions and precursors therefore all of them show different shape, size, particle distribution, structure, morphology, elemental configuration, even different practical applications. All these physical and chemical properties are crucial for designing a nanomaterial with required practical applications. Therefore characterization of synthesized nanomaterial is extremely important to find out its size, shape, morphology, chemical composition and all other properties for its complete study and application. Herein, we are going to review some important characterization techniques used for various reported TMONCs.

3.1. XRD (X-ray diffraction technique)

XRD technique is widely utilized for identification of phase and the crystallite size of each phase of nanocomposite. Almost all the researchers have reported XRD studies of their synthesized transition metal oxide based nanocoposites for their crystalline quality and phase study. Shengyan Pu et al. reported XRD data of their synthesized graphene oxide/Fe_3O_4 nanocomposite and the XRD data revealed successful synthesis of graphene oxide/Fe_3O_4 nanocomposite [29]. Hadi Eslami et al. has also done XRD data study for the verification of successful fabrication of Iron-Manganese metal oxide nanocomposite. He has also reported the calculated size of nanocomposite by using scherrer's equation.

$$D = 0.89\lambda(\beta\cos\theta)^{-1} \quad (1)$$

Here, λ = X-ray wavelength, D = Size of the crystal, β = the full-width at half-maximum and θ = Diffraction angle [26].

In a very similar manner many researchers synthesized transition metal oxide based nanocomposites and studied their XRD data as K. Egizbek reported XRD data of $NiFe_2O_4$ nanocomposite [27], Thongthai Witoon et al reported XrD data study of Cu/ZnO nanocomposite [28], Tariq Jan et al studied XRD data of synthesized ZnO-CuO nanocomposite [25] and many others. It is clear from above that XRD is basic and Crucial required characterization technique for nanomaterials.

3.2. SEM and TEM (Electron Microscopy)

The properties and application are very much depending on the morphology and particle size of nanomaterials. Researchers have being using Electron microscope for observing the morphology and size of their synthesized transition metal oxide as well as other nanomaterials. Transmission electron microscopy (TEM) and Scanning electron microscopy (SEM) are two type of electron microscopy which are widely used by researcher for this purpose.

Abdessalem Hamrouni et al. reported TEM image of ZnO-SnO_2 nanocomposite and Observed that the size of nanocomposite was in 30-80nm range [45]. Ardiansyah Tanfik et al studies FESEM image of CuO/TiO_2/ZnO and observed spherical and clew like CuO/TiO_2/ZnO nanocomposites were formed [35]. M. A. Ahmed synthesized NiO/TiO_2 with 1, 2, 5 and 10% of NiO. He analyzed TEM images and found out, reduction of particle size on increasing nickel oxide percentage. Also size range of nanocomposite was in 4-32nm [36]. Literature survey revealed that electron microscopy is widely used for nanomaterials study and very crucial characterization technique. Alexin Lavin et al synthesized ZnO/CuO nanocomposite and analyzed them under SEM and TEM for finding the SEM image that shows irregular shaped semispheroidal nanoparticles and TEM confirmed agglomeration of spheroidal particles (10nm) connected with the rod-shaped particles (50nm) [37]. Shahab Ansari Amin et al. used TEM for morphology and size study of his synthesized TiO_2- Ag nanocomposite [38]. Bin Gao et al synthesized CNTs/TiO_2 nanocomposites and studied its morphology and size through TEM images [39]. Above example explained that morphology is crucial parameter for specific application use of any nanomaterial.

3.3. EDX (Energy dispersive X-ray)

EDX is useful to indentifying elemental content present in synthesized transition metal oxide based nanocomposites also other nanomaterials. Susanta Kumar Biswal et al. synthesized Fe_2O_3-Ag nanocomposite and reported its EDX characterization which confirmed the presence of oxygen (weight % of 51.12), silver (weight % of 23.25) and iron (weight % of 25.63) elements [98]. Zahra Noohpisheh et al reported EDX which showed that the weight percentage of Ag was 5.60 and the weight of ZnO was 86.02 in the synthesized Ag-ZnO nanocomposite [97]. Yoki Yulizar et al synthesized Au/TiO_2 nanocomposite and indentified its elemental composition by using EDX. EDX showed the presence of Au (weight % of 9.53), Ti (weight % of 32.66), and O (weight % of 57.81) elements in Au/TiO_2 nanocomposite [96].

EDX is important characterization technique. Elemental composition determination by EDX together with other characterization technique such as XRD, FTIR, UV etc give complete structural information of any nanomaterial.

3.4. FTIR (Fourier transform infrared spectrometer)

Numerous research works have been published for synthesis and characterization of transition metal oxide based nanocomposites, where researcher investigated functional groups in their synthesized nanomaterials by using FTIR spectroscopy. K. Karthik studied FTIR spectra for CdO-ZnO nanocomposite. Where he concluded that 861cm^{-1} peak was for cubic CdO and 505cm^{-1} peak was for hexagonal ZnO [85]. Mohammadreza Mansournia et al. synthesized CuO-ZnO composites and studied their FTIR spectra, he observed that all 0.4% CuO-ZnO, 2% CuO-ZnO, 10% CuO-ZnO and 50% CuO-ZnO showed a band in the range of 400-600cm^{-1} for metal-oxygen (M-O) stretching vibrations [68]. Zahra Noohpisheh et al reported FTIR spectra with 520cm^{-1} which was related to the vibration of the Zn-O bond and 900 to 1500cm^{-1} belonged to extract components present on the surface of the Ag-ZnO nanocomposites [97]. Above example shows the importance of FTIR spectra, it is useful for complete structural studies along with other characterization technique discussed before.

3.5. Some other characterization techniques

Some other technique also have being popular for the characterization of transition metal oxide based nanocomposites and other nanomaterials such as UV-Visible spectroscopy, Atomic force microscopy (AFM), Brunauer-Emmett-Teller (BET), Thermogravimetric analysis (TGA), Differential thermal analysis (DTA) etc. Abdur Rahman et al reported fabrication of CdO-MgO-Fe_2O_3, he also reported UV-Visible spectra for the same. He observed a wide absorption band in UV-Visible range of 340-500nm. That nanocomposite showed red shift as compare to individual CdO, MgO and Fe_2O_3 nanoparticles. He had also calculated band gap which was 1.76eV for CdO-MgO-Fe_2O_3 nanocomposite [72]. AFM was also being used for imaging nanocomposite as K. Kaviyarasu et al synthesized MgO:CuO nanocomposite and seen its images through Atomic force microscopy [75]. The BET is useful for surface area observations and pore properties analysis. Ramin Mohammadi-Aloucheh et al reported BET studies of ZnO/CuO nanocomposite. He observed images with mesopores and calculated specific surface area for ZnO/CuO(5%) and ZnO/CuO(10%) were 36.0 and 26.0m^2g^{-1} respectively. Also he calculated pore volumes for ZnO/CuO(5%) and ZnO/CuO(10%) were 0.0939 and 0.0984cm^3g^{-1} respectively [94]. Researchers are using TGA/DTA for thermal performance studies of a sample. Mohammad A. Khalizadeh et al synthesized Fe_3O_4/CNC/Cu nanocomposite and studied its TGA/DTA results showing its thermal stability [106].

It is clear from above studies that all characterization are having its own role and they work together for complete study of structural and chemical properties of a transition metal oxide based nanocomposites. Figure.2 give a brief summary of characterization techniques used in transition metal oxide based nanocomposites.

3.6. Morphology

The morphologies of transition metal oxide based nanocomposites are playing extremely important role for their practical uses in various field of sciences or we can say that practical applications of transition metal oxide based nanocomposites and even other nanomaterials are strongly related to their morphologies. Various method of fabrication have been used to control morphologies of transition metal oxide based nanocomposites. Literature survey revealed variety of morphologies of transition metal oxide based nanocomposites for example, Sharma parveen et al synthesized $Al(OH)_3/MnO_3$ nanocomposite, and he also reported its FESEM images from where it was clear that $Al(OH)_3/MnO_3$

nanocomposite having dandelion-fiber flake-like surface morphology [107]. G. Jenita Rani et al. reported SEM images of her synthesized Fe_3O_4/rGO, and concluded that Fe_3O_4/rGO showed *Watsonia Meriana* flower like morphology [108]. Razieh Aladpoosh et al. reported star-like Ag/ZnO nanocomposites on cotton fabric [109].

Different transition metal oxide combinations have been reported with different application uses such as T. K. Jana et al reported fabricated of CdS-ZnO nanocomposites with flower shaped morphologies and its photocatalytic study for degradation of organic dye (rhodamine B (RB)) [111], P. Vennila et al synthesized flower like Ni-Co/Fe_3O_4 nanocomposite. And that nanocomposite was reported for glucose sensor application [112], Jun Geng et al. reported ZnO/Au with hollow doughnut like morphology. ZnO/Au showed biosensors and biomedical properties [113] and many others.

Above study revealed various unique morphologies were achieved in literature. Figure.3 shows some already reported SEM and FESEM images with different morphologies of transition metal oxide based nanocomposites.

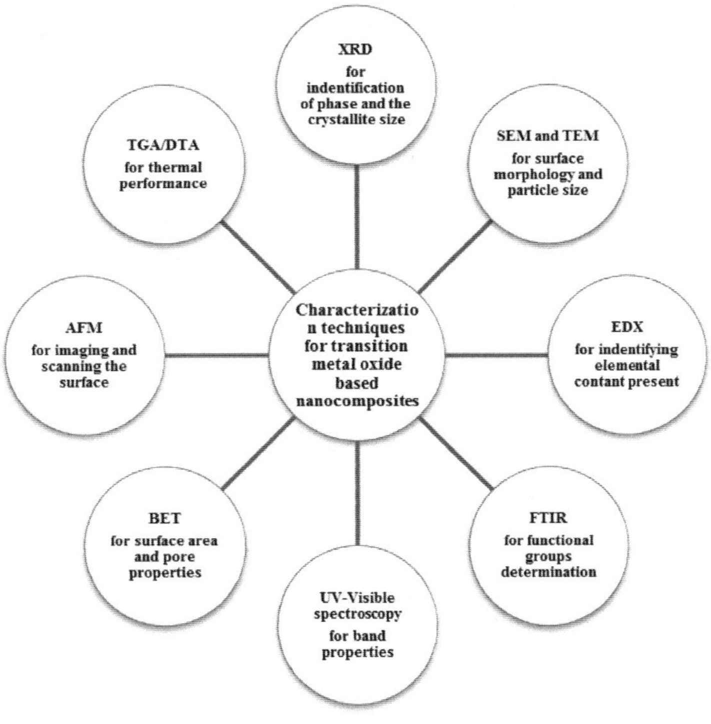

Figure 2. Simplified block diagram of the general characterization techniques for transition metal oxide based nanocomposite study using above literature survey [25, 27, 37, 45, 68, 72, 96, 98, 106].

Figure 3. Different morphologies of some reported transition metal oxide based nanocomposites, (a) [107], (b) [110], (c) [108], (d) [112], (f) [114] (copyrights reserved to the Elsevier) and (e) [113] (copyrights reserved to the American chemical society publications).

4. Various practical applications of Transition metal Oxide based nanocomposites

4.1. Supercapacitors

Supercapacitors are acting as a tie between dielectric capacitors and batteries by their higher power density and higher energy density. Electrode materials for supercapacitors application should have to be conducting and having high surface area. Therefore carbon based nanoparticles, TMONPs and TMONCs (hybridization of TMONPs with carbon nanomaterials or with other metal/metal oxide nanomaterials) attracts wide interest due to their greater specific capacitance and longer cycle life. Several researchers have utilized TMONCs for the ultracapacitor/supercapacitor application, for example, Guangyu et al. reported Co_3O_4/graphene nanocomposite with supercapacitor applications. He calculated maximum specific capacitance of Co_3O_4/graphene nanocomposite was 430F/g at 1 A/g (using three electrode system) and 215F/g at 0.4 A/g (using two electrode system). Specific capacitance obtained by two electrode system is less than three electrode system. He was able to achieve good cyclic stability of Co_3O_4/GO in both three electrode and two electrode systems [115]. D.V. Leontyeva et al synthesized carbon supported NiO (NiO/C) nanocomposite and studied its electrochemical behavior for supercapacitor applications. He calculated specific capacitance from cyclic voltammetry curve was 1100 to 777F/g with increase of scan rate from 5 to 40mV/s. And 970 F/g at 0.5A/g mass normalized current was calculated by galvanostatic charge-discharge curve. He performed thousand cycles and found out excellent cycle life (with $840Fg^{-1}$) [116]. Pintu Sen et al. fabricated Poly 3,4-ethylenedioxythiphene (PEDOT)-MnO_2 and Polyaniline (PANI)-MnO_2 nanocomposites. He studied their electrochemical properties and compared to pure MnO_2. Calculated higher specific capacitance for PEDOT-MnO_2 was 315F/g, for PANI-MnO_2 was 221F/g and for only MnO_2 was 158F/g. Where, nanocomposites showed higher specific capacitance because of their higher internal pore volume [117]. Anantha kumar Ramadoss et al synthesized graphene/ZnO nanocomposite with supercapacitor applications. He calculated maximum 109F/g capacitance at 5mV/s [118]. Thibearchews Prasankumar et al fabricated supercapacitor electrode (Polyaniline /Fe_3O_4 nanocomposite), and obtained topmost specific capacitance of 572F/g, 82% of capacitance retention after more than 5000 cycles at 1A/g [119].

Many other researchers also investigated the application of transition metal oxide based nanocomposites for supercapacitor applications such as, Sivalingam Ramesh et al. reported supercapacitor application of MWCNT/GO/$NiCo_2O_4$ nanocomposites [120], Ian Y.Y. Bu et al studied supercapacitor applications of ZnO/Reduced graphene oxide [121], Irum Shabeen et al investigated supercapacitor properties of MoO_3/$ZnMoO_4$ nanocomposite [122], Montree Sawangphruk et al also reported MnO_2-rGO for supercapacitor properties.

Above study shows transition metal oxide based nanocomposites are very good candidate for the use of supercapacitors electrode material. The possible reason of better supercapacitor performance is high conductivity and high specific surface area [123,124].

4.2. Hydrogen generation

Transition metal oxide based nanocomposites have acquired significant attention in recent time for their potential use for water splitting.

Moumita Chandra et al synthesized CuS/TiO_2 nanocomposite with varied TiO_2 contents. Where, CuS/TiO_2 with mole ratio 1:0.4 showed maximum H_2 production rate of 6310μmol g^{-1} in 5 hours or 1262μmol $g^{-1}h^{-1}$ [125]. Nagappagai Lakshmana Reddy et al reported synthesis of 1.3% Cu-TiO_2 (at reaction condition of 500°C/4h) and 1.5% Cu-TiO_2 (at reaction condition of 500°C/4h) nanocomposites, and revealed a highest rate of H_2 generation of 20.36 mmol h^{-1} g^{-1} (under LED) and 21.7mmol h^{-1} g^{-1} (under Solar light) [126]. Karim R. Diab et al synthesized $NiTiO_3$/rGO with 100:0, 99:1, 97:3, 95:5 and 93:7 weight ratios. Among all $NiTiO_4$/rGO with 95:5 showed maximum rate of H_2 generation (8383 μmol g^{-1} h^{-1}) also $NiTiO_4$/rGO with 95:5 weight ratio was showing maximum pore volume (0.176 mL/g) and maximum pore diameter (3.92nm) [127]. Several other transition metal oxide based nanocomposites also reported in literature for H_2 generation applications for example, CdO-CdS [129], Au/TiO_2 [130], reduced graphene oxide –TiO_2 [128], In and N co-doped TiO_4-Pd nanomaterial [131].

Hydrogen is an environmental friendly fuel which is promising alternative to fossil fuels. From above study, we can understand that transition metal oxide based nanocomposites are promising photocatalysts for Hydrogen production from water splitting because of their enhanced surface area and surface activity [125-127].

4.3. Solar cell applications

We have studied about H_2 as an alternative source of fossil fuel earlier in this article. Here, we are coming to review transition metal oxide based nanocomposites application as solar cells. Solar cell is most promising alternative of fossil fuels as solar energy is most abundant energy on earth also is most green and environmentally friendly energy.

Zohreh Dehghani Mahmoudabadi et al synthesized two CuO/TiO_2 nanocomposites by microplasma assisted electrochemical method. Where, CuO/TiO_2 (a) was synthesized by 15min irradiation with plasma and CuO/TiO_2 (b) by 25min irradiation with plasma. Reported conversion efficiency (CE), short-circuit current density (SCCD) and open circuit voltage (OCV) for CuO/TiO_2 (a) were 7.4%, 15.38mA/cm^2 and 0.70V respectively, and for CuO/TiO_2 (b) were 9.3%, 18.80mA/cm^2 and 0.68V respectively in order to dye sensitized photovoltaic cell application [132]. Gaurav K. Upadhyay et al reported fabrication of CdO:TiO_2 by adding various volume of the precursor solutions, Where 3:2 CdO/TiO_2 showed better CE, SCC and OCV with 3.23%, 7.6mA, 0.26V values respectively for solar cell applications [133]. Junling Song et al reported rGO-TiO_2 (coated on FTO (fluorine-doped tin oxide)) with solar cell application and studied its photovoltaic performance for solar cell applications.

He observed SCCD, OCv, fill factor and overall CE were 18.2mA/cm^2, 576mV, 57.9% and 6.06% respectively [134]. Abdul-Mojeed Ilyas et al. reported synthesis of TiO_2/CdO by coupling 10%, 20% and 40% weight of CdS with TiO_2. And observed photovoltaic studies for solar cell applications. He found that 10% CdS/TiO_2 was showing maximum CE among all with SCCD, OCV, fill factor and CE were 1.248mA/cm^2, 0.5581V, 0.4935% and 4.30% respectively [135]. Shahram Ghasemi et al synthesized Fe_3O_4/rGO nanocomposite and reported its properties for dye sensitized photovoltaic cell applications. Calculated values of short of SCCD, OCV, fill fector, and CE were 24.45mA/cm^2, 0.755V, 0.32% and 5.91% respectively which are better than only rGO [136].

Above study reveal that transition metal oxide based nanocomposites are widely used for solar cell application because when transition metal oxide semiconductors doped with other semiconductor or nanomaterials which can increase the charge transfer processes also decrease electron/hole recombination that enhances the performances of solar cell device [137-139].

4.4. Sensors

Sensitivity and selectivity are important and challenging in the wide field of sensors. To improve sensor performance different methods have been used such as doping, surface modification etc. TMONCs sensors were proved increased selectivity and sensitivity than individual transition metal oxide nanoparticles because of their enhanced surface area and surface activity or introduction of meso-porosity etc. [140-142]

Mehrnaz Joulazadeh et al fabricated Polypyrrole (PPy)/ZnO and PPy/SnO_2 nanocomposites sensors for ammonia detection. He reported that transition metal oxide based nanocomposite (PPy/ZnO) showed maximum response of ~34% towards ammonia than PPy/SnO_2 (~25%) and only PPy (~15%). He suggested possible reason for higher response of PPy/ZnO than PPy/SnO_2 was higher conductivity of PPy/ZnO due to presence of Zn^{2+} cations [143]. Meenakshi Dutt et al. used synthesized Li_2O-doped Fe_2O_3 and SnO_2- Fe_2O_3 nanocomposites as sensors in response to formaldehyde and ethanol in operating temperature range of 25-200°C. Where, reported response of 6.08 and 5.82 for HCHO and C_2H_2OH respectively at 25°C with Li_2O-doped Fe_2O_3 sensor, and SnO_2- Fe_2O_3 exhibits response of 6.44 and 4.93 for HCHO and C_2H_2OH respectively at temperature 150°C [140]. Ehab Salih et al fabricated ZnO/GO nanocomposites and studied its sensing properties towards H_2O_2 by electrochemical system. Where, cyclic voltammetry measurements confirmed oxidation peak currents of H_2O_2 at ~1.1V by using synthesized nanocomposite electrode due to direct electron transfer. That reflected sensing activity of ZnO/GO nanocomposites towards H_2O_2 [141]. Hung et al. synthesized series of WO_3/p-MWCNT nanocomposites with different % weight. Among all amount of MWCNT was 0.5% weight showed maximum response towards sensing NO_2 at its concentration of 5ppm with response of 18 and response time of 87s at 150°C [144].

Several other researchers also analyzed sensing capability of their synthesized transition metal oxide based nanocomposites for example, R. Sivasubramanian et al used synthesized copper (I) oxide-reduced graphene oxide nanocomposite for dopamine sensing [147], Bose Dinesh et al. fabricated rGO-Co_3O_4 for effective sensing of serotonin in the existence of dopamine and ascorbic acid [142], M.R. Mahmaudian et al reported sensing behavior of MnO_2/rGO nanocomposite for H_2O_2 [145]. Hyoun Woo Kim et al. synthesized ZnO nanoparticles and ZnO/Graphene nanocomposites and compared the sensing behavior ZnO-G which was synthesized by microwave irradiation show higher response than only ZnO nanoparticles towards NO_2, Ethanol, Acetone, Toluene, Benzene and CO gases [146].

Overall summery of above study is that, transition metal oxide based nanocomposite sensors mostly showed improved sensor performance than individual transition metal oxide nanoparticles.

4.5. Biomedical applications

Iman Gholamali et al prepared a series of oxidized starch/CuO nanocomposites (hydrogels with 0.00, 0.01, 0.02 and 0.03M of Cu_2). He studied the relation between encapsulation and release of ibuprofen drug in all oxidized starch/CuO nanocomposites at pH of 2.1 and 7.4. This study revealed a higher swelling capacity and controlled delivery of drug by oxidized starch/CuO nanocomposites than pure oxidized starch [148]. Dedhila Devadathan et al compared antifungal activities of their synthesized polyindole based nickel-zinc oxide nanocomposite, nickel-zinc oxide, polyindole, nickel oxide and zinc oxide and observed that polyindole based nickel-zinc oxide nanocomposites showed antifungal activity of 1cm, nickel-zinc oxide showed antifungal activity of 0.7cm and polyindole, nickel oxide and zinc oxide showed no antifungal activity [149]. Faisal Mukhtar et al reported antibacterial properties against *E. coli* of synthesized binary NiO-Fe_2O_3 and NiO-CdO nanocomposites and ternary NiO-Fe_2O_3-CdO nanocomposite. For that he prepared various concentrations (10, 20, 30 and 40g/mL) solutions of each and all nanocomposites. Among them 40g/mL solutions of each nanocomposite showed maximum antibacterial activities. He compared each nanocomposite (with 40g/mL concentration) and observed that NiO-Fe_2O_3-CdO showed maximum zone of inhibition (1.7mm) than NiO-Fe_2O_3 (~9mm) and NiO-CdO (~14mm). He suggested possible reason for antibacterial activity increase was contact stress or direct oxidation and penetration of heavy metals ions to negatively charged cell membrane [150]. Maqusood Ahamed et al synthesized ZnO nanoparticles, SnO_2-doped ZnO (SnO_2-ZnO) nanocomposite and SnO_2-ZnO/rGO (SnO_2-doped ZnO/reduced graphene oxide) nanocomposite and studied their anticancer activities via oxidation stress pathway. Fluorescent microscopic images revealed that ROS (Reactive Oxygen Species) and MDA (Malondialdehyde) levels were higher with all ZnO, SnO_2-ZnO and SnO_2-ZnO/rGO nanomaterials rather than control cells. Higher the ROS production more will be the oxidative damage to the cancer cells. They treated Michigan Cancer Foundation-7 (MCF-7 which are breast cancer cells) with certain amount of ZnO, SnO_2-ZnO and SnO_2-ZnO/rGO nanomaterials for 24h with and without of N-acetylcysteine (NAC) and found decrease in ROS generation in presence of NAC. Herein, SnO_2-ZnO/rGO nanocomposite reported with higher anticancer activity against MCF-7 cells in compared to pure ZnO nanoparticles [151].

It is clear from these examples that transition metal oxide based nanocomposites have been actively reported with several vital biomedical applications such as anticancer, biosensors, antifungal, antibacterial and in drug delivery. Further more such examples are reported in Table 2.

Table 2. Biomedical applications of various reported transition metal oxide based nanocomposites.

S.N.	Transition metal oxide Abased nanocomposites	Synthesis methods	Biomedical application uses	Ref.
1	Zeolite/zinc oxide-copper oxide	The facile chemical method	Antibacterial properties	152
2	WO_3-GO	Ultrasonic method	Antibacterial And Anticancer properties	153
3	Chitosan coated zinc oxide	Green chemistry approach	Antibacterial properties	154
4	GO-SnO_2-TiO_2	Solvothermal method	Biosensor for detection of acetone in diabetes mellitus patient's breath	155
5	ZnO/GO	Co-precipitation method	Antibacterial properties	156
6	rGO/NiO	Microwave-assisted hydrothermal	Glucose sensors in diabetes	157

		synthesis		
7	PbS-NiO	Chemical synthesis	Antifungal properties	158
8	Core-shell α-Fe_2O_3-ZnO	Wet chemical method	Antibacterial And Anticancer properties	159

4.6. Photocatalytic degradation of organic dyes

Discharge of industrial effluent streams is becomes environmental threats in modern world of the fast development of technology and world economy. Organic dyes one of the major pollutants which is causing the most serious water pollution. Where, Photocatalytic degradation of organic dyes is effective, efficient and low cost way to solve this major water pollution issue of dyes. In the last few decades, many researchers have been indentified transition metal oxide based nanocomposites as good photocatalysts over semiconductor transition metal oxide nanoparticles by overcoming of their limitation of wide band gap, fast electron hole recombination, slower reaction rate of photocatalysis etc. Many researchers have been focused on modifying and doping of transition metal oxide nanoparticles to achieve a nanocomposite with inhibition of electron hole recombination, fast reaction rate of photocatalysis [160-163]. Table 3. Presents various reported TMONCs for photocatalytic applications.

Table 3. Photocatalytic application of transition metal oxide based nanocomposites for different toxic dyes.

S.N.	Transition metal oxide nanocomposites	Synthesis techniques	Toxic dyes	Photocatalytic η (%)	Time	Light source	Ref.
1	Fe_2O_3/RGO	Green hydrothermal method	4-Nitrophenol	98%	50min	Visible light	164
2	Graphene-Zinc oxide nanocomposites	Chemical precipitation method	Rhodamine-B	100%	90min	Visible light	165
3	ZnO/Fe_2O_3	Solvothermal techniques	GRL dye	81.1%	200min	Ultra violet	166
4	Graphene-V_2O_5	Solution mixing method with hydrothermally grown V_2O_5	Methylene Blue	100%	90min	Sunlight	167
5	$LaFeO_3$/Ag_2CO_3	Co-precipitation method	Rhodamine-B And p-chlorophenol	99.5% And 59%	45min	Sunlight	168
6	Fe/Al/Ti oxide	Chemical routes	Methylene Blue	98.4%	10min	Visible light	169
7	Zinc Oxide activated charcoal Polyaniline	Simple precipitation method	Rhodamine-B	95%	120min	Visible light	170
8	Polyaniline/CdO	Chemical	Malachite	99%	4h	Sunlight	162

		oxidation polymerization	Green				
9	Ag-ZnO	Biogenic synthesis	Methyl Orange	~100%	5h	Visible light	160
10	ZnO/CuO (98.0/2.0 mass ratio)	Microwave assisted method	Acid Orange 7	80.5%	120min	Sunlight	171

5. Conclusion and future perspective

Many useful strategies have been investigated for the synthesis of several TMONCs such as sol-gel method by Taufik et al [35], hydrothermal techniques by Gao et al. [60], chemical vapor deposition by Ye et al. [50], wet chemical method by Collazos et al. [71], microwave assisted method by Karthik et al [85] and green methods by Elemike et al [95] respectively. Herein, the transition metal oxide based nanocomposites have been synthesized and modified by various technique of synthesis. As we know that on the basis of modified technique of synthesis, we can achieve unique sets of properties like, high surface area, enhanced surface activity/ more number of active site on surface, introduction of mesoporosity, decreasing electron hole recombination by doping and increasing charge transfer process. These characteristic properties have been investigated for different potential applications in the literature [24-25, 140-141]. Herein, we have also reviewed different reported characterization technique of synthesized metal oxide based nanocomposites like XRD, EDX, SEM, TEM and FTIR for structural and morphological studies. Herein, authors have investigated unique applications of transition metal oxide nanocomposites based on their different morphologies. Morphologies and surface properties play a vital role in practical application uses of these transition metal oxide based nanocomposites [108-111]. Nanocomposites are known as the materials of the era for their broad extend of applications in almost all fields of sciences such as physics, biology, environmental sciences chemistry and material sciences. Many potential real-life application of TMONCs have been reported such as sensors by Joulazadeh et al.[143], Photocatalytic degradation of toxic dyes by Liu et al. [163], biomedical applications like for drug delivery, antimicrobials , biosensors and anticancer by Alswat et al., Bharathi et al. and Kalidoss et al. respectively [152, 154, 155], in energy production by Mahmoudabadi and Dubey et al. [132, 128], solar cell and H_2 generation, supercapacitors for energy storage by Chen et al. [115] and various other.

Further more research is required in the area of fabrication and designing of new TMONCs with enhanced surface area, morphological control, size control, porosity control and homogeneity of all the homogenous dispersion of all materials in nanocomposites. Still, the researchers and scientists have to hard work for developing cost effective and environmental friendly routes of synthesis of transition metal oxide based nanocomposites with high rate of product at normal temperature and in moderate reaction conditions. Applications of many transition metal oxide based nanocomposites are somewhat limited to laboratory level only therefore further research should be focused towards industrial scale production and industrial applications of these nanocomposites.

Conflict of interest

No conflict of interest was declared by authors.

Abbreviations

TMONCs, Transition metal oxides based nanocomposites; CNTs, Carbon nanotubes; rGO, Reduced graphene oxide; PVA, Polyvinyl alcohol; E. coli, Escherichia coli; MWCNTs, Multi-wall carbon nanotubes; OiPr, Isopropoxy; SEM, Scanning electron microscopy; hfa, 1,1,1,5,5,5-hexafluoro-2,4-

pentanedionate; FESEM, Field emission scanning microscopy; dpm, 2,2,6,6,-tetramethyl3,5-heptanedionate; TEM, Transmission electron microscopy; EDX, Energy dispersive X-ray; TMEDA, N,N,N′,N′-tetramethylethylenediamine; FT-IR, Fourier transform infrared spectrometer; XRD, X-ray diffraction technique; AFM, Atomic force microscopy; BET, Brunauer-Emmett-Teller; TGA, Thermogravimetric analysis; DTA, Differential thermal analysis; PEDOT, Poly3-4-ethylene dioxythiphene; PANI, Polyaniline, ppy, Polypyrrole; ROS, Reactive oxygen species; MDA, Malondialdehyde; MCF-7, Michigan cancer foundation; NAC, N-acetylcysteine; CV, Conversion efficiency; SCCD, Short-circuit current density; OCV, open circuit voltage.

Acknowledgement

Authors are thankful to Miranda House, University of Delhi for lab facility. Sapna Yadav (CSIR, File No. 08/700/(0004)/2019-EMR-1) also want to thank CSIR, New Delhi for financial support.

References

[1] Tiwari S, Marathe S, Chandola M, Thakur N, Rathore R and Garg S 2016 *J. Polmer. Text. Eng.* **3** 01-07.
[2] Magro M and Vianello F 2019 *Nanomaterials* **9** 1608.
[3] Tyagi M and Tyagi D 2014 *Int. J. Electron. Electr. Eng.* **7** 603-608.
[4] Shyamala S and Muthuchudarkodi R R 2018 *Int. J. Sci. Eng. Manag.* **3** 588-595.
[5] Khorrami S, Abdollahi Z, Eshaghi G, Khosrani A, Bidram E and Zarrabi A 2019 *Sci. Rep.* **9** 9167.
[6] Sachinjith K R and Krishna S K R 2018 *Int. J. Adv. Res. Ideas Innov. Tech.* **4** 235-236.
[7] Camargo P H, Satyanarayana K G and Wypych F 2009 *Mater. Res.* **12** 1-39.
[8] Rathod V T, Kumar J S and Jain A 2017 *Appl. Nanosci.* **7** 519-548.
[9] Okpala C C 2013 *Int. J. Eng. Res. Dev.* **8** 17-23.
[10] Rani N, Mona S, Yadav S, Gupta K, Saini K and Khanuja M 2020 *AIP Conf. Proc.* **2276** 020042.
[11] Pino F, Mayorga-Martinez C C and Aerkoci A 2016 *Electrochem. Commun.* **71** 33-37.
[12] Aboulouard A, Gultekin B, Can M, Erol M, Jouaiti A, Elhadadi B et al. 2020 *J. Mater. Res. Technol.* **9** 1569-1577.
[13] Bhatia S and Verma N 2017 *Mater. Res. Bull.* **95** 468-476.
[14] Danish M S, Estrella L L, Alemaida I M, Lisin A, Moiseev N, Ahmadi M, et al. 2021 *Metals* **11** 1-25.
[15] Zhang Y, Sun X, Pan L, Li H, Sun Z, Sun C, et al. 2009 *Solid State Ion.* ***180*** 1525-1528.
[16] Widiarti N, Sae J and Wahyuni S 2017 *Mater. Sci. Eng.* **172** 012036.
[17] Ashok C, Rao K V and Chakra C S 2016 *J. Adv. Chem. Sci.* **2** 223-226.
[18] Durmus Z, Kurt B Z and Durmus A 2019 *Chemistry select* **4** 271-278.
[19] Mehto A, Mehto V R, Chauhan J, Singh I and Pandey R 2017 *J. Nanomed. Res.* **5** 1-8.
[20] Ghoreishi F, Ahmadi V and Samadpour M 2013 *J. Nanostruct.* **3** 453-459.
[21] Sedighi F, Esmaeilizare M, Sabhani-Nasab A and Behpour M 2018 *J. Mater. Sci. Mater. Electron.* **29** 13737-13745.
[22] Jazi F S, Parvin N, Rabiei M, Tahriri M, Shabestari Z M and Azadmehr A R 2012 *J. Ceram. Process. Res.* **13** 523-526.
[23] Somraksa W, Suwanboon S, Amornpitoksuk P and Randorn C 2020 *Mater. Res.* **23** e20190627.
[24] Thambidurai S, Gowthaman p, Venkatachalam M and Suresh S 2020 *J. Alloys Compd.* ***830*** 154642.
[25] Jan T, Azmat S, Mansoor Q, Waqas H, Adil M, Ilyas S et al. 2019 *Microb. Pathog.* **134** 103579.

[26] Eslami H, Ehrampoush M H, Esmaeili A, Ebrahimi A A, Ghameian M T, Falahzadeh H et al. 2019 *Mater. Chem. Phys.* ***224*** 65-72.

[27] Egizbek K, Kozlovskiy A, Ludzik K, Zdorovets M, Korolkov I, Marciniak B et al. 2020 *Ceram. Int.* **46** 16548-16555.

[28] Witoon T, Permsirivanich T, Donphai W, Jaree A and Chareonpanich M 2013 *Fuel Process. Techno.* **116** 72-78.

[29] Pu S, Xue S, Hou Z Y, Zhu R and Chu W 2018 *Environ. Sci. Pollut. Res.* **25** 17310-17320.

[30] Subhan M A, Uddin N, Sarker P, Azad A K and Begum K 2015 *Spectrochim. Acta A. Mol. and Biomol. Spectrosc.* **149** 839-850.

[31] Parashar M, Shukla V K and Singh R 2020 *J. Mater. Sci. Mater. Electron.* **31** 3729-3749.

[32] Tseng T K, Lin S Y, Chen Y J and Chu H 2010 *Int. J. Mol. Sci.* **11** 2336-2361.

[33] Ward D A & and Ko I 1995 *Ind. Eng. Chem. Res.* **34** 421-433.

[34] Hench L L and West J K 1990 *Chem. Rev.* **90** 33-72.

[35] Taufik A, Albert A & and Saleh R 2017 *J. Photochem. Photobiol. A.* **344** 149-162.

[36] Ahmed M A 2012 *J. Photochem. Photobiol. A* **238** 63-70.

[37] Lavin A, Sivasamy R, Mosquera E and Morel M J 2019 *Surf. Interfaces* **17** 100367.

[38] Amin S A, Pazouki M and Hosseinnia A 2009 *Powder Technol.* **196** 241-245.

[39] Gao B, Chen G Z and Puma G L 2009 *Appl. Catal. B.* **89** 503-509.

[40] Yang H, Shi R, Zhang K, Hu Y, Tang A and Li X 2005 *J. Alloys Compd.* **398** 200-202.

[41] Ghorbani M, Abdizadeh H, Taheri M and Golobostanfard M R 2018 *Int. J. Hydrog Energy* **43** 7754-7763.

[42] Grigorie A C, Muntean C, Vlase T, Locovei C and Stefanescu M 2017 *Mater. Chem. Phys.* **186** 399-406.

[43] Mosquera E, Pozo I d & and Morel M 2013 *J. Solid state Chem.* **206** 265-271.

[44] Ansari F, Sobhani A and Salavati-Niasari M 2018 *J. colloid Interface Sci.* **514** 723-732.

[45] Hamrouni A, Moussa N, Parrino F, Paola A D, Houas A and Palmisano L 2014 *J. Mol. Catal. A Chem* **390** 133-141.

[46] Puma G L, Bono A, Krishnaiah D and Collin J G 2008 *J. Hazard. Mater.* **157** 209-219.

[47] Rashid H U, Yu K, Umar M N, Anjum M N, Khan K, Ahmad N et al. 2015 *Rev. Adv. Mater. Sci.* **40** 235-248.

[48] Zhang Y, Zhang L Y and Zhou C 2013 *Acc. Chem. Res.* **46** 2329-2339.

[49] Gladfelter W L 1993 *Chem. Mater.* **5** 1372-1388.

[50] Ye J S, Cui H F, Liu X, Lim M T, Zhang W D and Shan F 2005 *small* **1** 560-565.

[51] Barreca D, Comini E, Ferrucci A P, Gasparatto A, Maccato C, Maragno C et al. 2007 *Chem. Mater.* **19** 5642-5649.

[52] Heale F L, Parkin I P and Carmalt C J 2019 *ACS Appl. Mater. Interfaces* **11** 41804-41812.

[53] Ma L, Chen A, Lu J, Zhang Z, He H and Li C 2014 *Particuology* **14** 24-32.

[54] Klanwan J, Akrapattangkul N, Pavarajarn V, Seto T, Otani Y and Charinpanitkul T 2010 *Mater. Lett.* **64** 80-82.

[55] Yuvaraj S, Fernandez A C, Sundararajan M, Dash C S and Sakthivel P 2020 *Ceram. Int.* **46** 391-402.

[56] Hessien M, Dana E and Tana A 2021 *Ceram. Int.* **47** 811-816.

[57] Feng S and Xu R 2001 *Acc. Chem. Res.* **34** 239-247.

[58] Morey G W 1953 *J. Am. Ceram. Soc* **36** 279-285.

[59] Somiya S and Roy R 2000 *Bull. Mater. Sci.* **23** 453-460.

[60] Gao Y, Wang L, Zhou A, Li Z and Chen J 2015 *Mater. Lett.* **150** 62-64.

[61] Zhang G Y, Sun Y Q, Gao D Z and Xu Y Y 2010 *Mater. Res. Bull.* **45** 755-760.

[62] Wang A, Yu W, Fang Y, Song Y, Jia D, Long L et al. 2015 *Carbon* **89** 130-141.

[63] Gan L, Shang S, Yuen C W, Jiang S X and Hu E 2015 *Appl. Surf. Sci.* **351** 140-147.

[64] Guan D, Gao Z, Yang W, Wang J, Yuan Y, Wang B et al. 2013 *Mater. Sci. Eng. B* **178** 736-743.

[65] Kaneti V Y, Moriceau J, Liu M, Yuan Y, Quadir Z, Jiang X et al. 2015 *Sens. Actuators* **209** 889-897.
[66] Teng F, Santhanagopalan S, Wang Y and Meng D D 2010 *J. Alloys compd.* **499** 259-264.
[67] Liang D, Cui C, Hu H, Wang Y, Xu S, Ying B et al 2014 *J. Alloys Compd.* **582** 236-240.
[68] Mansournia M and Ghaderi L 2017 *J. Alloys Compd.* **691** 171-177.
[69] Samanta P K 2017 *Journal of Tissue Science & Engineering* **8** 1000197.
[70] Zgura I, Preda N, Socol G, Ghica G, Enculescu M, Negut I et al. 2018 *Mater. Res. Bull.* **99** 174-181.
[71] Collazos A M, Gomer C D and Paez J R 2019 *Mater. Chem. Phys.* **222** 230-245.
[72] Rahman A, Sabeeh H, Zulfiqar S, Agboola P O, Shakir I and Warsi M F 2020 *Synth. Met.* **259** 116228.
[73] Zein S H, Yeoh L C, Chai S P, Mohamed A R and Mahayuddin M E 2007 *J. Mater. Process. Technol.* **190** 402-405.
[74] Rumyantseva M, Kovalenko V, Gaskov A, Pagnier T, Machon D, Arbiol J et al. 2005 *Sens. Actuators B* **109** 64-74.
[75] Kaviyarasu K, Magdolane C M, Anand K, Manikandan E and Maaza M 2015 *Spectrochim Acta A Mol. Biomol. Spectrosc* **142** 405-409.
[76] Wang Q, Geng B and Wang S 2009 *Environ. Sci. Technol.* **43** 8968-8973.
[77] Jiang T, Yin N, Bai Z, Dai P, Yu X, Wu M et al. 2018 *Appl. Surf. Sci.* **450** 219-227.
[78] She X, Zhang X, Liu J, Li L, Yu X and Huang Z 2015 *Mater. Res. Bull.***70** 945-950.
[79] Kannan K, Radhika D, Nikolova M P, Andal V, Sadasivuni K K and Krishna L S 2020 *Optik* **218** 165112.
[80] Wang Y and Lee J Y 2005 *J. power sources* **144** 220-225.
[81] Faraji S and Ani F N 2014 *J. Power Sources* **263** 338-360.
[82] Mirzaei A and Neri G 2016 *Sens. Actuators B* **237** 749-775.
[83] Xu X, Shen J, Li N and Ye M 2014 *J. Alloys Compd.* **616** 58-65.
[84] Karthik K, Dhanuskodi S, Gobinath C, Prabukumar S and Sivaramakrishnan S 2018 *J. Mater. Sci. Mater. Electron.* **29** 5459-5471.
[85] Karthik K, Dhanuskodi S, Gobinath C and Sivaramakrishnan S 2015 *Spectrochim. Acta Part A Mol. Biomol. Spectrosc.* **139** 7-12.
[86] Dou P, Tan F, Wang W, Sarreshteh A, Qiao X, Qiu X et al. 2015 *J. Photochem. Photobiol. A* **302** 17-22.
[87] Li L, Zhang X, Zhang W, Wang L, Chen X and Gao Y 2014 *Colloids Surf. A Physicochem. Eng. Asp.* **457** 134-141.
[88] Kumar R, Youssry S M, Ya K Z, Tan W K, Kawamura G and Matsuda A 2020 *Diam. Relat. Mater.* **101** 107622.
[89] Yulizar Y, Bakri R, Apriandanu D O and Hidayat T 2018 *Nano-Struct. Nano-Objects* ***16*** 300-305.
[90] Bharathi D, Ranjithkumar R, Chandarshekar B and Bhuvaneshwari V 2019 *Int. J. Biol. Macromol.* **141** 476-483.
[91] Matinise N, Kaviyarasu K, Ketsi N M, Khamlich S, Kotsedi L, Mayedwa N et al. 2018 *Appl. Surf. Sci.* **446** 66-73.
[92] Hessien M, Dana E and Taha A 2021 *Ceram. Int.* **47** 811-816.
[93] Liu Q, Ma P, Liu P, Li H, Han X, Liu L et al. 2020 *Water Sci. Technol.* **81** 694-708.
[94] Aloucheh R M, Yongjeh A H, Bayrami A, Navid S L and Asadi A 2018 *J. Mater. Sci. Mater. Electro.* **29** 13596-13605.
[95] Elemike E E, Onwudiwe D C and Singh M 2020 *J. Inorg. Organomet. P.* **30** 400-409.
[96] Yulizar Y, Sudirman Apriandanu D O and Wibowo A P 2019 *Compos. Commun.* **16** 50-56.
[97] Noohpisheh Z, Amiri H, Farhadi S and Gholami A M 2020 *Spectrochim. Acta A Mol. Biomol. Spectrosc.* **240** 118595.
[98] Biswal S K, Panigrahi G K and Sahoo S K 2020 *Biophys. Chem.* **263** 106392.

[99] Ji H, Hu S, Jiang Z, Shi S, Hou W and Yang G 2019 *Electrochim. Acta* **299** 143-151.
[100] Lonkar S P, Pillai V and Abdala A 2019 *Appl. Surf. Sci.* **465** 1107-1113.
[101] Abdellahi M, Bahmanpour H and Bahmanpour M 2014 *Ceram. Int.* **40** 9675-9692.
[102] Thankachan R M, Rahman M M, Sultana I, Glushenkov A M, Thomas S, Kalarikkal N et al. 2015 *J. Power Sources* **282** 462-470.
[103] Kahimibi H, Hong S B, Yong M and Chai G B 2017 *J. Electroanal. Chem.* **786** 14-19.
[104] Akgul F A 2016 *Compos. B. Eng.* **91** 589-594.
[105] Gu X, Tong C J, Wen B, Liu L, Lai C and Zhang S 2016 *Electrochim. Acta* **196** 369-376.
[106] Khalilzadeh M A, Tajik S, Beitollani H and Venditti R A 2020 *Ind. Eng. Chem. Res.* **59** 4219-4228.
[107] Parveen S, Kavyashree and Pandey S 2019 *Electrochim. Acta* **319** 832-842.
[108] Rani G J, Babu K J, Kumar G G and Rajan M J 2016 *J. Alloys Compd* **688** 500-512.
[109] Aladpoosh R and Montazer M 2016 *Carbohydr. Polym.* **141** 116-125.
[110] Fernander L M, Alkan G, Milosevic O, Rabanal M and Friedrich B 2019 *Catal. Today* **321-322** 26-33.
[111] Jana T K, Pal A and Chatterjee K 2014 *J. Alloys Compd* **583** 510-515.
[112] Vennila P, Yoo D J, Kim A R and Kumar G G 2017 Ni-Co/Fe_3O_4 *J. alloys Compd.* **703** 633-642.
[113] Geng J, Song G H, Jia X D, Cheng F F and Zhu J J 2012 *The Journal of Physical Chemistry* **116** 4517-4525.
[114] Bharathi D, Ranjithkumar R, Chandarshekar B and Bhuvaneshwari V 2019 *Int. J. Biol. Macromol.* **129** 989-996.
[115] He G, Li J, Chen H, Shi J, Sun X, Chen S et al. 2012 *Mater. Lett.* **82** 61-63.
[116] Leontyeva D V, Leontyeva I N, Avramenko M V, Yuzyuk Y I, KuKushkina Y A and Smirnova N V 2013 *Electrochim. Acta* **114** 356-362.
[117] Sen P, De A, Chowdhury A D, Bandyopadhyay S K, Agnihotri N and Mukherjee M 2013 *Electrochim. Acta* **108** 265-273.
[118] Ramadoss A and Kim S J 2013 *Mater. Chem. Phys.* **140** 405-411.
[119] Prasankumar T, Wiston B R, Gautam C R, Ilangovan R and Jose S P 2018 *J. Alloys Compd.* **757** 466-475.
[120] Ramesh S, Vikraman D, Kim H S, Kim H S and Kim J H 2018 *J. alloys Compd.* **765** 369-379.
[121] Bu I Y and Huang R 2015 *Mater. Sci. Semicond. Process.* **31** 131-138.
[122] Shaheen I, Ahmad K S, Zequine C, Gupta R K, Thomas A G and Malik M A 2020 *J. Mater. Sci.* **55** 7743-7759.
[123] Sawangphruk M, Srimuk P, Chiochan P, Krittayavathananon A, Luawuthi S and Limtrakul J 2013 *Carbon* **60** 109-116.
[124] Wang K, Dong X, Zhao C, Qian X and Xu Y 2015 *Electrochim. Acta* **152** 433-442.
[125] Chandra M, Bhunia K and Pradhan D 2018 *Inorg. Chem.* **57** 4524-4533.
[126] Reddy N L, Emin S, Kumari V D and Venkatakrishnan S M 2018 *Indust. Eng. Chem. Res.* **57** 568-577.
[127] Diab K R, El-Maghrabi H H, Nada A A, Youssef A M, Hamdy A, Roualdes S et al. 2018 *J. Photochem. Photobiol. A* **365** 86-93.
[128] Dubey P K, Tripathi P, Tiwari R S, Sinha A K and Srivastava O N 2014 *Inter. J. Hydrog. Energy* **39** 16282-16292.
[129] Kahane S V, Sasikala R, Vishwanath B, Sudaram V and Shailaja M 2013 *Inter. J. Hydrog. Energy* **38** 15012-15018.
[130] Mezni A, Ibrahim M M, El-kemary M, Shaltout A A, Mostafa N Y, Ryl J et al. 2018 *Electrochim. Acta* **290** 404-418.
[131] Sasikala R, Shirole A R, Sudarar V, Naik R, Rao R and Bharadwaj S R 2010 *Appl. Catal. A-Gen.* **377** 47-54.
[132] Mahmoudabadi Z D and Eslami E 2019 *J. Alloys Compd.* **793** 336-342.

[133] Upadhyay G K, Kumar V and Purohit L P 2021 Optimized CdO:TiO_2 *J. Alloys Compd.* **856** 157453.
[134] Song J, Yin Z, Yang Z, Amaladass P, Wu S, Ye J et al. 2011 *Chem. Eur. J.* **17** 10832-10837.
[135] Ilyas A M, Gondal M A, Yamani Z H and Baig U 2017 *Int. J. Energy Res.* **4** 1422-1435.
[136] Ghasemi S, Hosseini S R Kazemi Z 2018 *J. Solid State Electrochem.* **22** 245-253.
[137] Tarwal N L, Devan R S, Ma Y R, Patil R S, Karanjkar M M and Patil P S 2012 *Electrochim. Acta* **72** 32-39.
[138] Mahmoudabadi Z D, Eslami E and Narimisa M 2018 *J. Colloid Interface Sci.* **529** 538-546.
[139] El-Shafai N M, Shukry M, El-Mehasseb I M, Abdelfatah M, Ramadam M S, El-Shaer A et al. 2020 *Mater. Sci. Eng. B* **259** 114596.
[140] Dutt M, Ratan A, Tomar M, Gupta V and Singh V 2020 *J. Phys. Chem. Solid* **145** 109536.
[141] Salih E, Mekawy M, Hassan R Y and El-Sherbiny I M 2016 *J. Nanostructure Chem.* **6** 137-144.
[142] Dinesh B, Veeramani V, Chen S M and Saraswathi R 2017 *J. Electroanal. Chem.* **786** 169-176.
[143] Joulazadeh M and Navaechian A H 2015 *Synth. Met.* **210** 404-411.
[144] Hung N M, Chinh N D, Nguyen T D, Kim E T, Choi G, Kim C et al. 2020 *Ceram. Int.* **46** 29233-29243.
[145] Mahmoudian M R, Alias Y, Basirun W J, Woi P M and Sookhakian M 2014 *Sens. Actuators B* **201** 526-534.
[146] Kim H W, Kwon Y J, Mirzaei A, Kang S Y, Choi M S, Bang J H et al. 2017 *Sens. Actuators B* **249** 590-601.
[147] Sivasubramanian R and Biji P 2016 *Mater. Sci. Eng. B* **210** 10-18.
[148] Gholamali I, Hosseini S N, Alipour E and Yadollahi M 2019 *Starke* **71** 1800118.
[149] Devadathan D and Raveendran R 2014 *Int. J. Chem. Eng. Appl.* **5** 240-243.
[150] Mukhtar F, Munawar T, Nadeem M S, Hasan M, Hussain F, Nawaz M A et al. 2020 *Appl. Phys. A* **126** 588.
[151] Ahamed M, Akhtar M J, Khan M M and Alhadlaq H A 2021 *Int. J. Nanomed.* **16** 89-104.
[152] Alswat A A, Ahmad M B and Saleh T A 2017 *Colloid Interface Sci. Commun.* **16** 19-24.
[153] Jeevitha G, Abhinayaa R, Mangalaraj D and Ponpandian N 2018 *J. Phys. Chem. Solids* **116** 137-147.
[154] Bharathi D, Ranjithkumar R, Chandershekar B and Bhuvaneshwari V 2019 *Int. J. Biol. Macromol.* **129** 989-996.
[155] Kalidoss R, Umapathy S and Sivalingam Y 2018 *Appl. Surf. Sci.* **449** 677-684.
[156] Trinh L T, Quynh L A and Hieu N H 2018 *Int. J. Nanotechnol.* **15** 108-117.
[157] Alghazzawi W, Danish E, Alnahdi H and Salam M A 2020 *Synth. Met.* **267** 116401.
[158] Suganya M, Balu A R, Anitha S, Prabha D, Balamurugan S, Priyanka B et al. 2018 *Mater. Sci. Eng. B* **229** 118-125.
[159] Jana T K, Jana S K, Kumar A, De K, Maiti R, Mandal A K et al. 2019 *Colloids Surface B* **177** 512-519.
[160] Ansari S A, Khan M M, Ansari M O, Lee J and Cho M H 2013 *J. Phys. Chem.* **117** 27023-27030.
[161] Kumar A, Sharma G, Naushad M, Singh P and kalia S 2014 *Indust. Eng. Chem. Res.* **53** 15549-15560.
[162] Gulce H, Eskizeybek V, Haspulat B, Sari F and Avci A 2013 *Indust. Eng. Chem.Res.* **52** 10924-10934.
[163] Liu C, Yang D, Jiao Y, Tian Y, Wang Y and Jiang Z 2013 *ACS Appl. Mater. Interfaces* **5** 3824-3832.
[164] Mohan B S, Ravi K, Anjaneyulu R B, Sree G S and Basavaiah K 2019 *Physica B Condens. Matter* **553** 190-194.

[165] Maruthupandy M, Qin P, Muneeswaran T, Rajivgandhi G, Quero F and Song J M 2020 *Mater. Sci. Eng. B* **254** 114516.
[166] Hashim F S, Alkaim A F, Mahdi S M and Alkhayatt A H 2019 *Compos. Commun.* **16** 111-116.
[167] Shanmugam M, Alsalme A, Alghamdi A and Ramasamy J 2015 *ACS Appl. Mater. Interfaces* **7** 14905-14911.
[168] Pirazada B M, Pushpendra Kunchala R K and Naidu B S 2019 *ACS Omega* **4** 2618-2629.
[169] Mukherjee A, Adak M K, Upadhyay S, Khatun J, Dhak P, Khawas S et al. 2019 *ACS Omega* **4** 9686-9696.
[170] Selvin S P, Kumar A G, Sarala L, Rajaram R, Sathiyan A, Merlin J P et al. 2018 *ACS Sustain. Chem. Eng.* **6** 258-267.
[171] Ruan S, Huang W, Zhao M, Song H and Gao Z 2020 *Mater. Sci. Semicond. Process* **107** 104835.

Flexural and shear strength properties of unidirectional carbon fiber reinforced polymer composite interleaved with recycled carbon fiber and short virgin aramid fiber non-woven mats

C Y Attahu[1], J Yang[1*], KH Wong[1*] and C K Thein [1]

[1]Faculty of Science and Engineering, University of Nottingham Ningbo China, Ningbo 315100, China

[*]E-mail: kok-hoong.wong@nottingham.edu.cn; jian.yang@nottingham.edu.cn

Abstract. Carbon fiber reinforced polymer composites (CFRPs) are one of the most widely used composite types and wastes associated with them (CFRPs) get generated through either their manufacturing or end-of-service-life. Predominately due to environmental concerns and governmental regulations, recycling these CFRPs is needed and to make use of the recycled carbon fibers (rCFs), a wet paper-making technique was used to convert the rCFs into a 60 g/m^2 non-woven mat. For comparison purposes, the same technique was used to convert short virgin aramid fibers (vAFs) into a 60 g/m^2 non-woven mat. Each mat was sandwiched with two resin films and then interleaved with 12-ply unidirectional (UD) prepreg tapes (carbon/epoxy). The assemblage was molded into composite laminates using a vacuum bagging assisted compression molding technique, and the samples for the tests were cut using a waterjet machine accordingly. Compared with the control, the results indicate an increment in the flexural modulus, and the specific flexural modulus for the CFRPs with non-woven mats: the flexural modulus increased by approximately 8.2% and 12.0% for the CFRP with rCF and vAF mats, respectively; the specific flexural modulus increased around 9.5% and 13.3%, respectively for the CFRP with rCF and vAF mats. On the other hand, the shear strength approximately decreased by 6.4% and 6.0% for the CFRP with rCF and vAF mats, respectively. The negative shear strength performances of the composite laminates with non-woven mats reflected on their flexural strength performances: the flexural strength increased about 1.1% and decreased by approximately 7.9% for the CFRP with vAF and rCF mats, respectively. To resolve the negative shear strength performances, it is recommended that the surfaces of the mats be treated with a coupling agent to improve their interfacial adhesions.

Keywords: Recycled Carbon Fibers; Non-woven Mats; Aramid Fibers; Reuse

1. Introduction

Carbon fiber reinforced polymer (CFRP) composites are widely used in many industries because of their high specific stiffness and high specific strength [1-3]. These properties are quantitatively adequately documented by Khanna et al. [4]. Predominately driven by environmental/sustainability purposes [5], there is the need to develop sustainable, eco-friendly materials that could compete with other materials. For light-weight structural applications, including aerospace applications where minimum structural weight is required, a high specific modulus is essential since this translates into more significant weight savings at high stiffness [6]. Flexural modulus is a mechanical property that indicates the capacity of a material to bend. Shear strength is an indicator of the interfacial adhesion of the layers of the composite laminates: poor interfacial adhesion could be detrimental: -possible delamination of the composite laminates.

In a recent comprehensive review article by Pakdel et al.[7], and May et al. [8], it has been noted that highly energy-efficient and cost-effective recycling techniques which are industrially applicable have been developed for carbon fiber reinforced composites. In the same review articles, very positive environmental and waste management impacts were noted. Furthermore, Pakdel et al. indicated that despite all the positives associated with recycling these fibers, very little research has been done into the end usage by turning these recycled fibers into non-wovens for making composite laminates. In addition, Oliveux et al. [9]also conducted a comprehensive literature review of the various recycling technologies (mechanical recycling; thermal processes: pyrolysis, fluidized bed pyrolysis, micro-waves assisted pyrolysis; and solvolysis) that are in existence to aid in the recycling of fiber-reinforced polymer composites (fibers: glass, carbon, and aramid). The review also focused on reusing the reclaimed fibers: particularly, the way they have been reused in newer materials or applications and the major technical difficulties at hand. It was noted in the review article that recycled fibers can substitute little amounts of virgin fibers in products for structural applications but not at a high fiber loading amount. This means that only a certain percentage of recycled carbon fibers (rCFs) could be used in a material destined for structural applications-whether it is light-weight structural applications or not. And, also, the recovered carbon fibers from high-end technology applications cannot be reused in the same applications from which they were reclaimed since they suffer a slight decrement in fiber properties, so alternative applications must be found to reuse these fibers. Regarding the above observation, it can be concluded that, for example, fibers recovered from the wing-box of an aircraft cannot be used in any parts of the wing-box again but rather, such fibers could be used in other light-weight structural applications in the cabin of the aircraft. Applications could range from cockpit panels to cabin trays in the aircraft.

To develop sustainable, eco-friendly materials that could compete with other materials, it is essential to adopt the hybridization technique since there is a slight decrement in the properties of the involving rCFs. Hybridization is a method in which two or more fibers are combined to enhance the properties of their composites. However, it should be noted that hybridization has its shortcomings: there is either a negative or positive effect [10]. In hybridization, the final composite is expected to exhibit the advantages of involving fibers (positive hybridization), but that is always not the case. A scenario could arise where the 'wrong' fibers are combined to take the disadvantages of the involving fibers rather than the advantages of the involving fibers in the final composite product (negative hybridization).

Since there was a slight decrement in the properties of the reclaimed fibers of the carbon-reinforced composites, combining high strength, standard modulus fiber, T700 with high modulus fiber, M46J should take advantage of both properties fibers.

The use of aramid non-woven mats for structural applications has been in existence for a long time, and as seen in Figure 1 [11], such materials have been used in the construction of the Boeing 757 aircraft: it has been used in making the fixed trailing edge panels; wing-to-body fairings; and flap support fairings (the forward segment). Boeings 757 and 767 were simultaneously developed in the 1980s. According to the Boeing company, the 757 and 767 both had the same technological advances in materials, avionics, aerodynamics, and propulsions [12]. According to the above statement, although the 757 was phased out in early 2004, the 767 is still being constructed; hence materials made with aramid non-woven mats are still being used for structural applications. Although aramid non-woven mats in structural composites are known for toughness improvements such as compression after impact (CAI) and resistance to delamination [13-16], with regards to the fixed trailing edge panels, for example, during their operation, one of the loads they experience is bending loads which are characterized as flexural moduli hence the use of aramid non-woven mats for benchmarking purposes in the study of the flexural and shear strength properties of UD CFRP composite interleaved with recycled carbon fiber non-woven mat is not far-fetched.

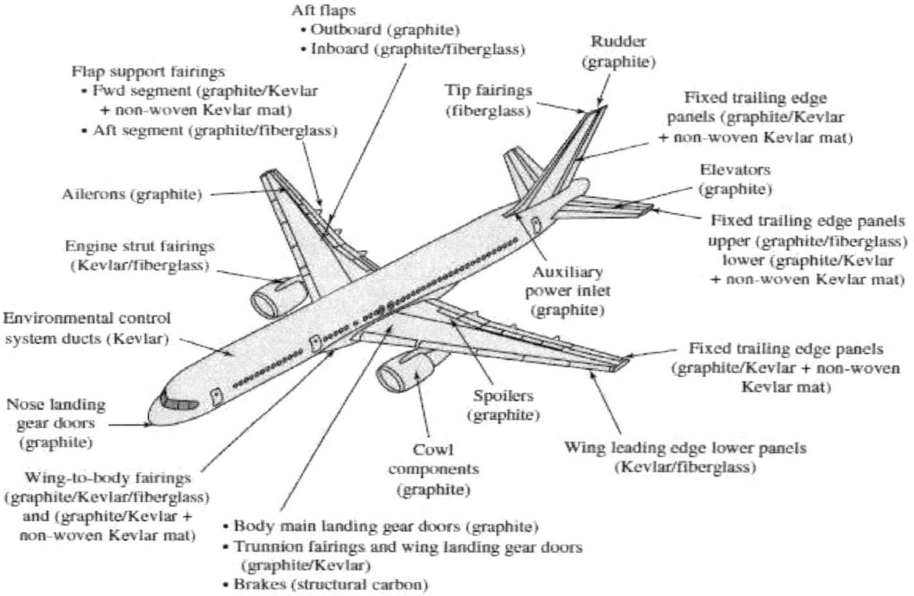

Figure 1. The main structure of the Boeing 757-200 aircraft showing composite parts used in its manufacturing [11]. Source: Copyright 2021, McGraw Hill LLC.

The composite laminates with the non-woven mats can be recycled despite the introduction of the vAFs. From the works of Asmatulu et al. [17], Okajima et al. [18], Yuyan et al. [19], and Chuang et al. [20], vAFs are recyclable. Although no literature has been sighted at the time of writing wherein mixed fibers(hybridized carbon and aramid fibers) have been recycled, with the current progress of recycling technologies, such materials can inevitably be recycled sooner than later. Regarding the configuration with the rCF mat only, to the best of the authors' knowledge by way of extensive literature, the methodology applied (the interleaving of UD fiber/epoxy prepreg with rCF non-woven mat) and the product developed(UD CFRP composite interleaved with recycled carbon fibers' non-woven mat) is the first of its kind. The UD CFRP composite interleaved with aramid fibers' non-woven mat is solely for comparison purposes.

Investigating the flexural and shear strength properties of unidirectional carbon/epoxy laminated composite interleaved with recycled carbon fibers non-woven mat for possible light-weight structural applications is the objective of this paper. Since the rCFs are fluffy and discontinuous; the best way to utilize them for potential light-weight structural applications is to convert them into non-woven mats using a paper-making technique and use the non-woven mats to interleave continuous fibers. With that, a closed-loop sustainable, environmentally friendly material can be developed for possible light-weight structural applications.

2. Materials and methods

2.1. Materials

The virgin aramid fibers, poly (p-phenylene terephthalamide) branded as Kevlar® 29 (K29) was supplied by DuPont in chopped strands with an average length of 3.0 mm. Kevlar® 29 (K29) is a para-amid with good mechanical and damping properties and is widely used for structural applications. The fiber has a diameter of 12 μm, a specific density of 1.44, ultimate tensile strengths of 2.92 GPa to 3.62 GPa, and a tensile modulus of 70.5 GPa. The picture of vAF is shown in Figure 3(a).

The recycled carbon fibers (rCFs) are standard modulus and high modulus Toray T700SC and M46J, respectively, recovered by the fluidized bed recycling process atthe University of Nottingham, UK. Although the properties of the rCFs were not measured in this project, it is expected that the rCFs will suffer a slight decrement in their tensile strengths.

The vAFs and rCFs in the form of 60 g/m^2 non-woven mats (mats),as seen in Figure 3(b) and Figure 4(c), respectively, were used as the interleaving materials (they were used to interleave a commercially available unidirectional carbon (UD) fiber/epoxy prepreg, Toray T700SC carbon fiber (12K) prepregs). The prepreg has a ply thickness of 0.20 mm,areal density of 200 g/m^2, and resin content of 35 wt.%. The resin filmhas an areal density of 125 g/m^2, a ply thickness of 0.10 mm and resin content of 100 wt.%. Both the UD carbon fiber prepreg and the resin film have the same resin system, YPH-42T, and were supplied by CA composites limited, Shanghai, PR China.

2.2. Experimental methods

2.2.1. Manufacture of recycled carbon fiber and chopped virgin aramid fibers non-woven mats.

Figure 2 shows the front view of the in-house rig used to make the non-woven mat using a paper-making technique. The wet paper-making technique was used to convert both recycled carbons and chopped virgin aramid fibers into rectangularly shaped non-woven mats (340 ×240 mm^2) with an areal density of 60 g/m^2 using the in-house rig. The technique consists of three stages: fiber dispersion, fiber filtration, and non-woven mat drying.

Firstly, an amount of water is determined using a set of parameters, and the resulting measurement is pumped into the dispersion tank. Next, the chemical compositions found in table 1 are measured and added to the water in the dispersion tank.

The chemicals are carefully added and thoroughly mixed for 8 hours (the resultant solution is used for a particular areal density as many times as possible) using a high-shear radial impeller at about 1200 ± 50 rpm. The resulting solution is pumped into the storage tank, and the desired amount is pumped from the storage tank into the dispersion for the mixing of the fibers and subsequently making the mats.

In reference to the steel mesh size on which the resulting mat will settle, the masses of rCFs and vAFs are measured in grams and added to the resultant solution in the dispersion tank separately. The rCFs measured are T700SC and M46J in the ratio of 75:25-the percentage ratio was because of the processing limitations associated with the M46J rCFs: the two were added to the solution simultaneously. Next, the fiber suspension is mixed for 30 ± 5 minutes with a high-shear radial impeller at about 1200 ± 50 rpm. The speed and time reported here were the best-compromised values that resulted in the making of good mats. The fiber-slurry is then pumped into the sheet former and the fiber-slurry sieve through a stainless-steel mesh with the aid of a vacuum pump. The resulting mat is wet and fragile at this stage; hence it is carefully removed from the stainless-steel mesh and dried at 100°C for 10 minutes.

2.2.2. Composite laminates' manufacturing.

The composites were fabricated using the vacuum bagging-assisted compression molding technique. Firstly, the Toray T700SC UD carbon fiber (12K) epoxy prepregs and mats were cut to 328 × 228 mm^2 using a ply cutter. The configurations with mats had a single mat sandwiched between two resin films (F_2) before being inserted into the middle of 12 layers of prepregs. With the arrangements with mats, six prepregs (P_6) were stacked in the UD fiber direction, followed by the sandwiched mat, then with another six prepregs. The layup sequence $[P_{12}]$, $[P_6F_1M_{rCF}F_1P_6]$, and $[P_6F_1M_{vAF}F_1P_6]$are configurations with no mat (the control laminate),100% rCF mat and 100% vAF, respectively.

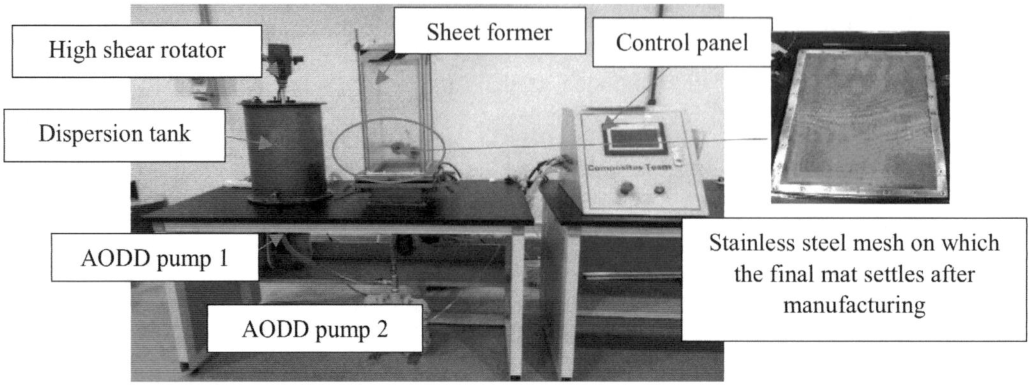

Figure 2. Front view of the non-woven mat rig.

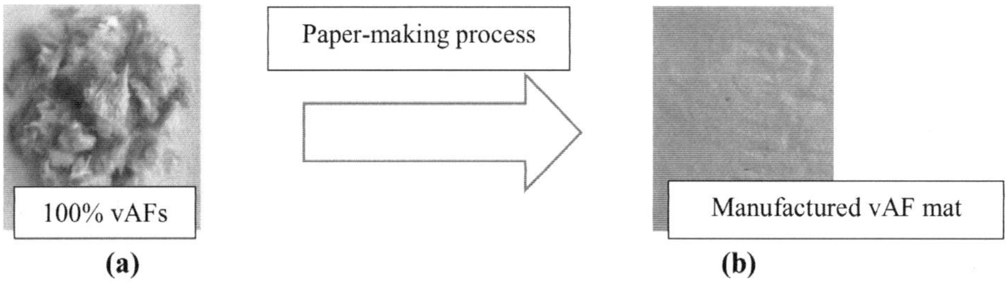

Figure 3.(a) vAFs and (b)manufactured vAF non-woven mat.

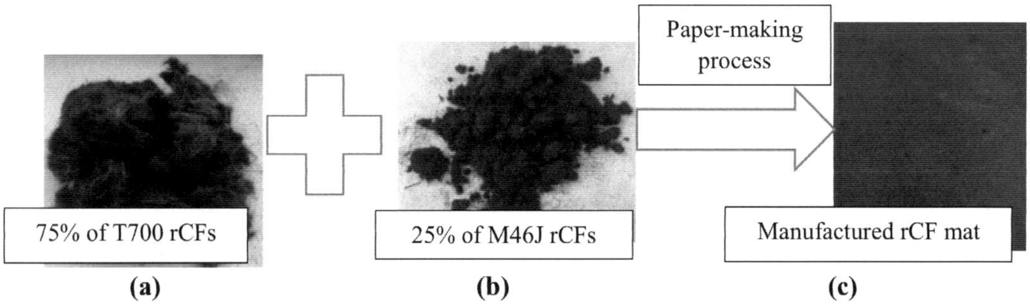

Figure 4. (a) 75% of T700 rCFs, (b) 25% of M46J rCFs and (c) manufactured 100% rCF non-woven mat.

Table 1. Chemical compositions.

Chemicals	Supplier	Functionality	Composition
Hydroxyethyl Cellulose	Macklin	Binding and Viscosity Modifier	5.5 g/l
Brij 35 non-ionic surfactant	Aladdin	Wetting and Dispersing Agent	1.0 g/l
Tributyl phosphate	Aladdin	Anti-foaming agent	0.5 g/l

The stainless steel mold and frame were cleaned with Acetone before being wiped with the mold releasing agent, Chemlease® PMR EZ, to remove the cured composite laminates. The preforms were then fitted into a 330 × 230 mm^2 (inner dimensions) stainless steel frame mold and two PTFE release films were used to cover the top and bottom of the steel plates. The assemblage was then placed on a 600 × 500 × 4 mm^3 stainless steel plate, with breather laid on top and vacuum, bagged using a sealant tape. A vacuum pump was connected, and the preforms were degassed at about room temperature till the vacuum pressure reaches 0.1MPa.Afterward, the assemblage was relocated to the hot press and dwelled at 80°C for about 30 min. At the about 30th minute, an additional 5 MPa pressure from the hot-press was applied, and the curing temperature was increased to 130°C in about 25 minutes and it dwelled at 130°C for 120 minutes. The cured composites were then taken out of the hot press at the 120th minute and allowed to cool down to about room temperature with the vacuum still applied (vacuum is applied throughout). Figure 5 shows the curing cycle. After the fabrication of the composite laminates, the samples for the various tests were cut using a water jet cutter. The samples were cleaned and dried in an oven at 100°C for 24 hours before testing.

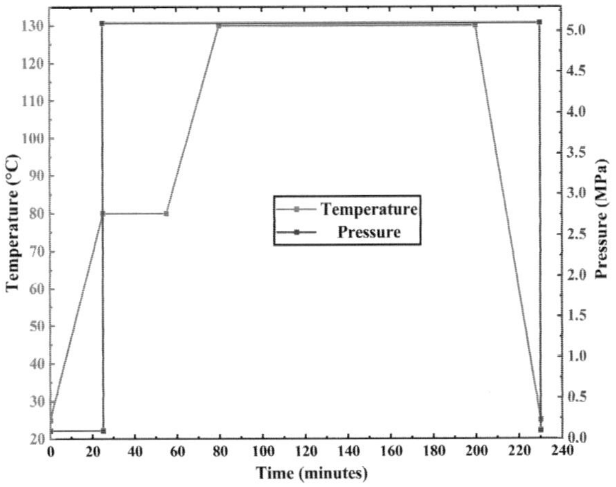

Figure 5. The curing cycle: temperature and pressure profile for the composite laminates.

2.3. Characterizations

2.3.1. *Flexural modulus and flexural strength.*

The flexural modulus and the flexural strength of the composite laminates were measured in a three-point bend configuration according to the standard, ASTM D790-2017 [21]. The flexural specimen had the dimension 130 mm × 12.7 mmwith an average thickness of 2.31 mm, 2.51mm, and 2.52 mm for the control, configurations with rCF and vAF mats, respectively. The support span-to-thickness ratio was 32, and the rate of crosshead motion was in the order of 4.0 mm/min, and 4.5 mm/min for the control and configurations with mats, respectively. The rate of crosshead motion for the configurations with mats falls within the ±10 % allowance as specified by the ASTM D790-17 standard. The tests were done at room temperature using the MTS UTM Exceed 45.105 with a load cell of 50 kN. The flexural modulus and flexural strength were computed using Equation (1) and Equation (3) respectively [21]:

$$E_f = \frac{L^3 m}{4wt^3} \quad (1)$$

Where: E_f = flexural modulus (MPa), L = support span (mm), w = width (mm), t = thickness(mm), and m = slope of the tangent to the initial straight-line section of the load-deflection curve, (N/mm of deflection); m is estimated by drawing a straight-line to the steepest initial section of the load-deflection curve.

With regards to the evaluation of the flexural strength, ordinarily, Equation (2) is used to compute the flexural strength but according to the standard, when higher support span-to-thickness ratios are used (ratios greater than 16) (32 was used in this project), significant-end forces get developed at the support noses. This will affect the moment in a simply supported beam. As such, Equation (2) includes an approximate correction parameter for the effect of these end forces hence Equation (3) being used to estimate the flexural strength.

$$\sigma_f = \frac{3F}{2wt^2} \tag{2}$$

where: σ_f = flexural strength (MPa), F = maximum load encountered before failure (N), L = support span (mm), w = width (mm), and t = thickness (mm).

$$\sigma_f = \frac{3F}{2wt^2}\left[1 + 6\left(\frac{D}{L}\right)^2 - 4\left(\frac{t}{L}\right)\left(\frac{D}{L}\right)\right] \tag{3}$$

where: σ_f, F, L, w, and t = the same as for Equation (2), and D = deflection of the centerline of the specimen at the middle of the support span or maximum deflection encountered before failure (mm).

2.3.2. *Short beam shear strength.*
In addition, to assess the composite laminates' interfacial adhesion, the short beam shear test was done according to the standard, ASTM D2344-2016 [22],wherein the interlaminar shear strength is calculated to quantify the interfacial adhesion. The specimen had the dimension, 8 mm × 6 mm with a thickness of 2.31 mm, 2.51 mm, and 2.52 mm for the control, configurations with rCF mat and vAF mats, respectively. The span length-to-thickness ratio was approximately 4. All the shear tests were done at room temperature with a crosshead rate of 1.0 mm/minusing the same universal testing machine utilized in the flexural tests. The short beam shear strength was computed using Equation (4) [22]:

$$P^{sbs} = 0.75 \times \frac{F_m}{w \times t} \tag{4}$$

where:P^{sbs} = short-beam shear strength (MPa); F_m = maximum load observed during the test (N); w = width (mm), and t = thickness (mm).

2.3.3. Density determinations.
The density was determined according to the ASTM D3171-15 (Test method II) standard [23] by using a set of vernier calipers and a Mettler Toledo laboratory scale (0.0001 g). All thickness variations of the tested samples were measured, and the average value was determined. A 20 mm × 30 mm sample size was suitable per the standard recommendations and available materials. The density in g/cm^3 was calculated using Equation (5) [23]:

$$\rho_c = \frac{M_i}{A \times t \times 100} \tag{5}$$

where: M_i = mass of the tested sample (g); A = area of the tested sample (m^2); andt = thickness (mm).

2.3.4. Surface electron microscopy (SEM).
The surface structure of the recycled M46J fibers was analyzed using a Zeiss Sigma VP scanning electron microscope with an accelerating voltage (EHT) of 5.0 kV. The working distance (WD) was 11.3 mm. Secondary electron detectors (SE2), and in-lens detectors were employed in the SEM imaging. Before the SEM imaging, a thin layer of gold was applied to the surfaces of the fibers using Leica SCD 500 gold sputter.

3. Results and discussion

3.1. Flexural modulus of the composite laminates
The flexural modulus of the control sample and the configurations with non-woven matsare summarized in table 2. From the table, the incorporation of the non-woven mats improved the flexural modulus.

The average flexural modulus in GPa for the control, CFRP with rCF non-woven mat and vAF non-woven mat is 125.1, 135.3, and 140.1, respectively.

Table 2. List of the average flexural modulus of the composite laminates.

Sample	Flexural modulus/GPa	Differences (%) compared with the control
$[P_{12}]$	125.1±3.1	0.0
$[P_6F_1M_{rCF}F_1P_6]$	135.3±5.1	8.2
$[P_6F_1M_{vAF}F_1P_6]$	140.1±3.3	12.0

Some of the factors affecting the flexural modulus are the properties of the fibers and fiber contents; hence the incorporation of the mats is expected to increase the flexural modulus because of the additional fibers from the non-woven mats. In addition, the 'somewhat' sandwich structure could explain further why the CFRP with non-woven mats had better flexural modulus compared with the control. The mats' configurations can be considered a sandwich structure with six skins (upper and bottom each) separated by a core material (the non-woven mats). A sandwich structure is made up of two skins separated by a core material inserted centrally. The core material behaves like the web in an I-beam which increases the moment of inertia without much increment in weight there by producing an efficient structure for sustaining bending and buckling loads [24]. This phenomenon is a possible explanation of why the configurations with mats had better flexural modulus values than the control sample. Regarding the vAF mat's performance, because it is slightly thicker than the rCF mat, this makes the moment of inertia significant, thereby increasing the flexural modulus; hence the CFRP with the vAF mat having better values compared with the configuration with the rCF mat. Notwithstanding the slight decrement in fiber properties of the reclaimed carbon fibers, to further suggest possible reasons for why the CFRP with rCF mathad slightly lower values compared with the CFRP with vAFmat, more studies must be done whiles considering other factors like the effect of rCF types and their micro-hybridization with vAF.

3.2. Flexural strength of the composite laminates

The flexural strength of the control sample and the configurations with non-woven mats are summarized in table 3. From table 3, the incorporation of the rCF non-woven mat had negative flexural strength performance, whiles the vAF non-woven mat had a marginal positive performance. The fundamental reason for the versions observed here is the correlation between the interfacial adhesion (see 3.4) and the flexural strength performances. Good interfacial adhesion generally leads to good flexural strength performances.

The average flexural strength in MPa for the control, CFRP with rCF non-woven mat and vAF non-woven mat is 1567.6, 1444.1, and 1585.3, respectively.

Table 3. List of the average flexural strength of the composite laminates.

Sample	Flexural strength/MPa	Differences (%) compared with the control
$[P_{12}]$	1567.6±74.3	0.0
$[P_6F_1M_{rCF}F_1P_6]$	1444.1±187.6	-7.9
$[P_6F_1M_{vAF}F_1P_6]$	1585.3±89.3	1.1

3.3. Specific flexural modulus

The specific flexural modulus of the control sample and the configurations with non-woven mats are summarized in table 4. From the table, the incorporation of the non-woven mats had a positive impact on the specific flexural modulus. Compared to the control, the specific flexural modulus increased by 9.5% and 13.3%, respectively, for the CFRP with rCF and vAF mats.

Table 4. List of the specific flexural modulus of the composite laminates.

Sample	Specific flexural modulus($\frac{GPa}{gcm^{-3}}$)	Differences (%) compared with the control
$[P_{12}]$	79.2±2.0	0.0
$[P_6F_1M_{rCF}F_1P_6]$	86.7±3.3	9.5
$[P_6F_1M_{vAF}F_1P_6]$	89.7±2.1	13.3

Compared with the control sample, the increments observed by the inclusion of the non-woven mats demonstrated that in applications where minimum structural weight is required at high flexural modulus, the configurations with the mats would be competitive materials.

3.4. Short beam shear strength

The short-beam strength of the control sample and the configurations with non-woven mats are summarized in table 5. From the table, the incorporation of the non-woven mats harmed the short-beam strength.

Chemically, aramids have inert surfaces that usually result in poor interfacial adhesion, but in this work, the vAF mats had better performance when compared with the rCF mats. Resin compatibility issues might explain this discrepancy. For example, taking the M46J rCFs, they have some resin residues on their fibers, as shown in Figure 6. These resin residues can result in resin compatibility issues that could lead to poor interfacial adhesions, as observed here.

The solution here is to either find an appropriate sizing agent and ratios to treat the surfaces of the manufactured mats or to strive to use cleaner rCFs. Improving interfacial adhesion is critical because poor interfacial adhesion could lead to poor flexural strength values and delamination issues with the final product.

Table 5. List of the average short-beam strength of the composite laminates.

Sample	Short-beam strength /MPa	Differences (%) compared with the control
$[P_{12}]$	87.1±5.3	0.0
$[P_6F_1M_{rCF}F_1P_6]$	81.5±3.4	-6.4
$[P_6F_1M_{vAF}F_1P_6]$	81.9±2.1	-6.0

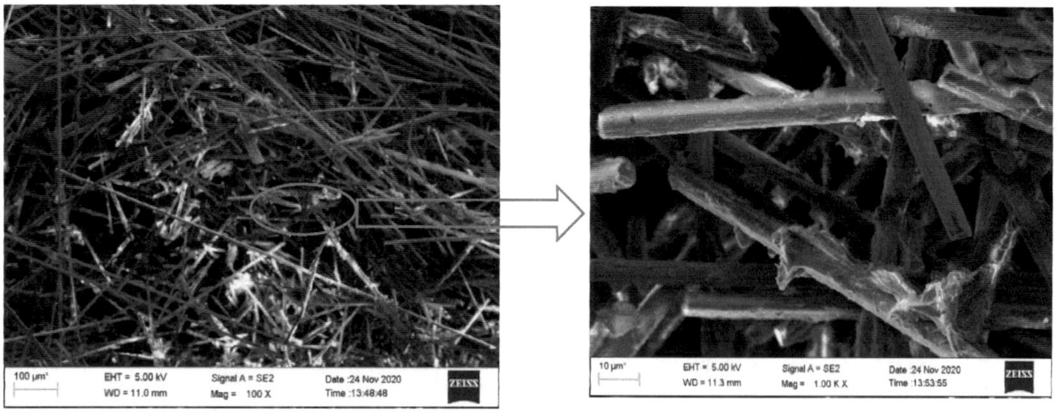

Figure 6. SEM images of the M46J rCFs.

4. Conclusion

This study evaluated the flexural and shear strength properties of a 12-ply unidirectional (UD) carbon/epoxy (T700/YPH-42T) laminated composite to incorporate recycled carbon fibers (rCFs) into carbon fiber reinforced composites for possible light-weight structural applications. The flexural and shear strength properties were evaluated according to the ASTM D790-17 and ASTM D2344-16 standards, respectively.

When compared with the control sample, it was found that the flexural modulus increased by approximately 8.2% and 12.0% for the CFRP with rCF and vAF mats, respectively; the specific flexural modulus increased approximately by 9.5% and 13.2% respectively for the CFRP with rCF and vAF mats; the flexural strength decreased by approximately 7.9% and increased by approximately 1.1% for the CFRP with rCF and vAF mats, respectively; and the shear strength approximately decreased by 6.4% and 6.0% for the CFRP with rCF and vAF mats, respectively. The correlation between the shear and flexural strengths means that good interfacial adhesion is necessary to improve the flexural strengths. Therefore, to resolve the negative shear strength performances, it is recommended that either cleaner rCFs should be used or the surfaces of their mats should be treated with a coupling agent to improve their interfacial adhesions.

Acknowledgment

This work was supported by the Ningbo Municipal Bureau of Science and Technology under the Natural Science Programme [Grant number 2019A610155].

References

[1] Attahu C Y and An L 2017 Influence of Assembly Gap and Shims on the Strain and Stress of Bolted Composite-Aluminum Structures ARPN *J. Eng. Appl. Sci.* **12(5)** 1593-1619

[2] Attahu C Y, An L, Li Z and Gao G 2018 Influence of Shim Layers on Progressive Failure of a Composite Component in Composite-Aluminum Bolted Joint in Aerospace Structural Assembly 2018 *Trans. Nanjing Univ. Aeronaut. Astronaut.* **35(1)** 188-202 DOI: 10.16356/j.1005-1120.2018.01.0188

[3] Zhu C, Yang J and Rudd C 2020 Vibration transmission and power flow of laminated composite plates with inerter-based suppression configurations *Int. J. Mech. Sci.* **190** 106012

[4] Khanna V, Kumar V and Bansal S A 2021 Aluminium-carbon fibre metal matrix composites: a reviewIOP Conf. Ser. *Mater. Sci. Eng.* **1033**, 012057 https://doi.org/10.1088/1757-899X/1033/1/012057

[5] Wong K H,Rudd C, Pickering Sand Liu X 2017 Composites recycling solutions for the aviation industry *Sci. China Technol. Sci.* **60** 1291–1300 doi:10.1007/s11431-016-9028-7

[6] Campbell F C 2010 *Structural Composite Materials* (Ohio: ASM International) **14**

[7] Pakdel E, Kashi S, Varley R and Wang X 2021 Recent progress in recycling carbon fibre reinforced composites and dry carbon fibre wastes *Resour. Conserv. Recycl.* **166** 105340

[8]May D, Goergen C and Friedrich K 2021 Multifunctionality of polymer composites based on recycled carbon fibers: A review *Adv. Ind. Eng. Polym. Res.* **4** 70-81

[9] Oliveux G, Dandy L O andLeekeG A 2015 Current status of recycling of fibre reinforced polymers: Review of technologies, reuse and resulting properties *Prog. Mater. Sci.* **72** 61–99 http://dx.doi.org/10.1016/j.pmatsci.2015.01.004

[10]Ebrahimnezhad-Khaljiri H, Eslami-Farsani R and Akbarzadeh E 2020 Effect of interlayer hybridization of carbon, Kevlar, and glass fibers with oxidized polyacrylonitrile fibers on the mechanical behaviors of hybrid composites *Proc. Inst. Mech. Eng. C J. Mech. Eng. Sci.* **234** 1823 - 1835

[11] Schaffer J P et al. 1998 *The Science and Design of Engineering Materials* (New York: McGraw-Hill) 580

[12] Boeing 2021 Boeing: Historical Snapshot: 757 Commercial Transport. [online] Available at: <https://www.boeing.com/history/products/757.page> [Accessed 10August 2021].

[13] Ramírez V, Hogg P and Sampson W 2015 The influence of the non-woven veil architectures on interlaminar fracture toughness of interleaved composites *Compos. Sci. Technol.* **110** 103-110
[14] Beylergil B,Tanoğlu M and Aktas E 2018 Effect of polyamide-6,6 (PA 66) non-woven veils on the mechanical performance of carbon fiber/epoxy composites *Compos. Struct.* **194** 21-35
[15] Beylergil B, Tanoğlu MandAktas E 2019 Mode-I fracture toughness of carbon fiber/epoxy composites interleaved by aramid non-woven veils *Steel and Compos. Struct.* **31** 113-123
[16] Ni N, Wen Y, He D, Yi X, Zhang T and Xu Y 2015 High damping and high stiffness CFRP composites with aramid non-woven fabric interlayers *Compos. Sci. Technol.* **117** 92-99
[17] Asmatulu E, Twomey J and OvercashM 2014 Recycling of fiber-reinforced composites and direct structural composite recycling concept *J. Compos. Mater.* **48(5)** 593–608 DOI: 10.1177/0021998313476325
[18] Okajima I, Okamoto H and T SakoT 2019 Recycling of aramid fiber using subcritical and supercritical water *Polym. Degrad. Stab.* **16** 222-28 https://doi.org/10.1016/j.polymdegradstab.2019.01.034
[19] Yuyan L, Li L and Linghui M 2009 The Experimental Research on Recycling of Aramid Fibers by Solvent Method *J. Reinf. Plast. Compos.* **28(18)** 2211-2220 DOI: 10.1177/0731684408092073
[20] Chuang Y, Bao L, Lou C and Lin J 2019 Hybrid-Fiber-Reinforced Composite Boards Made of Recycled Aramid Fibers: Preparation and Puncture Properties *Fibers Polym.* **20(2)** 398-405 DOI 10.1007/s12221-019-8868-1
[21] ASTM D790-20 2020 *Standard Test Methods for Flexural Properties of Unreinforced and Reinforced Plastics and Electrical Insulating Materials* (West Conshohocken, PA: ASTM International) DOI: 10.1520/D0790-17
[22] ASTM D2344 / D2344M-16 2016 *Standard Test Method for Short-Beam Strength of Polymer Matrix Composite Materials and Their Laminates* (West Conshohocken, PA: ASTM International) DOI: 10.1520/D2344_D2344M-16
[23] ASTM D3171-15 2015 *Standard Test Methods for Constituent Content of Composite Materials* (West Conshohocken, PA: ASTM International) DOI: 10.1520/D3171-15
[24] Gibson L J and Ashby M F 1997*Cellular Solids: Structure and Properties* 2nd ed. (Cambridge: Cambridge University Press) 345

The effect on the tool flank wear width VBmax during turning of Hybrid Mg/(8wt% SiC(p) +2wt% Al2O3(p) +1wt% Gr(p) MMC using PVD coated cermet tool

Jaspreet Hira[1], Alakesh Manna[2], Binit Kumar Jha[3], Rohit Sharma[4], Anup Malik[5], Deepak Bhutani[6]

[1]Associate Professor and Head, Department of Mechanical Engineering, Raffles University, Neemrana, Rajasthan, India
[2]Professor and Head, Department of Mechanical Engineering, PEC, University of Technology Chandigarh, India
[3]Assistant Professor, K.S.R. College Sarairanjan Samastipur, Bihar, India
[4]Assistant Professor, Department of Mechanical Engineering, Amity University Noida, India
[5]Assistant Professor, Department of Mechanical Engineering, Malaviya National Institute of Technology Jaipur
[6]Senior Researcher, Department of Production Engineering, PEC, University of Technology Chandigarh, India

*Corresponding Author mailing ID: jas.iitdelhi@gmail.com

Abstract. In this study Taguchi Optimization Technique was used, followed by ANOVA to evaluate the machinability i.e. turning of a hybrid Mg/(8wt%SiCp+2wt%Al2O3p+1wt%Grp) MMC using PVD coated cermet tool under the dry turning conditions. Several experiments were conducted using the L27 (313) orthogonal array on NH26 automated lather machine. ANOVA was used to determine the impression of the turning parameter on the VBmax of the tool flank. The finding of the research work investigated that the cutting speed in the dominant factor influence the tool wear width. The confirmation test revealed that the Taguchi technique has been successful in the optimization of the machining parameters for minimum tool wear. The optimal parameter arrangement for minimum tool wear width is at cutting speed of 120 m/min, with the feed rate of 0.4 mm/rev and depth of cut at 0.75 mm. The scanning electron microscope images revealed notches, groves and pits on the tool flank face that is clearly visible at different magnification units. Herein, this paper discuss about several reported synthesis techniques, various characterization techniques used for structural and surface properties identifications, different morphologies and various potential applications of transition metal oxide based nanocomposites.

Keywords: Hybrid MMC, Taguchi, Anova, S/N ratio, PVD coated cermet tool

1. Introduction

Hybrid metal matrix composite has gained tremendous attention in the manufacturing industry meeting the latest industrial requirements. These materials have inherited properties, such as high specific strength, higher rigidity and higher characteristics that the pure alloys. Despite having higher strength values, hybrid MMC are challenging to machine due to the presence of hard reinforced particles. Machining these metal matrix composites is a huge challenge and industries are working to develop an optimal solution in which machining parameters must be optimally suited to achieve the desired results. The physical interaction of hard reinforced particles on the tool causes a variety of tool failures, including high-temperature shearing, diffusion wear, abrasive wear, fatigue wear, oxidation effect, flank wear, crater wear, and chipping action. Holmberg et al [1] investigated the outstanding strains in TiN, DLC and MnS2 coated material surface and discovered that the PVD coated surface compressional residual stress range from 0.5 to 4 GPa and can reach upto 11 GPa. Yang et al [2] discovered that the carbon, nitride cermet coated material coated with TiAlN and TiAl/Cr AlN showed the improved adhesion, wear resistance and abrasive wear. During machining, Bushlya et al [3] discovered that the presence of SiCp in the Al-SiCp MMC deteriorated the protective layers on diamond and boron nitride tools. TiAlN coatings deposited on Si3N4 cutting inserts via PVD technique demonstrated higher adhesion strength than Cr-AlN coatings, according to the author [4]. Vaxiour et al [5], reviewed machinability of hybrid MMC, and observed that the presence of graphite material in the MMC enhances the surface finish and studied the improvement of the machinability due to its low heat expansion coefficient. Lim et al [6] explored the wear of the magnesium composite reinforced with nano sized Al particles and discovered the wear resistance increases with the increase in particles during higher machining operations. Rajesh et al [7] utilized L9 orthogonal array for conducting the machining on the red mud Al MMC, and this particular orthogonal proves to be beneficial in determining the best turning parameter combinations. Rubio et al. [8] used an L27 orthogonal array to optimise the parameters for the dry facing of magnesium in order to evaluate surface roughness, and they discovered that 0.04 mm/rev feed rate, 280 rpm spindle speed, and TP2500 tool coating are the best combinations for minimizing surface roughness. Hasan et al [10] examined the effect of the reinforcement, i.e. chopped fibres and woven roving, and discovered that as the particle strength and specimen hardness increased, so did the impact strength and specimen hardness. Barman et al [11] investigated the impact of the laser interferometry technique and demonstrated that these laser measurement systems provide higher accuracy and quality. Kumar et al found out the during the turning of nanocomposites Al2219 alloy matrix there is a formation of build up edge at lower cutting speed values. Kumar et al [13] investigated the effect of during parameters on the turning of Al 7075/10/SiCp and Al 7075 composite and found that the most significant parameters are the lower feed rate and high cutting speed. In view of the above research study an attempt has been done to find out the effect of machining parameters on hybrid MMC workpiece specimen using Taguchi design of experiments (DOE) and ANOVA methods.

2. Experimental Procedure

Hira et al [14], used the gas injection liquid stir casting to create an experimental workpiece specimen including the hard reinforced particles, which has been utilized for the work study. The dimension of the workpiece including the length 250 mm and diameter 90 mm respectively. The tool insert technical specifications has been inserted in the Table 1 and the chemical composition that is also listed in Table 2. Figure 1 (a) and (b) shows the schematic layout of the PVD coated cermet tool during turning and actual mounting of the tool on the holder. Figure 2(a) shows the schematic layout showing the instruments measuring the tool flank wear width VBmax and (b) instrument for measuring the value of the wear.

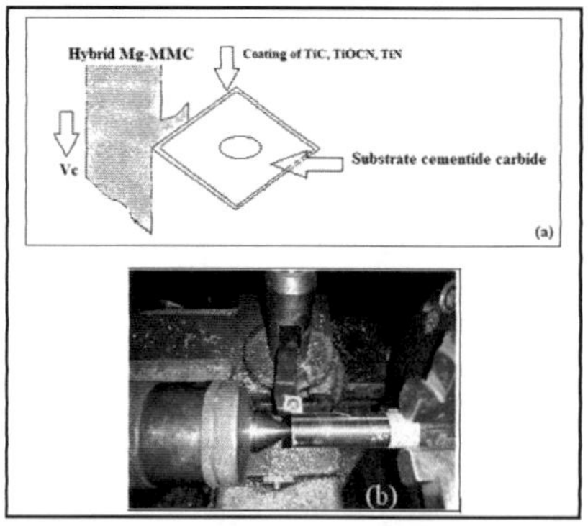

Figure 1. Schematic layout of the PVD coated cermet tool turning hybrid Mg-MMC; (b). Actual mounting of tool with tool holder.

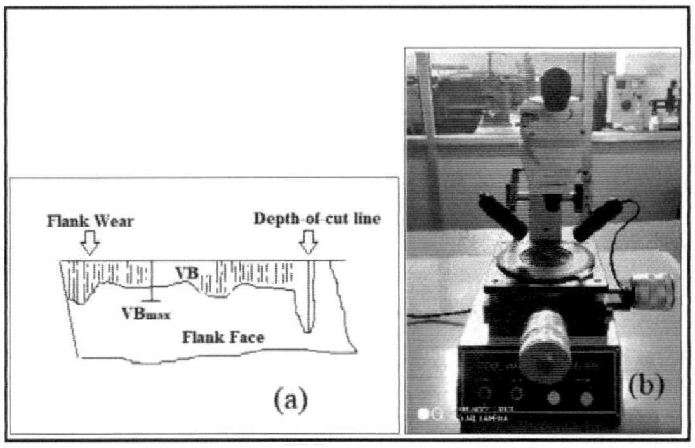

Figure 2. Schematic representation of the flank face of the PVD cermet tool with flank wear land VBmax and depth of cut line; (b) tool maker microscope for measurement of wear

Table 1. PVD coated cermet tool detail specifications

Tool shape	Tool Type	Cutting tool specification	Tool Material and Grade	Rake angle(⁰)	Clearance angle (⁰)	Cutting Edge (⁰)angle	Nose radius (mm)
Rhombic	TMax positive insert	CCMT09 T308WF1525	PVD coated cermet tool	6	7	80	0.8

Table 2. Chemical composition of hybrid Mg/(8wt%SiC(p)+2wt%Al2O3(p)+1wt%Gr(p)) MMC

Type of MMC	Size of reinforced particles (μm)	% $SiC_{(p)}$	% $Al_2O_{3(p)}$	% $Gr_{(p)}$	Mg
Discontinuous MMC	SiC(p)=23 Al2O3(p) =69 Gr (p)= 53	8	2	1	Balance

Table 3. Mechanical Properties

Material	Density (ρ)	Tensile strength (σ_t)	Yield strength (σ_y)	Elongation % (ϵ)	Hardness (HV)
Mg/(8wt%SiC(p)+2wt%Al2O3(p)+1wt%Gr (p))	2.9	230	210	5	63

3. Taguchi Design Approach

A strong experimental design tool, method based Taguchi design has been utilized for finding out effect of machining parameters on the workpiece specimen Mg/(8wt% SiCp+2wt% Al2O3+1wt% Gr) MMC. For a factorial design an orthogonal array L27 (313) has been utilized for the study and ANOVA has been utilized for important parameters. The Mathematical model has been developed considering the significant parameter and using multiple regression. Table 4 and table 5 shows the turning parameters and their level for detailing experimentation in order to optimize the significant parameters and also the result for the tool flank wear width VBmax.

Table 4. Turning (machining) parameters and their levels

Sr.No.	Machining Parameters	Unit	Level		
1.	A: Cutting Speed	m/min	40	80	120
2.	B: Feed	mm/rev	0.08	0.2	0.4
3.	C: Depth of cut	mm	0.2	0.5	0.75

Table 5. Test results for cutting tool flank wear width (VBmax) mm

Experiment No.	Y1	Y2	Y3	Test results for cutting tool flank wear width
1	0.078	0.079	0.083	0.08
2	0.089	0.087	0.094	0.09
3	0.09	0.091	0.119	0.1
4	0.1	0.11	0.12	0.11
5	0.13	0.139	0.151	0.14
6	0.129	0.13	0.131	0.13
7	0.149	0.15	0.151	0.15
8	0.148	0.149	0.153	0.15
9	0.19	0.2	0.21	0.2
10	0.2	0.21	0.22	0.21
11	0.189	0.19	0.191	0.19
12	0.187	0.19	0.193	0.19
13	0.239	0.24	0.241	0.24
14	0.269	0.27	0.271	0.27
15	0.289	0.29	0.291	0.29
16	0.3	0.31	0.32	0.31
17	0.32	0.33	0.34	0.33
18	0.31	0.319	0.331	0.32
19	0.19	0.21	0.23	0.21
20	0.289	0.29	0.291	0.24
21	0.259	0.26	0.261	0.26
22	0.341	0.35	0.359	0.35
23	0.369	0.37	0.371	0.37
24	0.389	0.39	0.391	0.39
25	0.38	0.42	0.43	0.41
26	0.489	0.49	0.491	0.49
27	0.49	0.5	0.51	0.5

4. Results and Discussion

4.1 ANOVA

Table 6 describes the ANOVA an F test results for the tool flank wear width (VBmax) it has been found that the percentage contribution of the cutting speed is approximately 32.78%, the feed rate contribution is 19.4% and the depth of cut is 23.05% which is equally responsible and has significant resulting in the performance of cutting tool flank wear width (VBmax). Through the Table 6 and analysis it is found that the highest contribution among various machining parameters is the cutting speed during the turning of hybrid Mg/MMC workpiece specimen.

Table 6. ANOVA and "f" test for tool flank wear width (VB_{max})

Sr.No.	Control Factor	DOF	Sum of Square	Variance	F Test	Percentage contribution
1.	A: cutting speed	2	0.14768	0.073484	25.8	32.78
2.	B: Feed rate	2	0.08625	0.043125	15.11	19.14
3.	C:Depthof cut	2	0.10375	0.051875	18.185	23.05
4.	AB Interaction	4	0.0281	0.007025	2.462	6.24
5.	AC Interaction	4	0.0468	0.0117	4.10	10.39
6.	BC Interaction	4	0.015	0.00375	1.314	3.34
7.	Error	8	0.02282	0.002853	-	5.06
8.	Total	26	0.4504	-	-	100.00

4.2 Signal to Noise Ratio (S/N)

Figure 3 shows the signal to noise ratio (dB) for the tool flank wear width (VBmax) and it can be seen through the figure that A3B3C3 are the optimizing parameters intended for minimum tool flank wear width (VBmax)

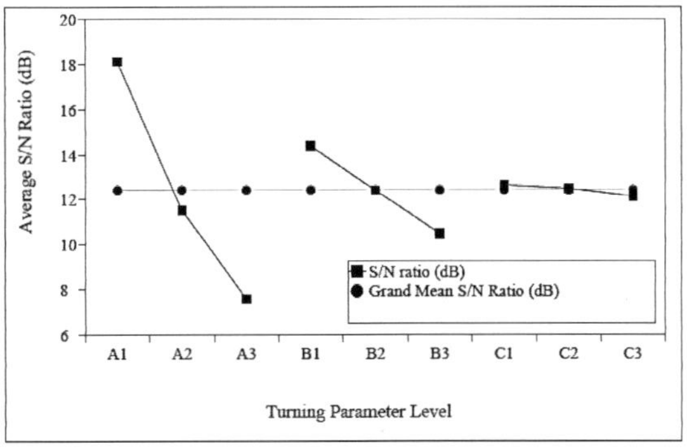

Figure 3. S/N ratio (dB) for tool flank wear width (VB_{max})

4.3 Development of Mathematical Model

Eq (1) depicts the developed mathematical model using the Gauss elimination method which is taking three parameters into consideration i.e. the cutting speed, feed rate and depth of cut:

YFlankwear=-0.07348+0.00289*A+0.24085*B+0.07041*C+0.0026*AB-0.00101*AC+0.06027*BC+0.000004*A2-0.2865*B2+0.00134*C2 (1)

Table 7 shows the assessment of the experimental investigation and the mathematical model, it has been found that the percentage error was discovered to be less than 10% which is indicating that the developed mathematical model described by Montgomery and Douglas was successful [15]. Through the confirmation and results its clear that the we can use this mathematical model to predict the machining parameters for obtaining preferred machining results in which the tool wear is minimum.

Table 7. Confirmation test results for cutting tool flank wear width VB_{max}

Confirmation testno.	Experimental results	Results as per developed mathematical model	% Error
1.	0.148	0.151968	2.6
2.	0.32	0.345589	7.9
3.	0.44	0.465806	5.8

4.4 Interaction plots of tool wear

Figure 4 (a), (b) and (c) demonstrate the different color region indicating in the figure which are representing the required machining groupings for achieving the minimum cutting tool flank value, the red color zone is representing the adverse conditions. The physical significance of the interaction plots are the interference values generated through the contact of tool tip which bears high temperature at high cutting speeds. The more the cutting speed the more the tool wear. The deformation in the tool is due to these high speeds and also interaction of hard reinforced particulates in the hybrid Mg/MMC workpiece specimen 5.

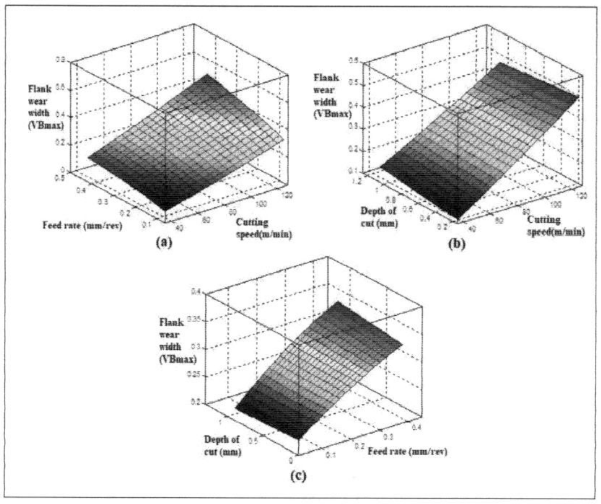

Figure 4. (a).Interaction effect tof feedrateandcutting speed on tool flankwear,(b).Interactioneffect of depthof cut and cuttingspeedon tool flankwear(c).Interaction effectof depthofcutandfeedrate ontool flank wearin machining(turning)hybridMg/(8wt%SiC(p)+2wt%Al2O3(p)+1wt%Gr(p))MMC

5. Scanning Electron Microscope of the tool

SEM of a PVD coated cermet tool was performed at various magnifications. The magnification depths were 3mm, 500µm, 50µm and 100 µm respectively. The effect of excessive forces, mechanical vibrations, intense stresses, and temperature rise on the PVD tool has been studied using SEM. At 3mm magnification, Figure 5(a) depicts the worn-out edge. The worn-out edge was caused by a decrease in yield intensity caused by high-temperature generation during turning. The cutting edge has softens and deformed resulting plastic deformation. This has been observed at high cutting speed. In the figure 5(b) the wear in the upper coated layer can been seen significantly.

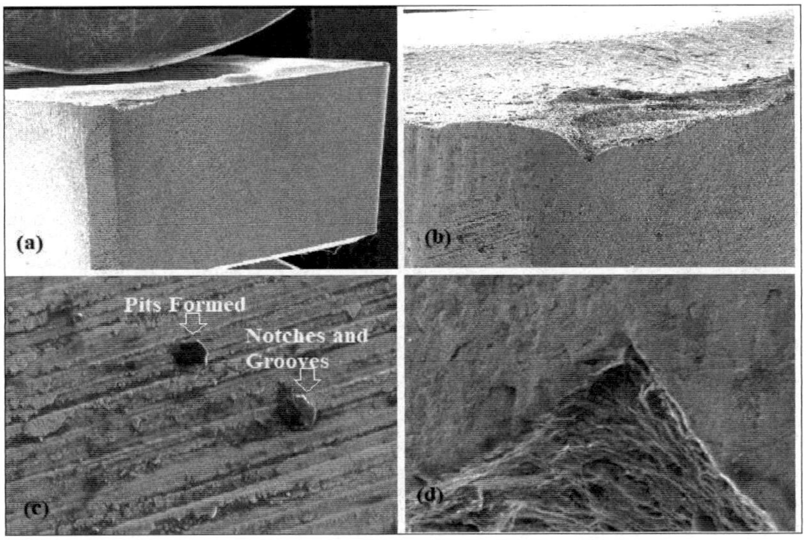

Figure 5. (a) SEM of the PVD tool showing tool wear on the cutting edge at 3mm; (b) Wear of coated surface 500µm; (c) Grooves and Notches wear at the tool rake face 100 µm and (d) Crater wear at 50 µm.

Figure5(d) depicts the tool's crater wear and discoloration patterns caused by high temperatures. This is the movement of atoms across the tool-workpiece chip interface via a diffusion mechanism. This crater wear significantly reduces the tool's strength, which may lead to further failure.The crater is formed as a result of adhesion and diffusion of hard reinforced particulates on the tool face flank. This crater wear significantly reduces the tool's strength, which may lead to further failure. The crater wear was formed at great cutting speed 120 m/min, higher feed rate 0.4 (mm/rev) and with lower depth of cut i.e. 0.2 mm during turning of Mg/MMC workpiece specimen.

6. Intensive Discussion

The current study is concerned with the wear of the PVD-coated cermet tool during hybrid Mg-MMC turning. Excessive forces and mechanical vibrational shocks are typically the primary causes of cutting tool failure. During machining, the intense stresses raise the temperature, causing plastic deformation. The machining parameters play a crucial role in turning as high velocity, are not favorable during turning of hybrid Mg-MMC as they add peak tool wear (edge chipping) later is the cutting depth and feed rate. The PVD cermet tool coating has the advantages for machining hybrid Mg-MMC. The tool wears enhanced with rises in cutting velocity, feed rate and cutting depth and with cutting speeds of 40 to 120 mm/min the tool wear increased as of 0.21 to 0.425 mm respectively. The inherent properties of hard reinforced particulates in the matrix magnesium resulting higher the weight fraction of particles resulting in the tool flank wear. The notches, groves, and pits formed on the tool flank face during machining at high cutting speeds. Mechanical wear of a cutting tool includes abrasion, chipping, delamination, adhesion, fracture, and flaking. The thermochemical wears on the tool are mass dissolution and atomic migration. The main factor in resisting damage such as grooving wear is chemical stability.

7. Conclusion

During machining, i.e., hybrid Mg-MMC turning under dry operating conditions, PVD- coated cermet instrument has been used effectively. The cutting speed (m/min) has the greatest percentage contribution in the cutting tool flank wear width while machining of hybrid Mg-MMC. The optimal parametric arrangement for minimum (VBmax) is A3B3C3 for minimum the tool wear width (VBmax). After turning at high cutting speeds, SEM revealed notches, grooves, and pits on the tool surface. The preceding study used standard tools, such as conventional High-Speed Steel tools, to machine hybrid MMC that would otherwise be difficult to machine. The PVD-coated cermet tool discovers the ability to cut materials with elevated hardness level values. The fields of tool geometry, difficult profile turning, and interrupted surfaces need more effort to fulfil future demand through turning.

References

[1] Holmberg, Kenneth, et al. 2009 Residual stresses in TiN, DLC and MoS2 coated surfaces with regard to their tribological fracture behaviour *Wear* **267** 2142 - 2156.

[2] Yang Wei et al. 2017 Structure and properties of PVD TiAlN and TiAlN/CrAlN coated Ti (C, N)- based cermets *Ceram. Int.* **43**, 1911-1915.

[3] Bushlya V et al. 2017 Performance and wear mechanisms of novel superhard diamond and boron nitride based tools in machining Al-SiCp metal matrix composite *Wear* **376** 152-164.

[4] Liu W et al. 2017 PVD-CrAlN and TiAlN coated Si3N4 ceramic cutting tools Microstructure, turning performance and Wear mechanism *Ceram. Int.* **43** 8999-9004.

[5] Anthony X M, and Ajith K J P 2017 Machinability of hybrid metal matrix composite-A review *Procedia Eng.* **174** 1110-1118.

[6] Lim C Y H et al. 2005 Wear of magnesium composites reinforced with nano-sized alumina particulates *Wear* **259** 1 620-625.

[7] Rajesh S et al. 2013 Multi-response optimization of machining parameters on red mud-based aluminum metal matrix composites in turning Int. J. Adv. Manuf. Syst. **67** 811-821.

[8] Rubio, Maria E et al. 2013 Experimental study of the dry facing of magnesium pieces based on the surface roughness *Int. J. Precis. Eng. Manuf.* **14** 995- 1001.

[9] Gnanavelbabu A, Kaliyamoorthy R and Saravanan P 2018 Investigation on the cutting quality characteristics of abrasive water jet machining of AA6061-B4C-hBN hybrid metal matrix composites *Mater. Manuf. Process.* **33** 1313-1323.

[10] Hasan E H, Khaled M S and Emam A A 2010 Study of impact energy and hardness on reinforced polymeric composites *Mapan* **25** 239-243.

[11] Barman S and Ranjan 2010 Enhancement of accuracy of multi- axis machine tools through error measurement and compensation of errors using laser interferometry technique *Mapan* **25**, 79-87.

[12] Siddesh K N G et al. 2017 Some studies on mechanical and machining characteristics of Al2219/n-B4C/MoS2 nano-hybrid metal matrix composites *Measurement* **10** 1-11

[13] Ravinder K and Chauhan S 2015 Study on surface roughness measurement for turning of Al 7075/10/SiCp and Al 7075 hybrid composites by using response surface methodology (RSM) and artificial neural networking (ANN) *Measurement* **65** 166-180.

[14] Hira J, Mangal S K and Manna A 2015 Fabrication of Hybrid Mg/(Al 2 O 3p+ SiC p+ Gr p) Metal Matrix Composite on Developed Gas Injection Liquid Stir Casting Setup *Arab. J. Sci. Eng.* **40** 2729 - 2738.

[15] Montgomery D C 2001 Design and analysis of experiments *John Wiley & Sons. Inc.* New York.

Performance optimization of operating parameters: Al 6061-Red Mud composite under sliding wear

N Panwar[1], S Bala[2], A Chauhan[3]

[1]Skill faculty of engineering and technology, Shri Vishwakarma Skill University, Dudhola, Palwal, Haryana-121102 (India)
[2]Department of Physics, SGGS College, Sector-26, Chandigarh-160019 (India)
[3]Department of Mechanical Engineering, University Institute of Engineering and Technology, Panjab University, Chandigarh-160014 (India).

[3]Email:drchauhan98@gmail.com, drchauhan98@pu.ac.in

Abstract. Aluminium 6061-red mud composite has been fabricated by using stir casting method. Uniform distribution of red-mud particles has been observed by Scanning electron microscopy. Energy dispersive spectroscopy confirms presence of constituent elements of red mud. Wear testing of samples has been done on pin-on-disk type wear tester. The weight loss due to wear first decrease then increase with increasing percentage reinforcement. From experimentation, authors have found that the weight loss first decreases and then increase with aging time. Also, weight loss increases with applied load, sliding speed and sliding distance. In ANOVA analysis, the variables: load and sliding distance are found significant while particle size, aging time, percentage reinforcement and sliding velocity are not found significant. The order of significance is load, sliding distance, percentage reinforcement, sliding speed, aging time and particle size. Optimum value of weight loss is predicted by using Taguchi technique and confirmed by experimentally obtained results.

1. Introduction

Metal matrix composites (MMCs) are the materials in which hard ceramics has been reinforced inside a matrix of metal. Matrix and reinforcement material of MMCs are non soluble in each other and chemically non reactive [1-2]. MMCs have mixture of good properties of constituent materials like good stiffness, better strength, higher corrosion prevention, and more hard, lighter in weight and superior wear resistance [2-5]. MMCs with Aluminium matrix has been utilized in manufacturing parts in auto, aircraft, spacecraft and marine industries because they are lighter in weight and have good heat conducting properties [6-10]. Aluminium metal matrix Composites (AMMCs) reinforced by various types ceramic particles has been highly utilized composite because they can be manufactured and machined similar to conventional metallic materials [11]. Reinforcements used as whiskers, fibres and particles [12]. Aging of MMCs may significantly increase hardness [13-14]. Mostly used ceramic reinforcements in manufacturing AMMCs are carbide, nitride, boride and oxide of different materials. AMMCs can be reinforced with silicon carbide, alumina, boron carbide, titanium carbide, titanium boride, magnesium oxide, TiO_2, zircon, graphite and boron nitride [15-18]. AMMCs has been manufactured by different processes such as stir casting, squeeze casting, liquid infiltration, rheocasting, compocasting and powder metallurgy technique etc. [19-21]. Stir casting has been very simple, cheaper and flexible process out of the discussed methods [22-24]. For microstructure

IOP Conf. Series: Materials Science and Engineering 1225 (2022) 012007 doi:10.1088/1757-899X/1225/1/012007

observation scanning electron microscopy (SEM) and X-ray diffraction can be used [25-26]. Different Design of experiments (DOE) and analysis techniques can be utilized for systematically execution of experiments and analysis [2, 27-30].
Author's finds in the literature that very few work studies on red mud reinforced aluminium matrix composite has been reported. Therefore the present work aims on the fabrication of desired composite with predefined input parameters. Taguchi method has been implemented for design the experiments incurrent wear study.

2. Experimental Procedure

2.1. Material Preparation
Al6061 has been chosen as matrix. Red Mud reinforcement has been brought from HINDALCO, Renukut plant situated in Utter Pradesh, India. XRF test of both material have been done to find out and verify constituent elements. Number of experiments has been decided by Taguchi Design of experiments technique. The input parameters/variables and their respective levels are given below in Table-1.

Table 1. Chosen variables and respective five levels for wear test

Sr. No.	Variables	L-1	L-2	L-3	L-4	L-5
1	Percentage Reinforcement (by wt.)	4	8	12	16	20
2	Particle Size (μm)	250	177	149	125	74
3	Ageing Time (hr)	0.5	6	12	18	24
4	Load applied (kg)	1	2	3	4	5
5	Sliding Distance (m)	1000	1500	2000	2500	3000
6	Sliding Speed (m/s)	0.2	0.4	0.6	0.8	1

2.2. Fabrication of composite
The Al6061 matrix based red-mud composite of current work has been prepared by stir casting process. Aluminium 6061 metal pieces has been heated to 800°C in a crucible made from graphite using electric muffle furnace Fig-1. Mechanical stirrer has been inserted in molten alloy for creating vortex. Red mud reinforcement particles preheated at 400°C in a separate furnace has been injected to vortex. Vortex has been created for obtaining uniform spreading of red mud in the matrix. Magnesium in a small quantity has been incorporated in the mixture for getting superior wettability of red-mud reinforcement with molten Al6061. The molten mixture has been poured inside preheated cylindrical mild steel moulds. Homogenizing of prepared composite has been performed to achieve uniformity the structure and better properties. While homogenizing samples has been heated to a temperature of 525°C for eight hours inside a electric furnace. Quenching of heated composite has been done by water. After those samples has been dried and used for further operation. Aging treatment of fabricated composite specimen has been performed inside furnace at a temperature of 175°C ±5°C. Aging has been done for 0.5, 6, 12, 18 and 24 hrs.

2.3. Wear Test
Wear tests have been done on wear testing machine (pin on disk) under sliding contact (Model: TR-20LE-PHM-CHM400 by DUCOM Bangalore). The equipment can perform wear testing a load up to 200N.The machine have working speed range from 200 to 2000 rpm. Tangential force acting and wear of sample under study has been observed using electronic sensors and displayed on computer screen. The parameter under study has been graphically drawn as function of applied load and rotational speed. The disc dimensions have been 100 mm diameter with a thickness of 8mm. The material of disk is EN-31 and hardness of 60 HRc (surface roughness at 1.6Ra). The samples have been cylindrical pin having dimensions 10 mm diameter with 30 mm length. The weight of the specimens has been taken before the conduction and after the conduction of wear tests by electronic weighing balance (least count = 0.0001 gm).

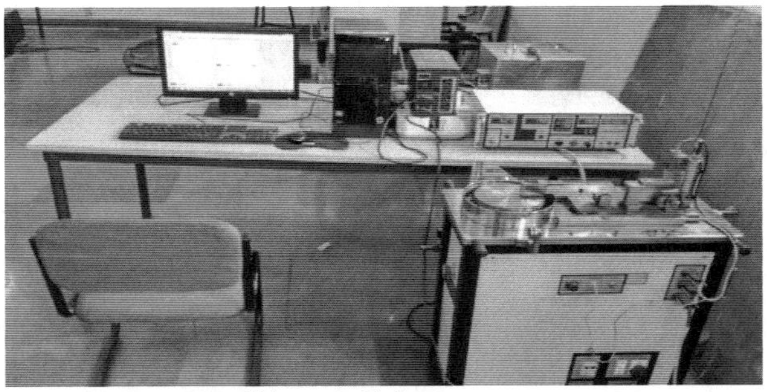

Figure 1. Pin on disk type wear tester

3. Results and discussion

3.1 Wear

To evaluate the impact of chosen parameters on output i.e. wear, five level of each input parameter have been taken. Wear testing have been conducted as per L_{25} orthogonal array. Fifty experiments have been carried out with a repetition of two and the outcome as weight loss due to wear (WLDTW) are given in Table-2. In the table, values are given for percentage reinforcement of red-mud i.e.PR, particle size of red-mud i.e. PS, aging time in the furnace i.e. AT, Load i.e. L in kilogram, Sliding distance i.e. SD, sliding speed i.e. SS, weight loss because of wear i.e. WL in grams and Signal to noise ratio i.e. S/N.

There have been three different type of quality characteristics in Taguchi technique given as: smaller the better, nominal the better and larger the better. Smaller the better criteria is chosen for current study out of the three quality characteristics because wear loss should be less for most of engineering application so we try to minimize the weight loss because of wear.

Mean value of WLDTW and signal to noise (S/N) has been evaluated by Minitab software. The calculated values of mean and S/N ratio has been shown in Table 2

Table 2. Mean WLDTW and S/N ratio

Sr. No.	PR (By wt)	PS (μm)	AT (hr)	L (Kgf)	SD (m)	SS (m/s)	WL mean (g)	S/N
1	4	250	0.5	1	1000	0.2	0.0025	52.0412
2	4	177	6	2	1500	0.4	0.0052	45.67993
3	4	149	12	3	2000	0.6	0.0133	37.52297
4	4	74	18	4	2500	0.8	0.03405	29.35766
5	4	125	24	5	3000	1	0.0574	24.82176
6	8	250	6	3	2500	1	0.0038	48.40433
7	8	177	12	4	3000	0.2	0.0182	34.79857
8	8	149	18	5	1000	0.4	0.0053	45.51448
9	8	74	24	1	1500	0.6	0.00295	50.60356
10	8	125	0.5	2	2000	0.8	0.0117	38.63628
11	12	250	12	5	1500	0.8	0.02045	33.78613
12	12	177	18	1	2000	1	0.0051	45.8486
13	12	149	24	2	2500	0.2	0.00475	46.46613
14	12	74	0.5	3	3000	0.4	0.01425	36.9237
15	12	125	6	4	1000	0.6	0.0073	42.73354
16	16	250	18	2	3000	0.6	0.0086	41.31003
17	16	177	24	3	1000	0.8	0.0079	42.04746
18	16	149	0.5	4	1500	1	0.0211	33.51435
19	16	74	6	5	2000	0.2	0.013	37.72113

20	16	125	12	1	2500	0.4	0.00705	43.03622
21	20	250	24	4	2000	0.4	0.0165	35.65032
22	20	177	0.5	5	2500	0.6	0.02695	31.38882
23	20	149	6	1	3000	0.8	0.0072	42.85335
24	20	74	12	2	1000	1	0.0015	56.4782
25	20	125	18	3	1500	0.2	0.00735	42.67425

3.1.1 Delta analysis for WLDTW

Delta analysis for the wear test results has been performed by Minitab software. In this average result of every variable at all five levels has been evaluated. After that the difference of highest mean value and lowest mean result has been calculated and has been called delta. More has been the value of delta higher is the impact of the input parameter on outcome. The Table 3, 4 displays the mean of S/N ratios and average result values for WLDTW of the developed Al red-mud composite at every level. The analysis concludes that the load has maximum impact on wear (rank 1) and particle size has least influence (rank 6). As from the analysis particle size has least influence on wear so not considered for further analysis.

Table 3. Response table for S/N ratios of WLDTW

Level	Percentage Reinforcement	Particle Size	Aging Time	Load	Sliding Distance	Speed
1	37.88	42.22	38.50	46.88	47.76	42.74
2	43.59	38.38	43.48	45.71	41.25	41.36
3	41.15	41.17	41.12	41.51	39.08	40.71
4	39.53	39.95	40.94	35.21	39.73	37.34
5	41.81	42.24	39.92	34.65	36.14	41.81
Delta	5.71	3.86	4.98	12.23	11.62	5.40
Rank	3	6	5	1	2	4

Table 4. Response table for Means of WLDTW

Level	Percentage Reinforcement	Particle Size	Aging Time	Load	Sliding Distance	Speed
1	0.022490	0.013150	0.015300	0.004960	0.004900	0.009160
2	0.008390	0.018160	0.007300	0.006350	0.011410	0.009660
3	0.010370	0.010330	0.012100	0.009320	0.011920	0.011820
4	0.011530	0.012670	0.012080	0.019430	0.015320	0.016260
5	0.011900	0.010370	0.017900	0.024620	0.021130	0.017780
Delta	0.014100	0.007830	0.010600	0.019660	0.016230	0.008620
Rank	3	6	4	1	2	5

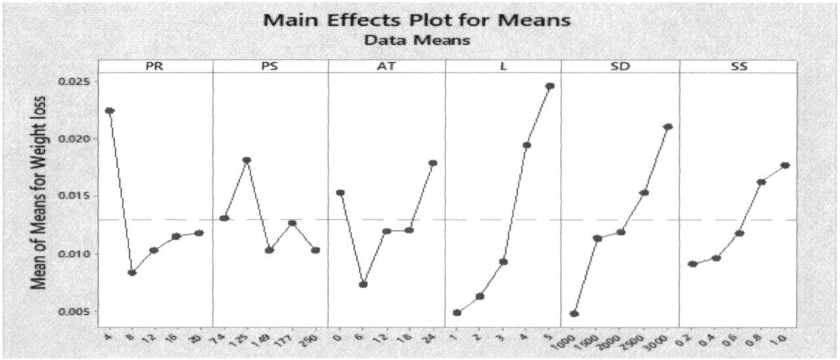

Figure 2. Mean values of parameters with respect to WLDTW.

3.1.2 Analysis of variance for WLDTW

The ANOVA analysis concluded that the input variables i.e. applied load and sliding distance has been come out significant but the effect of particle size, aging time, percentage reinforcement and sliding velocity are not found significant. The significance order of the variables is load, sliding distance, percentage reinforcement, sliding speed, aging time and particle size.

Table 5. ANOVA for Wear

Source	Degree of Freedom(DOF)	Sum of Squares(SS)	Mean Squares(MS)	F Ratio	P value	% Contribution
Percentage reinforcement	4	95.28	23.82	1.77	0.298	7.17
Aging Time	4	66.82	16.70	1.24	0.420	5.03
Load	4	653.44	163.36	12.11	0.017	49.22
Sliding Distance	4	372.52	93.13	6.91	0.044	28.06
Speed	4	85.56	21.39	1.59	0.333	6.44
Residual Error	4	53.94	13.49			4.06
Total	24	1327.56				
R-Sq = 95.9% R-Sq(adj) = 75.6%, Variance(S) = 3.672						

3.1.3 SEM analysis

At lower loads in figure 3e, 3d, 3g abrasive wear is seen more prominent. At higher applied loads in figure 3b, 3c, 3i, 3j, 3a mechanism of wear is adhesive type. At 8% reinforcement figure 3c, 3d least wear debris are visible shows minimum wear and abrasion wear dominates adhesion wear. At lower i.e. 4% and higher reinforcement percentage i.e.12, 16, 20% more amount of wear can be seen in the images 3a, 3b, 3e, 3f, 3g, 3h, 3i, 3j. At lower sliding distance abrasive grooves are less visible in figure c despite of higher load. AT higher sliding distance grooves are clearer fig 3b, 3f, 3g. Lower sliding speed have high tendency for adhesive wear figure 3i, 3f, 3c while the higher sliding speeds have tendency for abrasion wear figure 3b, 3h.

3.1.4 Estimation of optimum WLDTW

Here, the optimum value of WLDTW and confidence interval of it has been predicted theoretically using Taguchi's technique. By considering the influence of parameters optimum value of WLDTW have been found out.

The Significant process variables for WLDTW are load at 1st level (L_1), sliding distance at 1st level (SD_1) The mean value of calculated WLDTW have been expressed as [31]:

$W_{wl}=L_1+SD_1-T_{wl}$ ...1

Where, T_{wl}: is Average of all values of WLDTW: 0.012936 g

L_1: Average value of WLDTW at first level of applied load: 0.00374 g

SD_1: Average value of WLDTW at first level of sliding distance=0.00482 g

W_{wl}= 0.004376 g

Value of confidence interval has been predicted using below equation [31]

$$CI=\sqrt{F_\alpha\,(1,f_e)V_e\left[\frac{1}{n_{eff}}+\frac{1}{R}\right]} \quad ...2$$

Where, CI= Confidence interval; $F_\alpha(1,f_e)$ = The F-ratio at a confidence level of (1-α) against DOF=1 and error DOF (f_e); V_e= Variance of error; R= no. of samples for confirmation experiment; N_{eff}= Effective number of replications= N/ [1+total DOF in the estimation of mean]; N = Total experimental results= 50

by putting the below values:

N = Total number of experimental results= 50

V_e= 0.000051
The total DOF in estimation of mean=8 (Table 5)
n_{eff}= 8.33
$F_{0.05}(1, 8) = 5.32$
The confidence interval = 0.0001682 g
The CI of predicted WLDTW at 95% confidence Interval has been =0.004376±0.0001682 g (0.004544 g-0.004207 g)
The optimum level of input variable parameters for the predicted range of WLDTW are: Reinforcement: 8%, Grain size: 149 micron, Aging time: 6 hours, Load: 1 kilogram, Sliding Distance: 1000 meter, Sliding Speed: 0.2 meter/second.

3.1.5 Confirmation Experiments
On the basis of analysis presented in last section, estimated results have been verified experimentally by conformation experimentation. The casting was performed and testing has been done at optimum levels of significant variables. The average value of experimental outcome has been shown below in Table 6. The outcome of confirmation experiments was compared with theoretically estimated results. It can be concluded that confirmation experiments result have been close to the estimated values and are in between the limits of confidence interval estimated (± 0.0001682 g). As reported in Table 6 developed composite has less wear than the base alloy under same load, sliding speed and sliding distance.

3.1.6 Effect of selected Parameters on WLDTW
The pattern of WLDTW with variables such as percentage reinforcement, particle size, aging time, load, sliding distance and sliding speed has been discussed in detail with the help of graphical representation.

3.1.6.1 Effect of percentage reinforcement on WLDTW
The WLDTW first decrease then increase with increasing percentage reinforcement. Firstly, the WLDTW reduced that may be due to higher the red-mud content more hard is the material, hence lower wear. But after 8% reinforcement, WLDTW starts increasing that may be due to rubbing reinforcement particle comes out and act as wear catalyst causing three body wear. Suresha and Sridhara [32] have reported similar trend for wear variation in hybrid AMMC with reinforcement of graphite and silicon carbide particulates. Harder reinforcement particles act as abrasive particles between two bodies i.e. plate and the work-piece. In ANOVA analysis of the variables, impact of percentage reinforcement (P=0.298>0.05) has not been found significant. The percentage reinforcement (7.17%) has been more effective than the sliding speed (6.44%), aging time (5.04%) and particle size but less significant than the load (49.22%) and sliding distance (28.06%).

3.1.6.2 Effect of particle size on WLDTW
Figure 2 shows the variation of WLDTW at each level of particles size. In this figure WLDTW has been displayed at y axis and particle size has been represented at horizontal axis. Varying particle size shows arbitrary trend. This arbitrary trend may be due to dominance of different aspects such as three body wear may occur due to reduction in particle size as number of particles increases, increase in hardness with decrease in particles size may occur due to increase in wettability etc. Particle size of 149 micron has minimum weight loss, whereas particle size of 125 micron has maximum weight loss. From delta analysis it has also been found out that particle size (rank=6) has been least effective factor on WLDTW.

3.1.6.3 Effect of aging time on WLDTW
Graphical representation of variation of WLDTW with respect to aging time has been shown in Figure 2. WLDTW has been represented as ordinate axis and aging time has been represented as abscissa.

IOP Conf. Series: Materials Science and Engineering 1225 (2022) 012007 doi:10.1088/1757-899X/1225/1/012007

The WLDTW first decreases then increase with aging time. This may be due to increment in hardness of the composite with higher aging time.

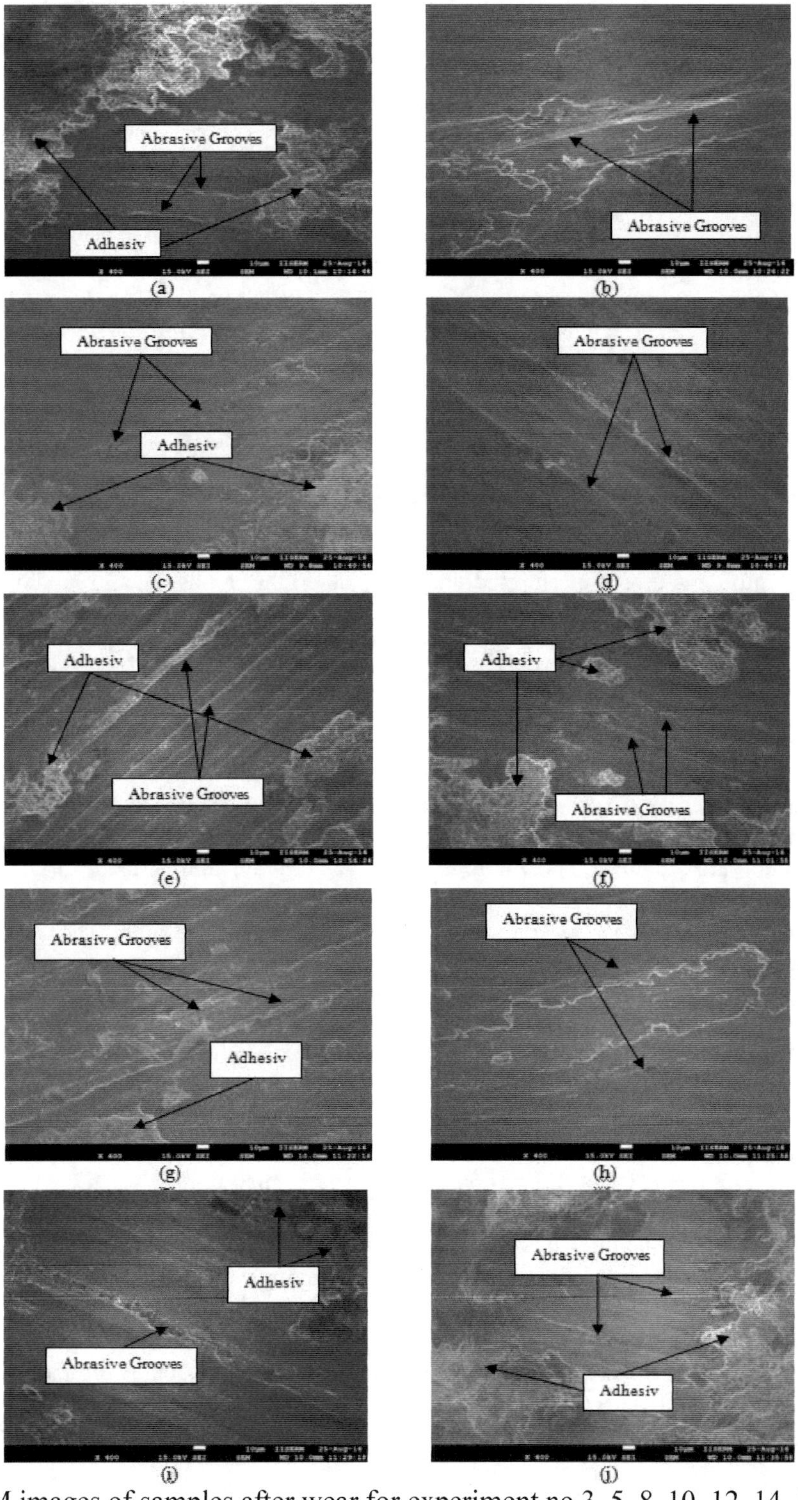

Figure 3. SEM images of samples after wear for experiment no 3, 5, 8, 10, 12, 14, 16, 18, 19, 22

Table 6. Confirmation experiments result for wear

Response Characteristics	Optimum level of parameters	Predicted Optimum Value of WLDTW	Confidence Interval at 95%	Experimental Value
WLDTW	A_5,B_5, C_5	0.004376 g	±0.0001682 g	0.0044 g
WLDTW (Base aluminum 6061)				0.0052g

But further increase in aging increases WLDTW that may be due to formation of precipitates of reinforcement causes three body wear along with sliding wear. The ANOVA of the parameters concluded that effect of aging time (P=0.420>0.05) on WLDTW has not been reported significant. When comparing the effect of all input parameters on WLDTW, the aging time (5.03%) has been more effective than the particle size (rank=6) but less effective than the load (49.22%), sliding distance (28.06%), percentage reinforcement (7.17%) and sliding speed (6.44%).

3.1.6.4 Effect of load on WLDTW

The variation of WLDTW with applied load has been shown in Figure 2 The weight loss has been displayed on vertical axis and load has been shown at horizontal axis. WLDTW is higher with higher load. This impact may be because of increase in friction due tolager value of normal load and hence high wear. Similar trend of results has been obtained by Sahin [33] and Adem [34] for aluminum composites. Significant effect of applied load (P=0.017≤0.05) has been found on WLDTW by ANOVA analysis of the variables. Further it has also been revealed by the analysis that the load (49.22%) has been most effective parameter among all selected input parameters.

3.1.6.5 Effect of sliding distance on WLDTW

Figure 2 shows the value of WLDTW at each level of sliding distance. Weight loss has been represented on y-axis and sliding distance has been represented on x-axis. By increasing sliding distance WLDTW increased. This may be because of larger is the sliding distance greater is the surface contact and hence higher erosion similar trend of results has been reported by Suresha and Sridhara [32] for aluminum composite reinforced by graphite and silicon carbide particles. The ANOVA of the variables on WLDTW reported that the influence of sliding distance has been significant. It has also reported by the analysis that the Sliding Distance has been second largest significant variable on weight loss after Load.

3.1.6.6 Effect of sliding speed on WLDTW

The value of weight loss at each level of sliding speed has been shown in Figure 2. The vertical axis in graph represents weight loss and horizontal axis represents sliding speed. By increasing sliding speed WLDTW is become highersimilar results has been obtained by Adem [32] for aluminum-copper alloy composite reinforced with silicon carbide. It may be because more is the velocity higher has been the surface area interaction between the sample and disc per unit time, hence higher weight loss. The ANOVA of the input parameters, reveals that the impact of sliding speed on weight loss has not been significant. The ANOVA analysis also reveals that the significance order of the factors has been load, sliding distance, percentage reinforcement, sliding speed, aging time and particle size.

4. Conclusions

By evaluating above experimentation and analysis below mentioned conclusions can be drawn:

1) The WLDTW first decrease then increase with increasing percentage reinforcement. The wear first decrease and then increase with aging time. Varying particle size shows arbitrary trend.
2) WLDTW has been higher with higher applied load, sliding speed and sliding distance. The ANOVA analysis concludes that the parameters i.e. load and sliding distance come out

significant whereas the influence of particle size, aging time, percentage reinforcement and sliding velocity are not found significant. The significance order of the inputvariables is load, sliding distance, percentage reinforcement, sliding speed, aging time and particle size.

3) ANOVA results reveals that the impact of load and sliding distance have been found significant whereas the impact of percentage reinforcement, particle size, aging time, sliding speed are not found significant. Load is most influencing parameter while particle size is least influencing parameter.
4) Optimum value of WLDTW is estimated with Taguchi method and confirmation has been done by experimentally obtained results.

References

[1] Chawla N and Chawla K K 2005 Metal Matrix Composites 1st *editionSpringer* New York.
[2] Basavarajappa S, Chandramohan G and Davim J P 2007 Application of Taguchi techniques to study dry sliding wear behaviour of metal matrix composites *Mater Des.* **28** 1393-98.
[3] Garcia-Cordovilla C, Narciso J and Louis E 1996 Abrasive wear resistance of aluminium alloy/ceramic particulate composites *Wear* **192** 170-77.
[4] Kok M 2005 Production and mechanical properties of Al_2O_3 particle-reinforced 2024 aluminium alloy composites J. Mater. Process. Technol. **161** 381-87.
[5] Panwar J and Chauhan A 2018Analyzing on the Effects of Process Parameters on Tensile strength for Al 6061-Red Mud Composite *J. of Sci. Ind. Res.* **77 (11)** 646-51.
[6] Li S, Su Y, Zhu X, Jin H, Ouyang Q and Di Z 2016 Enhanced mechanical behavior and fabrication of silicon carbide particles covered by in-situ carbon nanotube reinforced 6061 aluminum matrix composites *Mater Des.* **10** 7130-38.
[7] Miracle D B 2005 Metal matrix composites-From science to technological significance *Compos Sci Technol*. **65** 2526-40.
[8] Umanath K, Palanikumar K and Selvamani, S T 2013Analysis of dry sliding wear behaviour of $Al6061/SiC/Al_2O_3$ hybrid metal matrix composites *Compos. B. Eng.* **53** 159-68.
[9] Baradeswaran A and Perumal A E 2014 Study on mechanical and wear properties of Al 7075/Al2O3/graphitte hybrid composites *Compos. B. Eng.* **56** 464-471.
[10] Pramanik A 2016Effects of reinforcement on wear resistance of aluminum matrix composites *Trans. Nonferrous Met. Soc. China* **26** 348-358.
[11] Srivatsan T S, Al-Hajri M, Smith C and Petraroli M 2003 The tensile response and fracture behavior of 2009 aluminum alloy metal matrix composite *Mater. Sci. Eng.* **346** 91-100.
[12] Miyajima T and Iwai Y 2003 Effects of reinforcements on sliding wear behaviour of aluminum matrix composites *Wear* **255** 606-616.
[13] Panwar N and Chauhan A 2019 Parametric behaviour optimisation of macro and micro hardness for heat treated Al 6061-red mud composite *J. Mater. Res. Technol.* **8** 660-69.
[14] Torralba J M, Da Costa C E and Velasco F 2003 P/M aluminum matrix composites: an overview. Journal of Materials Processing Technology *J. Mater. Process. Technol.* **133** 203-06.
[15] Baradeswaran A and Perumal E A 2013 Influence of B_4C on the tribological and mechanical properties of Al 7075-B_4C composites *Compos. B. Eng.* **54** 146-52.
[16] Ramesh C S and Ahamed A 2011 Friction and wear behaviour of cast Al 6063 based in situ metal matrix composites *Wear* **271** 1928-39.
[17] Zhiqiang S, Di Z and Li G 2005 Evaluation of dry sliding wear behavior of silicon particles reinforced aluminum matrix composites *Mater Des.* **26** 454-458.
[18] Kumar T S, Subramanian R, Shalini S, Anburaj J and Angelo P C 2016 Synthesis, microstructure and mechanical properties of Al-Si-Mg alloy hybrid (zircon + alumina) composite *Indian J. Eng. Mater. Sci.* **23** 20-26.
[19] Raj R, Singh D and Thakur G 2016 Qualitative and quantitative assessment of microstructure in Al-B4C metal matrix composite processed by modified stir casting technique *Arch. Civ. Mech. Eng.* **16** 949-60.

[20] Kaczmar J W, Pietrzak K and Woosinaski W 2000The production and application of metal matrix composite materials *J. Mater. Process. Technol.* **106** 58-67.
[21] Deuis R L, Subramanian C and Yellup J M Dry sliding wear of aluminium composites-a review *Wear* **201** 132-144.
[22] Li Y, Li Q, Li D, Liu W and Guo-gang S H U 2016Fabrication and characterization of stir casting AA6061−31%B_4C composite *Trans. Nonferrous Met. Soc. China* **26** 2304-12.
[23] Gopalakrishnan S and Murugan N 2012 Production and wear characterisation of AA 6061 matrix titanium carbide particulate reinforced composite by enhanced stir casting method *Compos. B. Eng.* **43** 302-308.
[24] Moses J J, Dinaharan I and Sekhar J S 2016 Prediction of influence of process parameters on tensile strength of AA6061/TiC aluminum matrix composites produced using stir casting *Trans. Nonferrous Met. Soc. China* **26** 1498-11.
[25] Rajmohan T, Palanikumar K andRanganathan S 2013 Evaluation of mechanical and wear properties of hybrid aluminium matrix composites*Trans. Nonferrous Met. Soc. China* **23** 2509-17.
[26] Baskaran S, Anandakrishnan V andDuraiselvam M 2014Investigations on dry sliding wear behavior of in situ casted AA7075–TiC metal matrix composites by using Taguchi technique *Materials and Design* **60** 184-192.
[27] Prakash S K, Kanagaraj A and Gopal P M2015Dry sliding wear characterization of Al 6061/rock dust composite*Trans. Nonferrous Met. Soc. China* **25** 3893-03.
[28] Sahin Y 2005 Optimization of testing parameters on the wear behaviour of metal matrix composites based on the Taguchi method *Materials Science and Engineering A* **408**1-8
[29] Mahapatra S S and Patnaik A 2009 Study on mechanical and erosion wear behavior of hybrid composites using Taguchi experimental design*Materials and Design* **30** 2791-01.
[30] Baradeswaran A, Vettivel S C, Perumal A E, Selvakumar N and Issac R F 2014 Experimental investigation on mechanical behaviour, modelling and optimization of wear parameters of B4C and graphite reinforced aluminium hybrid composites *Mater. Des.* **63** 620-32.
[31] Ross P J 1998 Taguchi techniques for quality engineering *McGraw-Hill Book Company*, Singapore.
[32] Suresha S and Sridhara B K 2010 Wear characteristics of hybrid aluminium matrix composites reinforced with graphite and silicon carbide particulates *Compos Sci Technol* . **70** 1652-59.
[33] SahinY 2005 The prediction of wear resistance model for the metal matrix composites *Wear* **258** 1717-22.
[34] Onat A 2010 Mechanical and dry sliding wear properties of silicon carbide particulate reinforced aluminium–copper alloy matrix composites produced by direct squeeze casting method *J. Alloys Compd.* **489** 119-24.

A fuzzy logic approach for investigating the tribological behavior of polymer composite

G V Jagadeesh[a]*, T Teja Rakesh Kumar[a], T Sai Aneesh[a], N Dheeraj Srinivas[a], V Harish[a], A Ravi Kumar[a]

[a]Department of Mechanical Engineering, Seshadri Rao Gudlavalleru Engineering College, Gudlavalleru, Andhra Pradesh, India-521356

*E-mail:gvenkatajagadeesh@gmail.com

Abstract.The application spectrum of natural fiber reinforced polymer composites has been increasing tremendously because of enhanced tribologicalbehavior. The present study investigates the wear and coefficient of friction of castor oil fiber reinforced composites for developing as new tribo-material. Composites with 40 vol.% with short fiber length of 5mm were fabricated by hand lay-up method. The tribology tests were performed with pin on disc tribometer at normal loads (15N, 30N & 45N) and sliding distances (1000m, 2000m & 3000m). A fuzzy logic model has been developed to predict and analyze the wear and friction characteristics at unknown test cases. Each input variable was divided into three linguistic variables and each output variable was divided into five linguistic variables. A triangular membership function was used for defining all the variables. The capabilities of fuzzy logic model were tested by confirmatory experiments. The model predicted the wear results with an error of 3.12% and coefficient of friction by 2.19%.

1. Introduction

In recent years, polymer matrix composites [1] and metal matrix composites [2] are being extensively used in all engineering fields. Among the polymer matrix composites, natural fiber reinforced composites were preferred because of low cost, biodegradable nature, effective mechanical properties, low density, energy recovery, enhanced CO_2 sequestration, reduced respiratory irritation and good tribological behavior. A large number of natural fibers have been used to enhance the wear and friction characteristics of polymers. For instance, reinforcing epoxy polymer with oil palm fiber and kenaffiber improved the tribological behavior significantly [3]. At low temperature, fine grooves, debonding and micro-cracks were observed. However, at large temperature, predominant wear mechanisms such as delamination and fractured fibers were observed. Similarly, addition of short Agave Americana fiber to epoxy resin improved the wear resistance by 60% [4]. The best wear resistance was obtained at 3mm fiber length. In another research, adding chemically treated hemp fibers to polypropylene matrix improved the mechanical strength and wear resistance significantly [5]. In order to improve the tribological performance of synthetic fiber reinforced composites, natural fiber such as sisal was incorporated. The hybrid ratio of 25/75 wt% for sisal/carbon and sisal/glass hybrid composites has shown the best friction performance [6]. Addition of chopped castor oil fibers to epoxy matrix improved the amount of wear by 70%, CoF by 40%, and interfacial temperature by 24%. Also, the optimum fiber length was found to be 5mm. At 5mm fiber length, the lowest wear was found to be

2.05mg and CoF 0.239 [7]. As it was impossible to perform experimentation at every condition, soff computing techniques such as Fuzzy Interface system (FIS), Artificial Neural Networks (ANN) has been employed to predict and analyze the output responses at unknown test cases. This study aims at developing a fuzzy logic model to analyze the wear and coefficient of friction of chopped castor oil fiber reinforced epoxy composites.

2. Material & Methods

Castor oil plant stalk was dispersed in water for 9 days at ordinary temperature in order to separate xylem, cortex and pith. The cortex fiber was found to exhibit higher strength and rigidity than xylem fiber. Hence, cortex fibers having fiber length 5mm and diameter 0.3mm were used for fabrication of specimens. The matrix material was Epoxy (LY556 grade) and it was combined with hardener (HY 951grade) to form a resin. The composites with 60 vol. % resin as matrix material and 40 vol.% chopped castor oil fiber as reinforcement were fabricated by hand lay-up technique. The tribological characteristics were measured by pin on disc equipment. The tribometer set up was represented in Figure. 1. The tribological tests were performed as per ASTM G99 standard. EN-31 steel disc having hardness of 69 HRC was used as the contour surface. The track diameters were maintained between 100-120 mm such that sliding distances between 1000-3000m were achieved respectively and the normal load is varied at three levels (15,30&45N). The initial surface roughness of contact surfaces was maintained with the help of emery papers having grit sizes 220 & 600. The initial and final weights of specimens were measured to determine the amount of wear.

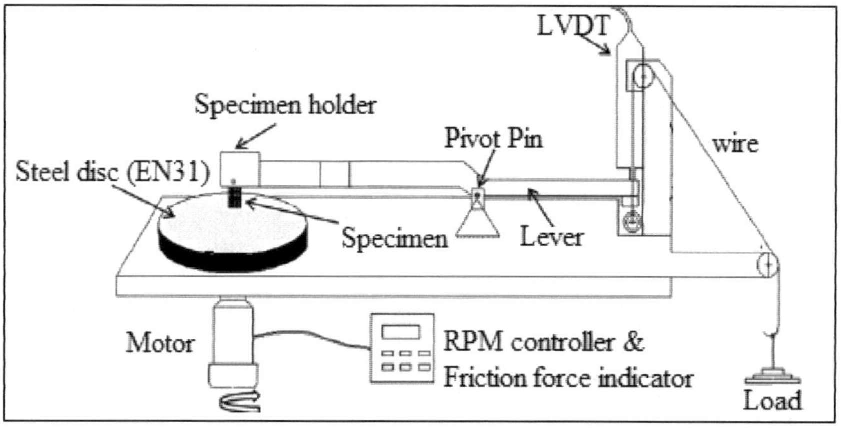

Figure 1. The schematic of tribometer equipment [7]

3. Results & discussion

3.1 Experimental results

The tribology tests were performed as per design teat conditions and the results were formulated in Table 1. Each experiment was conducted twice and average of two readings was taken for analysis. Each experiment has exhibited better accuracy and precision.

3.2 Fuzzy logic approach

A fuzzy logic system is an approach of mapping input data set nonlinearly to a scalar output [9]. Broadly, it comprises of five steps.

Step I: Deciding the descriptors of input and output variables

A crisp data was collected and transformed to fuzzy set with the help of linguistic variables. They were the input variables or output responses with values as words, instead of numerical values [10]. Three linguistic variables were chosen for each input variable and similarly, five linguistic variables were chosen for each output variable, as shown in Table 2.

Table 1. Tribological results of castor oil fiber reinforced epoxy composite [8]

SNo	Normal Load (N)	Sliding Distance (m)	Amount of Wear (mg)			CoF		
			Test 1	Test 2	Mean	Test 1	Test 2	Mean
1	15	1000	2.1	2.0	2.05	0.231	0.247	0.239
2	15	2000	2.7	2.5	2.60	0.243	0.258	0.250
3	15	3000	3.5	3.3	3.40	0.259	0.289	0.274
4	30	1000	2.9	2.7	2.80	0.299	0.311	0.305
5	30	2000	3.7	3.3	3.50	0.319	0.323	0.321
6	30	3000	4.2	4.0	4.10	0.337	0.345	0.341
7	45	1000	4.6	4.5	4.55	0.365	0.371	0.368
8	45	2000	5.3	4.8	5.05	0.388	0.402	0.395
9	45	3000	5.8	5.5	5.65	0.406	0.418	0.412

Table 2. Linguistic variables of input and output variables

Input variables		Output variables	
Normal Load (NL)	Sliding Distance (SD)	Amount of Wear	CoF
SL : Small NL ML : Medium NL LL : Large NL	SSD : Small SD MSD : Medium SD LSD : Large SD	VS : Very small wear S : Small wear M : Medium wear L : Large wear VL : Very Large wear	VS : Very small CoF S : Small CoF M : Medium CoF L : Large CoF VL : Very Large CoF

Step II: Defining the membership functions of each variable

Membership functions play a vital role in mapping the non-fuzzy input values to fuzzy linguistic terms during fuzzification and defuzzification respectively. A linguistic term was quantified by a membership function [11]. Triangular membership function having gradual increase and decrease characteristics has been used in this study. Membership function of each input and output variable was represented in Table 3.

Table 3. Membership function of each variable

Membership function of normal load	Membership function of sliding distance
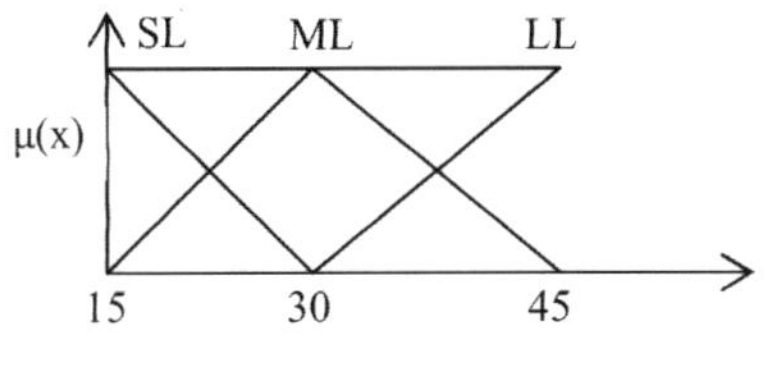	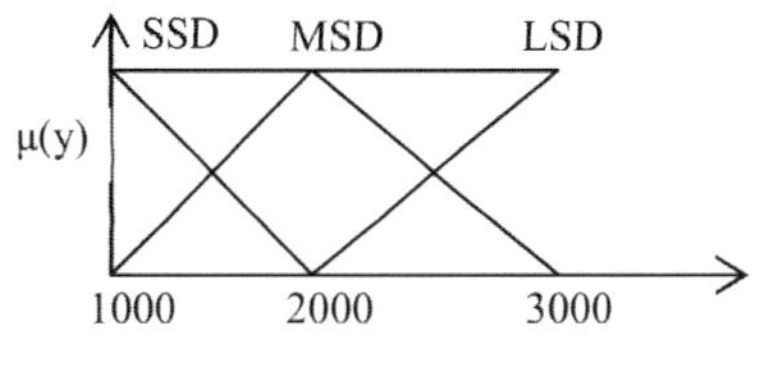
$\mu_{SL}(x)=\frac{30-x}{15}\quad 15\le x\le 30$ $\mu_{ML}(x)=\frac{x-15}{15}\quad 15\le x\le 30$ $=\frac{45-x}{15}\quad 30\le x\le 45$	$\mu_{SSD}(y)=\frac{2000-y}{1000}\quad 1000\le y\le 2000$ $\mu_{MSD}(y)=\frac{y-1000}{1000}\quad 1000\le y\le 2000$ $=\frac{3000-y}{1000}\quad 2000\le y\le 3000$

$\mu_{LL}(x)=\frac{x-30}{15}\quad 30\le x\le 45$	$\mu_{LSD}(y)=\frac{y-2000}{1000}\quad 2000\le y\le 3000$
Membership function of amount of wear	Membership function of CoF
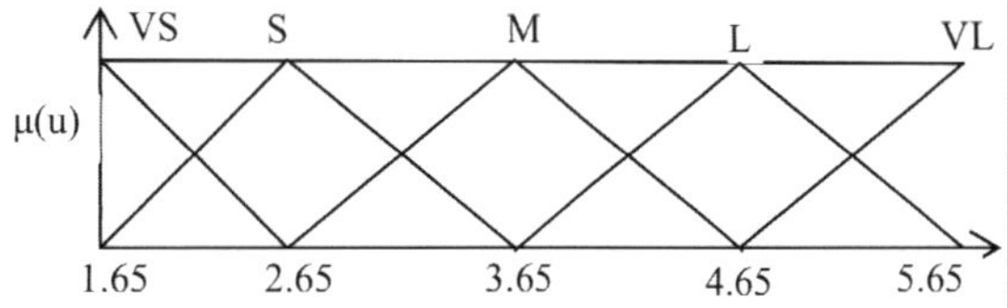	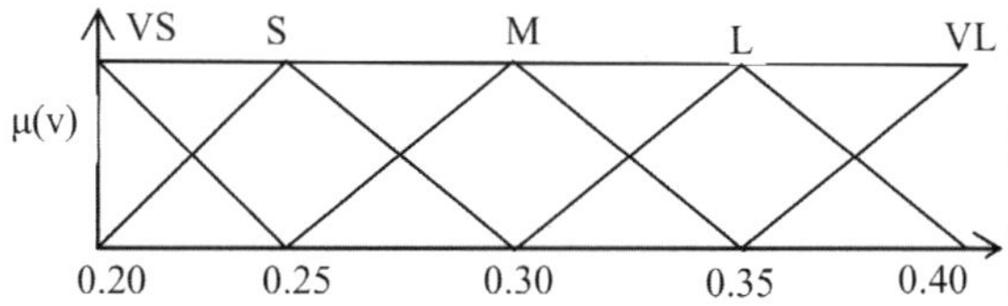
$\mu_{VS}(U)=2.65-U\quad 2.65\le U\le 3.65$ $\mu_{S}(U)=U-1.65\quad 1.65\le U\le 2.65$ $=3.65-U\quad 2.65\le U\le 3.65$ $\mu_{M}(U)=U-2.65\quad 2.65\le U\le 3.65$ $=4.65-U\quad 3.65\le U\le 4.65$ $\mu_{L}(U)=U-3.65\quad 3.65\le U\le 4.65$ $=5.65-U\quad 4.65\le U\le 5.65$ $\mu_{VL}(U)=U-4.65\quad 4.65\le U\le 5.65$	$\mu_{VS}(V)=\frac{0.25-V}{0.05}\quad 0.20\le V\le 0.25$ $\mu_{S}(V)=\frac{V-0.20}{0.05}\quad 0.20\le V\le 0.25$ $=\frac{0.30-V}{0.05}\quad 0.25\le V\le 0.30$ $\mu_{M}(V)=\frac{V-0.25}{0.05}\quad 0.25\le V\le 0.30$ $=\frac{0.35-V}{0.05}\quad 0.30\le V\le 0.35$ $\mu_{L}(V)=\frac{V-0.30}{0.05}\quad 0.30\le V\le 0.35$ $=\frac{0.40-V}{0.05}\quad 0.35\le V\le 0.40$ $\mu_{VL}(V)=\frac{V-0.35}{0.05}\quad 0.35\le V\le 0.40$

Step III: Forming a rule base

A rule base was formulated to regulate the output responses. A fuzzy rule consists of 'IF–THEN' rule with a condition and a conclusion [12]. The database and rule base was used to predict the wear and friction characteristics of chopped castor oil fiber composites. As there were two input variables with three linguistic variables, a total of nine rules were formulated for each output response, as depicted in Table 4.

Table 4. Fuzzy rules based on membership functions

Amount of Wear

	SSD	MSD	LSD
SL	VS	S	M
ML	S	M	M
LL	L	L	VL

CoF

	SSD	MSD	LSD
SL	VS	S	S
ML	M	M	L
LL	L	VL	VL

Step IV: Rule evaluation

The capability of developed fuzzy model can be determined through rule evaluation [13]. The fuzzy rules were evaluated for the condition: Normal load – 40N and Sliding distance – 2700m.Out of nine rules, four rules map to the condition, as represented in Table 5 and the corresponding fuzzified value was evaluated through rule map. Rule no. 2&4 found to exhibit higher strength than other two. Defuzzification was performed based on rule strength

Table 5. Rule evaluation for normal load 40N and Sliding distance 2700m

Normal load – 40N	Sliding distance – 2700m
$\mu_{ML}(x)=\frac{1}{3}\ \mu_{LL}(x)=\frac{2}{3}$	$\mu_{MSD}(y)=\frac{3}{10}\ \mu_{LSD}(y)=\frac{7}{10}$
Rule 1: If the Normal load is Medium & sliding distance is Medium than the amount of wear and CoF is Medium ***Rule 2:*** If the Normal load is Medium & sliding distance is Large than the amount of wear is medium and CoF is large ***Rule 3:*** If the Normal load is Large & sliding distance is Medium than the amount of wear is large and CoF is very large ***Rule 4:*** Normal load is Large & sliding distance is Large than the amount of wear and CoF is very large	
Strength of Rule 1	$S_1 = Min(\mu_{ML}(40), \mu_{MSD}(2700)) = \frac{1}{3}$
Strength of Rule 2	$S_2 = Min(\mu_{ML}(40), \mu_{LSD}(2700)) = \frac{7}{10}$
Strength of Rule 3	$S_3 = Min(\mu_{LL}(40), \mu_{MSD}(2700)) = \frac{2}{3}$
Strength of Rule 4	$S_4 = Min(\mu_{LL}(40), \mu_{LSD}(2700)) = \frac{7}{10}$

Step V: Defuzzification

The fuzzified data has been converted to crisp data by defuzzification technique. It executes the reverse phenomenon of the fuzzification process [13]. It transforms the fuzzy values of output responses into corresponding practical values of wear and CoF. In the present work, 'max-min' defuzzification technique has been used and the results obtained were tabulated in Table 6.

Table 6. Defuzzification of predicted output response

$$Max\,Strength = Max(S_1, S_2, S_3, S_4) = Max(\tfrac{1}{3}, \tfrac{7}{10}, \tfrac{2}{3}, \tfrac{7}{10}) = \tfrac{7}{10}$$

	Amount of Wear	CoF
Defuzzification of Rule 2	3.65	0.350
Defuzzification of Rule 4	5.35	0.385
Avg. Defuzzification value	4.50	0.367

4. Validation of Fuzzy Model

In order to determine the capabilities of developed fuzzy logic model, confirmatory tests were conducted at rule evaluation condition. The experimental results and fuzzy model results were depicted in Table 7. The results were in good accordance with each other. The fuzzy logic model predicted the wear results with an error of 3.12% and coefficient of friction by 2.19%. As the average error of fuzzy logic model was around 5%, the fuzzy interface system can be reliable technique for estimating the wear behavior at unknown test cases.

Table 7. Validity check of Fuzzy logic model

Output Variable	Experimental Value	Fuzzy Model Value	Error %	Avg. error %
Amount of Wear	4.80	4.65	3.12	2.65
CoF	0.378	0.367	2.19	

5. Conclusion

The fuzzy logic approach has been implemented in estimating the tribologicalbehavior of chopped castor oil fiber composites successfully and the following conclusions were drawn:

- The amount of wear and CoF increased with increase in normal load and sliding distance.
- The normal load has high impact on tribological characteristics than sliding distance.
- The triangular membership functions can be used to analyze the wear response of polymer composites.
- The fuzzy logic model estimated the amount of wear with 3.12% and CoF by 2.19% with an average error of 2.65%.
- The fuzzy interface system can be a reliable technique in estimating the wear behavior of natural fiber reinforced polymer composites.

References

[1] Omrani E, Menezes P L and Rohatgi P K 2016 State of the art on tribological behavior of polymer matrix composites reinforced with natural fibers in the green materials world *Eng. Sci. Technol. an Int. J.* **19(2)** 717–736.

[2] Jagadeesh G V and Gangi S S, 2020 A review on micromechanical methods for evaluation of mechanical behavior of particulate reinforced metal matrix composites *J. Mater. Sci.* **55(23)** 9848–9882.

[3] Shuhimi F F, Bin M, Abdollah F, Kalam M A, Hassan M, Mustafa A, and Amiruddin H 2016, Tribological characteristics comparison for oil palm fibre/epoxy and kenaf fibre/epoxy composites under dry sliding conditions *Tribol. Int.* **101** 247–254.

[4] Mylsamy K and Rajendran I 2011 Influence of fibre length on the wear behaviour of chopped agave americana fibre reinforced epoxy composites *Tribol. Lett.* **44(1)** 75–80,.

[5] Sullins T, Pillay S, Komus A, and Ning H 2017 Hemp fiber reinforced polypropylene composites: The effects of material treatments *Compos. Part B Eng.* **114** 15–22.

[6] Aslan M, Tufan M, and Küçükömeroğlu T 2018 Tribological and mechanical performance of sisal-filled waste carbon and glass fibre hybrid composites *Compos. Part B Eng.* 140 241–249.

[7] Egala R, Jagadeesh G V, and Setti S G 2021 Experimental investigation and prediction of tribological behavior of unidirectional short castor oil fiber reinforced epoxy composites *Friction* **9(2)** 250-272.

[8] Jagadeesh G V, Satish J, Egala R, and Srinivasu G 2019 Dry sliding wear of unidirectional short castor oil fiber reinforced epoxy composite *Mater. Sci. Forum* **969** 140–145.

[9] Heidarzadeh A, Testik O M, Guleryuz G, and Barenji R V 2020 Development of a fuzzy logic based model to elucidate the effect of FSW parameters on the ultimate tensile strength and elongation of pure copper joints *J. Manuf. Process.* **53** 250–259,.

[10] Leyva-Bravo J, Chiñas-Sanchez P, Hernandez-Rodriguez A, and Hernandez-Alba G G 2020 Electrochemical discharge machining modeling through different soft computing approaches *Int. J. Adv. Manuf. Technol.* **106** 3587–3596.

[11] Sinha A K, Narang H K, and Bhattacharya S 2020 A fuzzy logic approach for modelling and

prediction of mechanical properties of hybrid abaca-reinforced polymer composite *J. Brazilian Soc. Mech. Sci. Eng.* **42(6)** 1–11.
[12] Bhasker B, Seetharamaiah N, Ramesh Babu P and Gugulothu S K 2020 A novel assessment study on a dynamic analysis of hydrodynamic journal bearing performance: A Taguchi-fuzzy based approach optimization *Transp. Eng.* **2** 100033.
[13] Belloufi A, Mezoudj M, Abdelkrim M, Rezgui I and Chiba E 2020 Experimental and predictive study by multi-output fuzzy model of electrical discharge machining performances *Int. J. Adv. Manuf. Technol.* **109**, 2065–2093.

Print fidelity evaluation of PVA hydrogel using computational fluid dynamics for extrusion dependent 3D printing

Ratnesh Raj[1, *], Siriki Vamsi Venkata Krishna[1], Akshat Desai[1], Chintapalli Sachin[1,] and Amit Rai Dixit[1]

[1]Department of Mechanical Engineering, Indian Institute of Technology (ISM), Dhanbad, Jharkhand 826004, India.

E-mail: rtnsh07@gmail.com

Abstract. This study presents a practical method of print fidelity evaluation for an extrusion-dependent technique of 3D printing. Simulation through computational fluid dynamics (CFD) tool has been used for evaluating the fidelity of printing. The polyvinyl alcohol (PVA) based hydrogel was prepared with deionized (DI) water and PVA powder using a magnetic stirrer at 90 °C for 3D printing. Rheological tests were carried out for checking the viscosity at various shear rates. CFD simulation was done by employing the Bird-Carreau modelusing rheological values. Velocity, pressure, shearing rate, and viscosity distributions through nozzle were obtained. From the shear rate and viscosity results, the increase in shear rate and decrease in gel's viscosity for both the nozzles prove that the material can be extruded. It was seen that the nozzle with a diameter of 0.51mm shows better results than the 0.41mm diameter, which was concluded from the values of maximum shear rate at the edges of the nozzles. The maximum shear rate value has reached a maximum of up to $326.5102s^{-1}$ whereas for a 0.41 mm diameter nozzle, it is $623.8037s^{-1}$ increasing the chances of developing wavy edges in a 0.41mm diameter nozzle than a 0.51mm diameter nozzle concluding that the nozzle with 0.51mm diameter gives far better results than the 0.41mm diameter nozzle.

1. Introduction

3D printing involves the deposition of material in the form of layers under computer control and its solidification to generate a 3D functional object. This is an additionprocess and not the same as the traditional removal method of object formation. Therefore, it is also termed as Additive Manufacturing (AM) [1]. Many 3D printing processes like extrusion, light-based vat polymerization, ink-jetting, and laser-based powder bed fusion are available, each having its pros and cons. Out of these technologies, this study is based on direct-write extrusion-dependent 3D printing (ED3DP), in which a paste/gel-like material can be printed utilizing a syringe-like extruder without the involvement of any heating or fusion processes [2]. The ability for a gel/paste to get 3D printed depends on how easily the material (gel/paste) can be extruded through the micron-sized nozzle and how strongly it can bond and solidify [3]. The extruder design and the extrusion of material through the nozzle depend on the complex flow behavior of materials in the extruder during the printing given by its rheological properties.

ED3DP technique has emerged as the competitive alternative in the fields of defense [4], biomedical [5], aerospace [6], food [2], and electronics [7]. Various sectors require different types of materials to be 3D printed as required while using suitable extruders. The extruder designed or chosen for printing should bear the behavior of the material in the extruder. The selected material for 3D printing should have fundamental capabilities like pressure-driven extrusion at a required flow rate; post extrusion shape-holding capability; the extruded structure must span a given gap and function as a mechanically strong substrate; dimensional stability while curing [8]. The wide use of ED3DP in various sectors makes studying the printability of material and the system for achieving fine functional objects with less dimensional inaccuracy. Analysis of the flow through the micron-sized nozzles of the technique could help in predicting the print fidelity well.

Prediction of print fidelity plays a vital role in the ED3DP process for achieving a structure-dependent functionality in the field of food, medical, and electronics. Various researchers have adopted different ways of fidelity prediction for accomplishing their requirements. For predicting the print resolution of extrusion-dependent bioprinting, a simple yet precise mathematical model has been developed. Themodel generated was for understanding the domination of various printing parameters like gauge pressure, printing speed, and nozzle-diameter on the line width while printing a continuous strand of Pluronic F127 hydrogel. The experimental results agreed with the model barring a few exceptions [9]. Apparently, it is found that maximum researchers have performed prediction through experimentation only.For this type of prediction, keeping some parameters constant with varying others is widely accepted. Carbon nanotubes (CNT) and Nafion-based inks' print fidelity was predicted by keeping nozzle diameter (0.29 mm) and printing speed (12 m/s) constant with varying pressure and Nafion concentration [10].CNT/Polydimethylsiloxane composite ink's print stability has been improved by varying CNT loadings, pressure, and gap span of printed gap-spanning model through several printing trials [11]. However, researchers have adopted software CFD simulation tool also for predicting the print fidelity by analyzing the flow behavior of pastes/gels through the printing nozzle. Five different food grains: mung bean, black rice, buckwheat, job's tear seeds, and brown rice-based pastes' extrusion analysis was made on the nozzle by simulation. The Bird-Carreau model's fitted coefficients were determined using data from the rheological tests, and CFD simulations for viscosity, pressure, and shear rates were carried outto examine the extrusion behavior. It was observed that the simulated pressure values at the central axis being lesser than the maximum applicable pressure by the extruder is the fundamental factor influencing the print fidelity of materials [12].

Conclusively, the print fidelity of a material is determined in three different ways; experimental, mathematical, or analytical, and simulation [9-12]. In the experimental method, a number of printings using different nozzles are carried out for assessing the print quality with the help of specific measuring instruments and visual observations. The second one is the method in which an analytical model is created that describes the output dimensions with the printing parameters, which gives an idea about the print conformity of the material. In the simulation, a mathematical model describing the material's flow behavior is incorporated into the software to produce virtual mapping, showing the distribution of the various properties (viscosity, velocity, pressure, and shear rate), which helps predict the print fidelity [13].

Out of these methods, simulation is the most accurate and convenient method [12]. It helps to analyze the print fidelity of the material without performing the 3D printing process while covering a more extensive range of values of rheological properties compared to rheological tests, which cover a limited range of values. It takes less time and effort than a mathematical model while painting a more comprehensive picture without wasting material [14].

Recent progress on ED3DP has focused on the production of tunable strength synthetic hydrogels with customized structures, recoverable and self-healing properties [15]. Poly (ethylene glycol) [16] and polycaprolactone [17], polyvinyl alcohol (PVA) [15,17], and pluronic F127 [9] are some polymers widely used in the form of hydrogels. Among them, PVA hydrogels have gained more popularity because of their biocompatibility, non-toxicity, swelling stability, low cost, thermal stability, high chemical stability, and mechanical properties [15,18]. According to its potential and usage in various industries like; medical, textile, paper, and food packaging, the extrusion behavior needs to be optimized for flawless deposition over substrates which is a prime requirement for 3D structuring through ED3DP. While research regarding the perfect print material is presently quite prevalent in the biomedical field, the research about the optimum nozzle is relatively less explored. This serves as an inspiration to carry out suitable analysis to establish the effect of changing print parameters and nozzle geometries on print output quality.

This study determines an accurate method of print fidelity evaluation for ED3DP. For this, the analysis of the flow of visco-elastic PVA hydrogel in the syringe-nozzle system is done. The hydrogel of PVA is prepared by mixing PVA into deionized (DI) water. CFD simulation was carried out using the Bird-Carreau model for the static rheological properties of the printing material. Pressure, velocity, shear rate, and viscosity analysis of syringe-nozzle system are made for two different nozzle-diameters.

2. Materials and Methodology

2.1. Materials and hydrogel preparation

Polyvinyl alcohol (PVA) (Molecular Weight= 88000-97000 $g\cdot mol^{-1}$, 98-99% hydrolyzed) is bought from Alfa Aesar (Thermo Fisher Scientific India Private Limited) in the form of tiny pellets. Deionized (DI) water was prepared in the laboratory.

To prepare a 24 wt % PVA hydrogel, 12 grams of PVA were combined with 50 ml DI water at 90 degrees Celsius with vigorous stirring. Stirring was done using a magnetic stirrer (Tarsons digital spinot, Tarsons Products Pvt. Ltd., Kolkata, India) and magnetic bead at 1200 rpm for 10 hours (figure 1 (a)). The prepared sample is allowed to cool to room temperature (25°C).

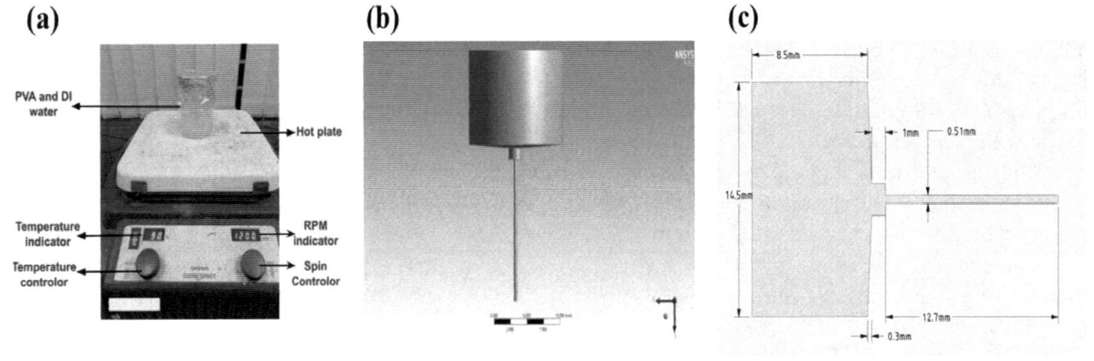

Figure 1. (a) Hydrogel preparation using a magnetic stirrer. **(b,c)** CAD designed model of a syringe-nozzle section through ANSYS with dimensions.

2.2. Rheological property determination

Hydrogel's rheological properties were calculated using Bohlin Gemini 2 (Malvern Instruments, UK), an advanced air-bearing rheometer. An oscillation test and a shear test were conducted. Oscillation tests give results about the dynamic rheological properties, while the shear test gives results about the static rheological properties. The change in viscosity with the shear rate was noted down.

The Bird-Carreau model was used for describing the alter in viscosity of the fluid with shearing rate. It was fit with the results obtained experimentally.

Bird-Carreau model equation:

$$\eta = \eta\infty + (\eta\infty - \eta 0)(1 + (\lambda\gamma)^2)^{(n-1)/2} \quad (1)$$

Here η = viscosity of the fluid (Pa.s); $\eta\infty$ = viscosity of the fluid at infinite shearing rate (Pa.s); γ = shearing rate (s^{-1}) λ = relaxation time(s); $\eta 0$ = viscosity of the fluid at zero shearing rate (Pa.s); n = rheological index [19].

3. Analytical Model

ED3DP CFD simulation was performed with the Mechanical Finite Element Analysis Software version-18.1 (ANSYS Inc, Canonsburg, PA, USA) in order to know the variable properties along the process.

3.1. Physical CAD model

The computational zone was restricted to the nozzle of the syringe-nozzle system. By this, the computational time for simulation was reduced. The simulated geometry was constructed in Design Modeler Geometry of Ansys 18.1 Version. The model has 0.41 mm and 0.51 mm nozzle diameters and 12.7 mm nozzle length for two different sized syringe-nozzle setups (figure 1(b,c)).

3.1.1. Model equations

The PVA hydrogel was laminar and presumed to be an incompressible fluid with a singlephase. ANSYS FLUENT solves the following momentum and continuity equations for an incompressible fluid undertaking isothermal flow via extrusion.

$$\frac{\partial}{\partial t}(\rho\vec{v}) + \nabla.(\rho\vec{v}\vec{v}) = -\nabla P + \nabla \cdot (\bar{\bar{\tau}}) \quad (2)$$

$$\nabla \cdot \vec{v} = 0 \quad (3)$$

Whereas $\bar{\bar{\tau}} = \mu[(\nabla\vec{v} + \nabla\vec{v}^T) - \frac{2}{3}\nabla \cdot \vec{v}I]$

Where, P is the static pressure (Pa), $\bar{\bar{\tau}}$ is the stress tensor (Pa), ρ is density (kg/m^3), and I is the unit tensor [20].

For solving the governing equations, the following conditions were used:

- There is no slip between the syringe wall and fluid, i.e., velocity near the wall is zero.
- Fluid's inlet velocity is presumed to be 5 x 10^{-5} m/s
- Outlet conditions gauge pressure is zero.
- Fluid has constant density, i.e., incompressible.
- Pressure is constant at a particular horizontal plane.
- There is viscosity change with respect to shear rate. For this condition, values mentioned in table 1 ($\eta 0$, $\eta\infty$, λ and n) have been used.

4. Results and Discussions

4.1. Hydrgel's Rheological properties:

The rheological properties of a material have been shown to be a significant predictor of its print fidelity in various studies. The data obtained from the experiment was plotted on the graph, as shown in figure 2(a). A shear-thinning behavior, i.e., when the shear rate is increased, the viscosity decreases, can be observed from the graph of viscosity vs. shear rate, which is a prime requirement for the ED3DP process.

By fitting the Bird-Carreau curve (figure 2(a)) in the data obtained, the coefficients of the Bird-Carreau equation were found by using ORIGIN software are listed in table 1.

Table 1. Co-efficient fitted in Bird- Carreau model

$\eta 0$ (Pa.s)	$\eta\infty$ (Pa.s)	λ (s)	n	R^2	ρ (kg/m^3)
2157.695	36.473	48.1997	0.1407	0.992	1057.098

The coefficient of determination (R^2) of fitted values was found to be 0.992, which indicates that the PVA Hydrogel viscosity change is accurately described by the Bird-Carreau model.

IOP Conf. Series: Materials Science and Engineering 1225 (2022) 012009 doi:10.1088/1757-899X/1225/1/012009

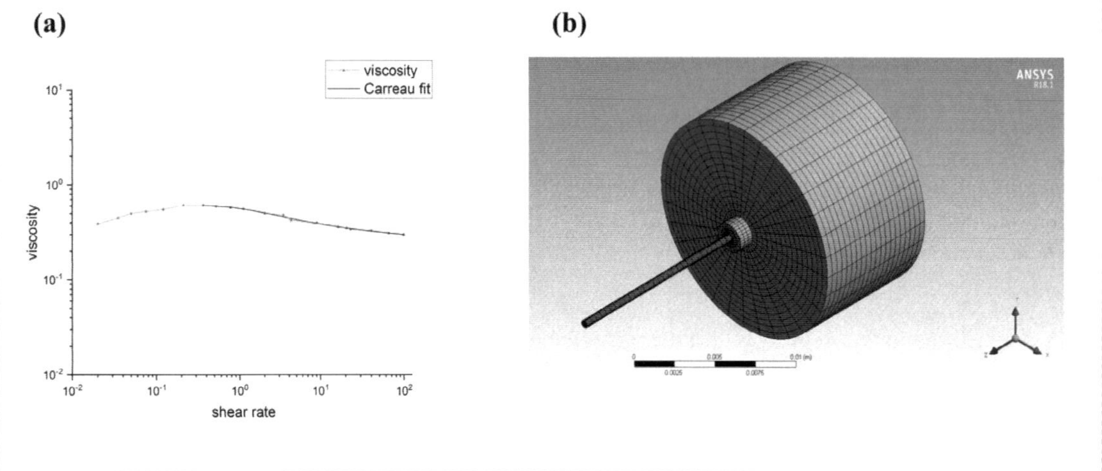

Figure 2. (a) Viscosity vs. Shear rate plot at different shearing rates with Carreau fit.**(b)** Mesh geometry of syringe-nozzle section.

4.2. CFD Simulation:

CFD simulation was carried out in software ANSYS R18.1 version for nozzles with two different diameters, 0.41mm and 0.51mm. The syringe-nozzle system was designed in the ANSYS drawing module.

Mesh independence tests were carried out to ensure no effect on simulated variables with mesh density. The design consists of 16,150 hexahedral mesh elements (figure 2(b)). The results are given in figures 3,4,5 and 6 for 0.41mm diameter nozzle followed by 0.51mm nozzle.

4.2.1. Simulated velocity distribution:

In the simulation (figure 3), it can be seen that the velocity gradually increases from the top to bottom towards the nozzle outlet. It can also be observed that velocity increases while moving from wall to center at the nozzle section.

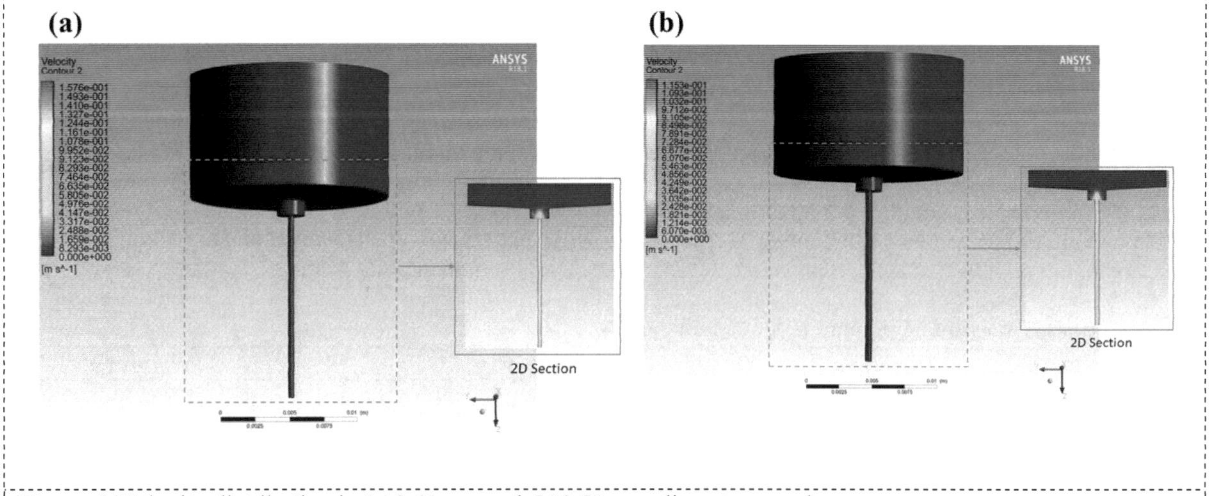

Figure 3. Velocity distribution in **(a)**0.41mm and **(b)**0.51 mm diameter nozzle.

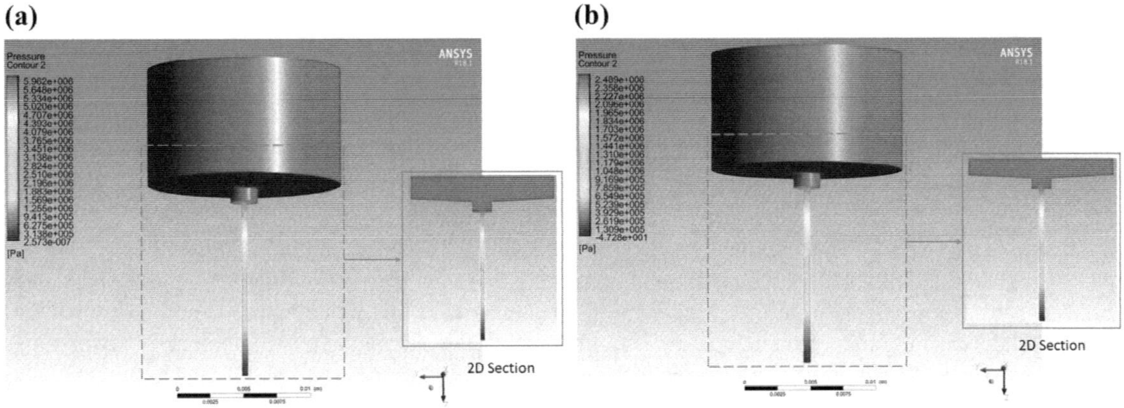

Figure 4. Pressuredistribution in **(a)**0.41mm and **(b)** 0.51 mm diameter nozzle.

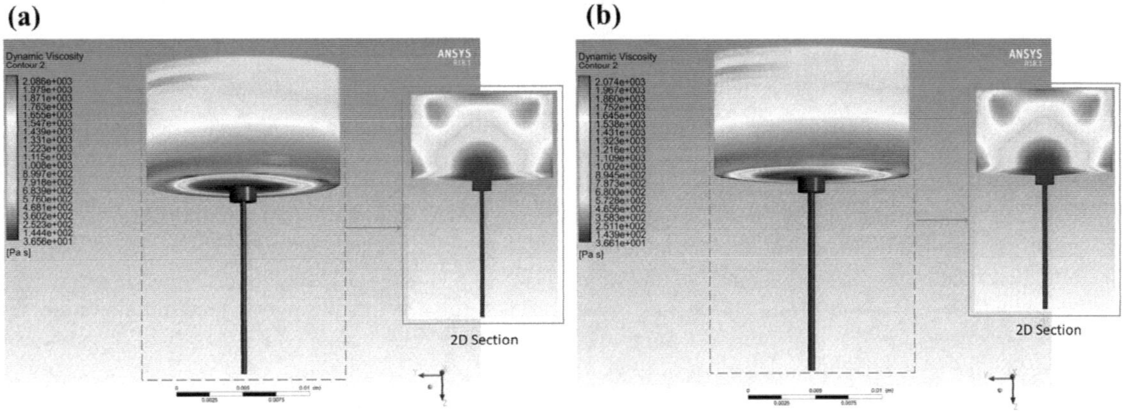

Figure 5. Viscosity distribution in **(a)** 0.41 mm and **(b)**0.51mm diameter nozzle.

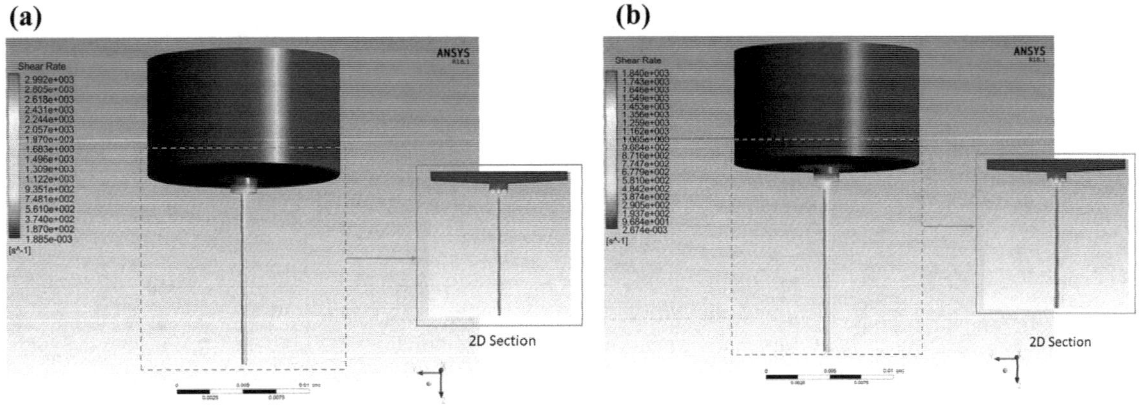

Figure 6. Shearing rate distribution in **(a)** 0.41 mm and **(b)**0.51mm diameter nozzle.

4.2.2. Simulated pressure distribution:
The syringe-nozzle system has given a pertinent level of applied pressure to assure a material. The pressure distribution of the material across the syringe is shown in figure 4. It is observed that with a decrease in pressure while approaching the nozzle, a smooth extrusion of the material occurs. As the material follows the continuity equation, pressure will genuinely decrease with the increase in velocity through the nozzle.

4.2.3. Simulated viscosity distribution:
The distributions of viscosities for both needles were similar, as shown in figure 5. The syringe had higher viscosities, particularly in the corners and top center, where shear rate values were lowest. As the hydrogel is shear thinning, shear rate increasing throughout the nozzle, viscosity decreases through the syringe, ensuring smooth flow of fluid layers one over another.

4.2.4. Simulated local shearing rate distribution:
According to the boundary conditions, the velocity at the walls is zero. Due to this, a different shear rate for the material occurs and increases with the cross-sectional area. It can be observed more shear at the nozzle tip, which ensures more deformation and suitable material extrusion,and alsothe shear is maximum at the walls and minimum at the center (figure 6).
From the distributions obtained through simulation, it has been observed that velocity, pressure, strain rate, and viscosity follow a similar trend in their change for both the nozzles.

5. Conclusions and outlook
Based on the observations following conclusions can be made:

- CFD simulation of the viscoelastic material employing the Bird-Carreau model shows the material's behavior across the syringe-nozzle system, which can estimate the print fidelity of the material.
- The increasing shear and decreasing viscosity through the nozzle show that the material is extrudable. Also, it was observed that velocity, pressure, shear rate, and viscosity show similar changes for different diameters of nozzles.
- It can be seen that the values of maximum shear rate at the edges of the 0.51mm nozzle have reached a maximum of up to $326.5102s^{-1}$ whereas for a 0.41 mm diameter nozzle, the maximum shear rate value is 623.8037 s^{-1}. So, there are more chances of developing wavy edges in a 0.41mm diameter nozzle than a 0.51mm diameter nozzle. Hence among the two nozzles, 0.51mm diameter gives far better results than the 0.41mm.

In addition, this study also plays a pivotal role in helping to identify the various irregularities in the understandings regarding this field that may exist at the moment. This study also provides an impetus for further exploration down this path to develop optimum parameters for achieving near-flawless ED3DP in the future. Further research should establish a standard procedure for finding out the minimum possible diameter of the nozzle used for the extrusion of a particular material.

6. References.
[1] Wu J 2018 Study on optimization of 3D printing parameters *In IOP conf series: Materials Science and Engineering* **392** 062050.
[2] Yang F, Guo C, Zhang M, Bhandari B, Liu Y 2019 Improving 3D printing process of lemon juice gel based on fluid flow numerical simulation *Lwt* **102** 89-99.
[3] Xia X, Duan G 2021 Effect of solid loading on properties of zirconia ceramic by direct ink writing *Mater. Res. Express* **8(1)** 015403.

[4] Chen H, Ma W, Huang Z, Zhang Y, Huang Y, Chen Y 2019 Graphene-based materials toward microwave and terahertz absorbing stealth technologies *Adv. Opt. Mater.* **7(8)** 1801318.
[5] Jakus A E, Secor E B, Rutz A L, Jordan S W, Hersam M C, Shah R N 2015 Three-dimensional printing of high-content graphene scaffolds for electronic and biomedical applications *ACS nano* **9(4)** 4636-48.
[6] Wu T, Jiang P, Zhang X, Guo Y, Ji Z, Jia X, Wang X, Zhou F, Liu W 2019 Additively manufacturing high-performance bismaleimide architectures with ultraviolet-assisted direct ink writing *Mater. Des.* **180** 107947.
[7] Zhang X, Hu H, Tang D, Zhang C, Fu J, Zhao P 2021 Magnetic Flexible Tactile Sensor via Direct Ink Writing *Sens. Actuators A: Phys* **20** 112753.
[8] Duty C, Ajinjeru C, Kishore V, Compton B, Hmeidat N, Chen X, Liu P, Hassen A A, Lindahl J, Kunc V 2018 What makes a material printable? A viscoelastic model for extrusion-based 3D printing of polymers *J. Manuf. Process.* **3** 526-37.
[9] Suntornnond R, Tan E Y, An J, Chua C K 2016 A mathematical model on the resolution of extrusion bioprinting for the development of new bioinks *Materials* **9(9)** 756.
[10] Luo B, Chen H, Zhu Z, Xie B, Bian C, Wang Y 2018 Printing single-walled carbon nanotube/Nafion composites by direct writing techniques *Mater. Des.* **155** 125-33.
[11] Luo B, Wei Y, Chen H, Zhu Z, Fan P, Xu X, Xie B 2018 Printing carbon nanotube-embedded silicone elastomers via direct writing ACS Appl. Mater. Interfaces **10(51)** 44796-802.
[12] Guo C, Zhang M, Devahastin S 2020 3D extrusion-based printability evaluation of selected cereal grains by computational fluid dynamic simulation *J. Food Eng.* **286** 110113.
[13] Norton T, Sun D W 2006 Computational fluid dynamics (CFD)–an effective and efficient design and analysis tool for the food industry: a review *Trends Food Sci Technol* **17(11)** 600-20.
[14] Zhu S, Stieger M A, van der Goot A J, Schutyser M A 2019 Extrusion-based 3D printing of food pastes: Correlating rheological properties with printing behaviour *Innov Food Sci Emerg Technol* **58** 102214.
[15] Jiang P, Yan C, Guo Y, Zhang X, Cai M, Jia X, Wang X, Zhou F 2019 Direct ink writing with high-strength and swelling-resistant biocompatible physically crosslinked hydrogels *Biomater. Sci* **7(5)**1805-14.
[16] Hong S, Sycks D, Chan HF, Lin S, Lopez GP, Guilak F, Leong KW, Zhao X 201 3D printing of highly stretchable and tough hydrogels into complex, cellularized structures *Adv. Mater.* **27(27)** 4035-40.
[17] Kim W, Kim M, Kim GH 2018 3D-printed biomimetic scaffold simulating microfibril muscle structure *Adv. Funct. Mater.* **28(26)** 1800405
[18] Gaaz T S, Sulong A B, Akhtar M N, Kadhum A A, Mohamad A B, Al-Amiery A A 2015 Properties and applications of polyvinyl alcohol, halloysite nanotubes and their nanocomposites *Molecules* **20(12)** 22833-47.
[19] Yu C, Gunasekaran S 2004 Modeling of melt conveying in a deep-channel single-screw cheese stretcher *J. Food Eng.* **61(2)** 241-51.
[20] Aloku G O, Yuan X F 2010 Numerical simulation of polymer foaming process in extrusion flow *Chem. Eng. Sci.* **65(12)** 3749-61.

Modelling, Simulation and Analysis of NPDA for Sub-retinal implant using Nano Materials

M. VijaiMeyyappan[1] and R. Joseph Daniel [2]
[1] Assistant Professor, M.I.E.T. Engineering College, Dept. of Electronics and Communication Engg., Gundur, Trichy, 620 007
[2] Professor, NPMaSS MEMS Design Centre, Dept. of Electronics and Instrumentation Engg., FEAT, Annamalai University, 608 002

corresponding author's e-mail address: vijaimeyyappan@gmail.com

Abstract. There are two primary illnesses that afflict the elderly worldwide those are Retinitis Pigmentosa (RP) and Age-related Macula Degeneration (AMD). Presently there has been a large research focus to develop treatment methods that would cure AMD induced vision disability and RP diseases. Silicon Micro Photo Diode Array, ARGUS II, and ALPHA IMS are a few of the latest therapeutic approaches suggested for curing these diseases (MPDA). These systems have led to many undesirable side effects in patients. The objective of this research work is to create an affordable device that improves the vision recovery. Photovoltaic cells composed of nanomaterials such as carbon nanotubes and graphene are known as Nano Photo Diode Arrays (NPDA). Under incident light, they may generate stimulating electrical pulses to restore vision, photoreceptor cells are stimulated in the retina.The authors of this study sought to optimise the several geometries of Graphene/SWCNT-C60/Al bilayer hetero junction photodiodes that would constitute the NPDA for subretinal implantation.

1. Introduction

The eye is a vital organ that provides us with the ability to see and understand much about our surrounding environment. The neural signals are decoded by the visual cortex into coherent image clarification. Signal processing and convergence is carried out by various neural cell layers in the retina. Rods are highly light-sensitive and responsible for peripheral vision. Color vision and sharp images are regulated by cones. Horizontal cells aid in the integration and regulation of multiple photoreceptor cell information. The electrical signal is converted into electrical pulse trains by the bipolar cells, which keep the spatiotemporal details. Via the optic nerve, electrical pulses are sent to the brain's visual cortex. Amacrine cells manipulate retinal signal processing in response to visual stimuli. Ganglion cells collect visual information from bipolar cells and amacrine cells in their dendrites, which they then transmit to the brain via axon [1]. Figure 1 (a) shows how light enters the retina neural layer before the photoreceptors.

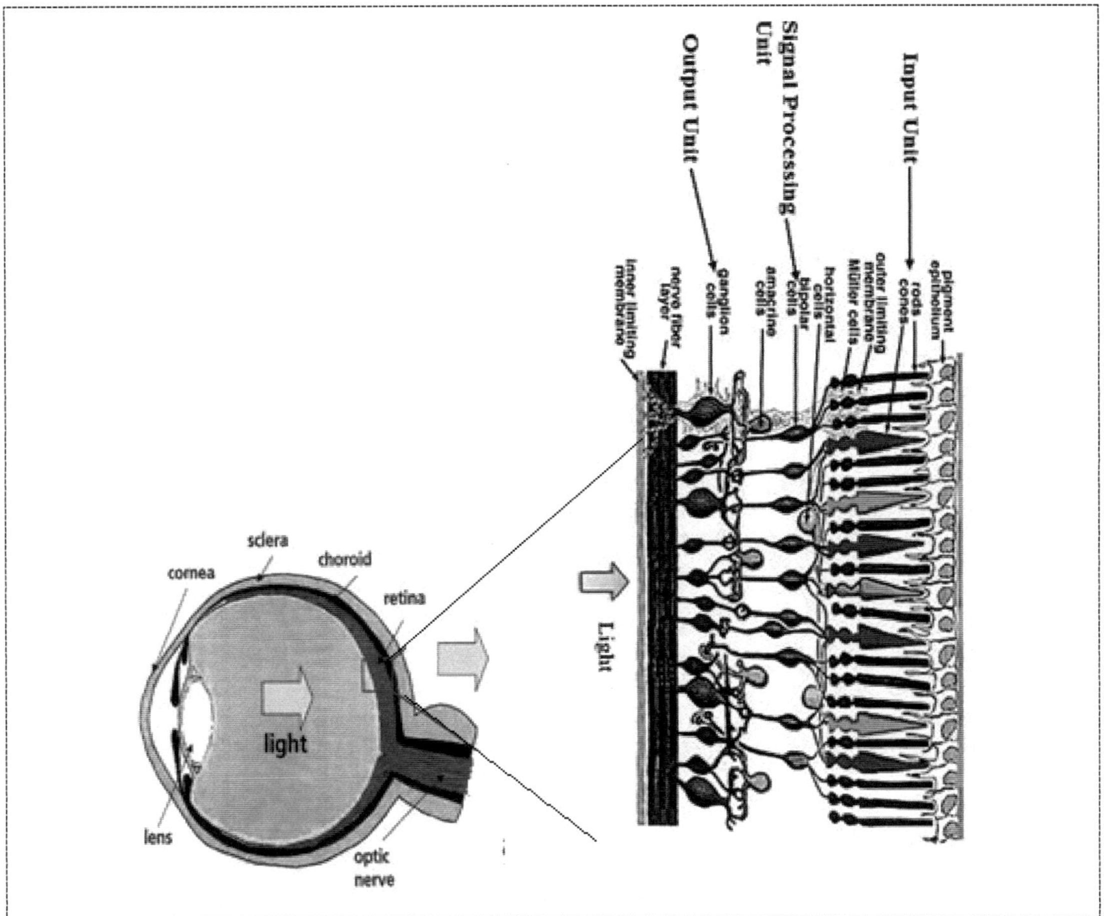

Figure 1(a). The light reaches the retina neural layer before the photoreceptors

Eye diseases including AMD and RP, which cause visual impairments by completely degenerating the retina, are affecting an increasing number of individuals worldwide.AMD is the most prevalent source of blindness for the persons above the age of 55 [2].It harms the macula, a tiny patch near the centre of the retina and the portion of the eye required for clear, central vision, which allows us to perceive objects front of you. A blurry region around the middle of vision is a frequent AMD symptom. Millions of light sensing cells make up the macula area, as shown in Figure 1(b), which provides bright, central vision. It's around 5.5 mm in diameter (0.22 in). It is situated in the rear of the eye the most sensitive part in retina.

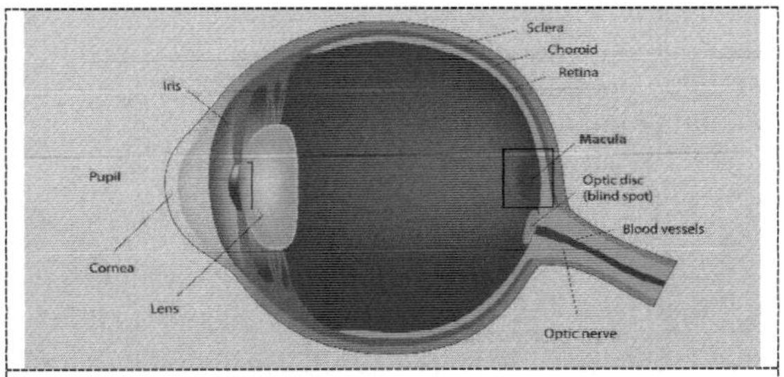

Figure 1(b). Macula region in an eye

If macula is damaged, the centre of angle of vision may look hazy, distorted, or dark.This disease will lead to progressive loss of central vision. Retinitis Pigmentosa is a common and debilitating cause of vision loss in both the central and peripheral vision. RP is an inherited disorder [3]. Outer retinal degeneration causes RP. This damages the other half of the inner anatomical retina, which is made up mainly of photoreceptors' outer and inner parts, as well as their cell bodies. RP is most often diagnosed in infants, teenagers, and young adults, with the disorder progressing over their lives. In the early stages of RP, people are likely to experience night blindness, and later on, their vision and field of view can gradually deteriorate. Others become fully blind as a result of RP, which may occur as early as infancy. The progression of RP can vary from case to case. The retinal cells gradually lose their ability to function and ultimately die. The peripheral rod cells are usually the first to be affected by RP, followed by the central cone cells [4]. In short, both RP and AMD cause photoreceptor cells to degenerate, rendering the visual system insensitive to light.

For those suffering from AMD or RP, there is no effective treatment available. They urgently need an alternate method for allowing the world's sights to reach the brain and be perceived. Professor Wilfried Mokwa of RWTH Aachen University claims that 30 percent of RP patients' retinal ganglion cells are still active after years of blindness [5]. As a result, the following two remedial techniques have been investigated all over the world.

i. Bypass the degenerated photoreceptors appears to be a viable option. As a result, electrical impulses are elicited in retinal ganglion cells, resulting in a visual sense. This concept is said to be Epi-retinal implant.

ii. A sub-retinal implant is the second choice. A chip replaces the deteriorated photoreceptors with a MPDA in this operation. A picture is recorded onto the MPDA using the normal optics of the eye.

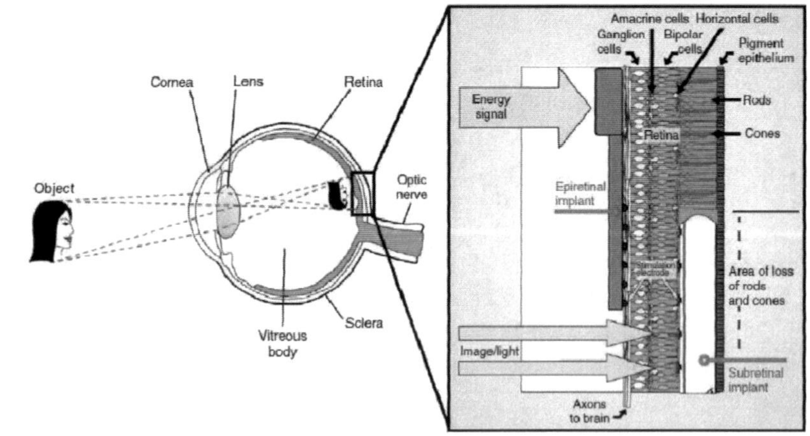

Figure 2. The light reaches the retina neural layer before the photoreceptors

Epi-retinal implants, on the other hand, with fewer electrodes, employ electrical stimulation to produce the sense of visual perception, and do all image processing outside. [5]. Figure 2 depicts a schematic diagram of Epi-retinal and Sub-retinal implants. Because of the numerous benefits, the sub-retinal implant is the most widely researched alternative. The sub-retinal implant works by using 1,500 electrodes to replace the role of degenerated photoreceptors and processing images internally. Only a few photodiode-based prostheses that have been developed and implemented successfully.

Regardless of the reality Argus® II device efficiently generates phosphenes in reaction to visual inputs, in a handful of seconds, the apparent picture might fade away, causing image persistence problems. The protocol for further implantation is much too complicated. They also need external devices.

In the 1980s, Chow and colleagues created the device called Artificial Silicone Retina (ASR)for subretinal[7] was the first application of this technique. However, after 9 months, there was evidence of corrosion [8]. Finally, because the MPDA was developed on stiff silicon substrates, implant of these devices proved challenging. As a result, developing a foldable flexible substrate sub-retinal implant device is critical.
In this study, we investigate the feasibility of successfully implementing Nano Photo Diode Array (NPDA) using nano organic materials such as graphene, Semiconducting (S-SWCNT), conducting polymers and fullerene C_{60}buck ball for sub-retinal implant. This proposed NPDA will stimulate the degenerated retina and restore the vision efficiently and it doesn't require any external equipment and power supply. Further the biocompatibility of the materials used in this NPDA is much superior. The biological responses that occur as a result of the surgery, foreign objects, and the effects of electrical stimulation all fall under the category of biocompatibility. Finally, relative to current computers, it is more adaptable.

2. Proposed Method

The authors intend to convert light intensity to stimulus current using a passive photovoltaic nano photo diode in this analysis. The primary goal of this project is to investigate the dependence of the photodiode performance on the layer thickness and material properties as well as to modify the design in order to fulfil the performance criteria for human subretinal implants. This section discusses the nano photodiode structure, operation and evaluation of photodiode performance using specific simulation tools.

2.1 Nano Photo Diode Array structure

NPDA structure includes polymer, carbon nanotubes, graphene, and aluminum material. This NPDA generates power under incident light to stimulate the retina for regaining lost vision function.

Figure 3 depicts the proposed Nano Photo Diode cell structure. The cathode is aluminium alloy, the electron donor is SWCNT, C_{60}bucky ball fullerence is the electron acceptor, and the anode is graphene. The performance and transparency of the NPDA are improved by processing these materials over a flexible substrate [10, 11].

In this structure as diagrammed in Figure 3, Between two electrodes, a donor and an acceptor with differing band gaps are utilized in this work are placed together having different work functions, graphene (5.3eV- Anode) and aluminium (4.2eV- Cathode).

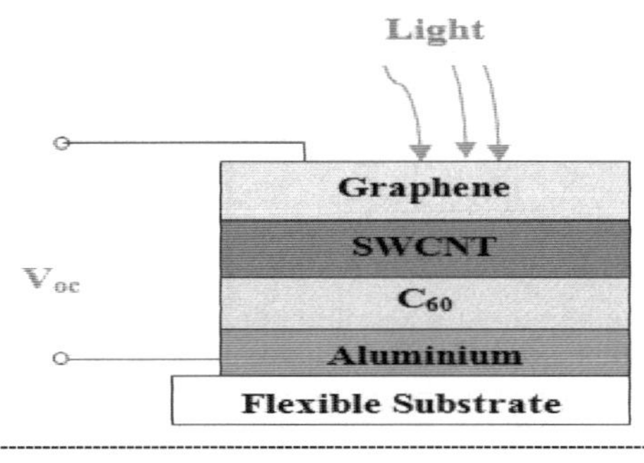

Figure 3. Structure of Single NPD cell

These two layers, graphene and aluminum electrodes, are associated with the formation of excitons. The work function of the electrode and semiconducting materials employed in this study are shown in Figure 4.

The transparent electrode is made of graphene. Graphene is a promising substrate in solar cell applications because of its strong and unique properties, huge area, water vapour, and lack of reactivity to oxygen. It is a suitable substitute for Indium Tin Oxide (ITO), which is costly and brittle in nature. ITO is also incompatible with foldable organic solar cells. Graphene is a feasible alternative for applications needing minimal cost and adaptability [12].

S-SWCNTs have a unique characteristics and combination of organic semiconductor. These characteristics make SWCNT the best candidate for use as a donor material in this study [13].

The electron acceptor C60 fullerence were chosen because it has better compatibility and allows for higher-quality blends with a donor material. Wet chemistry is also used to process it. Charge separation is possible when the donor's Low Unoccupied Molecular Orbit (LUMO) is higher than the acceptor's LUMO and the donor's High Occupied Molecular Orbit (HOMO) is greater than the acceptor's HOMO. It's a prerequisite, but not adequate. As can be seen in Figure 4, this condition is fulfilled by the materials chosen. For charge separation to occur, the combination of these differences in electron affinity and ionisation potential outcomes must be able to resolve the exciton binding energy [14].

Cathode metal should have a lower work function than anode metal. The work function of aluminium (4.2eV) is lower than that of graphene (5.3eV), so it meets the requirements.
The electrode materials and active layers are deposited one after the other in this method for forming a bond between two distinct materials. As a result, it is referred to as a bi-layer junction solar cell. Figure 5 depicts the proposed NPDA structure, it has a 3mm× 3mm size.

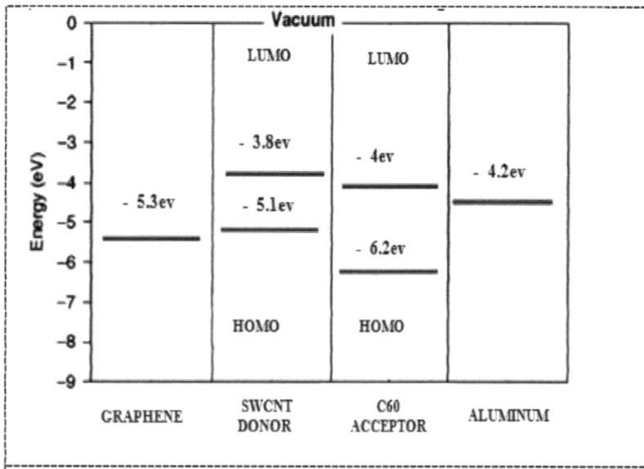

Figure 4. Shows the electrode materials and semiconducting materials work function employed in this study.

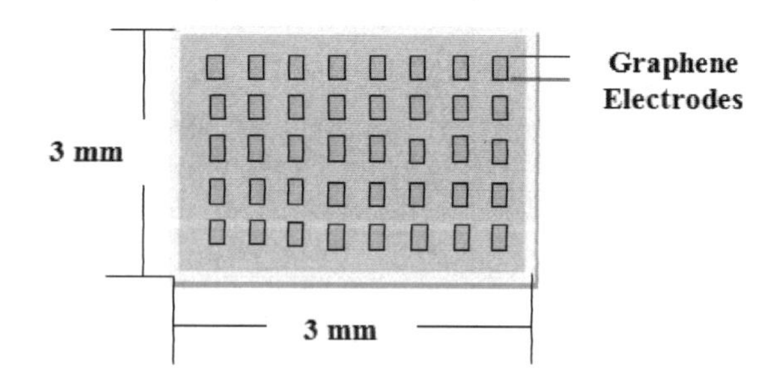

Figure 5. Future NPDA Structure Dimensions of 3mm×3mm area in top view.

2.2 Nano Photo Diode operation

2.2.1Photovoltaic impact and photovoltaic voltage

The photovoltaic impact is what allows photovoltaic (or solar) cells to convert light into electricity. Light, which is pure energy, enters a PV cell and gives certain electrons (negatively charged atomic particles) enough energy to make them free [15].

Depending on carbon atoms that have been sp2-hybridized, Organic solar cell materials are generally composed of alternating single bonds and double bonds (conjugated structure). The electrons

IOP Conf. Series: Materials Science and Engineering 1225 (2022) 012010 doi:10.1088/1757-899X/1225/1/012010

will reside in two energy states due to the π-π* interactions amongst the bonding and anti-bonding pz orbitals. The molecular structure of the organic semiconductor can be changed to adjust their location and separation. Figure 4 depicts the energy band diagram for various materials. The donor's relative HOMO (ionisation potential) energy level and the acceptor's LUMO (electron affinity) determine the photovoltage (V_{oc}).

$$V_{oc} = [E_{Donor}\,\text{HOMO} - E_{Acceptor}\,\text{LUMO}] \quad (1)$$

2.2.2 Device Physics of PHJ photo voltaic

A PHJ photovoltaic cell is made up of semiconductors. Planar heterojunction is another name for this structure. The photo - generated singlet excitons might permeate through the donor to the planar contact with the second material, which is typically very electronegative. That the electron will go to a lower energy state within the acceptor, the energy required to detach the singlet exciton is provided by the acceptor material. As a result of the charge transmission, the exciton is separated, and the electron transfers to the acceptor substrate while the hole remains on the donor. The C_{60}fullerence is well-known electron acceptor material.

Light absorption, excitons diffusion to an active interface, charge isolation, charge transfer, and charge collection are all phases in the functioning of an organic photovoltaic cell. Figure 6(a) – 6(d) depicts the stages involved in OPV energy conversion.

The first step is light absorption, which leads to the forming of exciton. When light strikes photosensitive semiconductor organic materials, electrons in the HOMO state are stimulated to the LUMO state, as a result, excitons arise in the donor material. The light absorption that leads to exciton formation is depicted in Figure 6(a). Excitons, or closely bonded electron hole pairs, develop and diffuse within an organic semiconductor that is unaffected by an applied electric field. The exciton diffusion is depicted in Figure 6(b). The excitons produced as a result of light absorption can contribute to the the development of free charge carriers in order to make an effective organic solar cell.

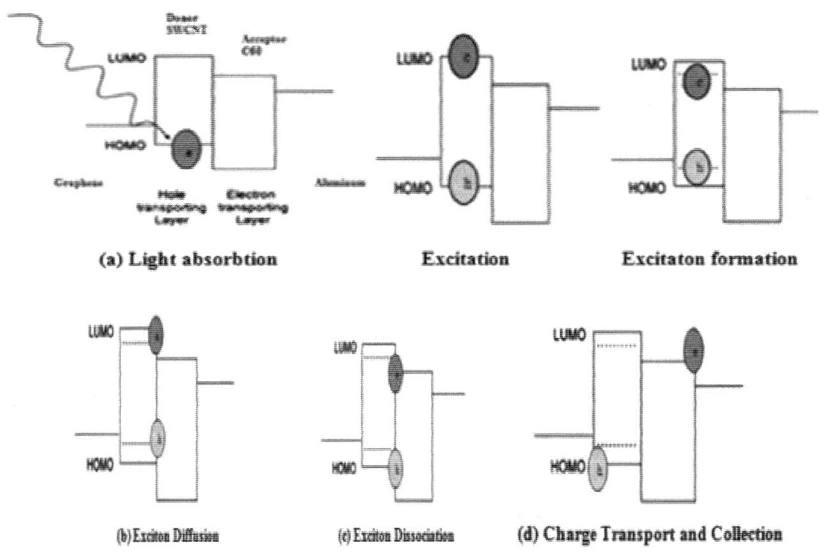

Figure 6. (a-d): Stages involved in OPV energy conversion

The exponential lifetime of an exciton (τ). Diffusion transports excitons, and distance (L) calculates an exciton as follows:

$$L = \sqrt{\tau \times D} \quad (2)$$

D denotes the excitons' diffusion coefficient. The resulting holes and electrons must migrate through the conductivity organic materials used to connect the device's exterior contacts once an exciton is divided into free charge carriers. The charge transport and set are depicted in Figure 6(c-d). The relationship between the velocity (V_{di}) The charge carriers that are acquired as a result of an electric field's influence (ξ) is:

$$Vdi = \mu_i \xi \quad (3)$$

in which μ_i is the mobility. Charges must frequently surpass the things in order to permeate a poor work capacity electrode material (Ca or Al).

3. Simulation studies

Organic Photo Voltaic (OPV) tool was used to model and simulate the modern hybrid design nano photodiode cell utilizing Nano materials in order to optimize NPDA device.

3.1 Simulation Software tool OPV

This tool is a universal learning resource[16]. The present work uses Graphene/SWCNT/C_{60}/Al cell structure as diagrammed in Figure 3 for NPDA. This structure was created for simulation by defining various thickness as and other physical parameters. These Nano Photodiode cell models are subjected to simulations in order to determine their parameters, the parameters are obtained from I-V characteristics as illustrated in Figure 7. A bias voltage of 0.0V to 0.9V with a 0.05 phase increase is provided to the anode contact.

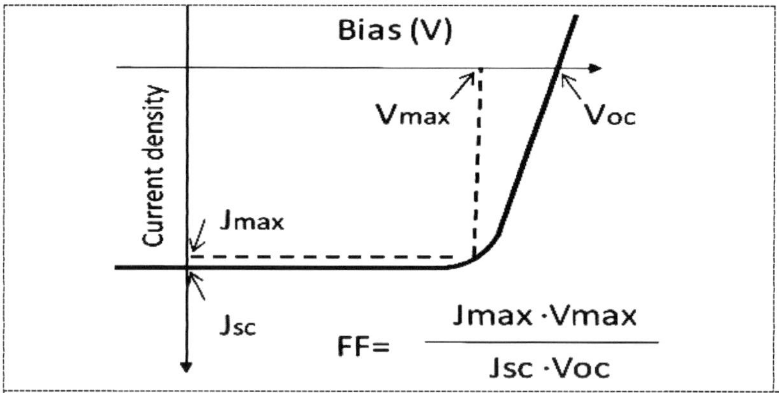

Figure 7. Organic solar cell I-V characteristics in the IV Quadrant.

$$\text{Jdark(V)} = J_o(e^{qV/K_BT} - 1) \quad (4)$$

The dark current density $J_{dark}(V)$is for an ideal organic solar cell, J_o denotes a constant, KB denotes Boltzmann's constant, and T denotes Kelvin temperature. The total number of short circuit photocurrents and dark currents can be used to approximate the cell's total current voltage response. Despite the fact that the reverse current flowing in an illuminated cell in response to voltage is not strictly equal to the current flowing in the dark, for certain photovoltaic materials, the approximation is rational. The sign standard for voltages and currents. The cell's net current density is:

$$J(V) = J_{dark}(V) - J_{sc} \quad (5)$$

which can be written as

$$J(V)= J_o(e^{\frac{qv}{KBT}}-1)-J_{sc} \tag{6}$$

for an ideal diode.

Open circuit voltage (Voc) is the potential difference reaches its highest value when the connections are disconnected. In order to create the optimal diode,

$$Voc = \frac{KT}{q}\ln\left(\frac{J_{sc}}{J_o}+1\right) \tag{7}$$

Equation (7) indicates a logarithmic rise in V_{oc} with the intensity of light. Figure 7 indicates that the current-voltage product is negative and that when the voltage is between 0 and V_{oc} the cell produces electricity. The illuminated device V<0 functions as a photodetector and consumes electricity to produce a light-dependent but partially independent photo current. The system absorbs power again at V> Voc. This is the area where diodes are emitted from light.

Fill Factor is calculated by

$$FF=\frac{I_{max} * V_{max}}{I_{sc} * V_{oc}} \tag{8}$$

Where I_{max} and V_{max} are maximal power points of current and voltage. The overall power conversion efficiency η is defined as

$$\eta = \frac{I_{max} * V_{max}}{P_{in}} = \frac{(FF * (I_{sc} * V_{oc}))}{P_{in}} \tag{9}$$

The short-circuit current density is calculated by

$$Jsc = qG(L_n + L_n) \tag{10}$$

Where generation rate is denoted by G, and electron lengths is denoted byL_n and lengths of the hole diffusion is denoted by L_P respectively.

3.2 Impact of Anode Layer thickness on performance

Graphene is used as an electrode in the nano photodiode. The impact of anode layer thickness on parameter diodes is defined in this section. The thickness of Graphene ranges from 25nm to 200nm though preserving donor is having constant thickness, receiver and aluminium layer thickness of about 50 nm. The anode layer thickness of 100 nm I-V curves is depicted in Figure 8. The thickness of donor, acceptor and cathode layers is 50 nm. The thick anode decreases the transmission of light and low sheet resistance.

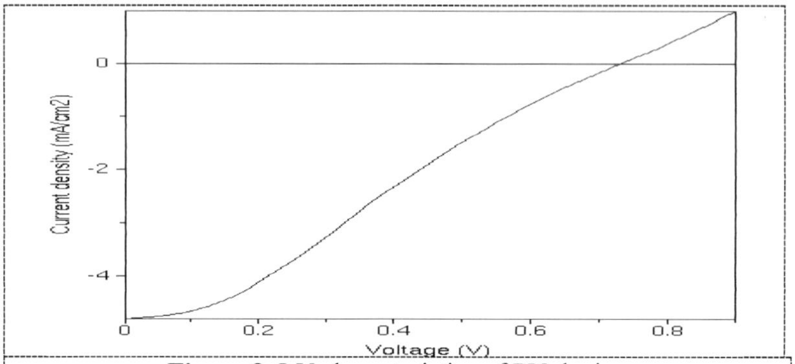

Figure 8. I-V characteristics of PV device

3.3 Impact of anode layer thickness on Voc

The variation of V_{oc} in relation to the anode layer thickness is seen in figure 9. From this figure it is evident that V_{oc} is not influenced by the anode layer thickness. This suggests that the graphene layer thickness doesn't play any role on the creation of excitons.

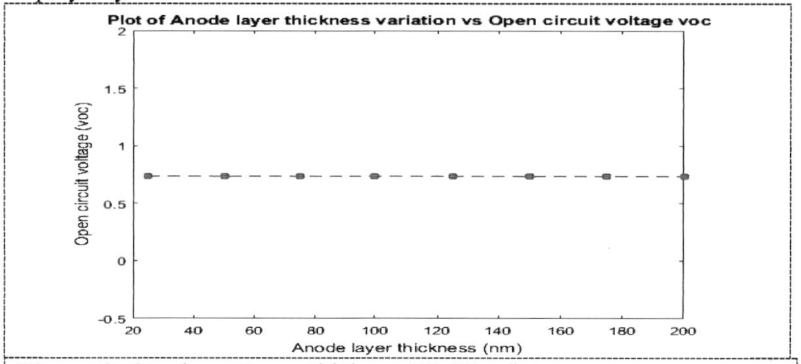

Figure 9. Graphene Layer thickness Vs V_{oc} plot

3.4 J_{sc} density is affected by the thickness of the anode layer

According to the simulation findings, the Figure 10 shows that the J_{sc} is improved due to high lights transmittance and low sheet resistance, if anode thickness rises between 25 and 100 nm. Graphene thickness increases leads to a falling current density owing to poor light transmission due to the thickness from 100 nm to 200 nm of the anode.

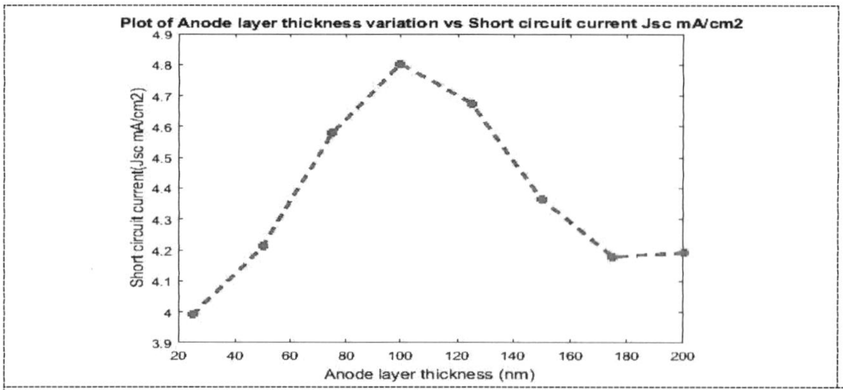

Figure 10. Plot of Graphene Layer thickness Vs J_{sc}

3.5 Impact of Graphene layer thickness on Efficiency

The ratio of output power and input power is defined as efficiency. Even with strong light transmission, when anode thickness grows, efficiency improves from15 nm to 110 nm. Low sheet resistance results in increase in anode-thickness. As the thickness of the graphene increases from 110 nm to 210 nm, the efficacy of anode thickness reduces owing to low light transmission. When all above-mentioned parameters are decreased, as seen in Figure 11, device efficiency also decreases.

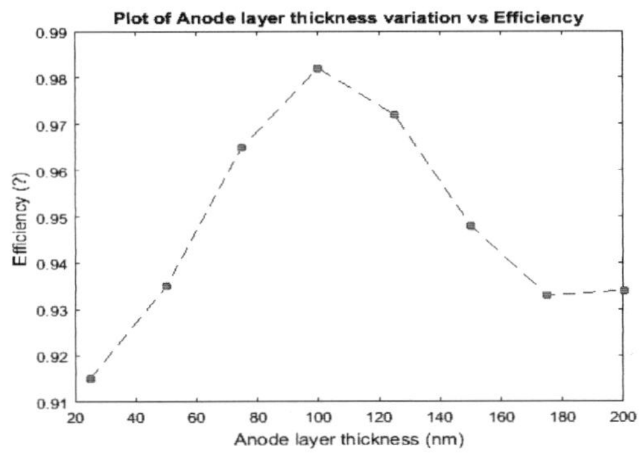

Figure 11. Anode Layer Thickness vs. Efficiency Plot.

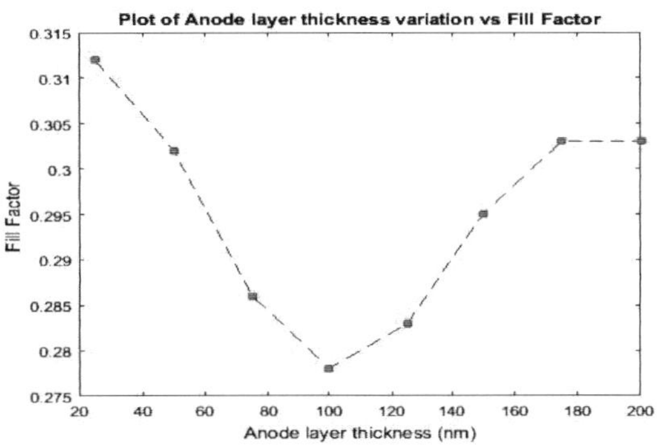

Figure 12. Graphene Layer thickness Vs Fill FactorPlot.

3.6 The thickness of the Graphene layer impact on the Fill Factor

Figure 12 shows that the FF is inverted in terms of J_{sc} and η. Diminishes the FF as J_{sc} and η increases.

3.7 Active Layer thickness Impacton performance

Active layer thickeness effects accomplished utilizing modelling and are compared to active layer thickness, on different parameters. In this analysis, the donor and acceptor thickness varies between 20 nm and 200 nm. The thickness of the anode and the cathode is stable. Figure 13 depicts the I-V curve with an Donor-acceptor layer thickness of 60 nm.

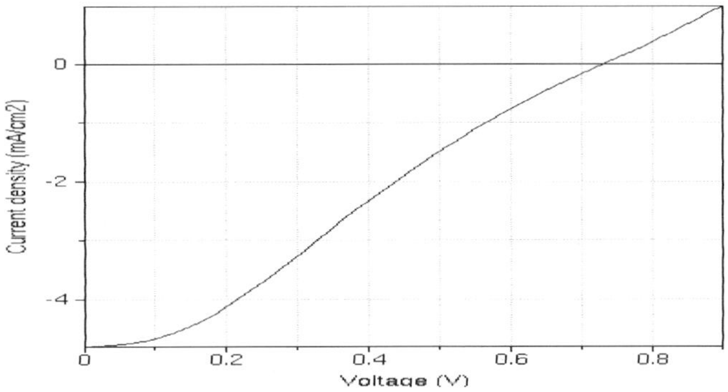

Figure 13. PV device I-V characteristics donor-acceptor layer thickness.

3.8 The impact of Donor-acceptor layer thickness on V_{oc}

Figure 14 shows that the V_{oc} value declines insignificantly. The operating circuit volt is solely determined by the difference in the equation (1) between HOMO donor and LUMO acceptor.

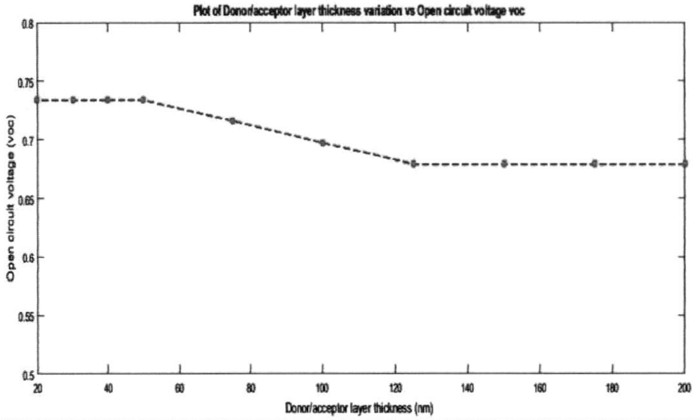

Figure 14. Donor/acceptor Layer Thickness vs. V_{oc} Plot.

3.9 J_{SC} effect on active layer thickness

Figure 16 depicts fluctuation of efficiency with donor/acceptor layer thickness. The device's efficiency declines when the parameters Voc, Jsc, and FF have been reduced. According to this study, efficiency of more than 1% may be impossible. As the active layer thickness increases, current density decreases.

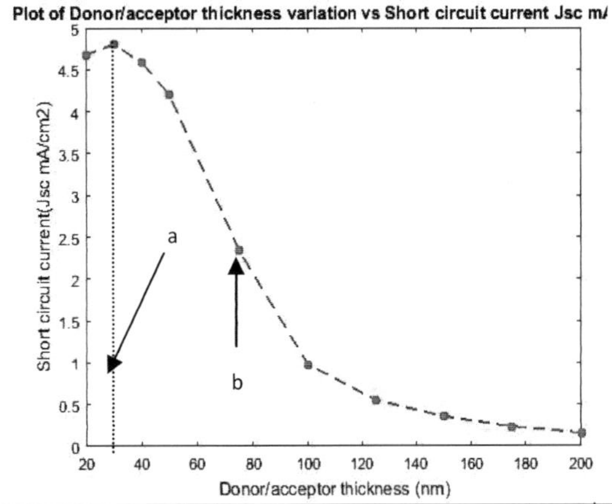

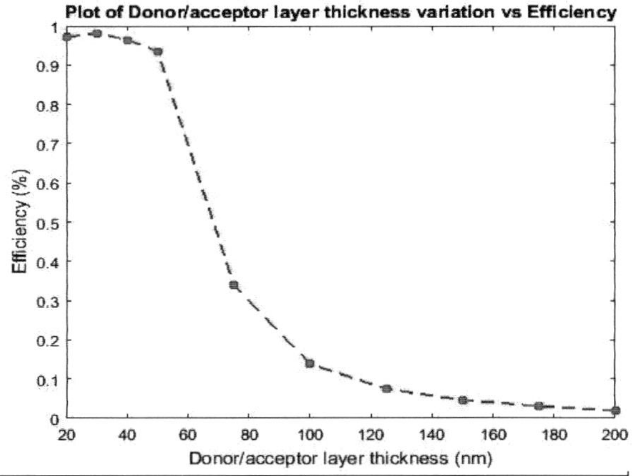

Figure 15. Plot of Donor/acceptor Layer thickness Vs J_{sc} short circuit current density.

Figure 16. Plot of Donor/Acceptor layer thickness vs efficiency

3.10 Impact of active layer thickness on Efficiency

Figure 16 depicts the Donor/Acceptor layer thickness impact on efficiency. The efficiency of the device decreases as the parameters V_{oc}, J_{sc}, FF are decreased. This analysis shows that more than % productivity might not be feasible.

3.11 Fill Factor Effect of Active Layer Thickness

The rise of Donor-Acceptor layer thickness over electrode thickness would marginally reduce the fill factor as seen in Figure 17. It indicates that under I-V characteristic curves the maximal power obtained decreases by an increasing thickness. As a result, thickness of the Donor-Acceptor layer increases it decreases.

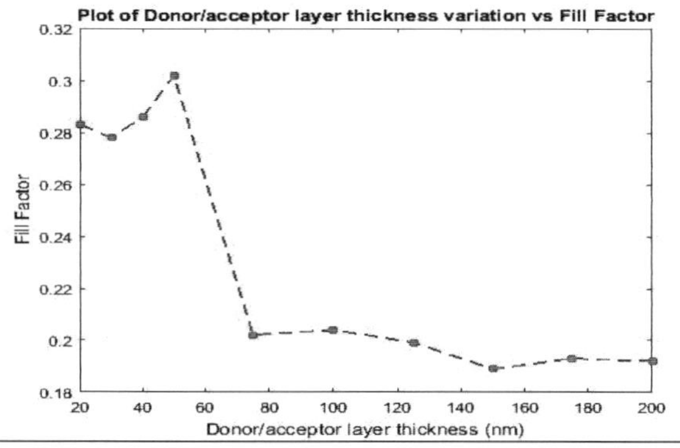

Figure 17. Donor/acceptor Layer Thickness vs FF Plot

Table 1. Optimum layer thicknesses.

Layer	Material	Layer thickness
Anode	Graphene	100 nm
Donor	SWCNT	50 nm
Acceptor	C60	50 nm
Cathode	Aluminum	50 m

3.12 Thickness of the cathode layer has an effect on performance

Simulation results clearly show that the performances of both the variation on cathode layer thickness and anode layer thickness resulting in similar output.

The increase in cathode thickness will increase the efficiency. The bottom of the solar cell is the Cathode. Therefore, in optical absorption, it has no function. The collection of charges is the responsibility of Cathode.

3.13 Design Optimization

The analysis above and the results of the different NPDA parameters show that the following are the ideal layer thicknesses to increase the efficiency. Now it is essential to design the geometries of layers for the PHJ devices to be fabricated. The simulation results form the basis for the design of optimum geometries for the the finest performance. It is worth noting that maximizing the efficiency is the key to achieve the best possible performance. This study present the simulation studies carried out for optimizing the Nano Photodiode (NPD) geometries using simulation tool. Several PHJ devices were simulated and compare their performance parameters. Based on the results the design guidelines for achieving optimum photocurrent density and photovoltage for subretinal stimulation applications were generated and optimum design was arrived at.

4. Conclusion

RP and AMD are the two main conditions leading to human vision loss. For retina implants in the past, MPDA was considered. The results discussed in the previous sections show that the transmittance of light reduces as the anode layer thickness is increased beyond 100 nm. It is also demonstrated that it is not able to stop the BHJ devices in producing higher efficiency even if the thickness of anode layer is more than 100 nm. The results further indicate that the efficiency of PHJ devices will be more as its thickness increases. Cathode layer thickness plays no role in efficiency improvement. Therefore, it is kept 50 nm layer thicknesses. The result of the single PHJ NPD optimum layer thickness presented in Table 1 indicates that it should be sufficient to produce the necessary photo voltage and photocurrent for stimulating a single nerve cell. More corrosive, though, leading to damage to the implant system itself [2]. This is also bio incompatible. A potential solution is NPDA nano-materials such as CNT and graphene made on bio-compatible versatile substrates. This NPDA, which produces power under light incident to stimulate retina, does not require an external power source to recover impaired vision function. This current structure of NPDA is more robust and can increase the level of vision dramatically. This work aims to optimize the diode configuration and geometry to maximize the efficiency of the photodiode array.

5. References

[1] JÖrg-Uwe M2002 Retinal Implant- a bioMEMS Challenge *Sens. Actuators* ***A*** **97**1-9.

[2] Van Lookeren Campagne M, Le Couter J, Yaspan B L and Ye W 2014 Mechanisms of age-related macular degeneration and therapeutic opportunities *J Pathol* **232** 151–164.

[3] Asher A, Segal W A, Baccus S A, Yaroslavsky L P, and Palanker DV 2007 Image processing for A High-Resolution Optoelectronic Retinal Prosthesis *IEEE. Trans. Biomed. Eng* **54** 993-1004.

[4] Alan Y Chow and Vincent Y Chow2004 The Artificial silicon Retina Microchip for the Treatment of Vision Loss From Retinitis Pigmentosa *Arch Ophthalmol* **122** 460-469.

[5] Mokwa W 2008Artificial Retinas *Comprehensive Microsystems* 3ISBN13:978-0-444-52194-1

[6] Luo YH, da Cruz L 2014 A review and update on the current status of retinal prostheses (bionic eye) *Med Bull.* **109** 31-44.

[7] Neal S Pand Alan Y C1999Subretinal implantation of semiconductor-based photodiodes: Progress and challenges *J Rehabil Res Dev* **36**4.

[8] Stett A, Barth W and Weiss S2000Electrical multisite stimulation of the isolated chicken retina *Vision Res* **40** 1785–95.

[9] Zrenner E, Bartz-Schmidt KU and Benav H2011 Subretinal electronic chips allow blind patients to read letters and combine them to words *ProcBiol Sci* **278** 1489–97.

[10] Swapna P and Srinivasa R Y 2014 Simulation Studies On Semiconducting Single Walled Carbon Nanotube Based Bilayer and Bulk Hetero Junction Organic Solar Cells *4th Int. Conf. on Mat Proc Char* **2** 1634-41.

[11] Ham MH, Paulus GL, Lee CY, Song C, Kalantar-zadeh K, Choi W, Han JH, Strano MS2010 Evidence for high-efficiency exciton dissociation at polymer/single-walled carbon nanotube interfaces in planar nano-heterojunction photovoltaics*ACS Nano* **4(10)** 6251-9.

[12] Kymakis E, Stratakis E, Stylianakis MM, Koudoumas E and Fotakis C2011Spin coated graphene films as the transparent electrode in organic photovoltaic devices*Thin Solid Films* **520** 1238–41.

[13] Meo Mand Rossi M 2006 Prediction of Young's modulus of single wall carbon nanotubes by molecular-mechanics based finite element modeling *Compos Sci Technol* **66** 1597-1605.

[14] Sariciftci N S, Smilowitz L, Heeger A J, and Wudl F1992Photo induced electron transfer from a conducting polymer to buckminster fullerene *Science* **258** 1474–76.

[15] Becquerel A E 1839Memory on the electrical effects produced under the influence of radiation solarire *C R Acad Sci Paris* **9** 561.

[16] Biswajit R, Camila G W, Mohammad R K, and Muhammad A A2014 OPV Lab Simulation Tool 1.2 www.nanohub.org

[17] Sebag J 2020 what-is-the-macula retrieved from*http://www.vmrinstitute.com/what-is-the-macula/.*

Instability in Casson nanofluids for Darcy-Brinkman model

M Devi [1], J Sharma [2,*], U Gupta [3]

[1]Energy Research Centre, Panjab University, Chandigarh-160014, India.
[2]University Institute of Engineering and Technology, Panjab University, Chandigarh, India.
[3]Dr. S.S. Bhatnagar University Institute of Chemical Engineering and Technology, Panjab University, Chandigarh, India.

*Corresponding author E-mail address: jyoti.maths@gmail.com

Abstract. The convective instability of Casson nanofluids saturating porous medium is investigated analytically and numerically. Darcy-Brinkman law is utilized to frame the momentum equation for the system. The governing equations are simplified using normal mode analysis and linear stability theory. The present study finds the expressions of thermal Rayleigh number for various boundaries and discusses the convective instability under various effects with the help of Wolfram Mathematica software. Out of all the boundaries, the instability of the layer is found to be least for both rigid boundaries. It is established that the convection in the fluid gets delayed due to porosity effects while non-Newtonian property and nano scale effects contribute significantly in making the layer unstable.

Keywords: Nanoparticles; Non-Newtonian fluid; Casson model; Brownian motion; Thermophoresis; Darcy Brinkman model.

Nomenclature

a	wave number
α_m	thermal diffusivity of fluid
c_p	specific heat capacity
d	thickness of the layer
C_B	diffusion coefficient due to Brownian motion
C_T	diffusion coefficient due to thermophoresis
e_{ij}	deformation rate
g	gravitational acceleration
k	thermal conductivity of nanofluid
k_m	overall thermal conductivity
K	Permeability
a_x	x-component of wave number
a_y	y-component of wave number
p	pressure
s	rate of growth
t	time

Greek-symbols

β	Casson parameter
β_t	thermal volumetric coefficient
γ	conductivity variation parameter
ε	porosity
$\hat{\mu}$	effective viscosity parameter
μ_d	dynamic viscosity
τ	stress tensor
ρ_p	nanoparticle density
ρ_0	fluid density at temperature T_0
$(\rho c)_m$	effective heat capacity
σ	thermal capacity
ϕ	Relative volume fraction $\dfrac{\phi-\phi_0}{\phi_1-\phi_0}$

Non-dimensional parameters

D_a	Darcian parameter

T	temperature	L_f	fluid Lewis number
T_0	temperature at top layer	N_a	diffusivity fraction
T_1	temperature at bottom layer	N_p	particle density increment
q_D	Darcian velocity ($q_D = (u,v,w)$)	p_r	Prandtl number
Y_y	yield stress for Casson fluid	R_a	Darcy-Rayleigh number
(x,y,z)	cartesian co-ordinate system	R_m	basic-density Rayleigh number
z	an axis of coordinate system	R_p	particle Rayleigh number

Superscripts

^	perturbed variable
*	non-dimensional variable

1. Introduction

Nowadays, there has been an increasing interest in nanofluids, which are a combination of ordinary fluid and a small number of suspended nanoparticles; the resulting fluid was coined as 'Nanofluid' by Choi [1]. Nanofluids transfer heat more efficiently than the conventional fluids and are widely utilized in a variety of applications in diverse fields. Buongiorno [2] made a significant contribution to nanofluid modeling by pointing out that that the base fluid velocity and the relative velocity contribute in absolute velocity of nanoparticles. The model was used by many researchers [3-6] to study the convective heat transfer in nanofluids and was further modified by Nield and Kuznetsov [7] to consider the solutal effects on nanofluid layer. The analytical investigations were made to establish the destabilizing influence of solute and nanoparticles on the system. Gupta et al. [8] considered the instability in a binary nanofluid layer subjected to magnetic field. Contrary to solutal and nano scale effects, magnetic field was observed to delay the instability of the layer. To consider the influence of density and conductivity of particles, steady state solution for particle volume fraction was assumed to be invariable [9] and the problem was studied numerically for various nanoparticles in water based nanofluid. Nield and Kuznetsov [10] considered the Brinkman instability in nanofluids for all three boundaries. Further it was established that rotation and magnetic field make the binary nanofluid layer more stable while porosity effects speed up the initiate the convection currents in the fluid layer [11,12]. Dhananjay et al. [13] considered the thermal instability in an internally heated porous layer and porosity effects were found to contribute in the stability while internal heat source and nanoparticles destabilize the system.

Casson model fits well to various non-Newtonian fluids [14]. The properties of blood coincide with that of Casson fluid in moderate shear rate flows according to Blair and Spanner [15], thus it is reasonable to use Casson model for blood flow. Casson's equation's success was investigated by Scott Blair [16]. The dual nature solution for the suggested model with thermal radiation impacts on steady as well as unsteady flow of Casson fluid was discovered by Hamid et al. [17]. Natural instability in a partially heated porous medium was studied by Aneja et al. [18] for Casson fluid using penalty finite element approach. Recently, Casson model has been used by researchers for various nanofluid flow problems to study the impact of nanoparticles on blood flow. Boundary layer MHD Casson nanofluid flow with variations in conductivity and viscosity was investigated by Gbadeyan et al. [19]. The convection currents in Casson nanofluids with internal heating effects were considered by Gupta et al. [20]. Recently, Gupta et al. [21] considered binary instability of Casson nanofluids analytically and numerically.

Unlike the previous works, the present paper studies the phenomenon of Darcy-Brinkman convection in Casson nanofluids. The equations based on conservation laws are simplified using normal mode technique [22] and linear stability theory is employed. Top layer of configuration is considered to have more nanoparticle volume fraction as compared to lower which assures the non-occurrence of oscillatory motions [7, 10]. The stationary convective instability is illustrated through graphs to depict the impact of nanofluid, non-Newtonian and porosity parameters on the Rayleigh number. It is established that Darcy number and porosity have stabilizing impact while Casson parameter, nanoparticle Rayleigh number and Lewis number show destabilizing influence. Also, diffusivity ratio and particle density increment do not show any impact on the instability of the system.

2. Conservation equations for the system

A fluid layer is considered (refer; Figure1). The temperature and nanoparticle concentration are T_1 *and* ϕ_1 at lower layer z=0 while T_0 *and* ϕ_0 at upper layer z=d (where $T_1 > T_0$ *and* $\phi_1 > \phi_0$).

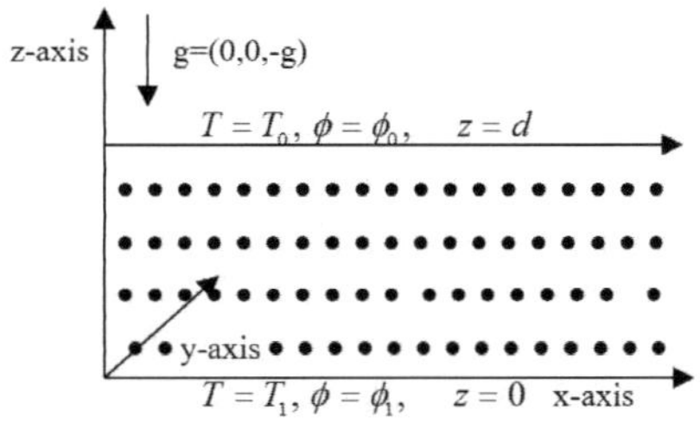

Figure 1. Sketch of the system.

The set of rheological equation for Casson fluid (refer; [23]) is

$$\tau_{ij} = \begin{Bmatrix} \mu_d + \left(\dfrac{Y_y}{\sqrt{2\pi}}\right) 2e_{ij},\ \pi > \pi_c \\ \mu_d + \left(\dfrac{Y_y}{\sqrt{2\pi_c}}\right) 2e_{ij},\ \pi < \pi_c \end{Bmatrix}, where\ Y_y = \frac{\mu_d\sqrt{2\pi}}{\beta}, \tag{1}$$

and $\pi = e_{ij}e_{ij}$, $e_{ij} = \dfrac{1}{2}\left(\dfrac{\partial v_i}{\partial x_j} + \dfrac{\partial v_j}{\partial x_i}\right)$.

The equations for fluid layer system are (refer; [2,5])

$$\nabla . q_D = 0, \tag{2}$$

$$\frac{\rho_0}{\varepsilon}\frac{\partial q_D}{\partial t} = -\nabla p + \tilde{\mu}\, div\, \tau - \frac{\mu}{K} q_D + \left[\phi \rho_P + (1-\phi)\{\rho_0 (1-\beta_t (T-T_0))\}\right] g. \tag{3}$$

Equations (1) and (3) give

$$\frac{\rho_0}{\varepsilon}\frac{\partial q_D}{\partial t} = -\nabla p + (1+1/\beta)\tilde{\mu}\nabla^2 q_D - \frac{\mu}{K} q_D + \left[\phi \rho_P + (1-\phi)\{\rho_0 (1-\beta_t (T-T_0))\}\right] g, \tag{4}$$

$$(\rho c)_m \frac{\partial T}{\partial t} + (\rho c)_f\, q_D . \nabla T = k_m \nabla^2 T + \varepsilon (\rho c)_p \left[C_B \nabla \phi . \nabla T + \frac{C_T}{T_0} \nabla T . \nabla T \right], \tag{5}$$

$$\frac{\partial \phi}{\partial t} + \frac{1}{\varepsilon} q_D . \nabla \phi = C_B \nabla^2 \phi + C_T \nabla^2 \frac{T}{T_0}. \tag{6}$$

At lower layer ($z = 0$);

$$w = 0, \frac{\partial w}{\partial z} + l_1\, d\, \frac{\partial^2 w}{\partial z^2} = 0, T = T_1, \phi = \phi_1,$$

and at upper layer ($z = d$);

$$w = 0, \frac{\partial w}{\partial z} - l_2\, d\, \frac{\partial^2 w}{\partial z^2} = 0, T = T_0, \phi = \phi_0, \tag{7}$$

where the parameters l_1 *and* l_2 approach to infinity for free and zero for rigid boundaries.
Now we non-dimensionlize the variables as

$$\left(x^*,y^*,z^*\right)=\left(x,y,z\right)/d,\ t^*=t\alpha_m/\sigma d^2,\ q_D^*=q_D d/\alpha_m,\ p^*=pd/\mu\alpha_m,$$

$$\phi^*=\left(\phi-\phi_1\right)/\left(\phi_0-\phi_1\right),\ T^*=\left(T-T_0\right)/\left(T_1-T_0\right),\ \text{where}\ \alpha_m=k_m/\left(\rho c\right)_f,\ \sigma=\left(\rho c\right)_m/\left(\rho c\right)_f. \tag{8}$$

Using Eqs. (8), Eqs. (2)-(7) (after dropping the asterisks) become

$$\nabla .q_D=0, \tag{9}$$

$$\frac{D_a}{P_r}\frac{\partial q_D}{\partial t}=-\nabla p+\left(1+1/\beta\right)D_a\nabla^2 q_D-q_D-R_m e_z+R_a Te_z-R_p\phi e_z, \tag{10}$$

$$\frac{\partial T}{\partial t}+q_D.\nabla T=\nabla^2 T+\frac{N_p}{L_f}\nabla\phi.\nabla T+\frac{N_a N_p}{L_f}\nabla T.\nabla T\ , \tag{11}$$

$$\frac{1}{\sigma}\frac{\partial\phi}{\partial t}+\frac{1}{\varepsilon}q_D.\nabla\phi=\frac{1}{L_f}\nabla^2\phi+\frac{N_a}{L_f}\nabla^2 T, \tag{12}$$

where

$$R_a=\rho_0 gKd\beta_t\frac{\left(T_1-T_0\right)}{\mu\alpha_m},\ L_f=\frac{\alpha_m}{C_B},\ P_r=\frac{\mu}{\rho_0\alpha_m},$$

$$N_p=\varepsilon\left(\rho c\right)_p\frac{\phi_1-\phi_0}{(\rho c)_f},\ N_a=\frac{C_T}{C_B}\frac{\left(T_1-T_0\right)}{T_1\left(\phi_1-\phi_0\right)},\ D_a=\frac{\tilde{\mu}K}{\mu d^2}, \tag{13}$$

$$R_m=(\rho_p\phi_1+\rho_0(1-\phi_1)gdK/\mu\alpha_m,\ R_p=\frac{\left(\rho_p-\rho_0\right)\left(\phi_1-\phi_0\right)}{\mu\alpha_m}gdK.$$

And boundary conditions for unit depth of layer are

$$w=0,\frac{\partial w}{\partial z}+l_1\frac{\partial^2 w}{\partial z^2}=0,T=1,\phi=0\ \ at\ z=0,$$

$$w=0,\frac{\partial w}{\partial z}-l_2\frac{\partial^2 w}{\partial z^2}=0,T=0,\phi=1\ \ at\ z=1. \tag{14}$$

3. Steady state solution and perturbation equations

At $t=0$, let velocity of the fluid is zero and variables vary along horizontal axis. Thus Eqs. (9)-(12) give

$$T(z)=1-z,\ \ \phi(z)=z. \tag{15}$$

Now we add perturbations to initial solutions and write

$$q_D=q_D+\hat{q}_D,\ p=p+\hat{p},\ \phi=\phi+\hat{\phi},\ T=T+\hat{T}. \tag{16}$$

Equations (9)-(12) with the help of Eqs. (15 and 16), give us a set of perturbed differential equations as

$$\nabla.\hat{q}_D=0, \tag{17}$$

$$\frac{D_a}{P_r}\left(\frac{\partial\hat{q}_D}{\partial t}\right)=-\nabla\hat{p}+D_a\left(1+1/\beta\right)\nabla^2\hat{q}_D-\hat{q}_D+R_a\hat{T}e_z-R_p\hat{\phi}e_z, \tag{18}$$

$$\frac{\partial\hat{T}}{\partial t}+\hat{w}=\nabla^2\hat{T}+\frac{N_p}{L_f}\frac{\partial\phi}{\partial z}\frac{\partial\hat{T}}{\partial z}+\frac{N_p}{L_f}\frac{\partial T}{\partial z}\frac{\partial\hat{\phi}}{\partial z}+2\frac{N_a N_p}{L_f}\frac{\partial\hat{T}}{\partial z}\frac{\partial T}{\partial z}, \tag{19}$$

$$\frac{1}{\sigma}\frac{\partial\hat{\phi}}{\partial t}+\frac{1}{\varepsilon}\frac{\partial\phi}{\partial z}\hat{w}=\frac{1}{L_f}\nabla^2\hat{\phi}+\frac{N_a}{L_f}\nabla^2\hat{T}, \tag{20}$$

Now on Eq. (18) together with Eq. (17) and using the relation $curlcurl=graddiv-\nabla^2$, we get

$$\frac{D_a}{P_r}\frac{\partial}{\partial t}\left(\nabla^2\hat{w}\right)+\nabla^2\hat{w}-\left(1+1/\beta\right)D_a\nabla^4\hat{w}=R_a\nabla_H^2\hat{T}-R_p\nabla_H^2\hat{\phi}, \tag{21}$$

where $\nabla^2=\frac{\partial^2}{\partial^2 x}+\frac{\partial^2}{\partial^2 y}+\frac{\partial^2}{\partial^2 z}$ and $\nabla_H^2=\frac{\partial^2}{\partial^2 x}+\frac{\partial^2}{\partial^2 y}$ (Laplacian operators).

To use normal mode technique on Eqs. (17)-(21), perturbed variables are written as [refer; 22]

$$\left(\hat{w},\hat{T},\hat{\phi}\right)=\left[W,T,\phi\right]\left(z\right)\exp\left(ia_x x+ia_y y+st\right). \tag{22}$$

Putting $s=0$ for stationary convection and using Eq.(22) in Eqs. (17)-(21); we get

$$\left((1+1/\beta)D_a\left(D^2-a^2\right)^2-\left(D^2-a^2\right)\right)W-a^2R_aT+a^2R_p\phi=0, \tag{23}$$

$$W+\left(D^2-a^2+\frac{N_p}{L_f}(D-2\frac{N_aN_p}{L_f}D\right)T-\frac{N_p}{L_f}D\phi=0, \tag{24}$$

$$-\frac{1}{\varepsilon}W+\frac{N_a}{L_f}(D^2-a^2)T+\frac{1}{L_f}(D^2-a^2)\phi=0, \tag{25}$$

with $D=\frac{d}{dz}$, $a=(a_x^2+a_y^2)^{1/2}$.

The boundary conditions (given by Eq. (14)) reduce to

$$W=T=\phi=0, \tag{26}$$

and

$$D^2W=0 \ \textit{for free boundary}, \tag{27}$$

$$DW=0 \ \textit{for rigid boundary}. \tag{28}$$

Further Eqs. (23)-(25) are solved to find the expressions of Darcy-Rayleigh number for boundary conditions given by Eqs. (26)-(28).

4. Expressions of Rayleigh number

4.1 Both-free case

The trial solutions satisfying Eqs. (26) and (27) are

$$(W,T,\phi)=(L,M,N)\sin\pi z, \tag{29}$$

The eigen value equation using Eq. (29) in Eqs. (23)-(25) and eliminating of unknowns L,M and N, gives

$$R_a=\frac{2D_a(1+1/\beta)J^4+2J^3-2a^2J\,L_f\,R_p/\varepsilon}{2a^2J}-R_p\,N_a, \tag{30}$$

where $J=\pi^2+a^2$.

4.2 Rigid-free case

The trial solutions satisfying Eqs. (26)-(28) are

$$W=Lz^2(1-z)(3-2z),\ T=M(z-z^2),\ \phi=N(z-z^2), \tag{31}$$

Using Eq. (31) in Eqs. (23)-(25) gives

$$R_a=\frac{28[D_a\varepsilon(1+1/\beta)(4536+432a^2+19a^4)+\varepsilon(216+19a^2))](10+a^2)^2-507a^2R_pL_f(10+a^2)}{507a^2\varepsilon(10+a^2)}-R_pN_a. \tag{32}$$

4.3. Both-rigid case

The trial solutions satisfying Eqs. (26) and (28) are

$$W=L(z^4-2z^3+z^2),\ T=M(z-z^2),\ \ \phi=N(z-z^2). \tag{33}$$

Equations (23)-(25) with Eq. (33) give

$$R_a = \frac{56[D_a\varepsilon(1+1/\beta)(504+24a^2+a^4)+\varepsilon(12+a^2))](10+a^2)^2 - 54a^2L_fR_p(10+a^2)}{54a^2\varepsilon(10+a^2)} - R_pN_a. \tag{34}$$

5. Approximate solution and results

5.1. Both-free case

When $\beta \to \infty$ in the Rayleigh number expression in Eq. (30) matches the one given by Nield and Kuznetsov [10]

$$R_a = \frac{D_aJ^3 + J^2}{a^2} - R_p(N_a + \frac{L_f}{\varepsilon}). \tag{35}$$

Also Eq. (30) can be written as

$$R_a + R_p(N_a + \frac{L_f}{\varepsilon}) = \frac{D_a(\pi^2+a^2)^3(1+\frac{1}{\beta}) + (\pi^2+a^2)^2}{a^2}. \tag{36}$$

When $D_a = 0$ then $R_a + R_p(N_a + \frac{L_f}{\varepsilon}) = 4\pi^2$, that is the lowest value attained at $a = \pi$.

When $D_a = 1$ then $R_a + R_p(N_a + \frac{L_f}{\varepsilon}) = \frac{27}{4}\pi^2$, that is the lowest value attained at $a = \frac{\pi}{\sqrt{2}}$.

Therefore, using the large value of D_a as unity, critical wave number for both-free boundaries is around 2.22 and critical Rayleigh number is around 657.5.

5.2. Rigid-free case

Equation (32) by putting $\beta \to \infty$, provides

$$R_a = \frac{28[D_a(4536+432a^2+19a^4)+216+19a^2](10+a^2)}{507a^2} - R_p(\frac{L_f}{\varepsilon} + N_a), \tag{37}$$

which is same as that of Nield and Kuznetsov [10] for rigid-free boundaries. Also, Eq. (37) can be written as

$$R_a + R_p(N_a + \frac{L_f}{\varepsilon}) = \frac{28[D_a(1+1/\beta)(4536+432a^2+19a^4)+216+19a^2](10+a^2)}{507a^2}. \tag{38}$$

When $D_a = 0$ then $R_a + R_p(N_a + \frac{L_f}{\varepsilon}) = 48.01$, which is minimum value obtained at $a = 3.27$.

When $D_a = 1$ and $\beta \to \infty$ then $R_a + R_p(N_a + \frac{L_f}{\varepsilon}) = 1139$, which is minimum value obtained at $a = 2.67$.

Hence, the critical wave number for the case is 2.67 and critical Rayleigh number as 1139 when D_a is as large as unity.

5.3. Both-rigid case

For $\beta \to \infty$ in Eq. (34), we get expression of Rayleigh number

$$R_a = \frac{28[D_a(504+24a^2+a^4)+(12+a^2))](10+a^2)}{27a^2} - R_p(N_a + \frac{L_f}{\varepsilon}). \tag{39}$$

This expression agrees with Nield and Kuznetsov [10] for porous layer only. Eq. (39) can also be written as

$$R_a + R_p(N_a + \frac{L_f}{\varepsilon}) = \frac{28[D_a(1+\frac{1}{\beta})(504+24a^2+a^4)+(12+a^2))](10+a^2)}{27a^2}. \tag{40}$$

When $D_a = 0$ then $R_a + R_p(N_a + \frac{L_f}{\varepsilon}) = 43.92$, which is minimum value obtained $a = 3.31$.

When $D_a = 1$ and $\beta \to \infty$ then $R_a + R_p(N_a + \frac{L_f}{\varepsilon}) = 1750$, which is minimum value obtained at $a = 3.12$.

Thus when D_a approaches to unity and in the absence of Casson parameter, the critical wave number is 3.12 and critical Rayleigh number 1750 approximately.
Here, Eqs. (36), (38), (40) reveal that rigid-rigid boundaries give the system most stability.

6. Numerical results using Wolfram Mathematica software

Equations (31), (34), and (37) are dependent on multiple parameters. Let us take fixed values of parameters as follows: $L_f = 100, N_a = 1, N_p = 0.01, R_p = 0.1, D_a = 0.5, \varepsilon = 0.9$ and vary one of the variables to analyze its impact on instability of the system. Figure 2 depicts the relationship between critical Rayleigh number and Casson parameter β for various values of R_p with all three boundaries. It is noted that when R_p increases, there is a fall in values of critical Rayleigh number with respect to Casson parameter β and hence R_p hastens the instability of the system. Figure 3 plots the graph of critical Rayleigh number versus Casson parameter β for various values of N_a. One can easily observe that critical Rayleigh number for all three boundaries do not get affected by increment in modified diffusivity ratio N_a and hence doesn't have any impact on the initiation of instability in the fluid.
There is no significant change in the stability curves for variations in N_p as interpreted in Figure4 and hence doesn't have much impact on the onset of instability of the fluid layer. Figure 5 shows the stability curves of critical Rayleigh-number in the presence of Casson parameter β for variations in L_f. The critical Rayleigh number falls for increasing values of L_f and therefore a destabilizing impact of Lewis number is established for all three boundaries.

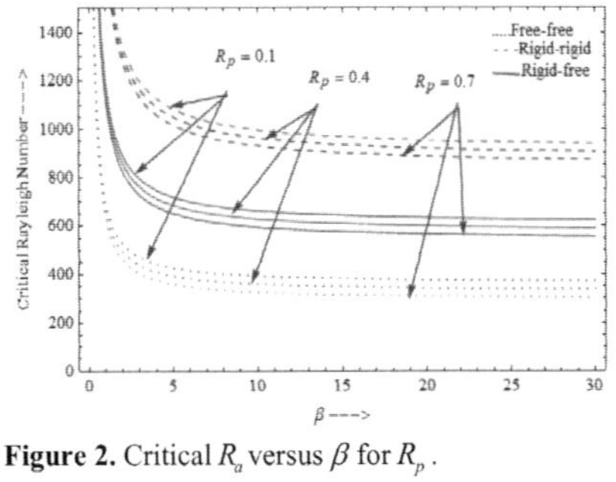

Figure 2. Critical R_a versus β for R_p.

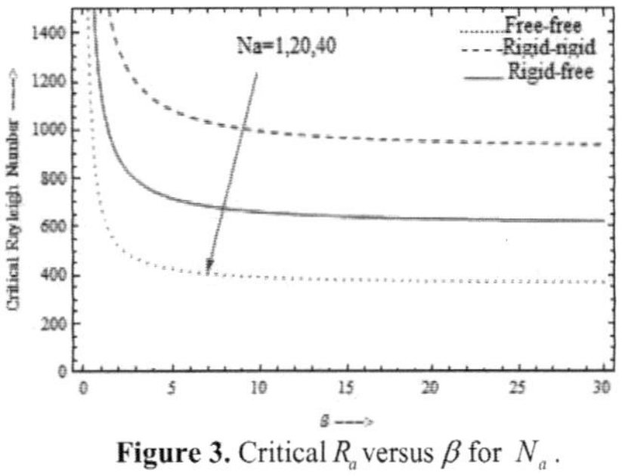

Figure 3. Critical R_a versus β for N_a.

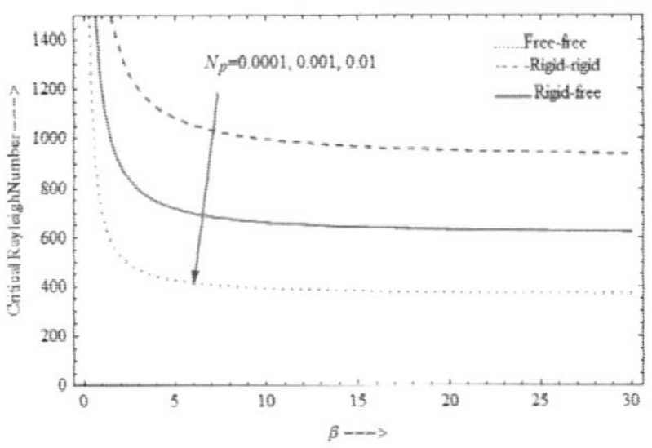

Figure 4. Critical R_a versus β for N_p.

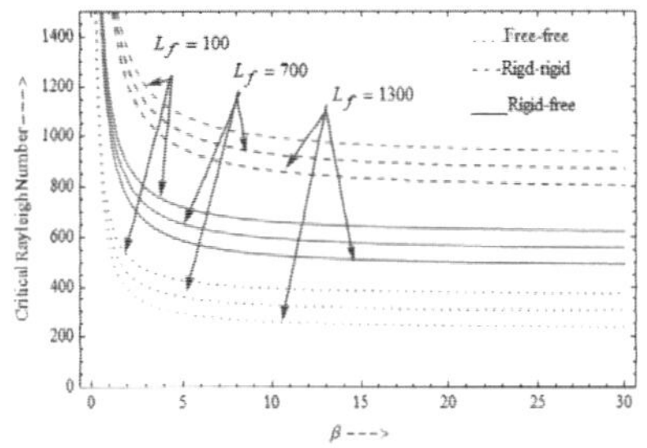

Figure 5. Critical R_a versus β for L_f.

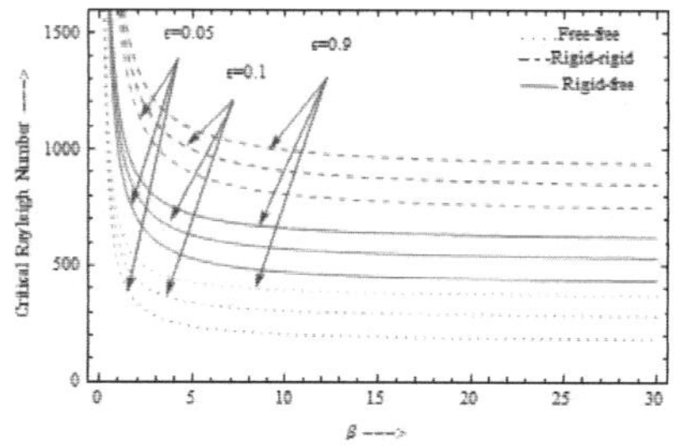

Figure 6. Critical R_a versus β for ε.

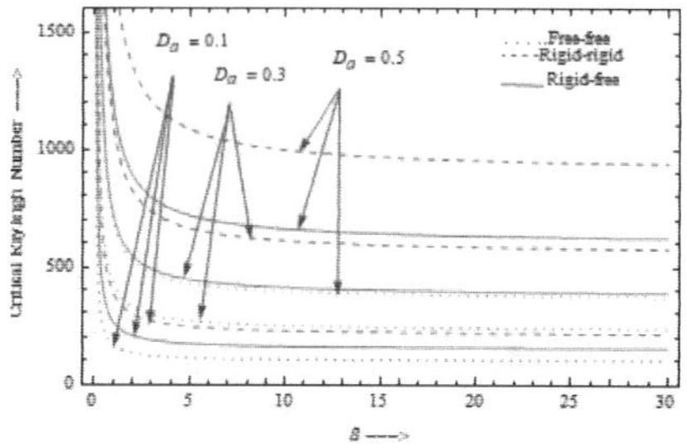

Figure 7. Critical R_a versus β for D_a.

Figure 6 depicts the impact of porosity ε on the onset of instability and it is noted that as ε rises, there is a rise in the value of critical Rayleigh number. So, it is indicating that porosity parameter delays the onset of convection currents in the layer. Figure 7 depicts the curves of critical Rayleigh number as function Casson parameter β for variations in Darcy number and it is shown to delay the convection in the layer. The reason is the increment in effective viscosity with the increment in Darcy parameter which retards the fluid flow. In Figs. 2-7, for smaller values of Casson parameter, critical Rayleigh number decreases quickly whereas for increasing values of β doesn't show much variation in its values.

To understand the complete effect of Darcy parameter, let us draw the graphs of $\log D_a$ versus critical Rayleigh number and $\log D_a$ versus critical wave number a_c. Here, one can observe that critical Rayleigh number increases when Darcy parameter increases in Figure 8 and hence has a delaying influence on the onset of instability. Figure 9 establishes that there is an increase in the size of convection cell due to presence of Darcy number which postpones the thermal instability of the system.

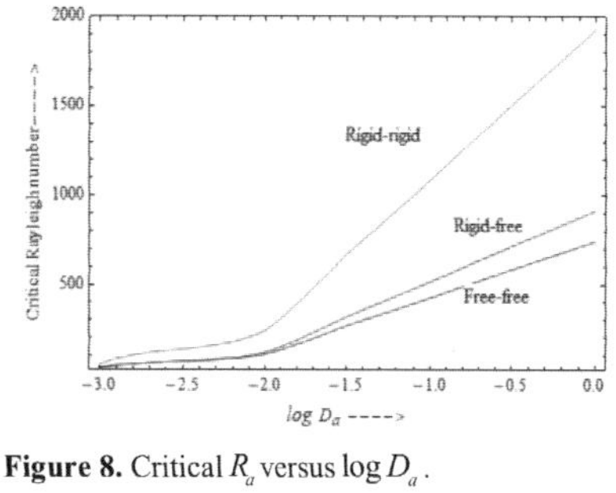

Figure 8. Critical R_a versus $\log D_a$.

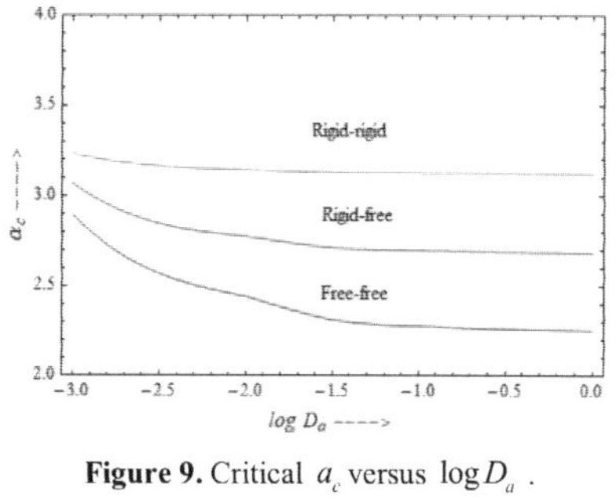

Figure 9. Critical a_c versus $\log D_a$.

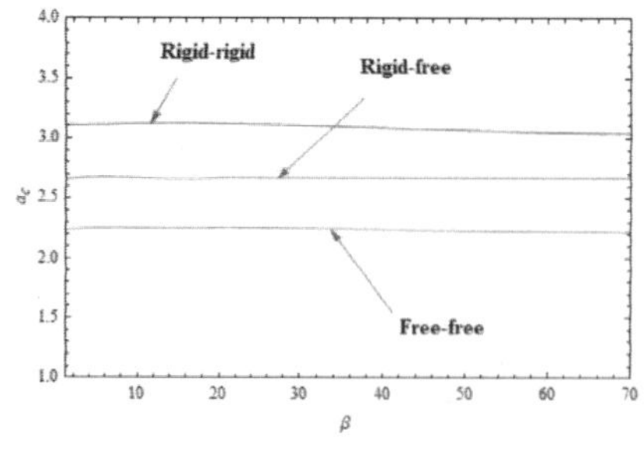

Figure 10. Critical wave number a_c versus β.

Impact of Casson parameter β on a_c is depicted in Figure 10. It is observed that the values of a_c do not vary much with the variation in Casson parameter β and only small values of Casson parameter has a slight impact on a_c. Thus, non-Newtonian behaviour largely doesn't show any impact on critical wave number.

7. Concluding remarks

The convective instability of a non-Newtonian nanofluid layer using Casson model is studied analytically and numerically using Darcy-Brinkman model. Method of normal modes is utilized to solve differential equations to get an eigen value equation for both free, rigid-free and both rigid cases. Effect of Casson parameter on critical Rayleigh number for various nanofluid and porosity parameters are illustrated graphically using the software Mathematica. It is established that Darcy number and porosity delay the convective instability of the system while other nanofluid parameters show a destabilizing effect on the fluid layer. Diffusivity ratio and particle density increment do not show any effect on the system. Non-Newtonian property destabilizes the system significantly for small values of Casson parameter which otherwise doesn't show much influence on the stability of the fluid. Out of all three boundaries, the layer is found to be most stable for realistic rigid-rigid boundaries.

Acknowledgements
One of the authors, Mamta Devi is thankful to Council of Scientific and Industrial Research, New Delhi-110012, India for the financial assistance in the form of JRF [Ref. No.: 09/135 (0895)/2019-EMR-I].

References

[1] Choi S U S, Eastman J A, Wang H.P. (Eds) 1995 Enhancing Thermal Conductivity of Fluid with Nanoparticle, Development and Application of Non-Newtonian Flow *ASME FED- 231/MD-**66** 99-105.*
[2] Buongiorno J 2006 Convective Transport in Nanofluid *ASME J. Heat Transfer. **128** 240-250.*
[3] Tzou D Y 2008 Instability of Nanofluids in Natural Convection *ASME J. Heat Trans. **130** 1-9.*
[4] Tzou D Y 2008 Thermal Instability of Nanofluids in Natural Convection *Int. J. Heat Mass Transf. **51** 2967-79.*
[5] Yadav D, Agrawal G S, Bhargava R 2011 Rayleigh-Benard Convection in Nanofluid *Int J. Appl. Math. Mech. **7(2)** 61-76.*
[6] Nield D A, Kuznetsov AV 2010 The Onset of Convection in a Horizontal Nanofluid Layer of Finite Depth *Euro. J. Mech. B/Fluids, **29** 217-223.*
[7] Nield D A, Kuznetsov A V 2011 The Onset of Double-Diffusive Convection in a Nanofluid Layer *Int. J. Heat and Fluid Flow. **32(4)** 771-776.*
[8] Gupta U, Sharma J, Sharma V 2015 Instability of Binary Nanofluid with Magnetic Field *Appl. Math. Mech. **6** 693-706.*
[9] Sharma J, Gupta U, Sharma V 2017 Modified Model for Binary Nanofluid Convection with Initial Constant Nanoparticle Volume Fraction *J. Appl. Fluid Mech. **10** 1387-95.*
[10] Nield D A, Kuznetsov A V 2010 Thermal Instability in a Porous Medium Layer Saturated by a Nanofluid: Brinkman Model *Trans. Porous Med. **81(13)** 409-422.*
[11] Sharma J, Gupta U, Wanchoo R K 2016 Numerical Study on Binary Nanofluid Convection in a Rotating Porous Layer *Differ. Equ. Dyn. Syst, doi.org/10.1007/s12591-015-0268-4.*
[12] Sharma J, Gupta U, Wanchoo R K 2016 Magneto Binary Nanofluid Convection in Porous Medium *Int. J. Chem. Engi., **2016** 9424036.*
[13] Yadav D, Agrawal G S, Bhargava R 2012 Boundary and Internal Heat Source Effects on the Onset of Darcy-Brinkman Convection in a Porous Layer Saturated by Nanofluid *Int J. Therm. Sci. **60** 244-254.*

[14] Copley A L, 1960 Apparent Viscosity and Wall Adherence of Blood Systems, *In Flow Properties of Blood and Other Biological Systems 97-121.*
[15] Scott Blair G W 1959 An Equation for the Flow of Blood Plasma and Serum Through Capillaries *Nature* ***183*** *613-614.*
[16] Scott Blair G W 1966 The Success of Casson's Equation *Rheo. Acta Sep.* ***5(3)*** *184-187.*
[17] Hamid M, Usman M, Khan Z H, Ahmad R, Wang W 2019 Dual Solutions and Stability of Flow and Heat Transfer of Casson Fluid Over a Stretching Sheet *Phys. Lett. A.* ***383*** *2400-2408.*
[18] Aneja M, Chandra A, Sharma S 2020 Natural Convection in a Partially Heated Porous Cavity to Casson Fluid *Int. Comm. Heat Mass Transf.* ***114*** *1045555.*
[19] Gbadeyan J A, Titiloye E O, Adeosun A T 2020 Effect of Thermal Variable Conductivity and Viscosity on Casson Nanofluid Flow with Convective Heating and Velocity Slip *Heliyon* ***6*** *E03076.*
[20] Gupta U, Sharma J, Devi M 2020 Casson Nanofluid Convection in an Internally Heated Layer *Mater. Today Proc.* ***28*** *1748-52.*
[21] Gupta U, Sharma J, Devi M 2021 Double-Diffusive Instability of Casson Nanofluids with Numerical Investigations for Blood-Based Fluid *Eur. Phys. J. Spec. Top.* ***9 (2021)*** *https://doi.org/1.1140/epjst/s11734-021-00053.*
[22] Chandrasekhar S 1981 Hydrodynamic and Hydromagnetic Stability *Dover Publications, New York, USA.*
[23] Mehmood S, Nawaz M, Ali A, 2018 Finite Volume Solution of Non-Newtonian Casson Fluid Flow in a Square Cavity *Comm. Math. Appl.* ***9(3)*** *459-474, https://doi.org/10.26713/cmav9i3.795.*

Simulation and Modelling of Triboelectric Nanogenerator for Self-powered Electronic Devices

Jashan Kumar Mainra[1], Akshpreet Kaur[1], Gaurav Sapra[1,*] and Parul Gaur[1]
[1]UIET, Panjab University, Chandigarh
*corresponding author e-mail:gaurav.sapra@pu.ac.in

Abstract Triboelectric Nanogenerators has revolutionised the area of energy harvesting and self-powered sensing. In recent years, variety of small scale applications of triboelectric nanogenerators have been explored extensively particularly in self powered electronics, wearable and implantable devices, self-powered biosensors, human motion monitoring, locationevaluation, air quality control etc. This paper discusses simulation and modelling of contact separation mode based triboelectric nanogenerator. In this work, triboelectric nanogenerators are simulated in COMSOL to compare the voltage profile of three different triboelectric materials – Kapton, Teflon and RTV Silicone with respect to Aluminium. Also, the effect of thickness of triboelectric layer on voltage profile is studied to optimize the thickness of the films.The output voltage recorded is 75 V, 60 V and 59 V for RTV Silicone, Teflon and Kapton respectively. It was observed that with increase in thickness of triboelectric layer, output voltagefirst increases linearly and then starts decreasing. The future research is directed towards fabricating a robust device for realising self – powered electronic devices.

Keywords: Simulation, Modelling, Triboelectric Nanogenerator

1. Introduction

With the advancement of technology, the energy demand is increasing over time globally and it has become essential to look out for renewable sources of energy. Most of the electronic devices usually utilise lithium ion batteries as a power source. Due to limited lifespan of the batteries, they tend to deplete over time and need to be replaced involving extra cost and complexities [1]. Therefore, harvesting energy is a promising technological solution to realise self – powered electronics. Various Energy harvesting technologies has been explored in recent years such as solar [2], pyroelectric [3], thermoelectric [4], electromagnetic [5], biofuel cell [6], piezoelectric [7], Triboelectric [8], hybrid [9],etc Among all the energy scavenging approaches, Triboelectric Nanogenerators (TENGs) has made asplash due to its low cost, simple design, high output performance and high energy conversion efficiency [10]. The first Triboelectric Nanogenerator was demonstrated in 2012 by Wang group [11].The Greek word "*tribo*" refers to friction. Triboelectric Nanogenerator basically operates on the amalgamated principle of triboelectric effect and electrostatic induction. Triboelectric Nanogenerator works in four working modes – Contact Separation Mode (CSM), Lateral Sliding Mode (LSM), Single Electrode Mode (SEM) and Free-Standing Mode (FSM). The contact separation mode is the most simple and widely used mode of operation. A typical Contact Separation Mode (CSM) based triboelectric nanogenerator comprises of two dissimilar triboelectric layers with two metal electrodes deposited on outer surfaces of triboelectric layers. With application of external force, two triboelectric layers make contact with each other leading to charge generation on opposite surfaces. When the two layers are separated due to removal of external force, electric potential is induced between the electrodes [12]. Therefore, Triboelectric nanogenerators has emerged as new era energy harvesting andsensing devices in variety of applications such as self-powered electronics [13], wearable and implantable devices [14], self-powered biosensors [15], human motion monitoring [16], location evaluation [17], environmental monitoring [18] etc.

This paper deals with design and simulation of triboelectric nanogenerators. In this work, three triboelectric nanogenerator with different triboelectric materials – Kapton, Teflon and RTV Silicone are designed and their output performance is compared in COMSOL. Also, the effect of thickness oftriboelectric layer on voltage profile is studied to optimize the thickness of the films. The output voltage recorded is 75 V, 60 V and 59 V for RTV Silicone, Teflon and Kapton respectively. It was observed that with increase in thickness of triboelectric layer, output voltage first increases linearly and then starts decreasing. The future research is directed towards fabricating a robust device to realise self – powered electronic devices.

2. Design of Triboelectric Nanogenerator

The output performance of triboelectric nanogenerator largely depends on the triboelectric materials. For maximum charge transfer, the materials for triboelectric layers are selected in such a way that they are located far apart in triboelectric series and are of opposite tribopolarity. In this work, thetriboelectric nanogenerator is designed to operate in contact separation mode due to its simple design.triboelectric nanogenerators (TENGs) are designed by using three different materials – Kapton, Teflon,RTV Silicone. These layers act as negative triboelectric layer. In all the three nanogenerators, Aluminium (Al) is used as positive triboelectric layer as well as second electrode. Another Aluminiumfilm is deposited on top surface of first triboelectric layer to form TENG device. Fig. 1 (a-c) shows three triboelectric nanogenerators.

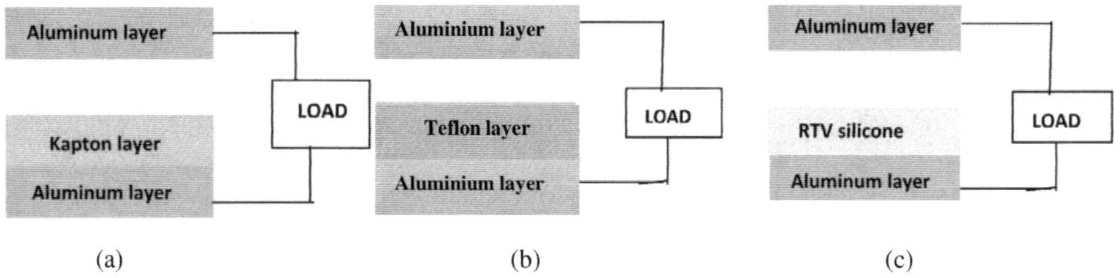

Figure 1. Design of three Triboelectric Nanogenerator using different negative triboelectric layer

3. Modelling and Simulation

The mathematical modelling of a typical TENG is performed that consists of two plates with opposite charges shuttling towards each other. The TENG shows capacitive behaviour with output voltage varying corresponding to change in distance between plates. The output voltage of TENG is inversely proportional to the capacitance whereas it is directly proportional to the corresponding distance between plates. The structure of a typical TENGwith two dielectric (triboelectric) layers and two metal electrodes is depicted in Fig. 2, where the distance 'D' is the distance between the two triboelectric layers.

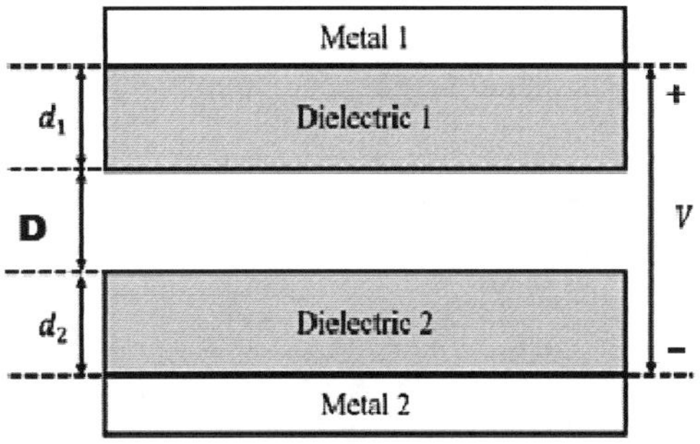

Figure 2. Structure of a typical TENG with two metal electrodes and two dielectric layers

The functioning of TENG involves coupling effect of triboelectrification and electrostatic induction, along with diffusion characteristics and charge transfer. With no external force initially, the Kapton, Teflon and RTV-siliconefilm will not be in contact with other triboelectric layer, thereby creating an air - gap between them. With application of external force, the two layers that are not absolutely smooth or flat come in contact with each otherresulting in generation and accumulation of charges due to deformations in the interfacial structure. As a result oftriboelectrification, inner surfaces of the two dielectric layers experience generation of equivalent positive and negative charges that get spread over the triboelectric surfaces because of the tunnel effect. The theory of work function states that the amount of charges on the surface are associated with the work function and the amount of the dielectric layers as per the following equation:

$$Q_t = \mathrm{A}\,(\hat{\mathrm{W}}_1 - \hat{\mathrm{W}}_2) \quad (1)$$

where Q_t, A, and $\hat{\mathrm{W}}$ are, respectively, denoted as the amount of tribo-charges formed on the dielectric surface, contact area of TENG, and work function of the dielectric materials. Work function mainly depends on the applied force and separation between the two dielectric layers.

$$\text{Work function} = \text{Force(F) x Distance (D)} \quad (2)$$

F = applied force

D= distance between two dielectric layers

During this period, polarization happens on the surface of the dielectric layer with a certain thickness, and a quite small amount of charges diffuse into the dielectrics to fill the inner charge hole or non-charged region, leading to spacer charges distributing with a distance from the interface layer. It is reasonable to assume that the tribo-charge density is uniform and the spacer charge distribution near the interface obeys the Poisson distribution; space charges per unit area can be given by -

$$\mathrm{Q(p)} = 2\Delta\hat{\mathrm{W}}\varepsilon\frac{\varepsilon_m}{xQ_t} \quad (3)$$

where, ε_m and ε are the permittivity of dielectric material and air respectively. Therefore, the dielectric surfaces accumulate equal and opposite charges and some charges remain dispersed inside the materials at the final contact.

With removal of external force, interfacial deformation between the triboelectric surfaces is released resulting in separation of two triboelectric materials. This causes generation of equal and opposite charges on the surface (i.e., Q_t) and some charges are dispersed inside the materials (denoted by Q_l). During the separation process of two triboelectric materials, an electric potential difference is developed between the Kapton/Teflon/RTV-silicone film and the Aluminium layer, which forms the dynamics of the whole TENG device prompting swift flow of electrons between the two metal electrodes. The charges transferred between the two electrodes can be obtained from Eqs. (1) and (2) .

$$Q_T = Q_t + Q_l$$

$$Q_T = Q_t + \sum_{x>0}^{d} Q(p) \quad (4)$$

When the distance between the two dielectric plates is maximum, an electrical equilibrium is attained and there is no flow of electrons. The maximum output voltage of the TENG device is attained from the following equation:

$$\mathrm{V} = -\frac{Q_T}{A\varepsilon}(\mathrm{d_t} + \mathrm{D}) + \frac{\sigma D}{\varepsilon} \quad (5)$$

Where, ε is the permittivity of free space, σ is the surface charge density, and D is the distance between the two dielectric plates. The total capacitance of the TENG is mainly influenced by two factors: the surface area of the plates (A) and the distance between the dielectric plates (D).

$$C_T = \frac{A\varepsilon}{d_t + \mathrm{D}} \quad (6)$$

where 'A' is the area of TENG, dt is the total thickness of the two plates i.e. summation of d_1 and d_2, and the relative permittivity of the dielectrics is denoted by ε_1 and ε_2, respectively. d_t is defined as:

$$\mathrm{d_t} = \frac{\mathrm{d_1}}{\varepsilon_1} + \frac{\mathrm{d_2}}{\varepsilon_2} \quad (7)$$

The material characteristics of a TENG are defined by Equations (1) (2), and (3). For generation of electricity from the TENG device and analyse its performance, it is necessary to perform modelling of output voltage of TENG. The output voltage V of the TENG device is defined as

$$\mathrm{V} = -\frac{Q_t}{C_T} + \frac{\sigma D}{\varepsilon} \quad (8)$$

From equation (4)

$$\mathrm{V} = -\frac{Q_t}{A\varepsilon}(\mathrm{d_t} + \mathrm{D}) + \frac{\sigma D}{\varepsilon} \quad (9)$$

4. Result and Discussion

In order to analyse the output performance of triboelectric nanogenerators and variation in voltage profile with change in thickness of negatively charged triboelectric (dielectric) layer, the three nanogenerators were modelled in COMSOL. Table 1 shows TENG modelling parameters. Fig. 3 compares voltage profiles of three nanogenerators with respect to variation in thickness and shows that for the same force applied (15N in each case) across the load resistance of 15 MΩ, the output voltage is observed to be highest in the case of RTV Silicone-Aluminium TENG i.e 75 V with dielectric layer thickness of 0.04 mm whereas the output voltage is 60 V and 59 V for Teflon-Aluminium and Kapton-Aluminium TENG device respectively.

All the three devices depicted an increase in the output voltage with increase in thickness of triboelectric layer varying from 0.04 mm to 0.12 mm. The simulated result shows that with further increase in thickness of triboelectric layer from 0.12mm to 0.2 mm, the output voltage starts decreasing and this decrease in voltage is less prominent in case of Kapton-Aluminium TENG device. Therefore, this indicates that 0.12 mm thickness is the optimum thickness of triboelectric layer for designing of TENG devices. The results are in congruence with the Distance Dependent Electric Field Model (DDEF) which states that with increase in thickness of dielectric layer, output voltage tends to decrease significantly.

Table 1. TENG modelling parameters

Variable	Value
A	10.16 x 7.62 $\times 10^{-4}$ seq. meter
ε	8.854×10^{-12} F/m
$\varepsilon 1$	2.1
$\varepsilon 2$	1.8
d_1	100×10^{-6} m
d_2	80×10^{-6} m
σ	50×10^{-6} C/m^2

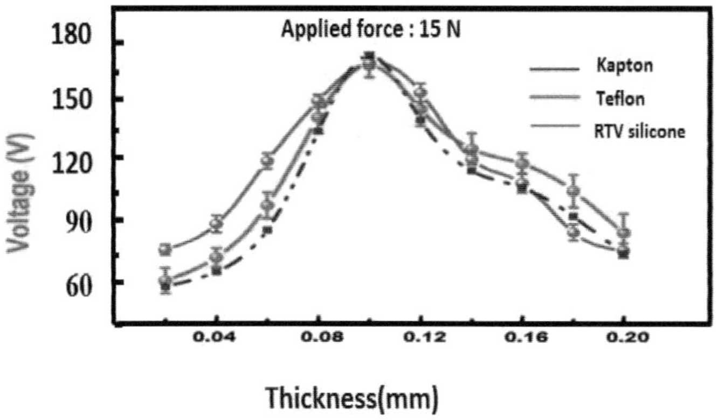

Figure 3. Simulation results comparing output performance of Kapton – Al TENG, Teflon – Al TENG, RTV Silicone TENG with variation in thickness of negative triboelectric layer

5. Conclusion

Triboelectric Nanogenerators have shown great potential for scavenging energy from ambient environment in variety of applications. This work successfully modelled three different TENG devices by varying negative triboelectric layer with different materials – Kapton, Teflon, RTV Silicone in order to compare their output performance. It is concluded that RTV silicon shows highest surface charge density among the three materials which is evident by its maximum output voltage of 75 V. The Thickness of Kapton, Teflon and RTV Silicone was varied from 0.04 mm to 0.20 mm and the effect on output voltage was studied. It is concluded that initially voltage increases with increase in thickness but thereafter output voltage tends to decrease with further increase in thickness. Therefore, 0.04 mm is the optimum thickness showing maximum output voltage for the TENG device. The future research direction could be fabrication of a robust device for self – powered electronic devices.

Acknowledgements

Akshpreet Kaur is thankful for financial support in the form of DST-INSPIRE Fellowship (DST/INSPIRE/Fellowship/2018/ IF190152) by Department of Science and Technology (DST), Ministry of Science and Technology, Government of India.

References

[1]. Liu X, Zhao K, Wang ZL, Yang Y 2017 Unity convoluted design of solid li-ion battery and triboelectric nanogenerator for self-powered wearable electronics *Adv. Energy Mater.* **7(22)** 1701629.

[2]. Guilar NJ, Kleeburg TJ, Chen A, Yankelevich DR, Amirtharajah R 2009 Integrated solar energy harvesting and storage *IEEE Trans. Very Large Scale Integr. (VLSI) Syst.* **17(5)** 627-37.

[3]. Sebald G, Lefeuvre E, Guyomar D 2008 Pyroelectric energy conversion: optimization principles *IEEE Trans Ultrason Ferroelectr Freq Control* **55** 538-51.

[4]. Ramadass YK, Chandrakasan AP A 2010 battery-less thermoelectric energy harvesting interface circuit with 35 mV startup voltage *IEEE J. Solid-State Circuits* **46(1)** 333-41.

[5]. Yan B, Zhang S, Zhang X, Wang K, Wu C 2017 Self-powered electromagnetic energy harvesting for the low power consumption electronics: Design and experiment *Int. J. Appl. Electromagn.* **54(2)** 165-75.

[6]. Gamella M, Koushanpour A, Katz E 2018 Biofuel cells–activation of micro-and macro-electronic devices*Bioelectrochemistry* **19** 33-42.

[7]. Sezer N, Koç M A 2020 comprehensive review on the state-of-the-art of piezoelectric energy harvesting *Nano Energy* 5105567.

[8]. Wu F, Li C, Yin Y, Cao R, Li H, Zhang X, Zhao S, Wang J, Wang B, Xing Y, Du X A 2019 Flexible, Lightweight,and Wearable Triboelectric Nanogenerator for Energy Harvesting and Self-Powered Sensing *Adv. Mater. Technol.* **4(1)** 1800216.

[9]. Qiu C, Wu F, Lee C, Yuce MR 2020Self-powered control interface based on Gray code with hybrid triboelectricand photovoltaics energy harvesting for IoT smart home and access control applications *Nano Energy* **70** 104456.

[10]. Xu C, Zi Y, Wang AC, Zou H, Dai Y, He X, Wang P, Wang YC, Feng P, Li D, Wang ZL 2018 On the electron-transfer mechanism in the contact-electrification effect *Adv. Mater.* **30(15)** 1706790.

[11]. Fan FR, Tian Z Q, Wang Z L 2012 Flexible triboelectric generator *Nano energy* **1(2)** 328-34.

[12]. Zhou L, Liu D, Wang J, Wang Z L 2020 Triboelectric nanogenerators: fundamental physics and potentialapplications *Friction* **8(3)** 481-506.

[13]. Hinchet R, Seung W, Kim SW 2015 Recent progress on flexible triboelectric nanogenerators for self powered electronics *ChemSusChem* **8(14)** 2327-44.

[14]. Kaur A, Sapra G, Gupta A 2021 Recent Progress on Energy Harvesters for Biomedical Applications Journal of Circuits, *Systems and Computers* **7** 2130010

[15]. Zhang H, Yang Y, Hou TC, Su Y, Hu C, Wang ZL 2013 Triboelectric nanogenerator built inside clothes for self-powered glucose biosensors *Nano Energy* **2(5)** 1019-24

[16]. Zhu J, Zhu P, Yang Q, Chen T, Wang J, Li J A 2020 fully stretchable textile-based triboelectric nanogenerator forhuman motion monitoring *Mater. Lett.* **280** 128568

[17]. Yu A, Song M, Zhang Y, Zhang Y, Chen L, Zhai J, Wang ZL 2015 Self-powered acoustic source locator in underwater environment based on organic film triboelectric nanogenerator *Nano Research* **8(3)** 765-73

[18]. Zhou Z, Li X, Wu Y, Zhang H, Lin Z, Meng K, Lin Z, He Q, Sun C, Yang J, Wang ZL 2018 Wireless self-poweredsensor networks driven by triboelectric nanogenerator for in-situ real time survey of environmental monitoring *Nano Energy* **53** 501-7

Investigation of Self-heating Effects in Ultrathin FDSOI MOSFETS

Vinod Kumar[1], **Ashwani K. Rana**[2]
[1,2] NIT Hamirpur, Hamirpur, India

[1]vinodsharma@nith.ac.in
[2]ashwani.rana@nith.ac.in

Abstract: This paper contains the design consideration of 18nm FDSOI MOSFET for mitigating the self-heating effects. As the device scaling reaches towards the nano regime, self-heating effects are quite significant. Increase in the lattice temperature will lead to decrease in the mobility and degradation of the drain current. Self-heating effects are investigated in FDSOI MOSFET by comparing different physical parameter variations like gate length, spacer thickness, buried oxide thickness and source/drain/metal contacts. It is being observed that the heat sinks used as metal contacts can reduce lattice temperature up to 62% and spacer size thickness can reduce it by 10%. However the buried oxide box thickness and gate length modifications do not have considerable impact on the self-heating.

1. Introduction

Silicon on insulator (SOI) MOSFET, as the name suggests has silicon layer on the top of insulator layer to build an active device. Insulator layer is usually made up of SiO2. FD-SOI MOSFET has several advantages over bulk MOSFET like the improved subthreshold slope, reduced parasitic capacitance and tolerant to radiations. Hence, FDSOI is always preferred for low-power digital applications along with high speed [1-2].

Concept of Self heating effect (SHE): From the figure 1, it is evident that primary path of energy transport is scattering between electron and optical phonon followed by optical phonon to the lattice. Physical sense is that on application of drain bias to the device, electron gain kinetic energy by lateral electric field and interacted with optical and acoustic bath.

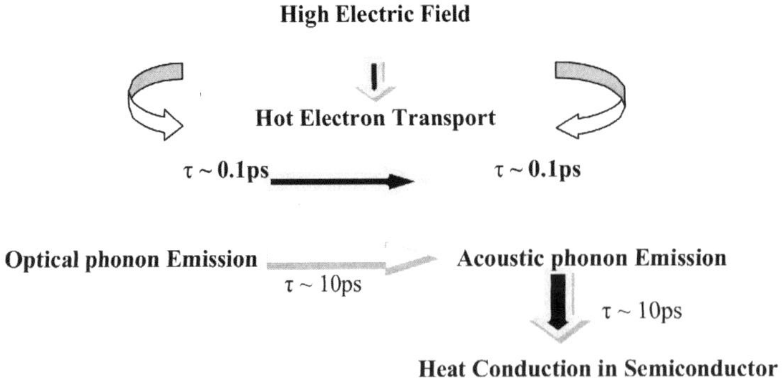

Figure 1. Generation of heat due to kinetic energy of particles in semiconductors

IOP Conf. Series: Materials Science and Engineering 1225 (2022) 012013 doi:10.1088/1757-899X/1225/1/012013

Most of the electron energy is stored as an optical phonon bath. The optical phonons have zero group velocity so heat is localized and transferred from optical phonon to acoustic phonon via a harmonic decay process. This process is relatively sluggish therefore a hot spot forms near the drain terminal of the device occurs. In active region the raised temperature leads to increase phonon scattering. Hence phonon scattering results in mobility reduction and subsequently current degradation. This phenomenon is called as self-heating. Majorityof the heat is absorbed by the lattice from the optical phonons.

Self-heating in a bulk MOSFET is not that much significant as heat generated near drain terminal is dissipated to the substrate. Whereas in SOI MOSFET, the buried SOI BOX acts as a barrier to the heat. Due to device scaling over the years, SHE becomes more prominent. The only way to reduce the self- heating effects is box or material engineering. The heat generated due to self-heating is dissipated from the device through the source, drain or gate contacts.

The SOI device is simulated using 2D hydrodynamic simulation along with thermodynamic simulation program to examine the device lattice temperature and drain current between terminals.

2. Device design and Simulation Set up

In this section, metal gate conventional FD SOI MOSFET is designed for high performance. The conventional metal gate SOI has physical gate length of 18nm, thickness of channel 10 nm, BOX thickness of 40 nm. Oxide thickness of 1nm is used. The maximum concentration of phosphorus in the S/D region is 10^{20} /cm^3. Channel region is doped with low boron concentration of 10^{17} /cm^3. All the simulations are performed using Sentaurus 2-D TCAD simulator. All the calculations are taken per unit channel width. The thermodynamic model, hydrodynamic models, drift-diffusion transport model and density gradient model are included for device simulations for calculating energy transport of the carrier in transport equation. In this paper the electron temperature is calculated differently than lattice temperature to investigate SHE. The physical dimensions and doping concentration values are chosen as per ITRS standards.

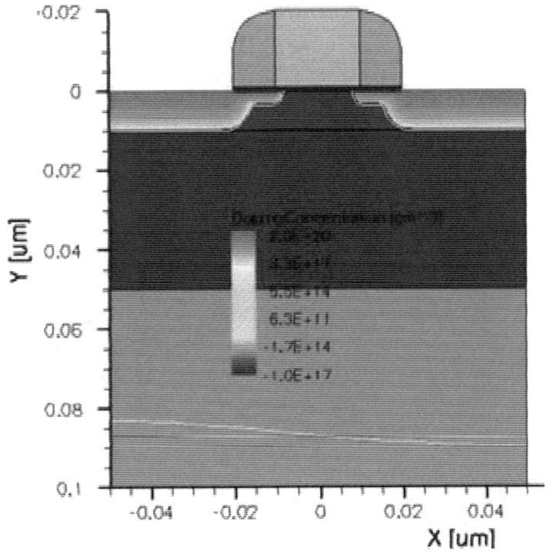

Figure 2. Doping concentration of an FD-SOI MOSFET.

Physical dimensions and doping concentration of FD-SOI MOSFET are given in table 1.

The basic model of electronics used for above device simulation consists of the Poisson and continuity equations. The Poisson equation is

$$\nabla.\varepsilon\nabla\varphi = - q(p - n + N_D - N_A)\ \rho_{-trap} \qquad (1)$$

Where ε is electrical permittivity, n and p are electron and hole densities, q is electronics charge.

Table 1. Structure of simulated FD-SOI MOSFET

Physical Parameter	Value
N^+ Source/drain doping	$1e^{20}/cm^3$
P-type Silicon film doping	$1e^{17}/cm^3$
P-type Silicon Substrate	$1.0e^{15}/cm^3$
Thickness of Source/Drain	10 nm
Buried oxide insulator thickness	40 nm
Channel length	18 nm
Gate oxide thickness	1 nm
Gate work function	4.5eV

N_D and N_A are ionized donors and ionized acceptor resectively. ρ_{-trap} is charge densities contributed by traps and fixed charges.
Electron and hole continuity equations are

$$\nabla.J_n = qR_{net} + q.(\partial n/\partial t) \quad (2)$$
$$-\nabla.J_p = qR_{net} + q.(\partial p/\partial t) \quad (3)$$

Where J_n and J_p is the electron and hole current densities and R_{net} = net electron hole recombination rate.
Shockley-read-hall recombination rate is given by

$$R_{net} = \frac{n.p - n_i,eff^2}{c_p(n+n_1)+c_n(p+p_1)} \quad (4)$$

where:

$$n_1 = n_{i,\,eff}\,.e^{(E_{trap}/KT)}$$

and

$$p_1 = ni,eff.\ e\text{-}^{(-E_{trap}/KT)}$$

The difference between defect level and intrinsic level is called as E_{trap}. And $n_{i,eff}$ is intrinsic carrier concentration. r_p and r_n is electron and hole carrier life time.
It is known that when thickness of material is smaller than phonon mean free path than thermal conductivity of material decreases drastically. Channel length of SOI is taken as 18nm and phonon mean free path is known to be 300nm in silicon material. Hence the hot spot temperature increases. In Monte Carlo simulation which solves BTE, this effect is taken but not in hydrodynamic approach. So modified version of approach of Sondheimer [8] is used.
For semiconductor of thickness t with z axis perpendicular to the plane of the film located between z=0 to z = t thermal conductivity as a function of t is given by

$$K(z) = K_0(T)\int_0^{\frac{\pi}{2}} \sin\text{^}3\,x \times (1 - e^{\left(-\frac{t}{2\Lambda(T)\cos x}\right)} \cosh(\frac{t-2z}{2\Lambda(T)\cos x}))dx \quad (5)$$

Where $\Lambda(T)$ is mean free path.

$$\Lambda(T) = \Lambda_0\,(\frac{300}{T})nm$$

With the room temperature mean free path Λ_0=290nm
Sentaurus does not provide distance dependent thermal conductivity inside material.

$$K_0(T)=\frac{1}{kappa_a+kappa_b\times T+kappa_c\times T^2} \tag{6}$$

Where default value at room temperature is

$kappa_a = 0.03 cmKW^{-1}$
$kappa_b = 1.56\times10^{-3} cmW^{-1}$
$kappa_c = 1.65\times10^{-6} cmK^{-1}W^{-1}$

Sentaurus uses above equation over the thickness of silicon film, in the case of our device thickness is 10nm, so modified value at room temperature as per the approach of Sondheimer are

$K_{av}(T) = 0.0664317 K_0(T)$
$kappa_a = 0.453311 cmKW^{-1}$
$kappa_b = 23.7307\times10^{-3} cmW^{-1}$
$kappa_c = 0.251\times10^{-6} cmK^{-1}W^{-1}$

Change these values of Kappa in sentaurus parameter file and perform the simulations for calculating the SHE.

3. RESULTS AND DISCUSSION

The self-heating effect of FD-SOI MOSFET for 18nm gate length is studied. The self-heating effects of the FDSOI are investigated for different gate lengths, spacer lengths and BOX thickness.

3.1 Quasi stationary Simulations

It is interesting to note that maximum lattice temperature and total heat generated due to reception of carriers at high speed is at the drain terminal. For estimation of lattice temperature,we have calculated surface resistance at gate and substrate contacts. The results shown by simulations are very much comparable to Vasileska et al. [9][10][11] and Etessam et al [12].

As we know that cost factor = "(maximum temperature rise)/(drain current). Also cost factor for heat generation = "(maxium heat generation)/(drain current). SO cost factor has almost linear variation with drain current for lattice temperature rise and heat generation.

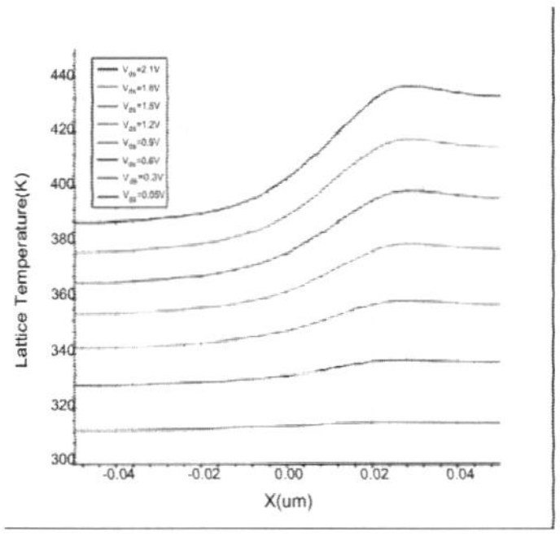

Figure 3. Lattice temperature in channel region near drain(y=-5nm) at Vgs=1.2V for varied drain voltage

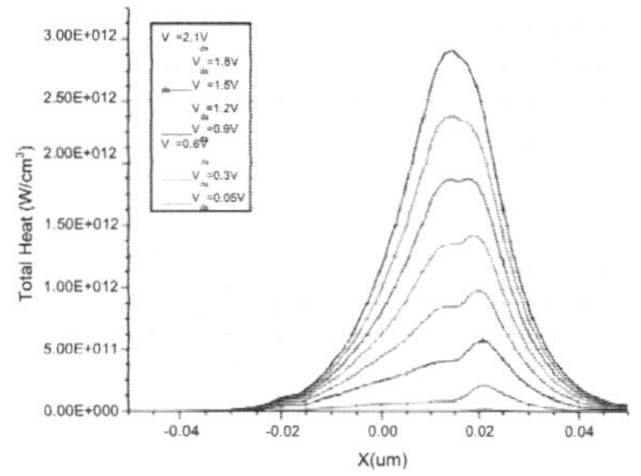

Figure 4. Total heat generated in channel region near drain(y=-5nm) at Vgs=1.2V for varied drain voltage

The reason behind the linear variation is the ballistic transport in the channel where hardly any electron phonon scattering occurs. This variation is more prevalent at lower voltages. For low lattice temperatures of the device we have to operate the device below 1V V_{DD}.

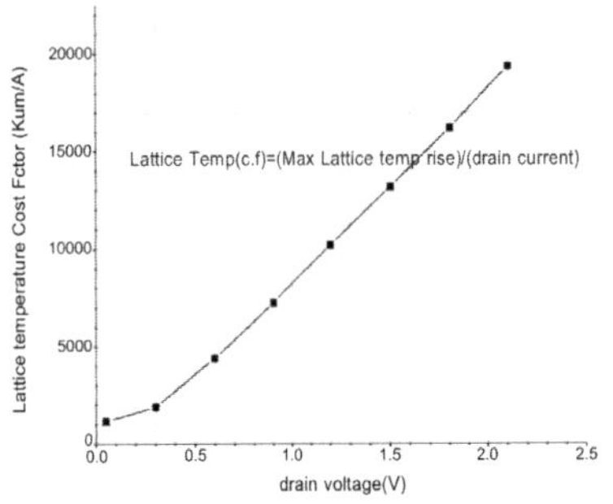

Figure 5. Variation of cost indexes temperature rise/drain current

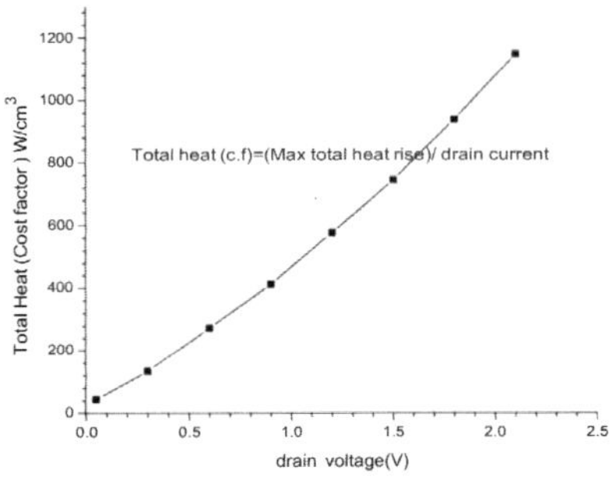

Figure 6. Heat generation/drain current with varied drain voltage at V_{GS}=1.2V

3.2 Thermal Boundary Conditions

Thermal boundary condition required for the device to be fixed in an LSI circuit. Substrate is considered to be an ideal heat sink at 300K.

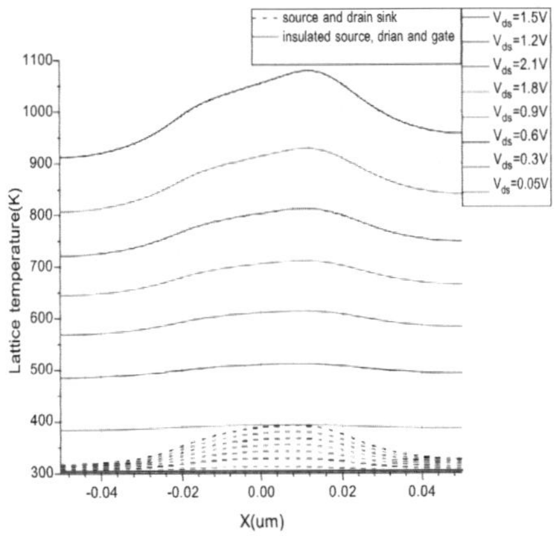

Figure 7. Impact of Insulating source ,drain and gate contacts on lattice temperature of the device

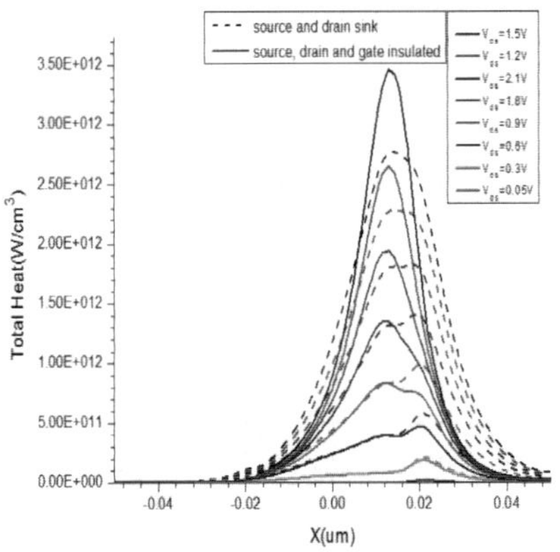

Figure 8. Impact of Insulating source ,drain and gate contacts on total heat generation of the device

Figure (7) shows maximum lattice temperature in the device when all the contacts are insulated. It is obvious because no heat sink path provided to the SOI device. As the voltage increases, the lattice temperature and total heat generation will increase. It is being observed that the lattice temperature can be reduced from 1100K fully insulated contacts to 400K without insulating contact. This reduction in the lattice temperature is 62% which is significantly very high. This is possible as the insulating material used near source/drain/gate contacts is having very low thermal conductivity as compared to semiconductors and metals. More insulation to the device contacts is the major cause of self-heating of the device.

3.3 Thermal resistance

Thermal resistance is well-defined as the ratio of the temperature difference between the two surfaces of a material to the rate of heat flow per unit area. Larger value of thermal resistance indicates smaller heat loss. Here thermal resistance

$$\text{Thermal resistance } (R_{th}) = \frac{Maximum\ Lattice\ temperature - 300K}{Id \times Vd}$$

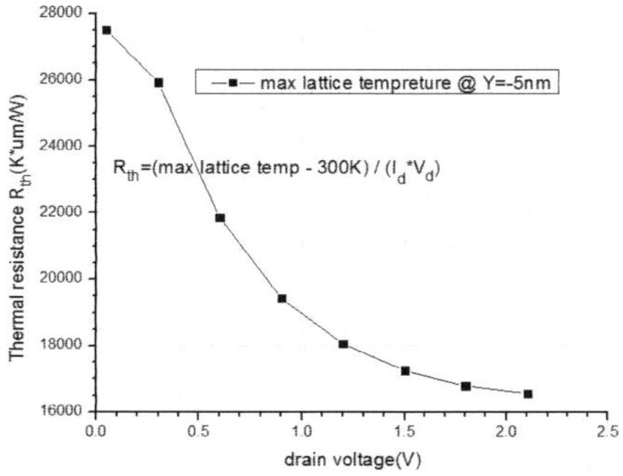

Figure 9. Thermal resistance variation due to varying V_{DS} and at fixed V_{GS}=1.2V

Simulations are performed for variation of drain voltage value from 0.05V to 2.1V at fixed gate voltage of 1.2V. From the figure 9 it can be noted that maximum thermal resistance decline is present in region ranging from 0.3-0.6V, which means escalation in drain current is maximum present at V_{DD}=0.3-0.6V.

3.4 Gate Length Variation

In this section we try to vary physical gate length of FDSOI MOSFET from 18nm to 200nm and simulate the device using hydrodynamic model. Simulation performed at a V_{DD}=V_{GS}=1V for analysing the electrical characteristic and thermal characteristic of the FDSOI MOSFET. For investigation of transfer characteristics of the device, drain voltage is fixed at 1V and varying gate voltage from 0.05V to 1V. The output characteristic are analysed at varying drain voltage from 0.05V to 1.5V and fixed V_{GS}.

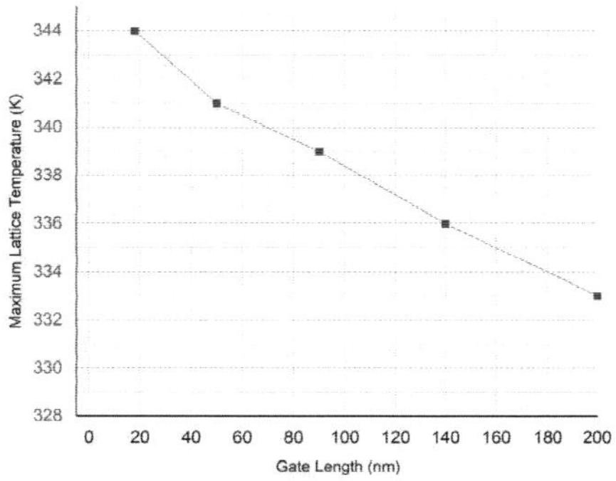

Figure 10. Impact of gate length on self-heating parameter of the ultrathinFDSOI

Figure 10 indicates that maximum lattice temperature is inversely proportional to the gate length which is due to the higher mean free path of the charge carriers.

3.5 BOX Thickness Variation

Optimizing the BOX thickness for utilizing substrate area for heat dissipation is often regarded the most prominent way to reduce the SHE. It is examined that the heat generation is almost identical in the range of 10 to 60nm thickness of the box. So it can be concluded that box thickness variation does not have impact on the SHE.

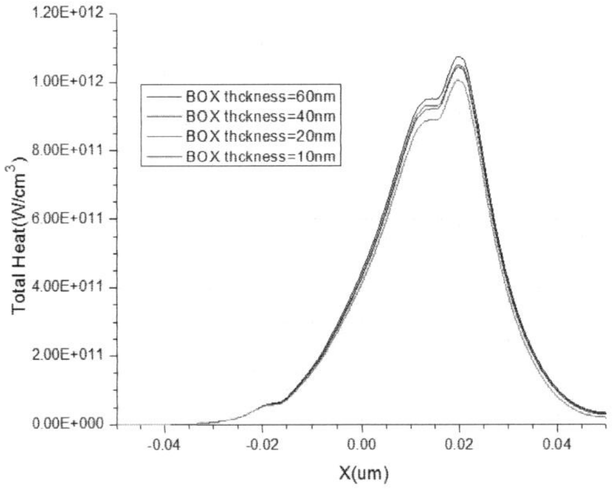

Figure 11. Impact of SOI box thickness on lattice temperature

3.6. Spacer Size Variation

The effect of variation of the spacer size is also studied to understand the thermal behaviour of the device. Hence it is analysed that smaller the size of the spacer, lower is self-heating. It can be reduced maximum upto approximately 10% by reducing spacer size from 40nm to 5nm.

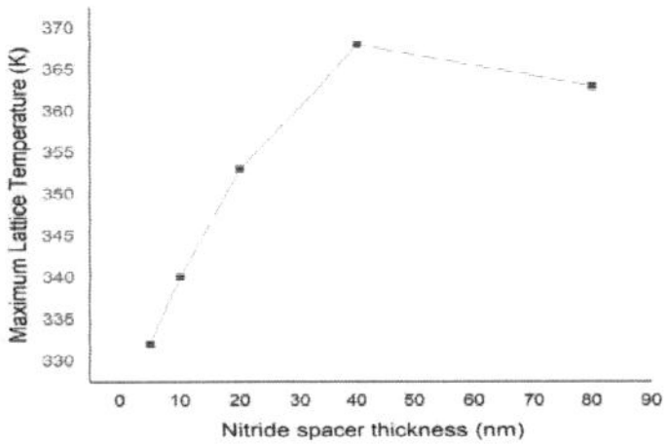

Figure 12. Maximum lattice temperature vs. nitride spacer thickness variation

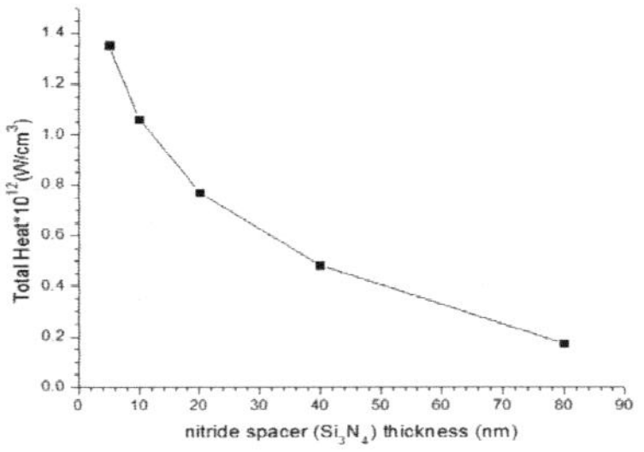

Figure 13. Total heat generation vs. nitride spacer thickness variation.

So we can say the heat generations inside the device will be lower for smaller size spacer but at the cost of higher fringing fields.

4. Conclusion and Future Scope

Detailed analysis of self-heating effect in FD-SOI MOSFET using Sentaurus TCAD hydrodynamic simulation under steady state condition is done. As the requirement of lessening of self-heating effects is increasing so it is suggested to use heat sinks at source, drain and gate terminals as metal contacts having higher thermal conductivity. The use of heat sinks can reduce the self-heating temperature up to 62% which is quite significant using metal contacts. We can reduce the self-heating by material or geometrical engineering inside the device. So scaling, spacer size, heat sinks are introduced to analyse the self-heating optimization. It is found that temperature inside the device is increased by 20K only for device scaling from 200nm to 18nm gate length which is insignificant. 5K increase in lattice temperature by varying the BOX thickness from 10nm to 60nm is observed. It is also investigated that the thermal resistance drops drastically whenever VDD applied is 0.3 to

0.6V. Cost factor of lattice temperature and total heat inside the device is directly proportional to applied drain voltage. The SOI box is blocking the heat dissipation unlike conventional MOSFET, so box engineering with higher conductive materials is required to be examined for reducing self-heating effects. Another solution to the problem is geometric variation of the buried oxide box which can also be helpful in exploring the optimization of self-heating effects

References

[1] Colinge J P 2004 *Silicon On Insulator Technology 2004: Materials to VLSI* (Kluwer Academic Publishers, London)
[2] Bernstein K and Rohrer N 2000 *SOI Circuit Design Concepts* (Kluwer Academic Publishers, London)
[3] Tien C L, Majumdar A, Gerner F M 1998 Eds., *Microscale Energy Transport* (Taylor & Francis,).
[4] Zhu M, Chen P, Ricky K Y, Zhenghua F, Lin C and Chu P K 2004 Numerical study of self-heating Effects of MOSFETs Fabricated on SOAN substrate *IEEE Trans. Electron Devices,*. **51** 901-906.
[5] Qing L, Zheng-Xuan Z, Ming Z, Xin-Yun X, Hua-Qing S, Cheng-Lu L 2003 A New Silicon-on-Insulator Structure of Metal- Oxide- Semiconductor Field Effect Transistor to reduce Self-Heating effect *Chin. Phys. Lett.* **20** 158-160.
[6] Kumar M J and Orouji A A 2006 Investigation of a New Modified Source/Drain for diminished Self–heating Effects in Nanoscale MOSFETs using Computer Simulation," *Physica E: low dimensional System and Nanostructures* **33** 134-138.
[7] Cole B and Parke S 2003 A Method to Overcome Self-Heating Effects in SOI MOSFETs *IEEE University/Government/ Industry Microelectronics Symposium, Proceeding of 15th Biennial* 295-297.
[8] Sondheimer E H 1952 The mean free path of electrons in metals *Adv. Phys* **1** 1–42 .
[9] Raleva V K, and Goodnick S M 2010 *Heating effects in nanoscale devices* (INTECH Open Access Publisher 2010)
[10] Raleva V K, and Goodnick S M 2009 Self-heating effects in nanoscale FDSOI devices:The role of the substrate, boundary conditions at various interfaces, and the dielectric material type for the box *Electron Devices, IEEE Transactions on* 563064–3071.
[11] Raleva K, Vasileska D, Hossain A, Yoo S K., and Goodnick S 2012 Study of self-heating effects in soi and conventional mosfets with electro thermal particle-based device simulator *Journal of Computational Electronics* **11** 106–117.
[12] Etessam-Yazdani K, Hussin R, and Asheghi M *2006* Impact of scaling on thermal behavior of silicon-on-insulator transistors *Thermal and Thermomechanical Phenomena in Electronics System ITHERM'06.* IEEE 1257–1264.

Eight Wave Mixing Process is suitable to make Single Photon Source for Quantum Cryptography

Priyanka[1] **and Savita Gill**[2*]

[1,2*] Department of Applied Science, University Institute of Engineering and Technology, Kurukshetra, 136119, India

[1] chauhan7101@gmail.com
[2*]savita2015@kuk.ac.in

Abstract. Single photon source for use in quantum cryptography should show higher order photon antibunching. We have analyzed non classical effect i.e. higher order photon antibunching in pump mode of eight-wave mixing non linear optical process using short time interaction technique and also the dependence of non classicality on the number of photons prior to interaction in non linear medium. Further we have reported that higher the order of photon antibunching, higher will be the probability to generate single photon source utilized in quantum cryptography.

Keywords: non-classicality, higher-order photon antibunching, quantum optics.

1. Introduction

It is conceivable to depict a state as a non classical state that doesn't have any classical analogue and generally negativity of Glauber Sudarshan P function explains the non classical state [1,2]. In particular, photon antibunching and squeezing have no classical equivalent [3,4]. Higher order squeezing [5-8] is one of these non-classical phenomena that has been well investigated, while higher order photon antibunching (HOA) has not been studied precisely. Perhaps the most important prerequisite for quantum information theory is a single photon source. Photon antibunching [9-15] is a non classical feature of radiation field that may be utilized to create single photon source [16,17]. Lee proposed the notion of HOA [18], which has been seen in the trio coherent state [19] and in several nonlinear optical processes [20,21].

The interest in non-classical states have been concentrated in latest past with the advent of numerous advanced applications in quantum information processing like quantum key distribution [22] , quantum cryptography [23,24], quantum teleportation [25,26], dense-coding [27,28]. As a result, investigations of non-classical characteristics of the radiation field are very significant in the circumstances of new developing subject of the quantum information theory.

Earlier the authors have studied the non classical effect i.e. photon antibunching up to second order in certain non linear optical processes [29]. In the present study, we will look into higher order photon antibunching and the dependence of the photon antibunching on photon number has also been studied. Section 2, underscoring the criteria of non-classicality of the non linear optical system. Besides, section 3 gives a second order solution of equation of motion of an eight wave mixing non linear optical process and shows the occurrence of HOA. In section 4 we have analyzed the results obtained using graphs and section 5 underscore on end.

2. Conditions of non-classicality i.e. antibunching of a non linear optical system

2.1. Condition for Higher order photon antibunching (HOA).

"HOA models was found by Lee" [18], which is

$$R(m,l) = \frac{\langle N_x^{m-1}\rangle\langle N_x^{l+1}\rangle}{\langle N_x^{m}\rangle\langle N_x^{l}\rangle} - 1 < 0 \tag{1}$$

where the number operator is addressed by N. $\langle N^{(y)}\rangle = \langle N(N-1)(N-2)............(N-y+1)\rangle$, $N^{(y)}$ is y^{th} factorial moment of the number operator. The two integers (l) and (m) fulfilling the condition $l \le m \le 1$ and addendum x connotes the specific mode. "$m=1$ is choosen by Ba An" [19]. The criterion of lth order photon antibunching is given as

$$A_{x,l} = \frac{\langle N_x^{l+1}\rangle}{\langle N_x^{l}\rangle\langle N_x\rangle} - 1 < 0 \tag{2}$$

and

$$\langle N_x^{l+1}\rangle < \langle N_x^{l}\rangle\langle N_x\rangle \tag{3}$$

A state that is antibunched in the higher order must be antibunched in the lower order physically. As a result, equation (3) may be simplified as follows:

$$\langle N_x^{l+1}\rangle < \langle N_x^{l}\rangle\langle N_x\rangle < \langle N_x^{l-1}\rangle\langle N_x\rangle^2 < \langle N_x^{l-2}\rangle\langle N_x\rangle^3 < \langle N_x^{l-3}\rangle\langle N_x\rangle^4 ....\langle N_x\rangle^{l+1} \tag{4}$$

by disentangling equation (3) and acquiring criteria for l^{th} order photon antibunching, which is

$$d(l) = \langle N_x^{l+1}\rangle - \langle N_x\rangle^{l+1} < 0 \tag{5}$$

Accordingly, in quantum cryptography, we can say that single photon source to be utilized must fulfill the criteria given in equation (5) of HOA.

3. Eight wave mixing process

The above process can be used to examine HOA in such a manner that three photons of frequency ω_1 are absorbed with four photons of frequency ω_2 and one photon of frequency ω_3 being expelled. For this form, the Hamiltonian is given as

$$H = \omega_1 a^\dagger a + \omega_2 b^\dagger b + \omega_3 c^\dagger c + g(a^3 b^{\dagger 4} c^\dagger + a^{\dagger 3} b^4 c) \tag{6}$$

where coupling constant is addressed by g, annihilation (creation) operators are $a(a^\dagger)$, $b(b^\dagger)$, $c(c^\dagger)$, respectively. $A = a\exp i\omega_1 t$, $B = b\exp i\omega_2 t$, $C = c\exp i\omega_3 t$ are the gradually varying operators wavering at frequencies ω_1, ω_2 and ω_3.

3.1. Time evolution of pump mode A.

Time evolution of operator A in pump mode is described as

$$\frac{dA}{dt} = \frac{\partial A}{\partial t} + i[H, A] \tag{7}$$

we have got

$$\dot{A} = -3igA^{+2}B^4C \tag{8}$$

$$\dot{B} = -4igA^3B^{+3}C^+ \tag{9}$$

and

$$\dot{C} = -igA^3B^{+4} \tag{10}$$

by applying short time approximation, expanding $A(t)$ in accordance with Taylor series and acquiring the terms up to second order as

$$A(t) = A - 3igtA^{\dagger 2}B^4C + \frac{3}{2}g^2t^2(6A^\dagger A^2N_B{}^4N_C + 6AN_B{}^4N_C - 16A^{\dagger 2}A^3N_B{}^3N_C$$
$$-72A^{\dagger 2}A^3N_B{}^2N_C - 96A^{\dagger 2}A^3N_BN_C - 24A^{\dagger 2}A^3N_C - A^{\dagger 2}A^3N_B{}^4 - 16A^{\dagger 2}A^3N_B{}^3$$
$$-72A^{\dagger 2}A^3N_B{}^2 - 96A^{\dagger 2}A^3N_B - 24A^{\dagger 2}A^3) \tag{11}$$

using equation (11), number operator $N_A(t)$ is given as

$$N_A(t) = A^\dagger A - 3igt(A^{\dagger 3}B^4C - A^3B^{\dagger 4}C^\dagger) + 3g^2t^2(9A^{\dagger 2}A^2N_B{}^4N_C + 18A^\dagger AN_B{}^4N_C$$
$$+6B^{\dagger 4}B^4N_C - 16A^{\dagger 3}A^3N_B{}^3N_C - 72A^{\dagger 3}A^3N_B{}^2N_C - 96A^{\dagger 3}A^3N_BN_C - 24A^{\dagger 3}A^3N_C$$
$$-A^{\dagger 3}A^3N_B{}^4 - 16A^{\dagger 3}A^3N_B{}^3 - 72A^{\dagger 3}A^3N_B{}^2 - 96A^{\dagger 3}A^3N_B - 24A^{\dagger 3}A^3) \tag{12}$$

To examine photon antibunching, at first we consider a quantum state $|\psi\rangle$ as

$$|\psi\rangle = |\alpha\rangle_A |0\rangle_B |0\rangle_C \tag{13}$$

where A mode is in coherent state i.e. $|\alpha\rangle$ and B and C mode are in vacuum state $|0\rangle$.
using equations (12) and (13) expectation value of photons number in pump mode A might be communicated as

$$\langle N_A(t)\rangle_\alpha = |\alpha|^2 - 72g^2t^2|\alpha|^6 \tag{14}$$

where $A|\alpha\rangle = \alpha|\alpha\rangle$. And value of $< N_A{}^4(t) >$ and $< N_A{}^5(t) >$ is given as

$$\langle N_A{}^4(t)\rangle_\alpha = |\alpha|^8 - 288g^2t^2(|\alpha|^{12} + 3|\alpha|^{10} + 2|\alpha|^8) \tag{15}$$

and

$$\langle N_A{}^5(t)\rangle_\alpha = |\alpha|^{10} - 360g^2t^2(|\alpha|^{14} + 4|\alpha|^{12} + 4|\alpha|^{10}) \tag{16}$$

now using equations (14-16) in equation (5), a direct however strenuous calculation yields

$$d_A(3)_\alpha = -576g^2t^2(|\alpha|^{10} + |\alpha|^8) \tag{17}$$

and

$$d_A(4)_\alpha = -288g^2t^2(2|\alpha|^{12} + 3|\alpha|^{10}) \tag{18}$$

equations (17) and (18) fulfill the criteria of higher order photon antibunching.

Moreover, to examine higher order photon antibunching, we have now taken the initial state $|\psi\rangle$ as

$$|\psi\rangle = |0\rangle_A |\beta\rangle_B |0\rangle_C \tag{19}$$

where A and C mode are in vacuum state i.e. $|0\rangle$ and B mode is in quantum state $|\beta\rangle$.

presently taking the expectation values of $N_A(t)$, $N_A^4(t)$ and $N_A^5(t)$ in pump mode A corresponding to condition (19) is

$$\langle N_A(t)\rangle_\beta = 0 \tag{20}$$

$$\langle N_A^4(t)\rangle_\beta = 0 \tag{21}$$

$$\langle N_A^5(t)\rangle_\beta = 0 \tag{22}$$

along these lines third and fourth order photon antibunching in mode A as for $|0\rangle|\beta\rangle|0\rangle$ is given as

$$d_A(3)_\beta = 0 \tag{23}$$

$$d_A(4)_\beta = 0 \tag{24}$$

Additionally we can contemplate the non classical effect (HOA) by assuming a quantum state $|\psi\rangle$ as

$$|\psi\rangle = |0\rangle_A |0\rangle_B |\gamma\rangle_C \tag{25}$$

where A and B mode are in vacuum state $|0\rangle$ and C mode is in quantum state i.e. $|\gamma\rangle$.

now taking expectation values of $N_A(t)$, $N_A^4(t)$ and $N_A^5(t)$ in pump mode A in relation to condition (25) is

$$\langle N_A(t)\rangle_\gamma = 0 \tag{26}$$

$$\langle N_A^4(t)\rangle_\gamma = 0 \tag{27}$$

$$\langle N_A^5(t)\rangle_\gamma = 0 \tag{28}$$

Therefore third and fourth order photon antibunching in mode A with respect to $|0\rangle|0\rangle|\gamma\rangle$ is given as

$$d_A(3)_\gamma = 0 \tag{29}$$

$$d_B(4)_\gamma = 0 \tag{30}$$

from equations (23, 24) and (29, 30) we can say that neither third order photon antibunching nor fourth order photon antibunching is existing in mode A with respect to $|0\rangle|\beta\rangle|0\rangle$ and $|0\rangle|0\rangle|\gamma\rangle$ respectively.

4. Result

The presence of third and fourth order photon antibunching is shown in equations (17) and (18) respectively. Taking $g^2t^2 = 10^{-4}$, the variations of third order photon antibunching as shown in equation (17) and fourth order photon antibunching as shown in equation (18) as a function of photons number in mode A i.e. $|\alpha|^2$ are shown in figures 1 and 2. It is obvious from figures that HOA increases non linearly with $|\alpha|^2$. This affirms that the antibunched state are related with the photons number in pump mode A i.e. $|\alpha|^2$. Further it is evident from the figures that non classicality increases as we go towards the higher order of photon antibuncing.

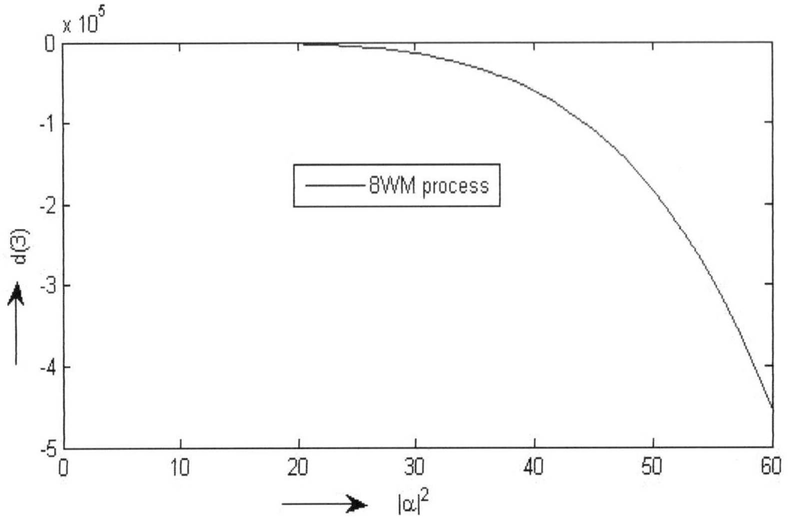

Figure 1. Dependence of third order photon antibunching d(3) on $|\alpha|^2$ (taking $g^2t^2 = 10^{-4}$).

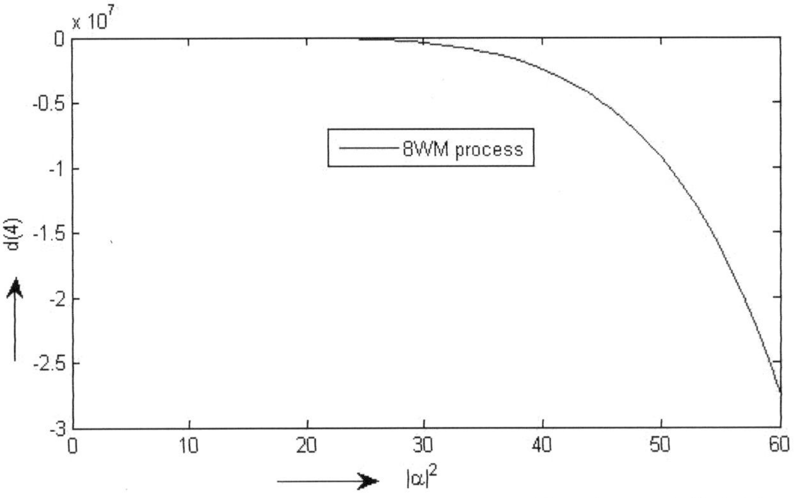

Figure 2. Dependence of fourth order photon antibunching d(4) on $|\alpha|^2$ (taking $g^2t^2 = 10^{-4}$).

5. Conclusion

We have reported non classicality in terms of HOA in pump mode of eight wave mixing non linear optical process. We have found that third and fourth order photon antibunching straightforwardly depends on the number of pump photons existing in the system and non classicality of a system increases as we move towards the higher order of photon antibunching. Thus we can conclude that fourth order photon antibunching is more suitable to generate single photon source utilized in quantum cryptography as compared to lower order photon antibunching.

REFERENCES

[1] Glauber R J 1963 *Phys. Rev.* **131** 2766-88

[2] Sudarshan E C G 1963 *Phys. Rev. Lett.* **10** 277-9.
[3] Hanbury Brown R and Twiss R Q 1956 *Nature* **177** 27-9
[4] Dodonov V V 2002 *J. Opt. B: Quantum Semiclass*. Opt. **4** R1-33
[5] Hong C K and Mandel L 1985 *Phys. Rev. Lett.* **54** 323-25
[6] Hong C K. and Mandel L 1985 *Phys. Rev. A* **32** 974-82
[7] Gill S 2017 *Inter. J. of Res. and Scient. Inno.* **4** 90-93
[8] Gill S 2018 *Intern. J. Of Advan. Res. in Sci. And Eng.* **7** 105-11
[9] Mandal S 2002 *Mod. Phys. Lett. B* **16** 963-65
[10] Pathak A and Mandal S 2003 *Mod. Phys. Lett. B* **17** 225-33
[11] Mandal S 2004 *Opt. Commun.* **240** 363
[12] Mandal S and Perina J 2004 Phys. Lett. A **328(2-3)** 144-56
[13] Kimble H J, Dagenais M and Mandel L 1977 *Phys. Rev. Lett.* **39** 691-5
[14] Diedrich F and Walther H 1987 *Phys. Rev. Lett.* **58** 203-6
[15] Short R and Mandel L 1983 *Phys. Rev. Lett.* **51** 384-7
[16] Pathak A and Verma A 2010 Ind. J. Phys. 84 1005-19
[17] Yuan Z, Kardynal B E, Stevenson R M, Shields a J, Lobo C J, Cooper K,Beattie N S, Ritchie D A and Pepper M 2002 *Science* **295** 102-5
[18] Lee C. K 1990 *Phys. Rev. A* **41** 1721-3
[19] An N B 2002 *J. Opt. B: Quant. and Semiclass. Opt.* **4** 222-7
[20] Gupta P, Pandey N P and Pathak A 2006 *J. Phys. B: At. Mol. Opt. Phys.* **39** 1137-43
[21] Sen B and Mandal S 2007 *J. Phys. B: At. Mol. Opt. Phys.* **40** 2901-16
[22] Acin A, Gisin N and Masanes 2006 *Phys. Rev. Lett.* **97** 120405(1-4)
[23] Hillery M 2000 *Phys. Rev. A* **61** 022309(1-8)
[24] Bennett C H, Brassard G, Mermin, N D 1992 *Phys. Rev. Lett. 68* 557–9
[25] Furusawa A, Sorensen L J, Braunstein L , Fuchs A C, Kimble J H 1998 *Plozik Science* **282** 706-9
[26] Bennett C H, brassard G, Crepeau C, Jozsa R, Peres A and Wootters W K 1993 *Phys. Rev. Lett.* **70** 1895-9
[27] Braunstein S L and Kimble H J 2000 *Phys. Rev. A.* **61** 042302(1-4)
[28] Bennett C H and Wiesner S J 1992 *Phys. Rev. Lett.* **69** 2881-4
[29] Verma A and Pathak A 2008 *Optics and Photonics for Information Processing II* **7072**

Revolution in electric vehicles

Amandeep Sagwal[1], Paras Agrawal[1], Randeep[1], Mangladeep Bhullar[1]
[1]Physics Department, DAV College, Sector 10, Chandigarh 160011, India.

Email: bhullarmangla@gmail.com

Abstract- Nowadays electric vehicles play an effective role in reducing greenhouse gases emission by the transportation sector. In recent times, electric vehicles such as battery, hybrid, plug-in hybrid, and fuel cell electric vehicles are becoming an advanced area of research related to the automobile industry. This article focuses on the latest emerging technologies in electric vehicles including various aspects of electric vehicles, their types, innovations in electric vehicle technology in terms of their motors, batteries as well as their applications, future perspectives, and different challenges associated with electric vehicles.

Keywords: Electric vehicles, Motor, Batteries, Ultracapacitors, Greenhouse gases

Abbreviations:

EVs	Electric vehicles	Ni-Zn	Nickel zinc
ICEV	Internal combustion engine vehicle	SRM	Switched reluctance motors
BEV	Battery electric vehicle	PM	Permanent magnet
FEV	Full electric vehicle	IM	Induction motor
HEV	Hybrid electric vehicle	BLDC	Brushless DC Motor
PHEV	Plug-in hybrid electric vehicles	SynRM	Synchronous reluctance machines
FCEV	Fuel cell electric vehicle	PMSM	Permanent magnet synchronous motor
NiMH	Nickel-metal hydride battery	GHG	Greenhouse gases

1. Introduction

As the number of vehicles keeps on filling in India, vehicles are resulting in being a wellspring of causing air contamination which is needed to be controlled down quickly [1]. The specialized measure that can be taken to reduce air pollution is the usage of electrically driven vehicles.

The concern towards the increasing global warming is not new to us. Back in 1996, the modern era began with the launch of EVs, such an EV was GM EV1 [2]. In comparison to ICEVs, EVs require new components such as supercapacitors, hydrogen fuel cells, hydrogen storage systems, automotive thermoelectric generators, charging technology for energy storage, etc. to provide smooth operating efficiency [3]. With time, we have witnessed the transition of motor design technology in electric vehicles. Moreover, future development trends are concentrated towards the improvement of motor performance by improving the insulation performance of magnet wires [4].

To meet the regulatory standards, the number of EVs such as BEV, HEV, PHEV, and FHEV has been introduced in the market. Though the aforementioned types have a significant role in reducing greenhouse gas emissions, each has its advantages and drawbacks. Understandably, FCEV has a much more promising potential in the transportation sector [1].

In near future electric vehicles will become more preferable than ICEVs, if research and development are successful in developing motors for EVs which are lighter, smaller, more reliable, have a more specific power, and are capable of offering maximum torque even at low speeds, etc.

This paper reviews the electrification of vehicles (cars) in the transportation sector. The paper discusses different models of electric cars in terms of components used. The review gives an understanding of the most recent achievements in terms of techniques and components used in the vehicles. Section 2 includes the various types of electric vehicles and cars models available in past and present days. Some futuristic trends in batteries and motors that are used and would be expected in the future in vehicles are reviewed in sections 3 and 4, followed by the conclusion.

2. Electric vehicles and their types

Landmarked reached after the appraisal of a survey done regarding the number of papers published on topics related to EVs during 1984 to 2020 is that continuous growth in development trends of various EVs has been marked. GMs EV1 was the first electric vehicle launched in 1996. Following this, Toyota Prius in 1997 was launched and witnessed a profuse response. Thereby, the increase in popularity of EVs as mass-produced cars was noticed. Amongst the prior labels mentioned, Tesla has been marked as the chief label since it has been extending the frontiers of innovation. Accordingly, the current report, EVs have reached the marked two million worldwide sales in the year 2019 and the growthin requirement for such cars is in continuity [2][5][6]. Also, a wide range of EVs is offered by automakers such as Tesla Roadster, Leaf, Mitsubishi i-MiEV, BMW i3, Renault Zoe, and many more [7]. The aforementioned models fit into the prime categories of EVs (figure 1) as follows:

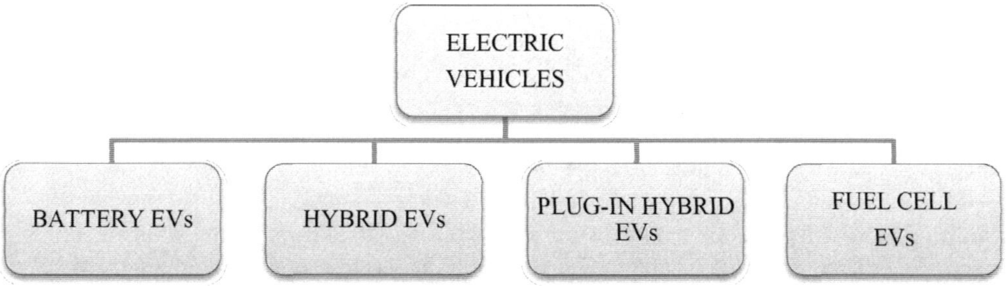

Figure 1. Various types of electric vehicles

i) Battery or Full electric vehicle (FEV or BEV)

BEVs exhibit an electric propulsion system. These are completely based upon electric motors or in other words, the power to drive train is provided with only the batteries [7]. Merits of such systems are their simple construction, operation, and convenience to be brought into utilization. The emission of GHG, as well as, noise is completely declined in such systems. Even at low speed, electric propulsion provides instant and high torque. The practice of such an approach can be marked in Teslas [8].

ii) Hybrid Electric Vehicle (HEV)

The concept of HEVs was introduced first in 1997. To power the vehicle, utilization of both ICE as well as electric power train is in requirement. During low power demand scenarios, HEV uses electric propulsion. During the rest period engine remains off and thereby reducing the fuel consumption like in case of traffic jams. In case of higher speed requirement HEV chop and change to ICE [3]. HEVs can be subcategorized into series, parallel, series-parallel, and complex hybrid HEVs and their properties are summarized in figure 2.

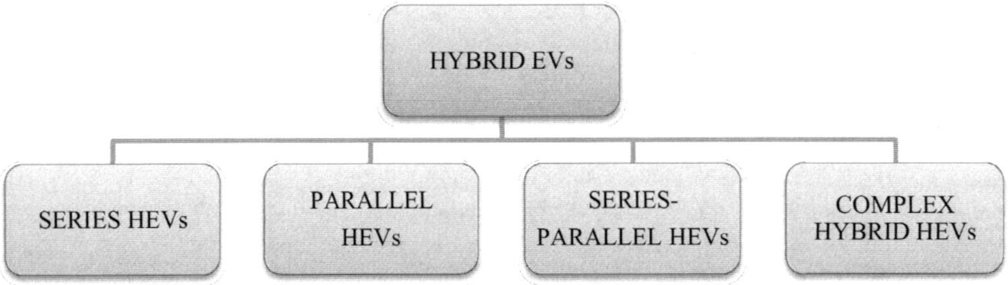

Figure 2. Types of Hybrid Electric Vehicles

a) Series HEVs: In series hybrid vehicle via transmission AC motor is contacted directly to the wheels or the differential. Into the bargain, DC/AC converter is needed to connect different power systems. Inthis, the DC link is responsible to decouple electrical machines and also, ICE from the wheels is mechanically decoupled with the help of an electrical system [7]. Due to the reason immensely large size of motor/generator, these find application in commercial heavy duty vehicles [1] [9].

b) Parallel HEVs: Since the electrical machine and ICE are connected to wheels/transmission, hence both contribute to the propulsion. Not only the reduction in fuel consumption in ICE is assisted by electrical machines but also it encourages regenerative functionand battery charging [7]. Due to its compact size, it can be brought into utilization in small vehicles such as Honda Insight, Ford Escape, and Honda Civic, etc [1] [9] [10].

c) Series-Parallel HEVs: Both series and parallel paths for power are developed by two electrical machines when combining with ICE. For the generation of electricity, for charging batteries ICE is coupled with an electrical generator. Regenerative function in the present case is viable too [7]. Toyota Prius was the first industrially available vehicle that uses this framework [1] [5] [11]. It provides only unidirectional power flow [8].

d) Complex hybrid HEVs: Such a system provides bidirectional power flow. It is an appropriate choice to be brought into use where vehicles require dual-axle propulsion. The only drawbacks which accompany such a system are their high complexity and cost [8] [12]. The use of such systems has been observed in certain Toyota and Ford models [8] [13].

iii) Plug-in Hybrid Electric Vehicle (PHEV)

Mechanically, PHEVs configuration is also similar to HEVs for its operation [8]. PHEVs are able to utilize fully electrically driven mode because such systems have adequate high energy density, energy storage systems capable of being charged externally [7]. Moreover, in contrast to the HEVs, PHEVs have high energy density capacity, so such vehicle uses electric propulsions as the main driving force [7] [8]. Hence, such systems are also termed as range-extended HEVs [7]. Examples of such systems are Chevrolet Volt and Toyota Prius [8].

iv) Fuel Cell Electric Vehicle (FCEV)

The Fuel cell is mounted in such vehicles to produce electricity which goes into the electric motor and drives the wheels. The production of electricity in the fuel cell is carried out by the chemical reaction (redox reaction) between high-pressure hydrogen and oxygen [14]. The source of hydrogen is high-pressure tanks carried out by the vehicle and the oxygen is obtained by the air sucked. FCEVs like Toyota Mirai or Honda Clarity are commercially available options [8]. These EVs are gaining more popularity in the transportation sector in recent times due to their properties [7] [8] (Table 1).

The electrified transport industry significantly relies on batteries and traction motors. The chief component in EVs and HEVs is the electric propulsion system and their energy storage system [1] [15]. The performance of electric machines in EVs can be considered up to par if it offers the needed high efficiency and high power density. To survive in the competitive world of automobiles sector the electric machine must not only fulfill the condition of high rated torque and wide speed range but also offer a fast dynamic response. Characteristics such as large capacity, reliability along with constant speed ability make such machines an enticing piece of technology [2] [16].

Table 1. Electric Vehicles and their characteristics

EVs	Advantages	Drawbacks
ICEV	Highly commercial Easy functioning	Economically unfriendly Emit green house gases Heavy framework [12]
BEV	Work smoothly Good Energy efficiency No greenhouse gas emitted	Long chargingperiod Limited charging sources [8] [12]
HEV	Economically friendly Durable Less pollution	High Cost Heavy Complex Structure [8] [12]
PHEV	Quite and smooth operation High fuel efficient Lower emission	Complex structure Initially expensive [8] [12]
FCEV	Maximum efficiency No petroleum products used Highly Reliable and durable Emission less	Costly Complex electronic structure [8] [12] [14]

3. Batteries

The most important component for an electric vehicle is the battery, serving as an energy source. The major concern regarding EVs is their driving range. An energy source is said to be perfect if it has maximum energy and power densities during the time of acceleration, fast charging, is inexpensive, and requires low maintenance [17]. The batteries with higher energy densities have the potential to drive vehicles a longer distance. The power to energy ratio is required more in HEV as compared to EV battery [18]. Many technological developments are being carried out to attain the desired goals. Some of the prominent types of batteries available are as follows [8]:

i) Lead-acid battery: One of the oldest batteries in use, for over 50 years. Their drawback includes the insufficient life cycle in case of deep rate of charge consumption with minimum power density. However, they are not highly expensive.

ii) Nickel-Metal Hydride battery: These batteries have almost double the energy density to what is in lead-acid batteries. At high voltages, NiMH batteries offer a safe operation. Unlike lead-acid batteries, NiMH batteries have a longer life cycle and a longer range of operating temperature. However, these batteries suffer a reduction in useful power.

iii) Lithium-Ion battery: These batteries have almost doubled the energy density than in nickel-metal hydride batteries. The reduction in usable power is less due to the low memory effect. Lithium-ion batteries offer good performance even at high temperatures. These batteries have a longer battery life which is around 1000 cycles and offer high specific energy and power. But, these batteries are expensive and have a longer recharging duration. Lithium-ion batteries have the highest energy densities among the available options.

iv) Nickel-Zinc battery: Not only these batteries have high power and energy densities butare also capable of offering deep cycle. Such batteries have a wide temperature operating range. But at the same time offers a disadvantage due to the fast growth of dendrite, which makes it unfit to be used in vehicle applications.

v) Nickel-Cadmium battery: The most significant advantage of these batteries is that they can discharge fully without any damage. Moreover, these batteries offer a long lifetime and are recyclable. Cadmium used in these batteries should be disposed of properly to avoid pollution. These batteries are too expensive to be brought into application in electric vehicles.

Not only batteries but supercapacitors also supply energy to drive an electric vehicle [7]. Supercapacitors or ultracapacitors help in achieving fast charging goals [7]. However, because supercapacitors have low specific energy hence, cannot be solely used as an energy source to drive an electric vehicle [12].

4. Motors

After compatible batteries, another important component is transition motors. In EVs new topologies such as Switched Reluctance Motors and Synchronous motors are used because the previously used DC motors have maintenance issues. Presently, induction motors and permanent magnet synchronous machines are said to be the best contenders to serve the purpose. Not only these but hybrid excited or less rare-earth magnet synchronous reluctance machines (SynRM) and PM assisted SynRM can also be considered for future perspectives. In the recent development of vehicles, PMSM specifically axial and radial flux is utilized. Synchronous PM type is the utmost and most efficient choice to meet the demands of the automotive [2] [16] and properties of these motors are characterized in Table 2. Figure3 shows different car models utilizing motors as their propulsion system. Depending upon the purpose of use, EVs are developed with a different number of motors as in Toyota Prius only a single motor is mounted whereas, in Acura NSX, three motors are fitted [8].

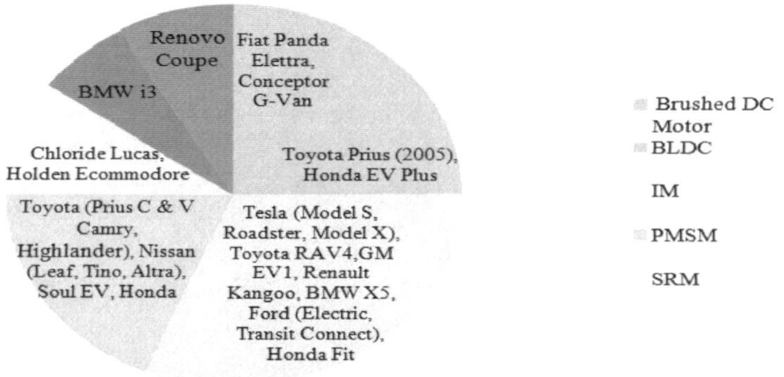

Figure 3. Electric Vehicles and their motors from ref [2] [8]

Table 2. Motors and their characteristics.

Type of Motor	Benefits	Drawbacks
Brushed DC Motor	Low speed offers high torque	Heavy Less efficient [8]
PM Brushless DC Motor	Negligible copper loss Efficient Lighter, compact, and Reliable	Expensive [8]
Permanent Magnet Synchronous Motor (PMSM)	Working in various ranges of speed Productive and Compact Low speed offers high torque	Maximum iron loss occurs at highspeeds [8]
Induction Motor (IM)	Can be used individually	High power losses Less efficient than PM [8]
Switched Reluctance Motor(SRM)	Inexpensive Hazardless	Less power factor Less efficient Bulky [8]
Synchronous Reluctance Motor (SynRM)	Highly efficient Compact Size	
PM assisted SynRM	Good power factor No demagnetization	[8]
Axial Flux Ironless PM Motor	Light in weight Efficient Minimal losses	[8]

Some recent problems and their solutions regarding motors are discussed as follows [19]:

i) Even though IM is generally acknowledged for utilization by the majority of EV/HEV makers, still it has poor efficiency and power factor in comparison to Permanent magnet motors [20-22]. Many developments are being made to increase efficiency by introducing a new converter topology or a new control scheme. IM drives are the most cost-effective among the compared motor drives.

ii) Though PM motors are the most popular choice for electric traction drive systems but the idea is not epoch. Due to the possibility of a shortage of rare-earth metals and their high cost we need to shift the approach towards IM and SRM as an alternative for EVs/HEVs applications. In case of permanent magnet motors improvement in performance cannot be achieved with inexpensive magnets such as Ferrites [19] [23]. Research and development of materials may improve the functioning and properties of permanent magnet motors.

iii) The presence of torque ripples and acoustic noise cause problems in functioning of switched reluctance motor [9] [20]. Nowadays, the most preferred choice is the axial flux PM brushless DC motor drives [24].

Besides the fact that electric vehicles promise to reduce GHG emissions and save us from climate changes, but such systems may lead to power instability. With the help of technological development, such an issue can be resolved which will make EVs more reliable [8].

The depletion and shortage of rare-earth materials will not allow us to use PM motors. Shifting the approach from PM type, IM with internal PM are being utilized by Tesla in its models presently. Great research has already been carried out and will be carried out in near future. Such technological initiatives will make EVs more affordable, efficient, and reliable.

Better charging technologies must also be brought into consideration for research. Wireless charging technologies will also be a conspicuous development if worked upon [8]. The trend and pattern of the sales of EVs across the globe is shown in figure 4.

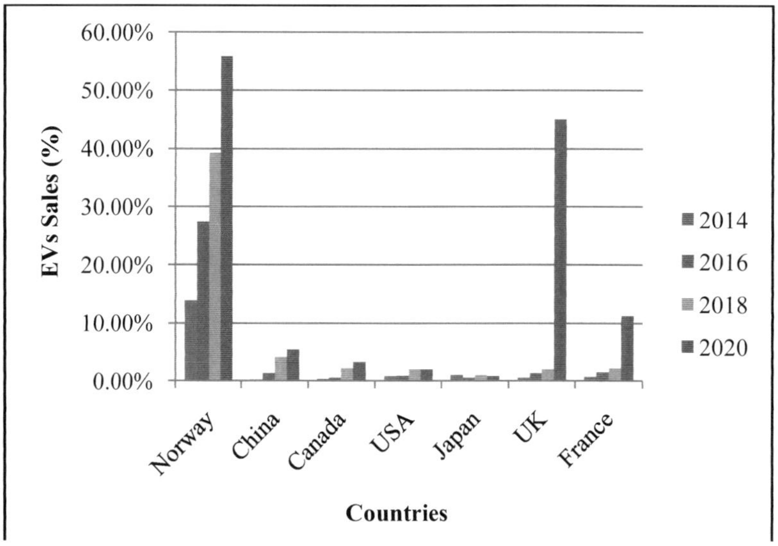

Figure. 4 Market shares of EVs car sales between 2014 and 2020 from ref [25]

5. Conclusion

The advancement of technology used in motors and batteries having good capacities will favor the use of EVs in other areas of transportation.

The use of superconductors and ultra-capacitors are however able to help in achieving quick charging objectives and provide high-power required during acceleration. The objective of achieving high energy densities is being fulfilled using Lithium ion batteries.

BEV and HEV are the latest innovations in the field of electric vehicles successfully growing at a faster pace. FCEVs have long-term potential, have negligible emissions, and have a good driving range making them a contender to be used as the standard vehicle in the future.

Along with DC and PMSM traction motors, SRM and IM are significant motors used in the electric vehicle sector. Every type of motor is used as far as cost, effectiveness, power density, and manufacturability are concerned. Presently, Synchronous permanent magnet motors are the most productive electric motor in electric vehicle application. The exhaustion of rare earth materials, has led to a change from the PM type motor to IM.

An enormous rise in production of EVs is underway in the coming years. Further upgrades in EVs are required and expected in upcoming years.

References

[1] Alam S Md and Khan A 2020 *IJSART* **6** 30-37
[2] Agamloh E, Jouanne A von and Yokochi A 2020 *machines* **8** 1-17
[3] Farfan-Cabrera L I 2019 *Tribol. Int.* **138** 473–86

[4] Sato A, Iiduka S and Kimura K 2020 *SEItech. rev.* **90** 17-21
[5] Husain, I 2011 *Electric and hybrid vehicles design fundamentals* 2nd ed. (CRC Press BocaRaton, FL, USA)
[6] Schefter K and Zaccagnino M 2019 *Electric vehicle sales facts and figures* (EEI PennsylvaniaAvenue, NW) 701
[7] Katić V A, Dumnić B, Čorba Z and Milićević D 2014 *Electronics and Energetics* **27** 299–316
[8] Un-Noor F, Padmanaban S, Mihet-Popa L, Mollah M N and Hossain E 2017 *Energies* **10** 1217
[9] Ehsani M, Gao Y and Miller J 2007 *Proc. IEEE* **95** 719-728
[10] A Emadi A, Lee Y J and Rajashekara K 2008 *IEEE Trans Ind. Electron.* **55** 2237 -2245
[11] Nagasaka A, Nada M, Hamada H, Shu H and Kikuchi Y 1998 *SAE trans.***107** 1721-1730
[12] Chan, C.C. 2002 *Proc. IEEE* **90** 247–275.
[13] Miller, J.M 2006 *IEEE Trans. Power Electron.* **21** 756–767
[14] EG & G Technical Services, Inc. 2002 *The Fuel Cell Handbook*, 6th ed.; U.S. Department of Energy (Morgantown, 2WV, USA)
[15] Finch J and Giaouris D 2008 *IEEE Trans. Ind. Electron.* **55** 481-491
[16] Sun X, Li Z, Wang X, Li C 2020 *Energies* **13** 90
[17] Deng J, Bae C, Denlinger A, and Miller T 2020 *Joule* **4** 1-5
[18] Baskar S., V.Vijayan, Isaac Prem kumar, I. J., Arun kumar D., and Thamaran D. 2020 *Mater. Today: Proc.* 1-3
[19] Rind S J, Ren Y, Hu Y, Wang J, Jiang L 2017 *Chin. J. Electr. Eng.* **3**
[20] Kumar L and Jain S 2014 *Renew. Sustain. Energy Rev.* **29** pp. 924-940
[21] Zhu Z, and Howe D 2007 *Proc. IEEE* **95** 746-765
[22] Zeraoulia M, Benbouzid M and Diallo D 2006 *IEEE Trans. Veh. Technol.***55** 1756-1764
[23] Riba J, Lpez-Torres C, Romeral L and Garcia A 2016 *Renew. Sustain. Energy Rev.* **57** 367-379
[24] Eldho Aliasand, A., and Josh, F. T. 2020 *Mater. Today: Proc.* **24** 1804–1815
[25] Sanguesa J.A., Torres-Sanz V., Garrido P., Martinez F.J. and Marquez-Barja J.M. 2021 *Smart Cities* **4** 372–404

Numerical Assessment of Highly Efficient Tin-Halide Perovskite Solar Cell

Ajay Kumar[1], Uddeshya Gupta[1], Neha Gupta[2] and Amit Kumar Goyal[1]

[1]Electronics and Communication Department, Jaypee Institute of Information Technology, Noida, India.
[2]Applied Science and Humanity Department ADGITM, New Delhi, India.

Email:ajay.kumar@jiit.ac.in

Abstract. Interface science of Perovskite Solar cells (PSCs) has a very crucial role in increasing the efficiency of solar cells. The process of numerical simulation has also helped the solar cell's efficiency to grow rapidly from 3.8% in 2009 to around 11% in 2010 in such a short period. This was a great achievement at that time and from then it kept on improving. Not only its rapid increase in efficiency attracted the researchers but also its low manufacturing cost and high V_{oc}. In this work, we will discuss the Lead-Free Perovskite Solar Cell (PSC) due to its non-toxic nature. A lot of research has already been done on Perovskite Solar Cell that has led to an increase in its efficiency. Despite the truth that they are toxic and can harm us anyway. The prime motive of producing electricity by using solar cells is to provide a cleaner way to produce electricity but if we are using toxic materials in the manufacturing of the cell, we are getting diverted. That's why this research paper will focus on the cell having Sn in its absorber layer. The model we are discussing here will have $CH_3NH_3SnI_3$ (Absorber) as the Perovskite Layer which is non-toxic also the following different layers were also used in simulating our device such as ITO (Indium Tin Oxide) as the front contact, ZnO (Zinc Oxide) as electron transporting layer (ETL), CuI is acting as Hole transport layer (HTL) and we are using Gold (Au) as a back contact. The motive of this research is to make it non-toxic that we have already achieved and to make it highly efficient and the efficiency of the proposed device is 24.72%.

1. Introduction

With the increasing modernization in this world, humans are tending to be more dependent on machines, and hence more energy resourcesare required to satisfy our needs. There are many ways to produce electricity some of them being non-renewable and some being renewable sources but since the past few years, there is a lot of focus on renewable sources because they will not emit harmful gases and harmful by-products [3]. Solar energy is a very famous renewable source of energy and being recently mostly discussed it has now many advancements that are making it successful. The rapid increase in efficiency in such a short span of time is also appreciable [4]. Since the advent of solar energy, the solar industry has seen many types of devices and the recent popular one is Perovskite Solar Cells (PSCs)[2, 8, 11, 16, 17]. Photovoltaic cells can directly transform the rays we get from the sun into electricity [6].

The discussion of a perovskite solar cell is done here in this paper. The Perovskite solar cell has played a pivotal role in the solar industry due to its low costing, high efficiency, and high V_{oc}. Lead-

Halide Perovskite solar cells are quite famous but the lead in them makes us shift to its alternative that is the Tin-Halide Perovskite Solar Cell [5, 12, 17]. So, in this proposed device we will use $CH_3NH_3SnI_3$ instead of $CH_3NH_3PbI_3$. The Tin-Halide Perovskite layer is not only non-toxic but also hasmany other benefits such as a small band gap that is of 1.3eV and high absorption coefficient (α)[13]. We successfully attained a highly efficient device because of the combination of layers used and also their properties. Figure 1. shown below depicts the necessary layers of the solar cell we will use in this device.

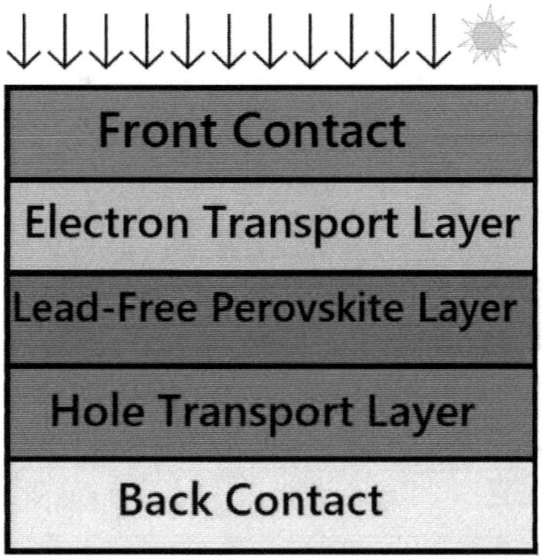

Figure 1. Layers used in the simulation of Tin-Halide Perovskitesolar cell.

In this work Tin-Halide, PSC will have the following five layers as Indium Tin Oxide/ Zinc Oxide/ Tin-Iodide Perovskite/ Copper Iodide/ Gold. In this proposed device ITO [7]is the front contact and sunlight will enter from here, the second layer is of ZnO which is the electron transport layer (ETL)[9], $CH_3NH_3SnI_3$ is a perovskite layer that behaves as an absorber, CuI and it is serving the purpose of the hole transport layer (HTL) and the role of back contact is fulfilled by Gold. A comparison is also done here with the other devices which will help us to assess our work.

2. Methodology and Simulation

The numerical simulation done here is done on Silvaco Technology CAD (Silvaco TCAD). The three-Dimensional device structure and the meshing of the solar cell obtained are shown in Figure 2(a) and Figure 2(b). All the layers are clearly visible and then a discussion will be done on the functions and the properties of all the materials used.

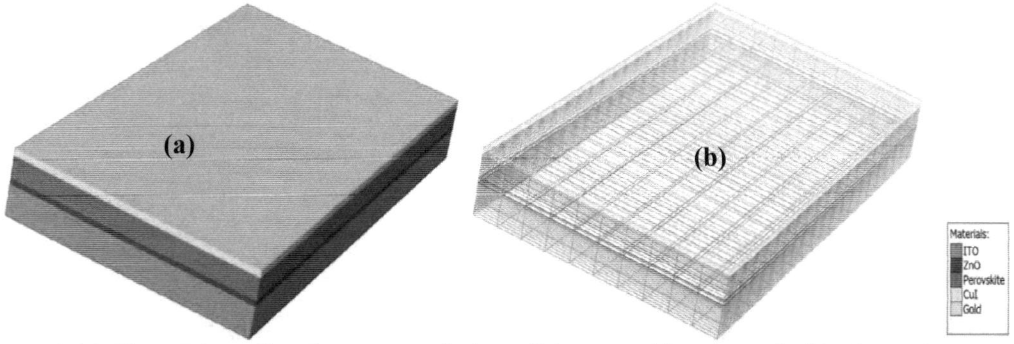

Figure 2.(a) Three-Dimensional structure of Tin-Halide Perovskitesolar cell; (b) Three-Dimensional meshed structure of Tin-Halide Perovskitesolar cell.

The simulated device that is presented in this paper hasthe following layers- Indium Tin Oxide (ITO) this layer is 0.500 µm thick and it acts as a passage for sunlight to enter the device it is the widest layer of this device so that maximum sunlight can enter the device. This layer is also known as front contact. The second layer is Zinc Oxide (ZnO), this layer serves as the electron transport layer (ETL) and its primary objective is to throw back holes towards HTL, this layer is 0.080 µm thick. The next layer is the Perovskite layer and which is 0.350 µm thick and is also known as the absorber layer. The material used in the next layer is CuI and it is serving as the hole transport layer. It is 0.100 µm thick. The final layer is of gold and it is 0.020 µm thick and it serves the purpose of back contact in this device. All the materials along with their electrical properties are shown in Table 1. given below.

Table 1. Parameters used in the simulation of Tin-Halide PSC.

Material	**Thickness**			**Doping Type**	**Permittivity ε_r (relative)**	**Band Gap (eV)**	**χ (eV)**	**μ_n ($cm^2/V/s$)**	**μ_p ($cm^2/V/s$)**
	x-axis (µm)	**y-axis (µm)**	**z-axis (µm)**						
ITO	10	0.500	5	-	-	-	-	-	-
ZnO	10	0.080	5	n-type	8.12 [1]	3.2 [15]	4.29 [1]	200	5
Perovskite	10	0.350	5	p-type	8.2 [18]	1.3 [18]	4.17 [18]	1.6	1.6
CuI	10	0.100	5	p-type	10 [19]	3.2 [19]	4.1 [19]	1.69e-4	1.69e-4
Gold	10	0.020	5	-	-	-	-	-	-

3. Results and Discussions

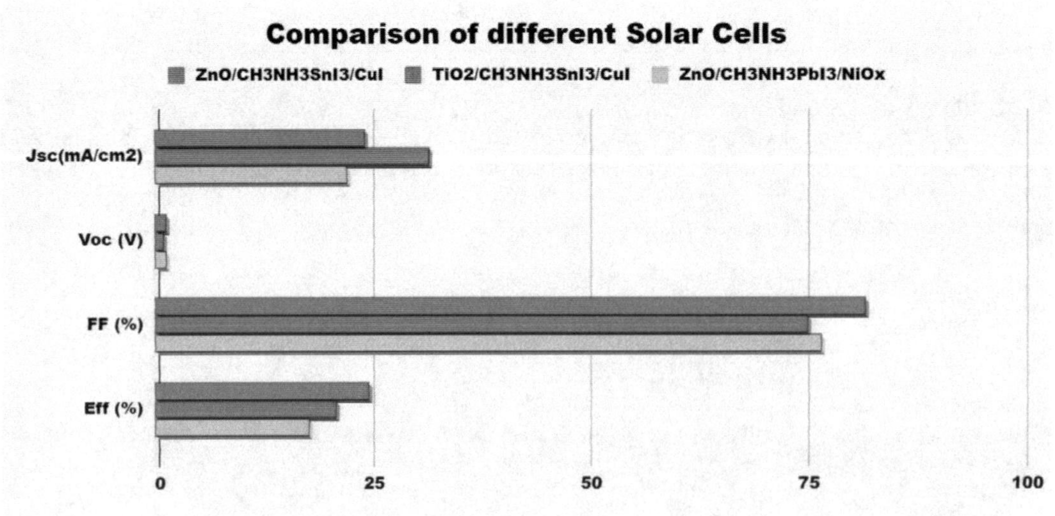

Figure 3. Comparison of different Solar Cells.

After simulating our device, the value of the following parameters was obtained J_{sc}, V_{oc}, FF, and efficiency. A comparison is also done with a device that had $CH_3NH_3PbI_3$ as the absorber layer instead of $CH_3NH_3SnI_3$ and also with a device having a different Electron Transport Layer (ETL). The comparison is also shown graphically. The graph is represented in Figure 3. Through this bar graph, it can be noticed that the parameters such as open-circuit voltage(V_{oc}), Fill Factor (%), and efficiency shoot up.Through the bar graph shown above, we can clearly interpret that the efficiency of the solar cell has been increased significantly and the combination of layers used is successfully increasing its efficiency. There is very little difference in the value of V_{oc} but still, the open-circuit voltage of the proposed device is better as compared to the other two devices and it is helping efficiency to increase[10]. The Value of the Fill Factor is also maximum in the device proposed having layers ($ZnO/CH_3NH_3SnI_3/CuI$) as compared to the other two. Optimization of thickness and other properties made it possible to achieve an efficiency of 24.72%. Values of various parameters calculated are shown in Table 2.

Table 2. Values of different parameters obtained.

Parameters	Values
J_{sc} (mA/cm^2)	24.2
V_{oc} (V)	1.06
FF (%)	81.8
Eff (%)	24.72

The architecture of Tin-Halide PSCs has exceptional features such as non-toxic, low-cost easy fabrication. The numerical simulation of the device helped us to observe the effect of different materials very easily and it also helps in the investigation of different attributes of different layers [14]. Many graphs are also obtained that verifies the credibility of this device. The graph of Wavelength vs Photon energy simulated by running it on Silvaco TCAD is clearly shown in Figure 4.andit can be clearly interpreted that upon decreasing wavelength of rays the potential starts to increase and then after a point that is around 1000 nm and the maximum potential obtained is 1.06 V. The potential started decreasing gradually. It is also verified theoretically as the wavelength is inversely proportional to the square root of potential.

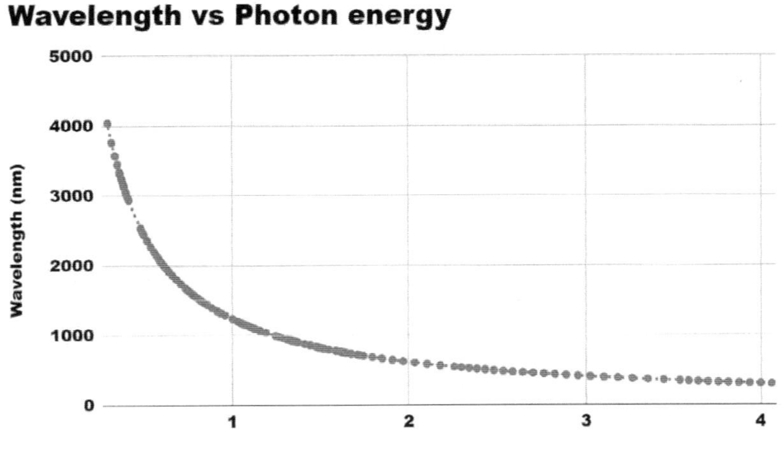

Figure 4. Wavelength as a function of photon energy.

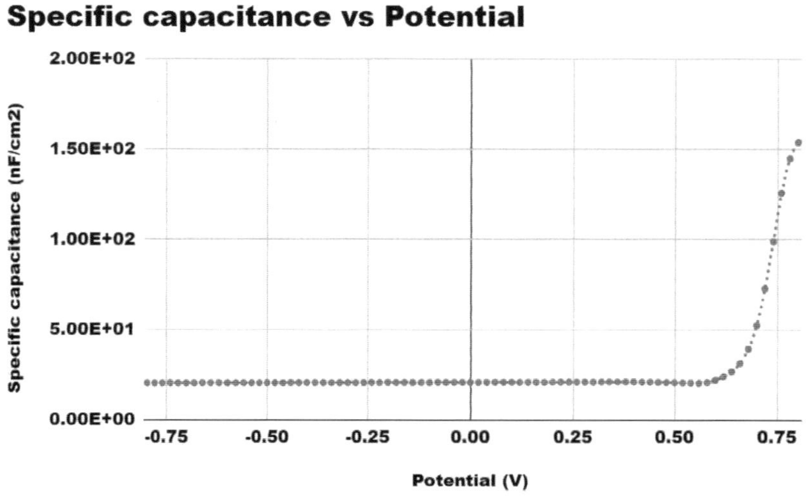

Figure 5. Specific capacitance vs potential.

The graph of specific capacitance vs potential is represented in the curve given in Figure 5. With the increasing potential, the value of capacitance is increasing and it is interpreted from it that the maximum capacitance is obtained at around 0.75 V and the capacitance obtained is about 158 nF/cm^2. Also, a curve between absorption constant i.e. α and optical wavelength is shown below in Figure 6. represents three different layers that areCuI, Perovskite, and ZnO through this curve we can interpret that the absorption constant of different layers is best in the range of 300nm to 900 nm and it is decreasing gradually with the increasing value of wavelength that is why the wavelength of light used for illumination is set to be at 0.350 µm. The graphs here portray the penetration of the rays in the cell.

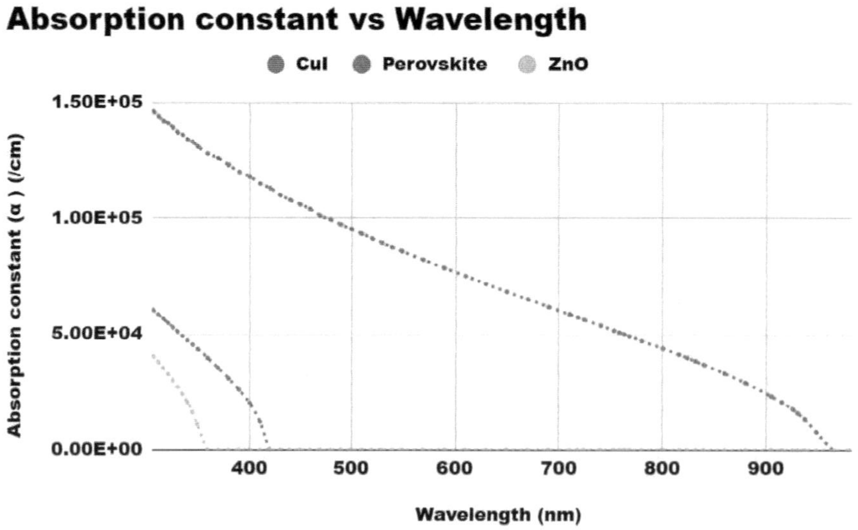

Figure 6. Absorption constant plots for CuI, Perovskite and ZnO with respect towavelength.

IOP Conf. Series: Materials Science and Engineering 1225 (2022) 012016 doi:10.1088/1757-899X/1225/1/012016

The next curve is shown in Figure 7, represents the variation of current density w.r.t potential at different values of thickness of absorber layer that is perovskite. It helps us in understanding that somewhere in between 0.233 μm and 0.700 μm we found the value of potential to be maximum and hence the value of 0.350 μm was used in the simulation. The different thicknesses of perovskite used in the simulation to find the optimization value were done and the graph is shown in Figure 8. This figureis showing that is the quantum efficiency will be maximum somewhere between 0.233 μm and 0.700 μm and hence the value of 0.350 μm was used and the maximum efficiency obtained after simulating the whole device was found to be 24.72%.

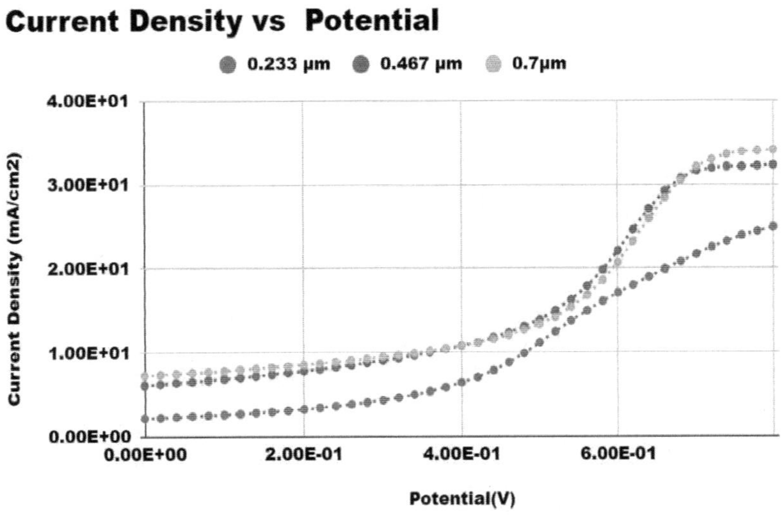

Figure 7. Current density vs potential.

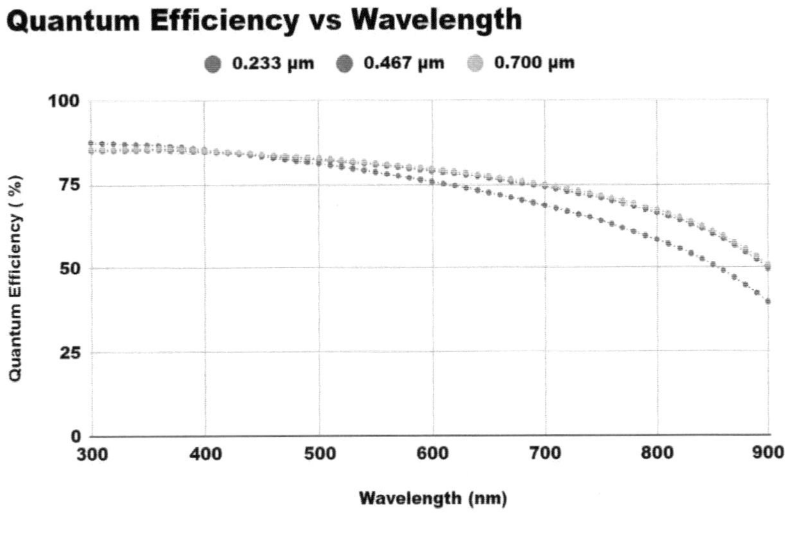

Figure 8. Quantum efficiency (QE%) vs wavelength.

The contour plots are shown in Figure 9 (a) and (b) represents the total current density and recombination rate. It can be observed that the perovskite solar cell with $CH_3NH_3SnI_3$ as the absorber layer and CuI as Hole Transport Layer (HTL) has a very high current density and also the recombination taking place in the device is decreasing and it is clearly visible in the absorber region as we all know more the recombination in absorber region lesser will be its efficiency. Hence a highly efficient solar cell is formed. Similarly, the graphs of hole concentration and electric field of perovskite solar cells having $CH_3NH_3SnI_3$ as its absorber layer and CuI as HTL are shown in Figure 9 (c) and (d). The contour plot of hole concentration is also supporting the increased efficiency of our device. Electric Field shown below is showing a varying electric field in the Hole Transport Layer and that is how the efficiency of 24.72% was achieved.

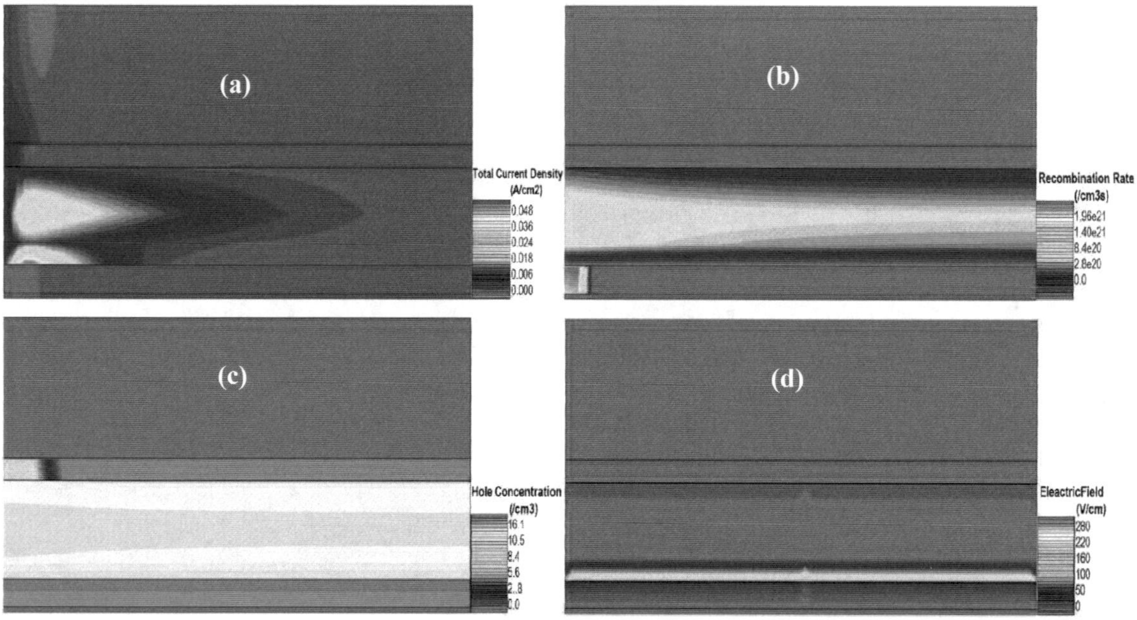

Figure 9. Contour plots of (a)Total Current Density, (b) Recombination rate, (c) Hole concentration, and (d) Electric Field of the proposed PSC.

4. Conclusion

Numerical optimization of layers was done such as their doping concentration and thickness hence we got to observe that the thickness of the layer plays a crucial role in increasing the Power Conversion Efficiency (PCE). The efficiency of every solar cell depends mostly on the thickness of its absorber layer as not only the absorption of sunlight takes place but the electron-hole pairs are also generated here. On decreasing the thickness of the absorber, the recombination of charges will increase thus decreasing the absorption and then efficiency. The results shown by the proposed device are satisfying and the comparison done was also supporting the work presented here. Different electrical parameters such as current density, open-circuit voltage, Fill Factor, and Efficiency were calculated efficiency of 24.72% was obtained. Perovskite solar cells can be seen as a promising candidate for future research as well because there is a huge scope of improvement such as we can further work on increasing its stability and also om the materials which can be used as its back contact instead of gold as it is expensive.

5. Acknowledgments

We are thankful to JIIT (Sec-62) for assisting us in resource providing.

References

[1] Chakrabarti T, Saha M, Khanda A and Sarkar S K 2018 *Advances in Communication, Devices and Networking*: Springer) 125-31.

[2] Du H J, Wang WC and Zhu J-Z 2016 Device simulation of lead-free CH3NH3SnI3 perovskite solar cells with high efficiency *Chin. Phys. B* **25** 108802.

[3] Ellabban O, Abu-Rub H and Blaabjerg F 2014 Renewable energy resources: Current status, future prospects and their enabling technology *Renew. Sust. Energ. Rev.***39** 748-64.

[4] Goswami D Y and Kreith F 2007 *Handbook of energy efficiency and renewable energy* CRC Press).

[5] Hao F, Stoumpos C C, Cao D H, Chang R P and Kanatzidis M G 2014 Lead-free solid-state organic–inorganic halide perovskite solar cells *Nat. Photonics* **8** 489-94.

[6] Jiang Q, Zhao Y, Zhang X, Yang X, Chen Y, Chu Z, Ye Q, Li X, Yin Z and You J 2019 Surface passivation of perovskite film for efficient solar cells *Nat. Photonics* **13** 460-6.

[7] Kumar A, Goyal A K, Gupta U, Gupta N and Chaujar R 2020 Increased efficiency of 23% for CIGS solar cell by using ITO as front contact *Mater. Today: Proc.* **28** 361-5.

[8] Kumar A and Gupta U 2021 Numerical assessment of high-efficiency lead-free perovskite solar cells *Mater. Today: Proc.* **45** 5041-5046.

[9] Kumar A, Gupta U, Chaujar R, Tripathi M and Gupta N 2020 Simulation of perovskite solar cell employing ZnO as electron transport layer (ETL) for improved efficiency *Mater. Today: Proc.* **28** 361-365.

[10] Kumari T, Jung S, Cho Y, Kim H-P, Lee J W, Oh J, Lee J, Lee S M, Jeong M and Baik J M 2020 A built-in electric field induced by ferroelectrics increases halogen-free organic solar cell efficiency in various device types *Nano Energy* **68** 104327.

[11] Lal N N, Dkhissi Y, Li W, Hou Q, Cheng Y B and Bach U 2017 Perovskite tandem solar cells *Adv. Energy Mater.* **7** 1602761.

[12] Löper P, Moon S-J, De Nicolas S M, Niesen B, Ledinsky M, Nicolay S, Bailat J, Yum J-H, De Wolf S and Ballif C 2015 Organic–inorganic halide perovskite/crystalline silicon four-terminal tandem solar cells *Phys. Chem. Chem. Phys.* **17** 1619-29.

[13] Lyu M, Yun J H, Chen P, Hao M and Wang L 2017 Addressing toxicity of lead: Progress and applications of low-toxic metal halide perovskites and their derivatives *Adv. Energy Mater.* **7** 1602512.

[14] Park N-G 2015 Perovskite solar cells: an emerging photovoltaic technology *Mater. Today* **18** 65-72.

[15] Rahman M S, Miah S, Marma M S W and Sabrina T 2019 Simulation based Investigation of Inverted Planar Perovskite Solar Cell with All Metal Oxide Inorganic Transport Layers Int. *Conf. Electr. Comput. Commun. Eng. IEEE* 1-6.

[16] Tonui P, Oseni S O, Sharma G, Yan Q and Mola G T 2018 Perovskites photovoltaic solar cells: An overview of current status *Renew. Sust. Energ. Rev.* **91** 1025-44.

[17] Tress W 2017 Perovskite solar cells on the way to their radiative efficiency limit–insights into a success story of high open-circuit voltage and low recombination *Adv. Energy Mater.* **7** 1602358.

[18] Umari P, Mosconi E and De Angelis F 2014 Relativistic GW calculations on CH 3 NH 3 PbI 3 and CH 3 NH 3 SnI 3 perovskites for solar cell applications *Sci. Rep.***4** 4467.

[19] Weiss M, Horn J, Richter C and Schlettwein D 2016 Preparation and characterization of methylammonium tin iodide layers as photovoltaic absorbers *physica status solidi* **213** 975-81.

Active Layer Thickness Optimization for Maximum Efficiency in Bulk Heterojunction Solar Cell

Susan G Daniel[1], B Devu[1] and C O Sreekala[1]

[1]Department of Physics, Amrita School of Arts and Sciences, Amritapuri, Amrita Vishwa Vidyapeetham, Kollam, India
Email address: sreekalaco@am.amrita.edu

Abstract. Organic solar cell (OSC) is one of the best promising candidate for harvesting energy mainly due to their simple and economic fabrication process, the reduced manufacturing cost, and easy integration to other products. Bulk heterojunction solar cell in which the active layer is a blend of donor –acceptor materials are one of the best organic photovoltaic device with highest efficiency and a significant improvement in the device performance occur over last years. Bulk heterojunction architecture gives a high interfacial surface area for efficient charge dissociation. In this study, bulk heterojunction solar cell is simulated using General Purpose Photovoltaic Device Model Software. A donor- acceptor blend of Zinc pthalocyanine (ZnPc) fullerene C60 is used as the active layer. The power conversion efficiency for various thickness of the active layer is studied. Optimization of active layer thickness for maximum power conversion efficiency are done. The dependence of various electrical parameters such as short circuit current density(J_{sc}), Open circuit voltage(V_{oc}),fill factor (FF),average carrier mobility on power conversion efficiency are also studied.

1. Introduction

Organic solar cells (OSCs) comes in the third generation solar cells where solar cells being categorised into first second and third generation solar cells and light absorbing layer is the carbon based (organic polymer) material [1, 2].Bulk heterojunction (BHJ) (organic) solar cells are applicable for the large scale device fabrication. Simple and economic fabrication process, device stability, mechanical flexibility, device performance makes the organic solar cell technologies scalable. The attractive unique properties of organic semiconductors including their low cost production, and lasers fabrication can be done using OSCs with high feasibility and efficiency. The organic molecules which is the building block of organic semiconductor are formed by π conjugated system [3]. Tang proposed the first two-component OSC in 1986 from the observation of photoelectric effect in the Tang cell [4]. The efficiency of two component OSC was very low [5]. This drawback was solved by the advent of bulk heterojunction (BHJ) solar cells in 1995 by Heeger and co-workers[6]. The active layer of a BHJ solar cell is a combination of a donor and acceptor material mixed in nanoscale [7, 8]. This gives required thickness for absorption and an appropriate diffusion distance across the active layer. The probability of charge dissociation is increased in bulk heterojunction architecture because of the increased interfacial surface area. Many researches have done since that time to discover new acceptors and donors and on their morphology.

This report consolidatesbulk heterojunction (BHJ) solar cell simulationalongwith ITO/PEDOT: PSS/ZnPc: C60/Al device structure. The active layer used is amixture of fullerene C-60 and Zinc

Pthalocyanine in the ratio 1:1. Furthermore the HTL (hole transporting layer) constitutes PEDOT: PSS, ITO (indium tin oxide) takes place of the transparent front contact and as the metal back contact Aluminium is used. The four isoindole units which provide conjugated 2 dimensional π electron system is the core structure of Zinc phthalocyanine (ZnPc) macrocycle. The trouble free synthesizing of ZnPc and its non-toxicity towards environment makes it a promising applicant in solar cell applications. ZnPc has high thermal and photo chemical stabilities. Phthalocyanine is enormously used as the donor material in the OSCs field [9]. Good carrier transport, high short circuit current density and efficiency is achieved using the best combination of C-60 as the electron acceptor. Organic solar cells with high power conversion efficiency can be fabricated using the blend ZnPc: C-60.In this study, thickness of the active layer ZnPc: C-60 is varied from 50nm to 500nm and the corresponding efficiencies arereported.

The dependence of power conversion efficiency on parameters like open circuit voltage (V_{oc}),short circuit current density (J_{sc}), fill factor (FF) of the bulk heterojunction (BHJ) solar cell with the variation in active layer thickness isdetermined using the electrical simulation in the General Purpose Photovoltaic Device Model (GPVDM) software. Such an analysis will help the experimental scientist to optimise the active layer thickness of their device.

2. GPVDM Software

General Purpose Photovoltaic Device Model (GPVDM) Software is ageneral-purpose tool (free), which is used in simulatingopto- electronic devices like solar cells, LEDs, diodes, FETs. During its initial time GPVDM is used to simulate the organic solar cell but now it is developed to simulate crystalline silicon solar cells, organic LEDs (OLEDs), a-Si solar cells, organic Field Effect Transistors (OFETs), CIGS solar cells [10]. GPVDM solves the poisson equation, carrier continuity equation using finite difference method. It solves the equation of drift diffusion and also solves the optical equation by applying the ray tracing or transfer matrix method [11]. The equations used to simulate the device is shown as equations 1-5.

$$\frac{d}{dx}\varepsilon_0\varepsilon_r\frac{\partial\varphi}{\partial x} = q(n-p) \quad (1)$$

$$J_n = q\mu_c n\frac{\partial E_c}{\partial x} + qD_n\frac{\partial n}{\partial x} \quad (2)$$

$$J_p = q\mu_n p\frac{\partial E_v}{\partial x} - qD_p\frac{\partial p}{\partial x} \quad (3)$$

$$\frac{\partial J_n}{\partial x} = q(R_n - G + \frac{\partial n}{\partial x}) \quad (4)$$

$$\frac{\partial J_P}{\partial x} = -q(R_P - G + \frac{\partial P}{\partial x}) \quad (5)$$

3. Electrical parameters

In this work, a bulk heterojunction (BHJ) solar cell having an activestructure ITO/PEDOT: PSS/ZnPc: C60/Al is simulated. This BHJ organic solar cell device contains 4 layers. The front contact transparent layer is ITO of thickness 140nm and the n-k values are also imported from the reference [12].The polymer contact layer is PEDOT: PSS of thickness of 100nm and the n-k values of PEDOT: PSS is extracted from the reference [13] and imported to GPVDM. The metal back contact layer is Aluminium of thickness 100nm and their n -k values are extracted from the reference [14]. The active layer is the mixture of fullerene C-60 and Zinc phthalocyanine (ZnPC)(1:1). The n-k values of ZnPc: C-60 is imported to the GPVDM from the reference [15].The thickness of the active layer ZnPc: C60 is ranged from 50nm to 500nm by a step of 50. The efficiencies corresponding to each thickness is noted [16]. The electrical parameters like open circuit voltage (V_{oc}), short circuit current density (J_{sc}), and fill factor (FF), average carrier mobility are to be determined.The device structure of simulated bulk heterojunction solar cell for 300 nm thickness ZnPc: C60 is shown in figure 1.

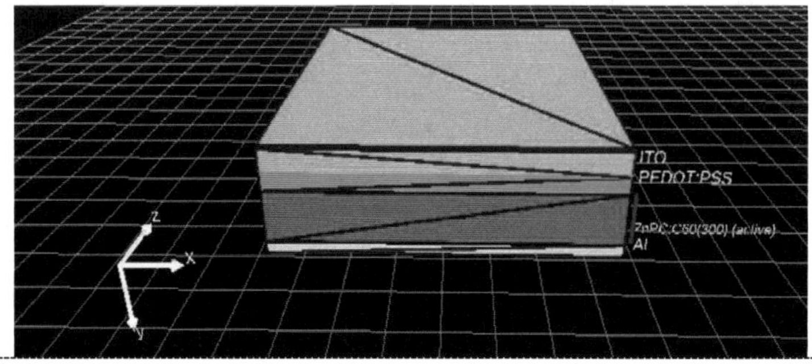

Figure 1. The device structure of simulated bulk heterojunction solar cell for active layer ZnPc: C-60 of 300 nm thickness

4. Results and Discussion

4.1. Electrical Parameters and Power Conversion Efficiency of Simulated BHJ Solar Cells

The power conversion efficiency and some of the electrical parameters such as short circuit current density (J_{sc}), open circuit voltage (V_{oc}), fill factor (FF), voltage (maximum power) (V_{max}), current density (maximum power) (J_{max}), maximum power (m_{pp}) and average carrier mobility of simulated ITO/PEDOT: PSS/ZnPc: C60/Al bulk heterojunction solar cells of different thickness of ZnPc: C-60 active layer is shown in table 1.

Table 1. Electrical parameters of the simulated bulk heterojunction organic solar cell devices.

Thickness (nm)	V_{oc} (V)	J_{sc} ($mAcm^{-2}$)	FF	V_{max} (V)	J_{max} $mAcm^{-2}$	Max power (m_{pp}) Watt m^{-2}	Average Carrier mobility($m^2V^{-1}s^{-1}$) (10^{-8})	Efficiency (%)
50	0.56	5.24	0.76	0.48	4.68	22.22	8.85	2.22
100	0.56	10.77	0.73	0.45	9.84	44.07	9.21	4.40
150	0.55	11.31	0.71	0.45	9.84	44.13	8.26	4.41
200	0.54	12.17	0.67	0.43	10.42	44.67	8.11	4.47
250	0.54	15.90	0.57	0.41	12.04	48.87	8.00	4.89
300	0.54	16.21	0.56	0.41	12.10	49.17	7.87	4.92
350	0.54	15.71	0.55	0.41	11.53	46.86	7.73	4.68
400	0.54	15.23	0.57	0.41	11.45	46.82	7.62	4.68
450	0.54	15.18	0.57	0.41	11.37	46.62	7.52	4.66
500	0.54	14.04	0.61	0.41	11.18	45.88	7.44	4.59

4.2. Variation of Average Carrier Mobility with Thickness

The carrier mobility increases from 50nm to 100nm then keeps on decreasing for all thickness. Figure 2 shows the variation of carrier mobility with thickness. The maximum value for average carrier mobility is obtained for the BHJ OSC simulated at 100nm thickness of ZnPc: C-60,which is equal to $9.21 x 10^{-8}$ m^2 $V^{-1}s^{-1}$. The decrease in carrier mobility is due to the increased blend thickness [17].

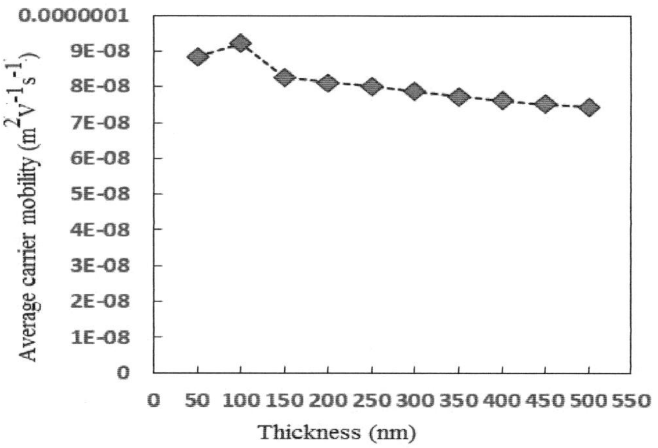

Figure 2. Average carrier mobility of ZnPc: C-60 in ITO/PEDOT: PSS/ZnPc: C60/Al bulk heterojunction solar cell device with varying thickness of active layer

4.3.Disparity of Open Circuit Voltage with Thickness

Varying thickness didnot affect the open circuit voltage much. Almost V_{oc} remains constant over the thickness only slight changes are noted before 100 nm thickness of the active layer. V_{oc} value shows notable variation up to 200 nm and thereafter remains constant. The slight decrease in V_{oc} is due to the decreased crystallinity which increases the defect density and hence decreased average carrier density at V_{oc} [10]. The figure 3 shows the active layer thickness dependence of open circuit voltage (V_{oc}) of simulated ITO/PEDOT: PSS/ZnPc: C60/Al BHJ OSC.

4.4.Disparity of Short Circuit Current Density (J_{sc}) with Thickness

Thickness variation makes a noteworthy change in short circuit current density (J_{sc}). As thickness increases from 50nm to 500nm, the short circuit current density also upsurges up to 300 nm and then as the thickness increases the short circuit current density gets decreased. The maximum short circuit current density obtained is at 300 nm of 16.21 mA/cm^2.The increase in J_{sc} from 50nm to 300nm is due to the increased charge carriers which are created by the absorption of light. The short circuit current variation (J_{sc}) with active layer thickness of ITO/PEDOT: PSS/ZnPc: C60/Al BHJ OSC is shown in the figure 4.

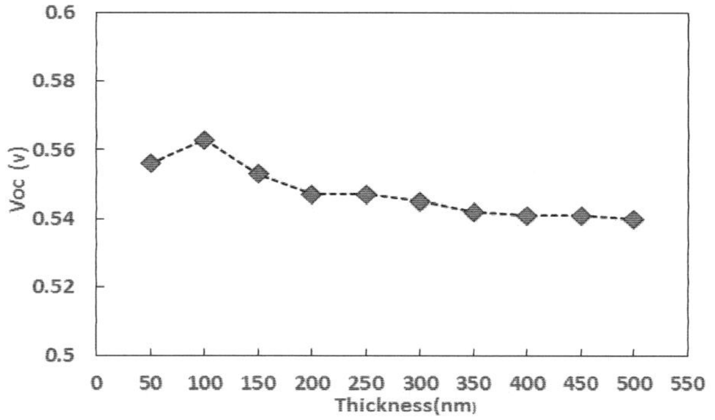

Figure 3. Disparity of open circuit voltage (V_{oc}) of ITO/PEDOT: PSS/ZnPc: C60/Al bulk heterojunction solar cell with varying thickness of active layer

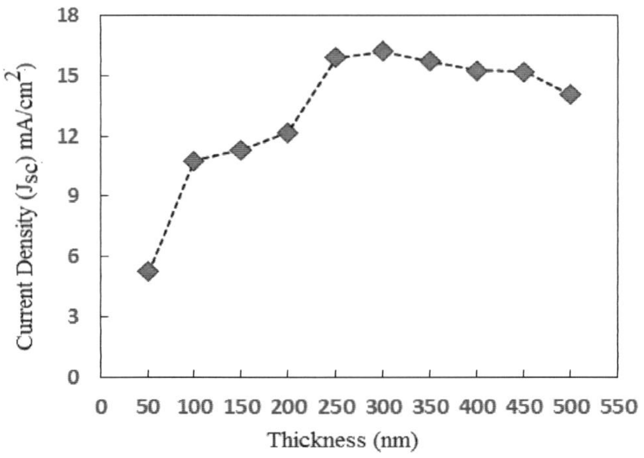

Figure 4. Disparity of Short circuit current density (J_{sc}) of ITO/PEDOT: PSS/ZnPc: C60/Al bulk heterojunction solar cell with varying thickness of active layer

4.5. Variation of Power Conversion Efficiency with Thickness

By simulating ITO/PEDOT:PSS/ZnPc:C60/Al BHJ organic solar cell for different thickness of ZnPc:C60, it is clearly understood that efficiency of the BHJ OSCs is dependent on thickness of the photo active layer (ZnPc:C60) .At 300nm efficiency is maximum (4.92%). Fig 5 shows that as the thickness of the active layer varies from 50 nm to 300nm the efficiency of BHJ OSCs increases and indicates maximum at 300 nm and again decreases as the thickness increases beyond 300nm. This may be due to the efficiency dependence on various parameters like open circuit voltage, short circuit current density, fill factor etc. The variation in efficiency with active layer thickness shows similar pattern to variation in efficiency of short circuit current density with thickness. Thus efficiency variation with thickness can be explained in terms of short circuit current density. It increases from 50nm to 300nm thus it increases the efficiency while after 300 nm the short circuit current density shows a decreasing pattern with increasing thickness which decreases the efficiency beyond 300 nm. The decrease in average mobility as shown in the figure 2 can also decreases the efficiency. The maximum short circuit current density (J_{sc}) is 16.21mA/cm^2 which corresponds to the maximum efficiency 4.92 % at 300nm. Thus the optimum thickness obtained is 300nm of active layer ZnPc: C60.

The increase in efficiency with variation of the thickness from 50 nm to 300nm is due to the larger absorption area of the photoactive layer which increases the incident photons and hence the generated excitons. After 300nm increase of thickness decreases the efficiency which is due to the decrease in short circuit current density, and the recombination of charge carriers. Also the V_{oc} value is slightly decreasing after 300nm thickness. V_{oc} shows the difference in energy between the HOMO of the donor and LUMO of the acceptor levels. Thus a decrease in V_{oc} value reduces the HOMO- LUMO energy levels which can reduce the exciton dissociation. This reduced exciton dissociation can reduce the carriers and hence decreases the efficiency. The decrease in V_{oc} can also increases the recombination. Thick absorbents can maximize the absorption, while the thickness should be less then the selected diffusion length for enhancing the charge collection. Therefore, the thickness greater than the optimum thickness can reduce the charge collection thus reduces the efficiency.

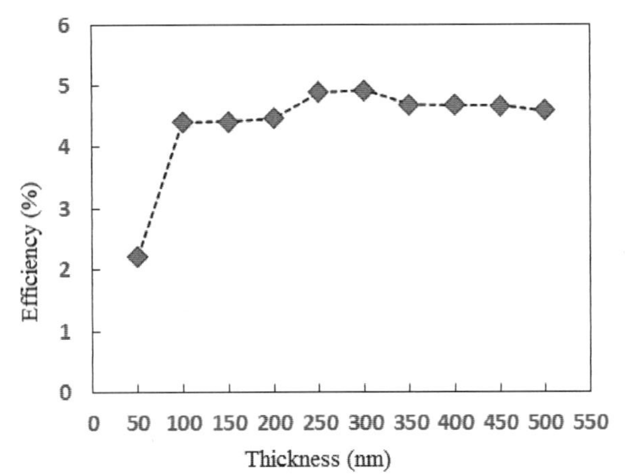

Figure 5. Variation of Power conversion efficiency (η) of ITO/PEDOT: PSS/ZnPc: C60/Al bulk heterojunction solar cell with varying thickness of active layer

4.6. Current Density -Voltage Curves (J-V Curves)

Current density voltage (J-V) plots of the devices for ZnPc: C-60 active layer is shown in figure 6. It is observed that the current density at zero voltage (J_{sc}) remains constant over a voltage range then increases and reaches at zero at a particular voltage (open circuit voltage). At 300nm, the current density at zero voltage is 16.21mAcm^{-2} and then it increases and reaches zero at 0.54V. At 50 nm the current density at zero voltage is 5.24 mAcm^{-2} and reaches zero at 0.56V. At 100nm the currentdensity at zero voltage is 10.77 mAcm^{-2} and reaches zero at 0.56V. At 150nm current density at zero voltage is 11.31 mAcm^{-2}and reaches zero at 0.55V. At 200nm current density at zero voltage is 12.17 mAcm^{-2} and reaches zero at 0.54V. At 250nm, current density at zero voltage is 15.9 mAcm^{-2} and reaches zero at 0.54 V. At 350nm current density at zero voltage is 15.71 mAcm^{-2} and reaches zero at 0.54V. At 400nm current density at zero voltage is 15.23 mAcm^{-2} and reaches zero at 0.54V. At 450nm current density at zero voltage is 15.18 mAcm^{-2} and reaches zero at 0.54 V. At 500nm current density at zero voltage is 14.04 mAcm^{-2} and reaches zero at 0.54 V. From the plot also J_{sc} is maximum for 300nm thickness of ZnPc: C-60 active layer of OSC.

4.7.Basic electrical parameters of the simulated BHJ Organic Solar Cell

The figure 7(a) shows the charge density voltage relation of ITO/PEDOT: PSS/ZnPc: C60/Al at 300nm. At 0V the charge density 1.71×10^{22} m^{-3} from 0.3V there is notable increase in charge density. Figure 7(b) shows the total charge density voltage relation. The total charge contributed by both electrons and holes. At 0V the total charge density is 2.22×10^{22} m^{-3}. The Figure 7(c) shows the current voltage relation of ITO/PEDOT: PSS/ZnPc: C60 /Al at 300 nm. At 0V the current is 0.97 mA. Beyond 0.4 V there occur a notable increase in the current and at 0.57V current becomes zero.

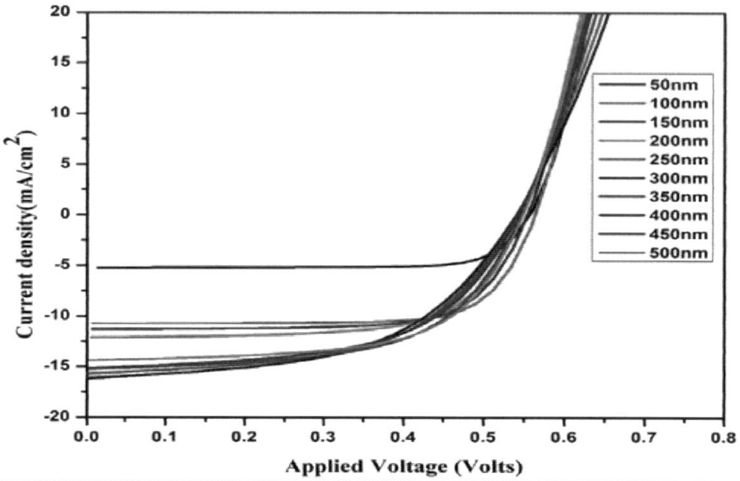

Figure 6. Current density-voltage graph of simulated ITO/PEDOT: PSS/ZnPc: C60/Al bulk heterojunction solar cells

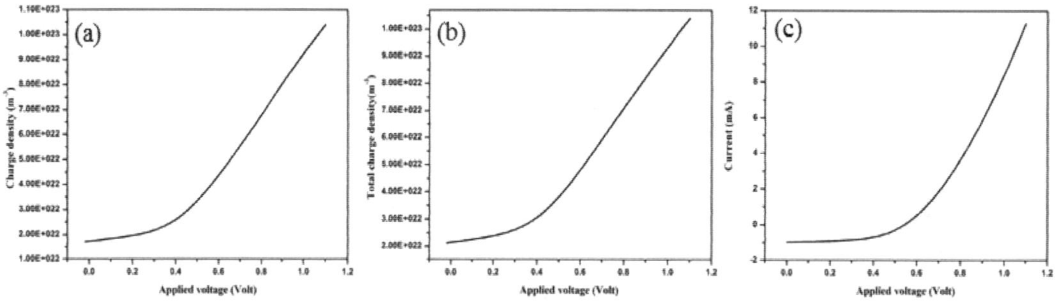

Figure 7. ITO/PEDOT: PSS/ZnPc: C-60/ Al bulk heterojunction OSC with 300nm thickness of ZnPc: C-60.(a)Charge density- Voltage plot. (b) Total charge density-Voltage plot. (c) Current-Voltage plot

5. Conclusion

ITO/PEDOT:PSS/ZnPc:C-60/Al bulk heterojunction solar cell is simulated for various active layer thickness (50 nm- 500nm with a difference of 50 nm).The power conversion efficiency along with the other electrical parameters such as open circuit voltage, short circuit current density (J_{sc}), fill factor, average carrier mobility are noted. The maximum obtained power conversion efficiency is 4.92% for 300nm optimum thickness of ZnPc: C-60 active layer of ITO/PEDOT: PSS/ZnPc: C-60/Al bulk heterojunction solar cell.The power conversion efficiency continuously increases from 2.22 % at 50 nm to 4.92% at 300nm and thereafter it decreases. The dependence of open circuit voltage(V_{oc}), Short circuit current density (J_{sc}), fill factor(FF), average carrier density is studied. It wasdetected that short circuit current density is growing with thickness up to 300nm of the active layer thickness in simulated bulk heterojunction (BHJ) solar cell and thereafter decreases. Showing a similar pattern as in the case of power conversion efficiency with thickness.The strong dependence of short circuit current density(J_{sc}) on power conversion efficiency is observed. The increase in short circuit current density with increase in thickness is due to higherlight absorption area. The decrease in short circuit current density after 300nm thickness is due to the increased recombination. A decreasing pattern of fill factor is observed previous to 400 nm thickness. A slight decrease in open circuit voltage (V_{oc}) is found below100nm.This decrease in V_{oc} is due to the reduced average carrier mobility. Average Carrier mobility get decreased with increasing thickness. Hopping mechanism of charge carriers reduces the carrier mobility at thick layers. This work is helpful for the experimental scientist to optimise the active layer thickness and thereby reducing their effort, time and materials.

Acknowledgement

The authors thank General Purpose Photovoltaic Device Model Software (GPVDM).

References

[1] Sreekala C O 2011 Studies on organic bulk heterojunction devices for photovoltaic solar cell applications.

[2] Dangate M, Munshi A, Sampath W S, Boltalina O V, Strauss S H, Saravanan C & Nair S 2016 Investigation of organic small molecules and polymer compounds for CdTe back contact. *In 2016 IEEE 43rd Photovoltaic Specialists Conference (PVSC)* 1438-1442.

[3] Tajik S et.al 2020 Recent developments in conducting polymers: applications for electrochemistry *RSC Advances.* **62** 37834-37856.

[4] Tang C W 1986 Two-layer organic photovoltaic cell *Appl. Phys. Lett.* **48** 183-185.

[5] Nath B G, Chinnusamy S, Anjali C M, Chandran S H, Dangate M S 2017 Density functional theory study of organic small molecules for renewable energy system *In Proc. 2017 IEEE Int. Conf. Technol. Adv. Power Energy* 1-6.

[6] Yu G, Gao J, Hummelen J C, Wudl F, Heeger A J 1995 Polymer Photovoltaic Cells: Enhanced Efficiencies via a Network of Internal Donor-Acceptor Heterojunctions. *Science* **270** 1789

[7] Tress W 2014 Organic Solar Cells: Theory, Experiment, and Device Simulation. Germany: *Springer* International Publishing.

[8] Sreekala C O, Saneesh P F, Sreelatha K S, Kishnashree A & Roy M S 2012 Organic bulk heterojunction solar cell based on rosebengal: ncTiO2 and parameter extraction by simulation. *In Adv Mat Res* . **403**, 4304 Trans Tech Publications Ltd.

[9] Ye Q, & Xu J W 2015 Bulk heterojunction solar cells—opportunities and challenges. *Solar cells—new approaches and reviews*. **359**.

[10] Gao Y, MacKenzie, R C, Liu Y, Xu B, Van L, Tian W 2015 Engineering ultra-long charge carrier lifetimes in organic electronic devices at room temperature *Adv. Mater. Interfaces* **2** 1400555.

[11] Mishra A K and Shukla R K 2020 Electrical and optical simulation of typical perovskite solar cell by GPVDM software *Materials Today: Proceedings.*

[12] Knipp D, Jovanov V, Tamang A, Wagner V and Salleo A 2017 Towards 3D organic solar cells. *Nano Energy* **31** 582.

[13] Lin Y, Zhao Y, Xin Q, Jiang C, and Song A, 2020 Electrical control of the optical dielectric properties of PEDOT: PSS thin films. *Optical Materials*, **108** 110435

[14] Lehmuskero Kuittinen M and Vahimaa, P 2007 Refractive index and extinction coefficient dependence of thin Al and Ir films on deposition technique and thickness. *Optics express*, **15** 10744

[15] Lenes M , Koster L J A , Mihailetchi V D and Blom P W.M., 2006. Thickness dependence of the efficiency of polymer: fullerene bulk heterojunction solar cells. *Appl. Phys. Lett.*, **88** 243502.

[16] Elumalai N K and Uddin A 2016 Open circuit voltage of organic solar cells: an in-depth review *Energy Environ. Sci*, **9** 391.

[17] Apaydın D H, Yıldız D E , Cirpan A and Toppare L 2013 Optimizing the organic solar cell efficiency: role of the active layer thickness *Sol. Energy Mater Sol. Cells* **113** 100.

MXene based Heterostructures for electrode materials of Batteries: A Review

M. A. H. Shanto[a], M. I. Chowdhury[a], A.B Antu[b], Naimur R. Niloy[c], N. Alam[a], M.A. Ullah[a], M.S. Anowar[a]

[a]Ahsanullah University of Science and Technology Dhaka, Bangladesh.
[b]Rajshashi University of Engineering and Technology,Rajshahi, Bangladesh.
[c]New Jersey Institute of Technology, New Jersey,USA.

[*]Corresponding author:abidhasan.aust@gmail.com

Abstract Growing concerns about fossil fuel's environmental impact, along with the recent breakthrough of electric vehicles, have turned research focus to energy storage solutions. Despite the fact that batteries were invented about 200 years ago, modern technologies are required to store energy in a larger grid with a high density. The electrode materials used in energy storage devices such as batteries and supercapacitors play a major role in their overall performance. A lot of materials have been explored but due to appealing electrical and electrochemical properties, MXene has received a lot of interest for energy storage devices. Because of their layered structure and high conductivity, MXenes are promising candidates for energy storage applications. Two-dimensional heterostructured materials are more advantageous than individual building blocks for batteries and supercapacitors. In this review work, we looked at different MXene based heterostructures and their electrochemical performance as electrode materials of batteries. A particular application of MXene in Lithium-ion batteries has been studied. Synthesis and characteristics of MXenes are briefly discussed here.Finally, future prospects and challenges are highlighted.

Keywords: MXene; Lithium-ion Batteries; Supercapacitors; Heterostructures.

1. Introduction

As the world's population grows, energy is becoming one of the most major challenges. Scientists are deeply concerned about the production, use, and storage of fuel because it is restricted beneath the earth's surface. If we can store renewable energy on a larger scale, it may be a viable option [1]. Batteries, capacitors, and supercapacitors are being used extensively as electrical energy storage devices [2-4]. Lithium-ion batteries and sodium-ion batteries are becoming very popular day by day because of their extensive uses in electric vehicles, laptops, digital cameras [5-8].Electrode materials are very important for these devices because outcome depends on materials of electrodes. Ru, Pt, Ir are common materials for electrodes that have been used for energy storage. But these materials have limitations such as availability;cost [9–15]. Graphene has been employed as anode material, although its specific capacity is restricted to 372 mAh/g [16]. MXenes, a new group of 2D materials consisting of transition metal nitrides, carbides, and carbonitrides, have received huge attention due to their outstanding properties. It is unique in the way that it has excellent electrical conductivity, easily accessible structure, fast diffusion, good thermal stability, large interlayer spacing, large surface area, and thickness controllability. Restacking in MXene sheets is one of the major challenges to design electrodes for energy storage devices but that can be avoided by introducing different spacers between two MXene layers [19]. In terms of energy storage technology, hybrid structures have a lot of potential. Researchers have recently introduced MXene/Graphene [73], Phosphorene/MXene [77], Bi_2MoO_6/MXene[76], MoS_2-on-MXene[72]. These heterostructures provide novel properties, enhance electrochemical performance, and also provide opportunities to design and manipulate nanoelectronic devices [20-22].Researchers discovered $Mo_2TiC_2T_x$ MXene in 2015 in which atomic layer of Titanium is inserted between two layers of Molybdenum. This arrangement produces features that differ from those previously reported for $Ti_3C_2T_x$ MXene [23]. They produced a MoS_2 /$Mo_2TiC_2T_x$ heterostructure that exhibits metallic characteristics due to the presence of MXene and exhibits high Coulombic efficiencies, high specific capacities, and great cycling stability when utilized as the electrode material of LIBs [23].

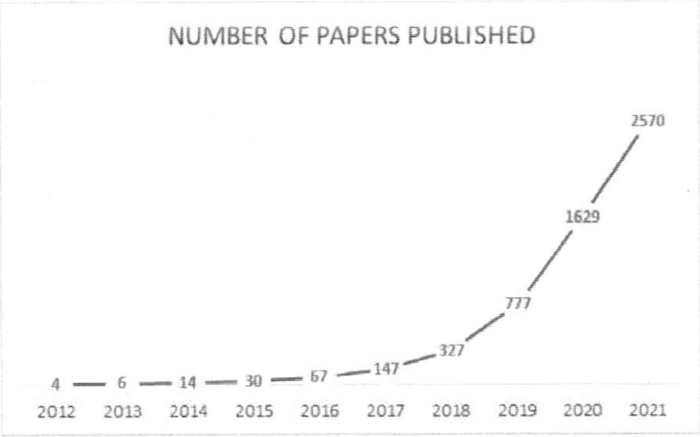

Figure 1. Number of papers on Mxene publishedper year (Data has been taken from sciencedirect.com)

MXene hybrid structures are vastly under research, and it will take a lot of effort to meet the energy storage needs. So, for the development and advancement of electrode materials for batteries, supercapacitors, and other energy storage devices, a highly focused evaluation is necessary. In this review, we looked at the most recent MXene synthesis technique, as well as the properties of MXene and their possible applications. It has been discussed how MXene can be used in lithium-ion batteries. The electrochemical performance of many heterostructures is investigated in depth in this work.

2. Synthesis & Preparation of MXenes

2.1 Synthesis

MXenes are synthesized from a specific etching of 'A' from $M_{n+1}AX_n$ (MAX) Phase. Here 'M' can be any initial transition metals, for example, Sc (21), Ti (22), V (23), Cr (24) etc. 'A' is a metal member of group 12-16 like Cd (48), Al (13), Si (14), P (15) etc. & 'X' is carbon (C-6), nitrogen (N-7) [24-28] .'A' is removed from MAX antecedent utilizing HF or NH_4HF_2provided in equation (1) & (4). The tough metallic bond between M and A is changed by weaker H-bond like fluorine (F), Hydroxyl (OH), Oxygen (O). Metallic hydrophilicity is present, and it reacts with various groups such as O, OH, and F, as shown in the equations. (2), (3) & (5) [17,26,27,29].

$$M_{n+1}AX_n + 3HF \rightarrow AF_3 + 3/2H_2 + M_{n+1}X_n \quad (1)$$
$$M_{n+1}X_n + 2H_2O \rightarrow M_{n+1}X_n(OH)_2 + H_2 \quad (2)$$
$$M_{n+1}X_n + 2HF \rightarrow M_{n+1}X_nF_2 + H_2 \quad (3)$$
$$M_{n+1}AX_n + 3\,NH_4HF_2 \rightarrow M_{n+1}X_n + (NH_4)AlF_6 + 3/2H_2 \quad (4)$$
$$M_{n+1}X_n + aNH_4HF_2 + bH_2O \rightarrow (NH_3)c\,(NH_4)dM_{n+1}X_n(OH)xFy \quad (5)$$

2.2 Preparation of Ti_3C_2

At room temperature aqueous HF is used to separate weakly bonded Al layers from Ti_3AlC_2 phase in order to preparing Ti_3C_2. Following are the reactions,

$$3HF + Ti_3AlC_2 \rightarrow Ti_3C_2 + AlF_3 + 3/2H_2 \quad (6)$$
$$Ti_3C_2 + 2\,H_2O \rightarrow Ti_3C_2\,(OH)_2 + H_2 \quad (7)$$
$$2HF + Ti_3C_2 \rightarrow Ti_3C_2F_2 + H_2 \quad (8)$$

Equations (7) and (8) show OH and F termination, respectively. To mitigate solids, centrifugation is utilized, followed by washing with deionized water. These MXenes are made up of several layers. Sonication is used to obtain a single structure [17, 46].

As an etchant, Halim [31] employed NH_4HF_2instead of the toxic HF. Because the nature of NH_4HF_2 is not as strong as that of HF, and because of the cation's simultaneous intercalation, NH_4HF_2 is more suited and safe for making MXenes. The reactions are as follows:

$$3NH_4HF_2 + Ti_3AlC_2 \rightarrow Ti_3C_2 + (NH_4)_3AlF_6 + 3/2H_2 \quad (9)$$
$$aNH_4HF_2 + Ti_3C_2 + bH_2O \rightarrow (NH_3)c\,(NH_4)dTi_3C_2(OH)xFy \quad (10)$$

3. Properties

3.1 Structural Properties

After extracting A from MAX phase MXenes are formed with various surface groups like F, O, and OH. MXenes with O/OH surface group are most stable. The surface group F can be switched with OH by rinsing or reserving in water [32]. High-temperature treatment or metal absorption process can be used to transform surface group OH into O. Bare MXenes can be found by contacting O terminated MXene with metals like Mg, Ca, Al [33].MXene's Structure modeling is very important to understand the structure. Previously it was assumed that the surface groups O, F, OH were located above the void among nearby three C atoms [34]. But later developments indicate position and arrangement of O, F, and OH groups are more complex and are not uniformly positioned. More studies are required to understand the coexistence and absorption of surface groups within MXenes including the factors interlayer interaction, Van der Waals force etc.

3.2 Stability

Bigger the negative worth of lattice energy, the stability of crystal is higher. By calculating structure of energy band using First principle, MXene's lattice energy is found negative so MXene can exist stably [18]. It is found that stability of $Ti_{n+1}C_n$ (MXene with Carbon) is more than that $Ti_{n+1}N_n$ of (MXene with nitrogen). It is also found that the stability and resistance of electron beam emission of Ti_3C_2film layer is more than that of graphene [29].

3.3 Metallic, Electronic and Electrical

Bare MXenes variants $Ti_{n+1}X_n$have metallic characteristic. Increasing of value of n decreases the metallic behavior because of additional M-X bond. $Ti_{n+1}N_n$ has more metallic characteristics than $Ti_{n+1}C_n$ as nitrogen atom has additional electron than carbon [37-38]. The density of $Ti_{n+1}C_n$ and $Ti_{n+1}N_n$ near Fermi energy states is 2.5-4.5 times than their antecedent MAX phase, according to the first principle. 3D state of Ti in the broken bond of Ti-Al transforms into the metal bond state of Ti-Ti, causing MXene to have metallic characteristics, which is compatible with the obtained results of [29]. MXene exhibits semiconductor properties depending on the type and order of surface groups (O, OH, Fetc.) [39].Using the first principle, M. khazaei [40]discovered M_2C (M = Ti, Zr, SC, V, Cr, Nb, HF, Ta) and M_2N (M = Ti,

Zr, Cr, Ta, Nb, HF) .They demonstrated that Sc_2CF_2, Hf_2CO_2 ,Sc_2CO_2, $Zr_2 CO_2$,Ti_2CO_2, $Sc_2C(OH)_2$ are semiconductor with a band gaps of 0.25 to 2.0 eV[40].Electric conductivities of MXenes pressed discs are alike to multi-layered graphene and greater than graphene oxide materials and carbon nanotubes [43-44] . Resistance increases with the number of layers and the different types of functional surface groups (OH, O, F) [29,46-47]. Because of variations in (i) delamination yield, (ii) d-spacing between MXenes flakes (iii) surface functional groups, (iv)lateral diameters (v) defect concentration generated by each etching technique, electrical conductivities of $Ti_3C_2T_x$ranged from 850 to 9880 Scm-1[57-60].

3.4 Magnetic

Various magnetic properties are found in several kinds of MXenes. For example,Ti_3CNT_x and$Ti_4C_3T_x$ are non-magnetic with the surface groups O, OH, F etc. [39,50,52], while at room temperature Cr_2CT_x and Cr_2NT_x are ferromagnetic with OH and F surface groups [18,40,54] and Mn_2NT_x is ferromagnetic independent of the surface groups [17]. These are calculated predictions and more experimental observations are needed.

3.5 Optical

For optoelectronic, photovoltaic, photocatalytic, and transparent conductive electrode technologies, UV and visible light absorption are important optical activities. Experiments on the $Ti_3C_2T_x$ (T: surface group F, OH, or O etc.) thin film demonstrate that it transmits 77 percent of visible light with a wavelength of 550nm [55]. The transmittance of a thin film of the $Ti_3C_2T_x$ (MAX) phase is 30%, which is not as transparent as MXene [31]. Intercalated $Ti_3C_2T_x$ with NH_4HF_2 has a transmittance of 90%. The absorbance of $Ti_3C_2T_x$ (T: surface group F, OH, or O, etc.) and intercalated films are not dependent on their thickness [31]. The optical characteristics of pure and functionalized $Ti_3C$and Ti_3C_2 with O, OH, F are presented by Wang[56]. Between infrared and ultraviolet, as well as visible light, MXenes containing F & OH functional groups have lower in-plane absorption coefficients than bare MXenes. OH and F surfaces clustered Ti_3C and Ti_3C_2should emit white color, according to refractivity analysis [56].

4. Application of MXenes in Lithium-ion Batteries

Lithium-ion batteries were first commercialized in 1990, and research has progressed at a rapid rate since then. Improved energy density and lightweight make it advantageous to conventional batteries in mobile phones, computers, and digital cameras. One of the most recent advancements (electric vehicles) needs a large amount of power at a high current[61]. Rapid charge/discharge rate degrades the electrochemical performance of lithium-ion batteries. As a result, new materials and technologies are necessary to enhance the performance of LIBs[62]. Graphite has been emerged as a promising alternative as an anode material for LIBs due to its good electronic conductivity, cost-effectiveness, and stability. However, when graphite is employed, a difficulty arises since it limits the specific capacity to 372 mAh/g[63].It is demonstrated from the theoretical calculations that MXenes, which have a wide range of structural and chemical properties, are one of the most attractive candidates for various energy storage applications[64].

Zhou reported 2D multilayered vanadium carbide (V_4C_3) MXene as an electrode material of LIBs. They synthesized this specific MXene by etching Al from V_4AlC_3 in HF solution (Figure-2). To get larger interlayer spacing and the specific surface they used Ball-milling treatment on V_4AlC [65].As anode material, it shows outstanding performance with a specific capacity of 225mAh/g after 300 cycles [65].Rao studied the interaction of lithium sulfide on Ti-based MXene and found that Ti-S bond dominates the interaction between MXene and Lithium sulfide which prevents the loss of sulfur from cathode of lithium-sulfur batteries [66].Recent advancements of MXene based heterostructures for Lithium-ion batteries will be reviewed in the latter part of this paper.

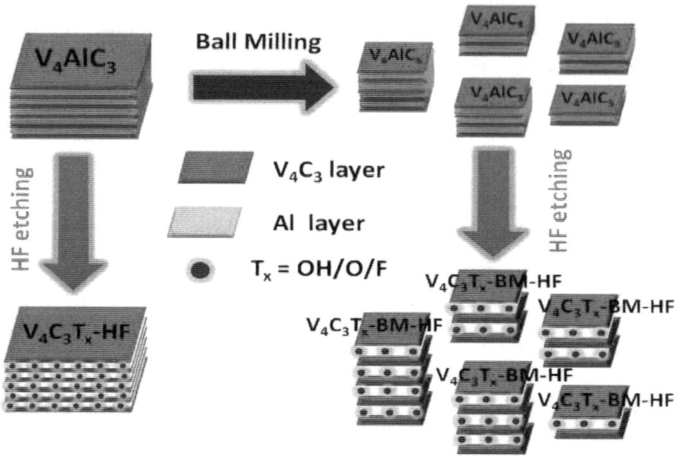

Figure 2. Preparation of $V_4C_3T_x$ MXene[65].@Copyright 2019 ELSEVIER.

5. MXene based Heterostructures for batteries

MXene-based heterostructures have paved the way for the development of high-capacity rechargeable batteries. Researchers have been paying close attention to it since it outperforms individual building blocks. They're working on new hybrid architectures that use a variety of catalysts to overcome the limitations of battery and other energy storage device design. [72-76].

Xu[67]fabricated an ultrafast kinetics electrode for sodium-ion batteries by synthesizing $MoSe_2$/MXene hybrid architecture. DFT calculation shows that diffusion barrier between Na+/MXene effectively reduced which enables high ion and electrical conductivity. This heterostructure shows high reversible capacities at high current (490 mAh/g at 1Ag−1,250mAh/g at 10 Ag−1), 99.8% coulombic efficiency indicates outstanding electrochemical performance[67]. Researchers design and fabricated a Co_3S_4@MXene hybrid architecture as an effective sulfur host for Mg-S batteries. DFT calculation and XPS result showed that polysulfide could be absorbed and it provides fast Mg ion transportation. A specific capacity of 1220 $mAhg^{-1}$, and after 100 cycles 528 $mAhg^{-1}$ capacity has been obtained in their work[68].

A representation of the 2D transition-metal dichalcogenide MoS_2 appears to be a strong contender for lithium-ion storage[69]. However, there are significant challenges, such as MoS_2's weak cycle stability and rate capabilities[70]. MXenes give an enhancement in electrochemical performance of MoS_2 based electrodes. MXene's layered structure and excellent conductivity make them ideal for energy storage, especially lithium-ion storage[71]. MoS_2 on Mo_2TiC_2Tx MXene heterostructures with a few layers optimize the potential of MXene as a support for MoS_2[72].

Obtained a-LPs from MoS_2 and $Mo_2TiC_2O_2/MoS_2$ heterostructures are 3.16 and 3.06 c, accordingly. In the heterostructure, MoS2 is such a semiconductor that has a bandgap of 1.7 eV and metallic characteristics. Higher conductivity is required for high-rate charging and discharging. It is concluded that the combination of MoS_2 and $Mo_2TiC_2O_2$ considerably improves the conductivity of MoS_2[72]. Initial charge, discharge capacities of pure Mo_2TiC_2Tx are 134 and 268 $mAhg^{-1}$. From Figure 3, we see MoS_2 /Mo_2TiC_2Tx - 500 gives a 4times improvement in charge, discharge capacities (554 and 646 $mAhg^{-1}$at 0.1A $^{-1}$). Charge capacity of the Mo_2TiC_2Tx/MoS_2 heterostructure is likewise significantly higher than that of multilayer $Mo_2TiC_2T_x$ previously reported [72].

MXene/graphene heterostructure functions exceptionally well as Li-ion battery's electrodes. The presence of graphene prevents MXene layers from restacking effects and improves electric conductivity, Li

adsorption strength, and mechanical stiffness [73]. Charge changes from graphene to MXene have been discovered in large quantities. Weakly coupled Ti_2CX_2/MoS_2heterostructures have already shown similar high charge transfer [74]. Single layers of Ti_2CO_2 are semiconducting, whereas, Ti_2CF_2 and $Ti_2C(OH)_2$ are known to be metallic[75]. The electrical conductivity of graphene states is higher than that of isolated MXenes, resulting in metallicity, as seen in Figures 2a-c of [73].MXene/Graphene heterostructures outperform Ti_2CX_2 electrodes in terms of electrical conductivity and electrolyte infiltration, resulting in improved electrochemical performance [73].
MXene nanocomposite in Plate-to-Layer Bi_2MoO_6/MXene-Heterostructure performs as a very conductive substrate for loading and anchoring Bi_2MoO_6nanoplates, resulting in a great improvement in electronic conductivity, as well as structural stability, is obtained [76]. Detailed electrochemical performance of (i)Bi_2MoO_6/MXene-50%(ii) Bi_2MoO_6/MXene-30% (iii)Bi_2MoO_6/MXene-10% and pristine Bi_2MoO_6 has been analyzed in the figure 5 of [76] to find out the best possible Bi_2MoO_6/MXeneheterostructure[76].

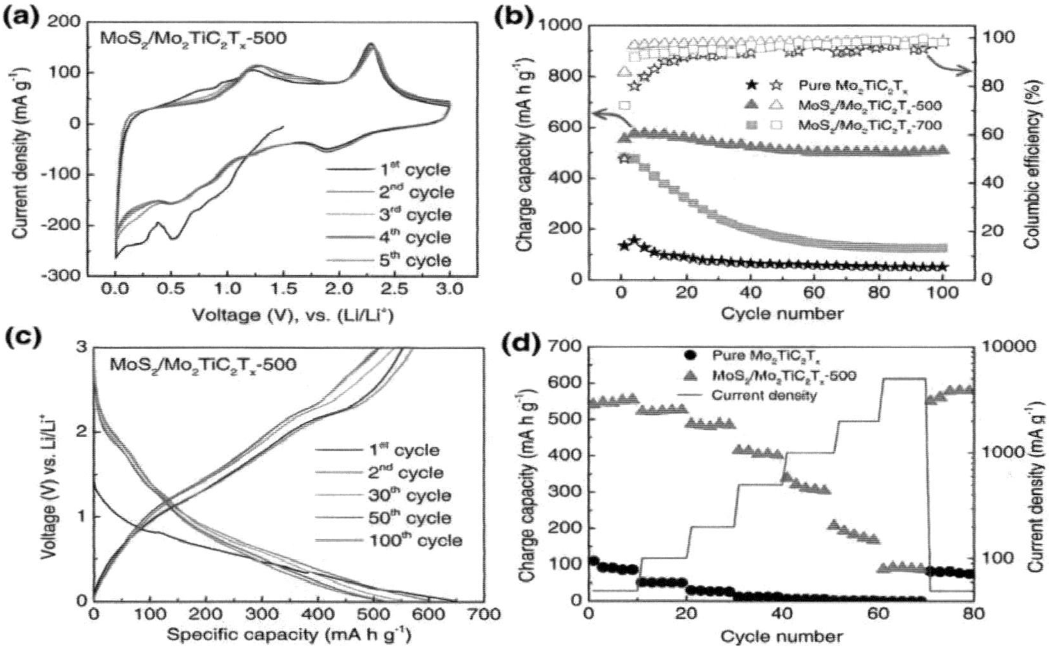

Figure 3. In LIBs, the electrochemical performance of MoS_2/Mo_2TiC_2Tx heterostructures[72].

The surface area of MXene increases as the MXene content increases, resulting in a more irreversible interaction between functional groups on the MXene surface and Li+ [76]. From figure 5 of [76] we see that MXene effectively enhances a composite electrode's rate performance. Figure 6 of [76] demonstrates Li storage behavior of Bi_2MoO_6/MXene-30%. During the lithiation/delithiation, Bi_2MoO_6/MXene-30% exhibits higher diffusion coefficients and lower over potential compared to Bi_2MoO_6, which implies better reaction kinetics. Because of their high conductive nature, MXenenanosheets considerably increase charge transmission[76].When MXene was used as a Bi_2MoO_6/MXene substrate during fabrication, the lithium storage capacity of Bi_2MoO_6 was considerably boosted. MXene improves charge transport and reduces Bi_2MoO_6 volume change, resulting in higher rate capability, long-term cycling stability, and high specific capacity.

Phosphorene/Ti_3C_2Tx MXene reduces structural expansion and allows electrons and sodium ions to migrate more easily, consequentlyimproving in the hybrid structured anode's cycling performance

[77].The increased radius of Na+ makes it difficult to construct traditional electrodes due to structural expansion[78-79].Based on the alloying reaction, phosphorus (P) has the largest theoretical capacity (2600 $mAhg^{-1}$). Among all the prospective candidates for Na+ storage, phosphorus has the largest capacity (2600 $mAhg^{-1}$). [80-82]. Scientists have looked into red phosphorus as a potential anode material for SIBs intensively. However, when used in batteries, itreduces ionic diffusion rate,and weak intrinsic conductivity hinder rapid charge/discharge performance and cycle life [83-84] .Figure 3 of [77] shows the initial 3 cyclic voltammetry of the phosphorene/MXeneheterostructure.The initial cathodic scan revealed a strong peak at 0V, indicating sodium intercalation, alloying processes, and SEI development. The anodic scan displays a peak around 0.55,0.7, and 0.92V, indicating a stepwise desodiation process of phosphorene[77]. Second cycle reveals a very reversible chemical process, as evidenced by the overlapping CV profile. The phosphorene/MXene hybrid anode's subsequent discharge and charge curves are nearly unchanged, which is compatible with the CV results.At a current density of 0.1Ag-1, cycling performances of (i) phosphorene/MXene, (ii)mixed BP/MXene, (iii)Ti_3C_2Tx MXene, and (iv) BP electrodes are presented in figure 3(c) of [77]. With a capacity of 467 mAh g^{-1}, the phosphorene/MXene heterostructure demonstrates good cycling stability after 100 cycles[77]. The volume expansion of the phosphorene/MXene heterostructured anode after 5 cycles is around 120 percent, demonstrating that the existence of MXene in the composite is well suited to supporting the phosphorene volume change (Figure 10)[77].Phosphorus electrode has faster diffusion of Sodium ion validated by low polarization change in galvanostatic titration(Figure-18)[85]. It is concluded that MXene's high electrical conductivity, together with higher Sodium diffusion kinetics and Sodium affinities in fluorine-rich phosphorene/MXene hybrids delivers high rate capability [77]

Jiao fabricated $TiO_2/Ti_3C_2T_x$heterostructure for Lithium-Sulfur Batteries. Tio2 is uniformly distributed on the MXene sheet and it works as capturing centers to polysulfides. Multifunctional catalyst Tio_2 improved the performance of Lithium-Sulfur batteries and it delivers 800 mAhg−1 at 2C. Very low capacity decay (0.028%) is found even after 1000 cycles at 2C and 93% of capacity retention is maintained after 200 cycles [86].The Co_3S_4@MXene electrode has a high initial capacity, but after 100 cycles, it loses half of its capacity. $MoS_2/Mo_2TiC_2T_x$ –500 has a smaller capacity than Co_3S_4@MXene but has a more stable performance. When current density rises, capacity decreases. To achieve large capacity and stable cycle performance, more analysis is needed.

Table 1. Specific capacity for different heterostructures in various batteries

Structure Name	Battery Name	Electrode type	Capacity (mAhg−1)	
Bi_2MoO_6/MXene-50%	LIBs	Anode	~500 at 0.1 Ag^{-1}	[76]
Bi_2MoO_6/MXene-30%	LIBs	Anode	~600 at 0.1 Ag^{-1}	[76]
Bi_2MoO_6/MXene-10%	LIBs	Anode	~700 at 0.1 Ag^{-1}	[76]
MXene/graphene	LIBs	Anode	426(O/G), 392(OH/G)	[73]
Phosphorene/MXene	SIBs	Anode	535 at 0.1 Ag^{-1}	[77]
TiO_2/Ti_3C_2Tx	LSBs	Anode	800 at 2C	[86]
Graphene/$ReSe_2/Ti_3C_2$ MXene	KIBs	Anode	395.3 at 0.1 Ag^{-1}	[35]
$MoSe_2$/MXene	SIBs	-	490 at 1 A g^{-1}	[67]
Co_3S_4@MXene	MG-SBs	Cathode	1220 at2C	[68]
Covalent triazine framework/Ti_3C_2MXenenanosheets	LSBs	Cathode	1200 at 0.5C	[36]
V_2O_5/MXene	LIBs	Cathode	~475 (discharge) at 0.5c	[45]
SnS_2/Sn_3S_4 on Ti_3C_2	LIBs	Anode	~1400(discharge)~800(charge) at 0.1 Ag^{-1}	[51]

6. Challenges and future prospects

It's necessary to make thin Mxene in order to enhance surface area by reducing thickness. The functional group of Mxene has a significant impact on the electrochemical performance of energy storage devices, although the relationship between functional group and Mxene features is still unknown. Researchers fabricated and analyzed electrochemical performance in great detail, but structural features of these heterostructures need to be studied further. Materials with a high dielectric strength are required to operate at high voltages. The dielectric strength of the above heterostructures is unknown. The resistance of a conductor changes as the temperature changes. So, the temperature Coefficient of Resistance needs to be calculated. The capacity of batteries to charge and discharge is determined by their electrode density. More physical and electrical features of electrode materials must be explored in order to arrive at a solution in battery design. Mxene will be a great material for future energy storage devices if the issues mentioned above can be resolved.

7. Conclusion

Energy storage is becoming a major concern due to recent advancement of electric vehicles. The high demand for electrochemical energy storage has led investigators to focus on the development of advanced electrode materials. MXene a new group of 2D materials is investigated in 2011. It has become an alternate material for electrodes to previously used materials due to its superior electrochemical characteristics. In this review paper, we discussed latest technology of MXene synthesis process and recent advancement of MXene heterostructures as electrode material of energy storage devices like lithium/sodium-ion batteries. MXene hybrid structures demonstrate superior performance compare to the individual building block that has been thoroughly discussed in this review work. We briefly discussed why MXene is promising in energy storage applications and many others as well due to its excellent properties. A specific application of MXene in lithium-ion batteries as electrode material has been reviewed. Because of several advantages like high electronic conductivity, improvement in ion transfer, good stability, eliminating one of the major problems in battery designing (restacking) MXene heterostructures opened a new door for research in the field of energy conversion and storage system. Hence it is reviewed here the advancement of MXene synthesis, specific application in Lithium-ion batteries, and heterostructures for electrode material in rechargeable batteries would be useful for further investigation of MXene in energy storage applications.

References

[1] Chu S and Majumdar A 2012 *nature* **488** 294-303
[2] Dunn B, Kamath H, and TarasconJ M 2011 *Science* **334** 928-935
[3] Westerlund S and Ekstam L 1994 *IEEE Trans Dielectr Electr Insul.* **1**826-839
[4]Pandolfo A andHollenkamp A F 2006 *J. Power Sources* **157** 11-27
[5] Niu S, Wang Z, Yu M, XiuL,Wang S, Wu X and Qiu J 2018 *ACS Nano* **12** 3928–3937
[6 Zhong Y, Li B, Li S, Xu S, Pan Z, Huang Q, Xing L, Wang C and Li W Nanomicro Lett **10** 1-14
[7] Sun N, Guan Z, Liu Y, Cao Y and Zhu Q 2019 *Adv. Energy Mater.* **9**
[8] Kundu D, Talaie E,Duffort Vand Nazar L F 2015 *Angew. Chem. Int. Ed.* **54** 3431-3448
[9] Danilovic N, Subbaraman R, Chang K C, Chang S H, Kang Y and Snyder J 2014 *Angew. Chem. Int. Ed.* **126** 14240-14245.
[10] Oh A, Sa Y J, Hwang H, Baik H, Kim J, Kim B, S. Joo S H and Lee K 2016 *Nanoscale* **8**16379-16386
[11] Yoon J, Park J, Sa Y J, Yang Y, Baik H, Joo S H and Lee K 2016 *CrystEngComm* **18**6002-6007
[12] YangL,Vukmirovic M B, Su D, Sasaki K, Herron J A, Mavrikakis M, Liao S and Adzic R R 2013 *J. Phys. Chem. A* **117** 1748-1753.

[13] Jin H, Lee K W, Khi N T, An H, Park J, Baik H, Kim J, Yang J H and Lee K 2015 *Small* **11**4462-4468
[14] Debe M K 2012 *Nature* **486** 43-51
[15]Gasteiger H A, Kocha S S, Sompalli B andWagner F T 2005 *Appl. Catal. B* **56** 9-35.
[16] Armand M and Tarascon J M 2008 *Nature* **451** 652-657
[17] Naguib M, Kurtoglu M, Presser V, Lu J,Niu J, Heon M, Hultman L and GogotsiY 2011 *Adv. Mater.* **23** 4248-4253
[18]Shein I R and Ivanovskii A L2013 *Micro Nano Lett* **8** 59-62
[19]Simon P 2017ACS *Nano* **11** 2393-2396
[20] Geim A K and Grigorieva I V 2013 *Nature* **499** 419-425
[21] Novoselov K S, Mishchenko A, Carvalho A and Neto A H C 2016 *Science* **353** 6298.
[22] Liu Y, Weiss N O, Duan X, Cheng H C, Huang Y andDuan X 2016 *Nat. Rev. Mater.* **1**1-17
[23] Anasori B, Xie Y, Beidaghi M, Lu J, Hosler B C, HultmanL,Kent P R C, Gogotsi Yand Barsoum M W 2015 *ACS Nano* **9** 9507-9516.
[24] She J W, Fredrickson K D, BAnasori B, Kibsgaard J, Strickler A L, Lukatskaya M R, Gogotsi Y,. Jaramillo T Fand Vojvodic A 2016 *ACS Energy Lett.* **1** 589-594.
[25] Ma T Y, Cao J L, Jaroniec M and Qiao S Z 2016 *Angewandte Chemie* **128** 1150-1154.
[26]Ghidiu M, Lukatskaya M R, Zhao M Q, Gogotsi Y and Barsoum M W 2014 *Nature* **516**78-81
[27]Khazaei M, Arai M, Sasaki T, Estili M and Sakka Y 2014 *Phys. Chem. Chem. Phys.* **16** 7841-7849.
[28]Xiao Y, Hwang J Y and Sun Y K 2016 *J. Mater. Chem.* **4** 10379-10393.
[29] Naguib M, Mashtalir O, Carle J, Presser V, Lu J, Hultman L, Gogotsi Y and Barsoum M W 2012 *ACS Nano* **6** 1322-1331.
[30]Aslam M K and Xu M 2020 *Nanoscale* **12** 15993-16007.
[31] Halim J, Lukatskaya M R, Cook K M, Lu J, Smith C R, Näslund L A and May S J 2014 *Chem. Mater.* **26** 2374-2381.
[32] Xie Y, Naguib M, Mochalin V N, Barsoum M W, Gogotsi Y, Yu X, Nam K W, Yang X Q , Kolesnikov A I and Kent P R C 2014 *J. Am. Chem. Soc.* **136** 6385-6394.
[33] Xie Y, Agnese Y D, Naguib M, Gogotsi Y, Barsoum M W, Zhuang H L and Kent P R C 2014 *ACS Nano* **8** 9606-9615.
[34] Hu T,Wang J, Zhang H , Li Z, Hu M and Wang X 2015 *Phys. Chem. Chem. Phys.* **17** 9997-10003.
[35] Xia Z, Chen X, Ci H, Fan Z, Yi Y, Yin W, Wei N, Cai J, Zhang Y and Sun J 2021 *J. Energy Chem.* **53** 155-162.
[36] Meng R, Deng Q, Peng C, Chen B, Liao K, Li Land Yang Z 2020 *Nano Today* **35**100991.
[37] Kurtoglu M, Naguib M, Gogotsi Y and Barsou M W 2012 *MRS Communications* **2**133-137
[38] Lashgari H, Abolhassani M R, Boochani A, Elahi S M and Khodadadi J 2014 *Solid State Commun.* **195** 61-69.
[39] Tang Q, Zhou Zand Shen P 2012 *Journal of the American Chemical Society* **134** 16909-16916.
[40] Khazaei M, Arai M, Sasaki T, Chung C Y, Venkataramanan N S, Estili M, Sakka Y and Kawazoe Y 2013 *Adv. Funct. Mater.* **23** 2185-2192.
[41] Guo L, Zhang Z, Li M, Kang R, Chen Y, Song G, Han S T, Lin C T, Jiang N and Yu J 2020 *Compos. Commun* **19** 134-141.
[42] Wu Y, Nie P, Jiang J, Ding B, Dou H and Zhang X 2017 *Chem Electro Chem* **4**1560-1565.
[43] Naguib M, Mochalin V N, Barsoum M W and Gogotsi Y 2014 *Adv Mater* **26** 992-1005.
[44] Li R, Zhang L, Shi L and Wang P 2017 ACS Nano**11** 3752-3759.

[45] Wang Y, Lubbers T, Xia R, Zhang Y Z, Mehrali M, Huijben M and Johan E 2021 *J. Electrochem. Soc.* **168** 020507.
[46] Mashtalir O, Naguib M, Mochalin V N, Agnese Y D, Heon M, Barsoum M W and Gogotsi Y 2013 *Nat. Commun.* **4** 1-7.
[47]Wang K ,Zhou Y, Xu W, Huang D, Wang Z and Hong M 2016 *Ceram. Int.* **42** 8419-8424.
[48] Wang H, Wu Y, Zhang J, Li G, Huang H, Zhang X and Jiang Q 2015 *Mater. Lett.* **160** 537-540.
[49] Zhang C, Anasori B, Seral-Ascaso A S, Park S H, McEvoy N, Shmeliov A, Duesberg G S, Coleman J N, Gogotsi Y and Nicolosi V 2017 *Adv Mater* **29**1702678.
[50] Urbankowski P, Anasori B, Makaryan T, Er D, Kota S, Walsh P L, Zhao M, Shenoy V B, Barsoum M W and Gogotsi Y 2016 *Nanoscale* **8** 11385-11391 .
[51] Li J, Han L, Li Y, Li J, Zhu G, Zhang X, LuTandPan L 2020 *Chem. Eng. J.* **380** 122590.
[52] Wu F, Luo K, Huang C, Wu W, Meng P, LiuYandKan E 2015 Solid State Commun. **222**9-13.
[53] Kumar H, Frey N C, Dong L, Anasori B, GogotsiYandShenoy V B 2017 *ACS Nano* **11**7648-7655.
[54] Ivanovskii A Land Enyashin A N 2013 Russ. Chem. Rev. **82** 735.
[55] Mariano M, Mashtalir O, Antonio F Q, Ryu W H, Deng B, Xia F, GogotsiY and Taylor A D 2016 *Nanoscale* **8**16371-16378.
[56] Bai Y, Zhou K, Srikanth N, Pang J H, He X and Wang R 2016 *RSC Adv.* **6** 35731-35739.
[57] Garg R, Agarwal A and Agarwal M 2020 *Mater. Res. Express* **7** 022001.
[58] Ronchi R M, Arantes J T andSantos S F 2019 *Ceram. Int.* **45** 18167-18188.
[59]TangX, Guo X, Wu Wand Wang G 2018 *Adv. Energy Mater.* **8**1801897.
[60]Maleski K, Ren C E, Zhao M Q, Anasori B and Gogotsi Y 2018 *ACS Appl. Mater. Interfaces* **10** 24491-24498.
[61] Lou X W, Deng D, Y.Lee J Y, Feng J and Archer L A 2008 *Adv. Mater.* **20**258-262.
[62] Li S, Zheng J, Zuo S, Wu Z, Yan PandPan F 2015 *RSC Adv.* **5** 46941-46946.
[63] Chaudhari N K, Jin H, Kim B, Baek D S, Joo S H and Lee K 2017 *J. Mater. Chem. A* **5** 24564-24579
[64] Lei J C, Zhang X and Zhou Z 2015 Front. Phys. **10** 276-286 .
[65] Zhou J, Lin S, Huang Y, Tong P, Zhao B, Zhu X and Sun Y 2019 *Chem. Eng. J.* **373** 203-212 .
[66] Rao D, Zhang L, Wang Y, Meng Z, Qian X, Liu J, Shen Z, Qiao G and Lu R 2017 *J. Phys. Chem. A* **121** 11047-11054.
[67] Xu E, Zhang Y, Wang H, Zhu Z, Quan J, Chang Y, Li P, Yu D and Jiang Y 2020 Chem. Eng. J. **385** 123839.
[68] Zhao Q ,Wang R, Zhang Y, Huang G, Jiang B, Xu C and Pan F 2021 *J. Magnes. Alloy.* **9** 78-89 .
[69] Stephenson T, Li Z, Olsen B and Mitlin D 2014 *Energy Environ. Sci.* **7** 209-231.
[70] Jing Y, Quiles E O, Cabrera C R, Chen ZandZhou Z 2014 *Electrochim. Acta* **147**392-400.
[71]AnasoriB,Lukatskaya M R and Gogotsi Y 2017 *Nat. Rev. Mater.* **2** 1-17.
[72] Chen C , Xie X, Anasori B, Sarycheva A, Makaryan T, Zhao M, Urbankowski P, Miao L, Jiang J andGogotsi Y 2018 *Angew. Chem. Int. Ed.* **57** 1846-1850.
[73] Du Y T, Kan X, Yang F, Gan L Y, Schwingenschlogl U 2018 *ACS ACS Appl. Mater. Interfaces* 10 32867-32873.
[74] Gan L Y, Zhao Y J, Huang D and SchwingenschlöglU 2013 *Phys. Rev.* B **87** 245307.
[75] Gan L Y, Huang DandSchwingenschlögl U 2013 *Journal of Materials Chemistry* A **1**13672-13678
[76] Zhang P, Wang D, Zhu Q, Sun N, Fu F andXu B 2019 *Nano-micro letters* **11**1-14

[77] Guo X, Zhang W, Zhang J, Zhou D, Tang X, Xu X, Li B, Liu H andWang G 2020 *ACS nano* **14**3651-3659

[78] WangT,Su D, Shanmukaraj D, Rojo T, Armand M and Wang G 2018 *Electrochemical Energy Reviews* 1 200-237

[79] J. Y. Hwang J Y, S. T. Myung S T, Y. K and Sun Y K 2017 *Chemical Society Reviews* **46** 3529-3614

[80] Tan H, Chen D, Rui X and Yu Y 2019 *Advanced Functional Materials* **29**1808745

[81]Yan F, Gao H, Chen J and Guo Z 2017 *Small Methods* **1** 1700216

[82] Ni J, Li L and Lu J 2018ACS *Energy Letters* **3** 1137-1144

[83]Gao H, Zhou T, Zheng Y, Liu Y, Chen J, Liu H and Guo Z 2016 *Advanced Energy Materials* **6** 1601037.

[84] Wu Y, Hu S, Xu R, Wang J, Peng Z, Zhang Q andYu Y 2019 *Nano letters* **19** 1351-1358.

[85] Dahbi M, Yabuuchi N, Fukunishi M, Kubota K, Chihara K, Tokiwa K and Yu X F *Chemistry of materials* **28**1625-1635

[86] Jiao L, Zhang C, Geng C, Wu S, Li H, Lv W and Tao Y 2019 *Advanced Energy Materials* **9** 1900219.

Exergy modeling and simulation of a horizontal concentric tube heat exchanger

Ankit Upadhyay[1], Sankalp Gour[2*], Ankul Kumar[2], Deepak Kumar[2], Aseem Chandra Tiwari[1]

[1]Department of Mechanical Engineering, UIT-Rajiv Gandhi Proudyogiki VishwavidyalayaBhopal, Madhya Pradesh, India-462033
[2]Department of Mechanical Engineering, Maulana Azad National Institute of TechnologyBhopal, Madhya Pradesh, India-462003
*E-mail: sankalpgour@gmail.com

Abstract. The present study deals with the exergy modeling and simulation of a horizontal concentric tube heat exchanger. In this context, a fundamental thermodynamics-based exergy model is developed here to predict the irreversibility and entropy generation numberfor various fluid flow arrangements of a concentric tube heat exchanger. To develop the same, a second law of thermodynamics-based approach is utilized. Additionally, a finite volume method-based simulation of the same heat exchanger is also performed using a computer-aided engineering software named ANSYS R15.0. At last, the analytical and simulation results are compared and validated with experimental results, and a good agreement is achieved.

1. Introduction

In general, the function of heat exchanger is the interaction of heat within the fluids due to change in temperature between them. Usually, they are based on the combination of conduction and convection without the interaction of external heat and work. The system can be taken as an isolated system, for example, Heat exchanger of concentric tube type, (i.e., neglecting the heat interaction and system work within the surrounding) [1]. However, heat transfer involves convection from hot water to an innermost layer of the tube wall, conduction within the tube wall (radially), and the tube wall's outer shell to the cold fluid through convection. Generally, heat exchangers are employed in engineering processes like the food industry, chemical industry, air conditioning, and power generation [2]. However, due to low energy sources and higher energy costs, many efforts were being put into enhancing the efficiency of heat exchangers. The first step in this process was to measure the fundamentals thermodynamics parameters, which deal with energy, transfer, and conversion of energy and entropy change at equilibrium condition. At the same time, exergy analysis of heat exchangers also became very useful in developing a better understanding of thermal performance than energy analysis [3]. Exergy analysis prevails over the major limitation in analysis of energy that generally ignores the losses and direction for a particular process [4, 5]. This analysis typically includes a system analysis that uses the thermodynamics based second law approach, composed of the principals of conservation mass and energy, to design and analyze energy systems.
To mention some earlier works on the exergy analysis in case of heat exchangers, Moran and Sciubba [6] discussed the importance of exergy analysis to evaluate the true amount (magnitude)of waste and the loss of non-renewable energy resources. In line with that, Ahmetcan and Doganeryener [7] experimented on a double tube type of heat exchanger to

evaluate the significance of process variables on the Irreversibility or exergy losses. They observed that the rise in effectiveness might result in the exergy loss in both the paths at constant temperature ratios. Further, Adrian Bejan [8] suggested the use of maximum exergy for a hot stream is compatible with the integrative theoretical design of energy flow systems in the case of aircraft. Paisarn Naphon [9] performed a study by using thermodynamics based second law approach to evaluate the different heat transfer rates in case of horizontal concentric tube type heat exchanger and validated it with the experimental work. Colorado and Velázquez [10] also performed the exergy modeling on the vapor absorption refrigeration on cascade system and found that, on operating the system at lower irreversibility in the generator as well as an absorber, it may reachthe maximum COP and energetic efficiency. Furthermore, Esfahani and Languri [11] performed the same exergy analysis by using graphene oxide nanofluids on a shell & tube type of heat exchanger and stated, On increasing hot fluid inlet temperature increases the exergy loss for turbulent flow & laminar flow regimes respectively.

In view of the simulation works, a CFD simulation is an efficient technique to design a heat exchanger system with all the modifications done in optimization. CFD possesses a fundamental methodology to resolve the complete system into small grids and proceeds with governing expressions on the discrete cells to obtain a numerical solution by validating the experimental work [12, 13]. In line with that, Correa et al. [14] suggested a multicell dynamicmodel that illustrates the transient behavior of the heat intraction in shell & tube type of heat exchangers. Further, Sunden [15] explained different feasible models that can be applied in the Computational fluid dynamics (CFD) analysis and performed a simulation on a plate type heatexchanger to show the prospect of CFD applications on various types of complications across heat exchangers. Huminic et al. [16] investigated that on using CuO and TiO_2 as nanoparticles with water as a base can increase the heat transfer in case of laminar flow region with increased water outlet temperature. At last, Congedo et al. [17] analyzed the configurations of three different geometry of ground heat exchangers with varying working conditions and learned that the velocity choice for the heat transfer fluid plays a vital role in the heat transfer performance for each arrangement.

2. Theoretical modeling

In this section, an analytical model based on horizontal concentric tube heat exchanger was developed to calculate the exergy loss and rate of entropy generation number. Further, To develop the same, a fundamental theory of heat transfer using thermodynamics based second law approach.

2.1 Problem definition

Consider a complete thermal unit composed of a test section having a concentric tube heat exchanger, refrigerant, cooling unit, heating unit, and data acquisition system shown in Figure 1. The water is considered a working fluid in the unit, and an R22 refrigerant is used tochilling the water.The piping and test connection system are connected for the easy removal or maintainace of parts at any time. The cooling unit includes reservoir tank , a voltage controlled electric heater, a stirrer , and a cooling coil immersed in the reservoir tank. For the working stage of the thermal unit, we may first adjust the temperature of water as required. Then, both the working fluids are pumped out of reservoir and allowed to flow through flow meter, testing unit and then collected back in the reservoir.

Next, we may control the water flow by regulating the valve and measuring the flow meters. Further, the temperature of water at the entry, exit, and mid section of heat exchanger as shown in Figure 1 and by using thermocouple probe we can estimate the inner tube wall temperature.

IOP Conf. Series: Materials Science and Engineering 1225 (2022) 012019 doi:10.1088/1757-899X/1225/1/012019

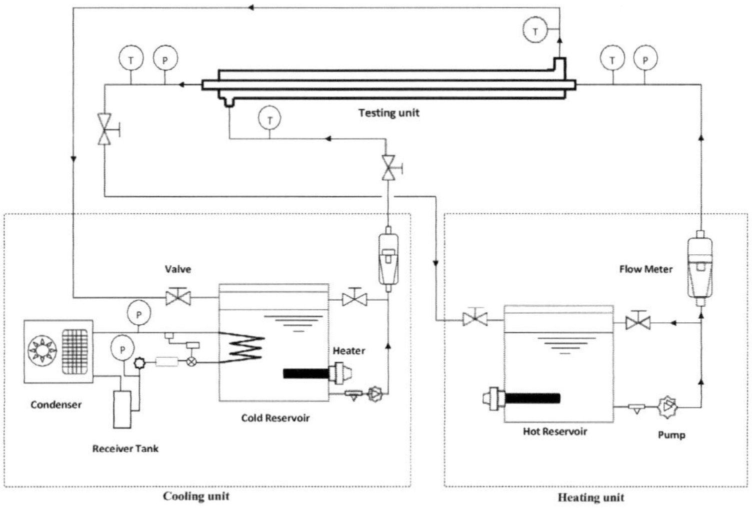

Figure 1. Schematic diagram of a thermal unit.

For a given working state of a thermal unit, the problem aims to derive an analytical expression of the exergy loss and entropy generation number following the thermodynamics based second law approach. The model is developed under the given assumptions

- Flow of hot fluid and cold fluid are assumed to be steady.
- The heat intraction between system and surrounding is assumed to be negligible.
- The convective heat transfer coefficients for annulus as well as tube sides are considered asa constant.
- Thermal conductivity for construction material of the tube is assumed to be constant.

The primary objective of the defined problem is to verify the second law based developed model for concentric tube heat exchanger with simulation findings and existing experimental results.

2.2 Exergy model

For a given horizontal concentric tube type heat exchanger wherein heat intraction take placein both the fluids, the total rate of entropy generation rate $\dot{S}_{gen}$ for the given system is defined as

$$\dot{S}_{gen} = \dot{S}_{gen_h} + \dot{S}_{gen} \quad , \tag{1}$$

where $\dot{S}gen_h$ and $\dot{S}gen_c$ denote the rate of entropy generation rates in hot and cold fluid, respectively.

From the theory of heat transfer [18], the above equation (1) may be rewritten as

$$\dot{S}_{gen} = \dot{m}_h C_h \ln\left(\frac{\theta_{h,2}}{\theta_{h,1}}\right) + \dot{m}_h C_c \ln\left(\frac{\theta_{c,2}}{\theta_{c,1}}\right), \tag{2}$$

where $\dot{m}_h$ represent the mass flow rate for hot-side and $\dot{m}_c$ represent the mass flow rate for cold-side, C_h and C_c represent the hot and cold water specific heat capacities, and $\theta_{h,2}$ and $\theta_{h,1}$ are the outlet & inlet hot and cold side water temperatures, respectively.

2.2.1 Case-A [$\dot{m}_h C_h < \dot{m}_c C_c$]: For a given minimum heat capacity rate in case of hot fluid $C_{min} = \dot{m}_h C_h$ than the cold fluid C_{max}, the effectiveness ϵ for the heat exchanger is defined as

$$\epsilon = \frac{\dot{m}_c C_c(\theta_{c,2} - \theta_{c,1})}{C_{min}(\theta_{h,1} - \theta_{c,1})} = \frac{(\theta_{h,1} - \theta_{h,2})}{(\theta_{h,1} - \theta_{c,1})} \tag{3}$$

On utilizing equation (3) in equation (2), we obtain

$$\dot{S}_{gen} = C_{min}\ln\left[1 - \epsilon\left(1 - \frac{1}{\theta_r}\right)\right] + C_{max}\ln\left[1 + \epsilon R(\theta_r - 1)\right] \tag{4}$$

where $\theta_r = \theta_{h,1}/\theta_{c,1}$ and $R = C_{min}/C_{max}$ denote the inlet temperature ratio and the heat capacity ratio, respectively. Following the Gouy-Stodola theorem [19, 20], an exergy loss or irreversibility I for a given working thermal system at a surrounding temperature T_0 is described by

$$I = T_0\dot{S}_{gen}. \tag{5}$$

Finally, For current case, the entropy generation number N_s, can be written as

$$N_s = \frac{\dot{S}_{gen}}{C_{min}} = \ln\left[1 - \epsilon\left(1 - \frac{1}{\theta_r}\right)\right] + \frac{1}{R}\ln\left[1 + \epsilon R(\theta_r - 1)\right] \tag{6}$$

2.2.2Case-B [$\dot{m}_c C_c < \dot{m}_h C_h$]: For a given minimum heat capacity rate in case of cold fluid $C_{min} = \dot{m}_c C_c$ than the hot fluid C_{max}, the effectiveness ϵ for the heat exchanger is defined as

$$\epsilon = \frac{\dot{m}_h C_h(\theta_{h,1} - \theta_{h,2})}{C_{min}(\theta_{h,1} - \theta_{c,1})} = \frac{(\theta_{c,2} - \theta_{c,1})}{(\theta_{h,1} - \theta_{c,1})} \tag{7}$$

On utilizing equation (7) in equation (2), we get

$$\dot{S}_{gen} = C_{min}\ln\left[1 + \epsilon(\theta_r - 1)\right] + C_{max}\ln\left[1 - \epsilon R\left(1 - \frac{1}{\theta_r}\right)\right] \tag{8}$$

At last, the entropy generation number N_s for the current case may be stated as

$$N_s = \frac{\dot{S}_{gen}}{C_{min}} = \ln\left[1 + \epsilon(\theta_r - 1)\right] + \frac{1}{R}\ln\left[1 - \epsilon R(1 - \frac{1}{\theta_r})\right] \tag{9}$$

3. Simulation

In this section, a simulational study for a horizontal concentric tube type heat exchanger calculates the exergy loss as well as rate of entropy generation number. A finite volume method-based ANSYS R15.0 software is utilized to simulate the heat exchanger. This methods of simulation based upon computational fluid dynamics is an essential tool to make applications related to fluid flow and heat and mass transfer. The methods used in current work for the flow of working fluid, i.e., water inside a concentric tube heat exchanger with counterflow arrangement. The methods include certain additional assumptions to reduce the complications of calculations.These additional assumptions are

- Water is assumed to be incompressible as well as a Newtonian fluid.
- Steady-state condition is assumed.
- The effect of thermal radiation and gravity are being neglected.

3.1 Geometry of concentric tube heat exchanger

Two concentric parallel tubes are taken in which the hot fluid flows inside the inner tube, and the cold fluid flows into the annulus between the two tubes with the counter flow direction. The specifications of geometry and other details of the simulated heat exchanger are demonstrated in Table 1.

Table 1. Geometrical specifications of concentric tube heat exchanger.

Dimensions	Outer tube	Inner tube
Tube material	Copper	Copper
Outer diameter	50 mm	22 mm
Inner diameter	44 mm	18 mm
Wall thickness	3 mm	2 mm
Tube length	2400 mm	2400 mm

3.2 Simulation methods and boundary conditions

For simulation of the problem, the finite volume method is adopted. A pressure and velocity coupling formulation is utilized by a Semi-Implicit Method for Pressure-Linked Equations algorithm (i.e.,SIMPLE algorithm) is utilized. An energy model is kept ON, and the Standard k-epsilon model is used with Standard wall functions . Water is considered as a working fluidfor both the hot & cold fluid in material section. The material of the outer and inner tubes is copper. In spatial discretization, for momentum computation, we adopted a second-order upwind scheme, and for evaluation of dissipation rate as well as kinetic energy in turbulent regime, first-order upwind method is used. The Hybrid-Initialization solution method is applied before running calculations. The considered number of nodes and elements information is shown in Table 2. The entry and exit boundary constraints are applied as the mass flow at entry and pressure at exit for the inner tube and the annulus side. It can be seen in Figure 2 containing temperature variation contours at the outlet sections of both the tubes.

Various data like an inlet temperature of the hot fluid $\theta_{h,1}$ = 40 °C and an inlet temperatureof the cold fluid $\theta_{c,1}$ = 20 °C are noted for a constant mass flow rates of cold m_c = 0.03 kg/s and hot m_h = 0.0329 kg/s fluids, respectively. Similar procedures are being adopted for all the parameters used in other conditions to get all the simulated plots of exergy loss and rate of entropy generation number.

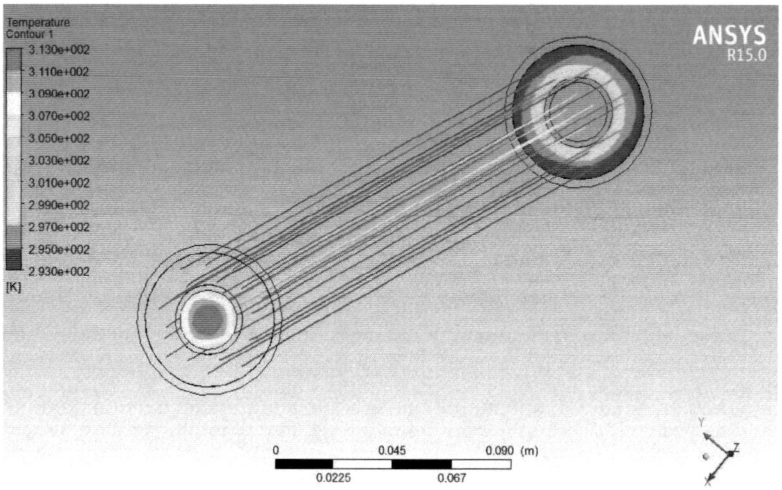

Figure 2. Temperature contours at inner tube outlet (hot fluid) and outer tube outlet (cold fluid) with $\theta_{h,1} = 40$ °C $\theta_{c,1} = 20$ °C, $m_h = 0.0329$ kg/s, and $m_c = 0.03$ kg/s.

Table 2. Nodes and elements specifications of concentric tube heat exchanger.

Domains	Nodes	Elements
Cold fluid	42688	31947
Hot fluid	41064	33360
Inner tube	31556	19180
Outer tube	12768	6666
All Domains	128076	91153

4. Results and discussion

This section presents, the analytical findings of the derived model and simulation results of ANSYS R15.0 software are analyzed and compared with the existing parallel experimental results [9] in the literature. In this context, Figures 3 and 4 illustrate the relationship between the rate of entropy generation and $\dot{m}_h$ at different inlet conditions. On increasing $\dot{m}_h$, we may observe the growth in the rate of entropy generation as expected and described via equation (4) or (8). The mass flow rate of hot water $\dot{m}_h C_h$ is having higher heat capacity rate than that of cold water $\dot{m}_c C_c$. Hence, the heat capacity in case of cold water is minimum, shown in equations (4) or (8).

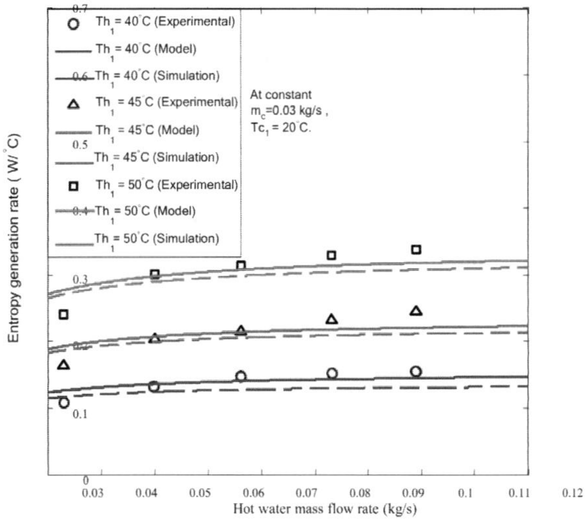

Figure 3. Effect of hot water mass flow rate ($\dot{m}_h$) on entropy generation rate ($\dot{S}_{gen}$) with varying inlet hot water temperatures.

Interestingly, a drop in this trend is observed at a higher value of the $\dot{m}_h$. Also, $\dot{S}_{gen}$ peaks constantly on the increase in the temperature at entry of hot water at steady temperature conditions for the cold water as well as the given $\dot{m}_h$. This observation shows that the $\dot{m}_h$ plays a fundamental role in the intensification of the rate of entropy generation . A similar observation may be noticed from figure 4. However, this grows closer for a lower value of the $\dot{m}_h$. One may also note that the theoretical and simulation findings agree well with the experimental data [9].

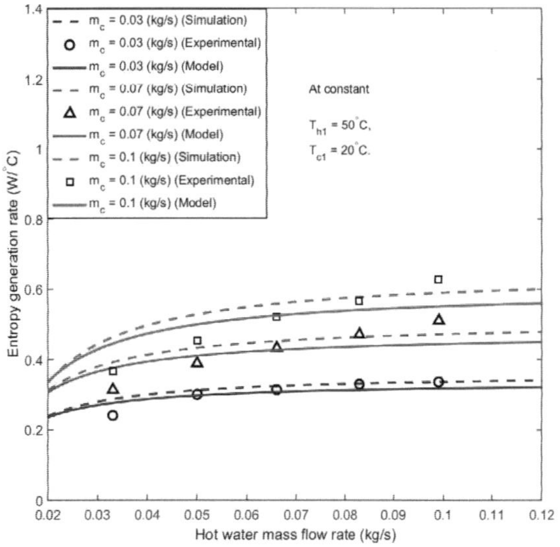

Figure 4. Effect of hot water mass flow rate ($\dot{m}_h$) on entropy generation rate ($\dot{S}_{gen}$) with varying cold water mass flow rates.

Figure 5 depicts the impact of hot water inlet temperature on the entropy generation number obtained from equations (6) or (9). Herein, increment in N_s is observed on increasing $\dot{m}_h$. However, the effect is more significant with higher inlet temperature for the same value of $\dot{m}_h$. According to the anticipated and experimental results, the theoretical and simulation findings somewhat here overpredict the measured results.

The consequence of the $\dot{m}_c$ on N_s is presented in Figure 6. One may here observe that for $\dot{m}_c = 0.03$ kg/s, with increment in $\dot{m}_h$ we observe a continuous rise in the rate of entropy generation number. This phenomenon is interpreted by equations (6) or (9) that recommends, the rate of heat capacity for cold water is minimum. It means that with increment in $\dot{m}_h$,the N_s also rises. However in case of $\dot{m}_c = 0.07$ kg/s, $\dot{m}_h$ is lower than that of 0.07 kg/s. Hence, The C_h is used in place of the minimum capacity rate. Therefore, one may identify a decreasing trend in the N_s concerning the $\dot{m}_h$. On the other hand, we recognize an increasing inclination of entropy generation number in regards to the $\dot{m}_h$ beyond $\dot{m}_h = 0.07$ kg/s. This behavior is obtained due to the change of the C_{min} to C_c. Similarly, entropy generation number(N_s) drops continuouslyat $\dot{m}_c = 0.10$ kg/s with increase in $\dot{m}_h$. Again, we can understand this phenomenon by help of equations (6) and (9), which show the minimum heat capacity rate is used as C_h. Hence,a decrease in N_s is observed as increase in $\dot{m}$. From both of the figures, the theoretical and simulation findings now reasonably validate the experimental results [9].

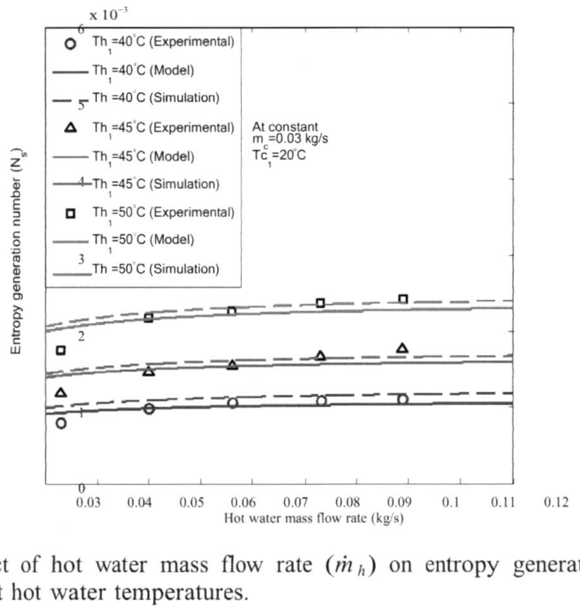

Figure 5: Effect of hot water mass flow rate ($\dot{m}_h$) on entropy generation number(N_s) with varying inlet hot water temperatures.

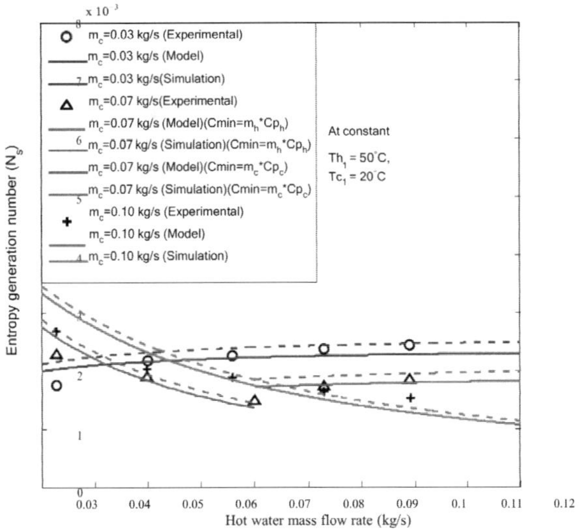

Figure 6. Effect of hot water mass flow rate ($\dot{m}_h$) on entropy generation numberN_s with varying cold water mass flow rates ($\dot{m}_c$).

Figures 7 and 8 depicts the dependence of the exergy loss on $\dot{m}_h$ for different entrytemperature conditions for hot water with specific values of mass flow rate for the cold water, respectively. One may observe that the nature of the curves for the I is identical to the $\dot{S}gen$, as displayed in Figures 3 and 4. Therefore, a similar description and interpretation may be followed for Figures 7 and 8 as well. Repeatedly, for both figures 7 and 8, the theoretical and simulation outcomes show a reasonable agreement with the experimental results [9].

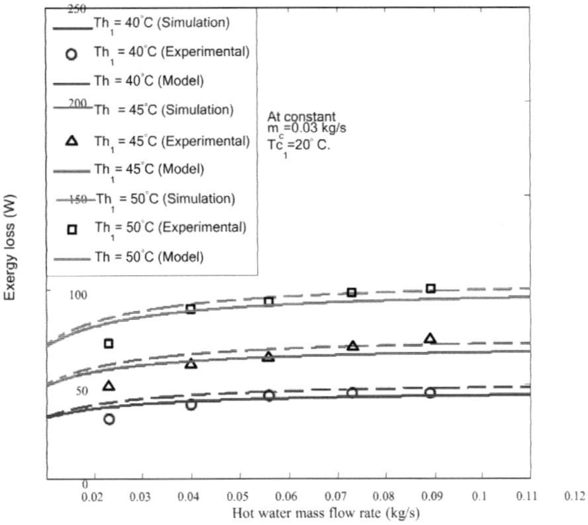

Figure 7. Effect of hot water mass flow rate ($\dot{m}_h$) on exergy loss/irreversibility (I) with varying inlet hot water temperatures.

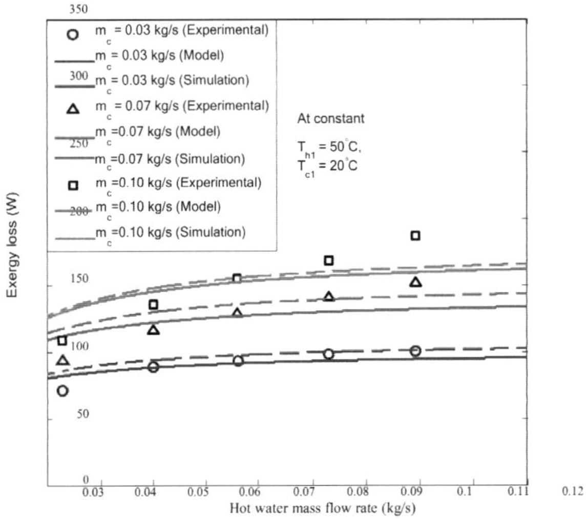

Figure 8. Effect of hot water mass flow rate ($\dot{m}_h$) on exergy loss/irreversibility I with varying cold water mass flow rates $\dot{m}_c$.

5. Concluding remarks

In the present work, an extensive study on the heat exchanger for horizontal concentric tube type has been performed following the thermodynamics based second law approach. The effect of exergy loss and entropy generation number is modeled and simulated for varying several parameters such as mass flow rates ($\dot{m}$) and inlet conditions for hot and cold water.The impact of these parameters upon the concentric tube heat exchanger design has been observed and analyzed. The theoretical and simulation findings of the current study are also successfully validated with the experimental results. Hence, the findings of the present work may help the researchers in designing the concentric tube type of other heat exchangers for various applications.

References

[1] Shah R K, Sekulic D P 2003 Fundamentals of heat exchanger design, *John Wiley & Sons.*

[2] Pacio J C, Dorao C A 2011 A review on heat exchanger thermal hydraulic models for cryogenic applications,Cryogenics **51 (7)** 366–379.

[3] Esen H, Inalli M, Esen M, Pihtili K 2007 Energy and exergy analysis of a ground-coupled heat pump system with two horizontal ground heat exchangers *Building and environment* **42(10)** 3606–3615.

[4] Prasad R, Shen J 1993 Performance evaluation of convective heat transfer enhancement devices using exergy analysis *Int. J. Heat Mass Transf.* **36(17)** 4193–4197.

[5] Singh N, Kumar D, Sarma A K, Jha M K 2017 Exergy conceptual-based study for comparative thermodynamic performance of ci engine fueled with petroleum diesel and biodiesel blends *Int. J. Power Energy Eng.* **143(1)** 04016030

[6] Moran M, Sciubba E 1994 Exergy analysis: principles and practice *J. Eng. Gas Turbines Power* **116(2)** 285-290.

[7] Can A, Eryener D, Buyruk E 1999 Experimental studies on influence of process variables to the exergy losses at the double tube heat exchanger, in: *Heat Transfer Enhancement of Heat Exchangers Springer* **641**–648

[8] Bejan A 2002 Fundamentals of exergy analysis, entropy generation minimization, and the generation of flow architecture *International journal of energy research* 26

(7).
[9] Naphon P 2006 Second law analysis on the heat transfer of the horizontal concentric tube heat exchanger *International communications in heat and mass transfer* **33(8)** 1029–1041.
[10] Colorado D, Velazquez V 2013 Exergy analysis of a compression–absorption cascade system for refrigeration Int. J. Energy Res. **37(14)** 1851–1865.
[11] Esfahani M R, Languri E M 2017 Exergy analysis of a shell-and-tube heat exchanger using graphene oxide nanofluids *Exp. Therm. Fluid Sci.* **83** 100–106
[12] Kumar V, Saini S, Sharma M, Nigam K 2006 Pressure drop and heat transfer study in tube-in-tube helicalheat exchanger *Chem. Eng. Sci.* **61(13)** 4403–4416.
[13] Kanaris A G, Mouza A A, Paras S V 2006 Flow and heat transfer prediction in a corrugated plate heat exchanger using a CFD code *Chem Eng Technol* **29(8)** 923–930
[14] Correa D J, Marchetti J L 1987 Dynamic simulation of shell-and-tube heat exchangers, Heat Transf. Eng. **8(1)** 50–59.
[15] Sundén B 2007 Computational fluid dynamics in research and design of heat exchangers, Heat Transf. Eng. **28(11)** 898–910.
[16] Huminic G, Huminic A 2011 Heat transfer characteristics in double tube helical heat exchangers using nanofluids,*Int. J. Heat Mass Transf.* **54** 4280–4287.
[17] Congedo P, Colangelo G, Starace G 2012 CFD simulations of horizontal ground heat exchangers A comparison among different configurations *Appl. Therm. Eng.* **33** 24–32.
[18] Bejan A 2016 Advanced engineering thermodynamics *John Wiley & Sons.*
[19] Pal R 2017 On the gouy–stodola theorem of thermodynamics for open systems, Int. J. Mech. Eng. Educ. **45(2)** 194–206.
[20] Reini M, Casisi M 2020 The gouy-stodola theorem and the derivation of exergy revised, Energy **210** 118486.

DFT Investigations of $BeSnN_2$ Chalcopyrite Compound for Optoelectronic Applications

Sanjay Lathwal[a], Aditi Gaur[b], Karina Khan[c], Sunil Kumar Goyal[d], Amit Soni[e*], Jagrati Sahariya[f*]

[a]Birla Institute of Technology & Science, Pilani-333031, Rajasthan, India
[b,d,e]Department of Electrical Engineering, Manipal University Jaipur, Jaipur-303007, Rajasthan, India
[c]Department of Physics, Manipal University Jaipur, Jaipur-303007, Rajasthan, India
[f]Department of Physics, National Institute of Technology, Srinagar (Garhwal)-246174, Uttarakhand, India

E-mail: **amitsoni_17@yahoo.co.in* ; **jagrati.sahariya@gmail.com*

Abstract. The ternary chalcopyrite compounds are a very renowned category to perform the theoretical investigation in order to find out a proper and apt compound for optoelectronic application. Solar cell is a very interesting field to compensate the energy supplying needs in place of other electricity generating sources. Several semiconductor compounds have been investigated and amongst them we have done a theoretical investigation of pure $BeSnN_2$ using DFT based computational tool i.e. Wien2k. The exchange correlation used for our study is Perdew Burke Ernzerhoff: Generalized Gradient Approximation (PBE-GGA). We have done electronic and optical investigation of the compound using the basic lattice parameters and other essential input parameters. The investigation has offered a bandgap of 1.005 eV which is suitable to quote for the optoelectronic applications. Optical properties like absorption, dielectric tensor (both real and imaginary), refraction and reflection have been investigated.

Keywords: Absorption, dielectric tensor, electronic, reflection, refraction, solar

1. Introduction

This work has been done to accomplish the investigation to know how successful BeSnN2 is as an optoelectronic material. The compound is a ternary pnictide semiconductor with a chalcopyrite-based structure which belongs to the base compound of group II-A. The motivation of this work is the theoretical investigation of several such similar compounds and to analyze their electronic and optical nature. As we all know solar cells have seen tremendous growth to meet the energy requirement efficiently in comparison to other energy generating sources [1]. With time we require to get a solar cell compound that offers more efficiency and less economical investment.

Many ternary compounds based on group II-A and II-B primary compounds have been studied and their brief literature has been mentioned here. A theoretical investigation in which $ZnSnN_2$ has been found to provide a better solution as a solar cell or photovoltaic application [2]. This work talks about the work and development that has taken place on this sample. They have discussed and compared the computational and theoretical results based on this sample. $ZnSiN_2$ was another compound along with $ZnGeN_2$ that has been studied due to its wide-bandgap nature and giving us an insight into the bandgap engineering option. The experimental work has been presented in which growth of $ZnSiN_2$ on c-sapphire & silicon substrates is done using a low-pressure technique i.e.

MOVPE [3]. The effect of doping $ZnSiP_2$ with Ge and Sn at the Si site has been identified using the structural, electronic, and optical aspects through DFT-based wien2k. The different doping percentages have shown decrement in bandgap with Sn doping and not much change has been observed in the case of Ge doping [4]. The doping of Ge at si site in $ZnSiP_2$ has also been investigated showcasing the comparison with pure and 12.5% of Ge in $ZNSiP_2$ [5]. This work deals with the wideband material (more than 2 eV) i.e. $MgSnN_2$ to provide an alternative to the light-emitting diode and other optoelectronic device applications [6]. This work depicts the computational study of $MgSnN_2$ offering a bandgap of 2.69 eV. The material has been quoted aptly for short-wavelength absorber material for photovoltaic applications [7]. With the use of metathesis reaction under extreme pressure, $MgSnN_2$ a unique ternary nitride semiconductor compound is obtained [8]. In another work, optoelectronic investigation of V doped $MgSiN_2$ has been done using the first principle method. Basic plots like Density of states (DOS) and band structures present half-metallic characteristics for V-doped $MgSiN_2$ along with half-metallic-based bandgap of 0.74 eV and 0.36 eV for V_{Mg} and V_{Si} respectively [9]. The lattice parameters and structural features of $MgSiN_2$ and $MgGeN_2$ which are orthorhombic structured and wide bandgap based semiconductors are investigated using DFT. $MgSiN_2$ powder's structural aspects and Raman spectra have been obtained [10]. The first principle calculation of 12.5 percent of Sr at the Mg site in the case of $MgGeN_2$ has been carried out to reveal the different features of the compound in optoelectronic applications [11]. Based on the first principle density functional calculations, the study of (BeSn, BeGe, MgGe)N_2 has been accomplished. The calculations depict that with $BeSnN_2$ and $BeGeN_2$, MnBe shows anti-ferromagnetism, whereas MnSn and MnGe show ferromagnetism [12]. Finally in one of the works,the band structures, equilibrium lattice inputs parameters and structural parameters, bulk moduli, and cohesive energies for chalcopyrite structure of $BeCN_2$, $MgCN_2$, $BeSiN_2$, $MgSiN_2$, and $MgSiP_2$ are calculated by the linear-muffin-tin-orbital method [13-15]. A theoretical study has been performed in one of the papers in which the FP-LAPW method has been adopted based on TB-mBJ approximations to investigate the chemical bond, structural aspect, electronic and optical features. The series undertook for investigation was $BeMN_{o2}$ (M = Si, Ge, Sn; N_o = P, As) in which $BeSiP_2$ has yielded the maximum and direct in nature bandgap [16]. The mechanical aspects to study the stability along with other features like optical and electronic to investigate the behavior of $BeSiP_2$ and $BeGeP_2$ [17]. $BeSiP_2$ offers direct and $BeGeP_2$ offers indirect bandgap obtained using the FP-LMTO potential embedded in the VASP package.

The outline of the paper goes as follows: Section I discusses the Computational method, then in section II results & discussion are presented with a sub-section of electronic property analysis and optical property analysis each. This research aims to produce an investigation for the compound i.e. suitable in optoelectronic applications specifically solar cells.

2. Computational Method

This work has employed density functional theory for the computational method in the Perdew Burke Ernzerhoff: Generalized Gradient Approximation (PBE-GGA) exchange-correlation. The tool utilized for this purpose is wien2k following a DFT-based approach. The wien2k is based on Augmented Plane Wave (APW) method and performs all-electron calculation rather than pseudopotential [18]. The lattice parameters used in the calculations are a=b= 4.689 Å & c= 9.890 Å and the radius of muffin tin spheres for Be, Sn, and N being 1.89 a.u., 2.08 a.u. & 1.7 a.u. respectively. The lattice constant a (which belongs to the Zinc blende's cubic lattice constant from where the chalcopyrite originates), q = c/a ratio, and the internal displacement parameter u are used as the structural parameters. With the use of 1000 k points in the irreducible part of the zone, well-converged results are produced. The k-mesh for the system consists of 10 x 10 x 10. The total energy's meaningful behavior was obtained through MT correction and is obtained as a function of the internal structural value u and c/a ratio. The spacegroup of our system is tetragonal in structure, numbered 122 (I-42D). The body-centered tetragonal (BCT) lattice consists of eight atoms per unit cell in the case of chalcopyrite structure ABC_2. The present compound belongs to the II-A group thus it has been compared with the compounds that contain other elements of this same group like Mg, Ca, etc.

3. Results & Discussion

a) Electronic Property Analysis

$BeSnN_2$ belongs to the tetragonal spacegroup and the crystal structure of the same briefly depicts the arrangement of Be, Sn, and N in the crystal cell. The top and bottom surface of the crystal is occupied by Be at all four corners along with their positions in the middle of the crystal as well. The Sn and N atoms are bonded together with a single bond. The representation of band structure comprises the high symmetrical and directional description of the Brillouin zone. The k points taken to plot the different bands ranged from -6.0 eV to 6.0 eV are Z, Γ, X, P, N & Γ. The Brillouin zone required to produce the band structure plot utilizes different positions of the k points and then presents the bandstructure representation. The Brillouin zone along with the crystal cell for $BeSnN_2$ is given here in figure1 (a) and figure 1 (b). Bandstructure has provided a bandgap placed in between the maxima of the valence band and minima of the conduction band. The calculated bandgap for $BeSnN_2$ is 1.005 eV as shown in given figure 2 (a). In figure (b), the density of states for $BeSnN_2$ is depicted explaining the different orbital's roles. The correlation between the band structure and DOS can be done to trace the band formation due to the orbital contribution [19]. The eigen-values were obtained from Kohn Sham and calculated over enough amount of fine k-grid existing in irreducible BZ. The density of states (DOS) is thus obtained by the modified tetrahedron method. The major contributions are as follows: in *Be,* the contribution is given by 2s orbitals, in *Sn,* the contribution is by 5p orbitals and 2p orbitals is the contributing orbital in N. The reduction of the bandgap is due to the major contribution by N – atom in BeSnX2. In addition, Be and Sn form the covalent bond as in the second layer of DOS were total of Be, Sn, N represented. The formation energy of a compound is a crucial feature that intends to judge the stability of a compound through its chemical composition [20-21]. The formation energy for the given pure compound is calculated to be -45.91 eV. The more negative value of calculated formation energy the more stable is the compound in terms of its existence.

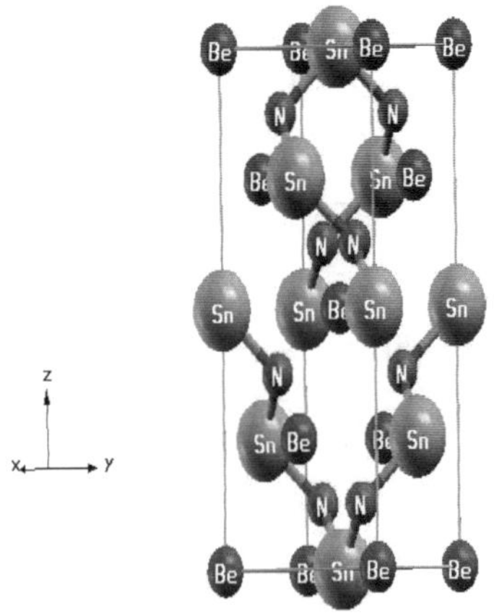

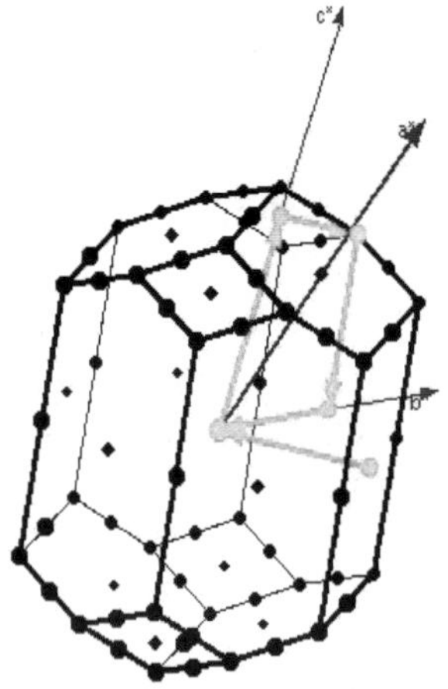

Figure 1. (a) Crystal structure and (b) Brillouin Zone for $BeSnN_2$ compound

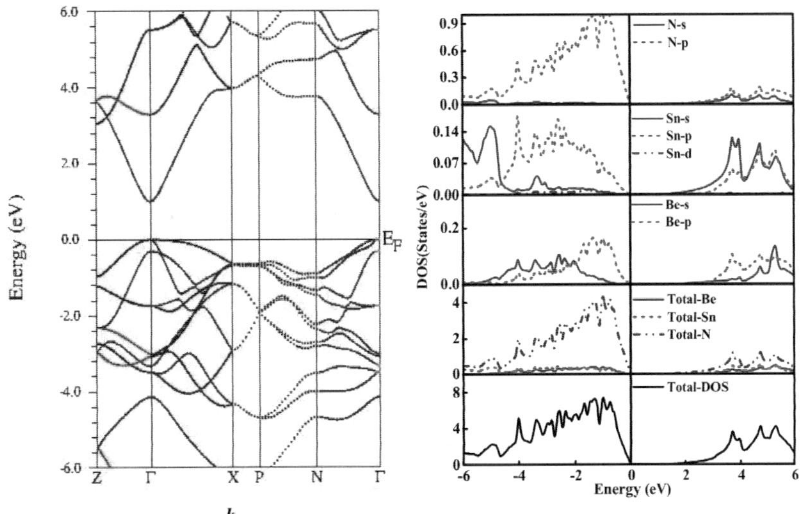

Figure 2. (a) Band structure and (b) DOS plot for $BeSnN_2$ compound

b) Optical property analysis

The optical properties of a semiconductor compound are determined by its absorption, dielectric tensor, reflection, and refraction features [22-23]. The dielectric tensor's real (ε_1) and imaginary part (ε_2) are determined by incident photon energy's function. The imaginary dielectric tensor part offers transition peaks that point out to the interband transitions that take place in the band structure plot of our compound under investigation as can be seen in figure 2 (a). The mathematical equations for the optical properties have been deduced from the dielectric function [24]. The equation for dielectric tensor consists of real as well as imaginary parts and is given as:

$$\varepsilon(\omega) = \varepsilon_1(\omega) + \varepsilon_2(\omega) \quad (1)$$

In the imaginary plot, these peaks are located at 4.74 eV and 5.56 eV as peak A and peak B respectively shown in figure 3 (b).

Also, the source of the different peaks is due to these interband transitions only. The static real dielectric tensor's value i.e. $\varepsilon_1(0)$is 6.20. The real dielectric tensor component is shown in figure 3(a).

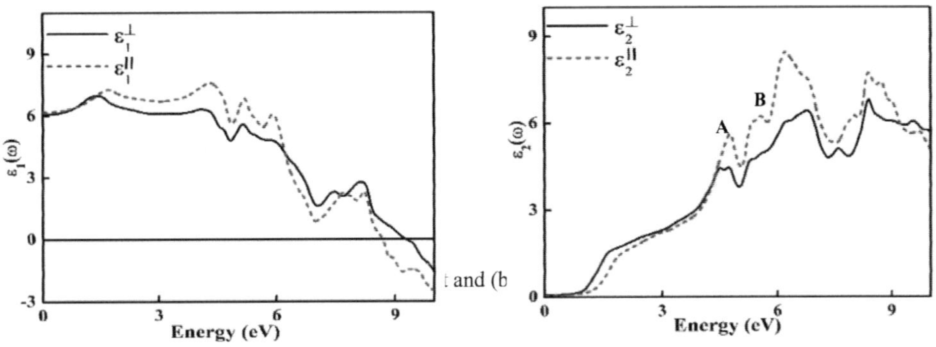

Figure 3. (a) real dielectric tensor component, (b) imaginary plot

IOP Conf. Series: Materials Science and Engineering 1225 (2022) 012020 doi:10.1088/1757-899X/1225/1/012020

The absorption coefficient α is the processing of area under the curve for analysis. Tauc's law states to determine the optical energy gap of a compound based on its absorption coefficient value. The absorption is determined by the expression:

$$\alpha(\omega) = \frac{4\pi}{\lambda} k(\omega) \quad (2)$$

This is defined by the interband transitions that are nearer to the bandgap. The absorption curve is shown in figure 4 below explaining the perpendicular and parallel components of absorption. As per the figure, we can see that the components show an isotropic nature for the (0 – 5) eV range, and the rest of the components show anisotropy.

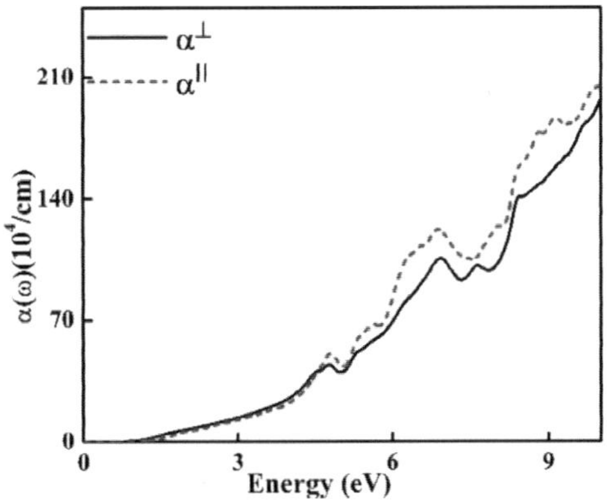

Figure 4. Absorption curve for the compound $BeSnN_2$

Refraction is the parameter that decides how a system behaves towards electromagnetic waves. The value of n (ω) i.e. refractive index is computed through the use of a real dielectric tensor [25]. It is given by the following equation:

$$n = \sqrt{\varepsilon_1} \quad (3)$$

The reflection parameter decides the optical nature of a semiconductor in terms of the solar cell. It is found to react more sensitively towards the dielectric tensor's real and imaginary parts [26]. The equation for reflection is given as follows:

$$R(\omega) = \frac{[n(\omega)-1]^2 + k(\omega)^2}{[n(\omega)+1]^2 + k(\omega)^2}, \quad (4)$$

The value of static reflectivity R(0) is 18% and the value of static refractivity is calculated to be 2.47. In this manner, all the parameters show the anisotropic nature of this semiconductor as per the whole investigation.

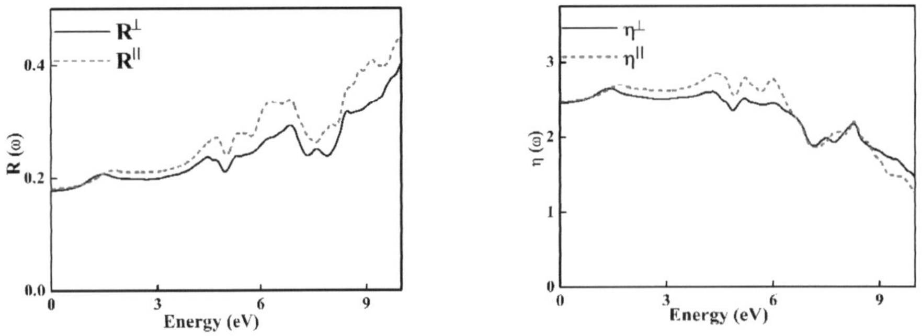

Figure 5. (a) Reflection plot and (b) Refraction plot for $BeSnN_2$

Additional optical characteristics like optical conductivity can be calculated with the use of dielectric function in particular energy ranges. The curve of optical conductivity depicts many transition peaks that correspond to the excitations aroused by the bulk plasmon. This in brief is the movement or shift of electrons from the valence band to the conduction band. The equation defining the optical conductivity is given by:

$$\sigma(\omega) = -\frac{i\omega}{4\pi}\varepsilon(\omega) \qquad (5)$$

We have made use of compounds with elements that offer no or least toxicity in their composition. Therefore our compound has not reported any toxic nature during its past experimental and theoretical work. This can be stated in the literature review from [3-17].

4. Conclusion

The compound has achieved a positive response in terms of optoelectronic application. This sample with the PBE-GGA exchange-correlation functional offers a bandgap of 1.005 eV which is suitable for applications in solar cells and other optoelectronic fields. For a system to stand out as a suitable optoelectronic device especially photovoltaics the requirement to be fulfilled is to deliver direct bandgap and also the value of bandgap should be around and above 1.2 eV. The present system though offering 1.005 eV needs a very less amount of suitable doping either with a similar isoelectronic element or with a p-type or n-type dopant. The electronic and optical analysis has been done with features explored such as band structure, DOS, absorption curve, dielectric tensor plot, reflection, and refraction. The optical properties have presented the anisotropic nature of the compound between perpendicular and parallel components. The compound offers 18% reflection concluding it achieves the aim to be an apt solar cell compound with lesser reflectivity feature.

5. Acknowledgment

We would like to express our gratitude towards DST-SERB project grant no. EMR/2017/005534 for providing financial support in the completion of such computational work. Along with this, we would like to thank sincerely, Manipal University Jaipur for providing the appropriate platform to carry out the research. At last, we would like to thank all our research team members for the completion of the computational tasks with precision.

References

[1] Gul M, Kotak Y, Muneer T 2016 Review on recent trend of solar photovoltaic technology *Energy Explor. Exploit.* **34(4)** 485-526.

[2] Khan I S et al 2020 Review of $ZnSnN_2$ semiconductor material *J Phys Energy* **2** 032007.

[3] Cloitre T et al 2004 Epitaxial growth of $ZnSiN_2$ single-crystalline films on sapphire substrate *Superlattices Microstruct.* **36** 377-383.

[4] Khan K et al 2020 Density functional investigations to study effect of M=(Ge, Sn) doping on opto-electronic response of $ZnSi_{(1-x)}M_xP_2$ *Optik* **208.**

[5] Khan K et al 2020 Investigation of structural and optoelectronic properties of $ZnSi_{1-x}Ge_xP_2$ (x = 0, 0125) compound using density functional theory *AIP Conference Proceedings* **2220.**

[6] York K R 2018 $MgSnN_2$: A New Eco-friendly Wide Band Gap Semiconductor *Western Michigan University* 1-56.

[7] Dumre B et al 2021 Stability, and electronic and optical properties of ternary nitride phases of $MgSnN_2$: A first-principles study *J. Phys. Chem. Solids* **153**, 110011.

[8] Kawamura F et al 2019 Synthesis of a Novel Rocksalt-Type Ternary Nitride Semiconductor $MgSnN_2$ Using the Metathesis Reaction under High Pressure *Eur. J. Inorg. Chem.* 1-17.
[9] Khan K et al 2021 Exploring Ab initio calculations of $Mg_{0875}Sr_{0125}GeN_2$ Alloy: A DFT Study theory *IEEE In: Int. Conf. Adv Comp Inno Techno Engg.*
[10] Huang H M et al 2013 First Principles Study of Half-Metallic and Magnetic Properties of V Doped $MgSiN_2$ *Chalcopyrite* **27** 257–261.
[11] Rasander M et al 2017 Structure and lattice dynamics of the wide band gap semiconductors $MgSiN_2$ and $MgGeN_2$ *Int. J. Appl. Phys.* **122** 1-9.
[12] Rufinus J and DeWinter J L 2008 Magnetic properties of Mn-doped chalcopyrites: $(BeSnBeGeMgGe)N_2$ *Journal of Applied Physics* **103.**
[13] Kocak B and Ciftci Y O 2016 Analysis of the structural electronic and optic properties of Ni doped MgSiP2 semiconductor chalcopyrite compound *AIP Conf Proc* **1722** 220013.
[14] Khan K et al 2020 Effect on structural and optoelectronic properties of Sn doping in MgSiP2-A DFT study *J Phys: Conf Ser* **1504** 012013.
[15] Basalaev Y et al 2005 Electronic structure of triple phosphides MgSiP2 ZnSiP2 and CdSiP2 *Russian Physics Journal* **48** 78-83.
[16] Fahad S et al 2015 Structural elastic electronic bonding and optical properties of $BeAZ_2$ (A = Si Ge Sn; Z = P As) chalcopyrites *J. Alloys Compd.* **646** 211-222.
[17] Gani Aa et al 2020 Mechanical stability and optoelectronic behavior of $BeXP_2$ (X=Si and Ge) chalcopyrite *Chin. J. Phys.* **64** 174-182.
[18] Blaha P et al 2020 WIEN2k: An APW+lo program for calculating the properties of solids *J Chem Phys* **152** 074101.
[19] Petukhov A G et al 1994 Electronic structure of wide-band-gap ternary pnictides with the chalcopyrite structure *Phys Rev B* **49** 4549-4558.
[20] Agrawal A et al 2016 A Formation Energy Predictor for Crystalline Materials Using Ensemble Data Mining IEEE 16th Int .Conf. on Data Mining Workshops 1276-1279.
[21] http://docsquantumatkcom/tutorials/formation_energies/formation_energieshtml.
[22] Terzzas E Y et al 2009 *Adv Mat Res* . **68** 89-95.
[23] Mousavi S J et al 2008 Calculation of the Structural Electrical and Optical Properties of κ-Al2O3 by Density Functional Theory *Chin. J. Phys.* **46** 170-180.
[24] Bennacer H et al 2014 First Principal Calculations of Optical Properties of InGaN2 Using in Solar Cells Applications *Conf on Multiphysics Modelling and Simulation for Systems Design and Monitoring. Applied Condition Monitoring* **2** 179-187.
[25] Berrah S et al 2009 Optical properties of the cubic alloy (In Ga) N Physica E **41** 701–704.
[26] Ziane MI 2015 et al First principles study of structural electronic and optical properties of indium gallium nitride arsenide lattice matched to gallium arsenide *Mater. Sci. Semicond. Process.* **30** 181–196.

Numerical and CFD analysis of a heat transfer enhancement in turbulent flow through a circular pipe using nanofluid

S Gupta [1*], J Singh [2], B K Gill[2]

[1]Dr. S. S. B. U. I. C. E. T, Panjab University, Chandigarh, Punjab, India
[2]Assistant Professor of Dr. S. S. B.U. I. C. E. T, Panjab University, Chandigarh, Punjab, India

*Corresponding author e-mail: surbhi1476@gmail.com

Abstract. This paper presents numerical and CFD simulation of an Al_2O_3-water nanofluid turbulent flow in a circular pipe which has a 36 mm diameter with a constant heat flux at the pipe wall using ANSYS FLUENT 2020 R2. The turbulent flow under different Reynolds numbers, from Re = 10000 to 50000 was used to process the numerical experiments. Different concentrations of nanoparticles, ranging from 1% to 4%, were used. The results were computed utilizing the single-phase approach. The coefficient of transfer of heat of nanofluids is found to be greater than that of base liquid. Enhancement in transfer of heat is also observed with rising in volume concentration of particles.

1. Introduction

In today's industry, heat exchangers are quite vital and their main role is to transfer heat at a maximum rate. Many various strategies are already being employedpreviously to improve the rate of transfer of heat[1] and achieve a desirable degree of thermal efficiency. Changes in flow design, boundary conditions, or thermophysical parameters, such as fluid thermal conductivity, can all passively improve heat transfer rates.

Traditional fluids for heat transfer, like ethylene glycol, water, and oil, have insufficient thermal properties [2]. To boost the overall heat transfer rate, a new novel fluid named Nanofluid has been first introduced by Choi et al. [3] which has piqued researchers' interest because their thermal conductivities are highwhen compared to the traditionalfluids for heat transfer. It is feasible to reduce the size of equipment for the transfer of heat by using nanofluids.

Nanofluids comprise nanoparticles ina usual fluidand these particles are in colloidal suspension. Carbides, metals, oxides, carbon nanotubes are nanoparticles that are most commonlyemployed in Nanofluids. TiO_2, CuO, Al_2O_3, SiO_2, and CeO_2 are examples of common nanoparticles.

In this field of heat transfer, there has been a significant rise in research activity recently. To explain the improvement of heat transfer, various models have been developed. Several publications published in the last few years have focused on the experimental investigation of the convection of nanofluid in both turbulent and laminar regimes.

Pak and Cho [4] reported the first research on heat transfer via turbulent convection of $TiO_2$and Al_2O_3 nanofluids within a circular tube, showing an increment in Nusselt number in association with a concentration of particles and developing a Nusselt number relation. Xuan and Li [5] did another key study onCu-based nanofluids undergoing turbulent convection, introducing a novel Nusselt number correlation that takes particle concentration into account directly. The forced convection in a laminar motion of nanofluids in a tube that is circular was experimentally evaluated by S. Zenali et al. [6]with a fixed temperature on the surface. The outcomes indicated that nanoparticles that are there in fluid enhanced transfer of heat.

Numerical simulations of turbulent and laminar convection of nanofluids have been the subject of numerous articles. To carry out the simulations, several methods are used.

The thermodynamic characteristics of nanofluid convectionwere first considered by Palm et al. [7], and they discovered significant differences between nanofluids with characteristics that are consistent and nanofluids with thermodynamic characteristics. Maiga et al. [8] investigated the feasibility of convection, bothturbulent and laminar, within a tube that is uniformly heated, finding that nanoparticles' existence improved transfer of heat. In a tube that is circular, Vincenzo et al. [9] tested a motion of turbulence nanofluids using single as well as mixture models.The results showed that conventional coefficient of heat transfer of base fluid is smaller than nanofluid and even Reynolds number as well as fraction of particle volume rises, enhancement in heat transfer increases.

In this current paper, turbulent transfer of heat flow inside a pipe which is circular using Al_2O_3/water nanofluid with constant heat flux onpipe wall is developed and explored numerically utilizing a single-phase technique. The results obtained are achieved in terms of volume concentration, temperature, length, Nusselt number, and Reynolds number profiles.

2. Mathematical modeling

2.1. Geometry

The two-dimensional geometrical configuration as shown in figure 1 consists of a circular pipe having D means diameter of about 0.01 meter and an L means length of about 1 meter. The nanofluid is composed of Al_2O_3 nanoparticles having a 36 nm diameter and water.

2.2. Numerical approach and Meshing

The current problem was solved using the code for CFD ANSYS FLUENT 20R2. The volume-control scheme wasutilized as a tool for resolving governing equations. Under this procedure, the conservation equations are transformed into a series of algebraic problems by spatial integration.For single-phase, pressure-velocity coupling was approached by the SIMPLE algorithm.

To ensure that numerical findings are accurate and consistent, meshing is done. Meshing divides the domain into parts, each representing a different element as shown in figure 2.A default meshing is also performed, which produces results that are not better than the one chosen. In the problem under consideration, the domain is divided into 400×22 elements.

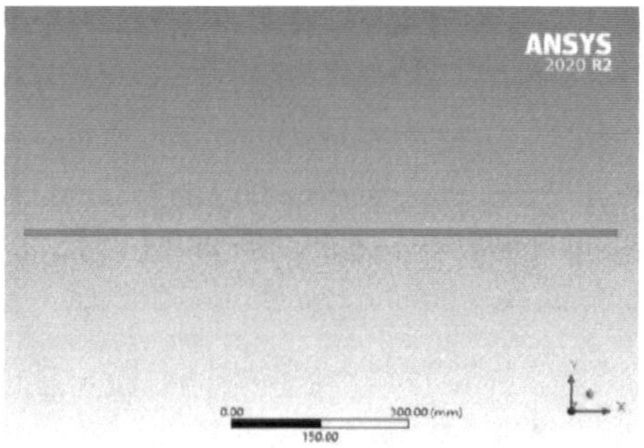

Figure 1. 2D Circular pipe geometry

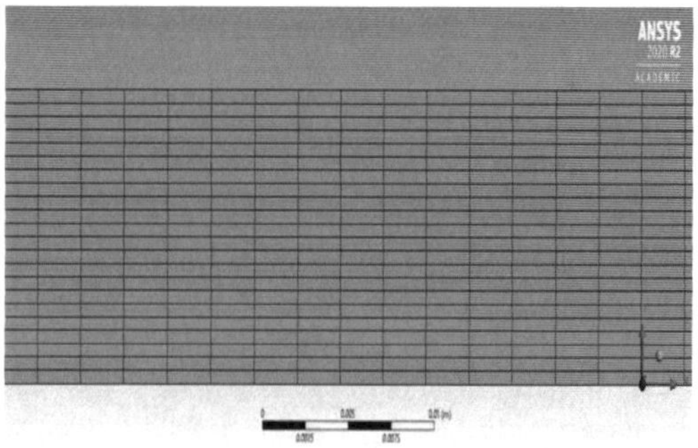

Figure 2. Meshing

2.3. Equationsthat directing

A dynamic fluid, as well as thermal behavior of investigated nanofluid, is studied utilizing a one-phase strategy as in this approach, the fundamental assumption behind it says combination acts as a one-

phase fluid(nanofluid behaves like a homogenous fluid), but thermophysical properties get enhancedowing to the presence of nanoparticles.
The directing equations of a one-phase strategy are mathematically described in the following way [10]

Mass conservation:

$$\nabla \cdot (\rho_m \vec{V}) = 0 \quad (1)$$

Momentum conservation:

$$\nabla \cdot (\rho_m \vec{V}\vec{V}) = -\nabla P + \nabla \cdot (\tau - \tau_t) \quad (2)$$

Energy conservation:

$$\nabla \cdot (\rho \vec{V} C_p T) = \nabla \cdot \left(\lambda \nabla T - c_p \rho_m \bar{vt}\right) \quad (3)$$

In the equation for energy, viscous dissipation and compression work are considered negligible.

2.4. Turbulence modeling

To account for turbulence processes, estimate models are required to complete the Thermo fluid dynamic field's governing equations. This work considers the k- ε model, which was developed by Launder et al. [11], which provides two equations.

For turbulent kinetic energy

$$\nabla \cdot (\rho_m \vec{V}_m k) = \nabla \cdot \left(\frac{\mu_{t,m}}{\sigma_k} \nabla k\right) + G_{k,m} - \rho_m \varepsilon \quad (4)$$

For dissipation rate

$$\nabla \cdot (\rho_m \vec{V}_m \varepsilon) = \nabla \cdot \left(\frac{\mu_{t,m}}{\sigma_k} \nabla \varepsilon\right) + \frac{\varepsilon}{\kappa}(C_1 G_{k,m} - C_2 \rho_m \varepsilon) \quad (5)$$

2.5. Thermophysical characteristics of nanofluids

The simulation technique necessitates the nanofluid thermophysical characteristicscalculationssince its outcomeshave a big impact. The thermophysical properties of nanofluids with various concentrationsof particle were calculated usingclassical theory of mixtures as shown in equations 6-9 and used for CFD simulation [10,12,13]. The following are the equations that were employed in this work.

For density

$$\rho_{nf} = (1 - \phi)\rho_f + \phi \rho_{np} \quad (6)$$

For specific heat

$$C_{p,nf} = \frac{\phi(\rho C_p)_{np} + (1 - \phi)(\rho C_p)_f}{\rho_{nf}} \quad (7)$$

For viscosity

$$\mu_{nf} = (123\phi^2 + 7.3\phi + 1)\mu_f \quad 8$$

For thermal conductivity

$$k_{nf} = \left[\frac{k_{np} + 2k_f + 2(k_{np} - k_f)\phi(1+\beta)^3}{k_{np} + 2k_f + (k_f - k_{np})\phi(1+\beta)^3}\right]k_f \quad 9$$

In the above equation,

$$\beta = \frac{t}{r_p} \quad 10$$

Table 1. ThermophysicalpropertiesatT=293k forAl_2O_3nanoparticles and water[14]

Material	ρ (kg/m³)	C_p (J/kgK)	μ (kg/ms)	k (W/mK)
Water, ϕ=0%	998.20	4182.0	9.98×10^{-4}	0.597
Al2O3	3880	773	-	36

Table 2. ThermophysicalpropertiesatT=293k forAl_2O_3/ water nanofluid

ϕ (%)	ρ (kg/m³)	C_p J/kgK)	μ (kg/ms)	k (W/mk)
1	1027.01	4053.21	1.023×10^{-3}	0.620
2	1055.83	3931.45	1.048×10^{-3}	0.643
3	1084.65	3816.16	1.073×10^{-3}	0.667
4	1113.47	3706.84	1.098×10^{-3}	0.692

2.6. A Boundary conditions

2.6.1. Inlet. The uniform axial velocity, V_0, for considered Reynold numbers are applied at the inlet. A temperature, $T_0 = 293$ K, and turbulence intensity, I, is imposed.

For initial velocity

$$V_0 = \frac{Re \times \mu_{nf}}{D \times \rho_{nf}} \quad 11$$

For the intensity of turbulence

$$I = 0.16 \times Re^{-1/8} \quad 12$$

*2.6.2. Pipe wall.*While kinetic energy of turbulence,as well asits dissipation, is zero, the non-slip requirements and consistent heat flux are enforced. The surface is heated to a constant temperature (100 °C).

3. Results

The transfer of heat in turbulent motion through a circular pipe utilizing Al_2O_3-water nanofluid was investigated using several numerical simulations in this paper. The results were computed using the single-phase approach for **ϕ = 0% to 4%, Re = 10 ×10³ to 50 × 10³, q = 500000W/m²**.

For all particle concentration, the case of Re = 20 × 10^3 and Re = 50 × 10^3 are considered. Figures3 and 4 show bulk and wall temperature curves along the pipe axis for both Reynolds numbers. Figure 3 illustrates that the bulk temperature increases along the pipe axis but when particle concentration rises, the result remains the same. Figure 4 shows that when the concentration of particles rises, the temperature of the wall goes down which indicates that adding nanoparticles has a favorable influence on the temperature of the wall. In comparison to base fluid, for a value of ϕ = 1%, the temperature of the wall is lower.

The temperature difference between base fluid and nanofluid at pipe exit is around 10 K, resulting in a significant particle effect.

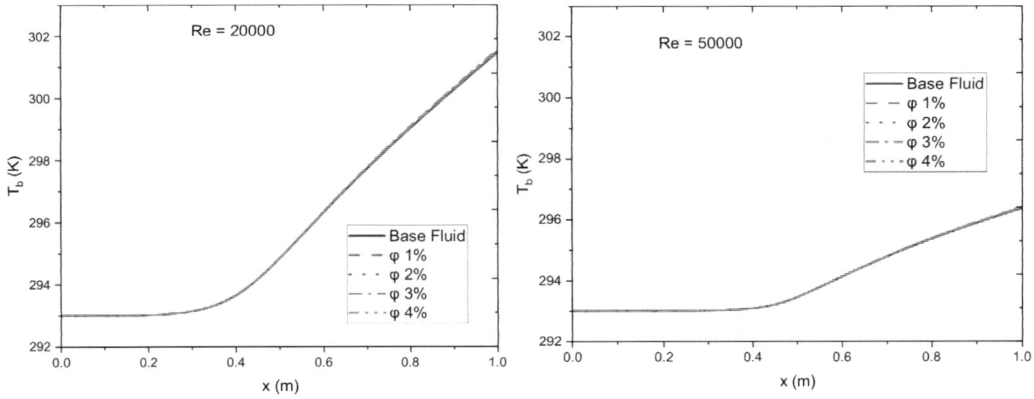

Figure 3. For i) Re = 20 × 10^3 and ii) Re = 50 × 10^3, the effect of particle concentration on bulk temperature along pipe axis.

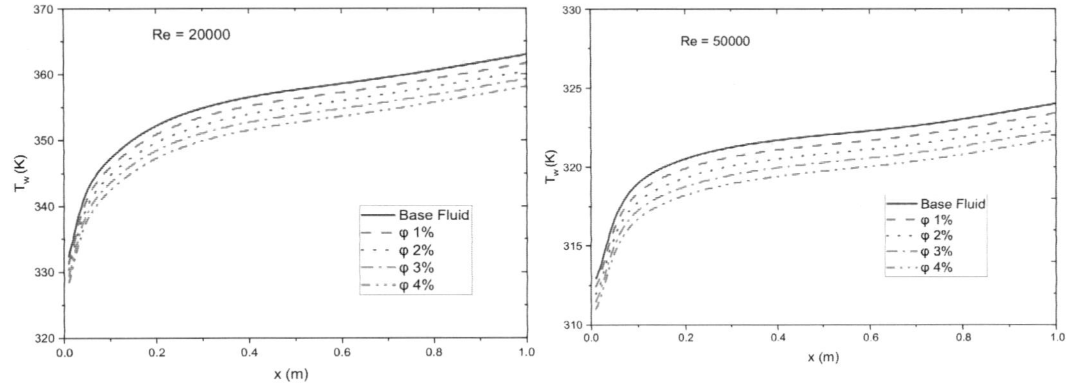

Figure 4. For i) Re = 20 × 10^3 and ii) Re = 50 × 10^3, the effect of particle concentration on wall temperature along pipe axis.

For Re = 10 × 10^3 to 50 × 10^3, the case of ϕ= 0% (base fluid) and ϕ = 4% are considered. Figures 5 and 6 show bulk and wall temperature curves along the pipe axis for both particle concentrations. Figure 5 illustrates that the bulk temperature increases along the pipe axis but drops as Reincreases. It's clear from figure 6 that when Re rises, temperature of wall reduces it depicts that changing Re has a positive impact on wallas well asbulk temperatures. In comparison to Re = 20 × 10^3, temperature of wall for Re = 50× 10^3 is lower.

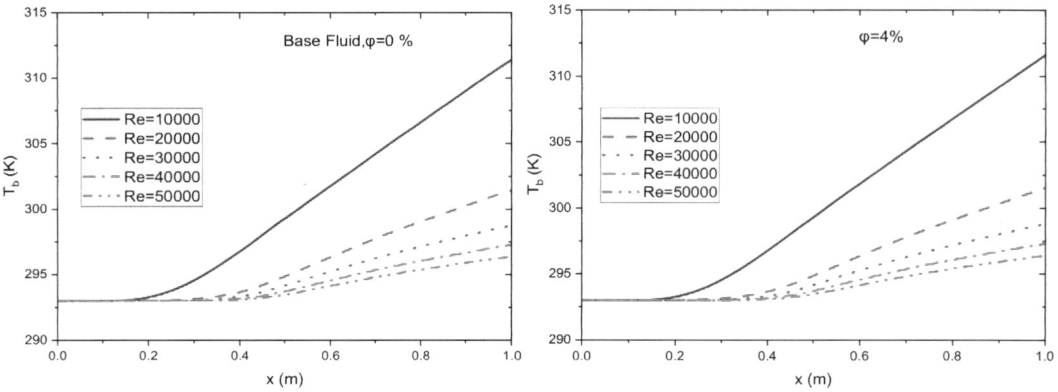

Figure5. For i) ϕ= 0% and ii) ϕ= 4%, the influence of varying Re on temperature of bulk along pipe axis.

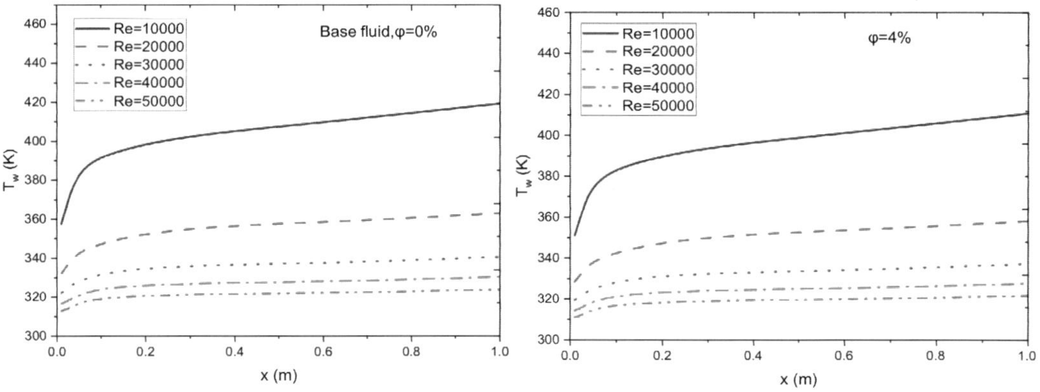

Figure 6. For i) ϕ= 0% and ii) ϕ= 4%, the influence of varying Re on wall temperature along pipe axis.

Influence of concentration of particles on coefficient of transfer of heat at surface throughout axis of tube for Re = 20×10^3 and Re = 50×10^3 is shown in figure 6. It reduces to a minimum value very immediately after entering the pipewhere temperature differentialexists amid bulk as well as wall is greatest and then slightly decreases along the pipe axis but with accordance to particle concentration, surface heat transfer coefficient increases.

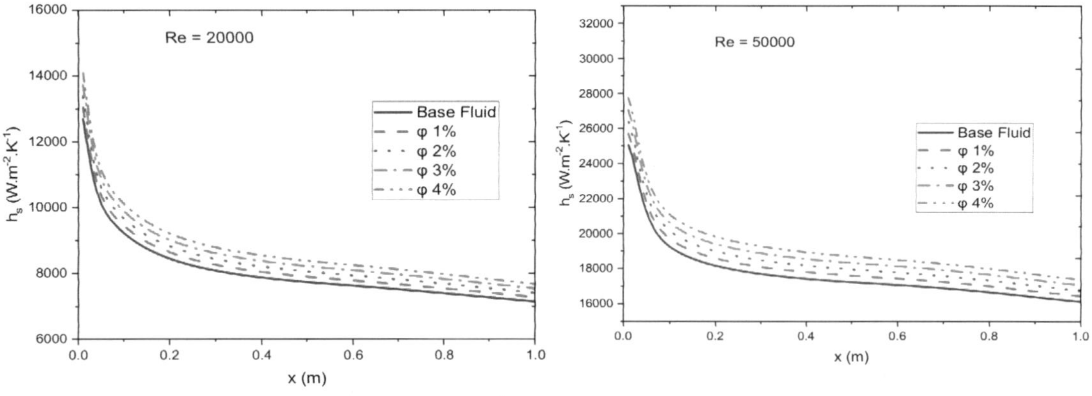

Figure 7. i) Re = 20×10^3 and ii) Re = 50×10^3, the influence of ϕ on coefficient of transfer of heat at surface on along pipe axis.

Average coefficient of transfer of heatrises as particle concentration increase for Re = 20 × 10^3 and Re = 50 × 10^3 as shown in figure 8. The increment in the ratios of the average coefficients of transfer of heat from nanofluid to base fluid is given in Tables 3 and 4. As a result, a nanofluid has a greater thermal conductivity, allowing heat transfer rates to also be expanded. But for increasing particle concentration and increasing average heat coefficient, the Nusselt number decreases.

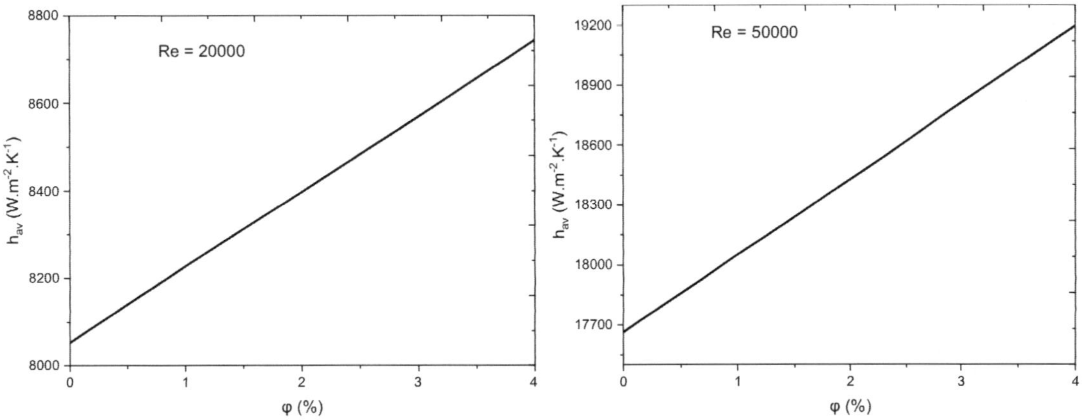

Figure 8. For i) Re = 20 × 10^3 and ii) Re = 50 × 10^3, the influence of ϕ on average coefficient of transfer of heat.

Table 3. Average Heat transfer coefficients and Nusselt Number for Re = 20000

ϕ (%)	h_{av}(W/mK)	h_{nf}/h_f	Nu_{av}	Nu_{nf}/Nu_f
0 (bf)	8052.95	1.00	134.89	1.00
1	8227.21	1.02	132.70	0.98
2	8396.93	1.04	130.59	0.97
3	8569.27	1.06	128.47	0.95
4	8743.95	1.09	126.36	0.94

Table 4. Average coefficients of heat transfer as well as Nusselt Number for Re = 50000

ϕ (%)	h_{av}(W/mK)	h_{nf}/h_f	Nu_{av}	Nu_{nf}/Nu_f
0 (bf)	17665.2	1.00	295.90	1.00
1	18051.2	1.02	291.15	0.98
2	18427	1.04	286.58	0.97
3	18808.7	1.06	281.99	0.95
4	19195.4	1.09	277.39	0.94

Similarly, influence of increasing Re on coefficient of heat transfer at surface along pipe axis as evidenced in figure 9. for ϕ = 0% and ϕ = 4%.

Table 5. shows average coefficient of transfer of heat at surface for all considered cases. The table clearly shows that averagecoefficient of transfer of heat increases as concentration of particles as well as Re rises. Therefore, the influence of particle concentration appears to be more significant at higher Re. Nusselt no. rises in parallel with coefficient of average heat transfer.

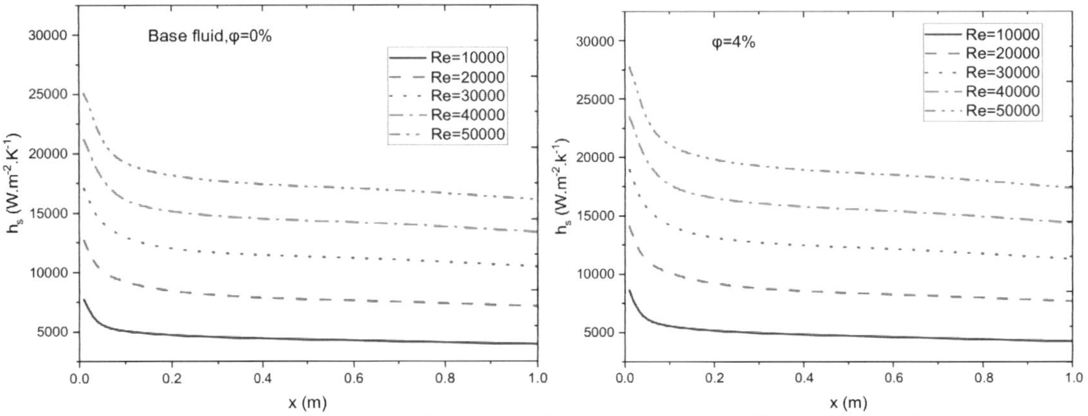

Figure 9. For i) ϕ= 0% and ii) ϕ = 4%, influence of different Re on coefficient of transfer of heat at surface along pipe axis

Table 5. Average coefficients of transfer of heat for all considered cases

Re	Base Fluid	ϕ = 1%	ϕ = 2%	ϕ = 3%	ϕ = 4%
10000	4522.58	4619.01	4712.96	4808.43	4905.26
20000	8052.95	8227.21	8396.93	8569.27	8743.95
30000	11636.7	11893	12142.5	12395.7	12652.1
40000	14732.9	15054.6	15367.9	15686.1	16008.6
50000	17665.2	18051.2	18427	18808.7	19195.4

Based on the simulation findings, the Nu numbers are computed as follows:

$$Nu = \frac{h \cdot D}{k} \tag{13}$$

The average Nufor consideredRe and concentration of particleas evidenced in Figure 10. Nu rises as Re number rises, but falls as concentration of particle rises.

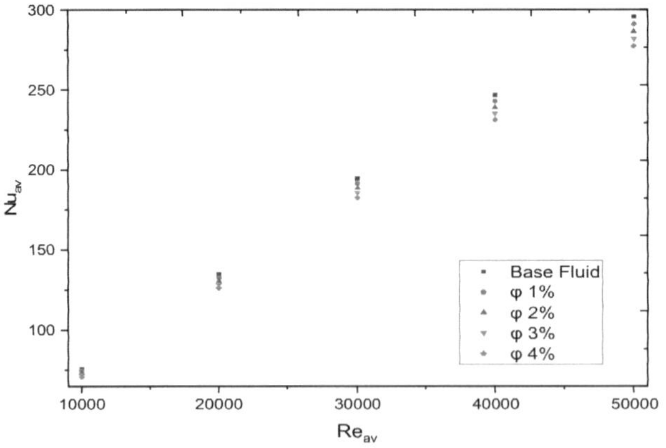

Figure 10. For varying ϕ, compare the Nusselt number to the Reynolds number.

4. Conclusion

The numerical and CFD simulation of coefficient of transfer of heat of turbulent convective motion of alumina-based nanofluid within a circular pipe is presented in this study using a single-phase model.

1. These results have demonstrated that in contrast tobase fluid, incorporating nanoparticles in a base fluid made a considerate contribution to coefficient of transfer of heat.
2. The data clearly illustrates that as concentration of particle as well asRe rises, so does coefficient of transfer of heat.
3. For all concentrations of particles, maximal rates of transfer of heat were found in correlation with the highest Reynolds.
4. Tables 3 and 4 exhibits average coefficient of transfer of heat& Nusselt number as a function of concentration of nanoparticle, with proportionate rises in total rates of heat transfer and decrease in Nusselt number but with rising Re, average coefficient of heat transfer and Nusselt number increases.

Nomenclatures

D diameter of the pipe, m
d diameter of nanoparticles, m
r radius of nanoparticle, m
V velocity offlow, m/s
C_p fluid's specific heat, J/kg K
L length of pipe, m
h coefficient of heat transfer, W/m2K
T temperature, K
k kinetic energy of turbulence, m^2/s^2
t thickness of the nanolayer
I turbulence intensity
Nu Nusselt number,
q heat flux of wall, W/m2
Re Reynolds number

Greek Symbols

ρ density of the fluid, kg/m3
ϕ volume concentration of particle
ε dissipation of turbulent kinetic energy, m^2/s^3
λ fluid thermal conductivity, W/mK
τ wall shear stress, Pa
μ dynamic viscosity of the fluid, kg/ms

Subscript

m mixture
t turbulent
b bulk
np nanoparticle
nf nanofluid
av average
s surface
0 inlet condition
f base fluid

References

[1] Ahmed H E, Mohammed H A and Yusoff M Z 2012 An overview on heat transfer augmentation using vortex generators and nanofluids: approaches and applications *Renew. Sust. Energ. Rev.* **16**(8) 5951-5993.
[2] Abouali O and Falahatpisheh A 2009 Numerical investigation of natural convection of Al_2O_3

nanofluid in vertical annuli. *Heat and mass transfer* **46**(1) 15-23.

[3] Choi SUS and Eastman JA 1995 Enhancing thermal conductivity of fluids with nanoparticles *ASME Int.Mech. Eng. Congress & Exposition* **66** 99-106.

[4] Pak B C and Cho Y I 1998 Hydrodynamic and heat transfer study of dispersed fluids with submicron metallic oxide particles *Exp. Heat. Transf.* **11** 151–70.

[5] Xuan Y and Li Q 2003 Investigation on convective heat transfer and flow features of nanofluids *J. Heat Transfer* ***125*** 151-155

[6] Heris S Z, Esfahany M N and Etemad S G 2007 Experimental investigation of convective heat transfer of Al2O3/water nanofluid in circular tube *Int. J. Heat Fluid Flow* **28** 203–10.

[7] Palm SJ, Roy G and Nguyen CT 2006 Heat transfer enhancement with the use of nanofluids in radial flow cooling systems considering temperature-dependent properties *Appl. Therm. Eng.* **26** 2209–18.

[8] Maiga S E B, Nguyen C T, Galanis N and Roy G 2004 Heat transfer behaviors of nanofluids in a uniformly heated tube *Superlattices and MicrostructuresAcademic Press* 543–57.

[9] Bianco V, Manca O and Nardini S 2011 Numerical investigation on nanofluids turbulent convection heat transfer inside a circular tube *Int. J. Therm. Sci.* **50** 341–9.

[10] Das S K, Choi S U S, Yu W and Pradeep T 2008 *Nanofluids Science and Technology*, John Wiley & Sons Hoboken NJ USA.

[11] Launder B E and Spalding D B 1972 Lectures in Mathematical Models of Turbulence *Academic Press,* London .

[12] Mansour R B, Galanis N and Nguyen C T 2007 Effect of uncertainties in physical properties on forced convection heat transfer with nanofluids *Appl. Therm. Eng.* **27** 240–9.

[13] Ravisankar B and Tara Chand V 2013 Influence of nanoparticle volume fraction, particle size and temperature on thermal conductivity and viscosity of nanofluids- A review *Int. J. Automot. Mech. Eng.* **8** 1316–38.

[14] Rohsenow W M and Hartnett J R 1999 Handbook of heat transfer **36** 36-3347

IOP Conf. Series: Materials Science and Engineering 1225 (2022) 012022 doi:10.1088/1757-899X/1225/1/012022

Effect of Alkyl Substitution to the Active Layer Material for Improved Efficiency in Bilayer Organic Solar Cell

S Rugma[1], B Devu[1] and C O Sreekala[1]

[1]Department of Physics, School of arts and sciences, Amritapuri, Amrita Vishwa Vidyapeetham, Kollam, Kerala.

Email adress:*sreekalaco@am.amrita.edu*

Abstract. Organic based thin film solar cells have achieved considerable power conversion efficiency (PCE). As there is a desire to the benefits of on-demand energy production, future world energy consumption will continue to rise. Bilayer organic solar cell has two active layers in between conductive electrodes. Present report, is a simulation study to compare the PCE of bilayer organic solar cells by varying one of its active layer material using General-Purpose Photovoltaic Device Model or GPVDM software. The materials used for active layer are MEH-M3EH-PPV, MEH-DOO-PPV, MEH-PPV and M3EH-PPV based on the degree of alkyl substitutions. Data of absorption spectrum, refractive index spectrum and photoluminescence spectrum are added as input and obtained the simulation results. Values of power conversion efficiencies, fill factor and other photovoltaic parameter values for each bilayer organic solar cell is obtained. Transmittance spectrum, reflectance spectrum, Current density- Voltage plot, variation of charge density and total charge density with applied voltage are attained from the simulation and compared. The results give out effect of side chain alkyl substitution on efficiency of bilayer organic solar cell. MEH-DOO-PPV with longest alkyl side chain substitution has maximum power conversion efficiency. This intuition helps to design suitable bilayer organic solar cell, which has got its relevant applications in the optoelectronics.

1. Introduction

One of the severe issues with the current world is the increase in energy consumption. So, it's very important to find new ways to generate energy in which the use of renewable energy generated by solar cell seems to be a solution. But the currentformed by silicon solar cell is at high price. Organic solar cells have gained the interest as they are low cost solar cells and provide the generation of inexpensive renewable energy [1, 2]. The light absorbing layer of organic cells are organic polymer material. In 1995, bulk polymer heterojunction was reported. After 2000, there was a rapid increase in efficiency from 1% to 12%(2013).So the density of photocurrent production by organic solar cell is good enough to be compared with that produced by inorganic solar cell [3].

For Bilayer cells, it has two layers in between the conductive electrodes [4]. The difference between ionization energies and electron affinity generates the electrostatic forces between the interfacesof the two layers. There is a need of exciton production by light in this charged regionfor efficient collection and charge separation [5]. In comparison with single layer photovoltaic cells, bilayer splits excitons much more efficiently because of the peculiarity of the materials chosen allowing the local electric fields to be strong [6]. The electron accepter layer has a higher ionization potential and electron affinity making the other layer the electron donor [7]. Recently, trial to setup bilayer solar cell is found in different devices but these cannot be compared with the bulk heterojunction equivalents [8]. Bilayer

solar cell have lower heterojunction interfacial area available. In order to overcome this issue, a modulation in the morphology of active layers is required. [9]
Organic Solar Cell consists of two electrodes with a photoactive layer in between. One of the electrodes is transparent and sputtered on a transparent substrate. The other one electrode improves the charge separation and transport in the range of the diffusion length of the excitons [11]. There is a buffer layer between active layer and the electrodes which acts like charge selective conductor by conducting holes and blocking electrons and vice versa. It is in the active layer, where light absorption and charge carrier generation happens. This metallic back electrode either acts as anode or cathode [12, 13]. In this simulation, side chain alkyl substitution effect is studied. This will help the experimental scientists to develop efficient solar cell device thereby saving the time, energy and materials.

2. GPVDM Software

General-Purpose Photovoltaic Device Model (GPVDM) refers to a tool, which is used for the simulation of photovoltaic devices and optoelectronics. The software is also used for the device stability and the clarification of power conversion efficiency (PCE). The software contains an electrical solver, an optical solver, and an easy-to-use graphical interface. The simulation of organic solar cells (OSCs) can be done including transfer matrix models and drift diffusion. The use of our own choice of material and laser spectra allowed both the stimulation and experiment to match [14]
GPVDM simulation software solve: Poisson equation (Equation 1), Bipolar drift–diffusion equations (Equations 2 and 3), Carrier continuity equations (Equations 4 and 5).

$$\frac{dy}{dx}\varepsilon_0.\varepsilon_T\frac{\partial\varphi}{\partial x} = q(n-p) \tag{1}$$

$$J_n = q\mu_c n\frac{\partial E_c}{\partial x} + qD_n\frac{\partial n}{\partial x} \tag{2}$$

$$J_p = q\mu_n p\frac{\partial E_v}{\partial x} - qD_p\frac{\partial p}{\partial x} \tag{3}$$

$$\frac{\partial J_n}{\partial x} = q(R_n - G + \frac{\partial n}{\partial x}) \tag{4}$$

$$\frac{\partial J_p}{\partial x} = -q(R_p - G + \frac{\partial p}{\partial x}) \tag{5}$$

3. Materials and Methodology

The device structures used areITO/PEDOT:PSS/M3EH-PPV/PCBM/Al, ITO/PEDOT:PSS/MEH-PPV/PCBM/Al, ITO/PEDOT:PSS/MEH-DOO-PPV/PCBM/Al and ITO/PEDOT:PSS/MEH-M3EH-PPV/PCBM/Al, where ITO is the front electrode, Al is used as back electrode, PEDOT:PSS as intermediate layer and M3EH-PPV, PCBM, MEH-PPV, MEH-DOO-PPV, MEH-M3EH-PPV,are active layers. Figure 1 depicts the structure of the active layer materials.

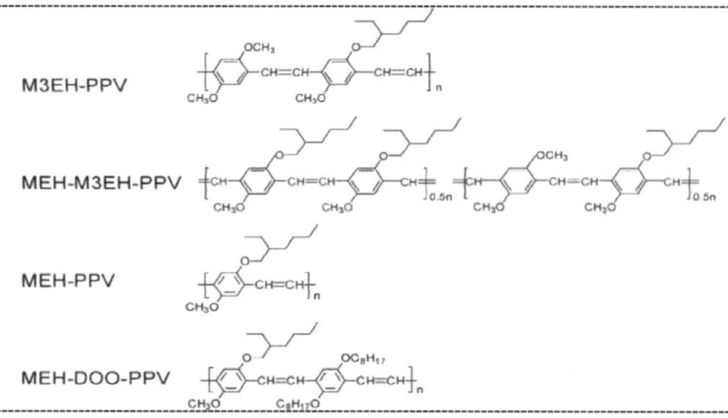

Figure 1. The molecular structure of the active layer material used in this report [15]

3.1. Simulation of Bilayer Organic Solar Cell

Aim of this series of work is to understand the PCE of organic bilayer solar cell on varying active layer material. For that the Organic Bilayer Solar cell with two contact electrodes, ITO and Al, the extinction coefficient *k* values and the refractive index *n* values are taken from the reported work [16]. There is an intermediate layer of material PEDOT: PSS and the *n-k* values are taken from the doctoral dissertation of Mengfang Li [17]. The simulated devices are having two active layers, one of which is made by the material PCBM, which is fixed in all the simulated devices and the *n-k* values are exported from the reported work [18]. The other active layer is varied and the materials are M3EH-PPV, MEH-PPV, MEH-DOO-PPV and MEH-M3EH-PPV whose molecular structures are shown in figure 1 and the n-k values are imported to the simulator from the reported work [15]. Using these data, organic bilayer solar cells with device structure ITO/PEDOT:PSS/M3EH-PPV/PCBM/Al, ITO/PEDOT:PSS/MEH-PPV/PCBM/Al, ITO/PEDOT:PSS/MEH-DOO-PPV/PCBM/Al and ITO/PEDOT:PSS/MEH-M3EH-PPV/PCBM/Al, are simulated in GPVDM software. The simulated device structure is shown in figure 2. Absorbance spectrum and refractive index - wavelength plot of ITO, PCBM,PEDOT:PSS, M3EH-PPV, MEH-PPV,MEH-DOO-PPV, MEH-M3EH-PPV and Al is obtained.

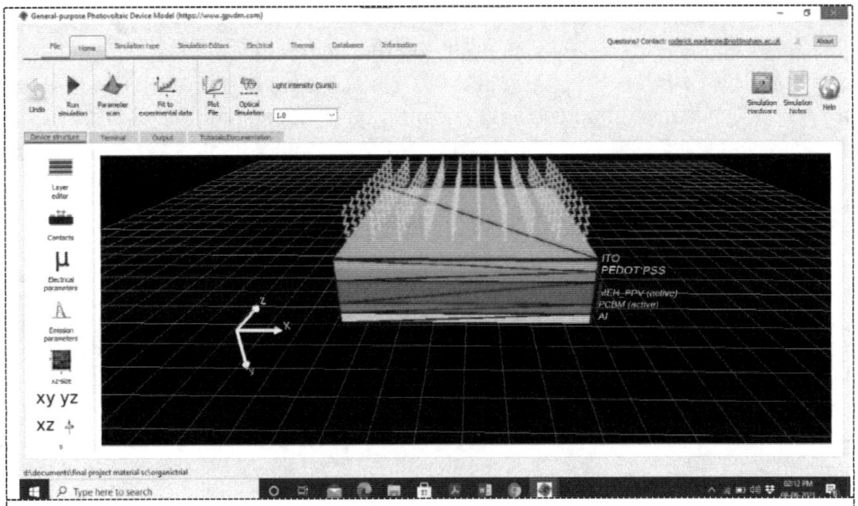

Figure 2. The device structure of simulated bilayer solar cell with active layer MEH-PPV/PCBM

The photoluminescence spectrum of active layer material, PCBM,M3EH-PPV,MEH-PPV, MEH-DOO-PPV and MEH-M3EH-PPVis also obtained. The refractive index spectrum and absorption spectrum of all materials are uploaded in the GPVDM software. Photoluminescence spectraof the active layers are also uploaded. Then the program is allowed to run. Efficiency values for each material, charge density plot, current density plot, photon density, generation rate are obtained as simulation result and it is compared.

4. Results and Discussion

4.1. Power conversion efficiency and electrical parameters of simulated bilayer solar cells

Efficiency of four different bilayer organic solar cells with device structureITO/PEDOT:PSS/M3EH-PPV/PCBM/Al, ITO/PEDOT:PSS/MEH-PPV/PCBM/Al, ITO/PEDOT:PSS/MEH-DOO-PPV/PCBM/Al and ITO/PEDOT:PSS/MEH-M3EH-PPV/PCBM/Alare examined using GPVDM software. In case of the materials, the degree of alkyl substitution increases respectively, M3EH-PPV, MEH-M3EH-PPV, MEH-PPV, and MEH-DOO-PPV. From GPVDM simulation results, the efficiencies of these four bilayer organic solar cells are compared. The photovoltaic parameters of the devices made from M3EH-PPV,MEH-PPV, MEH-M3EH-PPV and MEH-DOO-PPV as active layers are obtained from the simulation. The obtained results aregiven in table 1.

Table 1. Photovoltaic parameters of the simulated bulk bilayer organic solar cell devices

Active layer	V_{oc} (V)	J_{sc} ($mAcm^{-2}$)	FF	V_{max}(V)	J_{max} $mAcm^{-2}$	Max power (m_{pp}) Watt m^{-2}	Efficiency (%)
M3EH-PPV	0.52	4.80	0.64	0.369	4.32	15.98	1.59
MEH-M3EH-PPV	0.52	5.38	0.63	0.368	4.78	17.65	1.76
MEH- PPV	0.52	5.39	0.63	0.368	4.82	17.77	1.77
MEH- DOO-PPV	0.52	5.65	0.62	0.368	5.03	18.55	1.85

4.2. Variation of Transmitted Light with Wavelength

The absorption and refractive index spectrum of active layer materials is related to its transmittance. Figure 3 shows transmittance versus wavelength for the materials M3EH-PPV, MEH-M3EH-PPV, MEH-PPV and MEH-DOO-PPV. The simulation is carried out for the optical wavelength ranging from 350nm to 625nm and transmittance ranges from 0.0005 a.u to 0.0035a.u. Initially transmittance of the films are very less, after 500 nm, transmittance is very high which is very useful for a solar cell device.

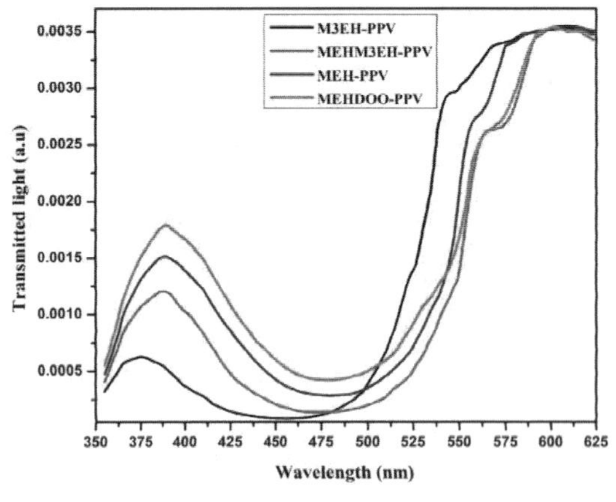

Figure 3. Transmittance spectrum of the active layers

It is observed that the first transmission peak is located between 350nm and 450nm. In this range, the highest peak is for MEH-DOO-PPV with transmission of 1.79×10^{-3}a.u at 391nm. The transmission is lowest for M3EH-PPV with a peak value of 1.51×10^{-3}a.u at 391nm. This finding is in accordance with the efficiency of the used material. That is, the value for transmittance is in the increasing order with increase in efficiency.

4.3. Variation of Reflectance light with Wavelength

The figure 4 shows the reflectance spectrum of MEH-DOO-PPV, MEH-M3EH-PPV, MEH-PPV and M3EH-PPV. There are three peaks, in the region of wavelength range from 350nm to 525nm, next from 525nm to 560nm and third one at 560nm to 600nm.

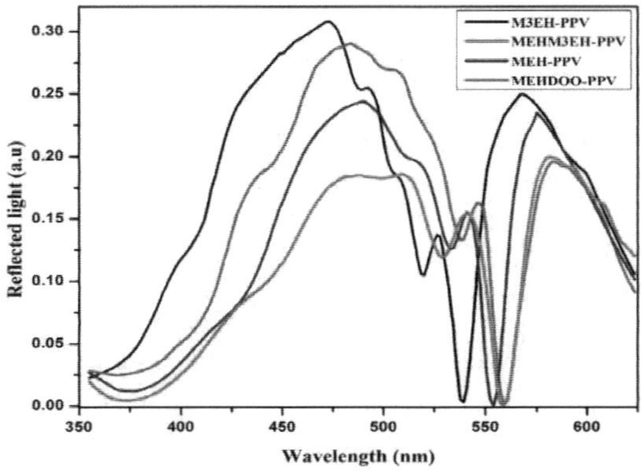

Figure 4. Variation of reflected light with wavelength for M3EH-PPV, MEH-PPV, MEH-DOO-PPV and MEH-M3EH- as active layers

The peak in the range, 350nm to 525nm has maximum reflectance value for M3EH-PPV. There is a peak in the range from 400nm to 600nm where the curvature increases, reaches a maximum peak value and then decreases. M3EH-PPV is having maximum peak value in both absorption spectrum and refractive index spectrum and it decreases respectively, M3EH-PPV, MEH-M3EH-PPV, MEH-PPV and MEH-DOO-PPV. Maximum absorption coefficient and refractive index decreases with increase in the degree of alkyl substitution, because increase in number of alkyl substitution leads to reduce in π-electrons per unit volume. In the transmittance spectrum obtained, for the same wavelength range, from 400nm, the transmittance value decreases, reaches a minimum value and then it increases. So, as absorption coefficient and refractive index increases, transmittance decreases. From 400nm, reflectance value increases, reaches minimum and then decreases. M3EH-PPV have maximum reflectance at this range. So, reflectance is directly proportional to absorption coefficient and refractive index.

4.4. Current density- Voltage Curve (J-V Curve)

Characteristics current density v/s applied voltage curve for MEH-M3EH-PPV, MEH-DOO-PPV, MEH-PPV and M3EH-PPV are shown in the figure 5. Short circuit current for the solar cell having the active layer M3EH-PPV is 4.8 mA/cm^2, 5.38 mA/cm^{2}for MEH-M3EH-PPV, 5.39 mA/cm^{2}for MEH-PPV and for MEH-DOO-PPV is 5.65 mA/cm^2. But the open circuit voltage (V_{oc}) remains the same as 0.52 V for all the simulated devices. The fill factor remains nearly same. At the same time, the efficiency of the device increases with the alkyl substitution.

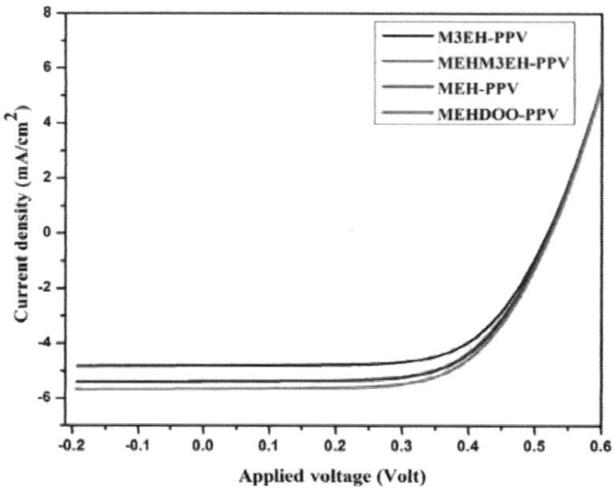

Figure 5. Variation of current density with applied voltage for the solar cells made from M3EH-PPV, MEH-PPV, MEH-DOO-PPV and MEH-M3EH-PPV

4.5. Basic Electrical Parameters of the Simulated Bilayer Organic Solar Cell

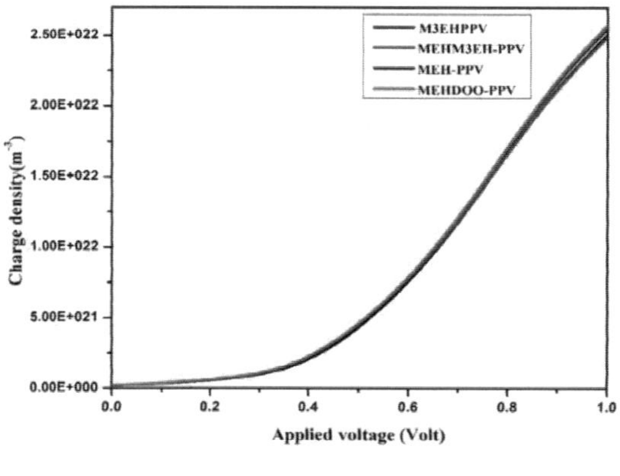

Figure 6. Variation of charge density with applied voltage for the solar cells with active layer asM3EH-PPV, MEH-M3EH-PPV, MEH-PPV and MEH-DOO-PPV.

The plot for charge density as the function of applied voltage is represented in the figure 6. The charge density strongly depends in on the applied voltage. It increases rapidly with the increase in the applied voltage. Charge density is almost same for all the solar cells made fromM3EH-PPV, MEH-M3EH-PPV, MEH-PPV and MEH-DOO-PPV.

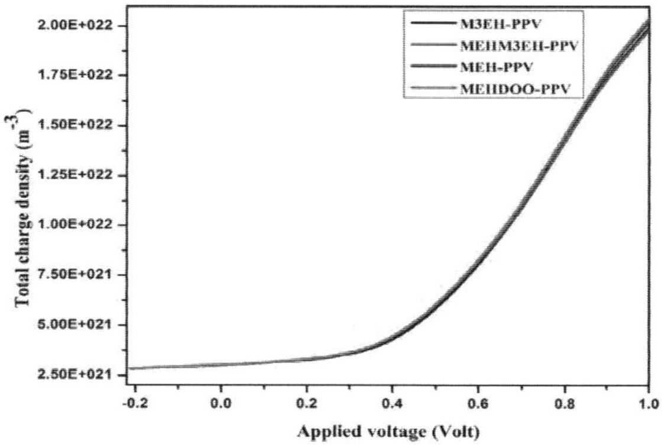

Figure 7. Variation of total charge density with applied voltage for the solar cells with M3EH-PPV,MEH-M3EH-PPV, MEH-PPV and MEH-DOO-PPV as the active layer.

Total charge density v/s applied voltage curve for the solar cells made from M3EH-PPV, MEH-M3EH-PPV, MEH-PPV and MEH-DOO-PPV as active layer is represented in figure 7. The total charge density is almost same for all the devices.

5. Conclusion

This report investigates the PCE of four different bilayer organic solar cell with device structureITO/PEDOT:PSS/M3EH-PPV/PCBM/Al, ITO/PEDOT:PSS/MEH-PPV/PCBM/Al, ITO/PEDOT:PSS/MEH-DOO-PPV/PCBM/Al and ITO/PEDOT:PSS/MEH-M3EH-PPV/PCBM/Al. Power Conversion Efficiencies of these four bilayer organic solar cells are obtained using GPVDM software. It is found that the higher efficiency is for ITO/PEDOT: PSS/MEH-DOO-PPV/PCBM/Al where, MEH-DOO-PPV is a PPV derivative which has a greater number of alkyl substitutions among them. From the simulation result, the power conversion efficiency of MEH-DOO-PPV is 1.85% with lowest fill factor of 62%. Transmitted light v/s wavelength plot, reflected light v/s wavelength plot, J-V plot and variation of charge density and total charge density with applied voltage are obtained. From the above simulation results, it is found that for a particular range of wavelength, MEH-DOO-PPV, shows the maximum transmittance and for the same range of wavelength, it shows minimum reflectance. The solar cells with MEH-DOO-PPV has minimum current density when voltage is applied. Beyond a particular voltage, current density of ITO/PEDOT:PSS/M3EH-PPV/PCBM/Al, ITO/PEDOT:PSS/MEH-PPV/PCBM/Al, ITO/PEDOT:PSS/MEH-DOO-PPV/PCBM/Al and ITO/PEDOT:PSS/MEH-M3EH-PPV/PCBM/Al is almost same. As per the simulation result, power conversion efficiency is directly proportional and fill factor is inversely proportional to the degree of alkyl substitutions. Among the simulated bilayer devices, the device with structure ITO/PEDOT: PSS/MEH-DOO-PPV/PCBM/Al is the most efficient one and is matching with the reported work.

Acknowledgement
The authors thank General Purpose Photovoltaic Device Model Software (GPVDM).

References

[1] Sreekala C O (2011) Studies on organic bulk heterojunction devices for photovoltaic solar cell applications

[2] Dangate M, Munshi A, Sampath, W S Boltalina, O V Strauss, S H Saravanan, C & Nair S 2016 Investigation of organic small molecules and polymer compounds for CdTe back contact *In 2016 IEEE 43rd Photovoltaic Specialists Conference (PVSC)* pp 1438-1442 IEEE.

[3] Hiramoto M & Shinmura Y 2017 Organic solar cells In Springer Handbook of Electronic and Photonic Materials (1-1) Springer, Cham.

[4] Tang C W 1986 Two-layer organic photovoltaic cell *Applied physics letters* **48** 183-185

[5] Kekuda D, Huang JH, Ho KC and Chu CW 2010 Modulation of donor– acceptor interface through thermal treatment for efficient bilayer organic solar cells *The Journal of Physical Chemistry C* **114** 2764-2768.

[6] Hösel M, Angmo D and Krebs FC, 2013 Organic solar cells (OSCs) *Handbook of Organic Materials for Optical and (Opto) electronic Devices*,473-507

[7] Nath B G, Chinnusamy S, Anjali C M , Chandran H, Shaikh N I & Dangate M S 2017 Density functional theory study of organic small molecules for renewable energy system *In 2017 International Conference on Technological Advancements in Power and Energy* (TAP Energy) 1-6 IEEE

[8] Sreekala C O, Saneesh P F, Sreelatha K, SKishnashree A & Roy M S 2012 Organic bulk heterojunction solar cell based on rosebengal: ncTiO2 and parameter extraction by simulation *In Advanced Materials Research* **403** 4304 Trans Tech Publications Ltd

[9] Mishra A K and Shukla R K 2020 Electrical and optical simulation of typical perovskite solar cell by GPVDM software *Materials Today: Proceedings*

[10] Ran N A, Love J A, Heiber M C, Jiao X, Hughes M P, Karki A Wang M Brus V V Wang H Neher D and Ade H 2018 Charge generation and recombination in an organic solar cell with low energetic offsets *Advanced Energy Materials 8*(5) p1701073

[11] Tajik S etal 2020 Recent developments in conducting polymers: applications for electrochemistry *RSC Advances* **62** 37834-37856

[12] G Yu J Gao J C Hummelen F Wudl A J Heeger 1995 Polymer Photovoltaic Cells: Enhanced Efficiencies via a Network of Internal Donor-Acceptor Heterojunctions *Science* **270** 1789

[13] Ye Q & Xu J W 2015 Bulk heterojunction solar cells—opportunities and challenges *Solar cells—new approaches and reviews* **359**

[14] Tress W 2014 Organic Solar Cells: Theory Experiment and Device Simulation Germany: Springer International Publishing

[15] Bahtiar A Koynov K Mardiyati Y Hörhold H H &Bubeck C 2009 Slab waveguides of poly (p-phenylenevinylene) s for all-optical switching: impact of side-chain substitution Journal of Materials Chemistry **19**(40) 7490-7497

[16] Ramsdale C M &Greenham N C 2003 The optical constants of emitter and electrode materials in polymer light-emitting diodes Journal of Physics D: Applied Physics **36**(4) L29

[17] Li M (2018) Conjugated Polymer-Based Functional Materials for Wearable Electronics: Electrochromism

[18] Gevaerts V S Koster L J A Wienk M M & Janssen R A 2011 Discriminating between bilayer and bulk heterojunction polymer: fullerene solar cells using the external quantum efficiency ACS applied materials & interfaces **3(9)** 3252-3255

[19] Bader M A Marowsky G Bahtiar A Koynov K Bubeck C Tillmann H & Pereira S 2002 Poly (p-phenylenevinylene) derivatives: new promising materials for nonlinear all-optical waveguide switching JOSA B **19(9)** 2250-2262

Raman Photoluminescence - An Efficient way to distinguish Natural diamonds and Lab Grown diamonds

Mr. Akesh Kumar G*[,a], Dr. N. Prabavathy[a], Dr. Laxmidhar Biswal[a],
[a]Centre for Innovation and Technology Excellence , Titan Company limited (Jewellery Division, Brand TANISHQ), Hosur, INDIA

*Corresponding author mail id:akeshkumarg@titan.co.in

Abstract. The advancements in Science and Technology introduced Lab grown diamonds (LGDs) that meet with diverse applications. The LGD's possess similar electrical, optical and mechanical characteristics compared to natural diamonds mined from earth. There is a possibility of LGDs infiltration in the Natural diamond pipeline in the jewellery industry. Hence, it becomes crucial to detect LGDs. The present article illustrates the importance of Raman Photoluminescence over absorption spectroscopy and Fluorescence/ Phosphorescence imaging for identification of diamond nature. In the current work, a single Laser source of 532 nm was used to obtain the distinct molecular information of Natural diamonds and Lab Grown diamonds.

Key words: Natural diamonds, Lab grown diamonds, Raman, Photoluminescence, Liquid Nitrogen, Laser.

1. Introduction

At present, the segregation of natural diamonds and Lab grown diamonds (LGD) is a huge challenge to the diamond Industry due to its similar optical, mechanical and electrical properties [1]. The natural diamonds are generally formed at pressures of 5.5 –8.0 GPa (55,000– 80,000 atmospheres) and temperatures of 1000°C – 1400°C at the hottest region of earth called mantel (depths of 140–240 km below earth surface). The oxidation or reduction of methane gas present at mantel, which depends on residence temperature and pressure, results in the growth of diamonds. They are brought to earth due to volcanic eruption [2]. The price of diamond is usually higher due its rarity and multiple processes involved in mining, processing, cutting & polishing and its exceptional diamond characteristics (high hardness, high electrical and thermal conductivity). The diamonds are used in the jewellery due to maximum Lustre (Specular Reflection) and Light Performance (Brilliance, Fire and Scintillation) that enhances its aesthetic look [3]. Additionally it contributes to wide applications such as diamond anvil cells, semiconductor devices, optical components, machine tools and abrasive wheels, etc [4].

Since diamonds are more valuable, that limited its application to be readily accessible. Hence the invention of Lab grown diamonds have reduced the demand of the diamond market. The LGDs are synthesized by High pressure and High Temperature (HPHT) and Chemical vapour deposition (CVD) techniques. The mechanical, electrical and optical characteristics of LGD are similar to natural diamonds except its origin. The LGD can also be manufactured basedon the application requirement. The HPHT synthesis consists of reaction chambers where a carbon source such as graphite or diamond powder is placed along with a catalyst to facilitate diamond growth on a diamond seed by temperature gradient at pressures of 5–6 GPa and at temperatures of 1300 °C – 1600°C [5]. CVD diamonds are produced by the ionization ofmethane gases to produce plasma discharge that deposits on the surface of the substrate, usually Silicon [6]. As technology advances in synthesizing LGD, there is the huge threat in infiltration/undisclosed mix-up of LGD into the Natural diamonds supply chain due its price advantage .The cost of LGD is approx.30-40 % of natural diamonds. In some HPHT diamonds a small magnetic flux due to the impurities added in the molten flux during their growth, fluorescence property, and stress pattern created on diamonds due to prolonged pressure is observed that distinguishes the natural diamonds from HPHT to an extent [7]. In a recent study the magnetic flux was also observed in natural diamonds, due to its formation from the melt containing magnetic materials [8]. And the most recent invention of CVD diamonds produces a similar stress pattern to natural diamonds. Henceforth, it is very crucial for the jewelry industryto confidently detect the LGD from natural diamonds [6]. However, as they are visually identical; the conventional tests are not 100% reliable to separate them. This article gives a glimpse about the equipment/technology available in the market and discusses their shortcomings, which can be overcome by Raman Photoluminescence [9]. The present research paper substantiates the detection of natural diamonds and Lab grown diamonds

2. Material and Characterization

The Absorbance studies in diamond were carried out using SHIMAZDU- UV Vis spectrophotometer. The Raman Photoluminescence (PL) was analysed using Enspectr Raman-R532. 1.5 L of Liquid Nitrogen was filled in a Cryoseal container (1.75L) was purchasedfrom INOXCVA. The Fluorescence images were carried out using IMAGER.

Totally 250 diamonds were analysed in the present study. Different grades (VVS-GH, VS-EF, SI2-IJ, I1/I2-GH) of natural, HPHT and CVD diamonds of various shapes and sizes were procured from markets across the globe and obtained from the preliminary screening equipment and from external grading. Among them, PL spectra of 50 diamonds have been reported in this work.

3. Results and Discussions

In General, diamonds are classified based on the presence of Nitrogen atoms as per Gemmological Institute of America (GIA) Type classification standards [10]. Figure 1 shows thelattice structure of type classified diamonds. Type I consists of Nitrogen atoms (> 5 ppm) either as aggregates (Type Ia) or isolated form (Type Ib). Type IIa consists of no nitrogen or boron impurities and Type IIb consists of boron impurities. Most natural diamonds fall in Type Ia category and few percent in Type 1b category. CVD diamonds fall in the Type II category. The presence/absence of nitrogen was one of the key factors to segregate diamonds [11]. The equipment faces limitations in identifying 100% diamonds at the first time, it refers to a certain percentage of diamonds to check in some other advanced technology. In some cases, equipment is not viable for detecting all categories of LGDs.

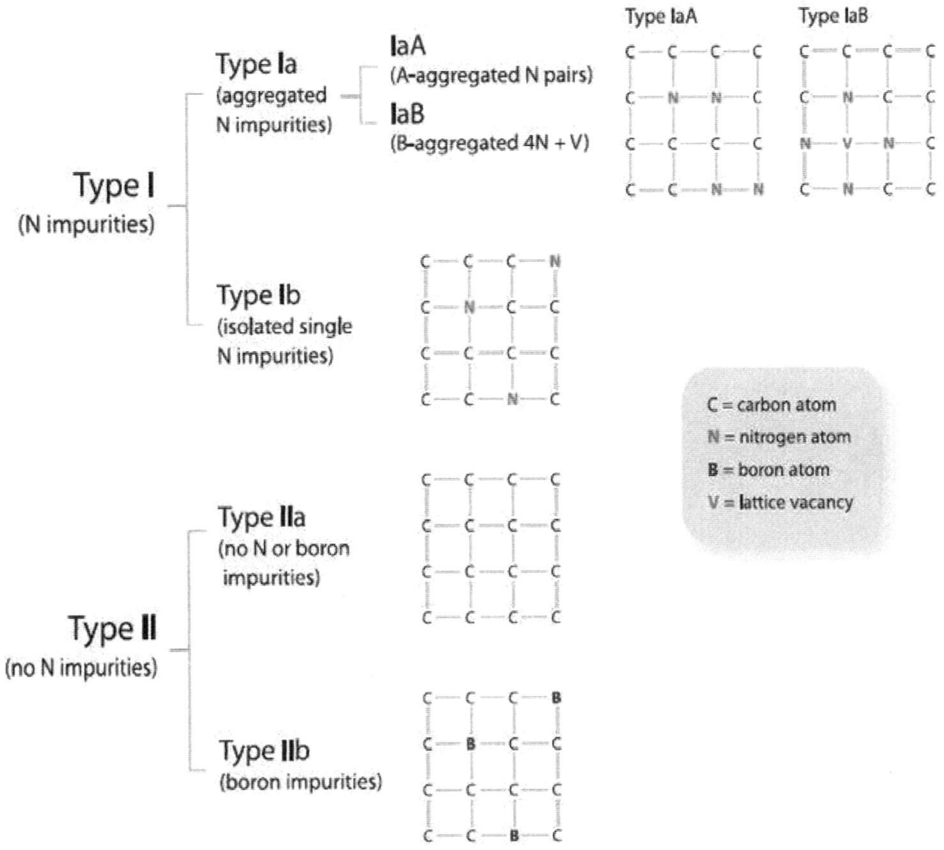

Figure 1. Type Classification of Diamonds
(https://4cs.gia.edu/wp-content/uploads/2014/06/Diamond-Type-Classification)

3.1 Absorbance spectroscopy

In UV Vis absorption spectra of diamonds, the presence of aggregated nitrogen atoms (Type 1aA) can be identified by the appearance of 415nm (N3 centre) peak[12] . This majorly distinguishes the natural diamond from LGD. Also, Natural diamonds of Type 1a show 480 nm band, H3 centre and flat absorption bands [13]. These diamonds are categorized as refer, since further screening is required on its authenticity. Figure 2 shows the UV- Vis spectra of natural diamond with N3 centres.

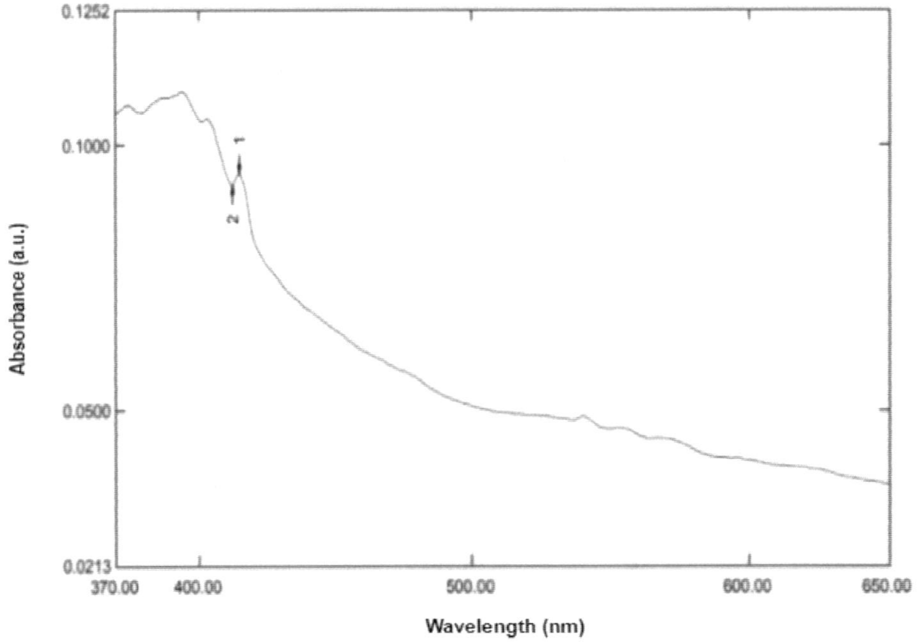

Figure 2. Absorbance spectra of natural diamond with N3 centres

3.2 Fluorescence and Phosphorescence imaging

The excitation time taken by the molecules of diamond leads to the change in color caused by fluorescence [14]. The Figure 3 shows the fluorescence and phosphorescence imaging of natural diamonds and LGD's. The natural diamonds give natural ink blue color in short phosphorescence. The HPHT diamonds show a Turquoise green color in long phosphorescence whereas CVD diamonds show colors ranging from brown, green, yellowish orange etc [15]. In some cases the presence of diamond color is distinguishable from the expected fluorescence. Those diamonds are considered as refer, which has to be tested further for its nature of origin.

3.3 Raman Photoluminescence

Photoluminescence analysis in Raman can provides information about the crystal defect, phase solid state structure of a material [16]. The high spectral resolution of Raman spectrometercan be useful in performing photoluminescence spectroscopy of solid state materials. Enspectr Raman (EnSpectr R532®) with 532 nm (green laser) is used for the photoluminescence peak analysis of diamonds. It combines the advantage of a portable probe system with the performance of a highly specified laboratory instrument. The peaks for diamond segregation fall in the excitation wavelength of green laser[17]. Hence it was used for the present study. Spectral peak of diamond is promising for the segregation of diamonds irrespective of diamond shape, fluorescence, color, carat, cut and clarity. Figure 4 (a,b,c,d) shows the PL spectra of Natural diamond with 4 different clarity and sizes. It was observed that with an excitation laser source of 532 nm, 572 nm was observed in all samples as diamond Peak. Some of the samples showed a peak of 656 nm and 773 nm and some with no distinct peaks that confirmed the origin of natural diamonds irrespective of diamond size, color and clarity [18].

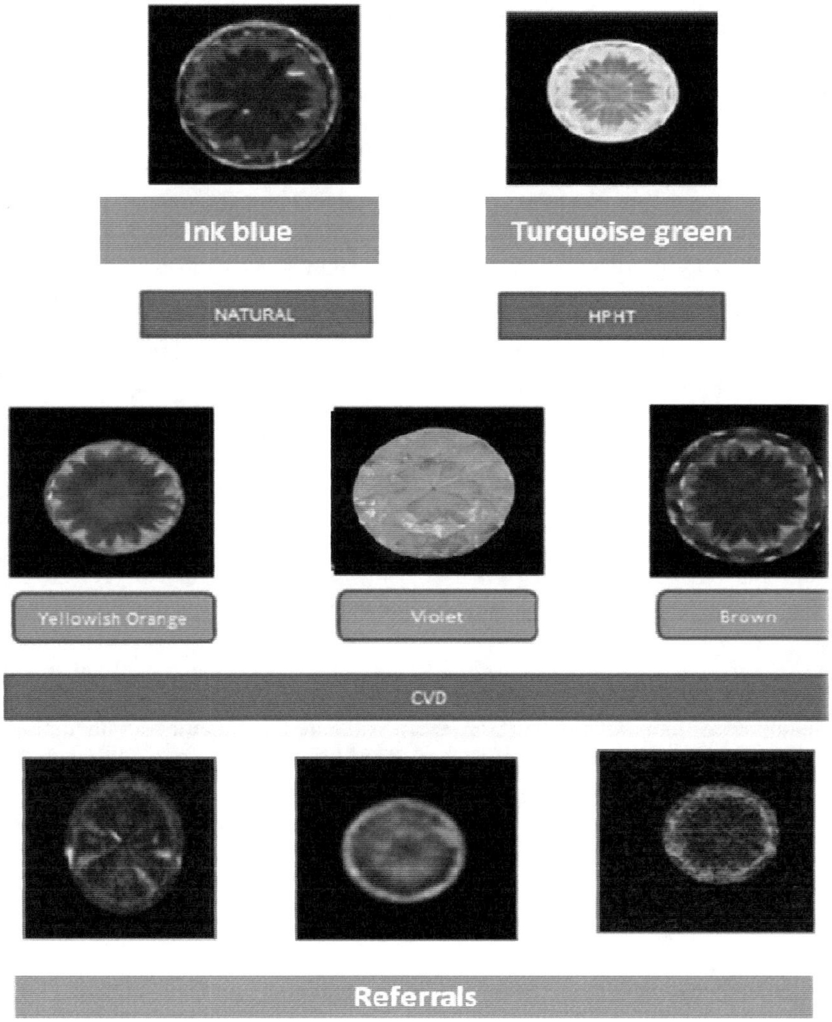

Figure 3. Fluorescence images of Natural, CVD, HPHT and Referral diamonds

IOP Conf. Series: Materials Science and Engineering 1225 (2022) 012023 doi:10.1088/1757-899X/1225/1/012023

Figure 5(a,b) and Figure (c) shows the PL spectra of CVD grown diamonds and HPHT diamonds respectively. The analysis of CVD diamonds showed a significant peak at 737 nm that corresponds to Silicon presence. It might be due to the plasma interaction with the Silicon substrate during their growth [19]. The presence of 596/597 nm was observed for the CVD diamonds with absence of Si peak. The natural diamonds and CVD diamonds are occasionally given High Pressure and High Temperature (annealing treatment) to remove or change its color and obtain a colorless one. Two peaks of 575 nm and 637 nm were prominent for the HPHT treated natural diamonds and CVD [20]. The presence of a peak at 880/883 nm evidently promises to be HPHT diamonds. The peak corresponds to the Nickel presence in the HPHT diamond, which was used as a catalyst in the molten flux during their diamond growth [21].

The samples that were reported as referral from Primary screening equipment wereanalyzed using Raman. Figure 6 (a) shows the PL spectra of diamonds that have been reported as referral diamonds. Most diamonds exhibited fluorescence that was observed from the hump in the PL spectra. The nature of origin was identified. In some referral diamonds, due to the interference of fluorescence and ro-vibrational hot bands, the peaks were hidden [22]. These diamonds required cooling treatment with liquid nitrogen to arrest the rapid vibrations of molecules that bring out the hidden peaks.

The PL spectra with hidden peaks are shown up only when the diamonds are directly cooled with liquid nitrogen on illumination with laser [23]. As the plunging pressure of Liquid nitrogen over diamond makes them roll over from the point of laser source, a small crucible set up with stainless steel and Teflon cup was designed to hold the diamond intact to the laser sourceas shown in Figure 7 (b). The liquid nitrogen was poured directly onto the diamond using a PET(Polyethylene Terephthalate) funnel. Figure 7(a) shows the Raman Enspectr with diamond cooling under Liquid Nitrogen.

Figure 6(b) shows the PL spectra of referral diamonds cooled under liquid nitrogen. The referral diamonds R3and R5 show hidden peaks of 881 nm, which confirms their HPHT origin. The 800 nm peak at R9 shows the presence of Nickel impurities in natural diamonds. It is likely to be natural. Further analysis (Growth Pattern/ absorption spectroscopy) is also required for its confirmation. The presence of 773 nm / 656 nm peaks in R4, R6, R7 confirm its natural origin. The main advantage of using Raman is the flexibility to analyze PL on diamond jewelryirrespective of its size and type. Figure 7(c) shows the PL analysis on the diamond jewelry.

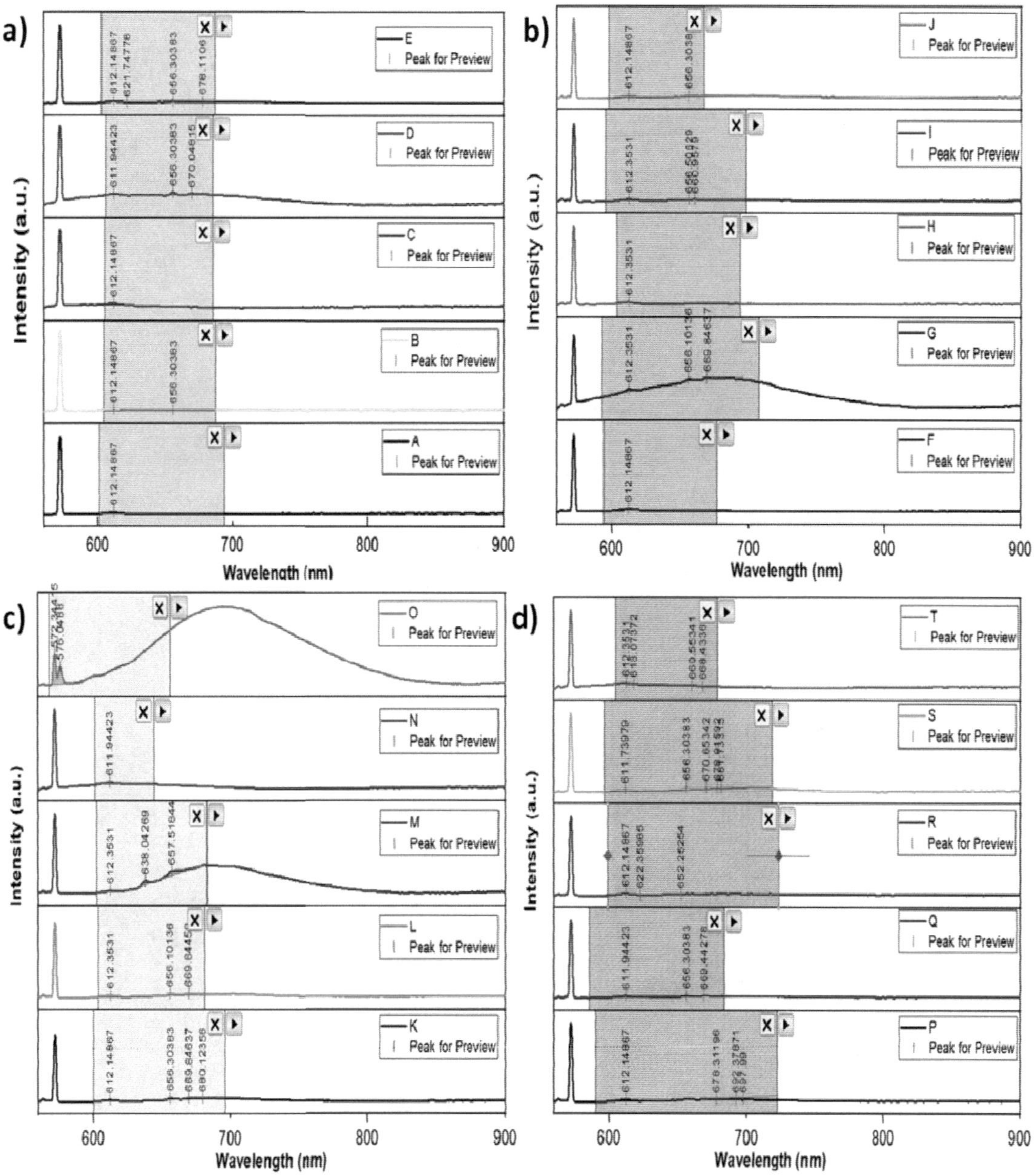

Figure 4. Photoluminescence spectra of Natural diamonds with Grade a) VVSGH; b) VSEF; c) SI2IJ; d) I1/I2 -IJ

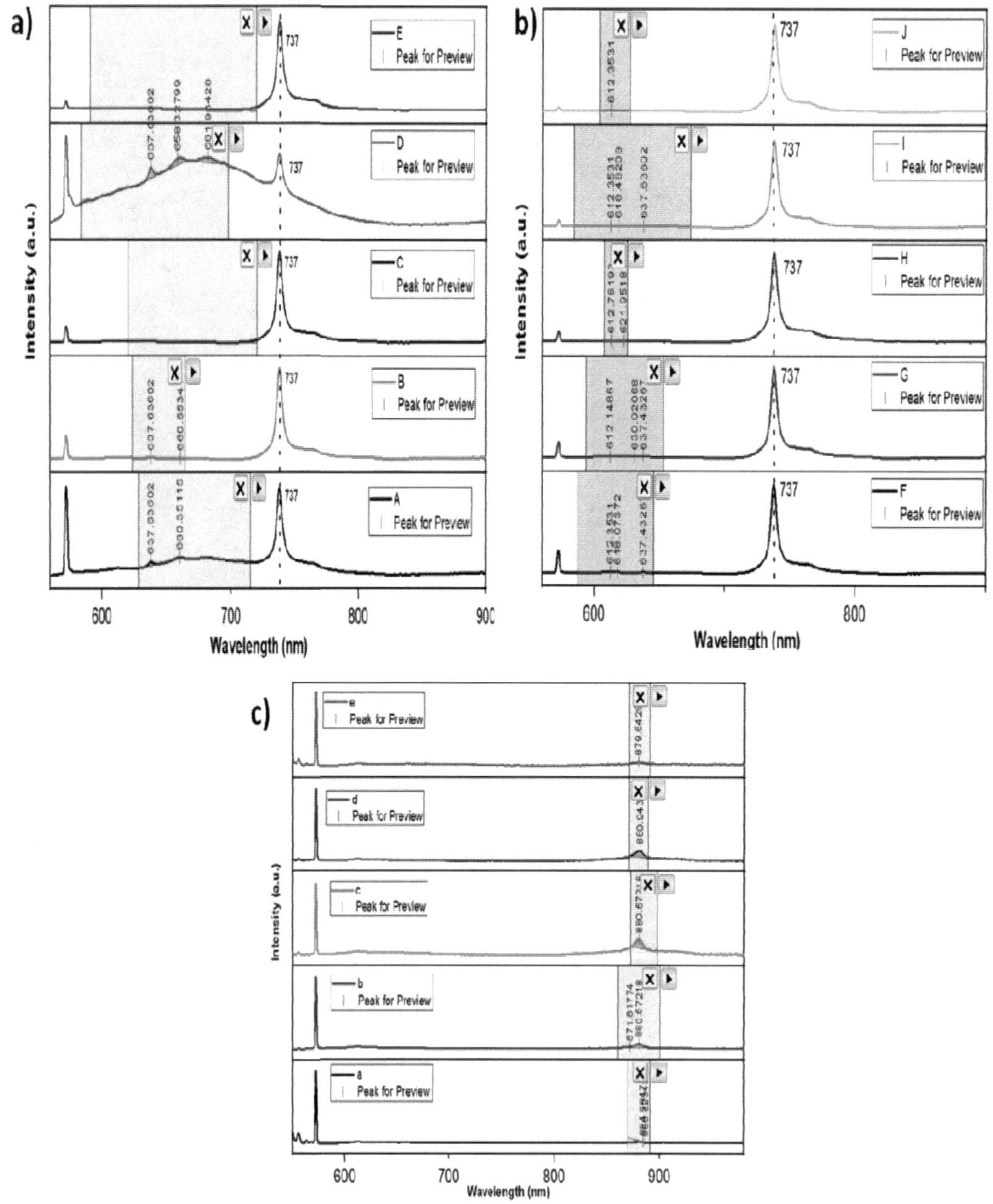

Figure 5. Photoluminescence spectra of (a, b) CVD; (c) HPHT diamonds

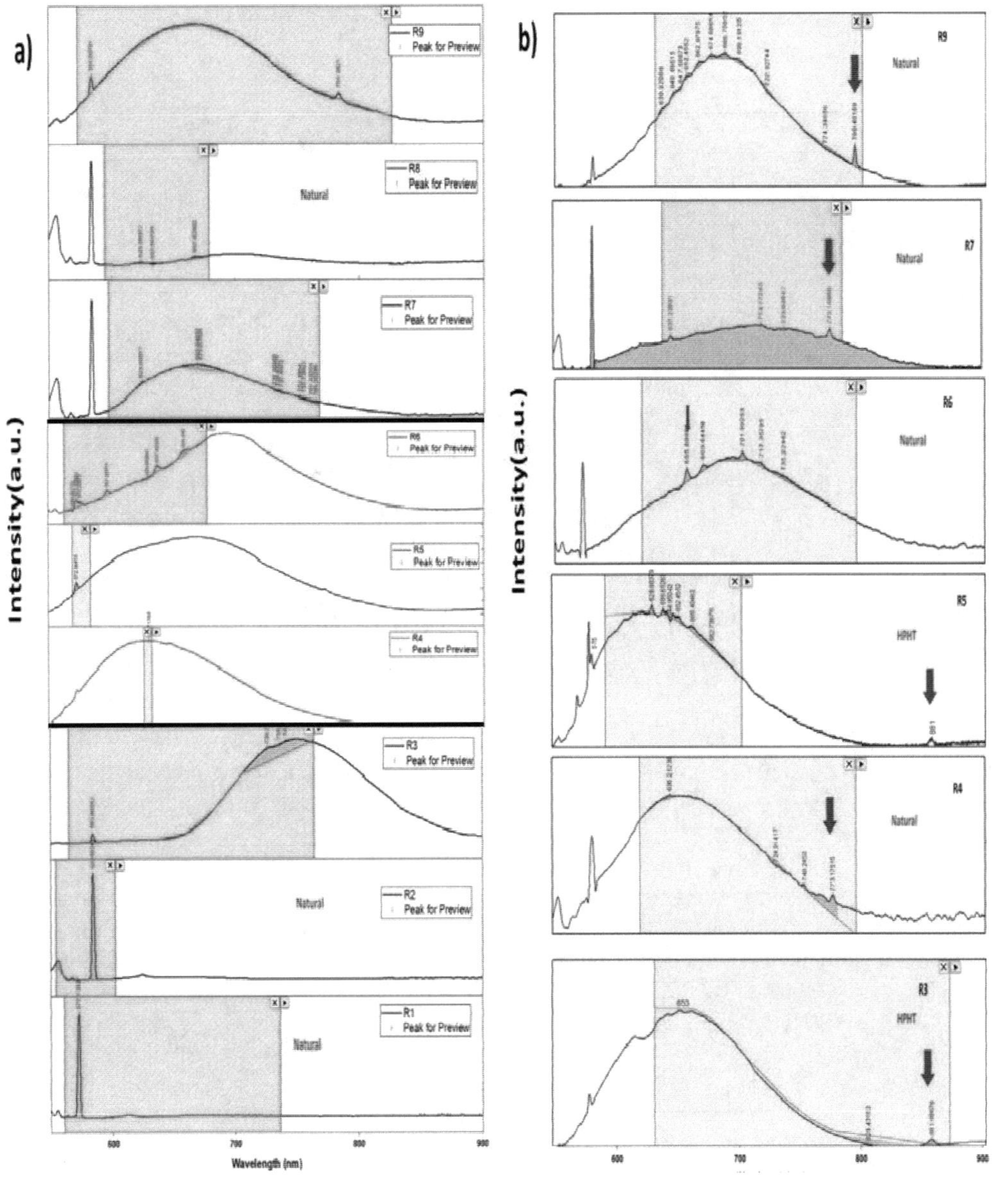

Figure 6. Photoluminescence Spectra of a) Referral diamonds; b) Referral diamond with Liquid Nitrogen

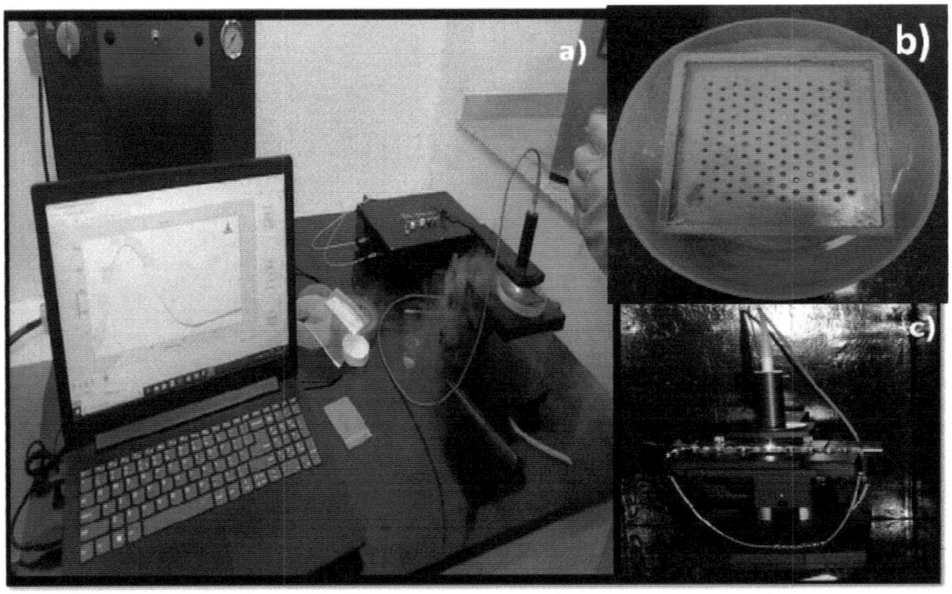

Figure 7. a) Photoluminescence measurement of diamond with Liquid Nitrogen; b) Crucible to hold diamond sample; c) PL analysis on diamond Product

4. Conclusion

Raman Photoluminescence with laser source 532 nm (green laser) was analysed for natural diamonds and Lab Grown diamonds. The spectral peaks give molecular information on the nature of diamond that forms a basis for effective separation. In addition, analysis was done on Referral diamonds, where the hidden peaks were identified by immersing the diamond in Liquid Nitrogen. Thus, Raman forms an authentic tool for separation of natural diamond and LGD.

References

[1] Brookes C A, Brookes E J 1991 Diamond in perspective : a review of mechanical properties of natural diamond *Diamond and Related Materials* **1** 13–17.

[2] Sorokhtin N O 2019 The Origins of Natural Diamonds *Scrivener Publishing*, Wiley, ISBN 978-1-119-59344-7.

[3] Shigley J E, Breeding C M 2015 Visible Absorption Spectra Of Colored Diamonds *GEMS GEMOL* **51(1)** 41–43.

[4] Raman C V 1942 The physics of the diamond *Curr Sci* **11** 261-268.

[5] Sivovolenko S, Zhulin S, Zipa K 2013 *The Australian Gemmologist* **25(2)** 82-121.

[6] Eaton-Magaña S, Shigley J E 2016 Observations on CVD-grown synthetic diamonds:a review *Gems & Gemology* 52(3) 222–245.

[7] Cotter P G , Kohn J A 1954 *J. Am. Ceram. Soc.* **37** 415-20.

[8] Eaton-Magaña S, Shigley J E, Breeding C M 2017 Observations on HPHT-grown synthetic diamonds: a review *Gems & Gemology* **53** 262–284.

[9] Scientific H 2015 Coloured Diamond Defect Identification by Raman Diffusion and Photoluminescence *Mineralogy Geology RA15.*

[10] Mildren R P 2013 Intrinsic Optical Properties of Diamond *Optical Engineering of Diamond (Wiley)* 1-34.

[11] Eaton-magaña S, Post J E, Heaney P J, Walters R A, Breeding C M, Butler J E 2007 Fluorescence Spectra of Colored Diamonds Using a Rapid, Mobile Spectrometer *Gems & Gemology* **43(4)** 332-51.

[12] Ramanathan B K G 1946 The absorption of ultraviolet radiation by diamond Proc. Indian Acad. Sci **24** 137.

[13] Eaton-maga S 2020 Naturally Colored Yellow and Orange Gem Diamonds: The Nitrogen Factor Gems & Gemology **56(2)** 194-219.

[14] Breeding C M, Eaton-magaña S 2019 Fluorescence of Natural and Synthetic Gem Diamond : Mechanism and Applications *In Encyclopedia of Analytical Chemistry, R.A. Meyers (Ed.)* https://doi.org/10.1002/9780470027318.a9670

[15] Shenderova OA, Shames AI, Nunn NA, Torelli MD, Vlasov I, Zaitsev A 2019 Review Article : Synthesis , properties , and applications of fluorescent diamond particles J. Vac. Sci. Technol. B 37 030802.

[16] Fish M L, Massler O, Reid J A, Macgregor R, Comins J D 1999 The application of photoluminescence and Raman spectroscopy of synthetic diamond *Diamond & Related Materials* **8** 1511–1514.

[17] Nagashima K, Nara M, Matsuda J 2012 Raman spectroscopic study of diamond and graphite in ureilites and the origin of diamonds *Meteoritics & Planetary Science* **1737** 1728–1737.

[18] Prawer S, Nemanich R J 2004 Raman spectroscopy of diamond and doped diamond *Philos Trans A Math Phys Eng Sci.* **362(1824)** 2537-65.

[19] Eaton-magaña S, Breeding C M 2016 An Introduction to Photoluminescence Spectroscopy for Diamond and Its Applications in Gemology *Gems & Gemology* **52(**1) 2–17.

[20] Nagashima K, Nara M, And Jun-Ichi M 2012 Raman spectroscopic study of diamond and graphite in ureilites and the origin of diamonds *Meteoritics & Planetary Science* **47** 1728-1737.

[21] Berg D M, Djemour R, Siebentritt L, Susanne S, Pérez-Rodriguez A 2012 Raman analysis of monoclinic Cu2SnS3 thin films *Appl. Phys. Lett.* **100** 192103.

[22] Donato M G, Faggio G, Marinelli M, Messina G, Milani E, Paoletti A, Santangelo S, 2001 High quality CVD diamond : a Raman scattering and photoluminescence study *Eur Phys J B* **20** 133–139.

[23] Wang W 2004 Identification of CVD Synthetic Gem Diamonds Using Raman Spectroscopy *Gems and Gemology* **40(2)** 2-25.

A study on bi-objective optimization for end milling of Aluminium based composite

Pardeep Saini, Deepak Kumar, Anuj K. Maurya, Pradeep K. Singh*

Department of Mechanical Engineering, Sant Longowal Institute of Engineering & Technology, Longowal, 148106, India

*Corresponding Author, E-mail: pkschauhan@yahoo.com

Abstract The composites are renowned for better properties than base alloy. In general, 2-8% weight fraction of reinforcement has been researched for various industrial applications. The addition of reinforcement gives improvement in mechanical properties but badly affects ductility. Liquid route method (casting) has been employed for fabrication of SiC reinforced aluminum composites. Machining has been performed by controlling input parameters such as cutting speed, feed rate and depth of cut. Roughness parameter (R_a) and material removal rate (MRR) have been considered for evaluation of surface quality and productivity respectively. Only MRR has been given consideration for rough machining conditions. Whereas, both surface roughness and MRR have been considered for the finish machining conditions. Standard L_9 orthogonal array have been employed for the experimentation. The occurrence of hard reinforcement in the base alloy creates the casted composite tough to machine. The confirmation experiments have been performed with the optimal settings of process parameters to confirm the output responses.

Keywords: Al-4032, End Milling, Composite, Roughness, Stir Casting, TGRA.

1. Introduction

Metallic and ceramics particles may be utilized for the fabrication of desired materials. El-Gallab and Sklad [1] have brought numerous functions of metal matrix composites (MMCs) such as chassis, drive shafts, pistons, and cylinder lining. The composite consists of a metal matrix and ceramic reinforcement [2-5]. MMCs fabricated by powder metallurgy and casting produces better quality than other methods. Although stir casting is normally favored, Hashim et al. [6,7] encountered a critical issue that influences the properties, and quality of MMCs is even allocation of ceramic particles. Excellence of fabricated products does not barely depend on fabrication process but also on machining performance. The machining of the composite is a difficult task for the research community [8,9]. The model obtained in present work has been tested through ANOVA to discover the statistically significant parameters. People have reported work on machining of metals, alloys, and composites over decades. Ozben et al. [10] observed improvement in the mechanical properties by altering the weight fraction of reinforcements such as SiC to Al base alloy. Results tell that both toughness and hardness expand with addition of SiC. Other literature related to fabrication and machining of aluminium based MMCs have been reported by different experts [11-17]. Machining of the MMCs is a key issue due to hard ceramic particles that results into low tool life, poor machined surface etc. [18]. Sahin [19] depicted few challenges during machining of AMCs such as less utilization of energy due to tool wear at the flank face. Surface roughness is one of the quality issues during machining of MMCs.

2. Materials and Methods

In the current study machining has been carried out for both the rough and finish machining. A logical use of the TGRA method has been described in Figure 1. Also, all other relevant details linked to applications of the TGRA in machining science have been testified in the literature [15, 20-24]. The metal matrix composite (MMC) has been casted taking Al-4032 as the base matrix and SiC (3% by weight, particle size up to 54μm) as reinforcement. All other relevant details regarding the fabrication of the composite sample have been provided in separate work carried out by the authors [18]. Milling experiments were performed on CNC vertical machining center. The process parameters are cutting speed (A): 15.70, 23.55, 31.40 m/min; depth of cut (C): 0.5, 1.0, 1.5 mm; feed rate (B): 40, 70, 100 mm/min (Table 1).

Table 1. Design matrix and experimental results

Exp. No.	Cutting Speed: A (m/min)	Feed Rate: B (mm/min)	Depth of Cut: C (mm)	Roughness R_a (μm)	MRR $(\frac{mm^3}{sec})$
1	15.70	40	0.5	3.36	03.33
2	15.70	70	1.0	3.62	11.66
3	15.70	100	1.5	3.42	25.00
4	23.55	40	1.0	2.99	06.66
5	23.55	70	1.5	2.75	17.50
6	23.55	100	0.5	3.04	08.33
7	31.40	40	1.5	2.91	10.00
8	31.40	70	0.5	3.18	05.83
9	31.40	100	1.0	3.07	16.66

The material removal rate (MRR) is calculated on the basis of material removed during by machining action and time consumed for the machining (Equations 1-3, Figure 2).

$$\text{Table Feed, } f_m = f_t \times N \times n \quad (1)$$

$$\text{Cutting time, } CT = \frac{L}{f_m} \quad (2)$$

$$MRR = \frac{volume\ removed}{cutting\ time} = \frac{L \times W \times t}{CT} = W \times t \times f_m \quad (3)$$

Researcher have preferred R_a as the roughness parameters due to its accuracy of representing the quality of the surface [25-28]. For the present work R_a parameter has been assessed with portable roughness tester (Taylor Hobson made, Figure 3).

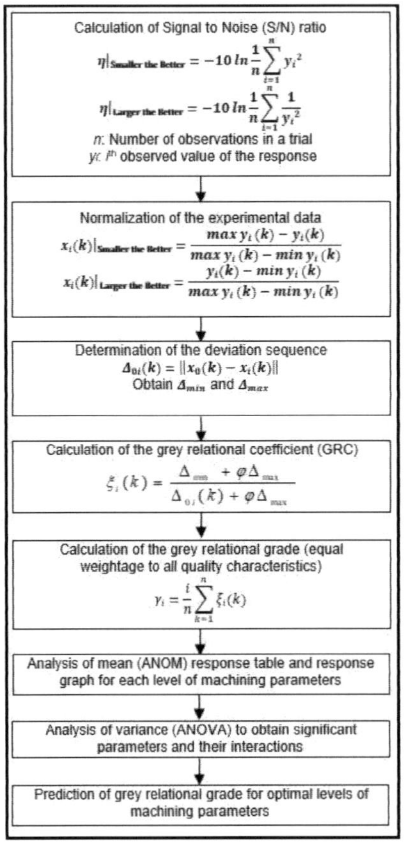

Figure 1. Block diagram of TGRA methodology

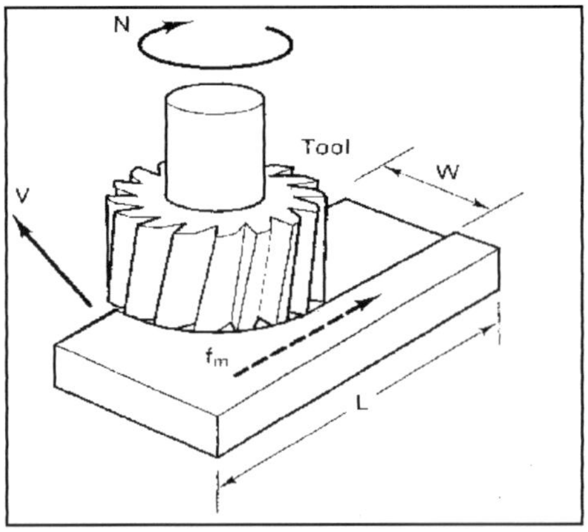

Figure 2. Schematic for calculation of material removal rate during end milling

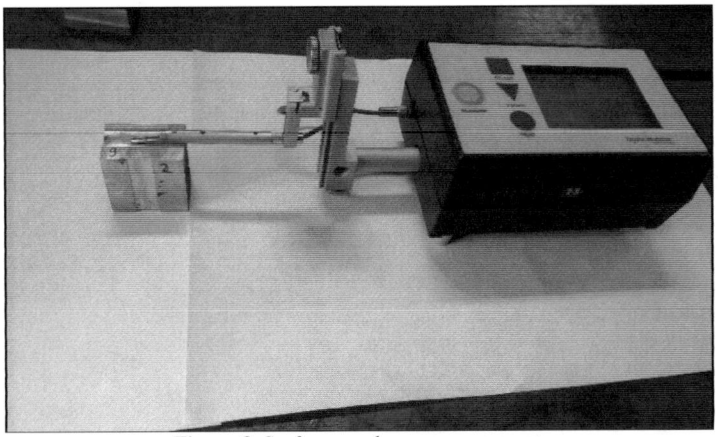

Figure 3. Surface roughness measurement

3. Results and Discussion

Total nine experiments have been performed based on Taguchi's L_9 orthogonal array. Three control parameters namely cutting speed, feed rate and depth of cut have been varied during the experimentation as described in design matrix (Table 1.)

3.1 Case I: Rough machining conditions

For rough machining, only MRR has been given weightage because for rough cut worker need not to pay any attention on the quality of surface produced after the cutting operation. The MRR increased linearly with increase in each input parameters as observed by other experts. Same could be understood from equations (1-3). In other words, the machining time decreases increase in each cutting parameter.

The theoretical MRR for all value of the cutting parameter was calculated applying analysis of mean (ANOM). DOC, FR, CS (i.e., 53.86 > 39.55 > 03.33) is the significance order for optimal MRR. The preferred settings of the cutting parameters turn out to be 'A_1 B_3 C_3' (Table 2 and Figure 4). Equation (4) shows the relationship between the productivity and the cutting parameters for rough machining conditions using the regression analysis of the results.

$$MRR = -7.92 - 0.1592\ CS + 0.1667\ FR + 11.67\ DOC \tag{4}$$

Table 2. Main effect for rough machining

Factor	A	B	C
1	13.33	06.66	05.83
2	10.83	11.66	11.66
3	10.83	16.66	17.50
Delta	02.50	10.00	11.67
Rank	**3**	**2**	**1**

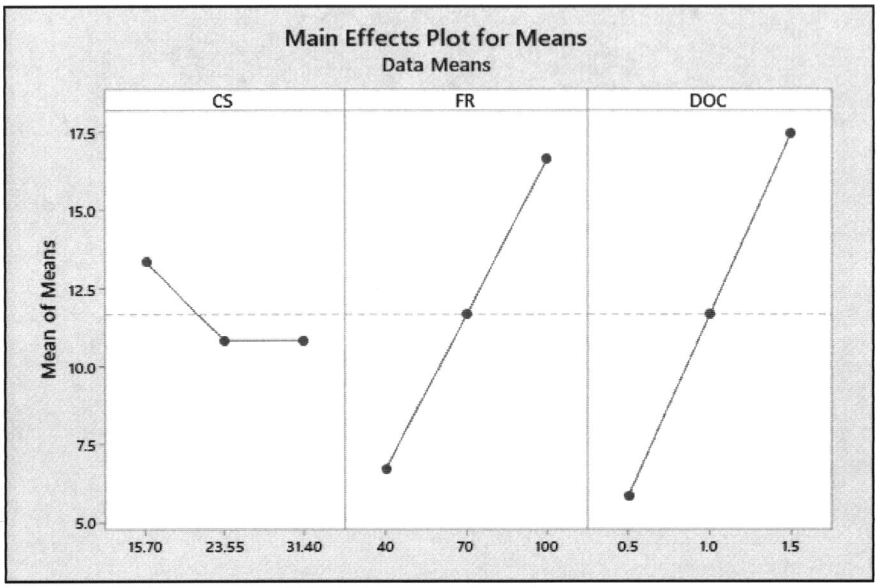

Figure 4. Main effects plot for rough machining

Analysis of variance (ANOVA)

Table 3 illustrates the concise results of ANOVA analysis. It appears that two process parameters DOC and FR are important parameter from the viewpoint of production rate. 53.86 %, 39.55 % and 03.30 % are the contribution % for DOC, FR and CS respectively.

Table 3. ANOVA table for rough machining

Parameter	Degree of Freedom	SS	MS	F Value	Percentage Contribution	F > F table
A	2	12.50	06.25	01.00	03.30 %	Insignificant
B	2	150.00	75.00	12.00	39.55 %	Significant
C	2	204.28	102.14	16.34	53.86 %	Significant
Error	2	12.50	06.25		03.30 %	
Total	8	379.28				
$F_{0.10\,(2,2)}$	9					

3.2 Case II: Finish machining conditions

The surface finish has been assumed more important compared to MRR considering finish machining conditions. 80% weightage to surface finish and 20% weightage to MRR has been given during the analysis for this case. Lower the better (L-B) criteria for surface roughness and higher the better (H-B) criteria for MRR have been used for conversion of response values into the S/N ratio. Moreover, it is essential to normalize /standardize the calculated S/N ratio as presented in Table 4.

Table 4. Signal to noise value for response parameters

Ex. No.	R_a Value			MRR Value		
	Calculated (μm)	S/N	Standardized	Calculated (mm^3/sec)	S/N	Standardized
1.	3.36	-10.52678555	0.728845609	03.33	10.45931191	0.000000000
2.	3.62	-11.17417141	1.000000000	11.66	21.33918396	0.621725155
3.	3.42	-10.68052212	0.793237418	25.00	27.95880017	1.000000000
4.	2.99	-9.513423766	0.304404001	06.66	16.47860910	0.343969898
5.	2.75	-8.786653877	0.000000000	17.50	24.86076097	0.822964012
6.	3.04	-9.657471672	0.364737759	08.33	18.41706992	0.454742327
7.	2.91	-9.277859780	0.205739181	10.00	20.00000000	0.545198119
8.	3.18	-10.04854240	0.528535814	05.83	15.31932850	0.277723355
9.	3.07	-9.742767510	0.400463502	16.66	24.43714871	0.798756889

After normalization/ standardization of the S/N ratio, next step is to calculate the grey relational grade (GRG) values as presented in Table 5. A_1 B_2 C_2 comes out to be the optimal parameters for the machining considering finish conditions. The end column in Table 6 grants the ranking for specific factor. Appropriately, the ranking order of input factor looks like CS, FR and DOC. In modern life the Al/SiC based composites are widely applied. The preferred MRR and roughness is accomplished by optimum machining of these MMCs. But then because of the excessive cutting action of reinforced SiC particulates into the Al matrix complications stand up during machining of these composites.

Figure 5 displays the GRGs regarding input machining factors. The response appears to diminish with FR throughout, but in relation to both CS and DOC, the response first grows up to level 2 and later drops. Also, both CS and FR appears to be major factors for surface machining requirements.

Equation (5) shows the relationship between the output response and the cutting parameters for finish machining conditions using the regression analysis of the results.

$$GRG = 0.867 - 0.02368\ CS + 0.003710\ FR + 0.0545\ DOC \qquad (5)$$

Table 5. GRC and GRG for finish machining

Expt. No.	L_9 Design Matrix			GRC		GRG for finish machining	Grade order
	A	B	C	R_a	MRR		
01	1	1	1	0.648378594	0.333333333	0.585369542	3
02	1	2	2	1.000000000	0.569297871	**0.913859574**	1
03	1	3	3	0.707451148	1.000000000	0.765960919	2
04	2	1	2	0.418201466	0.432514689	0.421064111	7
05	2	2	3	0.333333333	0.738513179	0.414369302	8
06	2	3	1	0.440426874	0.478350949	0.448011689	6
07	3	1	3	0.386320897	0.523668847	0.413790487	9
08	3	2	1	0.514687013	0.409072694	0.493564149	5
09	3	3	2	0.454737065	0.713019482	0.506393548	4

Table 6. Main effects on GRGs for finish machining

Factor	A	B	C
1	0.8071	0.5054	0.6154
2	0.6295	0.6385	0.5867
3	0.4354	0.7281	0.6699
Delta	0.3717	0.2226	0.0832
Rank	**1**	**2**	**3**

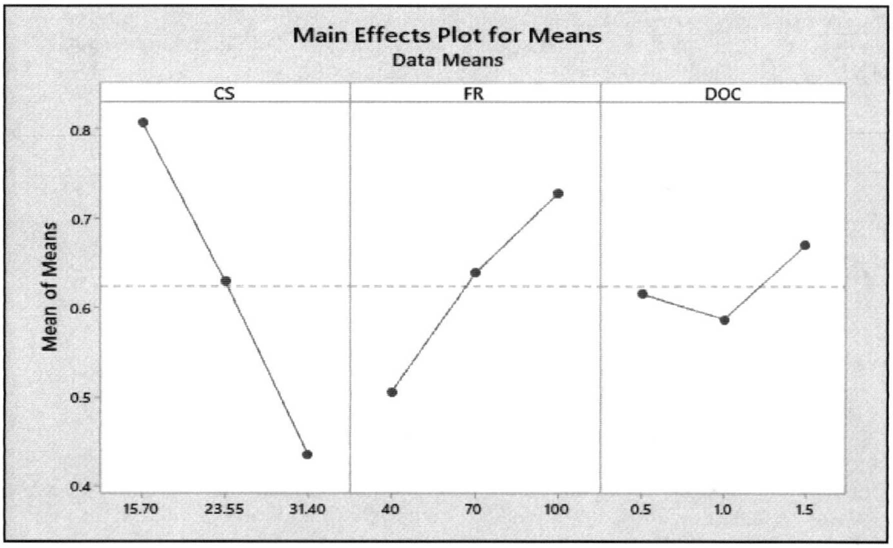

Figure 5. Main effects plot for finish machining

Analysis of variance (ANOVA)

Table 7 illustrates the concise results of ANOVA analysis. It appears that both CS and FR are important parameter from the viewpoint of finish machining conditions. 67.52%, 24.51 %, and 03.48 % are the contribution % for CS, FR and DOC respectively.

The result can be realised considering the process of machining operation. The elevated FR normally produce vibrations between the composite material and cutting tool. This causes an increase in intensity of surface roughness for the duration of cutting operation [29]. Likewise, expanding the FR results a rise in the machined sign owing to more contact area creation that leads to additional forces and consequently R_a parameter emerges considerably [30].

Table 7. ANOVA table for finish machining

Parameter	Degree of Freedom	SS	MS	F value	% Contribution	$F > F_{table}$
A	2	0.20743	0.103716	15.06	67.52 %	Significant
B	2	0.07529	0.037643	05.47	24.51 %	Significant
C	2	0.01070	0.005352	00.78	03.48 %	Insignificant
Error	2	0.01377	0.006887		04.48 %	
Total	8	0.30720				
$F_{0.05\,(2,8)}$	4.45					

The μ at the best combination is calculated applying Equation (6) where A_1, B_2 and C_2 denotes the mean GRG at the most favourable settings, and T_{gg} is the overall mean GRG [31]. The Confidence interval is found employing Equation (7).

$$\mu = A_1 + B_2 + C_2 - 2T_{gg} \tag{6}$$

$$CI = \sqrt{F_\alpha(1, f_t) \times \frac{V_e}{N_e}} \tag{7}$$

Where

$F_\alpha(1, f_t)$ = $F_{critical}$
f_t = 8
$F_{0.10}(1, f_t)$ = 4.45
V_e = 0.0068
N_e = Effective number of replications

$$= \frac{\text{Total number of result}}{\text{DOF of mean (=1,always)+DOF of all factors included in the estimate of the mean}} = N_e = \frac{9}{1+6} = 1.285$$

Validation through confirmation experiment

The validation is performed with the optimal settings of process parameters to confirm the output responses. The optimal solution for the forecasted machining parameters was tested experimentally in duplicate. The confirmed results are close with the expected ones (Table 8). The outcomes recommended that the bi-objective process has been adequately improved applying the Taguchi based grey relational analysis.

Table 8. Expected and validated scores at best setting for finish machining

Predicted		Confirmed		Range
Optimal level	GRG	Test level	GRG	
$A_1 B_2 C_2$	0.8733	$A_1 B_2 C_2$	0.9138	$0.7471 \leq \mu \leq 0.9995$

4. Conclusion

Optimization of machining of AMC has been studied for individual rough and finish machining conditions. The influence of three parameters has been investigated on the two output characteristics using L_9 orthogonal array. The occurrence of hard reinforcement in the base alloy creates this composite material difficult to cutting operation. But at high CS and low FR, machining turn out to be somewhat easier producing superior finish. Two parameters namely DOC (53.86%) and FR (39.55%) appear to be important for rough machining. Best level for the rough machining seems to be CS 15.7 m/min, FR 100 mm/min, and DOC 1.5 mm. Whereas, CS (67.52%) and FR (24.51%) looks to be the major influences for finish machining conditions. The most favorable level for finish machining found to be CS 15.7 m/min, FR 70 mm/min, and DOC 1 mm.

Conflict of Interest

The authors have no conflicts of interest to declare.

References

[1] El-Gallab M, Sklad M 1998 *J Mater Process Technol.* **83** 151
[2] Suthar J, Patel KM 2018 *Mater Manuf Process.* **33** 499
[3] Kumar D, Singh PK 2019 *Mater Res Express.* **6** 106543
[4] Kumar D, Singh PK 2019 *Mater Today Proc.* **18** 4201
[5] Kumar D, Singh PK 2019 *Mater Today Proc.* **18** 2563
[6] Hashim J, Looney L, Hashmi MSJ 2002 *J Mater Process Technol.* **123** 251

[7] Hashim J, Looney L, Hashmi MSJ. 2002 *J Mater Process Technol.* **123** 258
[8] Gururaja S, Ramulu M, Pedersen W 2013 *Mach Sci Technol.* **17** 41
[9] Umesh Khandey, Sudarsan Ghosh KH 2017 *Mater Manuf Process* **32** 1082
[10] Ozben T, Kilickap E, Çakir O 2008 *J Mater Process Technol.* **198** 220
[11] Muthukrishnan N, Murugan M, Rao KP 2008 *Int J Adv Manuf Technol.* **38** 447
[12] Seeman M, Ganesan G, Karthikeyan R, Velayudham A 2010 *Int J Adv Manuf Technol.* **48** 613
[13] Kuram E, Ozcelik B 2013 *Meas J Int Meas Confed.* **46** 1849
[14] Karabulut Ş, Karakoç H, Çıtak R 2016 *Compos Part B Eng.* **101** 87
[15] Lin K, Wang W, Jiang R, Xiong Y, Song G 2017 *Int J Adv Manuf Technol.* **88** 887
[16] Balasubramanian K, Nataraj M, Duraisamy P 2019 *Mater Manuf Process.* **34** 1389
[17] Palanikumar K, Karthikeyan R 2007 *Mater Des.* **28** 1584
[18] Saini, P, Singh PK 2021 *Eng Res Express.* **3** 045009.
[19] Sahin Y 2003 *Mater Des.* **24** 671
[20] Angappan P, Thangiah S, Subbarayan S 2017 *J Mech Sci Technol.* **31** 4159
[21] Mia M et al 2018 *Meas J Int Meas Confed.* **122** 380
[22] Singh PK, Saini P, Kumar D 2019 *Mater Today Proc.* **26** 2564
[23] Singh PK, Kumar K, Saini P 2019 *Mater Today Proc.* **26** 2961
[24] Singh PK, Saini P, Kumar AK 2019 *J Phys Conf Ser.* **1240** 012016
[25] Wang W, Kweon SH, Yang SH 2005 *J Mater Process Technol.* **162** 702
[26] Bhardwaj B, Kumar R, Singh PK 2014 *J Mech Sci Technol.* **28** 5149
[27] Bhardwaj B, Kumar R, Singh PK 2014 *Proc Inst Mech Eng Part B J Eng Manuf.* **228** 704
[28] Babu GHVP et al 2017 *Proc Inst Mech Eng Part B J Eng Manuf.* **231** 2084
[29] Okokpujie IP, Bolu CA, Ohunakin OS 2020 *Int J Adv Manuf Technol.* **108** 1473
[30] Shihab SK, Gattmah J, Kadhim HM 2021 *Silicon.* **13** 1403
[31] Trehan R, Singh S, Garg M 2015 *Proc Inst Mech Eng Part L J Mater Des Appl.* **229** 263

Up-conversion and Tunable Up-conversion Emission from Nitrogen Functionalized Graphene Quantum Dots

A B Siddique[1]*, K L V Prasad[1], S M Joseph[1], D Kishore[1], B Das[2,3], S Mubarak[4], K Mukhuti[5], M Ray[6‡]

[1]Aditya College of Engineering and Technology, Surampalem -533437. Andhra Pradesh, India.
[2]Sister Nibedita Govt. General Degree College for Girls, Hastings House, Kolkata, India,700027
[3]Department of Physics, Indian Institute of Engineering Science and Technology, Shibpur, PO. Botanic Garden, Howrah, India, 711103
[4] Department of Electronics and Communication Engineering, Saveetha School of Engineering, Saveetha Institute of Medical and Technical Sciences, India
[5]Indian Institute of Science Education and Research Kolkata, Mohanpur, Nadia – 741246, West Bengal, India
[6] School of Engineering and Sciences, Tecnológico de Monterrey, Monterrey, 64849, Nuevo Leon, Mexico

*Email id: abs.iiest@gmail.com

‡Email id: mallar.ray@tec.mx

Abstract. In amidst of conflicts of opinion regarding the presence of up-conversion in carbonaceous nanostructure. We have critically and precisely asserted that there is a real up-conversion in the nitrogen functionalized graphene quantum dots (NGQDs), which is found to be its inherent feature and not a measurement artefact. We have demonstrated that NGQDs exhibit real up-conversion in the blue- green region. To define the perspective of the study a variant of GQDs, NGQDs was synthesized using facile, one-pot synthetic methods. We have critically examined the up-conversion emission in NGQDs, interestingly, N-GQDs exhibited very superior optical properties such as high quantum yield of 97%, up-conversion emission at 535 nm and tunable up-conversion that exhibited a red shift of 12 nm. The two-photon excitation process is the governing phenomenon behind the real up-conversion in NGQDs which is suggested by the quadratic dependence of the integrated intensity on excitation power. Up-conversion can enhance the efficiency of the photovoltaic solar cell by even converting infrared region of sunlight to electricity. Moreover, tunable up-conversion open the door for promising applications of NGQDs in tunable fibre lasers, multicolour displays and multiphoton optical data storage.

Keywords:up-conversion, tunable up-conversion, UPCL, GQDs, NGQDs

1. The Introduction

Carbon quantum dots are one of the most sought carbon nanostructures for research, especially graphene quantum dots (GQDs). Recently, GQDs have appealed the global attention not only for their novel photo-physical features [1] but also for the versatile applications. A significant number of applications of GQDs structures have already been demonstrated in the past decade. Most of the demonstrated applications of GQDs relates to the field of biology due to the benign nature of these nanostructures. Thorough reviews of bioanalytical applications of GQDs can be found in the review articles by Zuo et al.,[2] Wang et al.,[3] and Namdari et al.[4] Luminescence quenching based detection/sensing of analytes is also widely reported for GQDs and NGQDs. Two recent reviews, by Li et al.[5] and Liu et al.[6] summarize the major advances in sensing applications using carbon and graphene based dots. Besides sensing and bio-related applications, GQDs, primarily the GQDs have been used in solar photovoltaics,[7]photodetectors,[8] and light emitting devices,[9-13]. Commencing from the initial reviews on GQDs to the most recent ones [14, 15] it has been noticed that fluorescence property has permanently remained the most striking feature of each and every variety of GQDs. All variants of GQDs display intense PL with extraordinarily high PL quantum yield (PLQY). Remarkably, despite the structural and compositional differences, several variants of GQDs exhibited similar photophysical properties.[16, 17] In addition to that there are significant number of reports that claim to observe up-conversion photoluminescence for a variety of carbon dots and GQDs.[1, 18-21] However, till date there are many contradictory arguments regarding the up-conversion PL in GQDs, in which one side of the researchers claim it to be the materials property, whereas counterparts argue it to be measurement artefact. In this paradigm, the reliability of up-conversion PL in GQDs has been debateable. Some of the claims of up-conversion PL measured with the white light in GQDs have been refuted by careful successive studies.[22-24] Wen et al. have carefully studied up-conversion in five variants of GQDs and considered this property to be measurement artefact.[25] However, there are significant reports that argue the presence of real up-conversion in GQDs that can be experimentally measured with pulsed laser excitation.[24, 26] This work not only critically examine and consolidate up-conversion in GQDs with the help of spatially resolved laser source but it also claims that the observed up-conversion is tunable with the excitation energy. To the best of our knowledge, there is no report on the tunable up-conversion PL in nitrogen functionalized GQDs. To define the perspective of the study a variant of GQDs—that is N-functionalized GQDs (NGQDs) was synthesized using facile, one-pot synthetic methods. We have critically examined the up-conversion PL in NGQDs, interestingly, N-GQDs exhibited very superior optical properties such as high quantum yield, up-conversion emission (two-photon process) and tunable up-conversion.

2. Experimental

2.1 Materials:
Citric acid (Merck, India) and ethylene diamine (SRL) were used for preparation of NGQDs. All chemicals were of high purity and suitable for the synthesis purpose. DI water having 18.2 MΩ.cm resistivity procured from Millipore was used for all experimental requirements.

2.2 Preparation of N-GQDs:
The N-GQDs were synthesized by mixing 2.1 g of CA and 1.8 g ethylene diamine in 50 ml DI water, the solution was mixed properly to make the clear solution before transferring to the Teflon lined 150 mL autoclave. The airtight closed autoclave was heated to 180°C in oil bath and kept for continous 8 hours. Finally, N-GQDs were collected and stored in refrigerator after being centrifuged at 5,000 rpm for 10 min.

2.3 Characterization:
The nano size of the synthesized material was examined by HR-TEM model JEM-2200 FS by JEOL which is armed with 200-kV field emission gun, energy filter (Omega) and Gatan software. In order to do the HR-TEM measurements, the NGQDs samples were diluted to deposit 3 μL of samples on the

carbon coated copper grids and dried. We used a GATAN energy filtered TEM integrated with HR-TEM for EELS study of our samples. The transition of carbon was identified from the spectrum obtained from electron energy loss spectroscopy of the NGQDs. The dried samples of NGQDs were used for doing Fourier transform infrared (FTIR) measurement by using JACSO 4700 LE in ATR mode.Omicron Multiprobe spectrometer (Omicron NanoTechnologyGmbh, UK) was employed to do the X-ray photoelectron spectroscopy (XPS) measurements of the dried sample. This unit is tailored with an EA125 (Omicron) hemispherical analyzer. It is worth noting here that any amount of moisture in the sample may lead to degassing in the vacuum chamber. The absolute PLQY was measured by Horiba made spectrofluorometer having model number as Nanolog model FL3-11. During the measurement the both the slits were fixed at 1nm and integrating time at 0.5 s. PL spectra were recorded using integrating sphere mounted that was mounted inside Nanolog-3 spectrofluorometer and absolute PLQY was estimated by Suzuki et al.[27]Pulsed excitation measurement was done by using a tunable Ti-Sapphire pulsed laser, named Tsunami from Spectra-Physics. It has maximum power of excitation of 2W and repetition rate of 80 MHz with pulse width ~100 femtoseconds. For spectroscopic measurement with pulsed laser, we have used an Andor Shamrock 303i spectrograph attached with an ultra-sensitive EM-CCD named iXon, also from Andor.

3. Results and Discussion

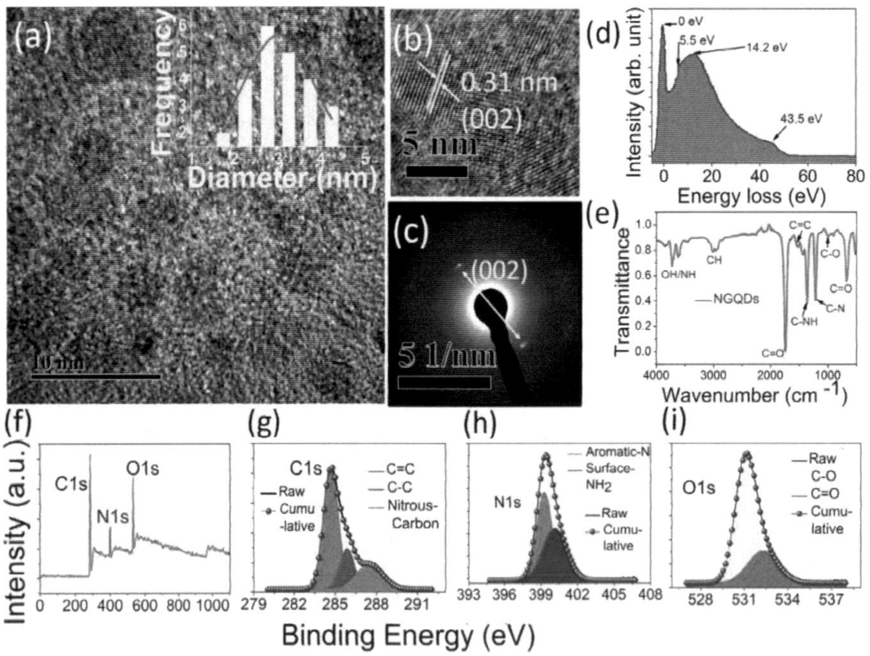

Figure 1. NGQDsstructural characterization: (a) bright field HRTEM image, inset: the size distribution histogram; (b) high magnification image; (c) SAEDP corresponding to the image in (a); (d) EELS profile of the NGQDs; (e) FTIR spectrum of NGQDs; (f) XPS wide scan spectrum; deconvoluted (g) C1s; (h) N1s spectra and (i) O1s.

The structural and surface features of the prepared NGQDs have been characterized by the above-mentioned techniques and the overall features were summarized in Figure 1. The localized clusters of carbon atoms are confirmed by the appearance of distinguishable dark spots shown in the bright field HR-TEM image of NGQDs in Figure 1a. The histogram shown as an inset exhibited the well distributed size variation of NGQDs from ~ 1-5 nm with average size of 2.75±0.4 nm. The crystalline nature of NGQD is revealed by the high magnification image of NGQD shown in Figure 1b, which clearly shows the fringes of 0.31 nm that corresponds to (002) graphitic planes. In addition to that SAEDP corresponding to Figure 1a, also consolidate on its crystallinity by the presence of bright spots that corresponds to the same (002) graphitic planes as shown in Figure 1c.

The electron energy loss spectroscopy (EELS) of the prepared sample, gave the clearer understanding about the NGQDs structure. The low-loss EELS spectrum revealed a π^* surface plasmon peak at 5.5 eV and exhibited a peak at 14.2 eV that corresponds to $\pi^* + \sigma^*$ bulk plasmon transition, showed in the Figure 1d. The occurrence of these peaks at obtained location direct towards the formation of graphene layers. Additionally, the peak at 14.2 eV even suggests the formation of 2D layers of graphene.[28, 29] Moreover, the broad but a weak peak at 43.5 nm manifest the incorporation of nitrogen in the graphene layer, confirms the functionalization of GQDs with nitrogen.[30] The outcomes of structural study that is HR-TEM and EELS consequently substantiate the formation of NGQDs. However, to get more insight into the surface of NGQDs, FTIR and XPS analysis was done.

The FTIR spectrum in Figure 1e, confirms the stretching vibrations of N-H and O-H exhibited by the band peaking between 3300-3600 cm $^{-1}$. The band peaking at 1730 cm $^{-1}$ and 1560 cm $^{-1}$corresponds to the vibrations of C=O and C=C respectively.[31, 32] The efficacious insertion of nitrogen atom in the graphitic structure were revealed by the presence of the bending vibration of C-NH at 1376 cm^{-1} and by the stretching vibrations of C-N at 1220 cm^{-1}.[33, 34]. To further understand the presence of various surface functional groups and chemical bonding, we have done the XPS measurement of the NGQD dry sample.

The wide scan XPS spectrum exhibited three strong peaks at 285 eV, 400 eV and 530 eV that correspond to C1s, N1s and O1s, respectively, shown in Figure 1f.[35] The peaks exhibited at 284.5eV, 285.8 eV and 287.55 eV correspond to C=C (sp^2 carbon), C-C (sp^3 carbon) and nitrous-carbon,[36]which were exhibited as the deconvoluted components of C1s as shown in Figure 1g. Similarly, XPS spectra of N1s deconvoluted to two components at 399.25 and 400.25 eV corresponds to aromatic-N and surface-NH_2 respectively shown in Figure 1h.[37] In addition to that the high resolution of O1s spectrum also de-convoluted to two components C-O and C=O peaking at 531.1 and 532.2 eV respectively.[9] The presence of Surface NH_2 in N1s peak of high resolution XPS and the peak of N-H bonding vibration in the FTIR could ensure the passivation and functionalization of nitrogen with GQD. The detailed description about the structural and surface features gives the direction to study its controversial property that is up-conversion PL in NGQD.

We have estimated the PLQY of NGQDs sample with help of integrating sphere mounted in the Nanolog spectrofluorometer. The absolute PL QY of the samples was 97% estimated by recording the PL spectra using an integrating sphere following the method proposed by Suzuki et al.[27] which is in agreement with other report.[38] Due to such a unprecedented PLQY and high absorption cross-section,[39]it is anticipated that leakage of even a weak magnitude of lower excitation wavelength is sufficient to produce normal emission output, which could be taken as the up-conversion emission. Thus, remarkably high photon conversion efficiency and unusually high absorption makes NGQDs vulnerable to wrong interpretation of up-conversion emission. Moreover, in order to proclaim on up-conversion emission even with laser excitations, it is necessary to negate the possibilities of exciting the samples with leaking of lower wavelengths. In order to repudiate any possibility of leakage, a spatially resolved excitation of 800 nm was used from Ti –Sapphire, tuneable femtosecond pulsed laser. To precisely negate any possibility of the presence of any other component except 800 nm in the excitation channel, the spatial distance between the 800 nm and any other component happened to be present was increased by using three mirrors as reflectors.

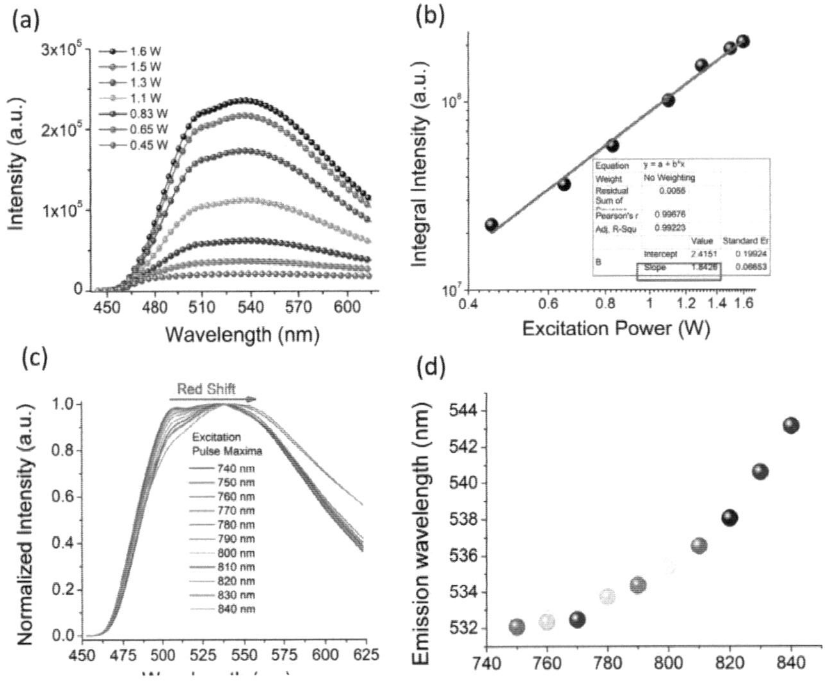

Figure 2. (a) Up-conversion spectra in NGQDs under excitation with the 800 nm femtosecond pulsed laser; (b) Integral intensity versus Excitation power (log-log scale); (c) Up-converted emission spectra of NGQDs measured with 800 nm femtosecond laser; (d) Excitation wavelength and up-converted emission peak position.

With that much precaution and steps, we were able to detect a significant up-conversion peak at around 535 nm with the 800 nm femtosecond laser excitation, even by the application of filter in the excitation channel as shown in the Figure 2a. It was observed that the real up-conversion emission hardly distinguishable with the bared eye, unlike flawed up-conversion which is intense. It is also important to add here that, the up-conversion signal was observed only with pulsed mode highly concentrated laser beam and not with same laser in continuous wave mode. In order to strengthen our claim, we have also varied the excitation power (P), and estimated the integral intensity to plot the graph of P (log scale) versus integral intensity (log scale) as shown in Figure 2b. The slope of this linearized plot came out tobe ~2, that strongly suggest the presence of up-conversion having a two-photon excitation process.[40] Remarkably, it was noticed that with the 800 nm pulsed excitation, NGQDs not only exhibit up-conversion emission but also these emission were found to be tunable with the excitation energy shown in Figure 2c. The up-converted peak at 535 nm clearly shows the red-shift with the increase in the excitation wavelength. The excitation wavelength and up-converted emission peak position is plotted in Figure 2d, that shows distinguishable red-shift and assert tunable up-conversion. This is for the first time that we have observed such interesting features in NGQDs, that open the new door for research in the most explored topic of the time. The outcome of this work is summarized and compared with other related work in the Table 1. Tunable up-conversion is promising for the potential application in tunable fiber lasers, multicolor displays and multiphoton optical data storage.However, in order to understand the mechanism responsible for up-conversion and tunable up-conversion observed in NGQDs warrant a separate investigation.

Table 1. Comparison of summary of present work with other related work

Sample	PLQY	Laser Excitation/ White light	Up-converted Emission	Up-conversion red shift	Ref
NGQDs	97%	Laser 800 nm	Up-converted peak535 nm	12 nm	This work
NGQDs	99%	-	No Upconversion	-	[38]
Carbon dots/ GQD	-	Laser and white light	Upconversion refuted	-	[22-25]
Carbon dots/ GQD	-	Laser	Real Upconversion reported	-	[24, 26]

4. Conclusions

In amidst of conflicts of opinion regarding the presence of up-conversion in carbonaceous nanostructure. We have critically and precisely asserted that there is a real up-conversion in the NGQDs, which is its inherent feature and not a measurement artefact. The prepared N-GQDs exhibited very superior optical properties such as high PL quantum yield of 97%, up-conversion emission at 535 nm and tunable up-conversion that exhibited a red shift of 12 nm. However, the up-conversion signal was observed only with pulsed mode highly concentrated laser beam and not with same laser in continuous wave mode. The two-photon excitation process is the governing phenomenon behind the real up-conversion in NGQDs which is suggested by the quadratic dependence of the integrated intensity on excitation power. To understand the mechanism responsible for up-conversion and tunable up-conversion observed in NGQDs warrant a separate investigation.Up-conversion can enhance the efficiency of the photovoltaic solar cell by even converting infrared region of sunlight to electricity. Moreover, tunable up-conversion open the door for promising applications of NGQDs in tunable fibre lasers, multicolour displays and multiphoton optical data storage.

Conflicts of interest
The authors declare that they have no known competing financial interests or personal relationships that could have appeared to influence the work reported in this paper.

Acknowledgments
ABS acknowledges MHRD, for PhD fellowship.

References

[1] Yan Y, Gong J, Chen J, Zeng Z, Huang W, Pu K, Liu J and Chen P 2019 Recent advances on graphene quantum dots: from chemistry and physics to applications *Adv. Mater.* **31** 1808283
[2] Zuo P, Lu X, Sun Z, Guo Y and He H 2016 A review on syntheses, properties, characterization and bioanalytical applications of fluorescent carbon dots *Microchim. Acta* **183** 519-42
[3] Wang J and Qiu J 2016 A review of carbon dots in biological applications *J. Mater. Sci.* **51** 4728-38
[4] Namdari P, Negahdari B and Eatemadi A 2017 Synthesis, properties and biomedical applications of carbon-based quantum dots: An updated review *Biomed. Pharmacother.* **87** 209-22
[5] Li M, Chen T, Gooding J J and Liu J 2019 Review of carbon and graphene quantum dots for sensing *ACS sensors* **4** 1732-48
[6] Liu M L, Chen B B, Li C M and Huang C Z 2019 Carbon dots: Synthesis, formation mechanism, fluorescence origin and sensing applications *Green Chem.* **21** 449-71

[7] Essner J B and Baker G A 2017 The emerging roles of carbon dots in solar photovoltaics: a critical review *Environ. Sci. Nano* **4** 1216-63
[8] Mihalache I, Radoi A, Pascu R, Romanitan C, Vasile E and Kusko M 2017 Engineering graphene quantum dots for enhanced ultraviolet and visible light p-Si nanowire-based photodetector *ACS Appl. Mater. Interfaces* **9** 29234-47
[9] Tang L, Ji R, Cao X, Lin J, Jiang H, Li X, Teng K S, Luk C M, Zeng S and Hao J 2012 Deep ultraviolet photoluminescence of water-soluble self-passivated graphene quantum dots *ACS nano* **6** 5102-10
[10] Zheng M, Liu S, Li J, Qu D, Zhao H, Guan X, Hu X, Xie Z, Jing X and Sun Z 2014 Integrating oxaliplatin with highly luminescent carbon dots: an unprecedented theranostic agent for personalized medicine *Adv. Mater.* **26** 3554-60
[11] Zhang X, Zhang Y, Wang Y, Kalytchuk S, Kershaw S V, Wang Y, Wang P, Zhang T, Zhao Y and Zhang H 2013 Color-switchable electroluminescence of carbon dot light-emitting diodes *ACS nano* **7** 11234-41
[12] Li C-X, Yu C, Wang C-F and Chen S 2013 Facile plasma-induced fabrication of fluorescent carbon dots toward high-performance white LEDs *J. Mater. Sci.* **48** 6307-11
[13] Wang F, Kreiter M, He B, Pang S and Liu C-y 2010 Synthesis of direct white-light emitting carbogenic quantum dots *Chem. Commun.* **46** 3309-11
[14] Baker S N and Baker G A 2010 Luminescent carbon nanodots: emergent nanolights *Angew. Chem. Int. Ed.* **49** 6726-44
[15] Mintz K J, Zhou Y and Leblanc R M 2019 Recent development of carbon quantum dots regarding their optical properties, photoluminescence mechanism, and core structure *Nanoscale* **11** 4634-52
[16] Cao L, Meziani M J, Sahu S and Sun Y-P 2012 Photoluminescence properties of graphene versus other carbon nanomaterials *Acc. Chem. Res.* **46** 171-80
[17] Siddique A B, Pramanick A K, Chatterjee S and Ray M 2018 Amorphous Carbon Dots and their Remarkable Ability to Detect 2, 4, 6-Trinitrophenol *Sci. Rep.* **8** 9770
[18] Cao L, Wang X, Meziani M J, Lu F, Wang H, Luo P G, Lin Y, Harruff B A, Veca L M and Murray D 2007 Carbon dots for multiphoton bioimaging *Journal of the J. Am. Chem. Soc.* **129** 11318-9
[19] Hutton G A, Martindale B C and Reisner E 2017 Carbon dots as photosensitisers for solar-driven catalysis *Chem. Soc. Rev.* **46** 6111-23
[20] Wang H, Revia R, Wang K, Kant R J, Mu Q, Gai Z, Hong K and Zhang M 2017 Paramagnetic properties of metal-free boron-doped graphene quantum dots and their application for safe magnetic resonance imaging *Adv Mater* **29** 1605416
[21] Li J-Y, Liu Y, Shu Q-W, Liang J-M, Zhang F, Chen X-P, Deng X-Y, Swihart M T and Tan K-J 2017 One-pot hydrothermal synthesis of carbon dots with efficient up-and down-converted photoluminescence for the sensitive detection of morin in a dual-readout assay *Langmuir* **33** 1043-50
[22] Zhuo S, Shao M and Lee S T 2012 Upconversion and downconversion fluorescent graphene quantum dots: ultrasonic preparation and photocatalysis *ACS nano* **6** 1059-64
[23] Tan D, Zhou S and Qiu J 2012 Comment on upconversion and downconversion fluorescent graphene quantum dots: ultrasonic preparation and photocatalysis *ACS nano* **6** 6530-1
[24] Shao M and Zhuo S 2012 Reply to Comment on Upconversion and Downconversion Fluorescent Graphene Quantum Dots: Ultrasonic Preparation and Photocatalysis *ACS nano* **6** 6532
[25] Wen X, Yu P, Toh Y-R, Ma X and Tang J 2014 On the upconversion fluorescence in carbon nanodots and graphene quantum dots *Chem Comm* .**5** 4703-6
[26] Gan Z, Wu X, Zhou G, Shen J and Chu P K 2013 Is there real upconversion photoluminescence from graphene quantum dots? *Adv. Opt. Mater.* **1** 554-8

[27] Suzuki K, Kobayashi A, Kaneko S, Takehira K, Yoshihara T, Ishida H, Shiina Y, Oishi S and Tobita S 2009 Reevaluation of absolute luminescence quantum yields of standard solutions using a spectrometer with an integrating sphere and a back-thinned CCD detector *Phys. Chem. Chem. Phys.* **11** 9850-60

[28] Gass M H, Bangert U, Bleloch A L, Wang P, Nair R R and Geim A 2008 Free-standing graphene at atomic resolution *NNat. Nanotechnol.* **3** 676

[29] Jiang D-e, Du M-H and Dai S 2009 First principles study of the graphene/Ru (0001) interface *J. Chem. Phys.* **130** 074705

[30] Tang L, Ji R, Li X, Teng K S and Lau S P 2013 Energy-level structure of nitrogen-doped graphene quantum dots *JJ. Mater. Chem. C* **1** 4908-15

[31] Tetsuka H, Nagoya A, Fukusumi T and Matsui T 2016 Molecularly designed, nitrogen-functionalized graphene quantum dots for optoelectronic devices *Adv Maters* **28** 4632-8

[32] Ţucureanu V, Matei A and Avram A M 2016 FTIR spectroscopy for carbon family study *Critical reviews in analytical chemistry* **46** 502-20

[33] Zhang H, Chen Y, Liang M, Xu L, Qi S, Chen H and Chen X 2014 Solid-phase synthesis of highly fluorescent nitrogen-doped carbon dots for sensitive and selective probing ferric ions in living cells *Anal. Chem.* **86** 9846-52

[34] Qu D, Zheng M, Du P, Zhou Y, Zhang L, Li D, Tan H, Zhao Z, Xie Z and Sun Z 2013 Highly luminescent S, N co-doped graphene quantum dots with broad visible absorption bands for visible light photocatalysts *Nanoscale* **5** 12272-7

[35] Qu D, Zheng M, Zhang L, Zhao H, Xie Z, Jing X, Haddad R E, Fan H and Sun Z 2014 Formation mechanism and optimization of highly luminescent N-doped graphene quantum dots *Sci. Rep.* **4** 5294

[36] Liu S, Tian J, Wang L, Zhang Y, Qin X, Luo Y, Asiri A M, Al-Youbi A O and Sun X 2012 Hydrothermal treatment of grass: a low-cost, green route to nitrogen-doped, carbon-rich, photoluminescent polymer nanodots as an effective fluorescent sensing platform for label-free detection of Cu (II) ions *Adv Mater* **24** 2037-41

[37] Yang Y, Cui J, Zheng M, Hu C, Tan S, Xiao Y, Yang Q and Liu Y 2012 One-step synthesis of amino-functionalized fluorescent carbon nanoparticles by hydrothermal carbonization of chitosan *ChemComm* . **48** 380-2

[38] Khan F and Kim J H 2018 N-Functionalized Graphene Quantum Dots with Ultrahigh Quantum Yield and Large Stokes Shift: Efficient Downconverters for CIGS Solar Cells *ACS Photonics* **5** 4637-43

[39] Sun J, Xin Q, Yang Y, Shah H, Cao H, Qi Y, Gong J R and Li J 2018 Nitrogen-doped graphene quantum dots coupled with photosensitizers for one-/two-photon activated photodynamic therapy based on a FRET mechanism *ChemComm.* **54** 715-8

[40] Pollnau M, Gamelin D R, Lüthi S, Güdel H and Hehlen M P 2000 Power dependence of upconversion luminescence in lanthanide and transition-metal-ion systems *Phys. Rev. B* **61** 3337

Electrochemical Micromachining of Linear Micropattern on Stainless Steel Surface

S. Kunar[1], V. Pragna[1], A.K. Singh[2], M. S. Reddy[1]

[1]Department of Mechanical Engineering, Aditya Engineering College, Surampalem, India
[2]Department of Mechanical Engineering, Aditya College of Engineering and Technology, Surampalem, India

Corresponding author's e-mail: sandip.sandip.kunar@gmail.com

Abstract. Maskless electrochemical micromachining process with SU-8 2150 masked tool has been introduced containing microstructures with higherprofile accuracy that are generated on SS-304. In this research work, to make maskless EMM method with more cost-effective, substantial adaptations are provided to this technique and an advanced version of EMM process is applied for microtexturing. In this modified technique, a special microtextured cell unit including (micromachining cell, electrical connection, and electrolyte flow scheme) is assembled by the mechanical force and perpendicular cross flow system has been developed inside the cell. Experimentation is conducted to explore the machining influence of the advanced maskless EMM process for fabricating linear micropattern. The machining outcomes of voltage, frequency, and duty ratio on MRR, dimensional accuracy, textured depth, and surface roughness are investigated during microtexturing. The investigational research reveals that this advanced maskless EMM technique can generate highly identical micropatterns having MRR of 2.2 mg/min, width overcut of 19.76μm, textured depth of 14.4μm and surface roughness of 0.0307μm.

Keywords: Maskless EMM, microtexturing, reused tool, profile accuracy, textured depth, roughness

1. Introduction

Surface microtexturing has performed an essential role in the progress of many sophisticated areas viz., biomedical, aerospace, optics, energy, etc. Microtexturing techniques will become still more significant in the future. Maskless EMM is the anodic dissolution process, which is the converse of electroplating. Maskless EMM has numerous advantages viz., no burr, stress, and tool wear, smooth microtextured surfaces, and the capability to machine complex shapes irrespective to material's hardness, etc. [1]. In maskless EMM, a photo resist patterned tool is prepared and the bare surface of masked tool flows the controlled current through textured masked tool and removes high amount of metal from the workpiece by anodic dissolution and forms replica of textured tool on work surface. Compared with through-mask EMM, maskless EMM provides higher output with low investment and

one patterned tool can generate many micropatterned samples over large area with high surface quality and high dimensional accuracy throughout the patterned area [2].
Different micropatterning methods are employed for advanced applications. Laser beam machining and abrasive jet machining (AJM) are the advanced techniques for surface micropatterning. In general, both processes have some limitations, viz. heat affected zone, low production efficiency, etc. AJM can fabricate the micropatterns on difficult-to-machine materials [3]. Abrasive jet machining produces micro dimples, which are suitable to reduce the friction coefficient [4]. Through-mask EMM is a micropatterning technique with comparatively moderate quality and high machining rate that generates good texture on job surface [5]. TMEMM is employed to generate micro dimples and the job surface is coated with an insulation, which covered a tool and a masked layer [6]. The micro morphology in the insulation is transferred to the job surface during electrochemical machining. This modified TMEMM have some advantages, viz. lower manufacturing cost and shorter time [7]. TMEMM is utilized to produce micro impressions on curved surface having diameter of 40μm [8]. Micropatterns are created on a tubular surface having 22.7μm depth and 94μm diameter using TMEMM [9]. It utilizes photolithography process to produce micromorphology on insulated layer. Though, the photolithography is a difficult method that significantly increases the production cost and decreases the machining productivity in TMEMM. EMM is applied to prepare good micro circular patterns with low cost in enclosed electrochemical cell [10]. Maskless EMM is used to create the microtexturesusing various wave shapes[11]. Micro circular impressions are made using maskless EMM [12].
In this study, an advanced method of EMM called maskless EMM, is developed to generate high-quality linear micropatterns and improve dimensional regularity in linear micropattern. In this method, the distribution of current flux is improved, which aids to attain a high dimensional accuracy. In modified method, linear micropattern is transferred to the work surface without photolithography of substrates. The setup is established with microtextured cell, electrode fixtures, flushing system, and power unit for carrying out the investigation. One patterned tool can generate many textured samples with high dimensional accuracy. The machining influence of significant variables on performance criteria is explored during micropatterning. An effort has been made to analyze the micrographs based on micrographs.

2. Experimental Procedure

Figure 1 demonstrates the developed system setup for conducting the experiments for producing linear micropatterns. The system has microtextured cell, flushing unit and pulsed DC power unit. The microtextured cell has tool and job fixtures, erect cross flow flushing system and connection facility of power supply. The machining unit i.e., cell is manufactured using Perspex material to avoid the rusting. The flow system of electrolyte is developed inside the machining unit. The electrolyte is supplied upright from below to above via the restricted direction of the cell between electrodes, as shown in Figure 1. This developed flow system is suitable for removing the sludges from the micromachining zone creating the extra back pressure at the upper side of the cell. The electrical connection is supplied to the electrodes from the pulsed DC unit. This power unit has inherent protection function and function generator with ultra-fast responses. LIGA process is employed for generating the linear micropatterned tool using SU-8 2150 mask having thickness of 203μm on stainless steel sheets. The width of linear slot is 445μm. Precision micrometer is used to maintain inter electrode gap settings.
Extensive pilot tests are conducted to finalize the influential process variables and ranges of variables for micropattern generation to perform the experimental investigation. Below and above ranges of these parameters have not acquired the satisfactory results with higher profile quality and dimensional accuracy. The voltage of 8 to 14V, duty ratio of 50 to 80%,and frequency of 10 to 25 kHzare the persuading process variables that are used to examine the machining effect on width overcut, material removal rate (MRR), machining depth, and surface roughness (R_a). Other attributes are static, viz. 1 minute machining time, 20g/l electrolyte concentration, 5.54m^3/h flow rate and 50μm inter electrode gap. The combined mixture of $NaNO_3$ (50%) and NaCl (50%) is selected as an electrolyte for experimentation. The output responses are monitored by Optical microscope, Atomic Force Microscope (AFM), and CCI Non-Contact Profilometer.

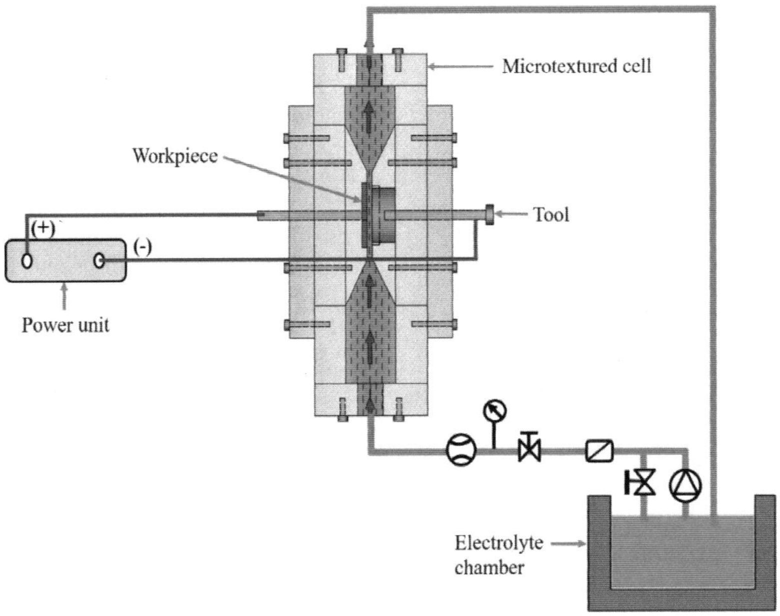

Figure 1. Maskless EMM setup

3. Results and discussion

The superiority of linear micropattern is controlled by machining rate, dimensional accuracy, textured depth, and surface quality. Investigational outcomes are plotted as various graphs and appraised the best micrograph with good quality.

Figure 2 displays the consequence of voltage on output measures with other staticattributes, such as 50% duty ratio and 25kHz frequency. With greater machining voltage, the MRR increases because of more current. The width overcut grows with greater voltage because the distribution of current flux is uncontrolled. For increasing the machining voltage, the depth rises for greater machining localization. The $R_a$increases with higher voltage owing to irregular etching for distribution of current flux. For high quality linear micropattern, lower voltage is proposed.

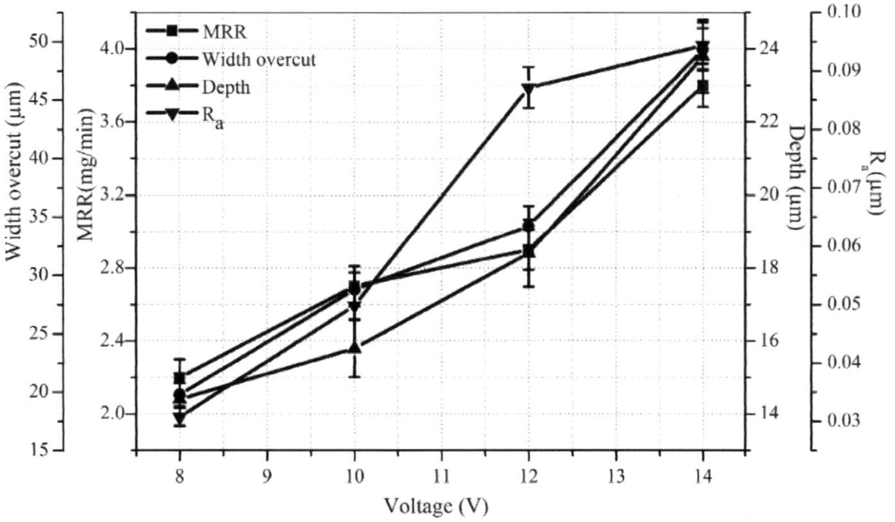

Figure 2. Variation of width overcut, MRR, depth and R_a with voltage

Figure 3 depicts the outcomes of duty ratio on performance criteria with other static parameters i.e., 25kHz frequency and 8V voltage. Because of the longer pulse on time (texturing time), the MRR rises as the duty ratio rises. Higher texturing time removes more material from machining area. With a higher duty ratio, machining accuracy lowers as texturing time increases, due touncontrolled distribution of current flux. The depth improves for greater duty ratio because of longer time, resulting in a greater localization effect. Because of the irregular etching, the surface finish reduces as the duty ratio increases. Lower duty ratio is suggested for better micropattern.

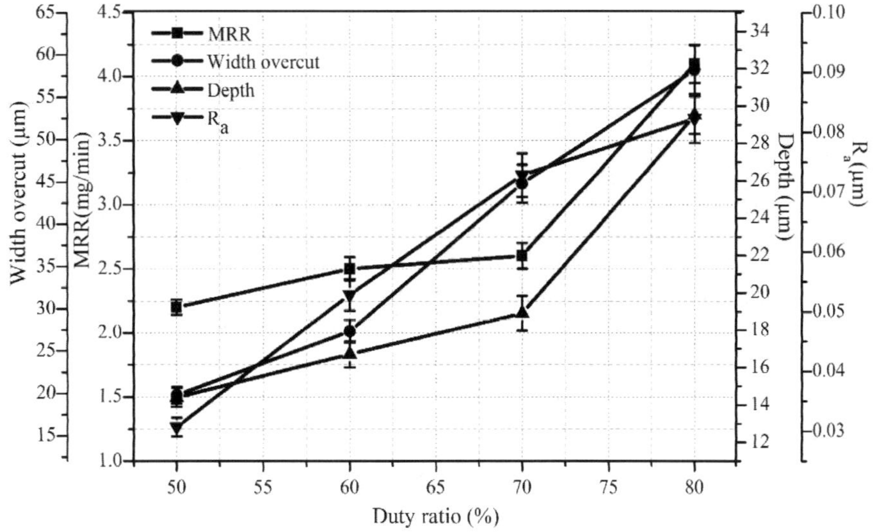

Figure 3. Variation of width overcut,MRR, depth and $R_a$with duty ratio

Experimentation is accomplished to assess the machining influence of frequency on output requirements with other constantvariables, such as 8V voltage and 50% duty ratio as exhibited in

Figure 4. With higher frequency, the MRR decreases because the repetition of the pulse on time decreases, resulting in less material removal. Since the texturing time decreases with higher pulsed frequency, the machining precision increases, and the stray current effect decreases. The micropatterned depth lowers with greater pulsed frequency because of lower machining time resulting lower controlled machining. Because of the shorter machining time, the surface finish improves with higher pulse frequency, which is responsible for regular etching. For précised micropattern, higher pulse frequency is suggested.

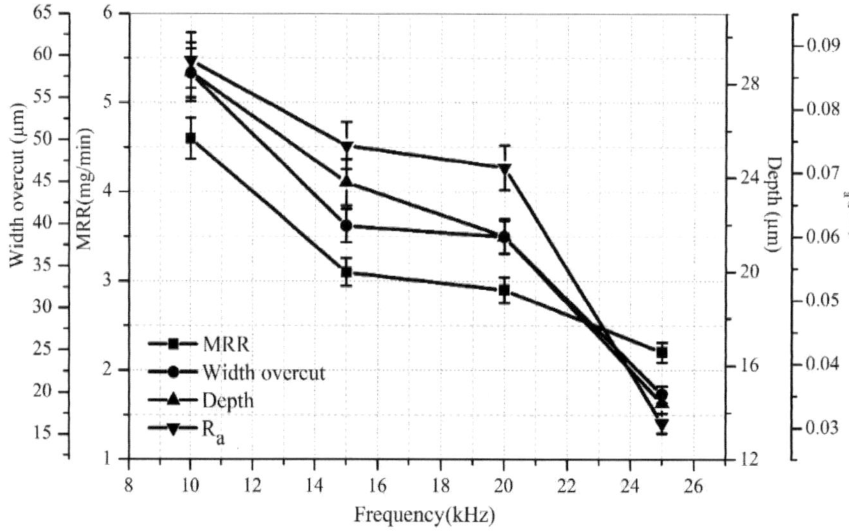

Figure 4. Variation of width overcut, MRR, depth and R_a with frequency

4. Analysis of micrographs

The micropatterned tool is utilized for producing many high quality linear micropatterns owing to higher strength, high resistance capability, etc. The masked tool is shown in Figure 5(a) before machining. It is applied up to fabrication of twenty-one high-quality micropatterned samples. The quality of patterned tool is deteriorated after fabrication of twenty-one samples. Figure 5(b) demonstrates the patterned tool after texturing. It is demonstrated that this mask has ability to produce many good quality samples.

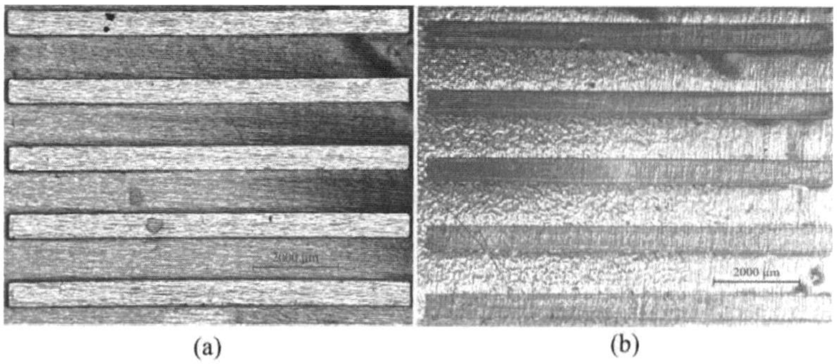

Figure 5. Textured tool (a) before texturing (b) after texturing

The micrograph of micropattern is fabricated at 8V voltage, 50% duty ratio and 25kHz frequency as shown in Figure 6. All impressions of linear micropattern are better in dimensional accuracy, size, and shape because of regular anodic dissolution and good flushing conditions. The textured depth is also unvarying all over the micropatterned zone owing to regular dissolution. Regulated etching also improves the consistency of the surface. The generated regular linear micropattern has MRR of 2.2 mg/min, width overcut of 19.76µm, textured depth of 14.4µm and surface roughness of 0.0307µm.Figure 7 shows a 3D view of a micro impression segment with a 2D depth profile having value of 16.1µm. Figure 8 shows the roughness profile having value of 0.0539 µm.

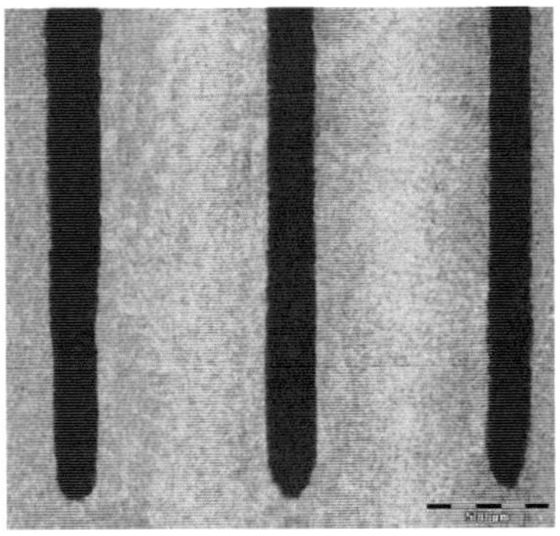

Figure 6. Regular linear micropattern

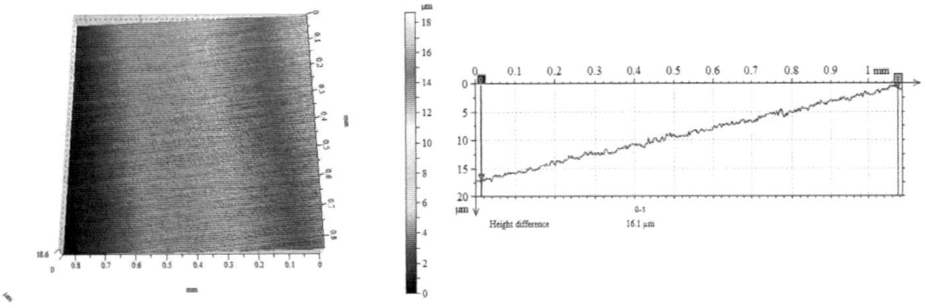

Figure 7. 3D view and 2D depth profile

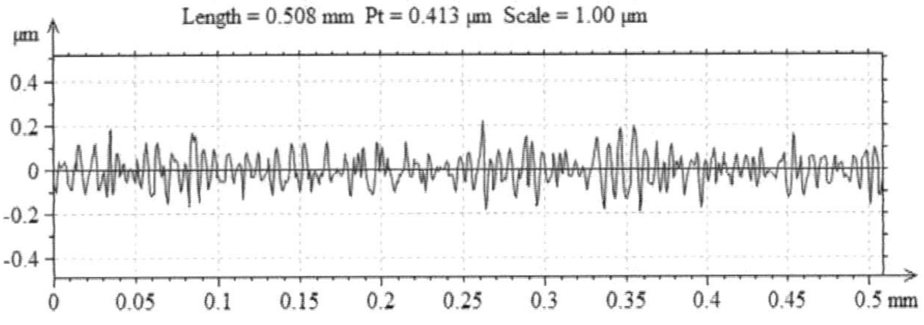

Figure 8. Roughness profile

AFM is utilized to take 3D view with 2D roughness profile of micro impression as shown in Figure 9 having the value of 190.89nm.

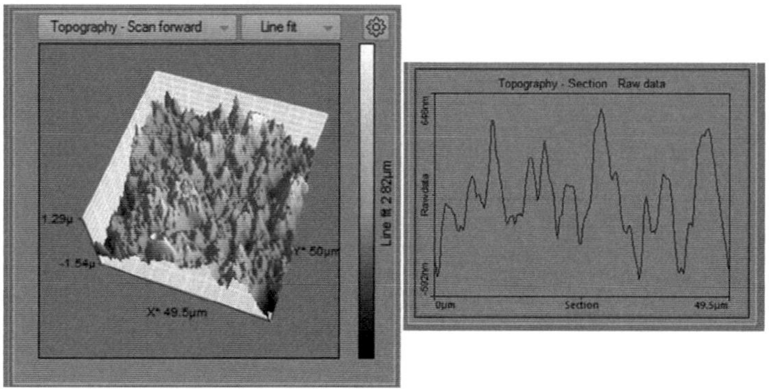

Figure 9. 3D view and 2D roughness profile

5. Conclusions

This advanced method is employed to produce high-quality linear micropatterns using developed electrolyte flow scheme and reusable masked tool. From the investigational analysis, the following deductions can be concise:

(i) With the developed setup and flushing scheme, an advanced EMM technique, maskless EMM, is used to generate good quality micro linear patterns.

(ii) One patterned tool can generate twenty-one high-quality micropatterned samples with high dimensional accuracy.

(iii) The best input parameters, viz. 8V voltage, 50% duty ratio and 25kHz frequency, can fabricate good quality micropatterns.

References

[1] Kunar S, Bhattacharyya B, Mahata S 2018 Influence of electrochemical micromachining process parameters during fabrication of varactor micropattern *Int J Adv Manuf Technol* 96 411-427.

[2] Kunar S, Bhattacharyya B 2019 Effect of various masked patterned tools during micro-electrochemical texturing *Microsyst Technol* **25** 1475-1492.

[3] Herbert D 1999 Blast finishing *Metal Finish* 97-1 93.

[4] Wakuda M, Yamauchi Y, Kanzaki S 2003 Material response to particle impact during abrasive jet machining of alumina ceramics *J Mater Process Technol* **132** 177–183.

[5] Madore C, Piotrowski O, Landolt D 1999 Through-mask electrochemical micromachining of titanium. *J Electrochem Soc* **146(7)** 2526–2532.

[6] Zhu D, Qu N S, Li H S, Zeng Y B, Li D L, Qian S Q 2009 Electrochemical micromachining of microstructures of micro hole and dimple array *CIRP Ann-Manuf Techn* **58** 177–180.

[7] Qian S Q, Zhu D, Qu N S, Li H S 2010 Generating micro-dimples array on the hard chrome coated surface by modified through mask electrochemical micromachining *Int J Adv Manuf Technol* **47 (12)** 1121–1127

[8] Hao X Q, Wang L,Wang Q D, Guo F L, Tang Y P, Ding Y C 2011 Surface micro-texturing of metallic cylindrical surface with proximity rolling-exposure lithography and electrochemical micromachining *App Surf Sci.* **2578** 906–89.

[9] Qu N S, Chen X L, Li H S, Zeng Y B 2014 Electrochemical micromachining of micro-dimple arrays on cylindrical inner surfaces using a dry-film photoresist *Chin J Aeronaut* 27

1030–1036.

[10] Kunar S, Bhattacharyya B 2018 Electrochemical microsurface texturing with reusable masked patterned tool *Eng Sci Technol Int J* **21** 1095-1103.

[11]Kunar S, Bhattacharyya B 2021 Influence of various pulse waveforms during fabrication of micro rectangular pattern by masked tool using electrochemical micromachining *J Adv Manuf Syst* **20(1)** 27-50.

[12] Kunar S, Rajkeerthi E, Mandal K, Bhattacharyya B 2020 Fabrication of array of micro circular impressions using different electrolytes by maskless electrochemical micromachining *Manuf Rev* **7** 15.

Focusing of Radially polarized Light Using Gaussian laser beam Near its focal length

Getu Endale[a,b,*], Devendra Mohan[a] and Sandeep Yadav[c]

[a]Physics Deptt, Guru Jambheshawar University of Science and Technology, Hisar, India
[b]Physics Deptt, Wolkite University, Wolkite, Ethiopia
[c]Physics Deptt, Gurugram University, India

E-mail: getuendale1@gmail.com

Abstract. The Richards-Wolf diffraction model is used to examine focusing of RPGB radially polarized Gaussian beam of low & large numerical aperture. The longitudinally polarized component becomes predominant at high numerical aperture and high radius. Numerical calculations for varied beam characteristics, numerical aperture (NA), and beam radius (θ_0) show the total optical intensity distribution in the focal region. The whole optical intensity distribution in the focal region is illustrated using numerical computations for various beam parameters, numerical aperture (NA), and beam radius (θ_0).The suggested technique appears to produce subwavelength focal spot of o.51λ& significant focal depth of 21λ. Optical trapping, data storage, biomedical imaging, laser cutting, microscopy, and the manipulation of optical traps with high refractive index particles can all benefit from such beams with a small spot size and large focal depth.

Keywords: Radial polarization, Focal length, Focal spot, Wavelength, Numerical aperture, Beam radius, Beam amplitude.

1. Introduction

One of the most important subjects in optical research and applications is focusing light on a very tight spot. Microscopy, data storage, imaging, and lithography are all applications that benefit from a small focal spot [1]. With the rising aperture angle and decreasing wavelength, the size of a focal spot decreases.As a result, the characteristics of laser beam focused by large NA lens have gotten lot of attention[2] - [4].Wang et al. [5]"use binary-phase and large NA lens to focus a radially polarized Bessel–Gaussian beam into a uniform and non-diffracting axial light beam with a sub-diffraction beam size of ~0.43λ".On focusing ofRPB, a sub-wavelength light needle with a depth of focus of around 4λ was recently created [6]. Another study used dual-beam focusing to create a sub-wavelength light needle with a greater depth of focus (over 9:5λ)[7].[8]showed an optical needle with a tight focal spotformed by the longitudinal electric field. The effects of pupil function [9] and apodizations [10] on the focal spot size are presented in a way that is near to reality and can be used as a guide for experimentation. Zha et al. [11] demonstrated that a radially polarized BG beam may be tightly focused by high NA lens &ternary optical element to produce longitudinally polarized beam. "A radially and azimuthally polarized Gaussian beam is proposed for illuminating the pupil plane of the objective lens to achieve a sub-wavelength longitudinal beam with a large focal depth [12]-[14]"."Laser beams with radial polarization are becoming more popular and its large NA focusing capabilities and applicability as large resolutionprobes have attracted researchers' attention"[15] -[18].An electric field with a radially polarized electric field is characterized by a strong axial component due to beam polarization symmetry. This characteristic can be utilized in various scientific fields like High-resolution microscopy, microlithography, metrology & nonlinear optics[2].

We look at how a radially polarized Gaussian laser beam behaves near its focal length in this work. We consider both low and high numerical aperture to analyze the field distributions of Gaussian laser beam near its focus and briefly discuss some of its recent applications. To evaluate the focusing of a Gaussian laser beam, we utilized the same Gaussian beam size, different beam diameters, and adjusted numerical aperture and radius throughout this study.

Theoretical modelling of electric field distribution for a RPGB presented insection 2. Section 3 covers numerical analysis for focusing radially polarized Gaussian beams usingdifferent parameters of beams. Section 4 discusses some of these beams' potential applications.

2. Theoretical Modelling

The Richards-Wolf diffraction method is used to investigate the field distribution around the focus of highly focused polarized beams[19] - [20]. Figure 1 shows the schematic diagram that corresponds to this situation.The unit vector $\boldsymbol{g}_0$canrepresented bythe following way:

$$g_0 = \cos\phi\,\hat{\imath} + \sin\phi\,\hat{\jmath} \tag{1}$$

g_0 representsradial component in the object space. The radial and azimuthal components of the incident field of the pupil plane can then be resolved:

$$\vec{E}_i(r,\varphi,\phi) = l_0\left[e_r^0 g_0 + e_\phi^0 (g_0 \times \hat{k})\right] \tag{2}$$

where$l_0$is the radially varying relative amplitude of the field.

According to the Richards-Wolf theory, $\vec{E}_t(r,\varphi,\phi)$ is[20].

$$\vec{E}_t(r,\varphi,\phi) = \frac{-ik}{2\pi}\int_0^{\theta_{max}} d\theta \int_0^{2\pi} \vec{a}(\theta,\varphi)\exp\left[ik(\vec{s}\cdot\vec{r})\right]\sin\theta\, d\varphi \tag{3}$$

Where $\theta_{max}$is maximum angle specified by objective lens of NA, k wavenumber, and $\vec{a}(\theta,\varphi)$ is the field strength factor which is given by:

$$\vec{a}(\theta,\varphi) = l_0 f P(\theta)\left[e_r^0 g_1 + e_\phi^0 (g_1 \times \vec{s})\right] \tag{4}$$

The polarization unit vector is also affected by the refraction of the focusing lens. The unit vector g_1 is perpendicular to $\hat{s}$, the ray's propagation direction. The radial polarization unit vector is written by[17]:

$$g_1 = \cos\theta\,(\cos\phi i + \sin\phi j) + \sin\theta\hat{k} \tag{5}$$

where θ indicates the polar angle. Then $\vec{s}\cdot\vec{r}$is:

$$\vec{s}\cdot\vec{r} = Z\ \cos\theta + r\sin\theta\cos(\varphi - \phi) \tag{6}$$

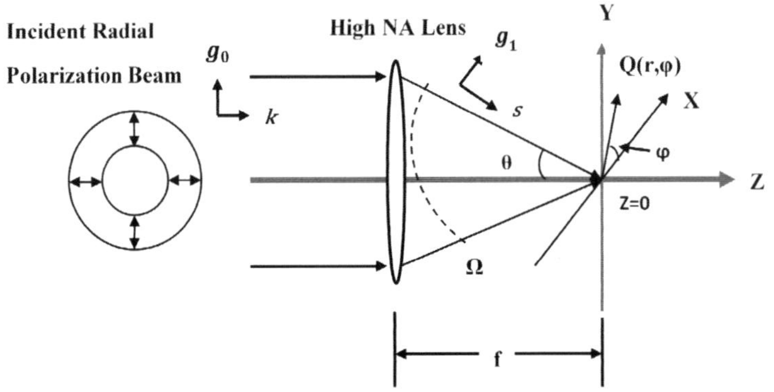

Figure 1. Schematic diagram of focused RPB.*f*objective focal length. Inplane, $Q(r,\varphi)$ is an observation point.

The electric field vector in the focus region is obtained by selecting a suitable field amplitude $\vec{\boldsymbol{a}}$.We use $e_{\phi}^{(0)}=0$ in equation (2) for radially polarized illumination. The electric field vector's Cartesiancomponents near the focus can therefore be represented as follows:

$$\vec{E}_t(r,\varphi,\phi)=\begin{bmatrix}\vec{E}_x(r,\varphi,\phi)\\ \vec{E}_y(r,\varphi,\phi)\\ \vec{E}_z(r,\varphi,\phi)\end{bmatrix} = \frac{-iA}{\pi}\int_0^{\theta_{max}} d\theta\int_0^{2\pi}\sin\theta\sqrt{\cos\theta}\,l_0\exp\left[ik(z\cos\theta + r\sin\theta\,\cos(\varphi-\phi))\right]\begin{bmatrix}\cos\theta\cos\phi\\ \cos\theta\sin\phi\\ \sin\theta\end{bmatrix}d\phi d\theta \quad 7$$

The following transformations are used to create the axis of $\vec{E}_t(r,\varphi,\phi)$:

$$\vec{E}_r = cos\ \phi\vec{e}_x + sin\ \phi\vec{e}_y \quad 8$$

and

$$\vec{E}_{\phi} = cos\ \phi\vec{e}_y - sin\ \phi\vec{e}_x \quad 9$$

The azimuthal axis is 0 in the image space. Then the two axis are given as:

$$\vec{E}_r(r,\varphi,\phi)=\frac{-iA}{\pi}\int_0^{\theta_{max}}\int_0^{2\pi}\sin\theta\sqrt{\cos\theta}\,l_0\cos\theta\cos(\varphi-\phi)\exp\left[ik(z\cos\theta + r\sin\theta\,\cos(\varphi-\phi))\right]d\phi d\theta \quad 10$$

And

$$\vec{E}_z(r,\varphi,\phi)=\frac{-iA}{\pi}\int_0^{\theta_{max}}\int_0^{2\pi}\sqrt{\cos\theta}\ (\sin\theta)^2 l_0\exp\left[\ ik(\ z\cos\theta + r\sin\theta\cos(\varphi-\phi))\right]d\phi d\theta \quad 11$$

The identity below can be used to complete the integration overϕ:

$$\int_0^{2\pi} \cos(n\varphi)\exp[ikr\sin\theta\cos\varphi]d\varphi = 2\pi i^n J_n(kr\sin\theta) \tag{12}$$

Where $J_n(kr\sin\theta)$is a first order Bessel function.Electric fields at the focus of a pupil illuminated by radial polarization take the following form:

$$\vec{E}_r(r,z) = A\int_0^{\theta_{max}} \sqrt{\cos\theta}\sin(2\theta\,)\, l_0 J_1(kr\ \sin\,\theta)\ \exp[ikz\cos\theta]d\theta \tag{13}$$

And

$$\vec{E}_z(r,z) = 2iA\int_0^{\theta_{max}} \sqrt{\cos\theta}\,(\sin\theta)^2 l_0 J_0(kr\ sin\ \theta)\exp\,[ikz\ cos\ \theta]d\theta \tag{14}$$

The constant A is given by:

$$A = \frac{\pi f l_0}{\lambda} \tag{15}$$

At all distances z, precisely as it does with the paraxial solutions[17].Equations 13 and 14 can be written in terms of $P(\theta)$ as follows:

$$\vec{E}_r(r,z) = A\int_0^{\theta_{max}} P(\theta)\ \sin(2\theta)J_1(kr\ sin\ \theta)\exp\,[ikz\ cos\ \theta]d\theta \tag{16}$$

And

$$\vec{E}_z(r,z) = 2iA\int_0^{\theta_{max}} P(\theta)\,(\sin\theta)^2 J_0(kr\ sin\ \theta)\exp\,[ikz\ cos\ \theta]d\theta \tag{17}$$

The radial & longitudinal components near the focus of z=0 are $\vec{E}_r(r,z)$ and $\vec{E}_z(r,z)$, respectively.The amplitude modulated annular multi-Gaussian beam is described by $P(\theta)$, which is given by [12]:

$$P(\theta) = \left(\frac{\theta}{\theta_0}\right)^m \sum_{-N}^{N} \exp\left[-\left(\frac{\theta-\theta_c-n\omega_0}{\omega_0}\right)^2\right] \tag{18}$$

$\theta$is the converging semiangle. The maximum converging semiangle is denoted by symbol θ_{max}, & related to NA by the formula $\alpha = arcsin\,(NA)$. The value of θ_0 is commonly selected to be slightly less thanα.The radial translation of P(θ) is determined byθ_c.ω_0 is given by:

$$\omega_0 = \frac{1}{2}\times\frac{\alpha}{N + \{\,1-\ln[\sum_{n=-N}^{N}\exp(-n^2)]\}^{\frac{1}{2}}} \tag{19}$$

The total field is expressed by using following equation[5], [7]:

$$\vec{E}_t(r,z) = E_r\vec{e}_r + E_z\vec{e}_z \quad (20)$$

The amplitudes of the two orthogonal components given in equations 13 and 14 are E_r and E_z, respectively,and $\vec{e}_r$ and $\vec{e}_z$are their corresponding unit vectors.

3. Numerical Analysis

Figure 2 depicts the amplitude profile of the Gaussian beam waist at $\omega_0 = 0.14$ beam size, as well as the different Gaussian beam radius, θ_0. Figures 2 (a) and 2(b) use a numerical aperture of 0.6, while figures 2 (c) and 2(d) use a numerical aperture of 0.95.The AD of Gaussian beam of numerical aperture 0.95 and 0.6 is shown to examine the Beam Amplitude profile of a Gaussian beam with beam size ω_0 and beam radius θ_0. The Gaussian beam width grows with increasing radius for both numerical apertures.

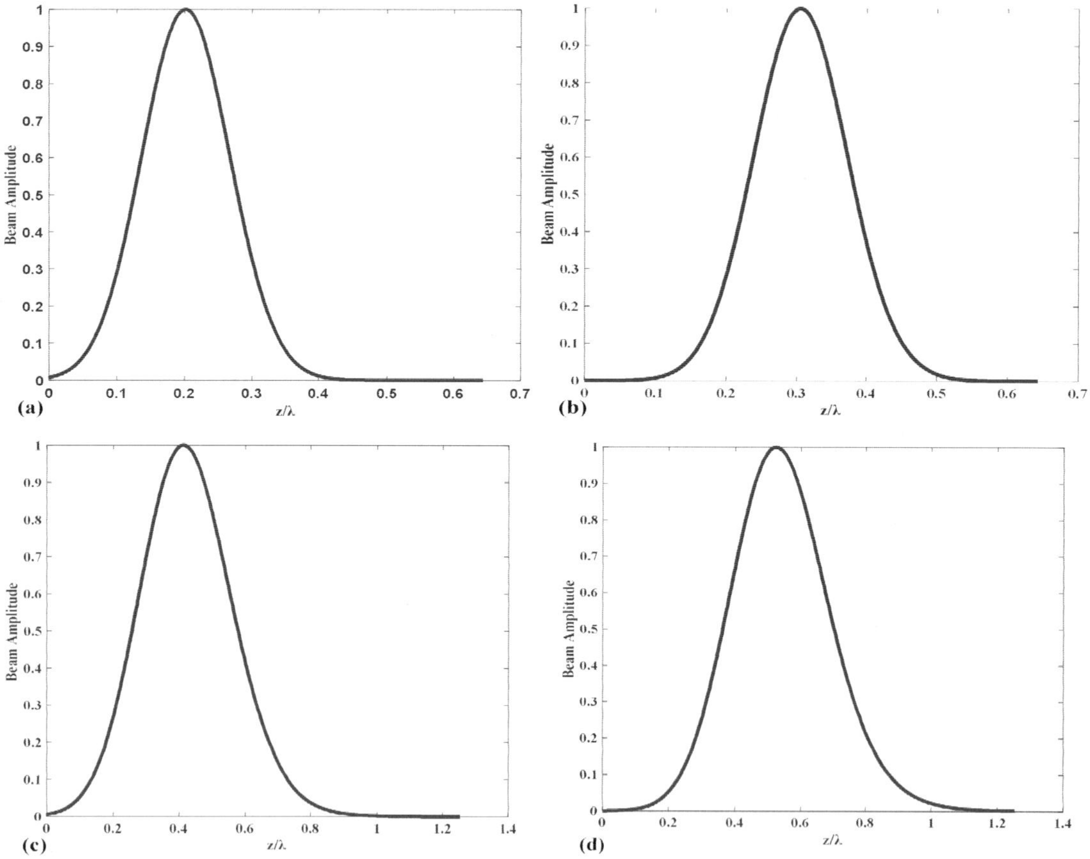

Figure 2. Amplitude profile of the Gaussian beam at the output pupil for $\omega_0 = 0.14$. (a) and (b) are for the numerical aperture, NA of 0.6 and (c) and (d) is for the NA0.95. θ_0 for (a), (b), (c)& (d) is $0.2, 0.3, 0.4$ and 0.5 respectively.

Figure 3 deals focusing of electric density distributions for RPGB focused byNA=0.6 and n=1, where NA and n stand for Numerical Aperture and Refractive Index, respectively.$\alpha = \text{asin}\left(\frac{\text{NA}}{\text{n}}\right) = 1.25$ rad. For $\theta_0 = \text{asin}\left(\frac{\text{NA}}{\text{n}}\right)$, $\theta_0$is no larger than α and assuming that the Gaussian profile is retained.$E_r(0,z)$ is zero due to the Bessel function's nature, although $J_1(0) = 0$ is also zero. As a result, for different z, the greatest field intensity occurs at location of$r \neq 0$. Indeed, the RP axis $E_r(r,z)$ widen the focusing spot's transversal width.It means that by lowering the high intensity of RP field& modifying point of highest intensities emerge, a small focusing spot obtained. $J_0(0)$ is 1 for $r = 0$, and $E_z(0,z)$ is high for a given z. In these scenario, RP axis has no effect on the focal spot minimization[21].

The radial axis dominates for the polarized Gaussian beam with $\theta_0 = 0.2$, as seen in Figure 3(a)-(d). The radially polarized component has a maximum intensitynearly 0.98 at $z = 1.12\lambda$, which is higher than the longitudinally polarized component, which has a high intensity of about 0.21at z = 0. Large total field intensity will appear at$z = 1.12\lambda$ because it cannot occur along the optical axis. The radial polarization axis dominates around z=o. As a result, the focusing spot at the specified focal point cannot be reached. It also shows that the 0.6 Numerical Aperture at $\theta_0 = 0.2$ fails to converge the PGB with small θ_0. Bright spots denote high electric intensity regions, while the dark areas denote low electric intensity regions.

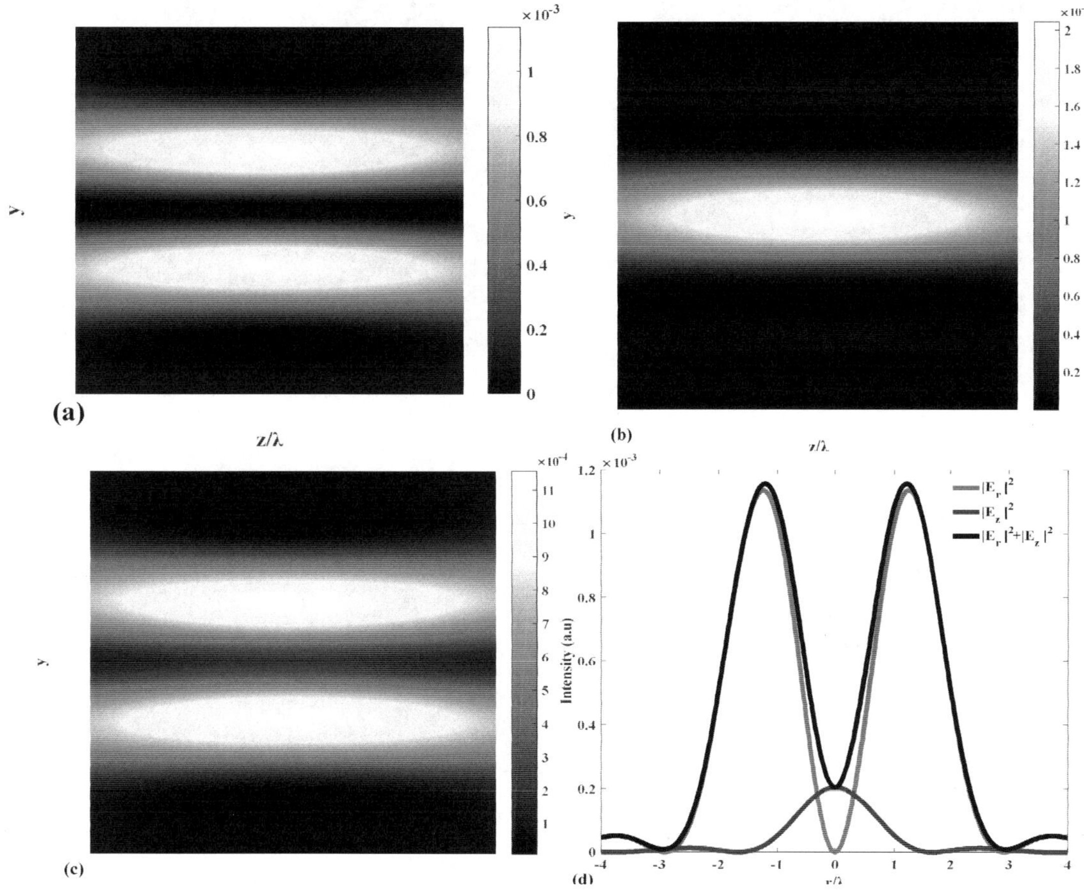

Figure 3: Contour plots of the focusing of electric density distributions in the *r–z* plane for GB that is radially polarized &used NA 0.6. $\omega_0 = 0.14$ and $\theta_0 = 0.2$ for (a) Radial component, $|E_r|^2$, (b) Longitudinally polarized component, $|E_z|^2$, and (c) Total electric density distribution, $|E_r|^2 + |E_z|^2$. (d) Focusing of electric density distributions in the plane focus z = 0 forGB that is radially polarized and focused by a lens for Radial component, $|E_r|^2$, Longitudinally polarized component, $|E_z|^2$,&Total electric density distribution, $|E_r|^2 + |E_z|^2$. Radial axis, longitudinal axis, & total fields are represented by red, blue, and black colors respectively.

Electric density distributions inr–z plane for Gaussian beam that is radially polarized and large NA 0.95, beam size ($\omega_0 = 0.14$), and Gaussian beam radius ($\theta_0 = 0.2$) are shown in Figure 4. The radially polarized component dominates the focusing field, hence the focusing spot is impossible to achieve at this focal point. The low radius of the Gaussian beam prevents a longitudinally polarized focusing field from being obtained, even if the numerical aperture is high.

Figures 5(a)-(d) illustrate the overall intensity distribution in r– z plane.We employed a high NA of 0.95& large GB radius of $\theta_0 = 0.8$. With the growth of θ_0, the maximum intensities longitudinal axis increases. As a result, longitudinally polarized component progressively becomes dominant. With rise of θ_0, the beam quality improves. It indicates using a radially polarized Gaussian beam of high θ_0, longitudinally polarized focusing field is generated. Figure 5(a)-(d) shows that for high NA & Gaussian beam radius, the produced focal field is focused.

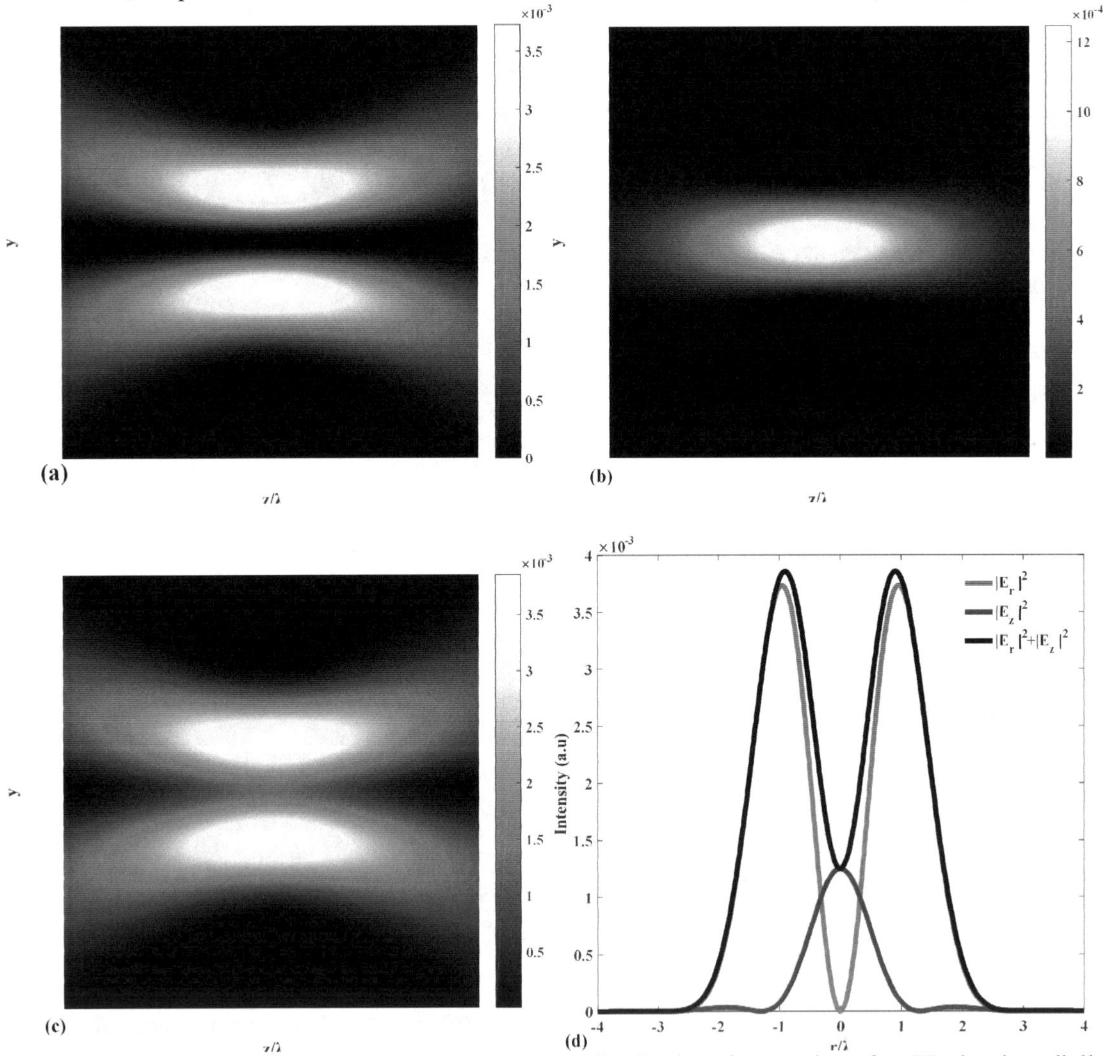

Figure 4. Contour plot of focusing of electric density distributions in r–z plane for GB that is radially polarized & focused bylens with NA of 0.95,$\omega_0 = 0.14$, & $\theta_0 = 0.2$ for (a) Radial component, $|E_r|^2$, (b) Longitudinally polarized component, $|E_z|^2$, and (c) Total electric density distribution, $|E_r|^2 + |E_z|^2$. (d) Focusing of electric density distributions in the plane of focus z = 0, for GB that is radially polarized and focused by a lensfor Radial component, $|E_r|^2$, Longitudinally polarized component, $|E_z|^2$,& total

electric density distribution, $|E_r|^2 + |E_z|^2$. Radial axis, longitudinal axis, & total fields are represented byred, blue, and blackcolors respectively.

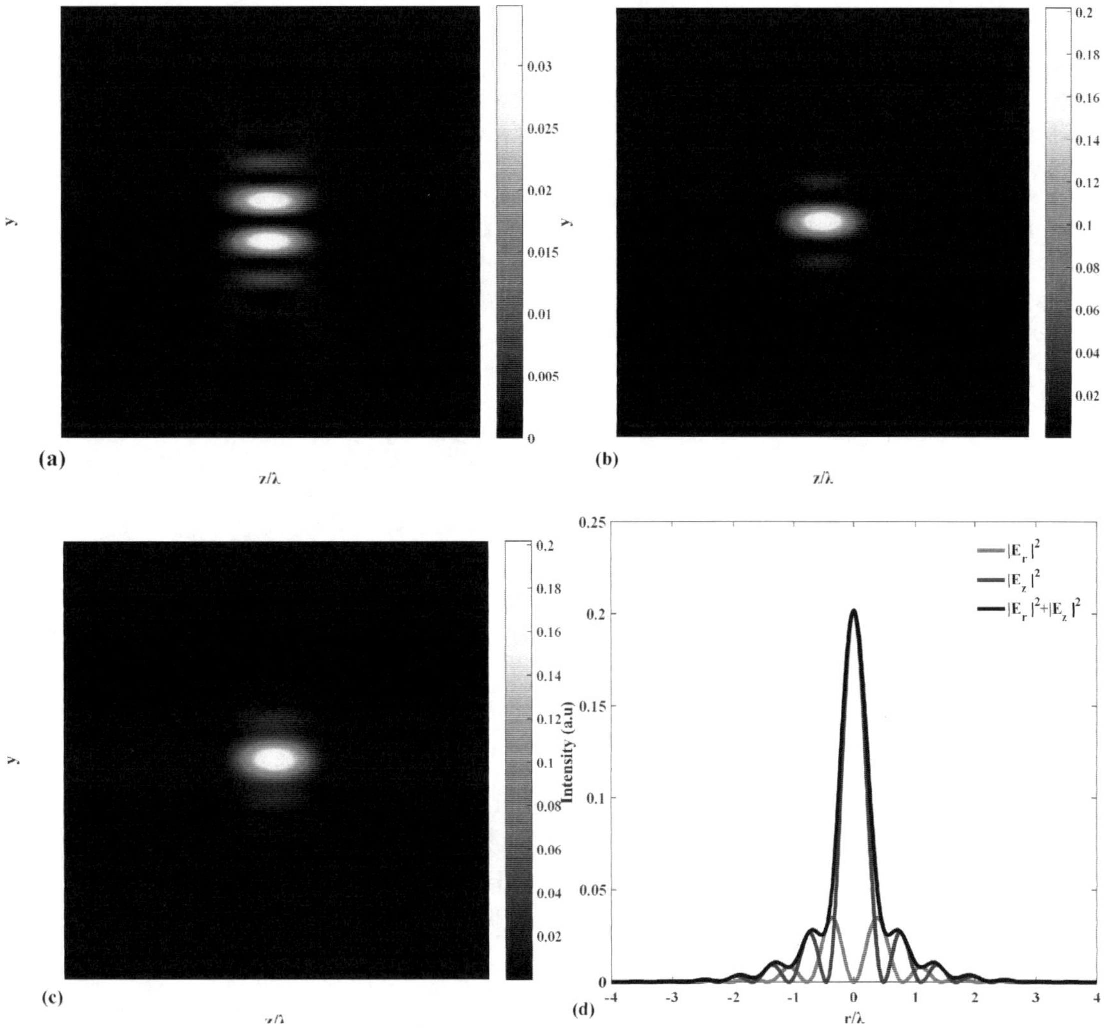

Figure 5: Contour plots of the focusing of electric density distributions $r-z$ plane of GB that is radially polarized & focused by lens &NA of 0.95,$\omega_0 = 0.14$,and $\theta_0 = 0.8$ for (a) Radial component, $|E_r|^2$, (b) Longitudinally polarized component, $|E_z|^2$, and (c) Total electric density distribution, $|E_r|^2 + |E_z|^2$. (d) Focusing of electric density distributions in the plane of focus z = 0 for GB that is radially polarized and focused by a lensfor Radial component, $|E_r|^2$, Longitudinally polarized component, $|E_z|^2$,& total electric density distribution, $|E_r|^2 + |E_z|^2$. Radial axis, longitudinal axis, and total field are represented by red, blue, and blackcolors respectively.

4. Conclusion

Finally, focusing of RPG beams is investigated using Richards' formulas at various numerical apertures and Gaussian beam radiuses. Numerical computations for varied beam parameters of NA, ω_0, and θ_0 showsthe overall optical intensity distribution. The suggested technique appears to produce a subwavelength focal spot (0.51λ)&significant focal depth (21λ).It believed that a beam with such asmall spot size and extended focaldepthwillfind widespread application in optical trapping,data storage, biomedical imaging, laser cutting, microscopy, & the control of high refractive index particle optical traps.

The Gaussian beam's Beam Amplitude profile is evaluated at low and high numerical apertures. For both low and large numerical apertures, it was demonstrated that as the Gaussian beam's radius increases, the width of the Gaussian beam also increases. The radially polarized component dominates the focusing fields in the focusing of electric field intensity, the polarized Gaussian beam at low NA and θ_0. The radialpolarization component dominates the focusing field around z=0. Due to the dominance of RP axis at this focal place, the focusing spot cannot be attained at low NA and θ_0.The longitudinally polarized component becomes prominent in the focusing field as the radius of the Gaussian beam increases, and the focusing spot occurs at a large numerical aperture and radius of GB.

Impact ofGaussian beam's radius on focusing was addressed and examined. Both the Gaussian beam radius and the Gaussian beam waist at beam size define the magnitude distribution of GB. Maximum intensity offield is determined by the radius of the Gaussian beam. Sub-wavelength focusing spot is obtained at high radius of the Gaussian beam, overcoming the diffraction limit.

Acknowledgement

I thank Professor Devendra Mohan for his invaluable aid in completing my paper. I also extend my thankfulness to Dr. Sandeep Yadav for his help.

References

[1] Chen Z and Zhao D 2012 4Pi focusing of spatially modulated radially polarized vortex beams *Opt Lett* **37** 8 1286 doi: 10 1364/ol 37 001286

[2] Zhan Q 2004 Trapping metallic Rayleigh particles with radial polarization *Opt Express*, **12** 1 3377 doi: 10 1364/opex 12 003377

[3] Yan S and Yao B 2007 Radiation forces of a highly focused radially polarized beam on spherical particles *Phys Rev A - At Mol Opt Phys* **76**(5) 1–6 doi: 10 1103/PhysRevA 76 053836

[4] Kawauchi H, Yonezawa K, Kozawa Y, and Sato S 2007 Calculation of optical trapping forces on a dielectric sphere in the ray optics regime produced by a radially polarized laser beam *Opt Lett* **32** 1839 doi: 10 1364/ol 32 001839

[5] Wang H, Shi L, Lukyanchuk B, Sheppard C and C Chong T 2008 Creation of a needle of longitudinally polarized light in vacuum using binary optics *Nat Photonics* **2** 501–505 doi: 10 1038/nphoton 127

[6] Huang K, Shi P, Kang X, Zhang X, and Li Y 2010 Longitudinally Polarized Field *Opt Lett* **35** 965–967

[7] Kuang C, Hao X, Liu X, Wang Tand Ku Y 2011 Formation of sub-half-wavelength focal spot with ultra long depth of focus *Opt Commun* **284** 1766–1769 doi: 10 1016/j optcom 2010 12 055

[8] Kitamura K, Sakai K, and Noda S 2010 Sub-wavelength focal spot with long depth of focus generated by radially polarized, narrow-width annular beam *Opt Express* **18** 4518 doi: 10

1364/oe 18 004518

[9] Sheppard C J R and Choudhury A 2004 Annular pupils, radial polarization, and superresolution *Appl Opt* **43** 4322–4327 doi: 10 1364/AO 43 004322

[10] Sheppard C J R and Yew E Y S 2008 Performance parameters for focusing of radial polarization *Opt Lett* **33** 497 doi: 10 1364/ol 33 000497

[11] Zha Y, Wei J, Wang H, and Gan F 2013 Creation of an ultra-long depth of focus super-resolution longitudinally polarized beam with a ternary optical element *J Opt (United Kingdom)* **15** 7 doi: 10 1088/2040-8978/15/7/075703

[12] Chen J N, Xu Q F, and Wang G 2011 Tight focus of a radially polarized and amplitude-modulated annular multi-Gaussian beam *Chinese Phys B* **20** 11 doi: 10 1088/1674-1056/20/11/114211

[13] Charles J W, Prabakaran K, Rajesh K B and Pandya H M 2014 Generation of sub wavelength super long dark channel using azimuthally polarized annular multi-Gaussian beam *Opt Quantum Electron* **46**, 8 1079–1086 doi: 10 1007/s11082-013-9827-5

[14] Charles J W, Prabakaran K, Parimaladevi R, Rajesh K B and Pillai T V S 2014 Tight focus of an azimuthally polarized and amplitude-modulated annular multi-Gaussian beam *Optik (Stuttg)* **125** 3023–3026doi: 10 1016/j ijleo 2013 12 076

[15] Dorn R, Quabis S, and Leuchs G 200Sharper focus for a radially polarized light beam, *Phys Rev Lett* **91** 2 1–4 doi: 10 1103/PhysRevLett 91 233901

[16] Quabis S, Dorn R, Eberler M, Gloclkl O and Leuchs G 2000 Focusing light to a tighter spot, *Opt Commun* **179** 1–7 https://www dropbox com/home/sem1 seminar?preview=1-s2 0-S0030401899007294-main pdf

[17] Youngworth K S and Brown T G 2000 Focusing of high numerical aperture cylindrical-vector beams *Opt Express* **7** 77 doi: 10 1364/oe 7 000077

[18] Kano H, Mizuguchi S, and Kawata S 1998 Excitation of surface-plasmon polaritons by a focused laser beam Calculation of Electric-Field, *America (NY)* **15** 1381–1386,

[19] Wolf E and P R S L A 1959 Electromagnetic diffraction in optical systems - I An integral representation of the image field *Proc R Soc London Ser A Math Phys Sci* **253** 1274 349–357 doi: 10 1098/rspa 1959 0199

[20] Richards B, Wolf E and P R S L A 1959 Electromagnetic diffraction in optical systems, II Structure of the image field in an aplanatic system *Proc R Soc London Ser A Math Phys Sci* **253** 1274 358–379 doi: 10 1098/rspa 1959 0200

[21] Lin J, Chen R , Yu H, Jin P , Cada M, and Ma Y 2014 Analysis of sub-wavelength focusing generated by radially polarized doughnut Gaussian beam *Opt Laser Technol* **64** 242–246 doi: 10 1016/j optlastec 2014 05 019

Integrating Silicon In Armchair Graphene Nanoribbon For Solar Cell Applications

Preetika Sharma[1*] and Gaurav Sapra[1]

[1]Assistant Professor, UIET, Panjab University, Sector 25, Chandigarh, INDIA-160025

* preetikamadhav@yahoo.co.in

Abstract. Solar power is not only free but also infinite and use of solar energy indeed has many advantages. Thus, solar technologies are greatly studied for producing low cost and better versions of existing devices that can be integrated in the existing photovoltaic (PV) systems. Many such devices like the solar powered curtains and clothes, cases for laptop and other electronic products are being produced. Silicon today is of course being used rapidly as a PV material for solar cell. However, a continuous effort has been made to use graphene for high efficiency solar cell applications and its use. Graphene on the other hand, is not very good at collecting the electrical current produced inside the solar cell but researchers are looking appropriately on this material. Many ways to modify graphene for this purpose are being studied. Solar cell requires semiconductor materials for exhibiting photovoltaic effect. As graphene is zero band gap material, hence, armchair graphene nanoribbons (AGNR) that are semiconducting are being looked up for solar cell applications. Therefore, in this present work, various attempts have been made to investigate the electronic bandgap by using AGNRs in solar cells. The in-depth analysis of electronic properties has been done where band structure, density of states and geometrical stability on the basis of transmission of energy has been examined. Further, AGNR doping with silicon has been performed in which we have replaced the carbon with silicon. Initially, we started with one-silicon atom as a dopant and then went upto 4 silicon as dopant and the variations has been compared. Thus, through this analysis, the use of doped AGNRs in solar cells is investigated.

Keywords: AGNR, Pristine Graphene, Solar Cell

1. Introduction

Solar energy is an economical, abundant and a sustainable source of energy. Solar energy systems rely on solar cells. Solar cells are made using pn junction made by a crystalline material such as silicon, germanium, gallium arsenide etc. These cells are connected in series to form an array of PV systems. They convert sun energy into electrical energy. These PV systems are a lot beneficial as they save a lot in electricity expenses as well as they are promoting green environment. This helps in reducing global warming. The materials required for solar cell must have high electrical conductivity and should be highly transparent. However, metals have high conductivity but they are not transparent. High transparency is needed to allow maximum light absorption so as to convert photons to electrons[1]. Semiconductors are the materials that are used in solar cell applications. These materials directly convert the light into electricity. The current is carried by either holes or electrons. Also highly doped semiconductors are transparent. They have bandgap. But for transmission of visible range, the value of bandgap should be greater than 3eV. This can be achieved in silicon. However, the transparency of silicon is a hindrance and hence, searches for better materials are needed. Thus, how good a semiconductor will perform as solar absorber or a collector, the study is governed by its bandgap.

Since last ten years, a lot of effort has been put on the PV applications. Much research and enhanced development has been focused to find new materials and enhanced cell structural designs. In particular, scientists and research organizations are searching materials that are similar or better performers than silicon semiconductors and its compounds(in its various types, namely, mono crystalline, polycrystalline, and amorphous silicon). However, as silicon is easily available in abundance and therefore its cheaper in terms of production and fabrication, there is a lot more to know about new materials that will replace them in PV industry. Some of the new materials searched include organic diodes, semiconductors, graphene and its composites, thin-film semiconducting materials that includeCdTe etc. The main area of focus is on the novel material called graphene. It is made of carbon atoms that are arranged in the form of hexagonal lattice [2,3]. It is not only a strong material but also completely transparent making it perfect for absorption of light and used for solar cell applications.

The evolution of solar cells technologies is usually divided into three generations [4]. The first generation consist of mono crystalline/ polycrystalline silicon wafers. The second generation is on thin-film semiconductors that include the amorphous silicon such as CdTe, CIGS etc. The third generation of solar cells consist of new materials that result in very cheap fabrication cost and production expenses. This generation thus yield high efficiencies and consist of dye sensitized solar cells (DSSCs), multijunction solar cells (MJSCs), organic solar cells (OSCs), quantum dot sensitized solar cells , perovskite solar cells (PSCs) etc. The classification of solar cells discussed is shown below in Fig. 1.

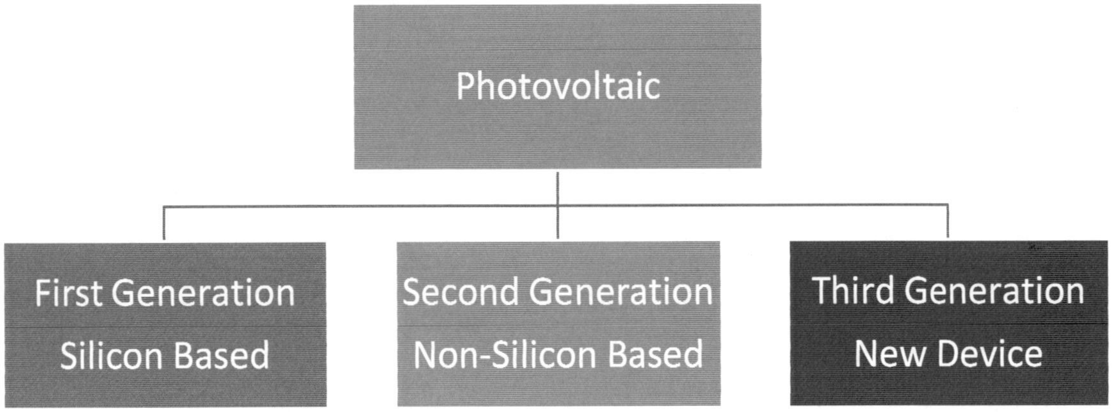

Figure 1. Classification of solar cell

Graphene Nanoribbons (GNR) based solar cells:

Graphene has good and potential applications in high-performance electronic devices, circuits such as photosensors, and smart applications. It has great electronic properties that include high carrier mobility, high carrier saturation velocity, excellent thermal conductivity, high mechanical strength, transparency, good thickness, and flexibility and versatility. This all makes graphene a great choice for advanced future applications. One such strongly needed application in future that whas become the need of the hour is the use of solar energy rather than conventional energy. For this, new solar cell technological design are needed [4]. In particular, as graphene has a unique integration of many properties such as high electrical conductivity and high optical transparency in visible and near-infrared regions, this gives it the tremendous ability to be used as the active layer for many metal and semiconductor junction based solar cells. Not only this, it is also reported that graphene solar cells have been also fabricated on various substrates. The commonly used are the silicon and compound semiconductors like cadmium disulphide. Further, many heterojunction photoelectric devices have also been fabricated in form of nanowires where good efficiency has been achieved. This has been made possible by using the combination of one-dimensional (1D) structure and physical characteristics of 1D nanowires. Graphene nano ribbons(GNRs) were also introduced as a theoretical model and it was examined through its edge and nanoscale size effect [5]. As electronic components are becoming ever smaller, the industry is gradually approaching the limits of what is achievable using the traditional approach with silicon as a semiconductor material.

Graphene nanoribbons (GNRs) are extremely amazing new materials that are being used for variety of applications. These applications range from electronic to thermoelectronic uses. The have tremendous role in mechanically robust thin films, high efficiencies in various devices, capacitors, nano batteries, photo detectors, diodes, transistors, integrated circuits and many more [6]. Among all the applications, the highest potential is seen in the feature of GNR is the tuneability. It can change its properties and can modify to perform more applications. These properties are a function of their precise atomic structure. Hence, it is important to exploit the GNRs to the fullest. Their two basic edge structures, namelyzigzag and armchair, and these edge structures have been intensively studied and it isseen that their electronic and magnetic depend on the structure. The difference between the electronic property of zigzag andarmchair GNRs has been reported. It is reported that the band gaps of zigzag GNRs are smaller than those of armchair GNRs, and the band gap increases with increasing size of the GNR [6]. Armchair nanoribbons show width-dependent band gap and also the band gap produced is inversely proportionalto the width of the GNR.

2. Computational Work

This work includes the calculation of the structural properties and equilibrium geometries of GNRs both prostine and doped. Thius study is based on the first principle calculation that is based on density functional theory[7-8]. An AGNR is taken first as pristine and then silicon as dopant is substituted. The concentration of silicon dopant is varied for 2%, 4%, 6% and 8%.The pristine structure and doped structures are shown in Fig 2(a-e). All geometrical structures are optimized by using density functional theory (DFT). Each carbon and silicon atom in the structure is relaxed at less than 0.05 eV/Å3. The electronic properties for pristine and doped AGNRs are studied using Atomistix Tool Kit (ATK). The local density approximation (LDA) is used in order to representthe exchange correlation potential. The Double Zeta polarized basis set has been chosen for all atoms when the geometries have been optimized. The structural systems are analysed with $1 \times 1 \times 10$ uniformly placed K points. The cut off energy is 75 Hartree which is set for all the structures. All the carbon atoms on the edges are saturatedwith H atoms for stability in structures.

3. Results And Discussions

3.1 Structures of pristine and doped armchair graphene nanoribbons are shown in Fig 2 (a-e)

In order to investigate the behavior of pure and doped AGNR, it is very important to see and study the electronic properties of structures. For this, optimization of all geometrical structure of pristine AGNR along with its electronic properties that include the band structure, DOS and transmission spectrum are calculated. Fig2(a-e) shows all geometrical structures in which the bond length between two carbon atoms is 1.42 Å and bond length between carbon and hydrogen atom is 1.10 A°. Further, the effect of silicon dopant on theelectronic properties of AGNRs is investigated. All the AGNRs with different silicon doping concentrations are studied. The doping concentration is varied so that the energy gap of AGNRs can be engineered for various applications. Due to doping phenomena, formation of the carbon–silicon bond occurs. The carbon atoms in the ribbon are then replaced randomly in the structure with silicon atoms for the study. The variation of band gap, DOS, transmission spectrum and total energy are observed and then compared with the pristine AGNR. It is reported that silicon dopingopens the band gap in graphene and increases its optical conductivity. Hence, AGNRs are seen promising materials for PV applications. .

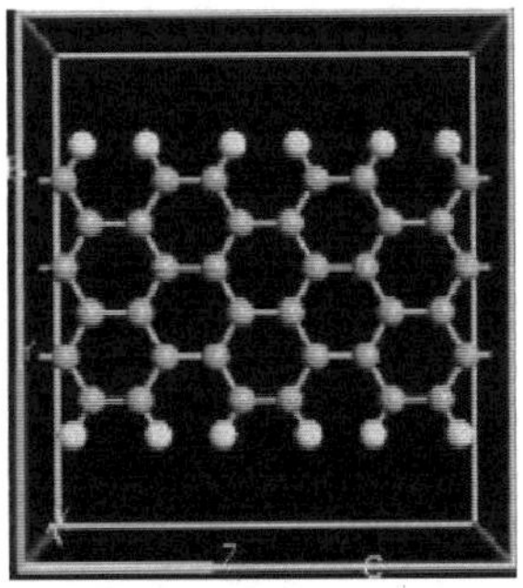

a)

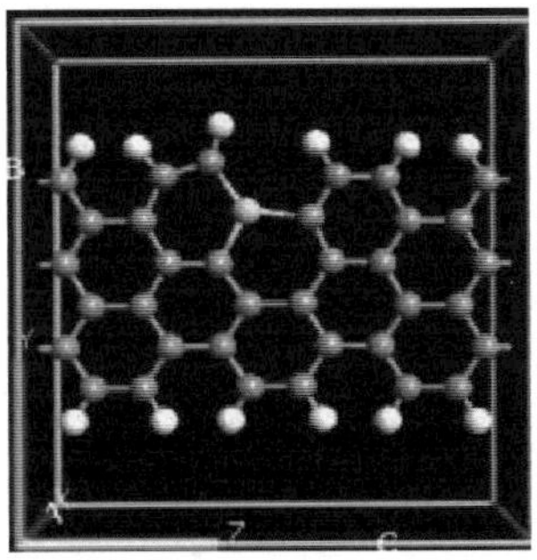

b)

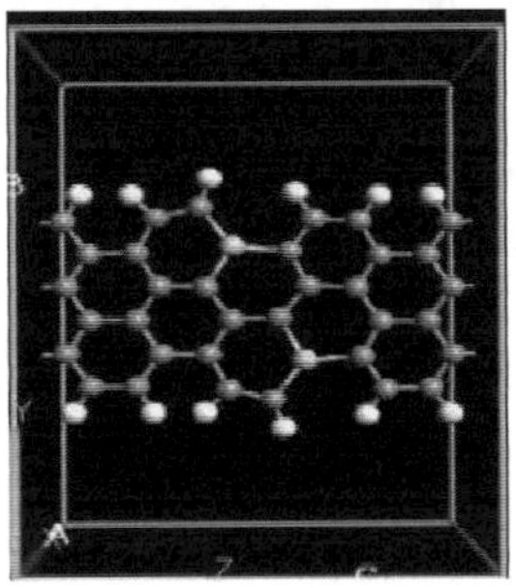

c)

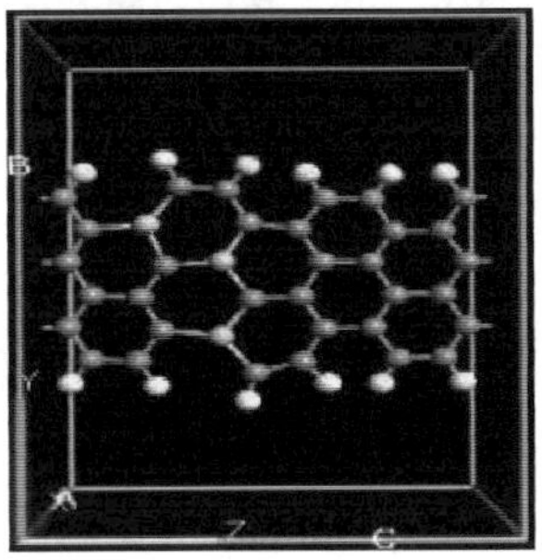

d)

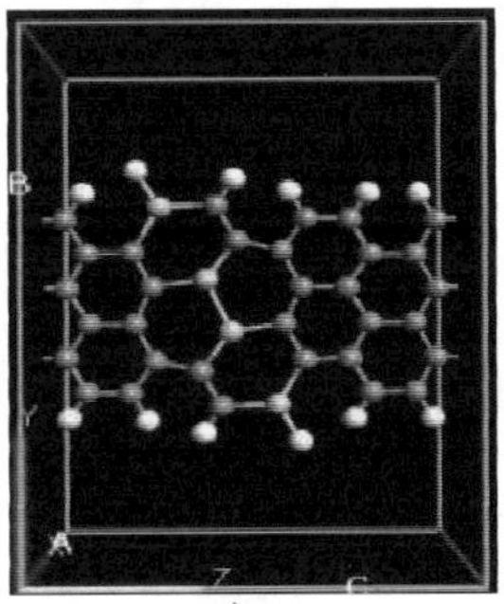

e)

Figure 2. Optimised geometrical structure of a) pristine armchair graphene nanoribbon, b) one doped silicon armchair graphene nanoribbon, c) two doped silicon armchair graphene nanoribbon, d) three doped siliconarmchair graphene nanoribbon and e) four doped silicon armchair graphene nanoribbon.

The geometrical stability of all the considered nanostructures of AGNR is investigated based on the total energy which is presented in Table 1.

Table 1. Calculation of total energy for various structures optimized.

S. No	Structures	Total energy
1.	Pristine	-5867.549304
2.	One doped	-5885.620916
3.	Two doped	-5903.905452
4.	Three doped	-5921.958995
5.	Four doped	-5939.494712

The total energy of the optimized structure of any combination on the different AGNR systems is calculated. Further, on doping, the total energy increases in negative side, confirming the stability of the structures. Also not much change in total energy is observed which shows the structures are not much disturbed due to doping. The total energy is calculated as the sum of kinetic and potential energy for the structure. Strong interaction between AGNR and dopant is attributed to more value of total energy (E total). In case of pristine, the totalenergy for pristine is -5867.549304eV/atom. The total energy calculations are done using the software VNL ATK. The pseudopotentials and basis set choosen is very crucial for energy calculations and it very much dependent on the elements included in your system whether its bulk or a molecule.

3.2 Band Structures

3.2.1 Pristine Structure

AGNR provide continued miniaturization and performance enhancement in devices. We have investigated the electronic band structure of an undoped GNR [9-10]. It shows that a nanoribbon always show a gapin its electronic spectrum. A band gap of 1.2eV is observed. It is reported that a bandgap of 1.1 eV is capable of absorbing about 75% of the solar rays. Many polymers have bandgaps greater than 2 eV and hence they only harness about 30% of the photons which can be converted into electrons. Therefore, attempts are done to observe the ideal band gap needed for photovoltaic phenomena. Another important requirement is the need of direct band gap materials. In case of a material which has an indirect bandgap, there a photon and phonon bothshould take part in transition from the valence band to the conduction band. Due to this, direct bandgap materials show stronger light emission and better absorption properties. In this paper, we are observing electronic bandgap which is not only ideal for solar cell but also is a direct bandgap. Upon studying the pristine, we see for silicon dopant in AGNRs. We can observe that as silicon dopant is added, the band gapstarts decreasing.

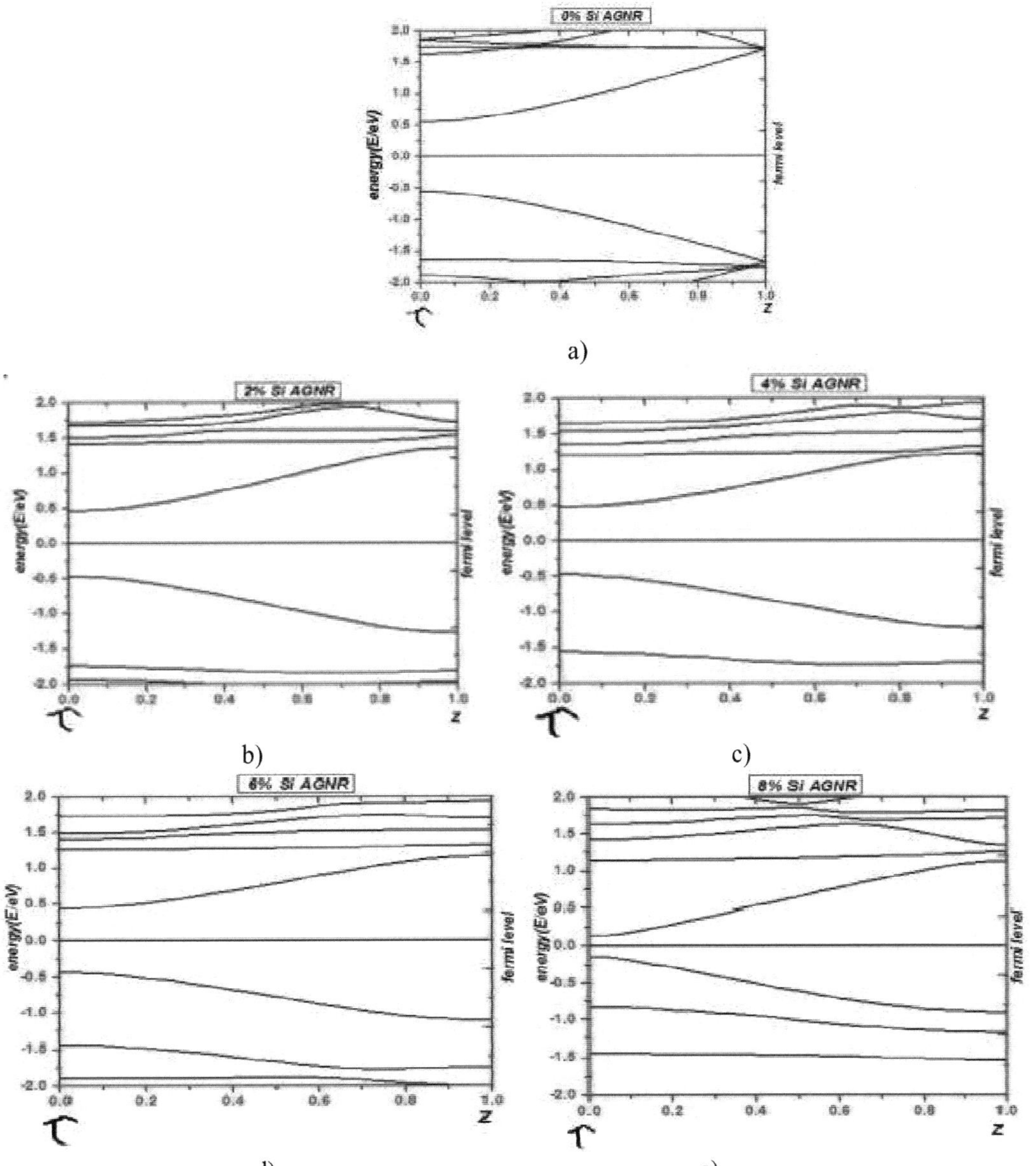

Figure 3. Band structure of a) pristine AGNR, b) 2% silicon doping in AGNR, c) 4% silicon doping inAGNR, d) 6% silicon doping in AGNR and e) 8% silicon doping in AGNR

3.3 Density of states

3.3.1 Pristine Structure

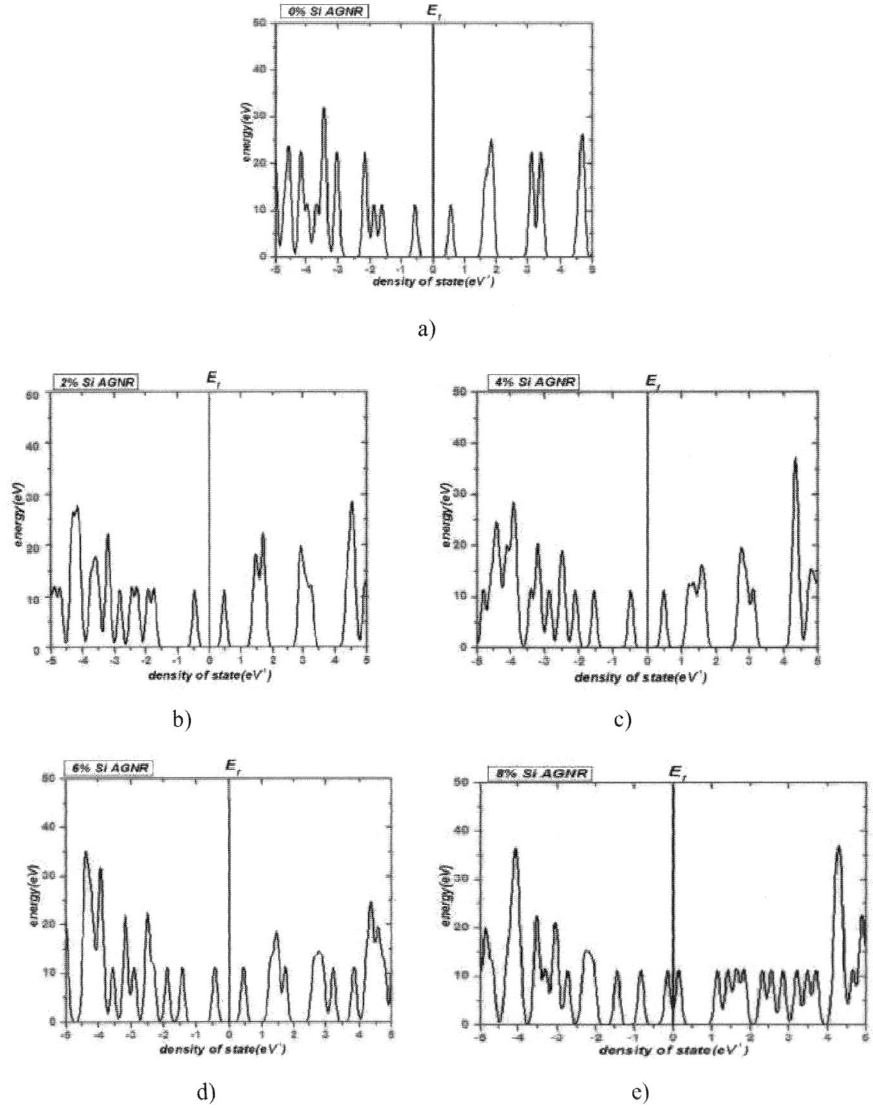

Figure 4. Plot shows density of states in structures w.r.t energy (eV) for a) pristine structure, b) 2% silicon doping in AGNR,c) 4%silicon doping in AGNR, d) 6% silicon doping in AGNR and e) 8% silicon doping in AGNR

Fig. 4(a) shows the calculation of density of states of AGNR for pristine structure. The plot show an energy gap around Fermi level. Introduction of silicon dopant in armchair graphene nanoribbon (AGNR) is done. As a result, metallic character is enhanced upon silicon doping which is seen in following plots. The plots shown below gives the presence of more conduction bands justifying the enhancement of metallic character upon doping.

Fig. 4(b) shows the calculation of density of states of AGNR for 2% silicon doped structure. Here the dopant is added with the concentration of 2% silicon. It is seen in the plot that as dopant is added, the metallicity has increased. The presence of more states andalso the widening of the states shows the increased conduction in the structure due to doping.

Fig. 4(c) shows the calculation of density of states of AGNR for 4% silicon doped structure. Here the dopant remains same but the concentration is now increased to 4%. It is seen in the plot that as dopant is increased from 2% to 4%, the metallicity also increases. Regarding the states and their widening, it has also conveyed the increase in the metallic character.

Fig. 4(d) shows the calculation of density of states of AGNR for 6% silicon doping. In this structure, the widening of the states is increased and also it is observed that the height of the transmission peak is also increased. From this observation, we can suggest increase in metallic character with additional doping concentration.

Fig. 4(e) shows the calculation of density of states of AGNR for 8% silicon doping. Here the dopant concentration is now increased to 8% . With widening of the states and increased height of the transmission peak, increased conduction in the structure can be seen. The high peak value shows highly transmission material structure.

3.4 Transmission Spectrum Analysis:

Finally, the transmission spectrum of pristine AGNR and silicon doped AGNR are obtained from DFT simulations. The transmission spectrum of the structures is shown in Fig. 5(a-e). As it can be seen from Fig. 5(a-e), the plots give variation of the transmission coefficient with respect to energy [10].

The transmission spectra are plotted to justify the level of transmission that can see in DOS plots. These transmission peaks indicate the transmission pathways that are available fortransmiossion. They show the path for the charge carriers at different applied voltages. Fig. 5(a) shows the transmission in a pristine AGNR. As it canbe compared with DOS, it is seen that a band gap is observed in the material.

IOP Conf. Series: Materials Science and Engineering 1225 (2022) 012028 doi:10.1088/1757-899X/1225/1/012028

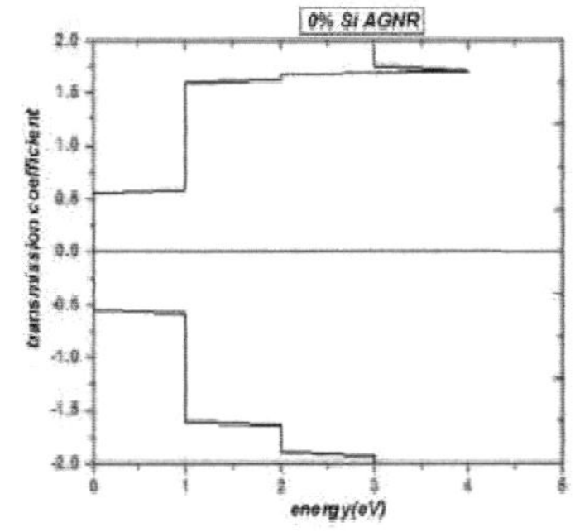

a)

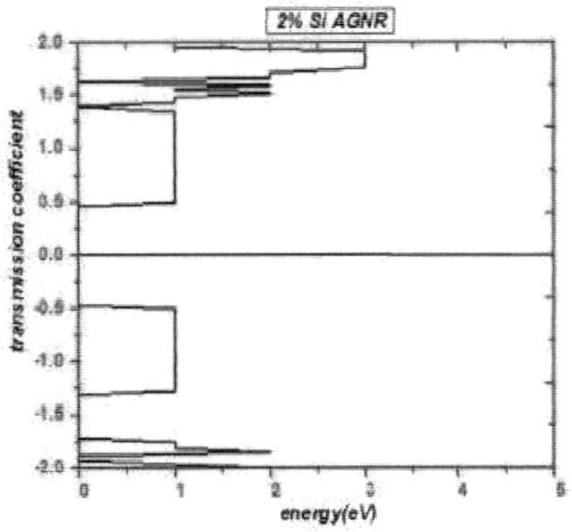

b)

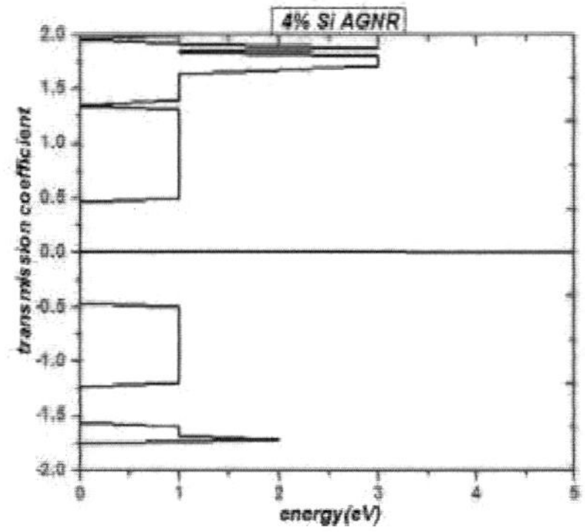

c)

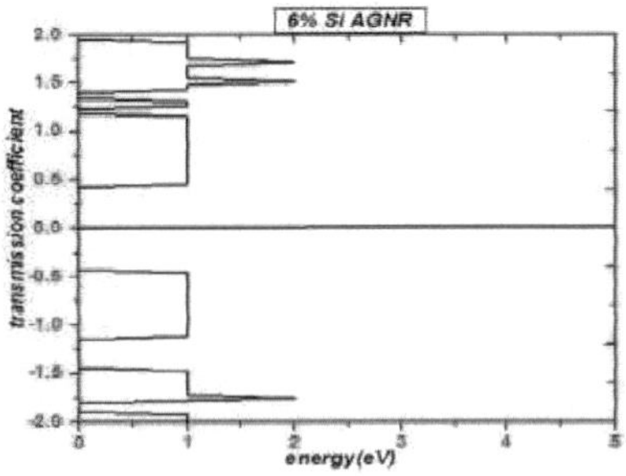

d)

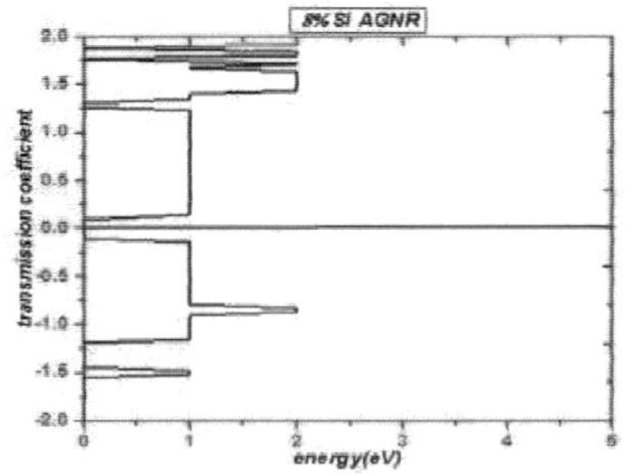

e)

Figure 5. Transmission coefficient w.r.t energy (eV) for a) pristine structure, b) 2% silicon doping in AGNR,c) 4% silicon doping in AGNR, d) 6% silicon doping in AGNR and e) 8% silicon doping in AGNR

Here, in Figure 5(b), through 2% silicon doping, the change in transmission has been obtained. The enhancedmetallic character is seen as the silicon dopant is added in pure structure of AGNR.

Figure 5(c) shows the transmission plot for 4% silicon dopant in AGNR. The change in transmission has beenobtained. The enhanced metallic character is seen as the silicon dopant concentration is increased.

Figure 5(d) shows 6% silicon doping in AGNR. The change in transmission obtained is seen.

Therefore, in all figures, we see that when AGNRs are doped with silicon, they do change their properties and characteristics. It can be noticed that passivated nanoribbons have more stable structure. The transmission clearly justifies the transition of semiconductor towards metallic state.

4. Conclusions and Future Scope

Solar cells appear to be more promising and show great potential while being used with new materials such as graphene or graphene composites. They are steadily being improved in terms of efficiency and parameters. The main challenge lies in finding suitable material which can provide ideal band gap as well as direct band gap for solar cell operation. So for this, the effect of silicon doping on band gap parameter is investigated. The variation in band gap parameter led to transition from semiconductor to metallic state. An ideal band gap is tried to achieve which is a direct band gap also thus fulfilling ideal conditions for solar cell operation. The electronic behaviour of graphene nanoribbon is seen through density of states and transmission spectrum. To make ideal solar cells, suitable materials are needed. Hence, the trajectory of these solar energy panels is upwards and the keep a room for further development.

References

1. Malik P, Kumari G Neelankshi et al. 2020 Regulating electronic properties of graphene sheet vian-type doping for solar cells applications J Mater Sci: Mater Electron 31 14306–14313.
2. Schwierz F 2010 Graphene transistors *Nat. Nanotechnol* **5(7)** 487.
3. Novoselov K S et al. 2012 A roadmap for grapheme *Nature* **490** 192–200 doi: 10.1038/nature11458.
4. Geim A K and Novoselov K S 2007 The rise of grapheme *Nat. Mater.* **6** 183–191.
5. Junfeng Y and Brian R S 2014 Third-generation solar cells: a review and comparison of polymer:fullerene, hybrid polymerand perovskite solar cells *RSC Adv.* **4** 43286.
6. Lukomskaya M V and Pavlovsky O V 2020 Electronic Properties of Graphene Nanoribbons *Phys. Atom. Nuclei* **83** 1611–1614.
6. Sharma P, Singh S, Gupta S et al. 2017 Enhancing linearity in I–V characteristics by B/N doping ingraphene for communication devices *J Mater Sci: Mater Electron* **28** 7668–7676 .
7. Sharma P, Gupta S, Kaur I, and Singh S 2016 Boron doped simulated graphene field effect transistor model AIP Conference Proceedings **1728**, 020110.
8. Lu G, Yu K, Wen Z, and Chen J 2013 Semiconducting graphene: converting graphene fromsemimetal to semiconductor *Nanoscale* **5(4)** 1353-1368.
9. Zhao T et al 2019 **J.** *Phys. D: Appl. Phys.* **52** 475301.
10. Xiao-Fei Li et al 2012 *J. Phys.: Condens. Matter* **24** 095801

Impact of 3% Molybdenum(Mo) nanoparticles on the interfacial and shear properties of lead-free Sn58Bi/Cu solder joint.

Amares Singh[1], Rajkumar Durairaj[2], K Ganesh Kumar[3] and Seng How Kuan [2]

[1] Faculty of Engineering Technology and Built Environment, UCSI Heights, 1, Jalan Puncak Menara Gading, Taman Connaught, 56000 Cheras, Wilayah Persekutuan Kuala Lumpur, Malaysia.

[2]Lee Kong Chian Faculty of Engineering and Science, Universiti Tunku Abdul Rahman, Jalan Sungai Long, Bandar Sungai Long, 43000 Kajang, Selangor, Malaysia.

[3]Faculty of Engineering & Technology (FET), Multimedia University,Jalan Ayer Keroh Lama, 75450, Melaka, Malaysia.

Email : amares@ucsiuniversity.edu.my

Abstract. The research investigates the impact of 3 % molybdenum (Mo) nanoparticles onthe interfacial properties and shear stress of the leadfree Sn58Bi (SB) solder joint. The research aims to providea resemblance between the interfacial and shear properties withthe additions of Mo nanoparticles. The interfacial between Mo added to the solder joint comprises of the common intermetallic (IMC) layer of Cu_6Sn_5andadditional IMCs of $MoSn_2$. that acted as further strengthening mechanism for the joint. Therefore, the shear strength of Mo reinforced solder joint was 32% higher compared to the pristine solder. The introduction of Mo into the SB solder resulted in an advantageous outcome.

1. Introduction

Inoculations of nanoparticles into the Sn-based solder alloys are predicted to provide better interconnection between the components and the printed circuit board (PCB) [1,2, 3]. Nanoparticleslike argentum (Ag) [4], zirconium oxide(ZrO_2) [5], cobalt (Co) [6], molybdenum (Mo) [7], aluminium oxide Al_2O_3 [8] and titanium dioxide (TiO_2) [9] are among the common nanoparticles added to the solder alloys. High melting temperature solders mainly the SnAgCu and SnAg solder system are vastly investigated in terms of its properties,and further studied by adding nanoparticles to the solders. Inclusion of nanoparticles did not affectthe melting temperature because the atomic configurationof the elements in the solder alloy is not disturbed [10, 11]. This is important to preserve low temperature soldering.Improving the mechanical properties oflow temperature solder alloys such as the SnBi solders will be useful for low temperature soldering. The study conducted by [12] shows an increase in the shear strength by 14.9 % upon addtions of 0.5 % TiO_2 compared to that of the non-reinforced Sn1.0Ag0.5Cu. In another work, the overall IMC layer thicknesses of the Sn3.8Ag0.7Cu solder joint was thinner after adding Mo nanoparticles. The relationship between the IMC layer and shear properties werecorelated[13]. The Cu_6Sn_5 and Cu_3Sn IMC layerswere present at the interfacial region of the titanium carbide (TiC) added SnAgCu solder joint. However, the growth rate of the layers was lower [14]. Thicknesses of IMC layers ranging between 1-5μm known to enhance the solder joint'sreliability. Few studies have concludedthe inability of nanoparticles toreactwith Sn and Cuand its presence as discrete particles obstructsthe formation of thick IMC layers [14, 15, 16]. Aided with the presence of the nanoparticles at the interfacial layer, the effects ofmaintaininga thinner IMC layer are more certain [17, 18].

As reported by many studies, minimal amount of the weight percentages of nanoparticles additions improves the properties of a solder alloy, while higher weight percentages of additions shows deterioration in the properties [19, 20]. A recent study on the SnBi solder alloy reinforced with 0.25 wt. % Mo nanoparticles showed improvement in the shear and tensile strength of the solder joint [21]. The presence of Mo nanoparticles as the load-bearing effect was noted to be the factor for the increment.[22] added 2 wt. % Al_2O_3 into the Sn58Bi solder and witnessed a hike of 6.7 % to the hardness value. However, up to now, not many comparisons have been used with different types of nanoparticles, especially involving the additions to low temperature solder alloys. Plenty workshavebeen done on high melting solders with minimal attention given to the low melting solders [23,24]. Nevertheless, low melting solders promotes soldering at much lower temperature and can provide protection to other electronic parts [3].Studies involving nanoparticles addition to the SnBi solder alloy are limited. This study focuses on the relation between the interfacial properties and shear strength of a low melting (139°C) SnBi solder added with 3% Mo nanoparticles.

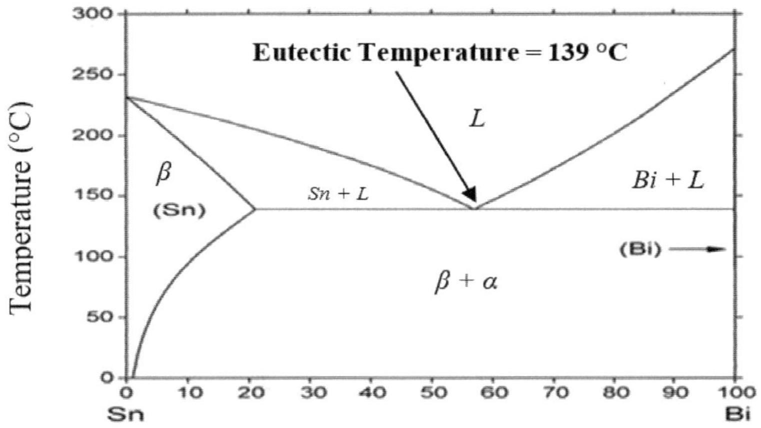

Figure 1. Phase diagram of Sn-Bi alloy [25].

2. Materials and experimental procedure

The Sn (Alfa Aesar-99.9%) and Bi (Alfa Aesar-99.9%) were weighed to the eutectic percentage as in figure 1, where the total weight of the Sn was 8.4 g and 11.6 g for the Bi respectively. These elements were mixed and melted at 600°C in a heat treatment furnace for 1 hour. The solder alloy was then let to solidify in the furnace. The Mo (Alfa Aesar-99.9%) nanoparticles were weighed according to 3 % from the total 20 g of the Sn58Bi. The high magnification image of Mo nanoparticles under scanning electron microscope (SEM)is shown in figure 2.

Figure 2. Mo nanoparticles at × 50 000 magnification.

The average size of the particles was measured to be 42 nm. The Mo nanoparticles were then added to the Sn58Bi solder alloy to the mixture. The Mo added solder is labelled as SB + 1 % Mo while the non-reinforced is termed as SB solder alloy onwards in this paper. Re-meltingusing the hot plate at 350°C for 15 minutes with stirring was done to enable the nanoparticles to disperse well on the solder. This process allows a proper dispersion of Mo nanoparticles. After solidifying, the solders were cut into 50 mm × 10 mm billets. The Cu substrates were cut into 40 mm × 5 mm × 10 mm following the ASTMD1002 standard to be used for the shear test.Soldering of the SB and SB + 3 % Mo solder billets on to the Cu substrates was done at 230°C for 1 minute. The single shear lap test method was usedto performthe shear test on the solder joints.

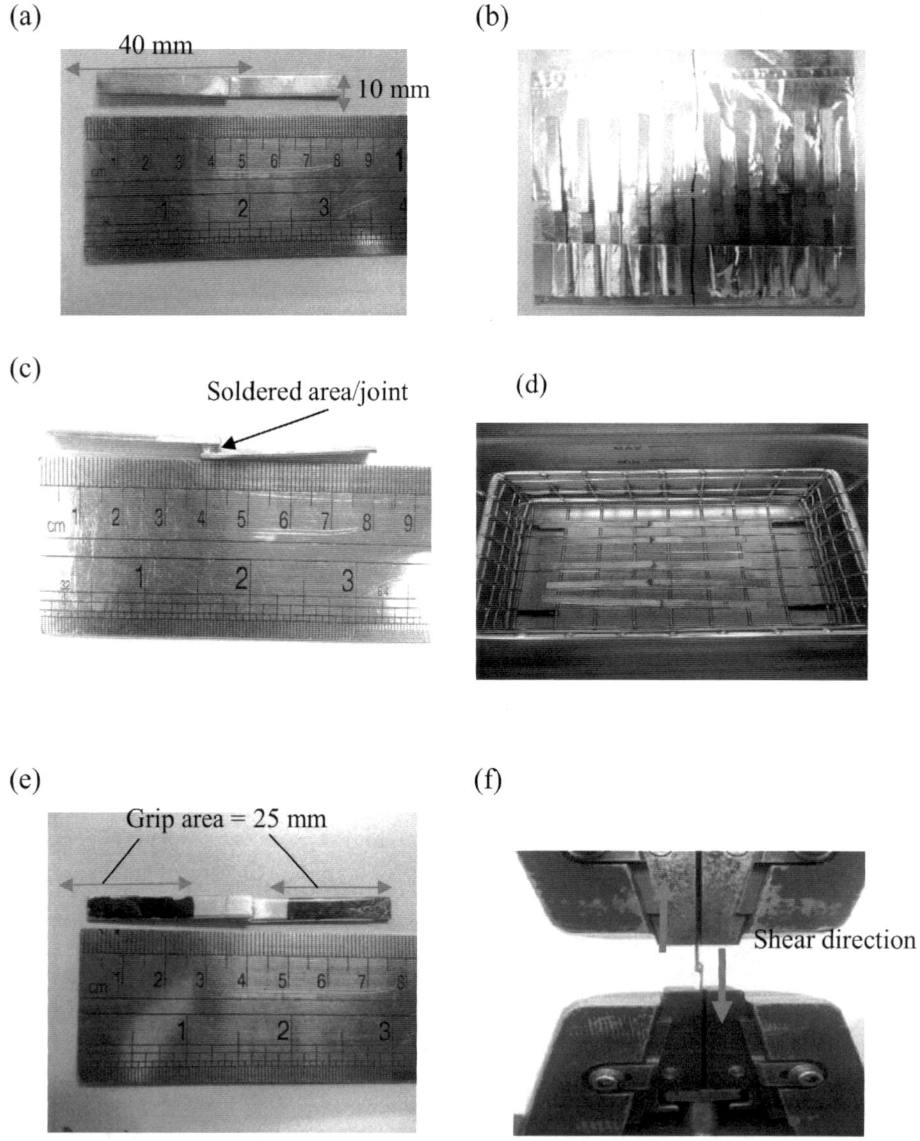

Figure 3. (a) Single shear lap specimen before soldering, (b) arrangement of samples prior soldering, (c) shear lap specimen after soldering (d) shear lap specimen cleaning with ultrasonic machine, (e) shear lap specimen prior testing and(f) shear test using universal tensile machine.

The specimens were then cleaned using ethanol and water. The cross sections of the solder jointswereviewed using the SEM (S3400N HITACHI) machine equipped with energy dispersive x-ray (EDX) which is used to detect and analyse the element at the interface.

Universal tensile machine (Instron 5582Q4970) is used to perform the shear test and the shearing of the joint was donewith a crosshead speed of 1.3mm/min. The shear stress was calculated and the data were analysed in relation to the interfacial properties. The shear test setup and machines are shown in figure 3. The overall flow chart of the methodology is shown in figure 4.

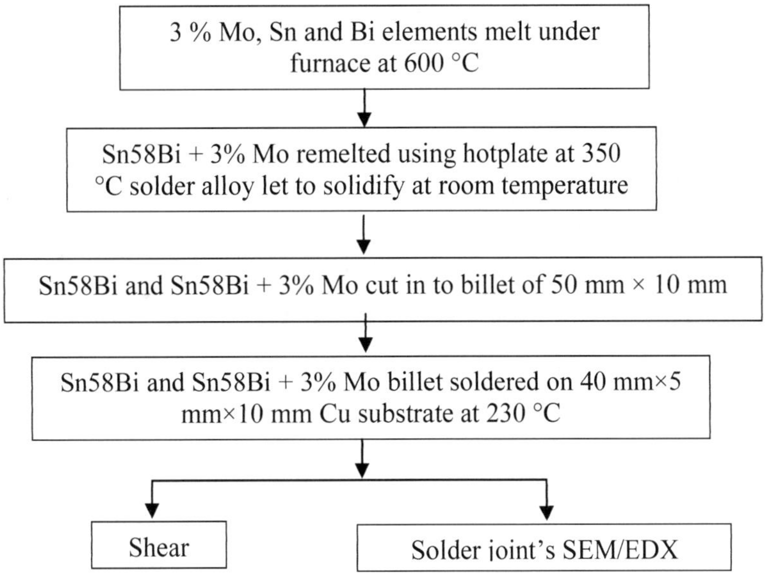

Figure 4. Flow chart of the methodology

3. Results and discussion

3.1 Interfacial properties of SB/Cu and SB + 3 % Mo/Cu joints

On the solder side, Bi (white phase) and the Sn (dark phase) were observed in a lamellar manner for the SB solder. Meanwhile, well-dispersed Mo nanoparticles were detected via EDX together with the Sn and Bi in the SB + 3 % Mo solder.

The regions at the interfacial layer analysed by SEM and EDX are shown in figure4. The reinforced solder joint shows also showed well dispersed Mo nanoparticles distribution. This was clarified with the EDX analyses.In a Sn-based solder joint, the common reaction between the Sn and the substrate forms a typical IMC layer of Cu_6Sn_5[26]. This layer would grow thicker with more presence of Sn and in many circumstances would react with Sn to produce another layer of Cu_3Sn [27]. These thicker Cu_6Sn_5 and Cu_3Sn contributes to the weakening of the joint [28].

Cross section images of Cu_6Sn_5 IMC layer in the bare SB/Cu joint are shown in figure 5 (a). In this system, Bi element plays the role of the non-reacting element and acts as a barrier for the diffusion between Sn and Cu. However, the Cu and Sn are more reactive with higher temperatures, the diffusion gets rapid and production of the Cu_6Sn_5 could not be restricted [16]. In fact, the presence of Cu_3Sn layer was found in the SB/Cu joint below the Cu_6Sn_5layer, with a thicker and non-uniformedlayer.Such condition of the IMC layer would be the weakening point for a solder joint [17,28]. Theseare few of many problems that needs to be sorted. Interestingly, with the reinforcement of the Mo nanoparticles, the Cu_6Sn_5 IMC layer was seen to be flatter and uniform as shown in figures5 (b).The Mo nanoparticles seen not to drastically hamper the production of the Cu_6Sn_5 layer. Nevertheless, thepresence of Mo nanoparticles allowed the Cu_6Sn_5 to form stably and blocked the formation of the Cu_3Sn IMC layer. Other IMC's, such as $MoSn_2$and $Cu_{10}Sn_3$, were also produced upon further XRD analyses. The Mo nanoparticles presence was not visible from the SEM imaging but the EDX and XRD analyses confirmed the existence of the nanoparticles.

TheIMC layer thicknesses of the SB/Cu and SB + 3 % Mo/Cu measures1.26822 μm and 0.8739 μm, respectively, as shown in figure 6.The reinforced solder joint produced a thinner IMC layer that would be more reliable towards shearing.The thickness measurementof less than 1 μm classifies this as a thin layer [29].Additionally,with existence of Mo_2Sn IMC, growth of the Cu_3Sn layer is reduced as Mo reacted with Sn and blocked the diffusion between Sn and Cu. The production of a thin IMC layer for the reinforced solder joint is attributed due to the blocking of the vast diffusion between Sn and Cu by the Mo nanoparticles that were found to bepresented at the top side of the IMC layer. The Mo nanoparticles were not visible from the SEM image due to theirnano-scaled size but with the EDX and XRD analyses, Mo nanoparticles was detected there. So, with these analyses, an illustration based on the real SEM image of the blocking process is shown in figure 7.

(a)

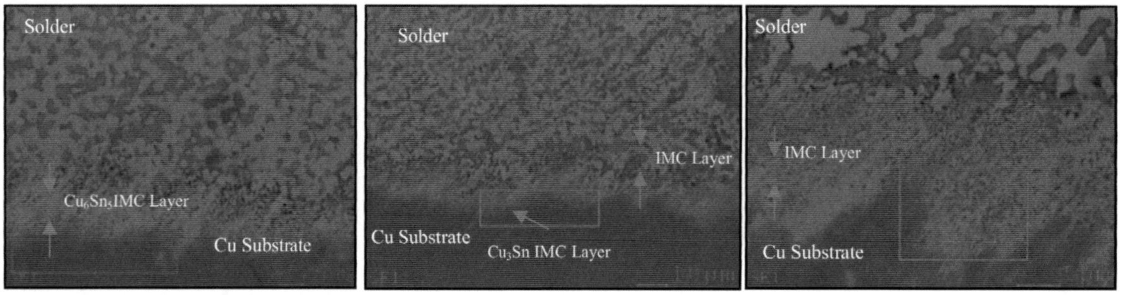

Element	Wt%	At%
Cu	63.00	77.70
Bi	07.45	02.79
Sn	29.54	19.51

Element	Wt%	At%
Cu	44.25	64.21
Bi	22.41	09.89
Sn	33.34	25.90

Element	Wt%	At%
Cu	02.02	05.98
Bi	89.25	80.21
Sn	08.73	13.81

(b)

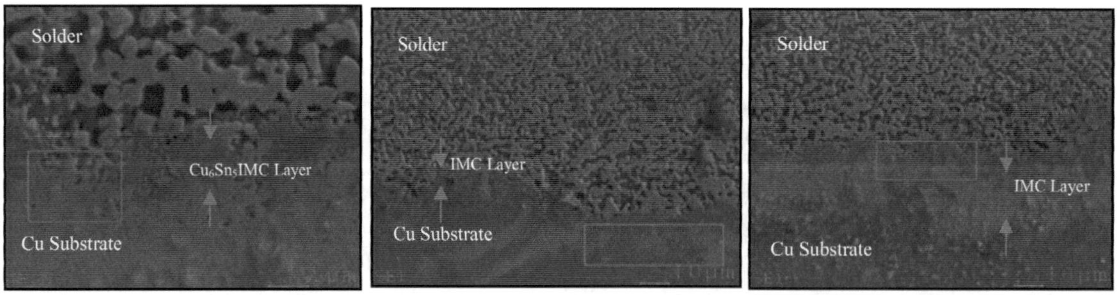

Element	Wt%	At%
Mo	03.63	04.90
Bi	54.08	33.54
Sn	26.08	28.48
Cu	16.21	33.07

Element	Wt%	At%
Mo	03.56	03.88
Bi	22.53	11.28
Sn	48.15	42.44
Cu	25.76	42.41

Element	Wt%	At%
Mo	03.42	03.68
Bi	31.58	15.58
Sn	32.83	28.53
Cu	32.17	52.21

Figure 5. SEM image and EDX of (a) SB/Cu and (b) SB + 3 % Mo/Cu solder joints.

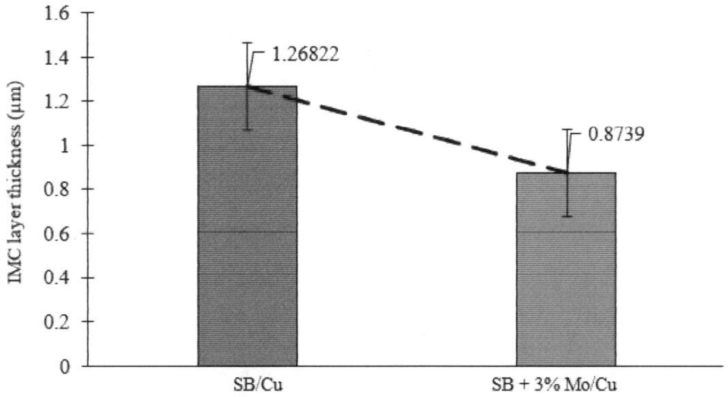

Figure 6. IMC layer thickness of SB/Cu and SB + 3 % Mo/Cu solder joints.

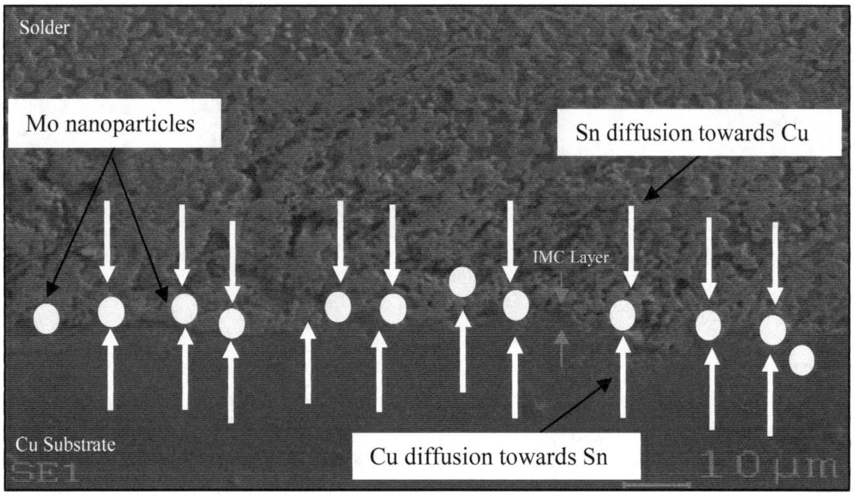

Figure 7. Blocking effect by Mo nanoparticles at the SB + 3 % Mo/Cu solder joints.

3.2 Shear properties of SB/Cu and SB + 3 % Mo/Cu joints

The single shear lap joint for the SB/Cu and SB + 3 % Mo/Cu was subjected to the shear test with the average maximum load and average maximum shear stress taken as the shear properties to be evaluated. The results for the SB/Cu and SB + 3 % Mo/Cu are shown in figures 8 (a) and (b). The average maximum load and average maximum shear stress of the SB + 3 % Mo/Cu joint were higher by 34 % than the non-reinforced solder joint. The increment is linked to the presence of the Mo nanoparticles, IMC layer and IMC compounds present at the interfacial layer as discussed in section 3.1.

As discrete and evenly dispersed particles, the Mo nanoparticles presented as a second phase strengthening mechanism at the solder joint. This simply means that the particles will be an added strengthening factor upon shearing. Hence, the shearingwill not occur with ease and the resistance toward deformation increases, resulting in higher shear stress. These particles are presented at the interfacial due to the high energy area at the region that attracts the active nanoparticles [30, 31]. This resistance from shearing is not found for the SB/Cu solder joint. Furthermore, Mo being hard unbreakable particles, the shear load will be absorbed by these particles that increase the shear stress to break the joint. This strengthening effect is termed as a load-bearing effect [24,32].

Presence of the Mo_2Sn IMCs along the interfacial at the solder joint acts as precipitation strengthening to the solder joint. Precipitation strengthening occurs when particles obstruct the dislocation of motion upon shearing [33]. The existence of more Mo_2Sn IMC at the interfacial layer creates more dislocation loops, which then increases the dislocation density. Hence, the shear stress for Mo reinforced solder joint increases as a result of the high dislocation densities. Contrariwise, the presence of Bi alone as the precipitation strengthening produced a lesser strengthening effect for the bare SB/Cu solder joints. Another reason for the increase in the average maximum load and average maximum shear stress is the thin IMC layer of Mo reinforced solder joints. Thinner IMC layers are less prone to crack initiation that resembles a stronger joint. This hypothesis is supported by many others too [24, 25, 32].

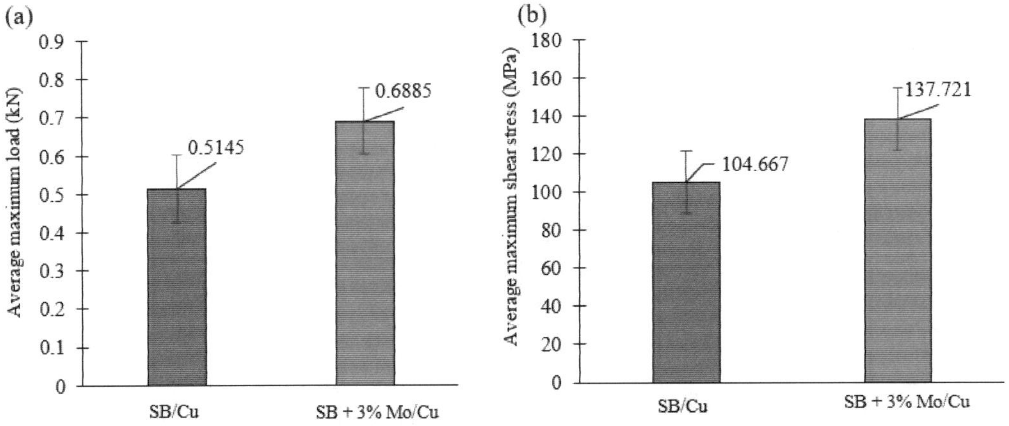

Figure 8. (a) Average maximum load and (b) Average maximum stress of SB/Cu and SB + 3 % Mo/Cu solder joints.

4. Conclusion

The impact of Mo nanoparticles on the SB solder alloy was studied in this work by investigating the interfacial and shear properties. The common Cu_6Sn_5 IMC layer was observed at both solder joints while the Cu_3Sn layer was only observed for the unreinforced solder joints.The presence of Mo as discrete particles together with the Mo_2Sn IMCs wasfound throughout the SB + 3 % Mo/Cu solder joints. Precipitation strengthening (Mo_2Sn), thin IMC layer and load bearing effect (Mo nanoparticles) influenced the increase in the shear properties of Mo reinforced solder joint. Reduction by 31 % in the IMC layer thickness observed upon addition of Mo nanoparticle. Increase in average maximum shear load and average maximum shear stress by 34 % was found for the reinforced solder joints. Certainly, the data of the shear properties corelated to the findings of the interfacial properties, showing the relevancy of this research with the hypothesis from other studies. These two properties are key in judging the credibility of a solder alloy in order to be implemented as the medium of interconnection between electronic components and printed circuit board of an electronic component at lower soldering temperature. Based on the primary results from this work, the SB + 3 % Mo had the potentiality.

Acknowledgments
The authors acknowledge the assistance provided by Universiti Tunku Abdul Rahman (UTAR) for the financial aid under the research fund IPSR/RMC/UTARRF/2017-C1/R01.

References

[1] Haseeb A S M A, Arafat M M, Tay S L, Leong Y M 2017 *J Electron Mater* **46** 5503–5518.
[2] Tang Y, Luo S M, Huang W F , Pan Y C , Li G Y 2017 *Alloys Compd* **719** 365-375
[3] Lobry P , Blaz L , Sugamata M , Kula A 2011 *Arch Mater Sci Eng* **49** 97-102
[4] Malgorzata J, Krystyna B , Marek K , Anna M , Wojciech N, Marcin S, , Janusz S 2010 *Electronics System Integration Technology Conf* 5642884
[5] Gain A K , Chan Y C , Yung K C 2011 *Microelectron Reliab* **51** 2306-2313

[6]Bashir M, Haseeb A S M A, Saliqur R A Z M , Fazal M A 2016 *J Mater Sci* **32** 1129-1136
[7]Haseeb A S M A , Arafat M M , Johan M R 2012 *Mater Charact* **64** 27-35
[8]Cheng T H , Tsao L C , Wang F , Kuo, W Y 2011 *12th Int Conf on Electronic Packaging Technology and High Density Packaging (Shanghai)*1-5
[9]Nasir S S M A, Mustafa E, Ahmad M , Yahaya M 2019 *BalázsIllés Ceram Int*, **45** 18563-18571
[10] Li, Yi, Chan Y C 2015 *J Alloys Compd* **645** 566-576
[11] Tu X , Yi D , Wu J , Wang B 2017 *J Alloys Compd* **698** 317-328
[12] Wen Y , Zhao X , Chen Z , Gu Y , Wang Y , Chen Z and Wang X 2017 *J Alloys Compd* **696** 799-807
[13] Yahaya M Z , Ani F C , Samsudin Z , Sahin S , Abdullah M Z , Mohamad A A 2016 *Mater Sci Eng A*, **669** 178-186
[14] Chen G , Peng H , Silberschmidt V V, Chan Y C , Liu C , Wu F 2016 *J Alloys Compd* **685** 680-689
[15] Han Y D, Gao Y, Zhang S T, Jing H Y, Wei J, Zhao L, Xu L Y 2019 *Mater Sci Eng A* **761** 138051
[16] Zhang L , Gao L 2015 *J Alloys Compd* **635** 55-60
[17] Tu X , Yi D , Wu J , Wang, B *J* 2017 *Alloys Compd* **698** 317-328
[18] Boettinger W, Kattner U, Moon K W 2006 Semiconductor Microelectronics and Nanoelectronics Programs *Solders and Solderability Measurements for MicroelectronicsNational Institute of Standards and Publication, Austin* 113-115
[19] Zhao Z, Lei L , Hyun S C , Jian C , Qian W, Yuming W , Guisheng Z 2016 *Microelectron Reliab* **60** 126-134
[20] Yang L , Shengli L , Wei S , Xuan W , Hao Z , Fenglian S 2019 *Mater Charact* **148** 280-291
[21] Yang L , Zhu L , Zhang Y , Liu P , Zhang N , Zhou S , Jiang L 2017 *Mater Sci Technol* **34** 992-1002
[22] Shen L , Foo A Q , Wang S , Chen Z 2017 *J Alloys Compd* **729** 498-506
[23] Hammad A E & Ibrahiem A A 2017 *Microelectron Reliab* **75** 187-194
[24] Singh A, Durairaj R, Seng H 2021 *J Adv Res Fluid Mech Therm Sci* **77** 69-87
[25] Okamoto H Desk Handbook: *Phase Diagrams for Binary Alloys* Second ed 2010 Ohio: Materials Park 44073-0002
[26] Chellvarajoo S & Abdullah M Z 2018 *Microelectron Reliab* **84** 230-237
[27] Ding M , Wenqing X , Xingye Y , Le M , Wei Z , Zhiqiang J 2018 *J Alloys Compd* **739** 481-488
[28] Eid E A , Fouda A N & Duraia M 2016 *Mater Sci Eng* **657** 104-114
[29] Ahmed M, Fouzder T, Sharif A, Gain A K, Chan Y C 2010 *Microelectron Reliab* **50** 1134-1141
[30] Li M , Zhang L , Jiang N , Zhang L , Zhong S J 2021 *Mater Des* **19** 710922
[31] Wang H , Hu X , Jiang X 2020 *Mater Charact* **163** 110287
[32] Amares S , Durairaj R , Kuan S H 2021 *Arch Metall Mater* **66** 407-418
[33] Efzan M N & Singh A 2014 *Solder Surf Mt Technol* **26** 147-161

FEA simulations of Lower Limb Prosthetics

Jashan Jyot Singh[1], Jaswinder Singh Mehta[1], Rajesh Kumar[1], Gaurav Sapra[1]

[1]UIET, Panjab University Chandigarh

Email: jashanjsingh0001@gmail.com

Abstract. Finite element analysis (FEA) is used to study the characteristics of various objects in different working condition for imitating real life like scenarios. The present work describes an overview of FEM simulations for lower limb prosthetics. The review will give a deep insight for stress interface and load distribution on the residual limb. It enables prosthetists to design the socket for effective functional performance. The study suggests that outcomes through FEA simulations under test conditions is needed It has been observed that predicted outcomes using FEA were in agreement with the experimental measurements. Further, the study suggests that outcomes of the FEA analysis under test conditions may be validated through GAIT cycle.

1. Introduction

Walking is the fundamental way to maneuver which reflects the locomotive characteristics of an individual[1]. Prosthesis stands for pro, meaning forward and thesis, as placement which comes from Greek language. It partially reinstate the ability of the biological lower limb[2]. There are various reasons for the prosthetics as a replacement in amputed limbs such as aesthetics and function, it helps in restoring the mobility of the amputee and increase self-esteem[3]. The different prosthetics for lower limb are transtibial and transfemoral prosthetics. The former is for amputation below the knee and the latter having amputation above the limb. Prosthetic foot, tube adaptor and transtibial socket are included in transtibial prosthetics and prosthetic foot, tube adaptor, prosthetic knee joint and transfemoral socket in the transfemoral prosthetics.

Lower limb amputations are generally much more prominent than upper limb. Amputations are generally in older adults than younger ones. Diabetic mellitus being the most common complication behind it, but vascular insufficiency, traumatic injury, cancer and congenital limb deficiency can be one of the many causes because of which the amputation can be carried out. While amputees undergo many challenges such as physical, social and emotional but it usually depends upon person to person[4]. Prosthetics can be classified into three categories with the engineering point of view i.e. passive devices which are primitive in nature having spring and damper system. Semi active system having microprocessor technology which can alter the behavior instantaneously. Power operated prosthesis works with the help of motor which help in mobility[5].

2. Finite Element Modelling of lower limb prosthetics

The part of the prosthetics which interacts with the remaining limb of the amputee is known as the prosthetic socket. The blueprint of the prosthetic transtibial socket depend on the individual requirement of the amputee as the size of the residual limb varies according to person to person. CAD FEA has been used to develop a FE model of the transtibial prosthetic socket. The results obtained through FEA were validated by experimentally measured stresses and the obtained results through FEA were found to be accurate [6].

The versatility of the CAD-FEA can be seen through the accurate results obtained through the process. While the geometry of the bones of the residual limbs are generally carried out through CT scanning and finite element analysis are used to obtain its 3D models. CT scans were used to obtain the 2D image of the limb while imaging and de-compiling helped in constructing the 3D solid model of the residual limb. The process of formation of 3D mode was found to be fairly accurate and less labor intensive without any need of formation of positive mold and manual intervention[7]. The designing process of an above knee prosthetic socket through computer aided design process has been described in the study. The fabricating through rectified cast is also seen. It was observed that the time spent over rectifying of residual limb mold could be reduced considerably with the help of computer aided designing[8]. Hence finite element analysis aid in reducing the time in the processing of design and analysis of the geometry.

As the size of prosthetic socket varies according to the need of the individual amputee so FEA analysis is a convenient process in its designing and analysis without any hassle of physical modelling reducing the time spent on physical model. FEA was used to predict the transtibial socket fit in the study. Two different pressure data were used to foresee the fitness of the transtibial socket with the help of FEM simulations. A new set of transtibial socket was then fabricated based on the computational results and fitted to the amputee. The computational analysis was found to be an efficient method in assessing socket fit which provided accurate evaluation and helped in redesigning of the prosthetic socket even before its actual fitting and fabrication.

In another study, the analysis of the prosthetic socket were carried out using FEA simulations through which the interface stresses, deformations in the socket due to applied force with the interaction of the residual limbs was analyzed. Parametric analysis was used to investigate the consequences of transtibial socket design and remaining leg geometry in the interface stresses between the remaining leg and prosthetic socket. It was found that the variation in the transtibial socket blueprint and remaining leg structure influenced the amount of interface stresses between socket and residual limb[9]. The sensitivity to variation of parameters were investigated using the CAD model in the study. It was observed that the degree of the rectification applied to the socket was directly proportional to the presumed values of the Young's modulus. The results obtained were in agreement as expected with increase in confluence pressure in the regions of higher positive rectification[10].

A nonlinear FE model was made to predict the stress w.r.t the actual geometry and the socket modification were predicted considering the slip boundary condition and large deformation[11]. The equivalent forces and moments at the knee joints at the time of movement were calculated. Three different compliant design of the prosthetic socket were observed and FEM was used to assess its structural integrity, the release of pressure during walking at fibula head was also seen. There was a reduction of interface pressure in a single leg stance at spiral slot. Largest pressure relief was found in one version of the spiral slot with reduction of 20-80% in the local interface pressure of the stance phase [12]. The non-linear FEA was used to examine the biomechanical response of stump slipping. Then a non-linear response was obtained in the biomechanical response in slump sliding. It was observed that stump sliding was one of the defining factors in socket evaluation and peak interface stress had a minor effect on the liner stiffness[13]. A FE Model was developed to study the interface pressure between the remaining leg and the transtibial socket. The friction state between the skin and socket was obtained through contact analysis. The influence of load transfer between transfemoral residual limb and the socket was explored in the study[14].

Saunders et al. desinged the solid ankle cushioned heel (SACH) model using computer modelling and finite element analysis were done based upon the loading conditions as presented upon the gait analysis of an amputee and further authenticated through mechanical testing. The system was viable in characterizing the overall performance of commercially available prosthetic feet [15]. The stress, strain and energy stored during the stance and fast walking speed were calculated with the gait data as well as the finite element modelling together in the prosthetic foot. The efficacy of the FEA simulations were validated with the gait data and then boundary conditions were tested using FEA analysis in the transtibial amputee. Therefore parameters which couldn't be calculated with gait data alone were obtained by the FEA and gait data together [16].

A combination of series and parallel of a damping element was introduced to a prosthetic foot design which was connected to a system of springs. The damping element having high damping constant was used to change the stiffness of the system under dynamic loading. The moment, displacement and mechanical reaction were predictable with the usage of the model and it could be useful in simulations of the response if any modifications were introduced to the original prosthetic design[17].

A flex foot was studied using FEM at three different curvature i.e. 20^0(small), 35^0(medium) and 50^0(large) degrees for a person standing or running. The result obtained were that there were higher stresses and larger displacement in big bending foot. The peak von Mises stresses were as maximum in big bending than in medium bending and least in small bending[18]. In another study, Roll over shape (ROS) with Ground Reaction Force (GRF) were clubbed together in the optimization of the prosthetic foot. It was found that desired deformation pattern in the foot could be obtained with the stiffness profile optimization while considering ROS corresponding to transient GRFs[19]. The anteroposterior was found to be having prominence in the figuring out of the peak von-mises stress, deformation values of mono-limbs. The optimization of the structural design of prosthesis was found to be ideal with FEM and Taguchi methods[20].

Two composite materials namely polyethylene epoxy and vinyl ester were used along with carbon fiber in the prosthetic legs blade which in turn reduced the number of layers of the carbon fiber ultimately reducing its weight and thickness. The addition of the aforesaid materials helped in reducing the strain in the existing prosthetic[21]. FE Model was used for the analysis of the structural behavior of the mono-limbs at different shank angles and the interactivity of the limb and socket were also seen. The amount of von mises stress values at different shank design at different walking phases were observed. It was observed that shank stiffness reduced with lower applied peak stress. The deformability of shank was predicted during walking without the actual pro fitting of the prosthetic and direct measurement. The effect of shank flexibility on socket-limb interaction was explored[22]. A FE model and a concurrent multibody of the femur, tibia, socket and ESAR prosthesis of a transtibial amputee was formed [23].The effect of friction on the residual limb of the transfemoral amputee was studied through FEM model. The friction had an effect on the soft tissues of the residual lower limb. The influence of coefficient of friction on the prosthesis, liner and soft tissues on the stress distribution between the complete system was analyzed with the FEA obtained results[24]. The study of the stresses appeared on the bone adjoining to the implant was carried out using FE Model. The amount and spread of bone stresses were compared for upper leg amputee with an Osseo-integrated prosthesis for 3 different loading conditions. A clear difference in the amount and stress distribution were observed for the 3 loading conditions [25].

Table 1. Summary of review articles based on optimization of process parameters

S.No	Author Name	Objective	Results
1	J.C.H. Goh et al[6].	A CAD-FEA prototype obtaining method was proposed from the geometry prevalent CAD system.	The interface pressures were predicted reasonably well with the developed system, it could be combined with prevalent system for obtaining numerical feedback in an automated process.
2	Zhang Shuxian et al.[7]	The limitation of the plaster casting in the traditional prosthetic socket fabrication was aimed to control and a method to form 3D models for the bones and skin of the residual limb are presented.	The construction of remaining leg and the prosthetic socket through CAD-FEA were found to be much better than previously done method, the downside was found to be the slightly higher price and processing time.
3	T.A. Krouskop et al.[8]	A Computer Aided Design process was described in the fabrication of the above knee prosthetic socket.	The results signified the use of computer aided design in the designing of the above knee sockets.
4	Ming Zhang et al.[9]	FEA was used to foresee the prosthetic socket fit in the study.	Two pressure data were used in predicting the fitness of prosthetic socket while using FEA.
5	D. P. Reynolds et al.[10]	The compression of tissues in the remaining leg within the prosthetic socket had been analyzed using FEA.	The results were obtained as expected with increase in interface pressure in areas of higher positive rectification.
6	XiaohongJia et al.[11]	The actual geometry of residual limb was used in the formation of its 3D nonlinear finite element model, the mechanical interplay between the socket and the remaining leg during the movement was studied with the formation of internal bones and socket liner.	The moments and equivalent forces during walking were calculated in the study.
7	Mario C. Faustini et al.[12]	The work carried about in the study was aimed at the reduction of the pressure when the remaining limb and the prosthetic socket were in contact.	It was observed from the results that when the acquiescent characteristics are merged together inside the walls of the socket then it can be helpful in reducing the pressure which arises due to the association of the remaining leg and the prosthetic socket, this would lead increased comfort for the amputees.
8	Chih-Chieh Lin et al.[13]	The biomechanical feedback of the stump gliding with the prosthetic socket were studied with the use of FE analysis , the effect of the stiffness of the liner on the transtibial prosthesis were given special attention.	It was seen from the solution that biomechanical feedback of the stump sliding were highly variating.
9	Linlin Zhang et al.[14]	The study of confluence pressure among the above knee remaining leg and the leg casing was carried out using nonlinear finite element model.	The effect of loads transfemoral remaining leg and the leg casing was analyzed in the study.

10	Marie M. Saunders et al.[15]	The motion analysis, computer analysis and mechanical testingwere used in developing a prosthetic foot design.	The utility of the approach to the parametric design was seen in the study, the prosthetic foot performance was analyzed numerically with inclusion of viscoelastic heel.
11	Xavier Bonnet et al.[16]	In the study the finite element analysis and gait data were clubbed together to obtain the stress, strain and the values of energy stored in the foot through the stance position at user defined walking motion.	The FEM and gait data combined assisted in obtaining the results which were not possible through kinematic analysis.
12	HeimirTrygggvason et al. [17]	The design modification of prosthetic foot was carried out with the study directed at the changeable stiffness of the object. The work is achieved by carrying out the design modification using finite element method.	The validation of the FE model was carried out using the calculated values of a prosthetic foot. It was also found to be useful in the analysis of the functional characteristics of the prosthetics.
13	Ming-Jen K et al.[18]	The interactivity among the remaining leg and socket were analyzed using FEA while using dynamic loads in the gait cycle.	It had been observed from the study that the highest results in the stresses gain in the dynamic model but the pressure areas may increase in either only size or both in size and magnitude.
14	M. Barbara Silver-Thorne et al. [19]	FEA was used to form the model of the amputed leg limb and its socket, the changes were done parametrically and the effects due to the stress dispersal were studied.	The solutions of the study concluded that variation in both the remaining leg geometry and the socket design have an effect on the interface stresses.
15	Winson C. C. Lee et al.[20]	The different design factor's significance were identified using Statistics- based Taguchi method in controlling of the stress and deformation within the mono-limbs.	It was concluded from the study that sagittal dimension was the most significant design factor in deciding the von Mises stress, deformation and dorsiflexion angle in the mono-limbs.
16	Linlin Zhang et al.[21]	The prosthetic racing legs [blades] were studied while using FEA process.	The value of stresses were found to be equivalent while changing the material for the same geometry, there was a growth in young's modulus with decrease in strain.
17	David A. Boone et al.[22]	The evolution of finite element model was described and theaction of mono-limbs structurally with various shank blueprints were analyzed, the interplay of the residual leg and the socket was also studied during regular movement such as while walking.	A FEM was formed to anticipate the shank deformation of the mono-limbs, higher understanding of the stress distribution was observed and the effect of shank flexibility on socket limb interaction.
18	Stacey M. Rigney et al. [23]	A FE model and concurrent multibody of femur, tibia, ESAR prosthesis and socket of a transtibial amputee athlete during sprinting were presented.	Thus, the model formulated an improvement in the calculation of energy storage presently with in vitro mechanical testing only as variance was observed with the true in vivo behavior.
19	Armando Ramalho et al.[24]	The study was aimed at development of FE Model to	Friction as a factor was evaluated using Finite Element Model in the

		estimate the effect of friction on the remaining leg of a person having amputation above the knee.	remaining leg of a person with above knee amputation.
20	Winson C. C. Lee et al[25]	CAD-FEA was developed to analyze the stresses in the bone adjoining the insert.	A comparison in the distribution and magnitude of bone stresses were carried out in the study for transfemoral amputees having Osseo integrated prosthesis at three different loading condition.

The above reviewed articles give an insight in the capabilities of Finite Element Analysis in the analysis of the lower limb prosthetics. It has also been seen from the literature that the mechanically conducted tests based upon ISO 16955 and the FE analyzed results were comparable with accuracy of around one percent while having maximum force of 824 N at the stance phase time of 600ms [19].

3. Conclusion

The study demonstrates the importance of FEA in the analysis of the lower limb prosthetics. It can be articulated from the analysis that FEA obtained results were near to practical measurements with socket modifications done based upon the FEA results. The effect of curvature of blade and material properties are the critical parameters for consideration while designing the lower limb prosthetics. FEA Simulations empowers to simulate different conditions which are otherwise difficult in physical conductions. Further, it helps in reducing the time in the rectification of the designed components hence improving its work efficiency. The workability of different prosthetics in similar working conditions can be checked through FEA but obtained results from simulations need to be appropriately close to the results obtained from physical testing. Further, testing conditions for simulation are required to be closer to the naturally occurring environment. GAIT analysis may provide significant data to validate these simulations.

Reference

[1]. Rajt'úková V, Michalíková M, Bednarcíková L, Balogová A, Živčák J 2014 Biomechanics of lower limb prostheses Procedia Eng **96** 382–91.

[2]. Dyer B T J, Sewell P, Noroozi S 2014 An investigation into the measurement and prediction of mechanical stiffness of lower limb prostheses used for running *Assist Technol.* **26(3)** 157–63.

[3]. Douglas T, Solomonidis S, Sandham W, Spence W 2002 Ultrasound imaging in lower limb prosthetics *IEEE Trans Neural Syst Rehabil Eng.* **10(1)** 11-21.

[4]. Murray C D, Fox J 2002 Body image and prosthesis satisfaction in the lower limb amputee. Disabil Rehabil **24(17)** 925-31.

[5]. Windrich M, Grimmer M, Christ O, Rinderknecht S, Beckerle P. 2016 Active lower limb prosthetics : a systematic review of design issues and solutions. *Biomed Eng Online.* **15(3)** 5–19.

[6]. Goh J C H, Lee P V S, Toh S L, Ooi C K 2005 Development of an integrated CAD-FEA process for below-knee prosthetic sockets *Clin Biomech* **20(6)** 623–9.

[7]. Shuxian Z, Wanhua Z, Bingheng L 2005 3D reconstruction of the structure of a residual limb for customising the design of a prosthetic socket *Med Eng Phys.* **27(1)** 67-74.

[8]. Krouskop T A, Muilenberg A L, Doughtery D R, Winningham D J 1987 Computer-aided design of a prosthetic socket for an above-knee amputee *J Rehabil Res Dev.* Spring **24(2)** 31-8.

[9]. Lee W C, Zhang M 2007 Using computational simulation to aid in the prediction of socket fit: a preliminary study. *Med Eng Phys.* **29(8)** 923-9.

[10]. Reynolds D P and Lord M1 1992 Interface Load Analysis for Computer-Aided Design of Below-Knee Prosthetic Sockets *Med. Biol. Eng. Comput.* **30 (4)**, 419–426.

[11]. Jia X, Zhang M, Lee WCC 2004 Load transfer mechanics between trans-tibial prosthetic socket and residual limb Dynamic effects *J Biomech* **37(9)**1371–7.

[12]. Faustini M C, Neptune R R, Crawford R H 2006 The quasi-static response of compliant prosthetic sockets for transtibial amputees using finite element methods *Med Eng Phys* **28(2)** 114–21.

[13]. Lin C C, Chang C H, Wu C L, Chung K C, Liao I C 2004 Effects of liner stiffness for transtibial prosthesis: A finite element contact model *Med Eng Phys.* **26** 1–9 .

[14]. Zhang L, Zhu M, Shen L, Zheng F 2013 Finite element analysis of the contact interface between trans-femoral stump and prosthetic socket *Annu Int Conf IEEE Eng Med Biol Soc* 1270-3.

[15]. Saunders M M, Schwentker E P, Kay D B, Bennett G, Jacobs C R, Verstraete M C et al. 2003 Finite element analysis as a tool for parametric prosthetic foot design and evaluation. Technique development in the solid ankle cushioned heel (SACH) foot *Comput Methods Biomech Biomed Engin.* **6(1)** 75-87.

[16]. Bonnet X, Pillet H, Fodé P, Lavaste F, Skalli W 2012 Finite element modelling of an energy-storing prosthetic foot during the stance phase of transtibial amputee gait *Proc Inst Mech Eng H.* **226(1)** 70-5.

[17]. Tryggvason H, Starker F, Lecomte C, Jonsdottir F 2020 Use of Dynamic FEA for Design Modification and Energy Analysis of a Variable Stiffness Prosthetic Foot *Appl. Sci.* **10** 650.

[18]. Ke M J, Huang K C, Lee C H, Chu H Y, Wu Y T, Chang S T et al. 2017 Influence of three different curvatures flex-foot prosthesis while single-leg standing or running: A finite element analysis study *J Mech Med Biol* **17(3)** 1–12.

[19]. Mahmoodi P, Aristodemou S, Ransing R S 2016 Prosthetic foot design optimisation based on roll-over shape and ground reaction force characteristics *Proc IMechE Part C: J Mechanical Engineering Science* **231(17)** 3093–3103

[20]. Lee W C C, Zhang M 2005 Design of monolimb using finite element modelling and statistics-based Taguchi method *Clin Biomech* **20(7)** 759–66.

[21]. Rahman M, Bennett T, Glisson D, Beckley D and Khan J 2014 Finite Element Analysis of Prosthetic Running Blades Using Different Composite Materials to Optimize Performance Int Mech Eng Congress Expo **10** 14-20

[22]. Lee W C, Zhang M, Boone D A, Contoyannis B 2004 Finite-element analysis to determine effect of monolimb flexibility on structural strength and interaction between residual limb and prosthetic socket *J Rehabil Res Dev.* **41** 775–86.

[23]. Rigney S M, Simmons A, Kark L 2015 Concurrent multibody and Finite Element analysis of the lower limb during amputee running Proc Annu Int Conf IEEE Eng Med Biol Soc EMBS 2434–7.

[24]. Ramalho A, Ferraz M, Gaspar M, Capela C 2020 Development of a preliminary finite element model to assess the effects of friction on the residual limb of a transfemoral amputee. Mater Today Proc **33** 1859–63.

[25]. Adam C J, Lee W C C, Doocey J M, Bra R, Evans J H, Pearcy M J et al 2008 FE stress analysis of the interface between the bone and an osseointegrated implant for amputees – Implications to refine the rehabilitation program **23** 1243–50.

IOP Conf. Series: Materials Science and Engineering 1225 (2022) 012031 doi:10.1088/1757-899X/1225/1/012031

Wearable Human Motion Monitoring Using Vertical Contact Separation Mode Triboelectric Nanogenerator

Akshpreet Kaur[1], Ankur Gupta[2], Cuifeng Ying[3], Mohsen Rahmani[3], Gaurav Sapra[1]

[1] UIET, Panjab University, Chandigarh, India
[2] Department of Cardiology, PGIMER, Chandigarh, India
[3]Advanced Optics and Photonics Laboratory, Department of Engineering, School of Science & Technology, Nottingham Trent University, Nottingham, United Kingdom
akshpreet9386@gmail.com[1]

Abstract. Wearable human motion monitoringhas become an integral part of the paradigm shift that is on the horizon in the field of healthcare for a wide variety of biomedical applications. In recent years, there has been incessant advancement in smart wearable technology capable of human motion monitoring ranging from strain sensors to piezoelectrics. This paper proposes to fabricate a Triboelectric Nanogenerator based on Vertical Contact Separation (VCS) mode with design characteristics offlexibility, low cost, simple and easy to fabricate design for real - time monitoring of human body movements. The output performance of fabricated VCS - TENG is analysed by finger tapping. In order to conduct human motion monitoring, VCS - TENG is affixed on different body parts, and the corresponding signal is analysed. The developed prototype can be further integrated with advanced electronics to deliver promising technology in the field of healthcare diagnostics and monitoring.

Keywords: Wearable devices, Human motion monitoring, Triboelectric nanogenerator, Vertical contact separation mode, Biomedical, Healthcare.

1. Introduction

Wearable sensors have emerged as a new platform for user-friendly, accurate, unobtrusive and non-invasive human motion monitoring. The movements of the body play a vital role in understanding the behavioural pattern and physiological conditions of human beings [1]. Wearable human motion monitoring is significant in a wide spectrum of healthcare applications [2], such as gesture recognition [3], posture correction [4], tactile and electronic skin [5], gait recognition [6], sleep quality assessment [7], physical rehabilitation and assess disease progression for patients suffering from 'Parkinson's [8], paralysis [9], multiple sclerosis [10], Huntington's disease[11], certain types of dementia [12]etc. The healthcare system has experienced a paradigm shift from standard clinical procedures to personalised assessment tools for automated health monitoring. This technological drift is accompanied by innovative technologies such as intelligent wearable devices. With this technological advancement, physicians can evaluate the dynamics of disease progression, early health risk prognosis, tailor spersonalisedtherapeutic tools, and in turn, effectively cut the expenses that are encumbering medical care systems globally [13]. There are several techniques that have been explored to monitor motions of the human body, such as strain sensing [14], ultrasound [15], doppler sensor [16], piezoelectricdevices [17], hybrid devices [18], triboelectric nanogenerators [19] etc. However, most of the methods used to date are quite expensive and complex. Therefore it is necessary to select easy fabrication methods and low cost materials to develop human motion monitoring devices. By using such technologies, a lot of useful information can be derived by analysing signals and transform future healthcare and lifestyle.

In this paper, a paper-based Vertical Contact Separation (VCS) - Triboelectric Nanogenerator (TENG) is developed with characteristics of flexibility, low cost, light-weight and easy fabrication for real – time human motion monitoring. The output performance of the VCS - TENG is measured by finger tapping with the force of approximately ~5N. The paper-based TENG is affixed on inner elbow and back of knee to monitor the corresponding human locomotion. The proposed work will open new routes for automated healthcare monitoring and lifestyle management.

2. Triboelectric Nanogenerator

Triboelectric nanogenerator was developed in the first place by Wang group in 2012[20]. A typical TENG consists of two triboelectric layers and two metal electrodes on each side of the triboelectric layers. The material for triboelectric layers is selected in such a way that they are located far apart in the triboelectric series and are of opposite tribopolarity. The device works on the amalgamated principle of triboelectrification and electrostatic induction. The friction ("*tribo*") caused by continuous contact and separation or relative sliding between the two triboelectric materials induces electrostatic charges on the surfaces of triboelectric layers resulting in electron flow between the metal electrode films. As shown in Figure 1, TENGs basically operates in 4 operating modes – Vertical Contact Separation mode (VCS), Single Electrode mode (SE), Lateral Sliding mode (LS), Free - Standing mode (FS).

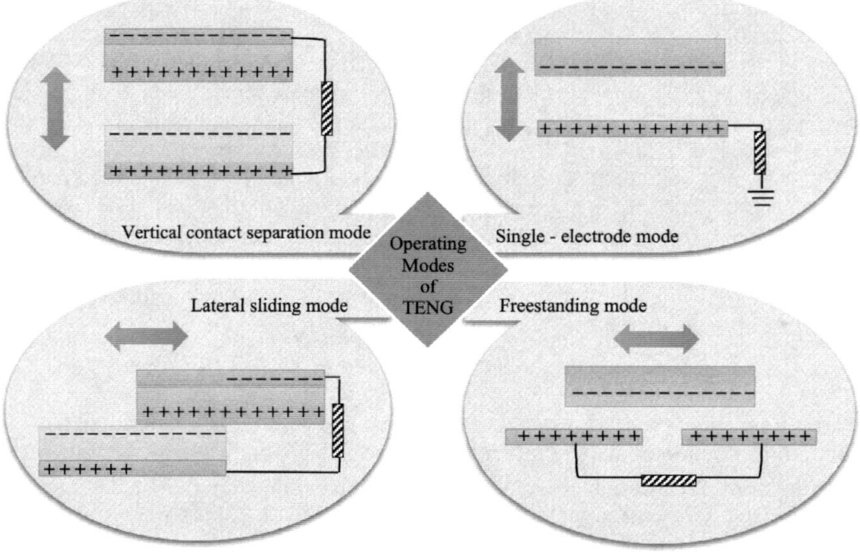

Figure 1. Operating modes of TENG [20]

2.1. Vertical contact separation mode

A typical vertical contact separation mode TENG consists of two dissimilar dielectric films acting as triboelectric layers separated by a certain distance and metal electrodes deposited on top and bottom surfaces of triboelectric layers. In this mode, the relative motion is perpendicular to the interface. With the application of external force, the two triboelectric surfaces come in contact with each other and opposite charges are created,i.e. one surface behaves as an anode and the other as a cathode. When the two layers are separated due to the release of external force, the surfaces maintain the charges and induces electric potential between the two triboelectric surfaces. These charges flow freely between the metal electrodes due to the electrostatic field. The continuous contact and separation generate potential drop, and current flows through the circuit, thereby generating AC power.

2.2. Single – electrode mode

The single-electrode modeinvolves the motion of only one material. In this mode, the ground is used as the reference electrode,which remains static and is used for electron injection to the mobile material. The potential drop will be created with contact and separation of mobile material with the static electrode such as human hand, human jogging, etc or vice-a-versa.

2.3. *Lateral sliding mode*

The lateral-sliding modeutilises two dissimilar triboelectric materials rubbing with each other in a lateral direction with different effective contact areas. In this mode, the relative displacement of frictional surfaces is in the direction parallel causing frictional sliding, thereby inducing potential difference between the two electrodes. The potential difference changes periodically due to change in the effective contact area and generates triboelectric power.

2.4. *Freestanding mode*

The freestanding triboelectric-layer mode comprises a dielectric film with a pair of symmetric electrodes under it with the thickness of electrodes of the same order as the gap with the moving film. The electron flow occurs due to asymmetric charge distribution arising from the movement of the mobile dielectric film[21-22].

3. Design and Fabrication of TENG

The proposed TENG is designed to operate in vertical contact separation mode. A typical TENG consists of two parts – a triboelectric layer of positive charge affinity with metal electrode film deposited on top and another triboelectric layer of negative charge affinity with metal electrode film deposited on the bottom. In this work, a very simple and easy to fabricate TENG with characteristics of flexibility is fabricated to monitor human motion in realtime. A strip of dimension 25 mm x 90 mm is cut from a 75 GSM copier paper (Trident Spectra), and aluminium film of 15 mm x 80 mm along with a peeled connecting wire is affixed on it using cello tape. Herein, two parts of TENG are fabricated to be identical to each other but stacked in such a way that the contact and separation occur between paper and cellotape resulting in charge transfer between the surface and electrons flow through the aluminium foil. The two flexible parts of TENG are folded in opposite directions from the middle in order to create an air gap between the two triboelectric layers, and the two edges are sealed with ordinary adhesive tape (Figure 2).

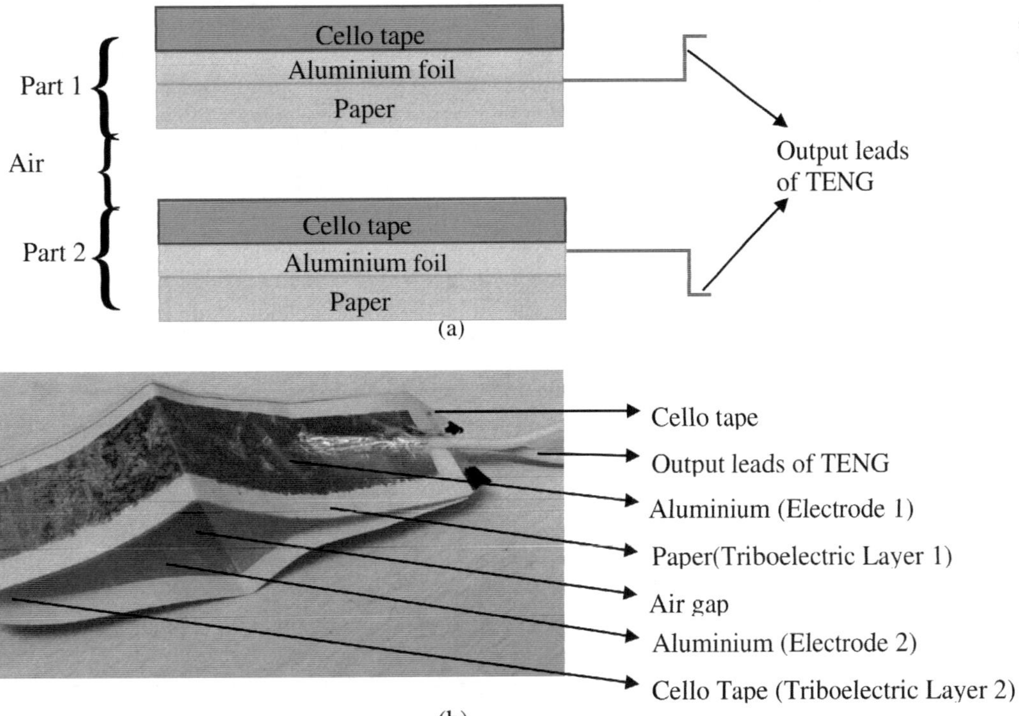

Figure 2. (a)Schematic of Paper based TENG (b) Photograph of fabricated Paper based TENG

4. Results and Discussion

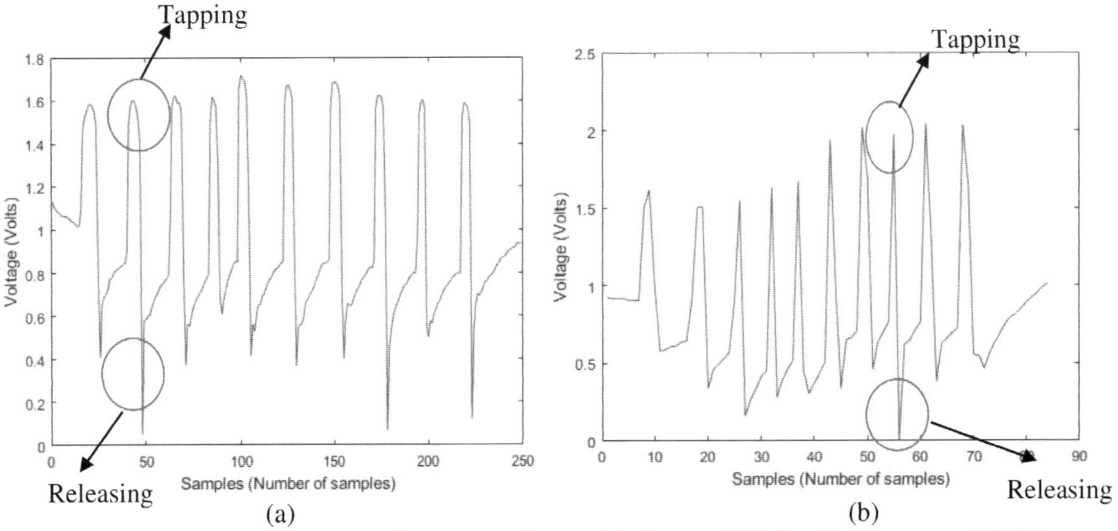

Figure 3. Voltage signal corresponding to (a) slow and (b) fast finger taps on TENG

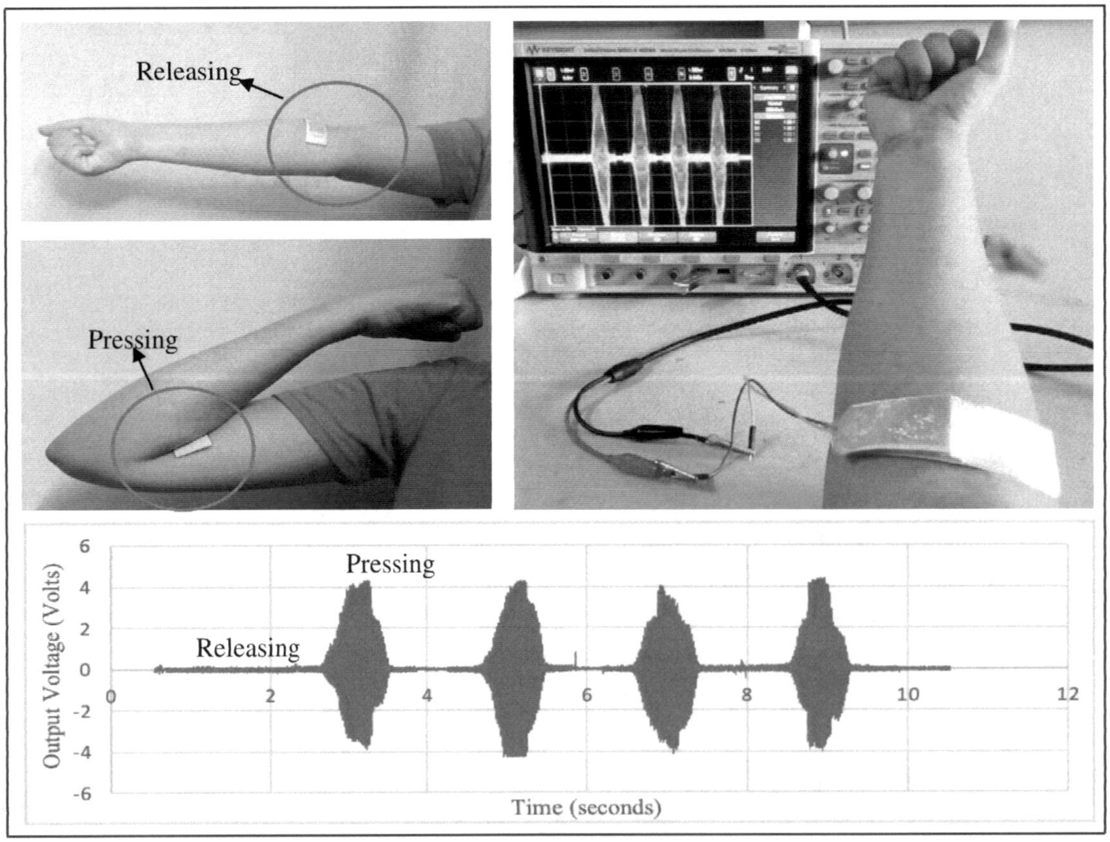

Figure 4. Real – time monitoring and corresponding voltage profile due to movement of elbow

The output performance of fabricated paper-based TENG was analysed by interfacing it with Arduino and MATLAB in order to acquire voltage signals. The contact and separation of TENG were achieved by tapping a finger on TENG.The voltage corresponding to 10 finger taps in slow motion and fast motion were recorded and are shown in Figure 3 (a) and (b), respectively. The maximum output peak to peak voltage recorded during slow finger tapping on TENG was 1.7 V whereas during fast tapping, it was recorded to be 2 V. The output voltage of TENG largely depends on the applied external pressure.

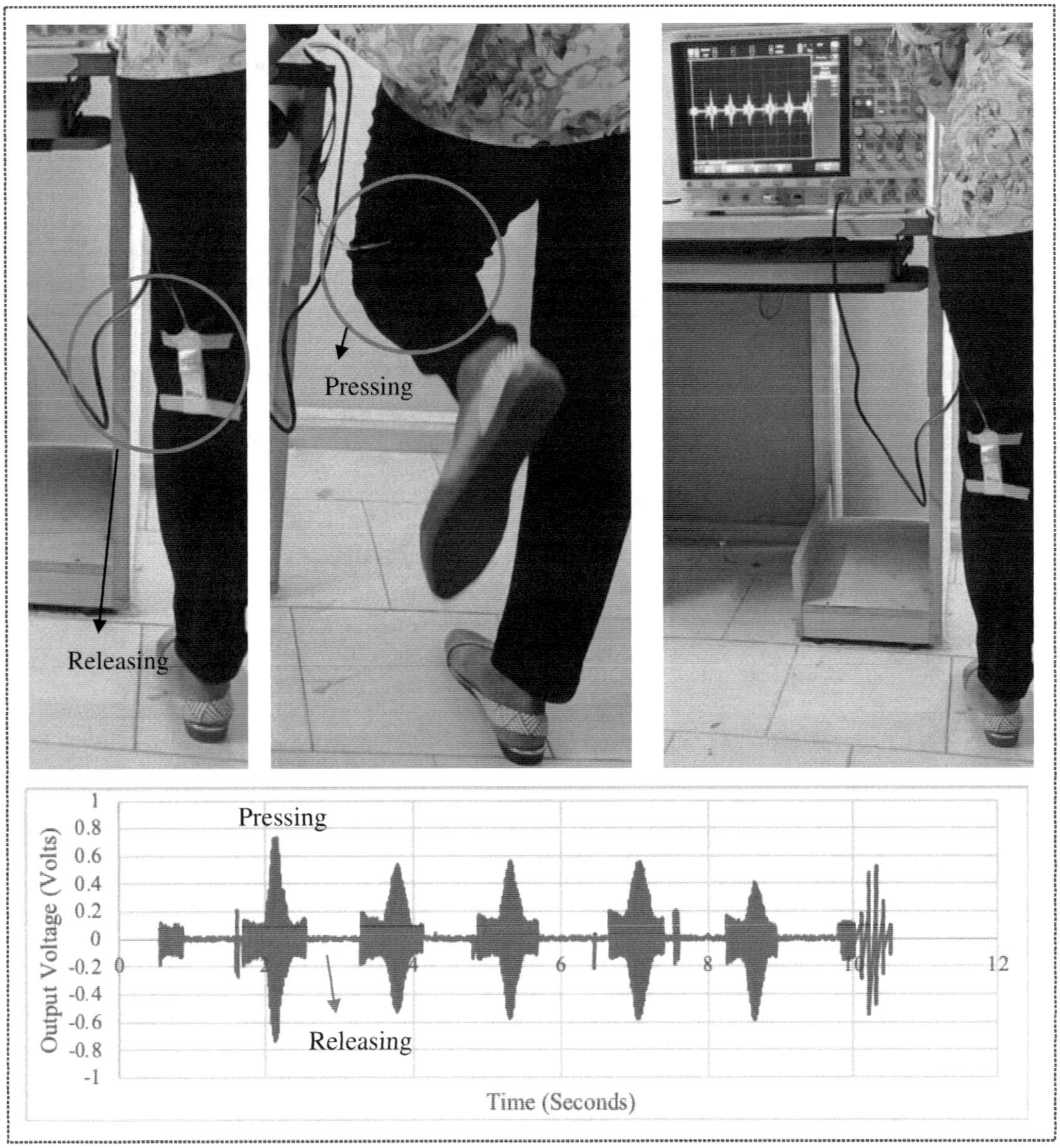

Figure 5. Real – time monitoring and corresponding voltage profile due to movement of knee

The fabricated TENG is affixed on different body parts to carry out human motion monitoring and interfaced with Mixed Signal Oscilloscope (Keysight Infinii Vision MSO-X 4024 A) for analysing the performance of TENG. The paper-based TENG is affixed on the inner elbow and back of the knee using double-sided tape to monitor the corresponding human locomotion. The continuous pressing and releasing between the triboelectric layers results in a transfer of charge and the development of electric potential between the aluminium electrodes. The corresponding voltage profiles are recorded using MSO as shown in Figure 4 and Figure 5. The maximum peak to peak voltage recorded due to movement of elbow and knee is 8 V and 1.4 V, respectively. The change in voltage is large enough to easily detect human motion. Therefore, the fabricated TENG device can successfully respond to body movements.

5. Conclusion

In this work, paperbased VCS - TENG of a very simple and lowcost design with characteristics of flexibility is developed for human motion monitoring. The performance of fabricated paper based TENG of size 25 mm x 90 mm is analysed by finger tapping and the corresponding recorded voltage is 1.7 V and 2 V for slow and fast tapping, respectively. The fabricated prototype is tested to monitor the movement of the arm and leg by affixing it on inner elbow and back of knee, respectively. The recorded voltage was observed to increase and subsequently decrease on continuous contact and separation of paper based TENG due to movement of the body part. Therefore, it is concluded that the paper based TENG is capable of human motion monitoring which is used in a wide range of healthcare and other applications.The TENG can be further used to monitor sleeping patterns and other human motion states like walking, stepping, jumping, cycling, etc. The prospective research direction could be an integration of the developed TENG with advanced electronics to realise smart wearable devices for remote healthcare monitoring and diagnostics.

Acknowledgement

Akshpreet Kaur is thankful for financial support in the form of DST-INSPIRE Fellowship (DST/INSPIRE/Fellowship/2018/IF190152) by the Department of Science and Technology (DST), Ministry of Science and Technology, Government of India. The authors appreciate the Nottingham Trent University – Panjab University strategic partnership funding. Mohsen Rahmani acknowledges the supports form The Royal Society and the Wolfson Foundation.

References

[1]. Homayounfar S Z, Andrew T L 2020 Wearable sensors for monitoring human motion: A review on mechanisms, materials, and challenges *SLAS TECHNOLOGY: Translating Life Sciences Innovation* **25**(1) 9-24

[2]. Kaur A, Sapra G, Gupta A 2021 Recent Progress on Energy Harvesters for Biomedical Applications *J. Circuits, Syst. Comput.* **7** 2130010

[3]. Huang Y, Gao L, Zhao Y, Guo X, Liu C, Liu P 2017 Highly flexible fabric strain sensor based on graphene nanoplatelet–polyaniline nanocomposites for human gesture J. Appl. Polym. Sci. **134** (39) 45340

[4]. Zhu J, Zhu P, Yang Q, Chen T, Wang J, Li J 2020 A fully stretchable textile-based triboelectric nanogenerator for human motion monitoring *Mater. Lett.* **280** 128568

[5]. Zhu M, Lou M, Abdalla I, Yu J, Li Z, Ding B 2020 Highly shape adaptive fiber based electronic skin for sensitive joint motion monitoring and tactile sensing *Nano Energy* **69** 104429

[6]. Vu CC, Kim J 2020 Highly elastic capacitive pressure sensor based on smart textiles for full-range human motion monitoring *Sens. Actuators A: Phys* **15** 112029

[7]. Li R, Wei X, Xu J, Chen J, Li B, Wu Z, Wang Z L 2021 Smart Wearable Sensors Based on Triboelectric Nanogenerator for Personal Healthcare Monitoring *Micromachines* **12**(4) 352

[8]. Albán-Cadena A C, Villalba-Meneses F, Pila-Varela K O, Moreno-Calvo A, Villalba-Meneses C P, Almeida-Galárraga D A 2021 Wearable sensors in the diagnosis and study of Parkinson's disease symptoms: a systematic review *J Med Eng Technol* . **30** 1-4

[9]. Rahman S S, Nusaka S S, Shezan F H, Sarkar M R 2015 The development of low cost exercise monitoring device for paralytic In 2015 *IEEE International WIE Conference on Electrical and Computer Engineering (WIECON-ECE)* Dec 19 530-533.

[10]. Sparaco M, Lavorgna L, Conforti R, Tedeschi G, Bonavita S 2018 The role of wearable devices in multiple sclerosis *Mult Scler Int.* **2018** 7627643

[11]. Tortelli R, Rodrigues F B, Wild E J 2021 The use of wearable/portable digital sensors in Huntington's disease A systematic review *Parkinsonism Relat Disord.* **83** 93-104.

[12]. Husebo B S, Heintz H L, Berge L I, Owoyemi P, Rahman A T, Vahia I V 2020 Sensing technology to monitor behavioral and psychological symptoms and to assess treatment response in people with dementia A systematic review *Front. Pharmacol.* **4**(10) 699.

[13]. Loncar-Turukalo T, Zdravevski E, da Silva JM, Chouvarda I, Trajkovik V 2019Literature on wearable technology for connected health: scoping review of research trends, advances, and barriers *J Med Internet Res* **21**(9) e14017

[14]. Song X, Liu X, Peng Y, Xu Z, Liu W, Pang K, Wang J, Zhong L, Yang Q, Meng J 2021 A graphene-coated silk-spandex fabric strain sensor for human movement monitoring and recognition *Nanotechnology* **32**(21) 215501

[15]. Summerhill E M, El-Sameed Y A, Glidden T J, McCool F D 2008 Monitoring recovery from diaphragm paralysis with ultrasound *Chest* **133**(3) 737-43

[16]. Kubo H, Mori T, Sato T 2010 Detection and measurement of human motion and respiration with microwave Doppler sensor *IEEE Int. Conf. Intell. Robots Syst.* **18** 5416-5423

[17]. Qian S, Qin L, He J, Zhang N, Qian J, Mu J, Geng W, Hou X, Chou X 2020 A lead-free stretchable piezoelectric composite for human motion monitoring *Mater. Lett.* **261** 127119

[18]. Li Y, Chen Z, Zheng G, Zhong W, Jiang L, Yang Y, Jiang L, Chen Y, Wong C P 2020 A magnetised microneedle-array based flexible triboelectric-electromagnetic hybrid generator for human motion monitoring *Nano Energy* **69** 104415

[19]. Shen X, Han W, Jiang Y, Ding Q, Li X, Zhao X, Li Z 2020 Punching pores on cellulose fiber paper as the spacer of triboelectric nanogenerator for monitoring human motion *Energy Rep.* . **6** 2851-60

[20]. Fan F R, Tian Z Q, Wang Z L 2012 Flexible triboelectric generator *Nano energy* **1**(2) 328-34

[21]. Wu C, Wang A C, Ding W, Guo H, Wang Z L 2019 Triboelectric nanogenerator: a foundation of the energy for the new era *Adv. Energy Mater.* **9**(1) 1802906

[22]. Zheng Q, Shi B, Li Z, Wang Z L 2017 Recent progress on piezoelectric and triboelectric energy harvesters in biomedical systems *Adv. Sci.* **4**(7) 1700029

Dual band compact antenna design using DCSRR

Ambika[a*], M. Mohamed Humaidh[a], S. Kowshic[a], S. Mohamed Afrath[a], M.Z. Abdul Rahman[a]

[a]ECE, BSACSIT, Chennai, Tamil Nadu, India
Email:[a*]ambika@crecent.education

Abstract. Wireless communication technology has revolutionised the way people interact all around the world. The rate of access and the accuracy of the information conveyed have increased as technology and gadgets have advanced. Multi-frequency systems and devices are the results of the requirement for mobility and the integration of several radio modules into this kind of device. Due to the limited area, integrating many antennas and a diplexer into a single device is very challenging. This means an advanced antenna design must be simple, small, and simple to incorporate in addition to being dual or multiband capable.Microstrip printed antennas are the most promising alternative for use in small systems due to its various features such as low profile, compact configuration, simplicity of integration, and low fabrication cost. There are many techniques to make an antenna operate in multiband such as by incorporating structures like complementary split ring resonator(CSRR), SRS (split ring slot), SRR (split ring resonator) etc., dual polarization method, loading of metamaterial. The DCSRR (D shape complementary split ring resonator) is inserted into the antenna patch to enable multiband operation. So, in this paper, to achieve dual band response using this method, which supports Wi-Fi and C band applications at frequencies of 5.8 GHz and 7.1 GHz, respectively. In the suggested work, a good gain of about 1.98 dB at 5.8 GHz and 2.27 dB at 7.1 GHz was obtained, as well as an efficiency of 80 percent and 81 percent at the relevant frequencies. This characteristic has been achieved without compromising on the compact miniaturized size of antenna, which is 12 x 12 x 1.6 mm^3.

Keywords:D-CSRR, compact antenna, dual band,s parameter

1. Introduction

To provide communication, GPS, and other services, modern wireless hand-held devices must operate on multiple bands [1].With a compact dimension and larger bandwidth, a particular Patch Antenna (MPA) can service entire frequency bands, including Wi-Fi, WLAN, WiMAX, and C band usages. This is the most common strategy for today's wireless handheld devices, rather than using various antennas for different working frequencies.In typical dual-band or multiband MPAs, the dimensions of the radiating patch are taken regarding the smallest frequency band, which requires further size increment. In the case of metamaterials based dual band antennas, the dimensions of the radiating patch are measured in terms of high frequency, which demands a smaller antenna.

The three categories of metamaterials are double negative materials (DNG) and single negative materials such as epsilon negative materials (ENG), permeability negative materials (MNG) [2, 3].Such type of materials are exhibited by SRR and CSRR structures. Several CSRR/SRR structures [4, 5] have investigated and examined the properties of CSRRsand SRRs. In particular, dual band antennas are provided in [6 - 7] by introducing CSRR into the radiating patch. Single band and multiple band antennas are designed in [8-11] by etching CSRR on the ground plane. The antennas outlined above take up a lot of space in regards of thickness or surface area for Wi-

Fi and WiMAX applications, and they additionally had a low bandwidth. Our goal is to maintain a good level of gain and bandwidth while decreasing the size by introducing a D shape CSRR (DCSRR)on antenna's construction. Suggested design antenna deals with single ring DCSRR created on patch antenna with compact size for 5.8GHz and 7 GHz dual frequencies.This antenna is suited for Wi-Fi and C band communication due to its compact size, low cost, and lightweight. This antenna's major benefit is its simplicity.

2. Antenna Design Structure

The design antenna is made up of three layers: patch, substrate, and ground. All dimensions are given in millimetres. The proposed antenna is built on a lossy dielectric sheet called FR-4 dielectric substrate ($\varepsilon_r = 4.4$) with height of 1.6mm. For better bandwidth partial ground is consider. Figure 1 a, b depicts a basic front and back of microstrip patch antenna with partial ground for 6.8 GHz resonance frequency. Based on transmission line model, the fundamental formula for determining patch antenna length and width is calculated [12] and the dimensions of antenna is given in Table 1.

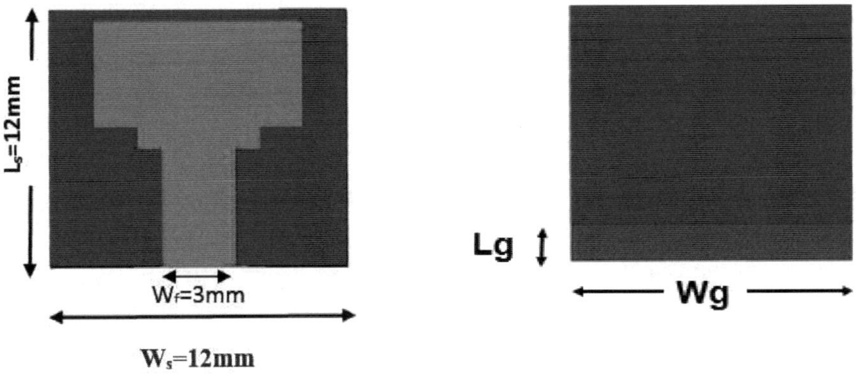

(a) (b)

Figure 1. Rear side and backside view of patch antenna

Table1. Dimensions of proposed antenna

Antenna parameter	dimension (mm)
Ws	12
Ls	12
Wf	3
Lf	7
Lslot	5
Wslot	1
Lg	12
Wg	1.5
Wp	8.5
Lp	5

2.1 D shape CSRR design

Figure 2 depicts the structure of the D Shape CSRR (DCSRR). Generally, the CSRR is the dual structure of SRR, according to the concept of Babinet's principle, wherein the functions of airand metal, as well as magnetic and electric field wherein the capabilities of metal and air, as well as electric and magnetic field characteristics are interchanged. The SRR may be energized by an external magnetic flux as a magnetic dipole, however, being an electric dipole, the CSRR may be activated by an external electric field.

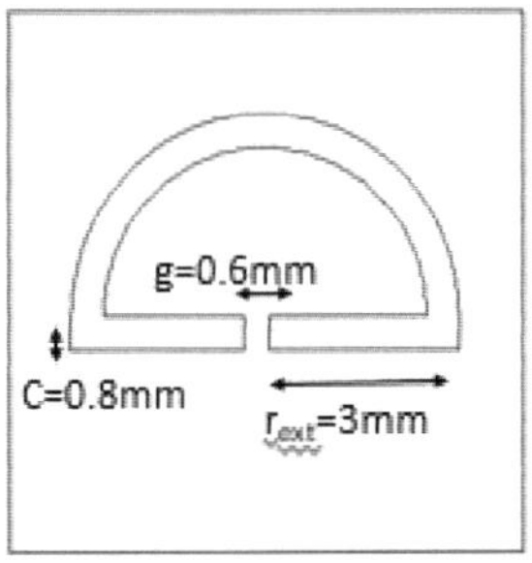

Figure 2. DCSRR structure

The proposed D shape CSRR resonant frequency is determined by the self-inductance and capacitance per unit length. The design equation of the proposed single D CSRR is calculated using mathematical calculation of resonant frequency,
L_t, C_t, f_{SRR} are calculated using the design equations given by (1), (2) and (6)
The formula for total Inductance,

$$L_t(\mu H) = [0.0002 * l * \left[2.303 * \log\left(N * \frac{l}{c+t}\right) - \frac{Y}{2}\right] * 10^{-6} \quad (1)$$

Where$l$ is per unit length inductance

C is strip width, t is thickness of strip,N is no of strip rings, constantY is 1.225 for D wire geometry.
Total Equivalent capacitance,

$$C_t = \frac{C_g}{2} \quad (2)$$

gap capacitance,

$$C_g = \frac{\varepsilon_0 * c * t}{g} \quad (3)$$

where g is split gap in ring, c is strip width, t is thickness of strip,

$$r_{avg} = r_{ext} - \frac{c}{2}, \quad (4)$$

Where r_{avg} is average D ring radius,
radius of D SRR ring, r_{ext} = 3 mm, c = 0.8 mm

The formula for resonance frequency, f_{SRR} is given in equation 6.

$$f_{SRR} = \frac{1}{2\pi\sqrt{C_t * L_t}} \quad (5)$$

The theoretical resonant frequency is 5.72 GHz.

$C_{SRR} = C_t$= 0.154 nF, $L_{SRR}= L_t$= 5.0215 pH,

The proposed D shape SRR design resonates at 5.8 GHz.

2.2. Waveguide Analysis of Single D Shape CSRR

The complex permittivity of DCSRR is calculated based on the time-domain transient response over a frequency range. Based on the Nicholson-Ross Weir (NRW) method [16], the parameter extraction is done. Figure.3 illustrates the waveguide setup using the HFSS tool for calculating real and imaginary S parameters of D CSRR. The D CSRR is placed within the waveguide and has perfect magnetic conductor (PMC) and perfect electric conductor (PEC) boundaryconditions.EM waves emerge via the input port and the values of S (Real & Imaginary) parameters from port 2 are computed. Extracted the real and imaginary of S parameters of a single D SRR resonating at 5.8 GHz from HFSS simulation.

The negative permittivity (ε) characteristics of DCSRR is extracted from these S-Parameter (S11 and S21) values by using equations (6&7).

$$\varepsilon_r = \frac{2}{jk_o d} * \frac{1-V_1}{1+V_1} \quad (6)$$

$$\mu_r = \frac{2}{jk_o d} * \frac{1-V_2}{1+V_2} \quad (7)$$

Where

V1 is S_{21}-S_{11},V2 is S_{21}-S_{11}
k_o is wave number of free space,d is substrate thickness

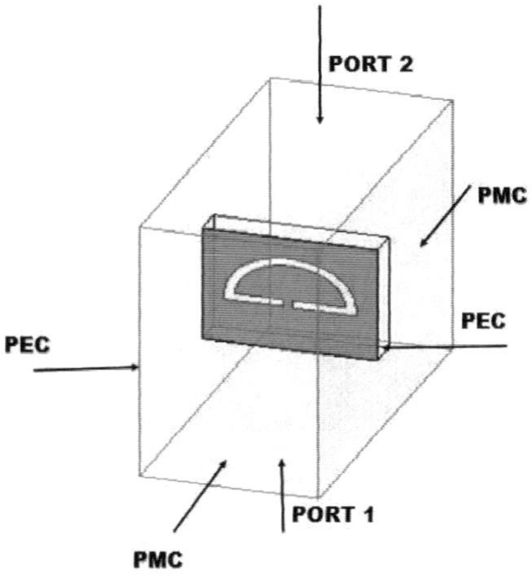

Figure 3. Unit Cell Design of Single D SRR Structure

The extraction of the single D CSRR unit cell of effective permittivity is obtained by the MATLAB tool. It can be seen in Figure 4 that the real permittivity value is negative at 5.8 GHz, which attributes to the D CSRR resonance behaviour.

2.3 Proposed antenna with DCSRR

The DCSRR structure is etched from the patch antenna after proper optimization based on mode characteristics which is shown in Figure 5. The proposed DCSRR structure creates additional resonance frequency at 5.8 GHz and patch antenna resonance frequency slightly shifted to 7 GHz. The antenna with and without DCSRR return loss response are simulated and plotted in Figure 6.

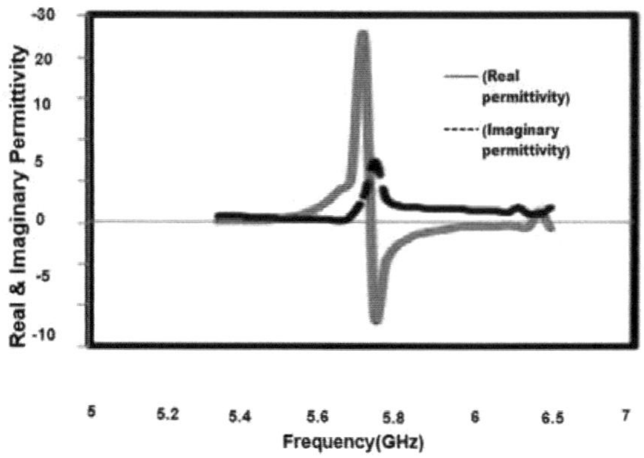

Figure 4. Extracted Effective Permeability ($\mathcal{E}_{eff}$) Single D CSRR Structure

With DCSRR provides better impedance matching at 5.8GHz around -27 dB after adding D structure. Two responses obtained at 5.8 GHz and 7 GHz, where 10 dB return loss is about -27 dB and -14.6 dB respectively. So, the obtained dual bands, i.e. one band is 5.8 GHz (Wi-Fi) and other is 7 GHz (upper C – Band). The optimization of antenna procedures discussed as follows:

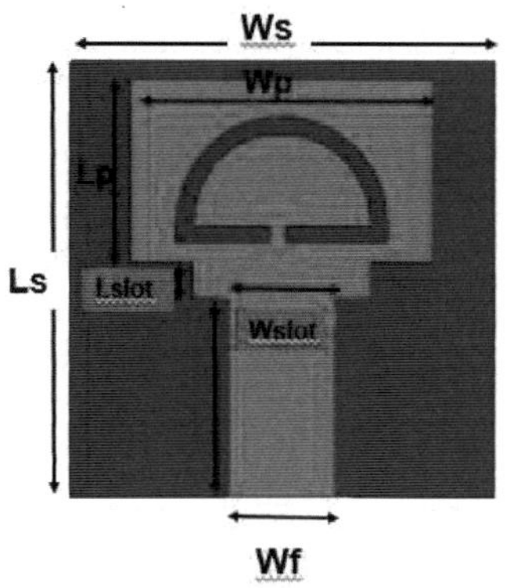

Figure 5. DCSRR Loaded Patch Antenna

With the patch antenna area dimension as 12 x 12=144 mm^2 without DCSRR, the antenna resonates at 6.8GHz after introducing DCSRR lower resonance frequency resonates at 5.8 GHz. Nearly 1GHz reduction in frequency response. Hence the proposed antenna achieved 14.7% reduction in size.

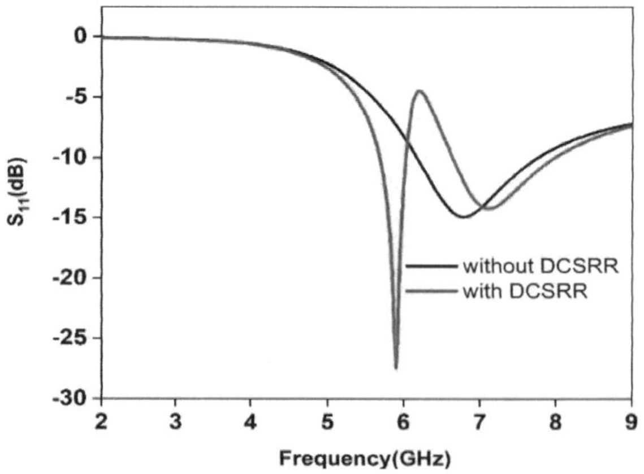

Figure 6. Simulated Return loss of patch antenna with a DCSRR

2.3.1. Parametric analysis

By adjusting and optimizing the width of the partial ground plane, in order to achieve good return loss at 6.8 GHz frequency which is shown in Figure.7.After introducing D shape CSRR(DCSRR) on the top of the patch the structure creates dual resonance frequencies at both 5.8GHz and 7 GHz. The dual resonance behaviour is created by negative permittivity of D shape CSRR

It was found that the obtained frequency with ground width of 1.5 mm is the best result as obtained a good return loss of -27 dB also exactly at desired frequency 5.8 GHz which is shown in Figure 7.

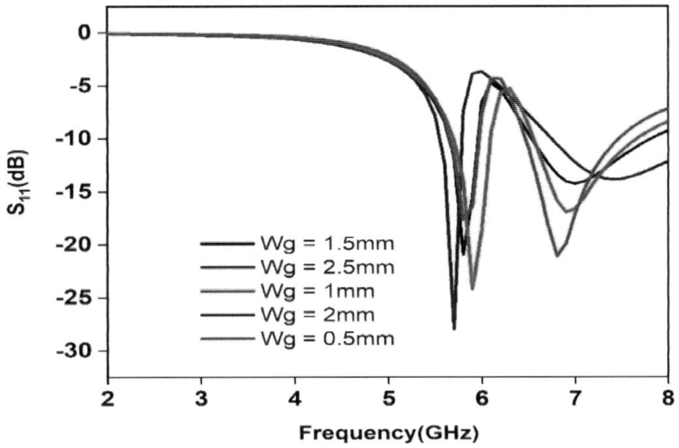

Figure 7. S_{11} responses for ground width variation

2.3.2. Surface current Analysis of proposed antenna

The mode of resonance can be determined by analysing the surface current distribution over the antenna surface. Field overlays are an option in HFSS for plotting the field distribution on a surface or object. Surface current analysis (J_{surf}), which is available under the Field overlays option, is used to calculate the current distribution on the radiating patch and ground patch. The patch under consideration is chosen, and the surface current distribution at a specific frequency is chosen. The study provides the amplitude and direction variation of the structure's surface current distribution. From Figure 8 ,clearly shows maximum current distributed near to the D shape strip which is responsible for 5.8 GHz but at 7 GHz the current distribution is very less near to D CSRR strip and maximum current distribution at patch feedlines .So 7GHz frequency response is almost similar to without DCSRR antenna design response .
From the ablove analysis, it was concluded that the low frequency band is mainly inspired by the proposed DCSRR structure in the patch antenna which is mainly controlled by the dimensions of the DCSRR.The second band is provided from patch itself with sligt disturbance from Split in DCSRR makes the small deviation in higher band frequency.

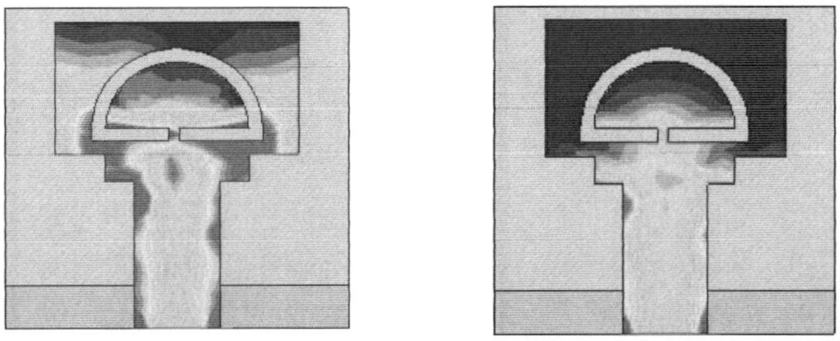

(a) at 5.8 GHz **(b) at 7 GH**

Figure 8. Surface current distribution of patch antenna with DCSRR

2.3.3. Radiation pattern of proposed antenna

The 2-D radiation pattern of proposed patch antenna with DCSRR is shown in Figure 9. BothE and H radiation pattern responses at dual band frequencies at 5.8 GHz and 7GHz.The 2-D radiation pattern of patch antenna with DCSRR provides figure of pattern for E plane and omnidirectional pattern obtained for H plane at 5.8 GHz and 7 GHz resonance frequency.

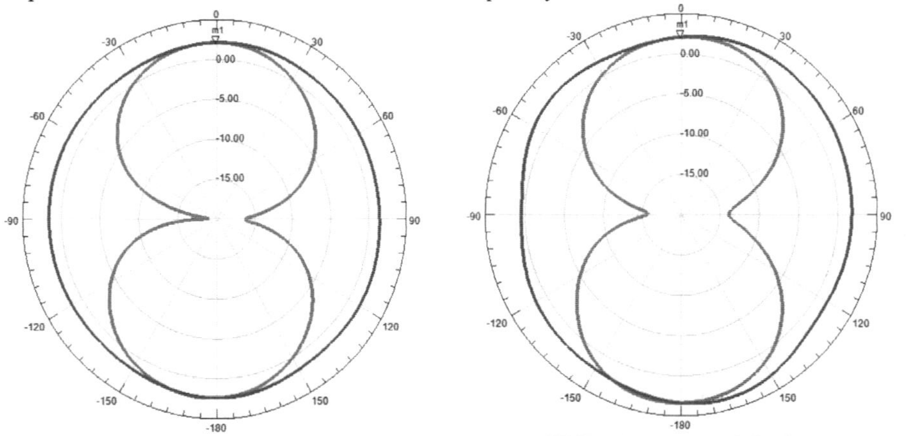

(a) Radiation plot at 5.5GHz (b)Radiation plot at 7.4 GHz

Figure 9. 2D radiation pattern of patch antenna with DCSRR

2.3.4 Gain and Efficiency of proposed antenna

The gain response of dual band frequencies is illustrated in Figure 10a. The maximum gain obtained at 5.8GHz and 7 GHz is 1.89 dB and 1.65 dB respectively.

(a) Gain plot(b) Efficiency plot

Figure 10. (a,b) Gain and Efficiency plot of proposed antenna with DCSRR

In the above Figure 10.b, efficiency at both the frequencies has been depicted. At 5.5 GHz, efficiency of about 0.95 (95%) is obtained and at 7.4 GHz, efficiency of about 0.919 (91.9%) is obtained. This proves that a good efficiency of antenna has been achieved at both the bands.

3. Conclusion

The proposed microstrip patch antenna has been incorporated with D structure in its patch, which is a structure similar to metamaterial and also such structures are capable of producing magnetic susceptibility, which referred as DSRR (Double Split Ring Resonator), this is used to achieve multiband operation. Using this technique, we have achieved dual band response to support Wi-Fi and C band application at frequencies 5.8 GHz (5.66-5.97) and 7 GHz (6.45-7.50) respectively. Also, a good gain of about 1.89 dB at 5.8 GHz and 1.65 dB at 7.4 GHz has been obtained along with the efficiency of 95% and 91.9% at the respective frequencies in the proposed work. This characteristic has been achieved with the compact miniaturized size of antenna 12 x 12 x 1.6 mm^3which is 14.7%. All the limitations such as size, gain, efficiency etc. identified in the existing structure and concept has been addressed and overcame in our proposed work. Therefore, the proposed antenna design has satisfactory characteristics for use in wireless communication system.

References

[1]. Garg R 2001 Microstrip antenna design handbook. *Artech House* ISBN 0-89006 513-6.
[2]. Caloz C and Tatsuo I2005 Electromagnetic metamaterials: transmission line theory and microwave applications *John Wiley & Sons.*
[3]. Eleftheriades G Vand Keith G B2005 Negative-refraction metamaterials: fundamental principles and applications *John Wiley & Sons*
[4]. Marqués et al. 2003 Comparative analysis of edge-and broadside-coupled split ring resonators for metamaterial design-theory and experiments *IEEE Trans. Antennas Propag.* **10** 2572-2581.
[5]. Lee Y and Yang H 2008 Characterization of microstrip patch antennas on metamaterial substrates loaded with complementary split-ring 45 resonators *Microw. Opt. Technol. Lett.* 50(8) 2131-2135.
[6]. Ortiz, N et al 2015 Design and implementation of dual-band antennas based on complementary split ring resonators *Waves in Random and Complex Media* 25(3) 309-322.
[7]. Ortiz N, Falcone F and Sorolla M 2011 Enhanced gain dual band patch antenna based on complementary rectangular split-ring resonators *Microw. Opt. Technol. Lett.* **53(3)** 590-594.
[8]. Ramzan Mand Topalli K 2015 A miniaturized patch antenna by using a CSRR loading plane *Int J Antennas Propag***2015**1-9.
[9]. Bilotti F, Andrea A, and Lucio V 2008 Design of Miniaturized Metamaterial Patch Antennas With μ-Negative Loading *IEEE Transactions on Antennas and Propagation* **56(6)** 1640-1647.
[10]. Tang MC et al 2016 A Compact Dual-Band Patch Antenna Design Based on Single-Ring Split Ring Resonator *Applied Computational Electromagnetics Society Journal* **31(3)**
[11]. Niu JX 2010 Dual-band dual-mode patch antenna based on resonant-type metamaterial transmission line *Electron. Lett.* **46(4)** 266-268.
[12]. Breed G 2009 The fundamentals of patch antenna design and performance *High Frequency Electronics* **3(12)** 49-51.

Characterization of Plastics and Polymers: A Comprehensive Study

Tarik Hassan, A K Srivastwa, Subhasis Sarkar and Gautam Majumdar

Department of Mechanical Engineering, Jadavpur University, Kolkata- 700032, India

Corresponding author's e-mail address- hassantarik9@gmail.com

Abstract. Since the beginning of polymer synthesis, a huge modification and development have taken place in this field which resulted in a world where people cannot think of a single day without a polymer product. They are light in weight, affordable and have the potential to provide a similar strength like the traditional metallic objects. As the demand for polymer products is increasing rapidly, its characterization and evaluation of mechanical properties become essential for making a reliable, scientific and cost-effective product design. Recently the incorporation of biopolymers in the global market has made this industry more attractive to consumers and government bodies. But the main challenge lies in its characterization as plastics and polymers found in nature or synthesized artificially have a wide range of physical and chemical properties. So, a particular set of instruments that can evaluate the properties of a group of polymers, may not be usable for a different group with the same accuracy and technique. Researchers are working incessantly to find out effective methods for this purpose and many of them are successful in finding some crucial properties of polymers like strength, elastic modulus, viscosity, hardness etc. The present work briefly discusses the superiority of polymer materials and the research work that has been carried out so far for the characterization of the same. Study on the biopolymer, its characterization and necessity in the context of environmental sustainability are also included in this literature.

1. Introduction

In the advancement of civilisation, the role of plastic and polymers is too much to say in a few sentences. These two terms are interrelated, plastics are a special type of polymers having huge applications as raw materials for the manufacturing of various modern appliances while polymers form a big umbrella under which plastic exists. So all polymers cannot be termed as plastics. As the name indicates, a polymer is formed by the combination of monomer units that have a wide range of physical and chemical properties. From the small household application to the heavy industrial equipment, its presence is very much dominant. The reason behind it is its favourable properties i.e. lightness, durability, chemically inertness (in most cases) and many other desired mechanical properties like hardness, strength per unit weight, scratch resistance etc[1,2]. Even our body is full of natural polymers i.e. Polysaccharides, Polynucleotides (DNA), Proteins etc[3]. The applicability of polymers has now expanded beyond the common domestic applications. Like the polyurethanes can be used in multipurpose areas viz. furniture, automobile industry, coating, construction and even in biomedical engineering due to its antimicrobial properties[4]. All these things made it impossible to think of a world without polymer and its products but it has a very adverse impact too on the environment. So, it is reasonable to invest effort not only in replacing plastics products with other biodegradable material but also in the recycling of the same by the maximum possible extent. That is why this is one of the most demanded topics of research nowadays. Several works have been done so far in this field and some of its fruitful results are advanced waste management processes, biopolymers, bioplastics (biodegradable), reinforced plastics and so on[5–7]. Polymers being a widely used material in many fields, their testing and characterization become an obvious thing for designing more reliable products with known or predictable material properties. The natures and properties of polymers are quite complicated to test and evaluate because of their wide range of physical states and unavailability of a unique testing methodology[8,9]. From a soft foam, viscous latex and rubber to a very hard and brittle thermoplast, all are different types and classes of polymers having different properties. Several unique physical testing methods are required for each unique class of polymers and that is the most challenging issue in this field of research[10]. Physical testing is mostly used to determine the mechanical properties of polymer namely hardness, tensile strength, flexural rigidity, scratch and wear resistance etc. Apart from that, a more detailed mechanical characterization can also be done for a sophisticated design of plastic products like creep, fatigue, fracture, impact strength, toughness, thermal stress etc and these data also plays a vital role in the failure analysis of the same[11,12]. Only the analysis of mechanical properties is not sufficient for the characterization of polymers as they can also be susceptible to chemical degradation, environmental corrosion, microbial(for biopolymers) and photolytic(due to sunlight) degradation, moisture-induced damage etc. These factors obviate the need for detailed chemical characterization of polymeric substances[13]. Properties of plastics and polymers are found to be highly influenced by functionalization, hydrocarbon chain length, incorporation of fibres, nanoparticles and different types of micro and nano fillers[14–18]. It has been found that the interfacial property of graphene oxide and a polymer matrix can be enhanced by controlling the functional group polarity of the polymer. Optimum polarity can increase the hydrogen bond density at the interface resulting in higher adhesion and superior strength[14]. Similarly, a study by Jian et al reveals that the interfacial shear strength and Young's modulus of CNT-epoxy nanocomposite polymer are improved by chemical functionalization on CNTs[19]. In modern applications, the use of polymer coating has drawn the attention of many engineering industries due to its favourable surface properties. A significant portion of polymer coatings are used to increase the durability of the surface of the substrate which resulted in a rigorous study and research for the evaluation of the important surface properties like scratch, wear, hardness, corrosion resistance etc in recent years[20–24].

Due to the unique properties of plastics and polymers and its constant worldwide demand, it will continue to be an integral part of the civilisation in future and the research in this field will also go on in the search of new material with more flexibility and advanced properties which will be able to serve the need of the respective engineering industries in a better way. Present work is focused on the analysis and comparative study of different types of plastics and polymers. Specifically, the characterization processes of synthetic and biopolymers and the path towards sustainability are discussed in detail.

2. About plastics and polymers

Polymerization is a basic chemical process that is made by combining the monomer units(Ethylene, Propylene etc) in a chain or a complex network with the help of a specific chemical reaction mechanism basically, addition and condensation[25] (Fig 1). Polyethene and polypropylene, the most used plastics in our everyday life are originated from crude oil and natural gas refinery. Firstly naphtha is separated from the by-products of the same and then passed through the cracking process at around 800 ^{0}C where the monomers i.e. ethylene and propylene are formed. It is then compressed and cooled to a liquid state and polymerisation is initiated by a free radical reaction with the help of an 'initiator' or 'catalyst' like organic peroxide.

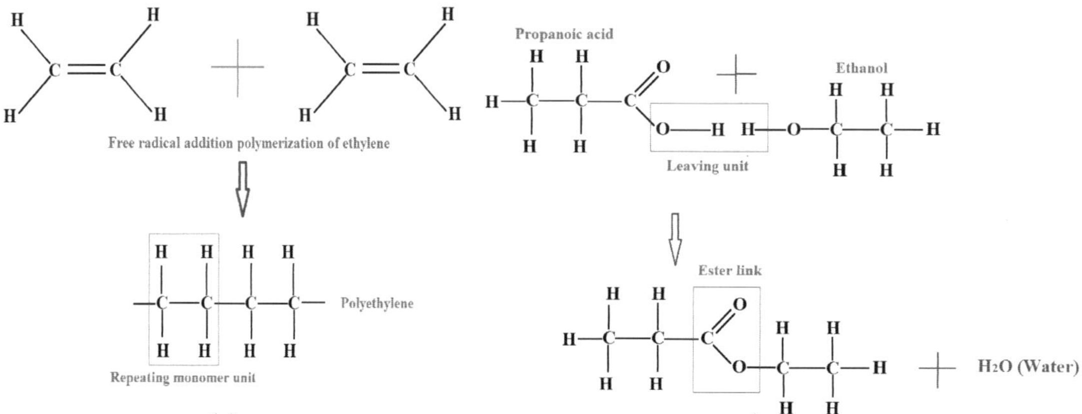

(a)

Figure 1. (a, b) Representation of addition and condensation polymerisation processes.

Finally, they are collected as plastic pellets and sent to factories for casting into different products. These two are the example of addition or chain growth polymerisation which mainly takes place by a free radical exchange while the condensation polymerisation is one which gives water as a by-product by the substitution of Hydroxyl (OH^-) radical. Polymerisation process is not limited to a small boundary. In fact, polymers were there from the creation of the universe and life and a wide range of natural polymerisation processes had taken place so far in our environment. In recent years, the increased demand for bio-based polymers has inspired chemists, microbiologists and process engineers to work on their synthesis in detail. Research has revealed that biopolymers can be produced from renewable resources or from organic substances like starch, cellulose, lactic acid, polysaccharide, polyamides etc which can be synthesized or degraded in the environment eco-friendly with the interaction of microorganisms[26,27]. Some basic types of biodegradable and non-biodegradable plastics are classified based on their source as shown in Table 1.

Table 1. Classification of some widely used plastics based on their source[28]

Plastic Type	Bio-based plastics (renewable resources)	Oil-based plastics (fossil resources)
Biodegradable plastics	poly(lactic acid) (PLA) polyhydroxyalkanoate (PHA) polysaccharide derivatives (low DS)* poly(amino acid)	poly(e-caprolactone) (PCL) poly(butylene succinate/adipate) (PBS/A) poly(butylene adipate-co-terephthalate) (PBA/T)
Non-biodegradable plastics	polysaccharide derivatives (high DS)* polyol–polyurethane bio-polyethylene (bio-PE) bio-poly(ethylene terephthalate) (bio-PET)	polyethylene (PE) polypropylene (PP) polystyrene (PS) poly(ethylene terephthalate) (PET)

*DS- Degree of substitution

3. Characterization of traditional polymers and innovations in this field

In polymer science, the role of synthetic polymers extracted from fossil fuels has been significant from its beginning due to its available production method and desirable material properties. For this reason, the modern polymers synthesized from renewable resources have not been able to replace their position completely till now. In our modern applications the fossil-based plastics i.e. PE (polyethylene), PP(polypropylene), PET(Polyethylene terephthalate), PS(polystyrene), PVC(polyvinyl chloride) etc. are still dominating as they are light, cost-effective, available in different forms and meet the design criteria properly. Among them PE and PP are crystalline at room temperature and can be used as a moulded object, especially the PET has glass transition temperature well above the room temperature which provides it higher hardness and dimensional stability. It is obvious that characterisation of any material requires proper testing methods to predict its physical and chemical properties which in fact, is the main requirement of engineering product design. There are several physical and chemical testing methods of polymers available today and many more are in the research stage[10,13]. Nowadays, as the use of recycled plastics has been emphasized worldwide as a part of sustainable development, its characterization becomes an important factor in design. As an example, the PET scrap is seen to be reusable after its separation from unwanted foreign particles, homogenization and heat treatment which gives it uniform mechanical properties. Its tensile properties, dynamic viscosity and thermo-oxidative stability are also studied for a detailed mechanical characterization[29]. PE and PP are widely abundant in domestic waste and there is a good possibility of recycling them if the waste is collected at the maximum extent by the municipality and other concerned authorities. There are many existing post-processing methods i.e. grinding, blending, homogenization, composite mixing, heat treatments which can make them reusable. It is found that the tensile properties improve if the plastic scrap mixture is finely milled to powder and also the possibility of alloy formation occurs among the polymers[30]. Researchers also suggest that the addition of waste rubber dust from the textile industry with the recycled PP blend elevates its bending, tensile and impact strength and damping characteristics[31]. Besides, PVC also contributes to plastic waste disposal significantly due to its heavy use in data, power transmission cable and fluid transmission pipes. A detailed mechanical characterization and composition analysis of the recycled PVC (r-PVC) reveals its reusability and processability after being separated from the waste. Due to some degradation in its mechanical properties over time researchers suggest the use of blended r-PVC with other recycled plastics or additives to improve its properties[32,33]. Research in the field of polymer science has been done extensively for the last few decades in the search of more effective polymers with superior mechanical and chemical properties and its demand is increasing day by day as it is benefitting the industries as well as our society a lot[34,35]. Polymer alloy can sometimes be a good replacement for the pure one i.e. PET/PP blend is more versatile than the individual polymers in many fields. It gives better stiffness, thermal and mechanical properties, favourable permeability etc[36]. Super polystyrene also is an example of modifying the existing polymer which is the copolymer of styrene and 1,1-Diphenylethylene. It exhibits long term service capability at elevated temperature and higher glass transition temperature due to the presence of bulky group DPE making it highly suitable in use as an insulating material, pump housing, fuse boxes, microwave dishes etc[2].

Recently a detailed characterization of Polyvinylpyrrolidone, a new generation polymer, is done which revealed its usefulness and versatility in biomedical applications as it is biocompatible, chemically stable, non-toxic, highly soluble in different organic solvents and capable to form complex with both hydrophilic and hydrophobic substances. It has also good electrical properties, adhesiveness and high solubility in water which makes it suitable for use in many beyond medical applications[1]. Menčík et al[37] carried a detailed study on the characterization of viscoelastic plastics viz. polymethyl methacrylate, epoxy resin and tooth enamel.They successfully performed nanoindentation test on the said polymers to estimate their material properties like elastic modulusand surface hardness. Besides, a separate creep test over a period of 3700 secondsis performed using a Berkovich indenter for evaluating the creep data.The average value of hardness and elastic modulus is found to be 0.205GPa and 3.66GPa respectively. Nowadays the study and characterization of polymer and polymer composites are being done extensively in various research and development sectors. Its application is found to be useful in almost every industry like renewable energy, aircraft, automobile and electric equipment, biomedical application and even in the modern drug synthesis process[38]. Carbon nanotube(CNT) is one of the most promising materials for future innovation because of its many exceptional properties which include excellent mechanical, thermal and electrical properties. Requirement of high strength material with lighter weight in various engineering industries like automobile, aerospace, biomedical, defence etc. has resulted in a rigorous study on CNT based composite and nanocomposite materials[39,40]. Its outer surface is hydrophobic in nature and chemically inert which makes it suitable in many applications like, CNT-Polymer based film electrode can be successfully utilized to detect glucose in an aqueous system[41,42]. Not only that, it is also found experimentally that anticancer drug-loaded multiwalled CNT has the ability to target and destroy the cancer cells more effectively[43]. Characterization of CNT reinforced epoxy composite is done by Dan et al using the finite element simulation to evaluate its mechanical properties like stress-strain relationship and tensile strength and the same has been validated by experimental data[44]. Advanced research suggests that the brake liner made of CNT reinforced epoxy and phenol-formaldehyde composite exhibits superior friction and wear characteristics as compared to the traditional asbestos liner[45]. Gupta et al conducted mechanical characterization on carbon fibre reinforced polycarbonate polymer composite using different techniques like universal testing machine, three point bending, compressive loading etc. They successfully found out the effect of carbon fibre on the tensile, compressive, flexural properties, toughness and ductility of the 3D printed polymer composite[11].Due to the increasing health concern in our modern era synthesis and characterization of antimicrobial plastics and polymers are being done widely especially for biomedical applications[46]. Researchers found that the existing synthetic polymers can be made antibacterial by applying a layer of antimicrobial peptide(APM) on their surfaces. As it is expensive and complex in production, the synthesis of other cost-effective polymers is also being studied[47,48].

In recent years, the use of wood plastic composite (WPC) in automobile, aerospace and building structure components has been incorporated significantly due to its low production cost, desired strength and sustainability as compared to glass fibre composites. WPC can be manufactured by the recycled plastic blend separated from construction and demolition waste[49]. It is found from the mechanical characterization that it is less strong than the reference material made with low-density PE(LDPE) but shows more stiffness than the reference. A similar kind of study and characterization is performed for the WPC where it is manufactured by recycled PE/PP blend reinforced with external fibre material which is a more effective and advanced method to meet the required strength criteria of design[50]. Modern high-performance engineering plastics can be a better option for the future because of their many favourable properties compared to the traditional ones but their synthesis cost is too high. Aromatic poly imino ether(PIE) is one of them which shows a promising future because of its low synthesis cost, high thermal stability and higher decomposition temperature[51]. Semicrystalline polymers also playing an important role in materials synthesis because its exceptional thermomechanical properties which are desirable in many applications. So its characterization also becomes necessary for further product development. Voyiadjiset al[52] conducted nanoindentation test on poly(ether-ether-ketone)(PEEK) to study its localized nanomechanical properties. Several other studies are also going on in this field like avoidance of density variation caused by the chain termination on the surface of semicrystalline polymers. Chain termination and dangling of fibres at the end surfaces can reduce the density by 17% of the average density[53]. Research done in this field is quite vast and multipurpose. Newer methods and techniques are yet to come for the study and characterization of plastics and polymers which will certainly enrich science and technology and consequently the human civilization.

4. Biopolymer, its characterization and sustainability

The journey of plastics and polymers started a long time ago in the second half of the 19th century when humans developed celluloid from cellulose fibre which is a natural polymer extracted from wood or straw[54]. From then a tremendous development has been done in this field for satisfying the need of our civilisation which resulted in a world full of plastics and polymers. Consequently, the dependence on artificial plastics has increased which has been affecting our environment and ecosystem for a long time. Microplastic is one of the most dangerous plastic derivatives found mainly in soil and water. They are tiny plastic particles produced by the gradual decay of plastic molecules from various sources viz tube, tire, waste container, textile, packaging industries etc[55]. Hence, the study, synthesis and characterization of alternate polymers i.e. the biopolymers has been the epicentre of research for the last few decades[56–58]. Biopolymers are those types of polymers that are produced from renewable and natural resources while biodegradable polymers are another type of biopolymer that dissolve or degrade easily in the environment producing non-toxic compounds or gases[26,59]. Bioplastics/polymers may be biobased(from biomass), biodegradable or can possess both the characters. So a fully biobased plastic may not be biodegradable and a fully oil-based(fossil) plastic may be completely biodegradable. Degradation of these polymers is also highly dependent on the sink it finally ends up in like seawater, marine water, soil etc. Haider et al[56] gave a detailed review on the biodegradation characterization of polymers and suggested the need for practical experimental study apart from laboratory experiments so that all the random factors in the environment can be taken into account. Several microorganisms like aerobes, anaerobes, photosynthetic bacteria, archaebacteria etc are responsible for the biodegradation of bioplastics which are mainly abundant in soil and compost[40]. Hydrolysis is one of the most important biodegradation processes which is capable of decomposing polyesters, polyethers, cellulose derivatives, starch etc[58]. It is already discussed that the demand for biopolymers has been increasing gradually and consequently the modification of existing biopolymers are also being done by composite blending, fibre reinforcement, addition of external functional groups etc in the existing hydrocarbon chain. These processes are seen to be effective in achieving superior properties as compared to the existing ones. When polyvinyl alcohol(PVA) is blended with corn starch and lignocellulosic fibre its thermal stability, water permeability and biodegradation rate improves which are suitable for specific application based product design[60]. Research suggests that potato and yam can be good raw materials for starch-based polymer synthesis due to their high starch content and excellent biodegradability in soil. Besides the resulting polymer shows moderate thermal stability which is confirmed by thermogravimetric analysis(TGA) and a workable tensile strength(0.6-1.9MPa) suitable for low strength applications[61,62].

Composite blending is a widely accepted modern polymerisation process and blending of synthetic polymers with biopolymers is a part of that. Its objective is to find the resulting characteristics of the composite polymer and whether it is more beneficial than the virgin ones or not. As an example, the mixture of low density polyethylene(LDPE) with corn starch has been studied which reveals that it exhibit reduced melt flow index(MFI) and increased elastic modulus[63]. Among biopolymers, the edible film is showing an increasing demand in the food processing industry as they are digestible and possess no threat to the human body. An edible film is a thin layer of edible polymer that can be placed over or between food components and plays a vital role in food preservation and distribution[64,65]. There are many processes of production of edible film and still, research is continuing for finding a more efficient and cost-effective synthesis process. Edible film produced from grass pea flour in the presence of transglutaminase enzyme has good potential because of its desired properties. It is found experimentally that the presence of microbial transglutaminase makes it mechanically resistant and gives a better digestive property. Besides, the scanning electron microscopy(SEM) study confirms its homogeneous structure obtained by enzyme treatment[66]. It's a matter of deep concern that the earth's mineral oil source is limited and it will be completely depleted in a few decades. So, the human civilisation cannot be fully dependent on traditional synthetic polymers for a long time. Researchers are trying to find alternate sustainable resources for its replacement and polyhydroxyalkanoates(PHAs) can be used quite satisfactorily for the same purpose. They are fully biodegradable, immunologically inert, highly suitable for biomedical applications and exhibits many desirable properties like mineral-based polymers[67]. PHAs are one type of polyesters that are produced from microbial fermentation. Because of its low production yield, the overall cost of PHAs increases causing a big barrier to its large scale use in the polymer industry[68]. Besides, proteins, lipids, fibres and polysaccharides are also widely available biopolymers obtained from plant and animal sources[69]. Efforts have been made to utilize these polymers in product synthesis by suitable modification and homogenization techniques. Among them, proteins have many available sources based on plants and animals like oilseeds, eggs, milk etc and most importantly, the wastes and surpluses from the food processing industry contribute the most. A detailed characterization of protein-based polymers reveals that the protein collected from these sources can be made usable by suitable mixing and other thermomechanical treatments. Experiments show that the addition of plasticizer, degree of blending and moulding process improve the value of young modulus, tensile strength, water uptake capacity and other properties of the polymer significantly[70]. The diverse properties of polymers and their composites along with the help of additive manufacturing have brought phenomenal changes in the biomedical field. They are capable of providing required strength, corrosion resistance and antimicrobial properties which make them the perfect raw material to produce customized biodevices, scaffolds for tissue culture, artificial bone replacements, heart valves, drug carriers and so on[71–73].The wide availability, easy processibility and many other impressive properties of plant-based cellulose fibre is also a good replacement for the synthetic polymer. It possesses higher mechanical strength as compared to the protein-based polymers and even, the natural fibre reinforced composite polymer has the potential to compete with the existing metals and ceramics[74].

5. Conclusion

The modern civilisation which we are living in is indeed nothing but the result of incessant innovation and advancement in the field of science and technology. The prime objective of any scientific research is to understand the way the universe works and to utilize our available resources for the welfare of mankind in the most efficient manner. It is discussed many times in this article that the role of plastics and polymers in our life is huge. From our body to every corner of the outer world, there is the existence of polymers that we often do not realize. Due to its immense importance, research in this field is still going on on a large scale. In most cases, plastics are proven to be more flexible, user-friendly, lightweight and have a higher strength to weight ratio as compared to metals. Not only that, plastic products are aesthetically and ergonomically far superior to traditional metals. The inventions of newer synthesis processes resulted in the creation of innovative polymers with excellent properties which had never existed before. Due to its increasing demand, there is tough competition among the manufacturers and day by day newer plastic and polymer products are being launched in the market with better performance and reliability. Characterization helps to know the actual nature and characteristics of the polymer which in fact, determines the capability of that material to perform in a specific application. Detailed knowledge about a particular material also improves the design methodologies and safety of the designed product. However, the characterization and testing procedures of polymers are not standardized, unlike the metals because of their widely available physical and chemical states. In the past few decades, due to the rising consciousness of the sustainable exploitation of natural resources, the emphasis on biopolymers have been increased significantly. Undoubtedly, the biopolymer is going to replace the traditional mineral oil-based polymer soon and it is the only way to make a habitable planet for our future generations. The advancement in the field of biopolymers done so far is promising and the future scope of study in this field is widening day by day as the whole world is trying to move towards a greener and cleaner environment with minimal wastage. The main advantages of biopolymers are that they are made from renewable resources and most of them can be made fully biodegradable in a suitable medium(sink). So finally, it can be concluded that the characterisation of plastics and polymers is one of the prime requirements for product design, development and quality assurance. More innovations are yet to occur in this field which will hopefully be able to serve the need of human civilisation in a better way.

References

[1] Teodorescu M and Bercea M 2015 Poly(vinylpyrrolidone) – A Versatile Polymer for Biomedical and Beyond Medical Applications *Polym. Plast. Technol. Eng.* **54** 923–43
[2] Gausepohl H, Oepen S, Knoll K, Schneider M, McKee G and Loth W 2000 Super-polystyrene: a new class of engineering plastics with versatile properties *Des. Monomers Polym.* **3** 299–315
[3] Olatunji O 2016 Classification of Natural Polymers *Natural Polymers* (Cham: Springer International Publishing) 1–17
[4] Rusu L C, Ardelean L C, Jitariu A A, Miu C A and Streian C G 2020 An Insight into the Structural Diversity and Clinical Applicability of Polyurethanes in Biomedicine *Polymers (Basel)* **12** 1197
[5] Siddique R, Khatib J and Kaur I 2008 Use of recycled plastic in concrete: A review *Waste Manag.* **28** 1835–52
[6] Sharma P V 2020 Polymers and Microplastics: Implications on Our Environment and

Sustainability *Emerging Technologies, Environment and Research for Sustainable Aquaculture* (IntechOpen)
[7] Folino A, Karageorgiou A, Calabrò P S and Komilis D 2020 Biodegradation of Wasted Bioplastics in Natural and Industrial Environments: A Review *Sustainability* **12** 6030
[8] Wanasinghe D, Aslani F, Ma G and Habibi D 2020 Review of Polymer Composites with Diverse Nanofillers for Electromagnetic Interference Shielding *Nanomaterials* **10** 541
[9] Yu Y, Yuk H, Parada G A, Wu Y, Liu X, Nabzdyk C S, Youcef-Toumi K, Zang J and Zhao X 2019 Multifunctional "Hydrogel Skins" on Diverse Polymers with Arbitrary Shapes *Adv. Mater.* **31** 1807101
[10] Brown R 1999 *Handbook of Polymer Testing: Physical Methods* (CRC press)
[11] Gupta A, Fidan I, Hasanov S and Nasirov A 2020 Processing, mechanical characterization, and micrography of 3D-printed short carbon fiber reinforced polycarbonate polymer matrix composite material *Int. J. Adv. Manuf. Technol.* **107** 3185–205
[12] Brandtner-Hafner M H 2020 Fracture Mechanical Characterization of Adhesively Bonded Wood-Ceramic Interfaces for Mode I Loading *J. Test. Eval.* **48** 20180256
[13] Lampman S 2003 *Characterization and Failure Analysis of Plastics* ISBN: 978-0-87170-789-5
[14] Zhang Y, Zhang Q, Hou D and Zhang J 2020 Tuning interfacial structure and mechanical properties of graphene oxide sheets/polymer nanocomposites by controlling functional groups of polymer *Appl. Surf. Sci.***504** 144152
[15] Ehlert G J, Lin Y and Sodano H A 2011 Carboxyl functionalization of carbon fibers through a grafting reaction that preserves fiber tensile strength *Carbon N. Y.* **49** 4246–55
[16] Nilagiri Balasubramanian K B and Ramesh T 2018 Role, effect, and influences of micro and nano-fillers on various properties of polymer matrix composites for microelectronics: A review *Polym. Adv. Technol.* **29** 1568–85
[17] Liu J, Shen J, Gao Y, Zhou H, Wu Y and Zhang L 2014 Detailed simulation of the role of functionalized polymer chains on the structural, dynamic and mechanical properties of polymer nanocomposites *Soft Matter***10** 8971–84
[18] Yan P, Zhao W, Zhang B, Jiang L, Petcher S, Smith J A, Parker D J, Cooper A I, Lei J and Hasell T 2020 Inverse Vulcanized Polymers with Shape Memory, Enhanced Mechanical Properties, and Vitrimer Behavior *Angew. Chemie Int. Ed.* **59** 13371–8
[19] Jian W and Lau D 2020 Understanding the effect of functionalization in CNT-epoxy nanocomposite from molecular level *Compos. Sci. Technol.* **191** 108076
[20] Bertrand-Lambotte P, Loubet J ., Verpy C and Pavan S 2001 Nano-indentation, scratching and atomic force microscopy for evaluating the mar resistance of automotive clearcoats: study of the ductile scratches *Thin Solid Films* **398–399** 306–12
[21] Wong M, Lim G ., Moyse A, Reddy J . and Sue H-J 2004 A new test methodology for evaluating scratch resistance of polymers *Wear* **256** 1214–27
[22] Charitidis C, Gioti M, Logothetidis S, Kassavetis S, Laskarakis A and Varsano I 2004 Comparison of the nanomechanical and nanoscratch performance of antiscratch layers on organic lenses *Surf. Coatings Technol.***180–181** 357–61
[23] Pelletier H, Mendibide C and Riche A 2008 Mechanical characterization of polymeric films using depth-sensing instrument: Correlation between viscoelastic-plastic properties and scratch resistance *Prog. Org. Coatings* **62** 162–78
[24] Baranwal R K, Hassan T, Agarwal G, Sarkar S and Majumdar G 2019 Optimization of the stress discontinuity value at the interface of a cylindrical stainless steel substrate and electroless Ni-P coating *Mater. Res. Express***6** 116421
[25] Flory P J 1946 Fundamental Principles of Condensation Polymerization. *Chem. Rev.***39** 137–97
[26] Breuer U 2009 Book Reviews: Microbial Production of Biopolymers and Polymer Precursors: Applications and Perspectives. Edited by Bernd H. A. Rehm *CLEAN - Soil, Air, Water* **37** 414
[27] Mohan S, Oluwafemi O S, Kalarikkal N, Thomas S and Songca S P 2016 Biopolymers – Application in Nanoscience and Nanotechnology *Recent Advances in Biopolymers* (InTech)
[28] Iwata T 2015 Biodegradable and Bio-Based Polymers: Future Prospects of Eco-Friendly Plastics *Angew. Chemie Int. Ed.* **54** 3210–5

[29] Pawlak A, Pluta M, Morawiec J, Galeski A and Pracella M 2000 Characterization of scrap poly(ethylene terephthalate) *Eur. Polym. J.* **36** 1875–84
[30] Jonna S and Lyons J 2005 Processing and properties of cryogenically milled post-consumer mixed plastic waste *Polym. Test.* **24** 428–34
[31] Jose J, Satapathy S, Nag A and Nando G B 2007 Modification of Waste Polypropylene with Waste Rubber Dust from Textile Cot Industry and its Characterization *Process Saf. Environ. Prot.* **85** 318–26
[32] Suresh S S, Mohanty S and Nayak S K 2017 Composition analysis and characterization of waste polyvinyl chloride (PVC) recovered from data cables *Waste Manag.* **60** 100–11
[33] Güney A, Özdilek C, Kangal M O and Burat F 2015 Flotation characterization of PET and PVC in the presence of different plasticizers *Sep. Purif. Technol.* **151** 47–56
[34] Hotta A, Cochran E, Ruokolainen J, Khanna V, Fredrickson G H, Kramer E J, Shin Y-W, Shimizu F, Cherian A E, Hustad P D, Rose J M and Coates G W 2006 Semicrystalline thermoplastic elastomeric polyolefins: Advances through catalyst development and macromolecular design *Proc. Natl. Acad. Sci.* **103** 15327–32
[35] Andrady A L and Neal M A 2009 Applications and societal benefits of plastics *Philos. Trans. R. Soc. B Biol. Sci.* **364** 1977–84
[36] Papadopoulou C P and Kalfoglou N K 2000 Comparison of compatibilizer effectiveness for PET/PP blends: their mechanical, thermal and morphology characterization *Polymer (Guildf).* **41** 2543–55
[37] Menčík J, He L H and Němeček J 2011 Characterization of viscoelastic-plastic properties of solid polymers by instrumented indentation *Polym. Test.* **30** 101–9
[38] Gupta B, Ghosh A K, Suzuki A and Rattan S 2018 *Advances in Polymer Sciences and Technology* (Singapore: Springer Singapore)
[39] Ali F, Ishfaq N, Said A, Nawaz Z, Ali Z, Ali N, Afzal A and Bilal M 2021 Fabrication, characterization, morphological and thermal investigations of functionalized multi-walled carbon nanotubes reinforced epoxy nanocomposites *Prog. Org. Coatings* **150** 105962
[40] Soni S K, Thomas B and Kar V R 2020 A Comprehensive Review on CNTs and CNT-Reinforced Composites: Syntheses, Characteristics and Applications *Mater. Today Commun.* **25** 101546
[41] Gopal N, Shukla P and Sahney R 2018 Development of CNT–Polymer Film-Based Electrode for the Detection of Glucose *Advances in Polymer Sciences and Technology* Springer 177–86
[42] Gergeroglu H, Yildirim S and Ebeoglugil M F 2020 Nano-carbons in biosensor applications: an overview of carbon nanotubes (CNTs) and fullerenes (C60) *SN Appl. Sci.* **2** 603
[43] Mangla B, Patel K S, Kumar P and Kohli K 2018 Anticancer Drug-Loaded Folate-Conjugated Multiwalled Carbon Nanotubes 197–210
[44] Li D, LiDing, Liu Z, Li Q, Guo K and Cao H 2020 Simulation Analysis on Mechanical Property Characterization of Carbon Nanotubes Reinforced Epoxy Composites *Comput. Model. Eng. Sci.* **125** 145–71
[45] Sridhar B S, George R and Shashikala A R 2021 Synthesis and characterization of carbon nanotube reinforced polymer brake liner material *Mater. Today Proc.*
[46] Anjum S, Gupta A, Kumari S and Gupta B 2020 Preparation and biological characterization of plasma functionalized poly(ethylene terephthalate) antimicrobial sutures *Int. J. Polym. Mater. Polym. Biomater.* **69** 1034–42
[47] Majhi S, Arora A and Mishra A 2018 Antibacterial Activity of Antimicrobial Peptide (AMP) Grafted Polystyrene Surface pp 39–46
[48] Arora A, Zheng W, Liang H and Mishra A 2018 Synthesis of Lysine Mimicking Membrane Active Antimicrobial Polymers 29–37
[49] Turku I, Keskisaari A, Kärki T, Puurtinen A and Marttila P 2017 Characterization of wood plastic composites manufactured from recycled plastic blends *Compos. Struct.* **161** 469–76
[50] Taufiq M J, Mansor M R and Mustafa Z 2018 Characterisation of wood plastic composite manufactured from kenaf fibre reinforced recycled-unused plastic blend *Compos. Struct.* **189** 510–5

[51] Al-Hussaini A S 2014 Synthesis and characterization of new thermally stable polymers as new high-performance engineering plastics *High Perform. Polym.* **26** 166–74
[52] Voyiadjis G Z, Samadi-Dooki A and Malekmotiei L 2017 Nanoindentation of high performance semicrystalline polymers: A case study on PEEK *Polym. Test.* **61** 57–64
[53] Fritzsching K J, Mao K and Schmidt-Rohr K 2017 Avoidance of Density Anomalies as a Structural Principle for Semicrystalline Polymers: The Importance of Chain Ends and Chain Tilt *Macromolecules* **50** 1521–40
[54] Trescott M M 1984 Pioneer Plastic: The Making and Selling of Celluloid. By Robert Friedel. (Madison: University of Wisconsin Press) *Bus. Hist. Rev.* **58** 284–5
[55] Rochman C M, Brookson C, Bikker J, Djuric N, Earn A, Bucci K, Athey S, Huntington A, McIlwraith H, Munno K, De Frond H, Kolomijeca A, Erdle L, Grbic J, Bayoumi M, Borrelle S B, Wu T, Santoro S, Werbowski L M, Zhu X, Giles R K, Hamilton B M, Thaysen C, Kaura A, Klasios N, Ead L, Kim J, Sherlock C, Ho A and Hung C 2019 Rethinking microplastics as a diverse contaminant suite *Environ. Toxicol. Chem.* **38** 703–11
[56] Haider T P, Völker C, Kramm J, Landfester K and Wurm F R 2019 Plastics of the Future? The Impact of Biodegradable Polymers on the Environment and on Society *Angew. Chemie Int. Ed.* **58** 50–62
[57] He D and Luo Y 2020 *Microplastics in Terrestrial Environments* 95 (Cham: Springer International Publishing)
[58] Amass W, Amass A and Tighe B 1998 A review of biodegradable polymers: uses, current developments in the synthesis and characterization of biodegradable polyesters, blends of biodegradable polymers and recent advances in biodegradation studies *Polym. Int.* **47** 89–144
[59] Bhagabati P 2020 Biopolymers and biocomposites-mediated sustainable high-performance materials for automobile applications *Sustainable Nanocellulose and Nanohydrogels from Natural Sources* (Elsevier) 197–216
[60] Imam S H, Cinelli P, Gordon S H and Chiellini E 2005 Characterization of Biodegradable Composite Films Prepared from Blends of Poly(Vinyl Alcohol), Cornstarch, and Lignocellulosic Fiber *J. Polym. Environ.* **13** 47–55
[61] Ismail N A, Mohd Tahir S, Norihan Y, Abdul Wahid M F, Khairuddin N E, Hashim I, Rosli N and Abdullah M A 2016 Synthesis and Characterization of Biodegradable Starch-Based Bioplastics *Mater. Sci. Forum* **846** 673–8
[62] Emadian S M, Onay T T and Demirel B 2017 Biodegradation of bioplastics in natural environments *Waste Manag.* **59** 526–36
[63] Rosa D S, Grillo D, Bardi M A G, Calil M R, Guedes C G F, Ramires E C and Frollini E 2009 Mechanical, thermal and morphological characterization of polypropylene/biodegradable polyester blends with additives *Polym. Test.* **28** 836–42
[64] Porta R, Di Pierro P, Rossi-Marquez G, Mariniello L, Kadivar M and Arabestani A 2015 Microstructure and properties of bitter vetch (Vicia ervilia) protein films reinforced by microbial transglutaminase *Food Hydrocoll.* **50** 102–7
[65] Falguera V, Quintero J P, Jiménez A, Muñoz J A and Ibarz A 2011 Edible films and coatings: Structures, active functions and trends in their use *Trends Food Sci. Technol.* **22** 292–303
[66] Giosafatto C, Al-Asmar A, D'Angelo A, Roviello V, Esposito M and Mariniello L 2018 Preparation and Characterization of Bioplastics from Grass Pea Flour Cast in the Presence of Microbial Transglutaminase *Coatings* **8** 435
[67] Keshavarz T and Roy I 2010 Polyhydroxyalkanoates: bioplastics with a green agenda *Curr. Opin. Microbiol.* **13** 321–6
[68] Amache R, Sukan A, Safari M, Roy I and Keshavarz T 2013 Advances in PHAs production *Chem. Eng. Trans.* **32** 931–6
[69] Avérous L 2004 Biodegradable Multiphase Systems Based on Plasticized Starch: A Review *J. Macromol. Sci. Part C Polym. Rev.* **44** 231–74
[70] Felix M, Perez-Puyana V, Romero A and Guerrero A 2017 Production and Characterization of Bioplastics Obtained by Injection Moulding of Various Protein Systems *J. Polym. Environ.* **25**

91–100
[71] Pekkanen A M, Mondschein R J, Williams C B and Long T E 2017 3D Printing Polymers with Supramolecular Functionality for Biological Applications *Biomacromolecules* **18** 2669–87
[72] Seppälä J, van Bochove B and Lendlein A 2020 Developing Advanced Functional Polymers for Biomedical Applications *Biomacromolecules* **21** 273–5
[73] Puertas-Bartolomé M, Mora-Boza A and García-Fernández L 2021 Emerging Biofabrication Techniques: A Review on Natural Polymers for Biomedical Applications *Polymers (Basel).* **13** 1209
[74] Thakur V K and Thakur M K 2014 Processing and characterization of natural cellulose fibers/thermoset polymer composites *Carbohydr. Polym.* **109** 102–17

Study of the Proton Irradiation Effects in n-FZ DSSD for the Si Tracker at R3B Experiment

Puspita Chatterjee[1], Nitu Saini[1], Shilpa Patyal[1], Balwinder Kaur[1], Thresia Michael[1], Sonia Raheja[1], Ajay K. Srivastava[1*]
[1]Department of Physics, University Institute of Sciences, Chandigarh University, 140413, Punjab, India.

*Corresponding author e-mail: kumar.uis@cumail.in

Abstract. The R3B collaboration aims to assemble an experimental setup with high resolution and efficiency to perform kinematically complete measurements of reactions with high-energy RIBs (radioactive ion beams). In the R3B experimental setup, the silicon tracker is positioned closest to the target region and can provide high-resolution position measurements of light-charged particles like protons. The three layers of the silicon tracker are constructed with a total of 30 Si double-sidedstrip detectors (DSSD). In this paper, as an option of 23 MeV proton irradiated n-type Float-Zone(Fz)Si double-sidedstrip detector (DSSD) havebeen considered, and by taking the two-trapproton irradiation damage model, the bulk damage effect and the macroscopic performance of the Sidouble-sidedstrip detector (DSSD) have been discussed. The detector is irradiated with three values of proton fluences (equivalent to 1Mev neutron fluence): 2×10^{14},5×10^{14} and 8×10^{14} cm^{-2}. By using the Shockley-Read-Hall recombination (SRH) formulation, the full depletion voltage and leakage current have been measured as a function of the irradiation dose, finally the device and process parameters and specifications for the Si double-sidedstrip detector (DSSD) used in the R3B silicon tracker experimenthavebeen proposed.

Keywords: Double sided strip detector, Float zone, Full depletion voltage, Leakage current, Bulk damage.

1. Introduction

FAIR (Facility for Antiproton and Ion Research) is an accelerator facility at GSI, Darmstadt, Germany [1-2]. FAIR can deliver ion beams of all elements up to Uranium. At FAIR we get high energy and high intensity of primary ion beams because of the addition of heavy-ion synchrotron(SIS 100/300) with previous GSI facility. After exiting from SIS 100/300 synchrotron, the primary ion beams will incident on a thin isotopic production target (positioned after the double ring synchrotron) can generate secondary rare isotopic ion beams up to 1.5AGeV [3],which will be delivered to R3B (Reactions with Radioactive Relativistic Beams) experimental setup at FAIR.In R3B, the high-energy rare isotopic ion beams /radioactive ion beams (RIBs) incident on a fixed target and will emit different particles like protons, neutrons, gamma rays, heavy ions, etc [4-6]. In the R3B experimental setup, the silicon tracker is positioned closest to the target region within a vacuum vessel [7]. When the Ion beams willpass through the three layers of the silicon tracker, it can provide high-resolution position measurement of light charged particles like protons [1]. The three layers of the silicon tracker are constructed with a total of 30 Si double-sided strip detectors (DSSD), 6 Si DSSD are required for the inner layer and 12 Si DSSD are required for eachouter layerof the Si tracker as shown in figure 1 [8-12].

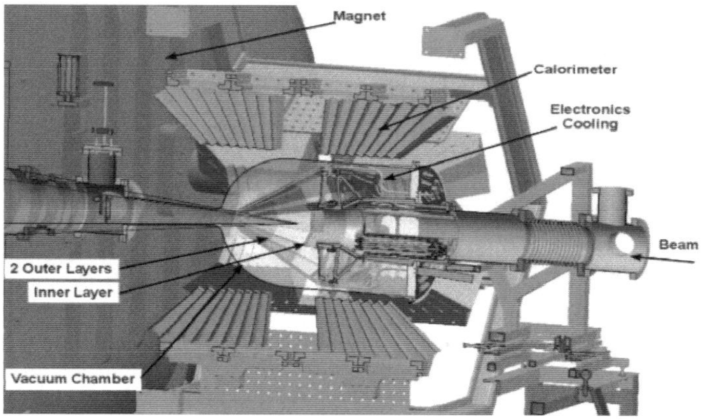

Figure 1. The position of the R3B silicon tracker in the R3B experimental setup [3].

The choice of the material for the DSSD is n-Fz (float-zone) silicon for the Si tracker[12]. The electrical performances of the detectors were needed tobe testedfor the designing, development, and optimization of the detectors for the high energy physics (HEP) experiments[13-14]. Within NUSTAR Collaboration for R3B silicon tracker, the n-FzDSSD is exposed by proton irradiation[1]. The macroscopic and microscopic performance of the detector can be measured using current-voltage (I/V), capacitance-voltage (C/V), Thermally Stimulated Current (TSC), Deep Level Transient Spectroscopy (DLTS) and Transient Current Technique (TCT). The induced radiation causes some damages, which overall affects the electrical performance of the detector. In order to study the radiation damages impact on the irradiated Si detectorsseveral research groups have proposed the n-Fz protonirradiation damage models [12].
In this paper, by using SRH (Shockley Read-Hall) calculations the bulk damage effect and the macroscopic performance of proton irradiated n-FzSi DSSD have been discussed for three different values of proton fluences. For the SRH calculations, the experimentally verified two trap n-Fz proton radiation damage model[12]have beenconsidered; this model is useful to explain the macroscopic performance of the n-Fz Si DSSD in the low irradiation environment. Finally design idea, process and device parameter for DC coupled Si DSSD and AC coupled Si DSSD have been proposed.

2. Double-sided silicon strip detector for R3B Silicon tracker

The Si double-sided strip detector is constructed by segmenting the electrodes on both sides of the bulk. Both sided strips are angled orthogonally with each other (see figure 2). A Si DSSD allows both types of charge carriers to be collected at the segmented electrode [9] and can provide two-dimensional position measurements of the incident particle. After the incidence of radiation electron-hole pair will generate in the active volume of the detector and will move towards the segmented electrodes under the influence of an external electric field and these charges can be collected and processed by the readout electronics [9].

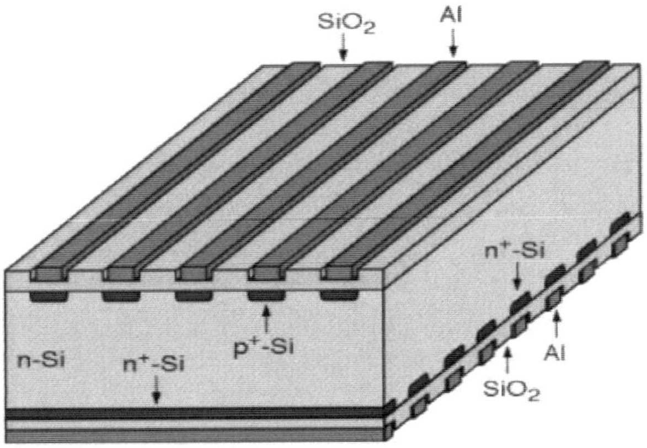

Figure 2. Cross-sectional view of a 3-D Si double-sided strip detector [10].

As we know from the basic principle of the detector, the incident radiation ionizes the charge particlein bulk as well as the silicon oxide layer of the detector.However, for the carriers in the insulating layer, some of them cannot escape from this layer and these remaining charges forms the fixed oxide charges. These positive charges attract electrons and form an accumulation layer of negative charges in the n-type bulk silicon. This electron accumulation layercan cause the shortening of the adjacent n^+strips. Hence, a medium doped p-type region was introduced between two n^+ strips to disrupt the electron accumulation layer to isolate the neighbouring strips [9]. This technique is called p-spray.An alternative technique to p-spray that is used for isolation is p-stop. A p-stop uses a highly doped p^+ implant positioned between adjacent n^+ strips to achieve strip isolation.

2.1. Bulk damage model

The incident radiation introduces some damage into the detector which results in degradation of detector performance. Bulk damage was incorporated by introducing some trap levels inside the forbidden energy levels having energy E_t [11]. Trap levels are characterized by some parameters like introduction rate, energy and capture cross-section. By the introduction of this trap level some macroscopic properties of the detector change likefull depletion voltage will increase, the leakage currentwill increase and the charge collection efficiency will decrease [11].

Here in this paper, as an option of 23MeV proton irradiated n-type float-zone Si double-sided strip detector have been considered. Typical linear dimensions are:55610μm × 29460μm × 200/250/ 300μm. The bulk damage effect induced by incident radiation as a function of three proton fluences2×10^{14},5×10^{14}, and $8\times10^{14}cm^{-2}$ (equivalent to 1MeV neutron fluence) are discussed here. Hence, an approximate bulk damage model incorporating just two trap levels, one acceptor and one donor trap, is devised for proton irradiated detectors are shown in table 1 [12]. By using SRH calculations the resulting bulk damage model can account for various macroscopic measured properties of proton irradiated silicon detectors i.e, increase of leakage current, increase of full depletion voltage and the decrease of charge collection efficiency with an increase of proton fluences.

Table1. Two deep trap proton irradiation damage model for n-Fz Si DSSD. Providing energy level distance from the conduction band edge E_c and valence band edge E_v. The fluence F is given in n_{eq} cm^{-2} [12].

Type of defect	Level (eV)	$\sigma_e(cm^2)$	$\sigma_h(cm^2)$	Concentration(cm^{-3})
Deep acceptor	E_C-0.525	1.0×10^{-14}	1.0×10^{-14}	$1.189cm^{-1} \times F + 0.65 \times 10^{14}cm^{-3}$
Deep donor	E_V+0.48	1.0×10^{-14}	1.0×10^{-14}	$5.589cm^{-1} \times F - 3.96 \times 10^{14}cm^{-3}$

This is an experimentally verified "two trap proton irradiation damage model" for n-Fz Si DSSD consisting one acceptor trap and one donor trap defect. The model is valid only for fluences higher than 10^{14} n_{eq} cm^{-2}[12].

2.2. Detector model and Shockley-Read-Hall theoretical calculations

In this section the proposed device model, specifications and the device and process parameters have been mentioned andby using the two-trap proton irradiation damage model the theoretical SRH calculations have been explained.

The size of Si DSSD that we have used in the calculations is $55610\mu m \times 29460\mu m \times 200/250/300\mu m$ (see figure 3) and the device and process parameters of the detector have been mentioned in table 2 and specifications have been mentioned in table 3. The Si DSSD that we are proposing is a DC-coupled and AC coupled sensor that consists of an n-strip, an n bulk, and a p-strip. Though the p-strips on the backside are placed orthogonal to the n-strip on the frontside, the direction of the readout strips is the same. This requires a double metal structure. The first metal and the second metal are electrically connected by a via-hole process [9]. The p-strips and the n bulk region just act as a p-n junction diode. The diode is operated in reverse bias condition. Figure 3(a) describes the two-dimensional view of a DC-coupled n-FzSi DSSD irradiated with proton irradiation and figure 3(b) describes the two-dimensional view of an AC coupled n-FzSi DSSD irradiated with proton irradiation. In AC coupled Si DSSD for electrical insulation, a very thick oxidation layer is deposited between two metals, and the electrical connection between the two metals is established by the via-hole process. There has a nitride layer after the oxide layer to prevent physical damages [9].

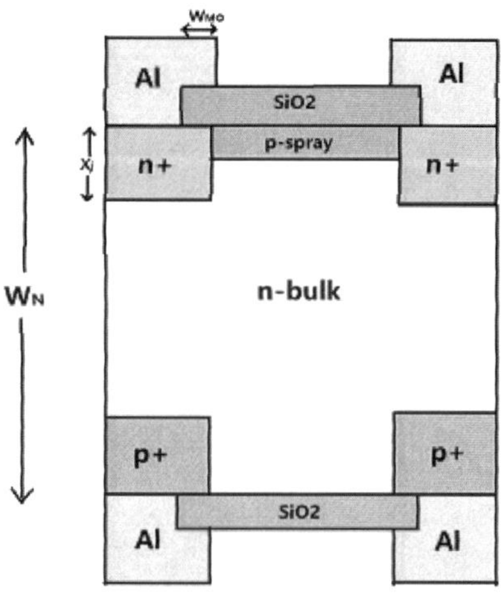

(a)

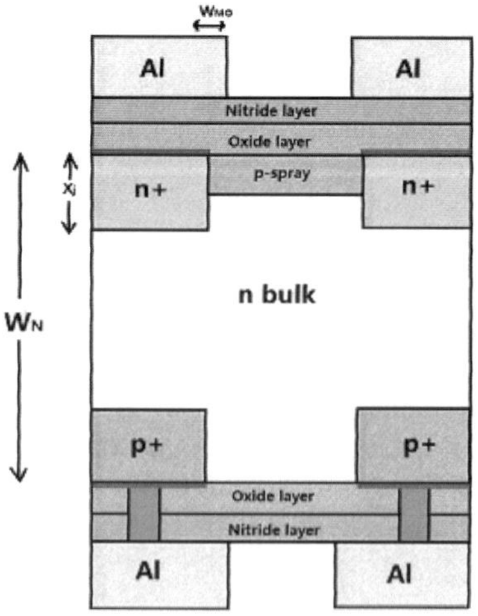

(b)

Figure 3. (a)Schematic of a two-dimensional view of single metal layer DC-coupled Si DSSD, (b) Schematic of a two-dimensional view of double metal layer AC-coupled Si DSSD.

Table 2. Device and process parameters of n-Fz Si DSSD for proton irradiation.

Sl. No	Physical parameters	Values
1	Doping concentration (N_D)	$1 \times 10^{12}, 3 \times 10^{12}, 5 \times 10^{12} cm^{-3}$
2	Device depth (W_N)	200μm, 250μm, 300μm
3	Junction depth (X_J)	2.5μm
4	Width of the metal over hang (W_{MO})	5μm
5	Thickness of the Al layer	1μm or 1.5μm
6	Thickness of the oxide layer	300 nm
7	Thickness of the nitride layer	50 nm
8	p-spray	10^{12}μm – 10^{15}μm

Table 3. Specifications of n-Fz Si DSSD for proton irradiation.

Sl. No.	Specifications	Values
1	Leakage current(I_L) per strip	<10 nA
2	Interstrip capacitance	< 1.2 pF
3	Breakdown voltage	< 1000 V

By using the Shockley Read Hall recombination formula the effective doping concentration(N_{eff}) /full depletion voltage(V_{FD}) and the leakage current(I_L) can easily be calculated;

$$N_{eff} = N_D + \sum n_T{}^{donor} - \sum n_T{}^{acceptor} \quad (1)$$

Here, $n_T = N_T \frac{e_{n,p}}{e_n+e_p}(2)$ (2)

$$e_p = C_p N_V exp^{\frac{-E_t-E_V}{kT}}$$

$$e_n = C_V N_C exp^{\frac{E_t-E_C}{kT}}$$

Here,N_D is the doping concentration, n_T is the steady state occupancy of defect level, N_T is the defect concentration, $e_{n,p}$ is the emission rate of electron or holes, $C_{n,p}$ is the capture coefficient of electron or holes, V_{th} is the thermal velocity.

After calculating the effective doping concentration, the V_{FD} can be calculated by using the formula;

$$V_{FD} = \frac{W_N{}^2 N_D q}{2\epsilon_{Si}} \quad (3)$$

Here,W_N is the thickness of the detector andϵ_{Si} is the dielectric constant of silicon.

By using Shockley Read Hall recombination formula the generation leakage current (I_L) can be calculated;

$$I_L = qAd\left(\sum n_T^{acceptor} e_n + \sum n_T^{donor} e_p\right) = \frac{qn_i A W_N}{\tau_g} \quad (4)$$

Here,I_L isgeneration leakage current, q is elementary charge, A is the area of the detector, d is the thickness of the detector, n_i is the intrinsic carrier concentration and τ_g is the generation lifetime.

Current damage constant (∝) can be calculated by using the formula:

$$\propto = \frac{A}{\emptyset_{eq} \times V} \quad (5)$$

Here, ∝ is current-damage constant, $\emptyset_{eq}$ is equivalent fluence and V is the volume of the detector.

3. TCAD device simulation for Double-Sided Strip Detector

To study the radiation damage effect in the double-sided silicon strip detector TCAD (technology computer aided design) device modelling approach can be used. The microscopic model can be fed into the TCAD program to simulate the macroscopic performance in the irradiated detectors. The Cogenda (Visual TCAD)[13] or SilvacoATLAS device simulator tool can be usedto optimize the electrical performance of the detector[14,15]. Atlas is a simulator tool that can perform DC, AC and transient analysis for semiconductor devices. This device takes a grid mesh made of discrete elements as an input structure and solves the Poisson's equation, carrier continuity, drift-diffusion equation for electrons and holes at every grid point of the mesh to calculate physical quantities such as electric field distribution and carrier mobility inside the device and electrical properties such as current and capacitance.The physical models that are used in the simulation are: Shockley-Read-Hall recombination, concentration dependent mobility, Auger recombination accounting for high-level injection effects, band gap narrowing and field dependent mobility[16].

4. Results and discussions

In this section, we have analysed the macroscopic performance of the irradiated n-FzSi DSSD for the R3B Si tracker. The n-FzSi DSSD detector is irradiated with the proton fluences (equivalent to 1Mev neutron fluence): 2×10^{14},5×10^{14} and 8×10^{14} $n_{eq.}$ cm^{-2}. By using the Shockley-Read-Hall (SRH) model, the full depletion voltage and leakage current have been measured as a function of the irradiation dose for the different doping concentrations and device depths, finally the device and process parameters and specifications for the DC/AC Si DSSD used in the R3B silicon tracker experiment are proposed. The macroscopic results on V_{FD} are shown in figure 4 for the different doping concentration in the irradiated n-FzSi DSSD.

In the irradiated detectors, it has been observed that the charge carriers (e/h) are trapped by the defects/deep-traps (Donor/Acceptor), leading to increase in the effective doping concentration, changing the electric field distribution and shifts the full depletion voltage. In figure 4, it has been shown that the full depletion voltage V_{FD} increases with the increase of proton irradiation fluences for three sets of n-Fz Si DSSD detectors having device depth 200μm, 250μm and 300μm for the different doping concentrations of the n-Fz bulk of the Si DSSD (see figure 4). All the irradiation proton doses have been normalized to 1MeV neutron equivalent doses using the appropriate hardness factor.

The generation leakage current is not a major issue in many radiation damage detection systems even it increases the overall noise of the full detector system. In irradiated detector, the leakage current can be controlled using cooling.

The leakage current at full depletion voltage is directly proportional to the fluence and it will increase with the increase of device depth. From the figure 5, we can see that the generation leakage current increases with the proton irradiation fluencesat 300 K. The lowest leakage current obtainedin 200μm thin DSSD irradiated with 2×10^{14} cm^{-2}.

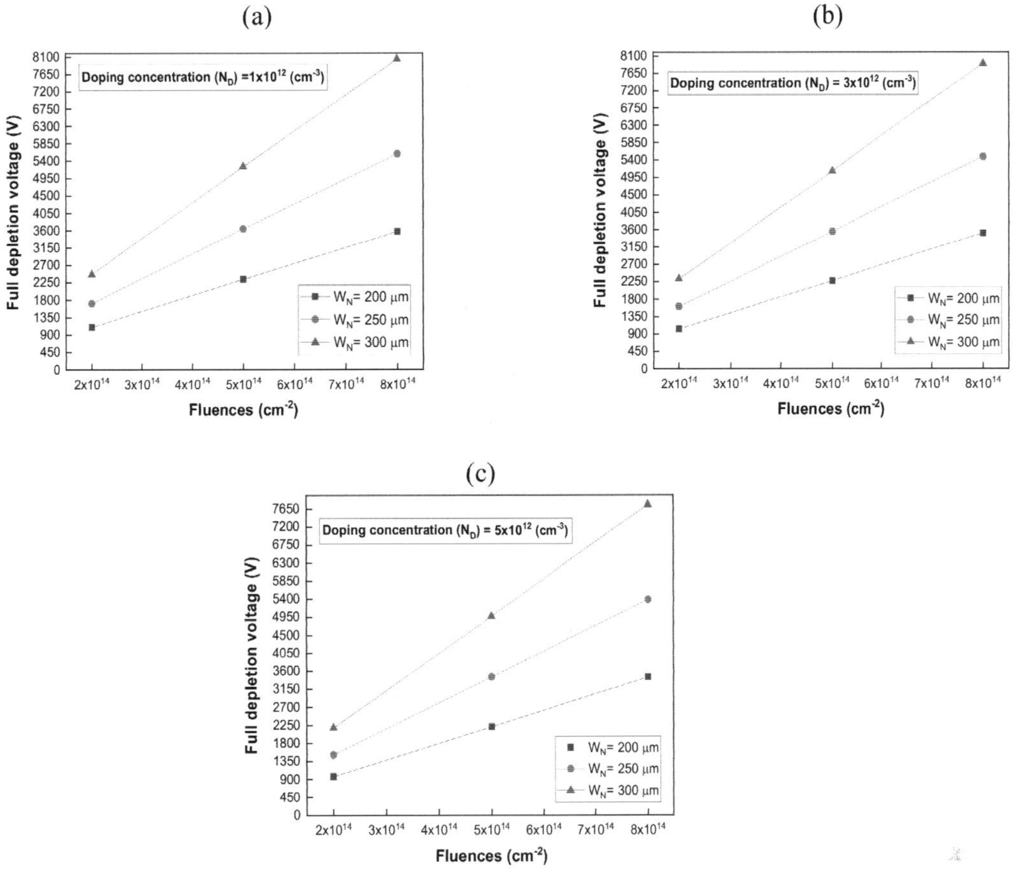

Figure 4. Full depletion voltages as function of equivalent (proton) fluences for the n-FzSi DSSD having (a) N_D=1x10^{12} cm^{-3}, (b) N_D=3x10^{12} cm^{-3}and (c) N_D=5x10^{12} cm^{-3} where device depths are as a running parameter.

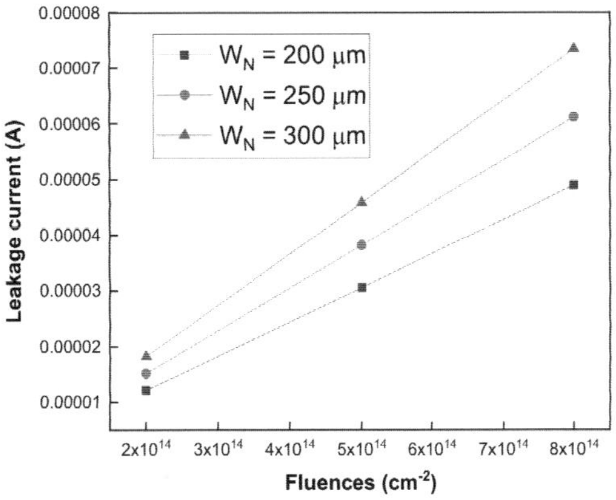

Figure 5. Leakage current as a function of equivalent (proton) fluences for the n-FzSi DSSD for N_D=5x10^{12} cm^{-3}, where device depths are as a running parameter at 300K.

5. Conclusions

In the R3B experiment at GSI, Darmstadt, Germany, a low proton dose (10^{14}cm^{-2}) single metal or double metal radiation hard (DC/AC coupled) Si DSSD requires for the Si Tracker. Here, we have

used a two-deep trap proton irradiation damage model to analyse the macroscopic performance of the Si DSSD in the Si tracker. The detectors irradiated with sufficiently high doses with a safety margin of neutron equivalent fluences; 2×10^{14},5×10^{14} and $8\times10^{14}cm^{-2}$. The variation of full depletion voltage and leakage current of the irradiated n-FzSi DSSD as a function of proton fluencesfor the different bulk doping concentrations and device depths showed. We have used SRH recombination formulation for the full depletion voltage and leakage current calculation at 300K.
It shows that for 200µm AC coupled irradiated n-Fz double metal layer Si DSSD with aproton fluence 2×10^{14} cm^{-2} shows the low full depletion voltage of 970 V at a doping concentration of 5 $x10^{12}cm^{-3}$. Finally, we have proposed the single metal layer DC coupled and double metal layer AC coupled Si DSSD design for the R3B Si tracker as a first option. The device, process parameters and specifications are shown for the proposed design. The TCAD device simulation using Silvaco ATLAS is underway to get the optimal design of DSSD for the Si tracker at R3B.

References

[1] Borri M, Lemmon R, Thornhill J, Bate R, Chartier M, Clague N, Herzberg R D, Labiche M, Lindsay S, Nolan P, Pearce F, Powell W and Wells D 2016 Detector production for the R3B Si-tracker *Nucl. Instruments Methods Phys. Res. Sect. A Accel. Spectrometers, Detect. Assoc. Equip.* **836** 105–12

[2] Henning W 2004 The GSI project: An international facility for ions and antiprotons *Nucl. Physics. A; ISSN 0375-9474; Worldcat; CODEN NUPABL 734 654-660*

[3] Powell W 2016 Development of a Silicon Tracker and Front-End *PhD thesis University of Liverpool*

[4] Kresan D, Heil M and Al-Turany M 2019 A High-Precision Tracking Algorithm for Mass reconstruction of Heavy-Ion Fragments in the R 3 B Experiment at FAIR *EPJ Web Conf.* **214** 02038

[5] Bertini D 2011 R3B Root, simulation and analysis framework for the R3B experiment at FAIR *J. Phys. Conf. Ser.* **331** 032036.

[6] Reifarth R, Altstadt S, Göbel K et al 2016 Nuclear astrophysics with radioactive ions at FAIR *J. Phys. Conf. Ser.* **665** 012044

[7] Lawrence Jones, Member, IEEE S 1985 A Readout ASIC for e R3 B Silicon tracker 9–13

[8] Douma C A, Boretzky K, Gasparic I, Kalantar-Nayestanaki N, Kresan D, Mayer J and Rigollet C 2018 Design studies for the NeuLAND VETO detector *J. Phys. Conf. Ser.* **1024** 012027

[9] Park H, Hyun H J, Kah D H, Kang H D, Kim H J, Kim Y I, Kim Y J, Shim D H, Choi Y I, Choi Y K, Lee J, Park J M and Park K S 2006 Design and fabrication of the double-sided silicon microstrip sensor *J. Korean Phys. Soc.* **49** 1401–6

[10] Ahmed M I 2015 New pixel detectors in SOI technology for particle physics applications *AGH University of Science and Technology* Ph.D. Thesis.

[11] Dalal R, Bhardwaj A, Ranjan K, Moll M and Elliott-Peisert A 2014 Combined effect of bulk and surface damage on strip insulation properties of proton irradiated n+-p silicon strip sensors *J. Instrum.* **9** P04007

[12] Eber R 2014 Development of radiation tolerant silicon sensors *J. Instrum.* **9** C02024

[13] Retrived from website: https://cogenda.com

[14] Retrived from website: https://www.silvaco.com

[15] Silvaco Inc., ATLAS User's Manual Device Simulation Software (Santa Clara, CA: 2010)

[16] Oussalah S, Mekheldi M, Filali W and Garoudja E 2021 A Comparative study by TCAD Simulation for two different planar n-on-p silicon particle detectors with different guard ring implant type *J. Phys. Conf. Ser.* **1766** 012015

Metamaterial-Based Miniaturized DGS Antenna for wireless Applications

[1]Jaget Singh, [2]Fateh Lal Lohar
[1]UIET, Panjab University, Chandigarh 160014, India
[2]Government Engineering College Jhalwar, 326001 Rajasthan, India

Email:[1]jaget_singh80@yahoo.co.in,[2]fatehudml@gmail.com,

Abstract. In this article, metamaterial based miniaturized DGS (Defective Ground Structure) antenna is presented. Initially conventional antenna is designed for 2.4 GHz and by using DGS method, the same conventional antenna size is reduced upto 50% of initial design. But antenna gain down due to DGS in antenna structure. By using metamaterial as a reflector, the antenna gain can be enhanced. The antenna and metamaterial are designed on Rogers RT5880 and Teflon Ceramic TF-1/2 ($\varepsilon_r = 10.2, \tan\delta = 0.001$) respectively. All simulation results are carried out by using EM simulator CST software and for the operating frequency of 2.415 GHz a gain of 6.28dB is observed.

Keywords: DGS, Miniaturized antenna, Metamaterial, and Wireless applications

1. Introduction

Nowadays in RFID tags, communication systems and radiating systems for radar and automotive applications integrated miniaturized antennas becomes relevant need. Since last decades efforts are being made to design low power RF components and miniaturized microstrip antennas which are not available readily without changing its features and size. Integration of radiating element like microstrip antenna along with remaining circuit on printed chip becomes the main challenge for the designer. In literature many researchers have used many techniques to reduce the size of antenna. For squeezing the dimension of the antenna, lumped elements, shorting post and suitable slots are used as a reactive load with patch of antenna. However, these techniques are not sufficient to provide the drastic and required reduction in dimension of the patch antenna used for various applications. Another method used to squeeze the dimension of patch is used by applying the dielectric substrate of high permittivity. But performance parameters like gain, efficiency bandwidth and shape of radiation pattern deteriorated due to increase in surface wave excitation between patch and substrate of antenna. The standard techniques available were not sufficient to provide the required miniaturized antenna. At this stage some other artificial materials like metamaterial seems to be much effective for reduction of dimension of antenna without affecting the other performance parameters [1-3].

Metamaterials have been discovered to exhibit some unique qualities not seen in natural materials. These artificial materials can be formed by arranging the periodic structure of metallic strip and rings at regular interval. First metamaterial concept was published by Victor Veselago a Russian physicist in 1968. Later on after thirty years, Sir John Pendry also presented that the geometry of conductor arranging in regular interval can produce the negative permeability and negative permittivity, which was also a another type of metamaterial. The negative permeability and permittivity of material can be used to design electrical small antenna by miniaturizing the antenna. Further by increasing the gain and directivity high directive antenna can be designed for various applications. Range of antenna array and beam scanning can be increased by using metamaterial to support communication link, navigation systems, surveillance sensor and command control systems [4-6].

Due to wide range of applications one of the most suitable techniques used for electromagnetic shielding and miniaturization of antenna is frequency selective surfaces (FSSs). An artificial substance called electromagnetic band gap (EBG) is utilised to provide the electromagnetic shielding required to minimise the specific absorption rate (SAR). FSS and EBG acts as bandpass or bandstop structure depending on the periodic structure at the required frequency [7-9]. EBG is also used to eliminate mutual coupling between antennas and increase microstrip patch antenna impedance matching and efficiency. A metamaterial loaded microstrip patch antenna reduces the resonance frequency by exciting the negative order resonance and reduces the actual size of antenna at the designed frequency. Reactive impedance surface (RIS) or artificial magnetic surface (AMS) is also another type of metamaterial which is also reported as effective technique used for antenna miniaturization by various researcher. There are various types of meta-surfaces available like periodic metallic patch and mushroom-like high-impedance surface etc [10]. Complementary split ring resonators (CSSRs) are also employed for miniaturisation and design of multiband antennas because of their ability to provide a quasi-static frequency that is much smaller than the guided wavelength [11].

Due to availability of unused bandwidth the interest is increasing towards the millimetre wave spectrum to meet the requirement of lower latency, high throughput and wider bandwidth for future wireless communication [12-14]. In this paper we have presented a metamaterial based antenna for wireless application. Initially a rectangular patch antenna is designed and simulated at 2.4 GHz. Then same antenna is miniaturized using DGS miniaturization technique and again simulated. By miniaturization technique the size of antenna is reduced approximately 50% of actual size of antenna at the same frequency. For the miniaturized antenna size has been reduced but gain has been decreased at the designed frequency. Further a FSS based (frequency selective surface) metamaterial has been used to improve the gain and other performance parameters.

2. Antenna Design

A microstrip patch antenna with dielectric constant $\varepsilon_r = 2.2$ and tangent loss 0.00009 is designed on Rogers RT5880 dielectric material. The height of substrate is 0.787 mm and length and width of the patch are calculated using the transmission line model.

Step 1: The width of the patch (W) is calculated as follows: The patch's width is determined by

$$W = \frac{c}{2f_r}\sqrt{\frac{2}{\varepsilon_r+1}} \quad (1)$$

Where, c denotes the speed of light in free space. For the current design, f_r stands for resonance frequency, which is 2.4 GHz. The substrate is 0.787 mm thick and has a relative permittivity of 2.2. The permittivity changes due to the presence of air between the patch and the dielectric, and this is known as effective permittivity (ε_{reff}) and is given by

$$\varepsilon_{reff} = \frac{\varepsilon_r+1}{2} + \frac{\varepsilon_r-1}{2}\left[1 + 12\frac{h}{w}\right]^{-\frac{1}{2}} \quad (2)$$

Where, h denotes the substrate's height or thickness.

Step 3: The length extension ΔL, which is given by

$$\frac{\Delta L}{h} = 0.412\frac{(\varepsilon_{reff}+0.3)\left(\frac{W}{h}+0.264\right)}{(\varepsilon_{reff}-0.2.58)\left(\frac{W}{h}+0.8\right)} \quad (3)$$

Step 4: The length of the patch must now be calculated.

$$\mathrm{L} = \frac{c}{2f_r\sqrt{\varepsilon_{reff}}} - 2\Delta\mathrm{L} \quad (4)$$

The extra length on both sides of the patch due to the fringing field is ΔL.

$$\mathrm{L_{eff}} = \mathrm{L} + 2\Delta\mathrm{L} \quad (5)$$

L_{eff} is the patch's overall length, including the length caused by the fringing effect..

2.1 Conventional microstrip patch antenna

The dimensions of conventional patch antenna using transmission line theory are calculated and shown in table 1.

Table 1. Antenna parameters

	Conventional Antenna(mm)	Miniaturized Antenna(mm)
W=L	80	40
W_p=L_p	42	20
L_f	14.50	7.50
W_f	0.50	0.50
L_1	-------	26.60
W_1	-------	0.80

Table 2. Metamaterial parameters

Parameters	Value (mm)
Ws=Ls	12
L2	3.90
L3	2.50
L4	3

A square substrate of 80 mm^2 and a patch of 42 mm^2 are taken at the designed frequency of 2.4 GHz. The Roger RT5880 of thickness 0.787 mm is used as a substrate. Figure 1 depicts the geometry of a constructed conventional antenna.

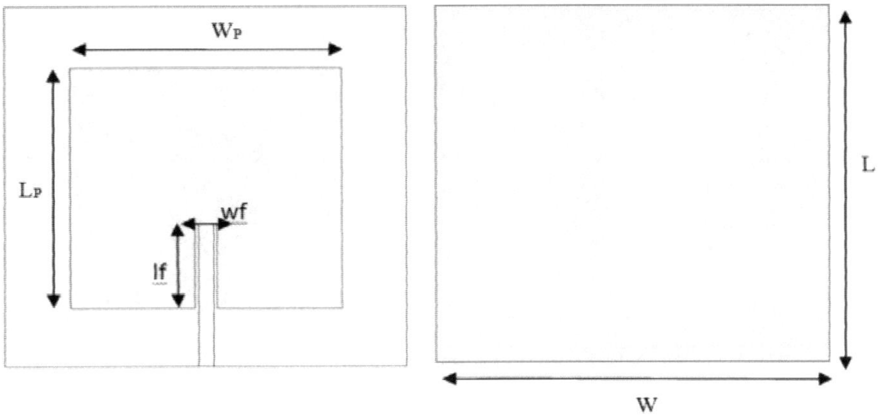

Figure 1. Conventional antenna at 2.4 GHz

2.2 DGS Miniaturized antenna

At low frequencies, it is always necessary to integrate and reduce the antenna size. At lower frequency antenna size become very high. As a result, numerous strategies are utilised to minimise the antenna at low frequencies. At the 2.4 GHz frequency, DGS (defective ground surface) is one of the most effective strategies for reducing antenna size. The concept behind DGS technique is increasing the current path, so that electrical length is increased and frequency down to shift in lower side. The slot L1*W1 is responsible to reduce antenna size by increasing electrical length of antenna, which is shown in fig.2. In DGS miniaturized antenna effective size of antenna has been reduced to 50% of its actual size at the same frequency.

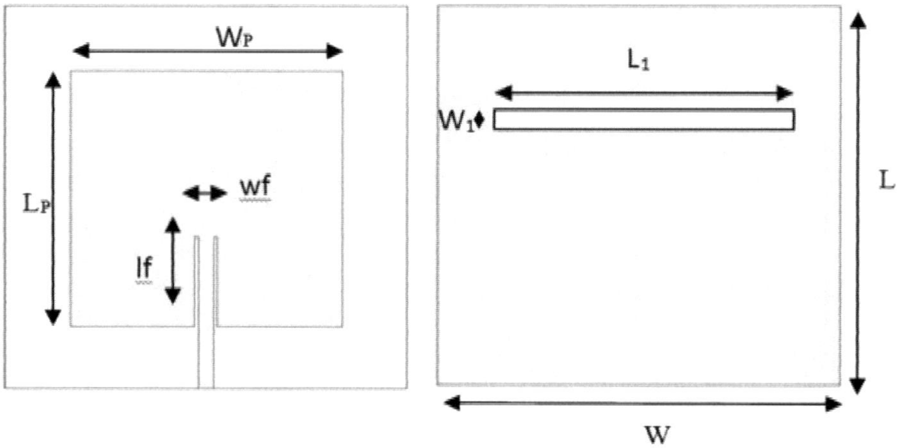

Figure 2. DGS Miniaturized antenna

2.3 DGS Miniaturized antenna loaded with metamaterial

Metamaterial does not exists in natural form, but they are created by arranging the metal strips, rings or rods in a regular intervals as shown in figure 3. In this figure an array of 4x4 metallic rings are arranged. The array of rings behaves as metamaterial and provides a band gap of frequency at which it provides high impedance. The dimensions of metamaterial are shown in table 2.

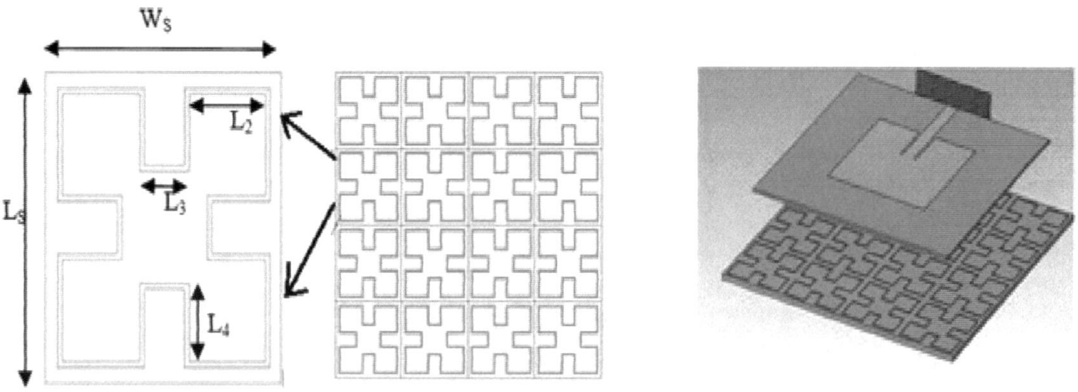

(a) (b) (c)

Figure 3. a) Metamaterial unit cell, b) Metamaterial array c) Antenna loaded with metamaterial

3 Result Analyses

The designed antenna is simulated using CST software. The operating frequency of a conventional antenna is 2.4 GHz, and the return loss is -21.36 dB, as shown in figure 4, and the corresponding radiation pattern is shown in figure 5, with 5.62 dB gain at the same resonating frequency and a maximum radiation direction of 2 degrees with an angular beam-width (3 dB) of 84.4 degrees. So antenna is able to radiate and intercept the energy from the desired direction.

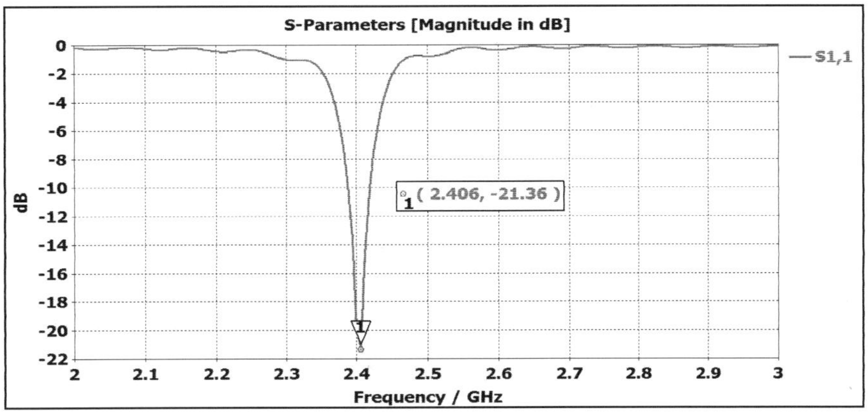

Figure 4. S11 Conventional antenna at 2.4 GHz

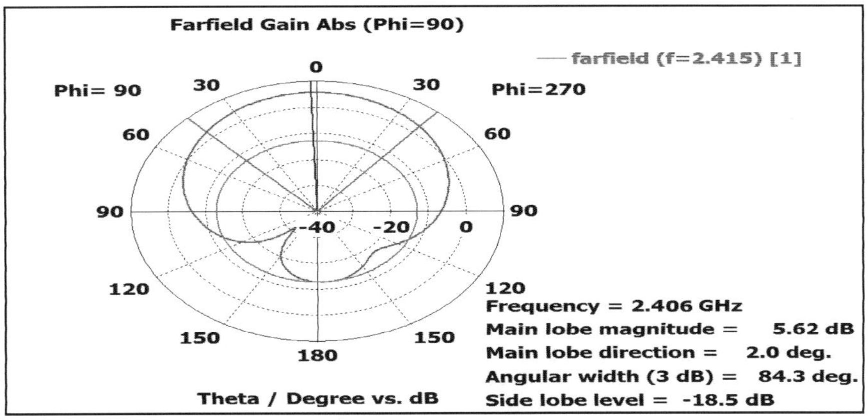

Figure 5. Gain Conventional antenna at 2.4 GHz

Figure 6 shows the return loss of DGS miniaturized antenna at 2.4 GHz, in this plot return loss is -33.988 dB, which is improved as compared to conventional antenna at same resonating frequency. But gain of the DGS antenna is reduce from 5.62 dB to 3.48 dB due to increasing the surface current at DGS slot and also increasing back side radiation shown in figure 7. To increase gain of DGS antenna, a reflector is required below the DGS antenna, which can improve gain by reflecting the back side radiation from DGS antenna and it add with main lobe of radiation pattern. So metamaterial is best option to work as reflector for this DGS antenna.

Figure 8 shows the reflection and transmission coefficient of proposed metamaterial, the forward transmission coefficient S21 and reverse isolation coefficient S12 are exactly same and traces the same curve, further input reflection coefficient S11 is out of phase with S12 and S21 at the designed frequency of 2.4 GHz. So metamaterial is acting as high impedance surface at this frequency and proves the metamaterial behaviour. DGS miniaturized antenna effectively reduced the size of antenna but at the same time gain has been also reduced. So another technique used to improve the gain of DGS miniaturized antenna is loaded with the metamaterial. Metamaterial loaded antenna is shown in figure 3 (c).

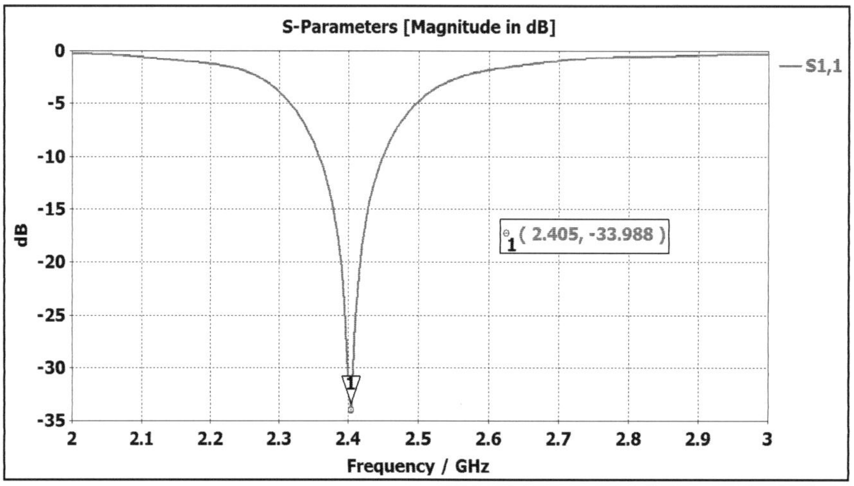

Figure 6. S11 DGS Miniaturized antenna

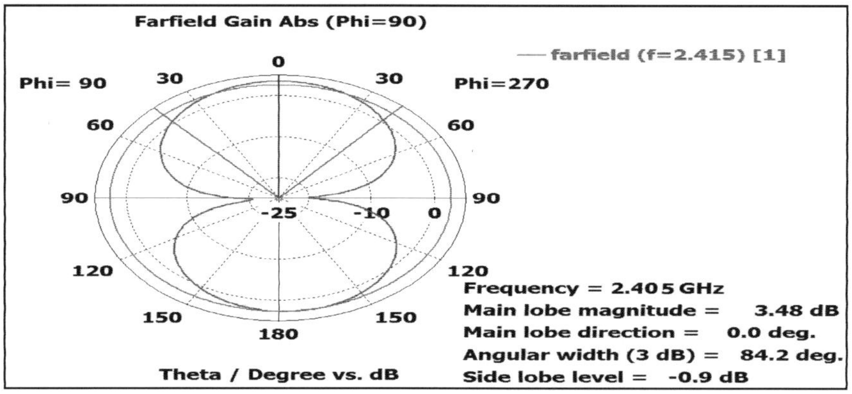

Figure 7. Gain DGS Miniaturized antenna

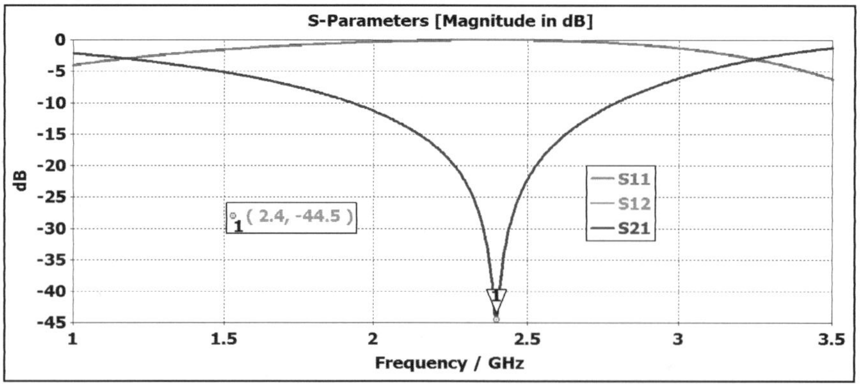

Figure 8. S11 and S21of metamaterial at 2.4 GHz

Figures 9 and 10 depict the S11 parameter and metamaterial loaded antenna gain, respectively. The The metamaterial loaded antenna's gain has been improved, because at this frequency metamaterial is providing high impedance and decreasing the spatial wave and directs all the power in desired direction of maximum radiations. So gain of antenna has been increased to 6.28 dB which is higher than conventional and DGS miniaturized antenna gain. Figure 11 shows radiation efficiency of DGS antenna with and without metamaterial, in which the radiation efficiency of DGS antenna at 2.41 GHz is 92.07 % and for DGS antenna with metamaterial at same frequency is 94.66 %. Figure 12 illustrates the VSWR for both situations and indicates that the antenna is perfectly matched. Figure 13 shows the realised gain for both situations, which is 3.44dB for the DGS antenna and 6.14dB for the DGS antenna with metamaterial at 2.41 GHz. The Surface Currents of a DGS antenna with and without metamaterial are shown in Figure 14.

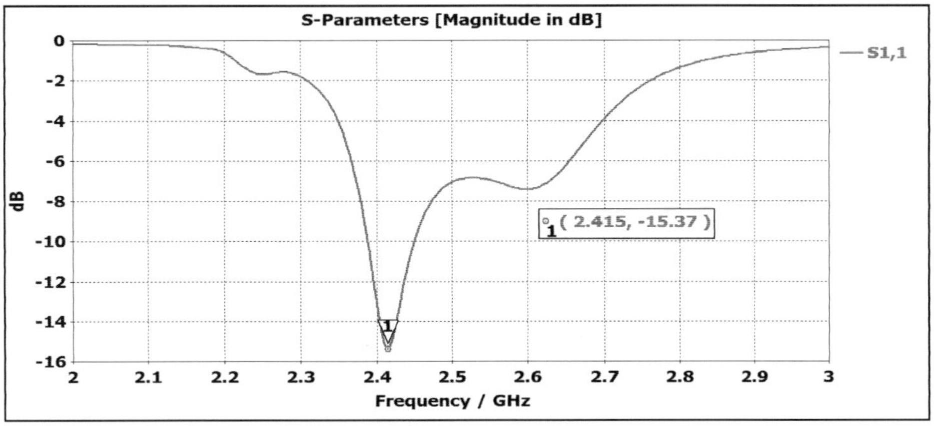

Figure 9. S11 and S21of metamaterial at 2.4 GHz

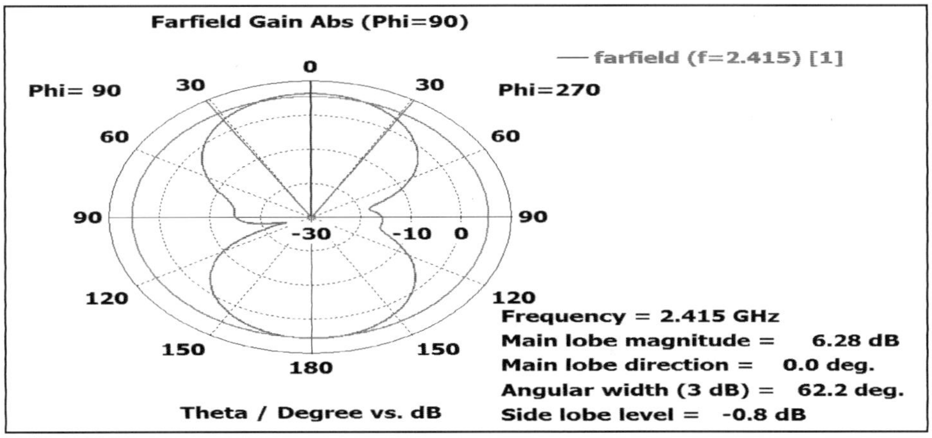

Figure 10. S11 and S21of metamaterial at 2.4 GHz

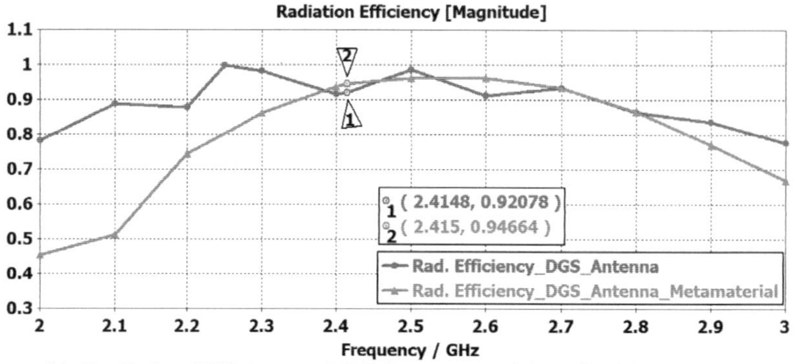

Figure11. Radiation Efficiency of DGS antenna with and without metamaterial

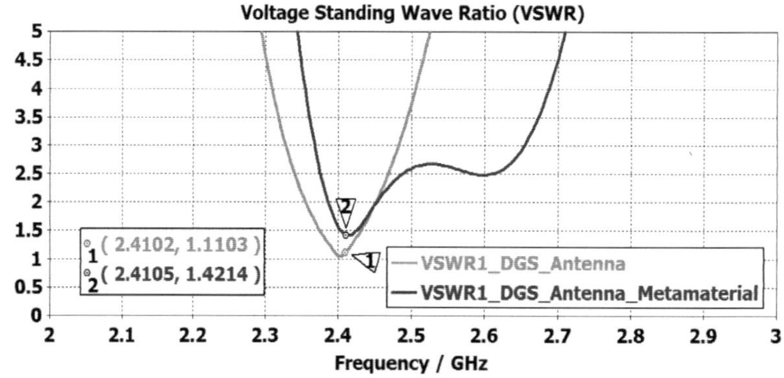

Figure12. VSWR of DGS antenna with and without metamaterial

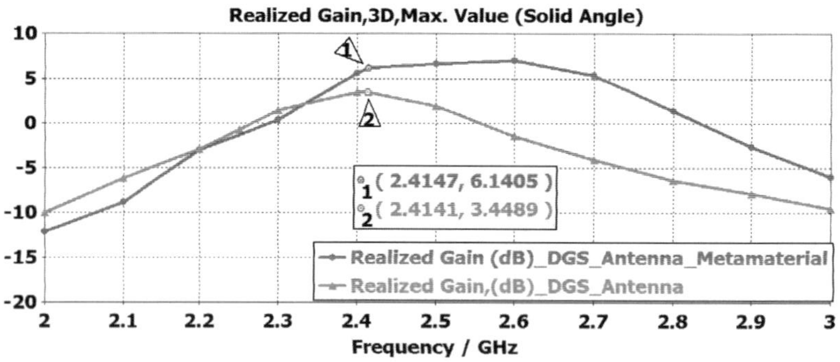

Figure13. Realized Gain of DGS antenna with and without metamaterial

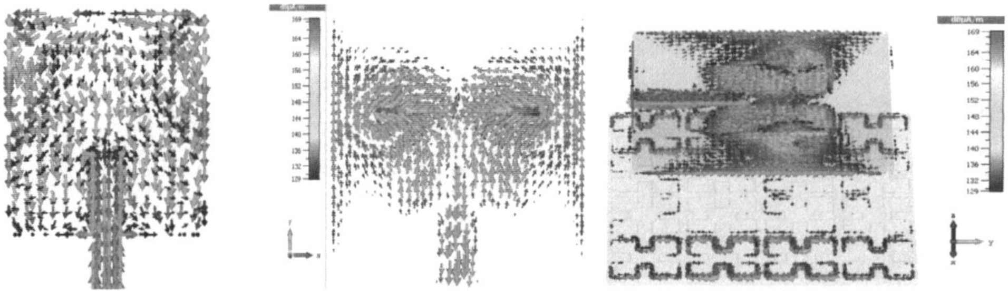

DGS_Antenna Top View | DGS_Antenna Bottom View | DGS_Antenna_Metamaterial 3D View

Figure14. Surface Currents of DGS antenna with and without metamaterial

4 Conclusions

A miniaturized antenna loaded with metamaterial operating at 2.4 GHz is designed and presented in this paper. Initially, a conventional 2.4 GHz antenna is developed by estimating the dimensions using transmission line model theory. The length and width of square patch is 42 mm at the designed frequency. When same antenna is designed using miniaturization technique the length and width has been reduced to 20 mm, which is less than 50% of original one. So using miniaturization technique the actual size of antenna has been reduced upto much extent but at the same time gain of antenna has been decreased at the designed frequency. A metamaterial array of 4x4 has been used to improve the gain of miniaturized antenna. At the desired frequency, Metamaterial has enhanced the antenna gain without increasing the antenna size.

Refrences

[1] Bilotti F, Alu A and Vegni L 2008 Design of Miniaturized Metamaterial Patch Antennas With μ-Negative Loading *IEEE Trans. Antennas Propag.* **56** 1640–1647.

[2] Sharma N, Gangwar R and Arya A K 2018 Metamaterial based compact super wide band microstrip antenna with dgs for wireless applications *Proc. - IEEE 2018 Int. Conf. Adv. Comput. Commun. Control Netw. ICACCCN 2018* 1016–1021.

[3] Dicandia F A and Genovesi S *2020* Characteristic Modes Analysis of Non-Uniform Metasurface Superstrate for Nanosatellite Antenna Design *IEEE Access* **8** 176050–176061.

[4] Anand M *2014* Applications of metamaterial in antenna engineering *Int J Eng Res Appl.* 2 49–52.

[5] Al-Bawri S S, Islam M T, Shabbir T, Muhammad G, Islam M S and Wong H Y 2020 Hexagonal Shaped Near Zero Index (NZI) Metamaterial Based MIMO Antenna for Millimeter-Wave Application *IEEE Access* **8** 181003–181013

[6] Alibakhshikenari M, Babaeian F, Virdee B S, A¨ıssa S, Azpilicueta L, See C H, Althuwayb A A, Huynen I, Abd-Alhameed R A, Falcone F et al. 2020 A Comprehensive Survey on Various Decoupling Mechanisms With Focus on Metamaterial and Metasurface Principles Applicable to SAR and MIMO Antenna Systems *IEEE Access* **8** 192965–193004

[7] Alonso-Gonz´alez L, Ver-Hoeye S, Fern´andez-Garc´ıa M and Andr´es F L H 2018 Broadband Flexible Fully Textile-Integrated Bandstop Frequency Selective Surface *IEEE Trans. Antennas Propag.* **66** 5291–5299

[8] Oesterling A X, Imani M F, Mizrahi O S, Gollub J N and Smith D R 2020 Detecting Motion in a Room Using a Dynamic Metasurface Antenna *IEEE Access* **8** 222496–222505

[9] Islam M T, Samsuzzaman M, Yahya I, Kogut A, Misran N and Islam M T 2019 Metamaterial inspired high gain antenna for microwave breast imaging *IEEE Asia-Pac. Conf. Appl. Electromagn. Proc.* 1–4

[10] Dong Y, Toyao H and Itoh T 2011 Compact Circularly-Polarized Patch Antenna Loaded With Metamaterial Structures *IEEE Trans. Antennas Propag.* **59** 4329–4333

[11] Li K, Zhu C, Li L, Cai Y M and Liang C H 2013 Design of Electrically Small Metamaterial Antenna With ELC and EBG Loading *IEEE Antennas Wirel. Propag. Lett.* **12** 678–681

[12] Lin X, Seet B C, Joseph F and Li E 2018 Flexible Fractal Electromagnetic Bandgap for Millimeter-Wave Wearable Antennas *IEEE Antennas Wirel. Propag. Lett.* **17** 1281–1285

[13] Sakli H, Chafai A, Chaker E and Sakli N 2021 Metamaterial-Based Antenna Performance Enhancement for MIMO System Applications *IEEE Access* **9**,38546-38556

[14] Suvarna K, Ramamurthy N, & Vardhan D V 2021 Miniaturized and Gain Enhancement of Tapered Patch Antenna Using Defected Ground Structure and Metamaterial Superstrate for GPS Applications *Prog. Electromagn. Res. C* ***108*** 187-200

Analysis of Effects of Change of Gear Parameter Module on Transmission Error in Spur Gear using Interference Volume Method

V Karma [1,4], G Maheshwari [2] and S K Somani [3]

[1,2] Mechanical Engineering Department, Institute of Engineering & Technology, Devi Ahilya University, Indore, Madhya Pradesh, 452001, India [3] Vice Chancellor, Oriental University, Indore, Madhya Pradesh, India

vkarma@ietdavv.edu.in

Abstract. Gear drives are the most used elements in power transmission systems. Any defect in the gear of these drives leads to noise and vibrations, which affect power/torque transmission. Transmission Error (TE) is one of the critical causes that arises due to the tooth geometry error, cracks in gear, profile error etc. Our work demonstrates, the consequences occurring due to the changes in gear parameter module on TE for spur gear pair of 1:1 gear ratio with pitch error, using the interference volume method. The spur gear without pitch error and with pitch error having the involute profile of standard tooth dimension system and stub tooth dimension system are modelled and assembled in CAD software SOLIDWORKS. The interference volume values obtained from the gear pairs are recorded for one mesh cycle. Various graphs plotted between the angle of rotation and interference volume values for one mesh cycle. It is found from the analysis that as the module and pitch error increase TE also increases. The effect of variation of module and pitch error is more in the standard tooth dimension system than the stub tooth dimension system.

Keywords: CAD, Gear Ratio, Pitch Error, No of Teeth, SOLIDWORKS, Geometry Error

1. Introduction

Gears drives are the most used elements employed to transfer power/torque. The condition to accurately transfer power/torque is that the drive elements such as gears should be free from any error else it could lead to noise and vibrations, which results in the development of Transmission Error (TE) despite of availability of world class manufacturing and design facilities. TE is a show of reading measured of angular or linear displacement by the side of activity on the base circle [1]. Hotait and Kahraman [2] demonstrated experimentally the association connecting dynamic factor and transmission error measurement. The values were obtained from unabated and adapted spur gears using a gear dynamic test set-up. Kohler and Regan [3] concluded that TE has harmonic components of significant amplitude at tooth contact frequency spectrum. Chang and Tang [4] conducted theoretical analysis under the influence of single and cumulative pitch error to investigate the nonlinear vibration response in a double-helical gear set. Velex et al [5] established theoretically a relationship in the middle of active mesh agitation and transmission errors using 3D analysis modal of

multi mesh gears. They demonstrated that the local confined TE is analogous with each individual mesh controls the dynamic mesh forces and provided a design criterion to minimize dynamic tooth loads by defining tooth shape modifications. Nina [6] developed equations of motion based on time varying mesh stiffness function, loaded quasistatic TE, and no-load or kinematic TE. Studied the effect of tooth pitch error and load on dynamic behavior of gear transmission. She also investigated the relationship between dynamic TE and mesh force or root stress. Sánchez et al [7] work is done to study the contact conditions of modified teeth under load. They widely explained the results by seeing the changes of various parameters of profile modification on actual load sharing quotient and TE. Palermo et al [8] in their work tried to calculate TE in comparison to the direct process which uses low-cost digital encoders, which are embedded on an accuracy gear pair test. Shweiki et al [9] does work to find the change in conduct when a quality and light weight cylindrical gear meshes. An elastic multi body expressions which has hybrid approach in FE method in the background representation is used by them. Zhiying and Pengfei [10] proposed and validated, analytical mesh stiffness and quasi static TE models with finite element method to find the inner working of spur gear meshing pairs. Pleguezuelos et al [11] found out the effect of changes in symmetric long on high contact ratio spur. The change is represented empirically in the optimum length as a function of the contact ratio. Oswald [12] works found out the efficacy in modification of profile in ever changing reduced loads in gears having spacing errors. The analysis considers the modifications of both linear and parabolic profiles. Bruyere et al [13] presented a closed-form analytical formula of TE for helical and solid spur gears for minimizing time varying amplitudes of TE amalgamation of lead crouch and profile relief. They showed that the optimum tooth modifications depend on ratio of profile Vs face contact, the normalized depth and number of modifications in profile in continuation with the normalized lead crown amplitude. Handschuh [14] experimentally, empirically, and mathematically find out the consequences of tooth spacing error on TE and root stresses in spur gear pairs. Bartosova et al [15] uses finite element analysis to find out TE at variety of combination of load states, axial distant variation, and tooth shape elevation modification. Park [16] find out friction in tooth and TE in spur gears which occurs due to result of friction sliding in quasi static condition. Somani [17] worked in his doctorate thesis on studying the effect of changing the various gear parameters on TE in spur gear.
Transmission through gears is used in every area of industry to convey power coming from shaft. A lot of research work is there in public domain in this area. Bruzzone et al [18] constructed model using multi-layered process in quasi-static state to find degree of deflection, tooth deformation and stiffness. Experiments results in terms of STE with different parameters has been found. TE is the main cause of vibration in gear box. Duan et al [19] established a test rig and made comparison between the result obtained theoretically and experimentally. Results shows that under heavy load conditions TE increases, but with the use of thin-walled housing it decreases considerably. A lot of study about the TE in gears is being done. Benaicha et al [20] in his work tries to develop a model which gives the insight efficient inaccurate working of gear TE. Results were compared with the earlier results obtained from classical techniques. Superiority of the model which is made was confirmed in his work. Backlash is also a type of TE which occurs due to the gap between the teeth at the pitch circles. This error is unavoidable as the time passes by. Ambaye et al [21] uses a Gear Trax software to model a gear with backlash with different degrees. In his experiments he uses plane strain analysis and finite element analysis to find mesh convergence for different contact pressure and stress. In this process he was able to predict backlash with a high degree of accuracy. Gearbox is used to transmit power from shaft to rotor in various machines. In case of wind turbine, it is important link to transmit power. Tao et al [22] uses Flank Pitch Error model to find gear faults. He proposed time varying meshing stiffness model to simulate meshing frequency and find severity of gear box vibration. Delay in drive which is popularly known as loaded TE is caused due to tooth deflection, manufacturing defects and assembly error in spur gear. Miguel et al [23] proposed a mathematical model to consider of meshing stiffness, load sharing ratio and TE for spur gears under minimal or varying load conditions. Gear TE causes vibration in moving parts. Chin et al [24] in his work find TE. His work is validated by readings

obtained from encoder and tachometer from a drive gear wear experimental setup. Model prepared by him prove to be very robust under sever conditions also.
From the available literature, it is observed that the researchers are working towards the determination of TE analytically, experimentally and or using finite element analysis for the gears having defects due to manufacturing, geometry error, crack, tooth breakage etc., in the standard tooth dimension system. In today's scenario there is an urgent need of more simplified methods/procedure using CAD software to analyze the various manufacturing defects, geometry errors etc., in spur gears. A lot of work is to be done by considering the gears in both standard and stub tooth dimension system. Hence, the objective of the current work is to analyze the TE using interference volume method in stub tooth dimension system and standard tooth dimension system for spur gear having pitch error in a simplified procedure. The gears are modelled in SOLIDWORKS software. Various gear parameters used are the modules 3, 3.5, 4, 4.5, and 5 mm, pressure angle 20°, and the number of teeth 18. The pitch error of magnitudes 1% and 2% is deliberately introduced in all the teeth of gear. The gear without pitch error is meshed with a gear having pitch error using the assembly module of SOLIDWORKS software. The interference volume values are then recorded for one mesh angle of the gear pair. The graphs are plotted between angles turned vs. interference volume values, and for each change in module and pitch error, the effect on the TE is predicted for both standard as well as stub tooth dimension system. It is observed that TE is affected when module and pitch error magnitude is change but the effect is more in case of standard tooth dimension system than in stub tooth dimension system. The analysis is done using the procedure as described by the authors in [25]. The work of the paper/outcomes are in line with the work of researchers presented in [3, 4, 6, 12, 14, 17].

2. Modelling of Master Gear and Erred Gear

The spur gears are modelled with true involute curve from mathematical equation of involute curve in CAD software SOLIDWORKS. The following spur gears are modelled for the analysis. The designation of gears and their pairs are described below.

a) Spur Gear of standard tooth dimension system and stub tooth dimension system without any defect or error are designated as Master Gear [MG] and Stub Master Gear [SMG] respectively.
b) Spur Gear of standard tooth dimension system and stub tooth dimension system with pitch error are designated as Erred Gear [E] and Stub Erred Gear [SE] respectively. Gear with 1% and 2% pitch error are designated as E1 and E2 in standard tooth dimension system, SE1 and SE2 in stub tooth dimension system, respectively.

After modelling of the master gear, erred gears are modelled by inducing pitch error of 1% and 2% magnitude. The pitch error is created by increasing the tooth thickness to 1% and 2% of its original value. These deviations are sufficient deviation from the original value in manufacturing of any gear. A nominal face width of 10 mm is considered for the gears. The gears with and without pitch error in standard tooth dimension system and stub tooth dimension system of module 3, 3.5, 4, 4.5 and 5 mm, number of teeth 18 and pressure angle 20° are designated in table 1.
For example as in table1, a gear of module 3 mm, number of teeth 18, pressure angle 20°, and without pitch error in standard tooth dimension system is designated as MG_m3_z18_phi20, and with pitch error of 1% and 2% is designated as E1_m3_z18_phi20 and E2_m3_z18_phi20 respectively (where MG is Master Gear, E1 is Erred Gear (1% pitch error), E2 is Erred Gear (2% pitch error), m is module, z is number of teeth and phi is pressure angle). Similarly a gear of module 3 mm, number of teeth 18, pressure angle 20°, and without pitch error in stub tooth dimension system is designated as SMG_m3_z18_phi20, and with pitch error of 1% and 2% is designated as SE1_m3_z18_phi20 and SE2_m3_z18_phi20 respectively (where SMG is Master Gear, SE1 is Erred Gear (1% pitch error), SE2 is Erred Gear (2% pitch error), m is module, z is number of teeth and phi is pressure angle). The various values of the spur gear parameters for the two tooth dimension systems are given in table 2 and table 3.

Table 1. Designation of various spur gears in two tooth dimension systems

Standard Tooth Dimension System		Stub Tooth Dimension System	
Master Gear / Gear without Pitch Error	Erred Gear / Gear with Pitch Error of 1% & 2%)	Master Gear / Gear without Pitch Error	Erred Gear / Gear with Pitch Error of 1% & 2%)
MG_m3_z18_phi20	E1_m3_z18_phi20 E2_m3_z18_phi20	SMG_m3_z18_phi20	SE1_m3_z18_phi20 SE2_m3_z18_phi20
MG_m3.5_z18_phi20	E1_m3.5_z18_phi20 E2_m3.5_z18_phi20	SMG_m3.5_z18_phi20	SE1_m3.5_z18_phi20 SE2_m3.5_z18_phi20
MG_m4_z18_phi20	E1_m4_z18_phi20 E2_m4_z18_phi20	SMG_m4_z18_phi20	SE1_m4_z18_phi20 SE2_m4_z18_phi20
MG_m4.5_z18_phi20	E1_m4.5_z18_phi20 E2_m4.5_z18_phi20	SMG_m4.5_z18_phi20	SE1_m4.5_z18_phi20 SE2_m4.5_z18_phi20
MG_m5_z18_phi20	E1_m5_z18_phi20 E2_m5_z18_phi20	SMG_m5_z18_phi20	SE1_m5_z18_phi20 SE2_m5_z18_phi20

3. Meshing and Analysis of Gear Pair of Standard Tooth Dimension System

After modelling of various gears, it is required to form assemblies/pairs (meshing of Master and Master or Master and Erred Gear) for doing the analysis. The various assemblies or combinations formed, and their designations are shown in table 4. Each of these assemblies or combinations are taken one by one for determining the TE. The interference volume is determined for all these gear pairs/assemblies. From the values of the interference volume obtained, the TE variation is determined for all these spur gear pairs. The methodology to determine the TE is verified from the pair of master gear with master gear and then the master gear and erred gear pair is taken for the analysis. The two gears are meshed using the assembly module of SOLIDWORKS and interference volume between the two gears in mesh is checked for initial meshed position by invoking the command of finding interference volume and then the value of interference volume is recorded.

Table 2. Values of Spur Gear parameters in Standard Tooth dimension System

Parameters [26]	Module, mm (m)	No. of Teeth (z)	Pitch Circle Diameter, mm (m*z)	Pressure Angle, ° (Phi)	Addendum, mm (1*m)	Dedendum, mm (1.25*m)	Base Circle Diameter, mm (m*z*cos(phi))	Circular pitch, mm (πm)	Tooth Thickness, mm (πm/2)
	3.0	18	54	20	3.0	3.750	50.7434	9.4248	4.7124
Master	3.5	18	63	20	3.5	4.375	59.2006	10.9956	5.4978
Gear	4.0	18	72	20	4.0	5.000	67.6578	12.5664	6.2832
(MG)	4.5	18	81	20	4.5	5.625	76.1151	14.1372	7.0686
	5.0	18	90	20	5.0	6.250	84.5723	15.7080	7.8540
	3.0	18	54	20	3.0	3.750	50.7434	9.5190	4.7595
Erred	3.5	18	63	20	3.5	4.375	59.2006	11.1055	5.5528
Gear	4.0	18	72	20	4.0	5.000	67.6578	12.6920	6.3460
(E1)	4.5	18	81	20	4.5	5.625	76.1151	14.2785	7.1393
	5.0	18	90	20	5.0	6.250	84.5723	15.8650	7.9325
	3.0	18	54	20	3.0	3.750	50.7434	9.6133	4.8066
Erred	3.5	18	63	20	3.5	4.375	59.2006	11.2155	5.6077
Gear	4.0	18	72	20	4.0	5.000	67.6578	12.8177	6.4088
(E2)	4.5	18	81	20	4.5	5.625	76.1151	14.4199	7.2100
	5.0	18	90	20	5.0	6.250	84.5723	16.0221	8.0111

For obtaining the next value of interference volume, one of the gears of gear pair is rotated by 1° and due to the meshing, the other gear is also rotated by 1° in opposite direction. The interference volume between the two gears is checked again and the value is noted. In this way for 21 such positions the two gears are rotated and at every instance the interference volume is checked and noted (The interference volume cycle repeats after 20° of rotation for 18 number of teeth). All the interference volume values are recorded in table 5. Figure 1 shows one of the gear pair of master gear and erred gear 1 (MG_E1_m3.5_z18_phi20).

Table 3. Dimensions of Spur Gear in Stub Tooth dimension System

Parameters [26]	Module, mm (m)	No. of Teeth (z)	Pitch Circle Diameter, mm (m*z)	Pressure Angle, ° (Phi)	Addendum, mm (0.8*m)	Dedendum, mm (1.0*m)	Base Circle Diameter, mm (m*z*cos(phi))	Circular pitch, mm (πm)	Tooth Thickness, mm (πm/2)
	3.0	18	54	20	2.4	3.0	50.7434	9.4248	4.7124
Master	3.5	18	63	20	2.8	3.5	59.2006	10.9956	5.4978
Gear	4.0	18	72	20	3.2	4.0	67.6578	12.5664	6.2832
(SMG)	4.5	18	81	20	3.6	4.5	76.1151	14.1372	7.0686
	5.0	18	90	20	4.0	5.0	84.5723	15.7080	7.8540
	3.0	18	54	20	2.4	3.0	50.7434	9.5190	4.7595
Erred	3.5	18	63	20	2.8	3.5	59.2006	11.1055	5.5528
Gear	4.0	18	72	20	3.2	4.0	67.6578	12.6920	6.3460
(SE1)	4.5	18	81	20	3.6	4.5	76.1151	14.2785	7.1393
	5.0	18	90	20	4.0	5.0	84.5723	15.8650	7.9325
	3.0	18	54	20	2.4	3.0	50.7434	9.6133	4.8066
Erred	3.5	18	63	20	2.8	3.5	59.2006	11.2155	5.6077
Gear	4.0	18	72	20	3.2	4.0	67.6578	12.8177	6.4088
(SE2)	4.5	18	81	20	3.6	4.5	76.1151	14.4199	7.2100
	5.0	18	90	20	4.0	5.0	84.5723	16.0221	8.0111

Table 4. Combinations/Assemblies of Various Gears (Standard Tooth Dimension System)

Module, mm (m)	No. of Teeth, (z)	Pressure Angle, ° (phi)	Meshing of Master Gear with Master Gear	Meshing of Master Gear with Erred Gear 1	Meshing of Master Gear with Erred Gear 2
3			MG_MG_m3_z18_phi20	MG_E1_m3_z18_phi20	MG_E2_m3_z18_phi20
3.5			MG_MG_m3.5_z18_phi20	MG_E1_m3.5_z18_phi20	MG_E2_m3.5_z18_phi20
4	18	20	MG_MG_m4_z18_phi20	MG_E1_m4_z18_phi20	MG_E2_m4_z18_phi20
4.5			MG_MG_m4.5_z18_phi20	MG_E1_m4.5_z18_phi20	MG_E2_m4.5_z18_phi20
5			MG_MG_m5_z18_phi20	MG_E1_m5_z18_phi20	MG_E2_m5_z18_phi20

The detailed procedure for determining the TE for all the combinations is explained in the sections from 3.1 to 3.3

Figure 1. Assembly/Meshing of Master Gear and Erred Gear 1 of standard tooth dimension system having module 3.5 mm, number of teeth 18 and pressure angle 20° (MG_E1_m3.5_z18_phi20).

3.1. Meshing of Master Gear with Master Gear (MG_MG)

Two gears forming a pair with combinations as shown in table 4 are taken one by one for analysis. The two gears are meshed and for this initial position interference between them is checked. Then gear pairs are rotated for contact angle (20° in case of 18 number of teeth). It is observed that the two gears are just touching each other, indicating that there is no interference occurring between them and hence the interference volume is zero. This also confirm the correctness of procedure.

3.2. Meshing of Master Gear with Erred Gear 1 (MG_E1)

In this case master gear is meshed with erred gear 1 (gear with 1% pitch error). The various combinations formed are represented in table 4. The same procedure as discussed in section 3.1 above is followed to determine the interference volume. Here it is observed that there is interference occurring in the two gears in mesh during rotation of contact angle (20° in case of 18 number of teeth). The values of interference volume are noted and tabulated in table 5 for plotting graph between interference volume vs rotation angle.

3.3. Meshing of Master Gear with Erred Gear 2 (MG_E2)

Now finally the master gear is meshed with erred gear 2 (gear with 2% pitch error). The various combinations are given in table 4. Here also similar procedure as discussed in section 3.1 is used to check the interference between the gears. It is again observed that there is an interference occurring in gears in mesh. The interference volume values for contact angle of 20° rotation is noted in table 5. Based on the interference volume values of table 5, various graphs are plotted between angle of rotation and interference volume values. These graphs are shown in the figure 2 and figure 3, respectively.

Table 5. Interference Volume values (mm^3) when Master Gear meshes with the Master Gear, Erred Gear 1, and Erred Gear 2 for Standard Tooth Dimension System

Angle of Rotation (°)	Master Gears [MG]	MG_ E1_ m3_ z18_ phi 20	MG_ E1_ m3.5_ z18_ phi 20	MG_ E1_ m4_ z18_ phi 20	MG_ E1_ m4.5_ z18_ phi 20	MG_ E1_ m5_ z18_ phi 20	MG_ E2_ m3_ z18_ phi 20	MG_ E2_ m3.5_ z18_ phi 20	MG_ E2_ m4_ z18_ phi 20	MG_ E2_ m4.5_ z18_ phi 20	MG_ E2_ m5_ z18_ phi 20
0	0	0.4000	0.5400	0.7000	0.8800	1.1000	1.1000	1.4800	1.9400	2.4400	3.0200
1	0	0.3700	0.5000	0.6500	0.8200	1.0000	1.0500	1.4300	1.8700	2.3600	2.9300
2	0	0.3600	0.5000	0.6300	0.8100	1.0000	1.0100	1.3800	1.8000	2.2900	2.8200
3	0	0.3600	0.4900	0.6500	0.8200	1.0100	1.0300	1.3900	1.8200	2.3000	2.8300
4	0	0.3600	0.5000	0.6500	0.8200	1.0100	1.0300	1.4000	1.8300	2.3100	2.8600
5	0	0.3700	0.4900	0.6600	0.8200	1.0100	1.0400	1.4000	1.8300	2.3200	2.8700
6	0	0.3600	0.5000	0.6500	0.8200	1.0100	1.0300	1.4000	1.8300	2.3200	2.8600
7	0	0.3600	0.4900	0.6500	0.8200	1.0100	1.0200	1.3900	1.8200	2.3000	2.8400
8	0	0.3600	0.4900	0.6300	0.8100	1.0000	1.0100	1.3800	1.8000	2.2900	2.8200
9	0	0.3700	0.5000	0.6500	0.8200	1.0100	1.0600	1.4500	1.9000	2.3900	2.9700
10	0	0.4000	0.5500	0.7000	0.9000	1.1000	1.1000	1.5200	1.9800	2.5000	3.0800
11	0	0.3700	0.5000	0.6500	0.8200	1.0100	1.0600	1.4500	1.9000	2.3900	2.9700
12	0	0.3600	0.4900	0.6300	0.8100	1.0000	1.0100	1.3800	1.8000	2.2900	2.8200
13	0	0.3600	0.5000	0.6500	0.8200	1.0100	1.0200	1.3900	1.8200	2.3000	2.8400
14	0	0.3600	0.4900	0.6500	0.8200	1.0100	1.0300	1.4000	1.8300	2.3200	2.8600
15	0	0.3700	0.5000	0.6600	0.8200	1.0100	1.0400	1.4000	1.8300	2.3200	2.8700
16	0	0.3600	0.4900	0.6500	0.8200	1.0100	1.0300	1.4000	1.8300	2.3100	2.8600
17	0	0.3600	0.5000	0.6500	0.8200	1.0100	1.0300	1.3900	1.8200	2.3000	2.8300
18	0	0.3600	0.4900	0.6300	0.8100	1.0000	1.0100	1.3800	1.8000	2.2900	2.8200
19	0	0.3700	0.4900	0.6500	0.8200	1.0000	1.0500	1.4300	1.8700	2.3600	2.9300
20	0	0.4000	0.5400	0.7000	0.8800	1.1000	1.1000	1.4800	1.9400	2.4400	3.0200

4. Meshing and Analysis of Gear Pair of Stub Tooth Dimension System

In this case the various gear pairs formed are listed in table 6. First the master gear is meshed with the master gear and the interference volume values obtained by rotation of gear pair for one mesh cycle is listed in table 7. Next using the same procedure, the interference volume values are obtained for master gear - erred gear 1 and master gear - erred gear 2 pair. All these values are listed in Table 7. Using these values various graphs are plotted between the interference volume values and angle of rotation. These are represented in figure 4 and figure 5 respectively.

Table 6. Combinations/Assemblies of Gears (Stub Tooth Dimension System)

Module, mm (m)	No. of Teeth, (z)	Pressure Angle, ° (phi)	Meshing of Master Gear with Master Gear	Meshing of Master Gear with Erred Gear 1	Meshing of Master Gear with Erred Gear 2
3			SMG_SMG_m3_z18_phi20	SMG_SE1_m3_z18_phi20	SMG_SE2_m3_z18_phi20
3.5			SMG_SMG_m3.5_z18_phi20	SMG_SE1_m3.5_z18_phi20	SMG_SE2_m3.5_z18_phi20
4	18	20	SMG_SMG_m4_z18_phi20	SMG_SE1_m4_z18_phi20	SMG_SE2_m4_z18_phi20
4.5			SMG_SMG_m4.5_z18_phi20	SMG_SE1_m4.5_z18_phi20	SMG_SE2_m4.5_z18_phi20
5			SMG_SMG_m5_z18_phi20	SMG_SE1_m5_z18_phi20	SMG_SE2_m5_z18_phi20

Table 7. Interference Volume values (mm^3) when Master Gear meshes with Master Gear, Erred Gear 1, and Erred Gear 2 for Stub Tooth Dimension System

Angle of Rotation (°)	Master Gears [SMG]	SMG_SE1_m3_z18_phi 20	SMG_SE1_m3.5_z18_phi 20	SMG_SE1_m4_z18_phi 20	SMG_SE1_m4.5_z18_phi 20	SMG_SE1_m5_z18_phi 20	SMG_SE2_m3_z18_phi 20	SMG_SE2_m3.5_z18_phi 20	SMG_SE2_m4_z18_phi 20	SMG_SE2_m4.5_z18_phi 20	SMG_SE2_m5_z18_phi 20
0	0	0.2590	0.3526	0.4604	0.5828	0.7194	0.7310	0.9950	1.2994	1.6446	2.0304
1	0	0.2583	0.3517	0.4594	0.5814	0.7178	0.7495	1.0202	1.3324	1.6864	2.0819
2	0	0.2935	0.3995	0.5219	0.6605	0.8154	0.8449	1.1499	1.5020	1.9009	2.3468
3	0	0.3476	0.4731	0.6181	0.7822	0.9658	0.9557	1.2989	1.6992	2.1506	2.6550
4	0	0.3651	0.4969	0.6492	0.8215	1.0143	1.0238	1.3937	1.8201	2.3036	2.8439
5	0	0.3665	0.4988	0.6516	0.8246	1.0181	1.0306	1.4028	1.8323	2.3189	2.8629
6	0	0.3651	0.4971	0.6492	0.8217	1.0144	1.0261	1.3965	1.8242	2.3087	2.8504
7	0	0.3497	0.4759	0.6216	0.7867	0.9713	0.9646	1.3124	1.7150	2.1705	2.6796
8	0	0.2963	0.4032	0.5267	0.6666	0.8229	0.8558	1.1641	1.5215	1.9256	2.3774
9	0	0.2585	0.3518	0.4592	0.5816	0.7181	0.7556	1.0279	1.3433	1.7001	2.0989
10	0	0.2588	0.3522	0.4602	0.5824	0.7190	0.7300	0.9936	1.2978	1.6426	2.0638
11	0	0.2585	0.3518	0.4624	0.5816	0.7181	0.7556	1.0289	1.3677	1.7001	2.0989
12	0	0.2963	0.4032	0.5402	0.6666	0.8229	0.8558	1.1656	1.5215	1.9256	2.3774
13	0	0.3497	0.4759	0.6303	0.7867	0.9713	0.9646	1.3137	1.7150	2.1705	2.6796
14	0	0.3650	0.4971	0.6495	0.8217	1.0144	1.0261	1.3968	1.8242	2.3087	2.8504
15	0	0.3664	0.4988	0.6514	0.8246	1.0181	1.0306	1.4029	1.8323	2.3189	2.8629
16	0	0.3655	0.4969	0.6490	0.8215	1.0143	1.0238	1.3934	1.8201	2.3036	2.8439
17	0	0.3548	0.4731	0.6074	0.7832	0.9658	0.9557	1.3003	1.6992	2.1506	2.6550
18	0	0.3043	0.3995	0.5090	0.6605	0.8154	0.8449	1.1493	1.5020	1.9009	2.3468
19	0	0.2602	0.3517	0.4595	0.5814	0.7178	0.7495	1.0198	1.3324	1.6864	2.0819
20	0	0.2590	0.3526	0.4604	0.5828	0.7195	0.7310	0.9950	1.2994	1.6446	2.0304

5. Result and Discussions

Based on the values of interference volume values for different cases of change of module and pitch error as reported in table 5 and table 7 for standard tooth dimension system and stub tooth dimension system respectively, various graphs are plotted between the angle of rotation on x axis and the interference volume values on y axis, respectively.

a. Figure 2 and Figure 3 shows the graph between interference volume values and rotation angle of gear pair for 1% and 2% pitch error cases with module changes from 3 to 5 mm respectively for standard tooth dimension system. The x axis contains the rotation angle in degree (°) and the y axis represents the interference volume values in cubic mm.
b. Similarly, the graph between interference volume values and rotation angle of gear pair for 1% and 2% pitch error cases with module changes from 3 to 5 mm for stub tooth dimension system is shown in Figure 4 and Figure 5, respectively. The x axis contains the rotation angle in degree (°) and the y axis represents the interference volume values in cubic mm. The angle of rotation is 20° since the curve repeats after 20° for 18 number of teeth.
c. In this paper the effects of variation of module on TE along with pitch error in gear is considered for the analysis purpose.

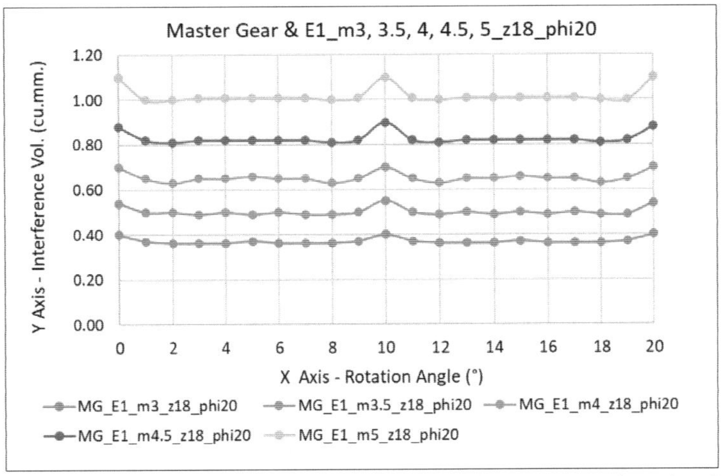

Figure 2. Graph between interference volume in cu. mm and angle of rotation (°) for analysing the effects of change of Module (3, 3.5, 4, 4.5 and 5) on TE with pitch error of 1% (Master Gear & Erred Gear 1). [Standard Tooth Dimension System]

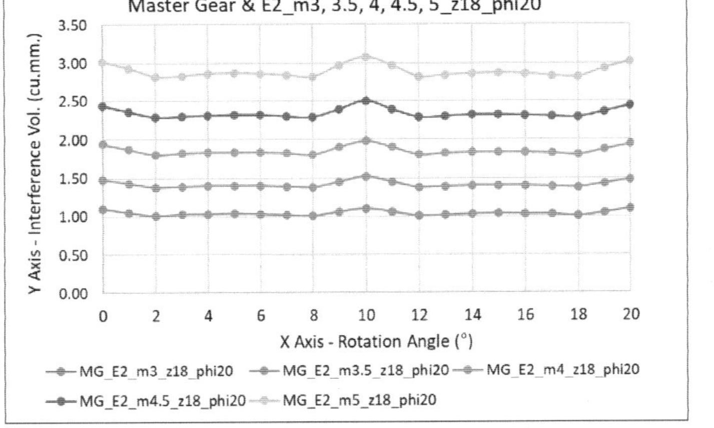

Figure 3. Graph between interference volume in cu. mm and angle of rotation (°) for analysing the effect of change of Module (3, 3.5, 4, 4.5 and 5) on TE with pitch error of 2% (Master Gear & Erred Gear 2). [Standard Tooth Dimension System]

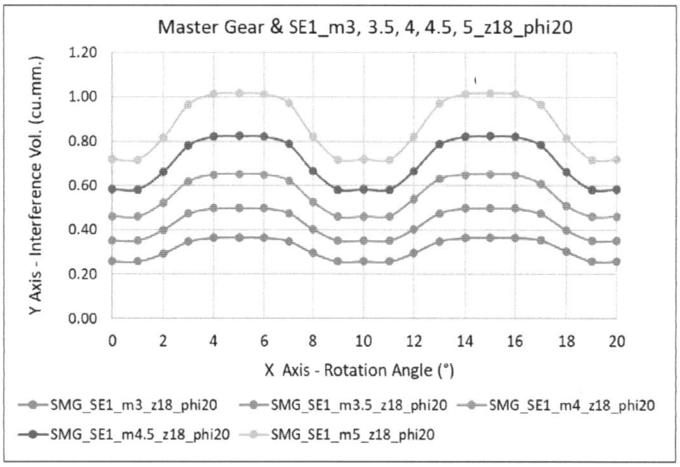

Figure 4. Graph between interference volume in cu. mm and angle of rotation (°) for analysing the effects of change of Module (3, 3.5, 4, 4.5 and 5) on TE with pitch error of 1% (Master Gear & Erred Gear 1). [Stub Tooth Dimension System]

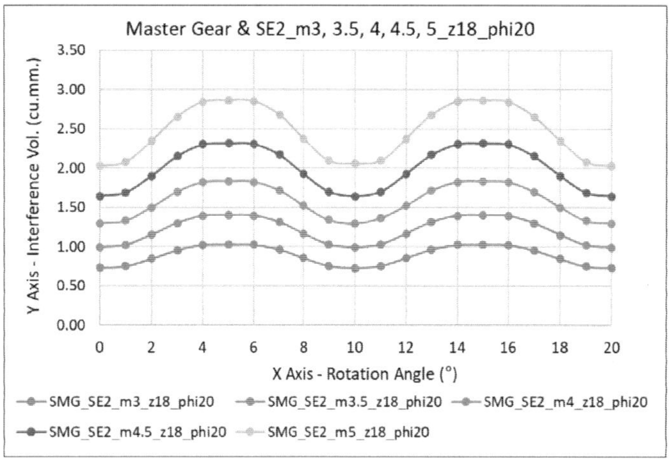

Figure 5. Graph between interference volume in cu. mm and angle of rotation (°) for analysing the effects of change of Module (3, 3.5, 4, 4.5 and 5) on TE with pitch error of 2% (Master Gear & Erred Gear 2). [Stub Tooth Dimension System]

Following are the observations drawn from the graphs of interference volume values and angle of rotation (figure 2 to figure 5)

1. There is a fluctuation in the values of interference volume, when angle of rotation changes from 0 to 20° (start and end of mesh cycle), there are some points having peak values of interference volume. This means that the TE is not constant for angle of rotation. The points having peak values are the point of concerned for the noise and vibrations in gears. This is observed for both the tooth dimension system i.e., standard tooth dimension system and stub tooth dimension system.
2. As the module changes from 3 to 5 mm, the interference volume values increases and hence the TE increases in standard as well as in stub tooth dimension system.
3. The interference volume values are found to be more in case of 2% pitch error than 1% pitch error and consequently the TE is more in 2% pitch error case. This is also observed in both tooth dimension system.

4. The graph is identical about its mid-point. i.e., it repeats after 10° for 18 number of teeth for both standard tooth dimension and stub tooth dimension system.
5. It is also observed that the interference volume or TE is more in standard tooth dimension system than stub tooth dimension system.

The work of the paper/outcomes are in line with the work of researchers presented in [3, 4, 6, 12, 14, 17]

6. Conclusions

In this work the consequences occurring due to the changes in gear parameter module on TE for spur gear pair with pitch error, using the interference volume method is presented. The spur gear without pitch error and with pitch error consisting of involute profile for standard tooth dimension and stub tooth dimension systems are modelled and meshed/assembled in SOLIDWORKS software. The interference volume values obtained from the meshing of gear pairs are recorded for one mesh cycle. Various graphs were plotted between the angle of rotation and interference volume values. It is found from the analysis that, as the module increases the TE also increases, TE also increases as the pitch error increase from 0 (no pitch error) to 2%. The effect of variation of module and pitch error is more in the standard tooth dimension system than that in the stub tooth dimension system.

References

[1] Beghini M, Presicce F and Santus C 2004 A method to define profile modification of spur gear and minimize the transmission error *Technical Paper American Gear Manufacturers Association* (Alexandria, Virginia) pp 1-9

[2] Hotait M A and Kahraman A 2013 Experiments on the relationship between the dynamic transmission error and the dynamic stress factor of spur gear pairs *Mech Mach Theory* **70** 116–128

[3] Kohler H and Regan R 1985 The Derivation of Gear Transmission Error from Pitch Error Records *Proc. Inst. Mech. Eng.* Part C **199** 195-01.

[4] Siyu C. and Jinyuan T. 2017. Effects of staggering and pitch error on the dynamic response of a double-helical gear set *J. Vib. Control* **23** 1844-56.

[5] Velex Ph, Chapron M, Fakhfakh H, Bruyère J and Becquerelle S 2016 On transmission errors and profile modifications minimising dynamic tooth loads in multi-mesh gears *J. Sound Vib.* **379** 28–52.

[6] Nina Sainte-Marie 2016 *A transmission-error-based gear dynamic model: Applications to single- and multi-mesh transmissions* (Doctoral dissertation, Université de Lyon)

[7] Sánchez Miryam B, Pleguezuelos M and Pedrero J 2019 Influence of profile modifications on meshing stiffness, load sharing, and transmission error of involute spur gears *Mech Mach Theory* **139** 506–25

[8] Palermo A, Britte L, Janssens K, Mundo D and Desmet W 2018 The measurement of Gear Transmission Error as an NVH indicator: Theoretical discussion and industrial application via low-cost digital encoders to an all-electric vehicle gearbox *Mech Syst Signal Process* **110** 368–89

[9] Shweiki S, Rezayat A, Tamarozzi T and Mundo D 2019 Transmission Error and strain analysis of lightweight gears by using a hybrid FE-analytical gear contact model *Mech Syst Signal Process* **123** 573–90

[10] Zhiying C and Pengfei J 2020 Research on the variation of mesh stiffness and transmission error for spur gear with tooth profile modification and wear fault *Eng. Fail. Anal.* **122** 2020 105184

[11] Pleguezuelos M, Sánchez Miryam B, and Pedrero JoséI 2020 Control of transmission error of high contact ratio spur gears with symmetric profile modifications *Mech Mach Theory* **149** 103839

[12] Padmasolala G., Lin H.H., Oswald F.B. 2000 *Influence of tooth spacing error on gears with and without profile modifications* (NASA/TM-2000-210061 PTG14436)

[13] Bruyere J, Velex P, Guilbert B, and Houser D R 2019 An analytical study on the combination of profile relief and lead crown minimizing transmission error in narrow-faced helical gears *Mech Mach Theory* **136** 224-43

[14] Handschuh M J 2013 *An Investigation into the impact of random spacing errors on static transmission error and root stresses of spur gear pairs* (Master of Science Thesis, The Ohio State University)

[15] Bartosova D, Otipka V and Rehak K 2018 Determination of transmission error in spur gear by numerical approach *Vib. Proced.* **19** 284-88

[16] Park C I 2019 Tooth friction force and transmission error of spur gears due to sliding friction *J. Mech. Sci. Technol* **33** 1311-19

[17] Somani S K 2004 *Transmission Error in Spur Gear* (PhD Thesis, Institute of Engineering & Technology, Devi Ahilya University, Indore)

[18] Bruzzone F, Maggi T, Marcellini C and Rosso C 2021 2D nonlinear and non-Hertzian gear teeth deflection model for static transmission error calculation *Mech Mach Theory* **166** 104471

[19] Duan T, Wei J, Zhang A, Xu Z and Lim T C 2021 Transmission error investigation of gearbox using rigid-flexible coupling dynamic model: Theoretical analysis and experiments *Mech Mach Theory* **157** 104213

[20] Benaïcha Y, Joël P-L, Beley J-D, Rigaud E, Thouverez F 2021 On a flexible multibody modelling approach using FE-based contact formulation for describing gear transmission *Mech Mach Theory* **167** 104505.

[21] Ambaye, G.A. and Lemu, H.G. 2020 Effect of backlash on transmission error and time varying mesh stiffness *Proc. Int. Work. of Advanced Manufacturing and Automation* (Springer, Singapore) pp18-28.

[22] Tao L, Tian D, Tang S, Wu X and Li B 2021 Dynamical modelling and simulation of spur gears with flank pitch error *Res. Sq.* **1** (*Preprint* rs.3.rs-534359/v1)

[23] Pleguezuelos M, Sánchez M B and Pedrero J I 2021 Analytical model for meshing stiffness, load sharing, and transmission error for spur gears with profile modification under non-nominal load conditions *Appl. Math. Model.* **97** 344–65

[24] Chin Z Y, Smith W A, Borghesani P, Randall R B and Peng Z 2021 Absolute transmission error: A simple new tool for assessing gear wear *Mech Syst Signal Process* **146** 107070

[25] Karma V K, Maheshwari G and Somani S K 2020 Analysis of effects of pressure angle and pitch error variation on transmission error in spur gear using interference volume method *Int. J. Adv. Eng. Sci. Technol.*15-22

[26] Maitra Gitin M 2001 *Handbook of Gear Design* 2nd Edn (Tata McGraw Hill Publishing Company Limited New Delhi)

Finite Element Modelling for Bending and Vibration Analysis of Composite and Sandwich Spherical Shells

Vikram Parmar[1*], **Najeeb ur Rahman**[1], **Naushad Alam**[1], **Varun Sharma**[1] **and Aftab Alam**[2]

[1]Department of Mechanical Engineering, ZHCET, AMU, Aligarh, UP-202002, India.
[2]Department of Business Administration, AMU, Aligarh UP-202002, India.
*vikramparmar@zhcet.ac.in

Abstract. Finite element modelling is used to illustrate the bending and vibration response of composite and sandwich spherical shells. ABAQUS is used to model the shell as a three-dimensional deformable solid part. Composites are modelled using mixed modelling technique. The quadrilateral continuum shell element, which has eight nodes, is used for finite element meshing. For a three-layered composite spherical shell with simply supported boundary conditions, the accuracy of the employed element is first evaluated. After that, three potential end conditions are evaluated for further analysis: cantilever (CFFF), two opposing sides clamped (CFCF), and all sides clamped (CCCC). Bending response is produced by applying two types of pressure loads to the exterior surfaces of composite and sandwich spherical shells: uniformly distributed load (UDL) and sinusoidal load (SINO). For various span to thickness ratios, free vibration response and forced vibration response (Time History) are provided under step pressure (UDL and SINO) loading. The element type being used is found to be quite precise and sturdy.

Keywords: FEM, ABAQUS, composite, sandwich, spherical shell, vibration, sinusoidal.

1. Introduction

Many mechanical and structural systems comprise of shell type geometrical features. There are different ways of manufacturing these shells. It depends upon their use and practical application. The Piezo laminated plates and shells are very common in smart structural applications. Composites and sandwich, as laminated structures, have received enormous research attention in present days. Such structures can be exposed to a variety of static and dynamics loads under different boundary and environmental conditions. A large variety of structural systems like aircraft wings, fuselage and empennage structures are generally modeled as shells, beams and panels. The composite and sandwich shell and beams also realize utilization as micro positioners, micro actuators and micro resonators that are used in, measuring apparatus, medical devices, robotics etc. These structures are generally made up of sandwich and composite laminates due to their light weight, high stiffness and good strength. They are often subjected to dynamic loading. Composite laminates are gradually emerging as a substitute to conventional materials and are nowadays frequently used in various applications subjected to dynamic loading such as heavy machines and vehicles, defense equipment, pressure vessels, satellites and aerospace structures. Vibration phenomenon may lead to resonance condition in mechanical structures. In addition to this it may also result in fatigue and fracture failures. Therefore, it is quite important to carry out vibration analysis of these structures in order to make them safe while in use. A number of FEM software packages are commercially available for modelling and analysis of composite and sandwich structures, such as

ABAQUS, NASTRAN, ANSYS etc. ABAQUS is well recognized among academic, research and development organizations due to ease in use, more user-friendly menu driven interface, wide material modeling capability and variety of available modules. Kumar et al. [1] investigated the response of laminated composite shells, combined with piezoelectric devices and actuators, which are exposed to mechanical, electrical and thermal loads using FE formulation based on FSDT. They used spherical and cylindrical shell geometries and carried out both static and dynamic analysis in their study. Yasin and Kapuria [2] used a new efficient four node finite element based on ZIGT for shallow shells with laminated composite and piezoelectric layers. They take into account layer wise mechanics and electromechanical coupling in their formulation. At each node of the new element, there are seven degrees of freedom. Hossain et al [3] used an improved FEM using FSDT for the linear investigation of spherical and cylindrical shells/shell-panels and uses the approach of Mixed Interpolation of Tensorial Components (MITC). Vel et al. [4] found a precise solution for the three-dimensional deformation of thick plates subjected to surface mechanical and electrical loads. They considered laminated plates of rectangular planform that consists of embedded actuators. The plates are assumed in simply supported boundary condition at the edges. Yasin and Kapuria [5] extended, an earlier developed quadrilateral element for laminated composite plates, to laminated shells for the static and free vibration response. The results obtained using the new element for shallow spherical and cylindrical shells are well compared with the available elasticity solutions and finite element solutions. The element's performance is demonstrated to be straightforward, precise, and productive. Qatu et al. [6-7] reviewed the recent literature on static as well as dynamic behavior of composite shells and the theory applied for the analysis. Biswal, D. K., & Mohanty, S. C. [8] used first order shear deformation theory for vibration analysis of sandwich panels with spherical geometry. The configuration of sandwich panel has viscoelastic core and elastic face sheets. To arrive at the equation of motion, Hamilton's principle is used. Pang et al. [9] presented an analytical model based on multi-segment partitioning strategy and FSDT. Their results show that the methodology used yield a stable and rapid convergence. Tang et al [10] developed a new procedure for getting vibration response of stepped cylindrical shell with circular cross section under various boundary conditions. They utilized reverberation-ray matrix method in their formulation to determine frequencies of vibration and steady state responses. The method used is proved to be precise and accurate for finding vibration response of stepped right circular cylindrical shells. Panda and Singh [11] presented a nonlinear finite element solution for vibration response of composite curved shells with large amplitudes of vibration. The higher order theory is used for laminate mechanics. To obtain the concerned governing equations, the variational principle is used. Shaukat Rafi S.M et al [12] performed FEM for vibration analysis of skew sandwich hybrid plates with piezoelectric top and bottom layers. Rafi et al [13] presented a detailed study on hybrid sandwich plate using commercial FE package ANSYS.

Civalek [14] performed static and dynamic analysis of shallow shells sitting on an elastic foundation. Donnell theory is used for formulation. Nonlinearity is considered while deriving the governing equations. Discrete convolution method and differential quadrature method are utilized to discretize the equations. Behera et al [15] examined the free vibration of different geometric aspects and cut-out or without central on spherical shell laminated composite panels using ABAQUS commercial finite element tool. Ma et al [16] presented two reinforcement methods to be applied at openings in spherical shells, to minimize effects of buckling and local weakening, when subjected to an external pressure. Their study includes both numerical analysis and experimentation. A good comparison of theoretical results with experimental data has been shown. Zhu et al [17] optimized the parameters of opening reinforcements used in spherical shells to have minimum instability in buckling. The optimization results were then used to fabricate experimental test models. Simulated buckling behaviour of shells was also obtained based on experimental data. Malavika et al [18] analyzed cracks in plates and evaluated the stress intensity factor using ABAQUS. They considered surface cracks as well as embedded cracks in their study. Under simply supported boundary conditions, Fantuzzi et al [19] investigated free vibration characteristics of spherical and cylindrical shells made-up of functionally graded material. Two types of FGM material configurations are considered for both shell geometries. A comparison of free vibration results is shown for exact and FE models.

Using finite element modelling, the bending and vibration response of composite and sandwich spherical shells is discussed in this paper. FE package ABAQUS is used to model the shell as a three-dimensional deformable solid part. A quadrilateral continuum shell element with 8 nodes is used for finite element meshing. First, the accuracy of the used element is evaluated for a three-layered composite spherical shell with clamped boundary conditions, and the findings are compared to those found in the literature. The results of free vibration, forced vibration and bending response are then presented for clamped-clamped (CFCF), cantilever (CFFF) and all side clamped (CCCC) boundary conditions for two different types of loads (UDL and sinusoidal load).

2. Finite Element Modeling using ABAQUS

2.1 Specifications of composite and sandwich spherical shells

Span to thickness ratio, $S = a/b = (10, 20,100)$.
Aspect ratio = $a/b = 1$, $Rx / a = 5$, $Ry / b = 5$
where the dimensions in the *x*- and *y*-axes are '*a*' and '*b*', respectively.
h (=1) is the dimension in *z* - direction.
The radii of curvatures in the *y* and *x* directions, respectively, are *Ry* and *Rx*.

Figure. 1 shows configuration of shallow composite and sandwich spherical shells. A lay-up of (0°/90°/0°) is considered for the composite laminae. The layup order and thicknesses of the layers are mentioned in the figure. Table-1 shows the properties of the materials used in the configuration of shells. The finite element mesh is created in ABAQUS using a quad in-plane general-purpose continuum shell element (SC8R) with eight nodes. The geometric order is considered to be linear. The used elements have only displacement degrees of freedom and no rotational degrees of freedom. Upon modelling these elements look like 3-D continuum solids however their response, both kinematic and constitutive, is similar to conventional shell elements. The shallow composite spherical shell finite element model created in ABAQUS is shown in Figure 2.

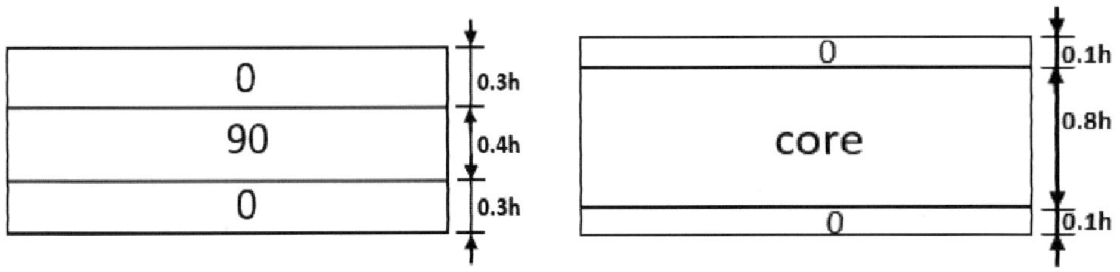

Figure. 1. Composite lamina configuration for shallow composite (0°/90°/0°) and sandwich spherical shells.

Table 1. The properties of materials used [5]

Properties of materials	Y_1 (GPa)	Y_2 (GPa)	Y_3 (GPa)	G_{12} (GPa)	G_{23} (GPa)	G_{31} (GPa)	υ_{12}	υ_{23}	υ_{31}	ρ (kg/m^3)
Composite	172.5	6.9	6.9	3.45	1.38	3.45	0.25	0.25	0.25	1600
Core	0.276	0.276	3.45	0.1104	0.414	0.414	0.02	0.02	0.25	1600

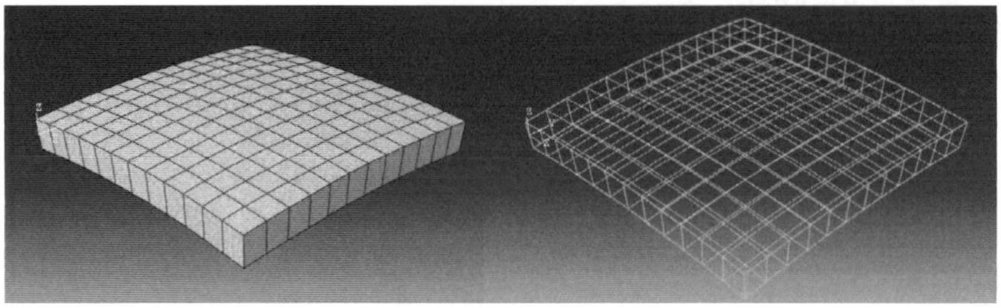

Figure. 2. FE model of shallow composite spherical shell using continuum shell element (SC8R).

Each elastic substrate mechanical node has three translational degrees of freedom: *x*, *y*, and *z* translations. The results are presented with a converged 12 x 12 mesh size. The boundary conditions used are cantilever at one edge (CFFF), clamped at opposite edges (CFCF) and clamped at all four edges (CCCC).

3. Results and discussion

3.1 Validation

For a three-layer composite spherical shell of square planform with layup (0°/90°/0°), the correctness of the employed element for static and dynamic response is first tested. [5]. Linear perturbation frequency procedure is used for obtaining dynamic free vibration response. Table 2 shows the comparison of non-dimensional natural frequency associated with bi-symmetric modes of composite spherical panel with all sides clamped (CCCC) and Table 3 shows the comparison of non-dimensional central deflection for simply supported (SSSS) boundary condition under uniformly distributed load. The non-dimensionalization is done using Eqs (1) and (2) respectively.

$$\bar{\omega}_n = \omega_n * a * S * \sqrt{\frac{\rho}{E_2}} \quad (1)$$

$$\hat{w} = 1000wY_0\,/hS^4P_0 \quad (2)$$

where, $S = a/h$, $P_0 = 1$ for UDL and Y_2 is equal to E_2 (MPa).
Both the results are found in accordance with the published reference which shows the correctness of the used modelling technique and robustness of the used element.

Table 2. Non-dimensional natural frequencies for all sides clamped boundary Condition.

Natural frequency ($\tilde{\omega}_n$) [S = 10, a/b =1, R/a = 5]			
Three-layer composite spherical shell (0°/90°/0°)			
S	Mode	Ref [5]	3D FE (Abaqus)
10	1	17.382	18.232
	2	-	25.274
	3	-	28.537
	4	-	33.361

Table 3. Non-dimensional central deflection for simply supported boundary Condition.

Non dimensional Central deflection ($\hat{w}$) [S = 10, a/b = 1, R/a = 5]		
Three-layer composite spherical shell (0°/90°/0°)		
S	Ref [5]	3D FE (Abaqus)
10	11.145	11.157

3.2 Free vibration response

3D FE model for dynamic response is developed using ABAQUS for composite spherical shell. The natural frequencies have been obtained for span to thickness ratio, S = 10, 20 and 100 for cantilever and clamped boundary conditions. These results have been tabulated in Table 4 for first five modes. As observed from the table, the non-dimensional natural frequencies for all sides clamped (CCCC) boundary condition are higher than corresponding frequencies for cantilever (CFFF and CFCF) boundary condition. Further the Non-dimensional natural frequencies increase as the shell becomes thinner. The frequencies of vibration of thick composite spherical shells are smaller in comparison to their corresponding values in thick sandwich spherical shells for all the considered boundary conditions. This is due to higher stiffness to mass ratio in sandwich structures. However reverse is true in thin shells where the effect of span to thickness ratio becomes more dominant.

Table 4. Non-dimensional natural frequencies for different boundary conditions.

		CFFF		CFCF		CCCC	
S	Mode	Composite	Sandwich	Composite	Sandwich	Composite	Sandwich
10	1	4.1472	13.2227	15.0845	41.6402	18.2392	46.4546
	2	4.6105	14.5456	15.1601	42.0954	25.2748	63.4697
	3	10.2822	24.8051	18.0076	48.0286	28.5378	75.3438
	4	11.2318	27.7264	21.7825	51.9055	33.3618	86.9291
	5	15.7915	45.2056	26.7275	68.3596	37.4141	91.4909
20	1	4.7787	8.1756	25.6014	37.6061	29.5139	39.4342
	2	5.3546	9.0783	25.6213	37.6439	36.4591	47.8699
	3	12.6256	16.2424	27.4926	39.3102	48.8149	66.9353
	4	20.2699	24.6766	36.5619	47.7807	53.2585	67.0411
	5	23.4403	33.1214	43.5814	51.9388	54.6447	72.4134
100	1	5.8909	2.0147	70.7124	21.1278	77.4289	21.9927
	2	6.2695	2.0679	70.7245	21.1429	84.6020	22.2724
	3	17.5155	4.4112	73.3191	21.6827	94.6783	24.6101
	4	30.6673	8.4067	74.5832	21.6843	98.6973	29.7223
	5	35.5603	10.9812	85.3429	23.3263	98.7669	30.0247

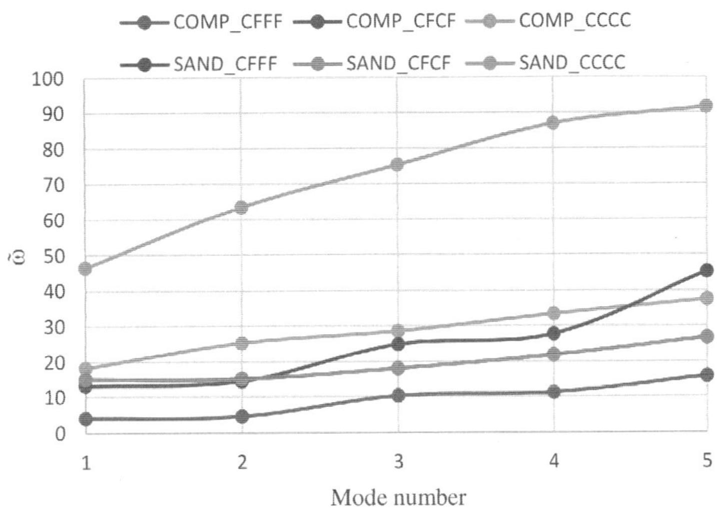

Figure 3. The first five non dimensional frequencies for cantilever composite and sandwich spherical (S = 10, a/b = 1, a/h = 10) shells.

3.3 Bending analysis

The bending study considered composite and sandwich spherical shells with cantilever (CFFF) and clamped boundary conditions (CFCF and CCCC) exposed to two load variations (UDL and SINO). Shells are modelled with span to thickness ratio, $S = 10$, 20 and 100. The loads are applied on the outer surfaces of the shells. Central deflection is recorded for the clamped shells and the tip deflection is recorded for the cantilever shells under both type of loads. The non-dimensional deflections have been tabulated in Table 5. A general decreasing trend of deflection is observed as the span to thickness ratio increases and the deflection due to UDL is found to be greater than deflection due to sinusoidal load.

Table 5. Non-dimensional deflection for different boundary Conditions.

		CFFF		CFCF		CCCC	
LOAD	S	composite	sandwich	Composite	Sandwich	Composite	Sandwich
UDL	10	55.6347	11.9646	1.21371	0.31271	0.32340	0.07030
	20	20.8251	5.3734	0.29472	0.11204	0.29334	0.01173
	100	0.8294	0.3453	0.00451	0.00526	0.00151	0.00027
SINO	10	20.9001	4.4878	0.69897	0.17079	0.32340	0.05112
	20	7.9307	2.0338	0.17263	0.06039	0.06447	0.00876
	100	0.3252	0.1326	0.00324	0.00262	0.00147	0.00023

3.4 Forced vibration Response

Forced vibration analysis is performed on composite spherical shells of span to thickness ratio, $S = 10$. The following load cases are considered.

1. Uniform Pressure $P_z = - P_0 F\,(t)$ on outer surface.
2. Sinusoidal pressure $P_s = -P_0 F\,(t)$ on outer surface.

For unit step load $F\,(t)$ is taken as 1.The deflection results for the load cases are non dimensionalized as;

$$\hat{w} = 100wY_0 \,/hS^4P_0 \tag{3}$$

where, $S = a/h$ and Y_0 is taken as the value of Y_2.

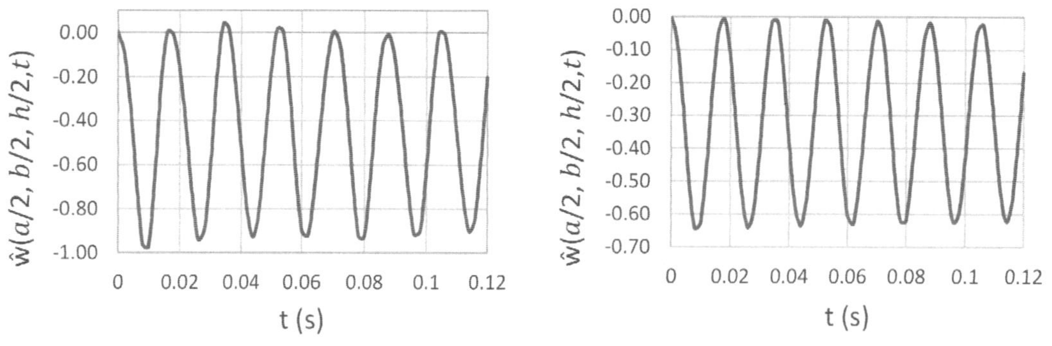

(a) (b)

Figure 4. Time history of central deflection of all sides clamped (CCCC) composite spherical shell for (a) UDL and (b) SINO loads.

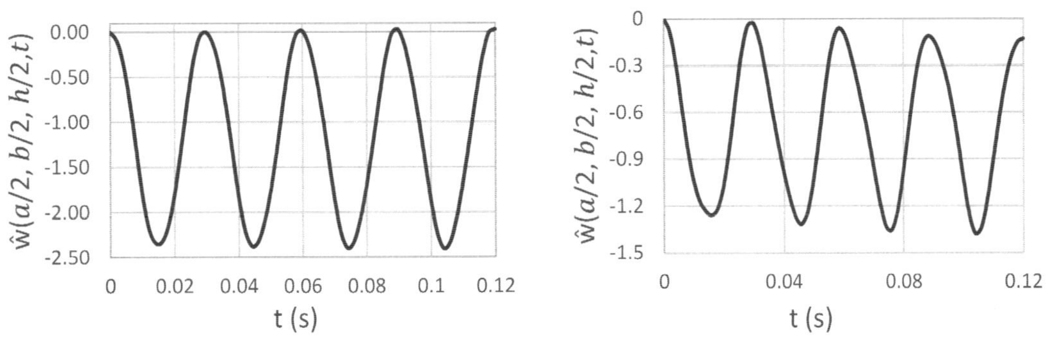

Figure 5. Time history of central deflection of opposite two sides clamped (CFCF) composite spherical shell for (a) UDL and (b) SINO loads.

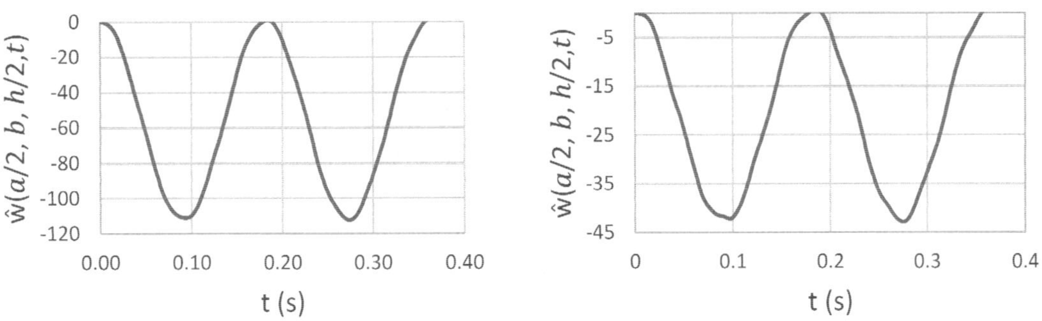

Figure 6. Time history of the tip deflection of only one side clamped (CFFF) composite spherical shell for (a) UDL and (b) SINO loads.

Forced vibration response is captured for 0.12 seconds for clamped (CCCC and CFCF) spherical shells under both uniform and sinusoidal loads (Figures 4, 5, 7 and 8) whereas for cantilever (CFFF) shells it is captured for 0.4 sec (figure 6 and figure 9). These figures show undamped time history response of the tip deflection, $\hat{w}(a/2, b, h/2, t)$, for cantilever and central deflection, $\hat{w}(a/2, b/2, h/2, t)$, for clamped boundary conditions. Almost 7 cycles are completed in 0.12 seconds under CCCC boundary conditions and 4 cycles are completed under CFCF boundary conditions for composite spherical shells. On the other hand in sandwich spherical shells, the number of cycles completed in the said time duration is almost 3.5 and 1.5 respectively. The time period for CFFF boundary condition is maximum for both composite as well as sandwich spherical shells. The amplitude of vibration is larger in case of UDL for all the boundary conditions.

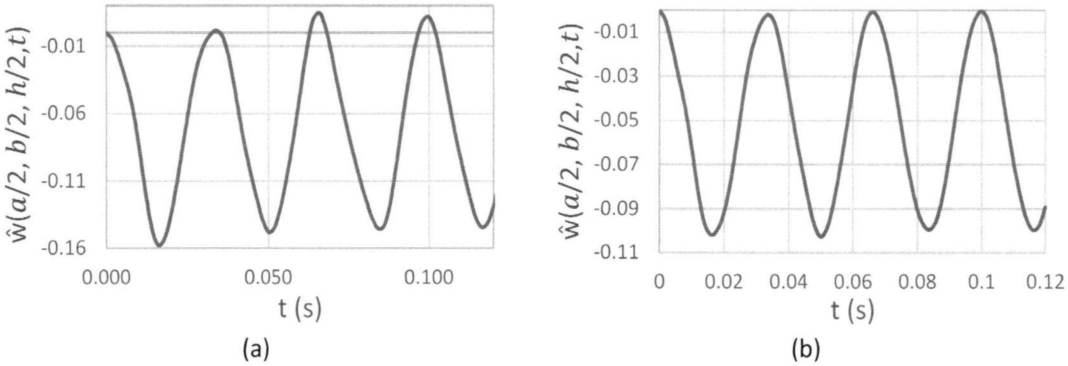

Figure 7. Time history of the central deflection of all sides clamped (CCCC) sandwich spherical shell for (a) UDL and (b) SINO loads.

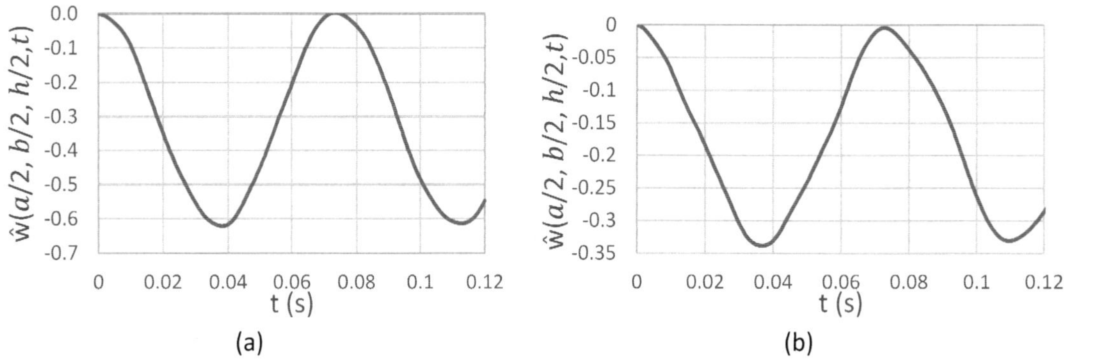

Figure 8. Time history of the central deflection of opposite two sides clamped (CFCF) sandwich spherical shell for (a) UDL and (b) SINO loads.

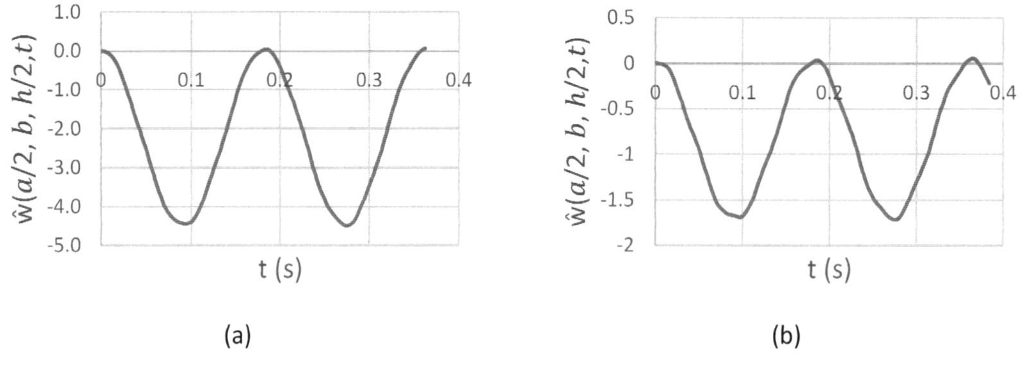

Figure 9. Time history of the tip deflection of only side is clamped (CFFF) Sandwich spherical shell for UDL and SINO loads from left to right.

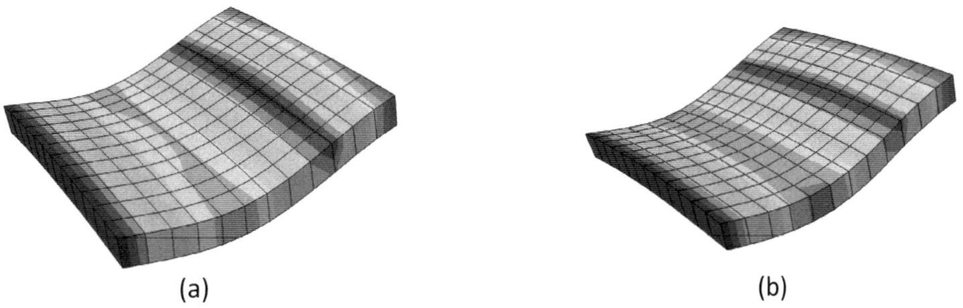

Figure 10. Fifth mode of vibration for cantilever (CFFF) (a) composite and (b) sandwich spherical shells.

Figure 10 shows the fifth mode shape for cantilever composite and sandwich spherical shells.

4. Conclusion

Using finite modelling in the commercial FE tool ABAQUS, the static and dynamic behaviour of composite and sandwich spherical shells is examined. The developed 3D FE model yield fairly accurate results for bending, free and forced vibration response. The used element is found to be quite accurate and robust. The natural frequencies under various boundary conditions increase as the shell becomes thinner. A general decreasing trend of deflection is observed as the span to thickness ratio increase and the deflection due to UDL is found to be greater than deflection due to sinusoidal load. Different layup configuration provide different stiffness and resistance to deflection under a given load type and boundary condition. Thus a suitable configuration can be optimized depending upon the application of shell.

5. References

[1] Kumar R, Mishra B K and Jain S C 2008 Static and dynamic analysis of smart cylindrical shell *Finite Elem Anal Des.* **45** 13-24
[2] Yasin M Y and Kapuria S 2014 An efficient finite element with layer wise mechanics for smart piezoelectric composite and sandwich shallow shells *Comput. Mech.* **53** 101-24
[3] Hossain S J, Sinha P K and Sheikh A H 2004 A finite element formulation for the analysis of laminated composite shells *Comput Struct.* **82** 1623-38
[4] Vel S S and Batra R C 2001 Exact Solution for Rectangular Sandwich Plates with Embedded Piezoelectric Shear Actuators *AIAA J.* **39** 1363-73
[5] Yasin M Y and Kapuria S, 2013 An efficient layer wise finite element for shallow composite and sandwich shells *Comput Struct.* **98** 202-14
[6] Qatu M S, Asadi E and Wang W 2012 Review of recent literature on static analysis of composite shells: 2000-2010 *Open J. Compos. Mater.* **2** 61–86
[7] Qatu M S, Sullivan R W and Wang W 2010 Recent research advances in dynamic behaviour of composite shells: 2000-2009 *Comput Struct.* **93** 14–31
[8] Biswal D K and Mohanty S C 2019 Free vibration study of multilayer sandwich spherical shell panels with viscoelastic core and isotropic/laminated face layers *Compos. B. Eng.* **159** 72-85.
[9] Pang F, Li H, Chen, H and Shan, Y 2021 Free vibration analysis of combined composite laminated cylindrical and spherical shells with arbitrary boundary conditions *Mech. Adv. Mater. Struct.* **28** 182-99.
[10]Tang D, Yao X, Wu G and Peng Y 2017 Free and forced vibration analysis of multi-stepped circular cylindrical shells with arbitrary boundary conditions by the method of reverberation-ray matrix *Thin-Walled Struct.* **116** 154-68

[11]Panda S K and Singh B N 2009 Nonlinear free vibration of spherical shell panel using higher order shear deformation theory–a finite element approach *Int. J. Press. Vessel. Pip.* **86** 373-83

[12] Rafi S S, Alam, M N and Rahman N 2020. Modelling and Analysis of Skew Hybrid Sandwich Plate. *In Proceeds. of Inter. Conf. in Mechanical and Energy Technology Springer Singapore* 589-96

[13]Rafi S S, Alam M N and Rahman N 2021 Dynamic analysis of hybrid sandwich plate *Mater. Today: Proc.* **46** 10009-14

[14]Civalek Ö 2014 Geometrically nonlinear dynamic and static analysis of shallow spherical shell resting on two-parameters elastic foundations *Int. J. Press. Vessel. Pip.* **113** 1-9

[15]Behera R K, Garg K, Patro S S and Sharma N 2018 Eigen-frequency analysis of spherical shell laminated composite plates with and without central cutouts using finite elements. *IOP Conf. Ser.: Mater. Sci. Eng.* **390** 012043

[16]Ma Q, Zhu Y, Liang W and Zhang J 2018 Buckling and strength of an externally pressurized spherical shell with reinforced opening *Int. J. Press. Vessel. Pip.* **165** 11-19

[17] Zhu Y, Ma Q, Zhang J, Tang W and Dai Y. 2017 Opening reinforcement design and buckling of spherical shell subjected to external pressure *Int. J. Press. Vessel. Pip.* **158** 29-36

[18]Malavika V A, Asraff A K, Kumar M and Sofi A 2021 Fracture analysis of plates and shells using FEM and XFEM *Innov. Infrastruct. Solut.* **6** 1-10

[19]Fantuzzi N, Brischetto S, Tornabene F and Viola E 2016 2D and 3D shell models for the free vibration investigation of functionally graded cylindrical and spherical panels *Comput Struct.* **154** 573-590.

Experimental investigation and optimization of weld bead shape factor and form factor using recycled steel slag as a flux in submerged arc welding process

Sumit Saini[1], Kulwant Singh[1], Jatinder Garg[2]

[1]Department of Mechanical Engineering, SLIET Longowal, Punjab, India-148106
[2]Department of Mechanical Engineering, BHSBIET, Lehragaga, Punjab, India-148031

sumitmeghraj@gmail.com

Abstract. The quality and productivity of welds produced depend upon the weld bead profile which is represented by weld penetration shape factor (WPSF) and weld reinforcement form factor (WRFF). The weld profile is further dependent upon welding parameters used hence optimization of welding parameters is essential. In this research work, the steel slag has been recycled to act as a flux in submerged arc welding. The effect of process parameters on WPSF and WRFF using recycled slag has been evaluated. The mathematical models have been developed from the data created. It was found that the weld deposited with recycled steel slag performs well as the WPSF and WRFF are comparable with that deposited with fresh flux. The welding parameters have been optimized using the desirability approach and genetic algorithm. The genetic algorithm yielded more accurate results as compared to the desirability approach.

Keywords: Recycling, steel slag, welding, bead geometry

1. Introduction

Steel slag is normally thrown as a waste which creates soil and air pollution apart from the need for landfill space for dumping. The chemicals, oxides, and minerals present in it are also wasted. There is a need for a methodology to trap these oxides, minerals going waste along with slag and convert them into a useful product. It will not only mitigate the dumping of slag but also save non-renewable resources. In present research work, a methodology was developed to convert steel slag into useful flux required for submerged arc welding process. The performance of flux produced in terms of WPSF and WRFF has been evaluated and presented in this paper. The chemistry of weld metal deposited with recycled has already been evaluated and published by the authors [1].

The load-bearing capacity of the welds produced depends upon the bead profile [2]. The bead profile can be represented by penetration, weld width and reinforcement. The weld penetration shape factor (WPSF) is well-defined as the ratio of bead width to weld penetration. Similarly, bead width to weld reinforcement is termed as weld reinforcement form factor (WRFF). The elements representing bead profile have been shown in Figure 1. The welds produced having WPSF less than one are prone to centerline solidification crack which is termed as ingotism. The problem of ingotism can be tackled by changing the welding procedure to obtain columnar grain structure [3]. Reference [4] found that welds produced having shape factor in the range of 1.3 to 2.0 provided improved mechanical properties due to freedom from ingotism defect. To avoid centerline cracking, it is necessary to keep the weld penetration shape factor from 1.0 to 1.4 [5]. The smoothness of the weld improves with an increase in the weld reinforcement form factor. This is due to fact that an increase in form factor, angle of convexity (α) decreases. For better mechanical properties, the value of the weld reinforcement form factor should vary between 2.0 to 7.0. The properties of the molten pool, slag and the gaseous phases under the arc have a pronounced effect on the weld form factor. Researcher obtained the form factor value of 2.0 at 800 amperes and 35 volts using a 5.0 mm diameter electrode [4]. The lower form factor is undesirable because it causes stress concentration at the weld toe where reinforcement merges with the top surface of the workpiece [6]. The mechanical properties of welded joints depend upon welding parameters [7]. The weld shape factor and form factor affect the performance of weld during service particularly the fatigue strength, hence optimization of welding parameters is essential.

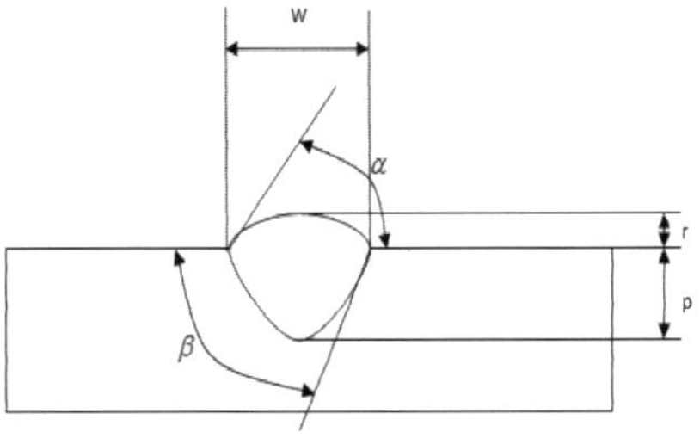

Figure 1. Elements of weld bead geometry

WPSF = w/p
WRFF = w/r

where,
p = Penetration
r = Reinforcement
w = weld width
α = Angle of convexity
β = Angle of entry

2. Methodology

The following steps were adopted to complete the research work:

2.1. Recycling of steel slag

The slag generated by the steel plant was collected from their dumping yard free of cost as it is a waste for them. The collected slag was crushed in a mechanical crusher and milled in a ball mill to convert it into powdered form (30 mesh size). Then the necessary alloying elements and deoxidizers in powdered form were added and dry mixed for 30 minutes to get a homogeneous mixture. Then liquid potassium silicate was added as a binder to get the wet mixture [13]. Then, this wet mixture was passed through the 10-mesh size sieve to obtain pellets. These pellets of the wet mixture were air-dried for 24 hours followed by baking in a muffle furnace at 800°C for 4 hours. After baking the baked mass was passed through a sieve of 10 mesh size to get granular flux known as "Recycled slag". Then the recycled slag in combination with EH 14 filler wire having a diameter of 3.14 mm was used to deposit beads on plates.

2.2. Identification of welding parameters and their working range:

To determine the welding parameters and their working range, extensive test runs were carried out, which are listed in Table 1.

Table 1. Input process parameters having lower and higher values

Parameter	Units	Symbols	Levels				
			-2	-1	0	+1	+2
Welding current	A	I	250	350	450	550	650
Welding speed	m/h	S	16	20	24	28	32
Arc voltage	V	V	20	23	26	29	32
Nozzle to Plate distance (NPD)	mm	N	20	22	24	26	28

2.3. Development of design matrix:
The central composite rotatable design has been adopted for this study. As dictated by this technique, the design matrix developed is indicated in Table 2.

Table 2. Design matrix along with responses

S.N o.	Factors				Responses				
	I (A)	S(m/h)	V(V)	N (mm)	p (mm)	w (mm)	r(mm)	WPSF	WRFF
1	350	20	23	22	5.67	6.77	6.23	1.194	1.086
2	550	20	23	22	8.85	9.68	4.96	1.093	1.951
3	350	28	23	22	6.00	8.5	4.56	1.416	1.864
4	550	28	23	22	6.97	6.62	5.00	0.949	1.324
5	350	20	29	22	6.28	15.11	3.78	2.406	3.997
6	550	20	29	22	10.95	13.25	4.37	1.210	3.032
7	350	28	29	22	5.20	11.82	2.99	2.273	3.953
8	550	28	29	22	9.49	7.26	5.06	0.765	1.434
9	350	20	23	26	5.97	6.95	6.11	1.164	1.137
10	550	20	23	26	7.21	7.31	6.5	1.013	1.124
11	350	28	23	26	6.33	10.71	4.05	1.691	2.644
12	550	28	23	26	6.29	8.31	6.57	1.321	1.264
13	350	20	29	26	5.32	14.57	3.04	2.738	4.792
14	550	20	29	26	9.37	9.56	5.51	1.022	1.735
15	350	28	29	26	4.73	13.21	2.67	2.792	4.947
16	550	28	29	26	7.66	8.89	6.7	1.160	1.326
17	250	24	26	24	4.21	11.19	2.52	2.657	4.440
18	650	24	26	24	9.77	7.78	5.92	0.796	1.314
19	450	16	26	24	7.67	11.23	6.84	1.464	1.641
20	450	32	26	24	6.43	9.09	5.23	1.413	1.738
21	450	24	20	24	4.43	8.85	6.05	1.997	1.462
22	450	24	32	24	6.04	17.11	4.12	2.832	4.152
23	450	24	26	20	7.91	10.12	2.85	1.279	3.550
24	450	24	26	28	7.87	9.22	5.11	1.171	1.804
25	450	24	26	24	8.33	8.36	5.95	1.003	1.405
26	450	24	26	24	8.04	7.13	6.01	0.886	1.186
27	450	24	26	24	8.48	6.48	6.46	0.764	1.003
28	450	24	26	24	7.87	7.05	6.11	0.895	1.153
29	450	24	26	24	8.33	6.55	6.35	0.786	1.031
30	450	24	26	24	9.55	7.35	6.56	0.769	1.120
31	450	24	26	24	7.34	7.55	6.23	1.028	1.211

2.4. Conducting the experiments:
A total no. of 31 beads on plates were deposited as dictated by the design matrix using Ador Fontech make submerged arc welding machine. The SA 516 grade 70 with a thickness of 12 mm was used as the base

material. This material is extensively used for the fabrication of boilers, pressure vessels and heat exchangers [8]. The filler wire EH-14 was used in combination with flux manufactured using steel slag as raw material. The chemical composition of base material and filler wire is given in Table 3.

Table 3. Chemical composition of base metal and filler wire

Material used	C	Mn	Si	P	S
SA 516 grade 70	0.10	1.17	0.6	0.03	0.03
EH 14 filler wire	0.14	1.70	0.10	0.03	0.03

2.5. Recording the responses:
A specimen of 20 mm width (Figure 2) was removed from the middle of the weld bead, which was polished, lapped and etched with 2% NITAL etchant. The beads dimensions such as weld penetration, weld width and reinforcement were measured and then WPSF and WRFF were calculated and presented in Table 2.

2.6. Developing models:
The mathematical models were developed from data generated using Design Expert-11 software. The developed models after dropping insignificant coefficients have been presented in Equations 1 & 2.

Figure 2. Specimen removed from the welded plate.

2.7. Checking the adequacy:
The 'F' test was applied to check the adequacy of developed models. It has been found that developed models are adequate at a 95% confidence level. The lack of fit test was used to evaluate the goodness of fit of the fitted quadratic models.

3. Results And Discussion

3.1. Effect of welding parameters on WPSF
The influence of welding parameters on WPSF has been presented in Figure 3 graphically. From the figure and Eq. 1, it is very clear that WPSF decreases exponentially with increasing welding current. This is due to the fact that the heat input increases with increasing welding current, which causes deep penetration resulting in decreased WPSF [9,12]. It is further observed that WPSF slightly decreases initially and then suddenly rises with an increment in arc voltage. The arc length increases with increasing arc voltage, which means that the arc hits a larger surface area of the workpiece. A larger spread of arc leads to the increased width of weld. Thus, increased width is responsible for increased WPSF. No significant effect of NPD and welding speed on WPSF has been observed.

3.2. Effect of welding parameters on WRFF
From Figure 4 and Eq. 2, it can be seen that the WRFF decreases exponentially with increasing welding current; the reason behind this is that increased welding current increases the melting rate of filler wire leading to increased reinforcement. Moreover, increased current provides digging arc and reduced width. Increased reinforcement is responsible for the decrease in WRFF. It was further noticed from Figure 4 that the WRFF increases with increasing arc voltage since the arc voltage increases the bead width as discussed earlier. The equivalent volume of molten filler metal is distributed over a larger surface area, hence reinforcement

decreases. Due to which WRFF increased. These results agree well with the results of [10]. The effect of NPD and welding speed on WRFF is insignificant.

$$WPSF = 0.87 - 0.45A + 0.017B + 0.257C + 0.057D - 0.050AB - 0.310AC - 0.037AD - 0.081BC + 0.095BD + 0.032CD + 0.181A^2 + 0.109B^2 + 0.353C^2 + 0.055D^2 \quad (1)$$

$$WRFF = 1.158 - 0.728A + 0.004B + 0.758C - 0.131D - 0.305AB - 0.568AC - 0.307AD - 0.230BC + 0.180BD + 0.027CD + 0.402A^2 + 0.105B^2 + 0.385C^2 + 0.352D^2 \quad (2)$$

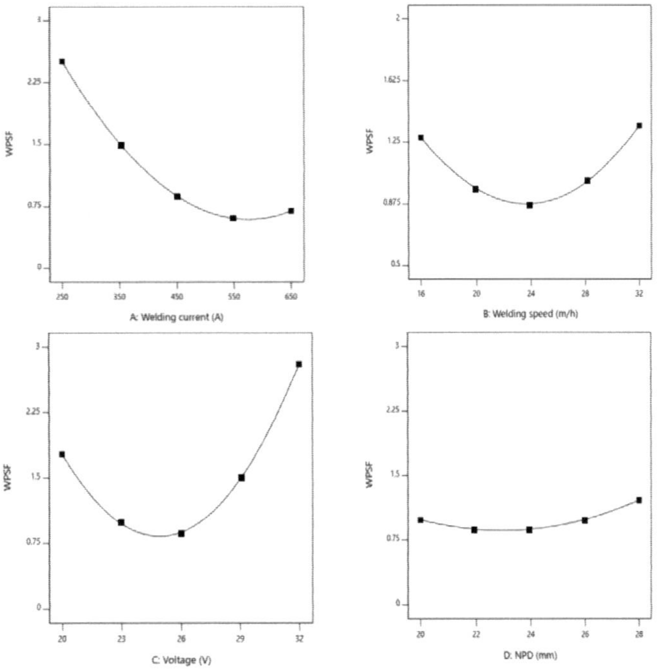

Figure 3. Direct effect of parameters on WPSF

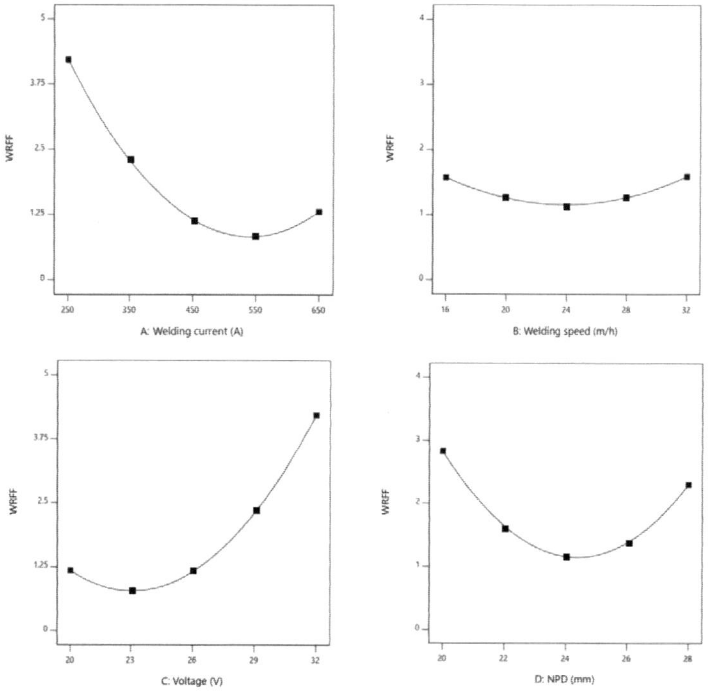

Figure 4. Direct effect of parameters on WRFF

3.3. *Interactive effects of welding parameters on WPSF and WRFF*

The interactive effects of several parameters on WPSF and WRFF have been presented in Figure 5. Figure 5 (a) indicates that WPSF decreases with increasing welding current at the lower and upper levels of welding speed. The reason behind this trend is that as the welding current increases, the size of the weld pool increases as a result of deeper penetration. This increase in penetration is responsible for the decrease in WPSF. Figure 5 (b) shows that the WPSF increases abruptly with increasing welding voltage at a lower level of welding current. Since the welding voltage increase the weld width, and low current decreases the penetration which reflects the increasing trend of WPSF. At a higher level of arc voltage, the WPSF decreases with increasing welding current due to the intense heat input which results in higher penetration.

From Figure 5 (c), it can be seen that the WRFF increases with decreasing the welding current at a higher welding speed. The reason behind this is the same as discussed above.

The WRFF also increases with decreasing welding current at a higher level of voltage since at higher voltage the arc length is larger which deposited a wider bead that's why an increase in WRFF was observed as shown in Figure 5 (d).

3.4. *Optimization of welding parameters*

The welds produced must be sound and free from flaws as well as economical. The quality and productivity depend upon welding parameters. Therefore, the optimization of process parameters is the need of the hour. In the present study, the optimization was done by using the desirability approach and genetic algorithm (GA) optimization. The desirability approach is widely employed in industries and research for the optimization of process parameters. It is quite simple as compared to other optimization techniques since it does not require any complicated algorithm or programming [11]. The desirability approach was done in Design Expert 11 software and the genetic algorithm was carried out in MATLAB 2020 software. The correct tuning of GA parameters influences the performance of optimization.

The optimum results given by both approaches are different and shown in Table 4. By comparing the percentage error given by these techniques, it was observed that the genetic algorithm gives better optimum results as compared to the desirability approach.

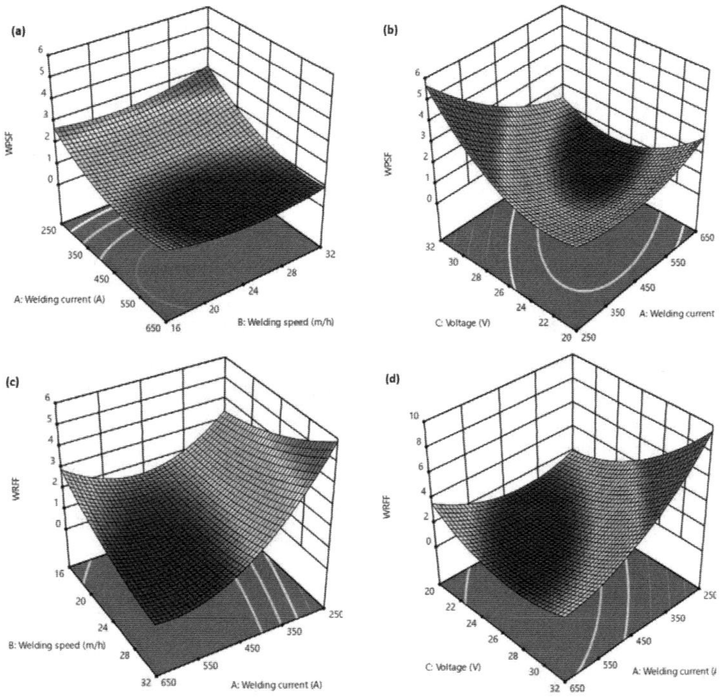

Figure 5. Contour plots of WPSF (a , b) and WRFF (c , d)

Table 4. Validation of models

Method	I (A)	S(m/h)	V (V)	NPD (mm)		WPSF	WRFF
Desirability	415	21	28	23	Predicted	1.66	2.80
					Confirmation	1.72	2.68
					% error	3.48	4.47
Genetic	442	20	30	26	Predicted	1.82	3.12
					Confirmation	1.78	3.05
					% error	2.24	2.29

For comparison of the WPSF and WRFF of welds deposited using recycled slag and fresh flux. A bead was deposited using fresh flux with optimized parameters obtained from the genetic algorithm technique as indicated in Table 5.

Table 5. Comparison of WPSF and WRFF of welds deposited using fresh commercial flux with recycled slag

Method	I (A)	S(m/h)	V (V)	NPD (mm)	WPSF	WRFF
Recycled slag	442	20	30	26	1.78	3.05
Fresh flux	442	20	30	26	1.66	3.25

From the above Table 5, it can be concluded that the WPSF and WRFF of the welds deposited with fresh commercial flux are comparable to those of weld deposited with recycled steel slag.

4. Conclusions

- The behaviour demonstrated by recycled slag in terms of weld shape factor and form factor is similar to that of fresh flux.
- The minimum and maximum WPSF attained with recycled slag are 0.764 and 2.83 respectively.
- The desirable WPSF (1.78) was provided by recycled slag to have good mechanical properties.
- The desirable WRFF (3.05) was achieved with recycled slag to get improved mechanical properties of weld produced.
- The welding parameters have been optimized to achieve desired shape factor and form factor using recycled slag and fresh flux.
- The optimized parameters obtained using the desirability approach are welding current 415 A, welding speed 21 m/hr, Arc voltage 28 V and nozzle-to-plate distance 23 mm.
- The optimized parameters obtained using the Genetic Algorithm approach are welding current 442 A, welding speed 20 m/hr, Arc voltage 30 V, and nozzle-to-plate distance 26 mm.
- The Genetic Algorithm has provided better optimization results having low percentage error as compared to the Desirability Approach.

References

[1] Saini S and Singh K 2021 Recycling of steel slag as a flux for submerged arc welding and its effects on chemistry and performance of welds *Int. J. Adv. Manuf. Syst.* **114** 1165-77.

[2] Smati Z 1986 *Automatic pulsed MIG welding* (British Maritime Technology)

[3] Weisman C and Kearns W H eds., 1976. *Welding Handbook: Welding processes: arc and gas welding and cutting, brazing and soldering* (Vol. 2) (American Welding Society)

[4] Srinath H 1975 Some Aspects of Submerged Arc Welding Process. *Indian Weld. J.* **7** 67.

[5] Dermott R G 1961 Goal: crackfree welds in high-strength steels *Weld. met. fabr.* ***49*** 61-73.

[6] Kou S 2020 *Welding Metallurgy* 3rd edn (Wiley)

[7] Sharma L and Chhibber R 2020 Study of weld bead chemical, microhardness & microstructural analysis using submerged arc welding fluxes for linepipe steel applications *Ceram. Int.* **46** 24615–23.

[8] Arya H K, Singh K and Saxena R K 2018 Effect of weld cooling rates on mechanical and metallurgical properties of submerged arc welded pressure vessel steel *J. Pressure Vessel Technol.* **140** 041406.

[9] Singh K and Pandey S 2009 Recycling of slag to act as a flux in submerged arc welding *Resour. Conserv. Recycl.* **53** 552-58.

[10] Murugun N and Gunaraj V 1999 Prediction and Comparison of area of heat affected zone for

bead on plate and bead on joint in submerged arc welding of pipes *J.Mater.Process.Technol.* **95** 246-61.

[11] Choudhary A, Kumar M and Unune D R 2019 Experimental investigation and optimization of weld bead characteristics during submerged arc welding of AISI 1023 steel *Def. Technol.* **15** 72–82.

[12] Ramirez M F, Hernández J W, Ladino D H, Masoumi M. and Goldenstein H. 2021 Effects of different cooling rates on the microstructure, crystallographic features, and hydrogen induced cracking of API X80 pipeline steel. *J. Mater. Res. Technol.* **14** 1848-61.

[13] Hayi, B A A 2019 Novel Production of Granular Flux for Welding Carbon Steel Plates *IOP Conf. Ser.: Mater. Sci. Eng.* **571** 012079.

IOP Conf. Series: Materials Science and Engineering 1225 (2022) 012039 doi:10.1088/1757-899X/1225/1/012039

Numerical Studies on Mode I Delamination and its Effect on the Vibrational Characteristics in Fibre Metal Laminates

Naresh Kali and Srikanth Korla

National institute of Technology Warangal, Warangal, India

Email: nareshk@student.nitw.ac.in

Abstract. Fibre metal laminates (FMLs) afford the notable advances over ongoing composite materials for aerospace and automotive applications due to their low weight and outstanding mechanical properties. Nevertheless, FMLs are prone to damages during manufacturing and loading conditions. Double cantilever beam (DCB) and vibration test are the commonly used tools to assess fracture energy values and the level of damage influence on the material properties respectively. Therefore, this paper aims at correlating the numerical validation of mode I delamination with already published experimental data by Y.Pan et.al and study the influence of delamination under free vibration analysis of Magnesium (Mg AZ31) alloy based fibre metal laminates. For the presented model, the numerical values showed good acceptance with the experimental values of DCB test. It was also further observed that there is significant reduction in natural frequency due to delamination in the fibre metal laminates.

Nomenclature

FMLs= fibre metal laminates,

CZM= cohesive zone method,

DCB= double cantilever beam,

G_c= fracture energy,

δ_{fail}= the point at which the elements fail completely,

K_{nn}, K_{ss}, K_{tt}= stiffness parameters,

N_{max},S_{max},T_{max}= strength parameters,

G_{IC},G_{IIC} = the mode I and mode II fracture energy values respectively,

ΔL_C= characteristic length,

OMF =Over meshing factor.

1. INTRODUCTION

Fibre metal laminates (FMLs) are the hybrid composite materials in which metal and fibre layers are arranged, alternatively. In recent years, the fibre metal laminates applicability is increasing rapidly in aerospace and automotive applications due to their low weight, fatigue resistance, outstanding mechanical properties, damage resistance and good energy absorbing capacity [1][2][3]. The first fibre metal laminates (ARALL) materials were originated at the Delft University of Technology (DUT) in the year of 1978 [4]. Even though Aluminium alloy based fibre metal laminates are exhibiting excellent performance in structural applications but when it comes to weight reduction in structures; the magnesium alloys may be the alternative material to aluminium alloys due to its light weight [5]. Magnesium alloy based FMLs provide corrosion resistance, damage resistance, and higher damping properties over the GLARE[6][7]. The lower stiffness of the magnesium alloy limits the structural applicability but offer good mechanical properties over monolithic magnesium alloy when integrated with the higher stiffness fibres like carbon/glass [8].

Fibre metal laminates are prone to damages during manufacturing and loading conditions. So among all damages, it has been stated that delamination is the most impact failure in the FMLs [9].It not only reduces the stiffness but also decreases the load bearing capacity of the material [10]. In specific, the delamination may occur at the interface region of the composite material [11].In materials the delamination may propagate in three modes namely mode I, mode II, mode III. In general, there may be a chance of occur mixed mode behaviour. DCB, ENF and MMB tests are available as the standard tests to evaluate the these modes [12]. Non-destructive methods are accessible to assess the influence of damage levels on the mechanical behaviour of FMLs. Vibration analysis is the commonly method to understand the behaviour of the material under loading conditions among all non-destructive methods [13].

Finite element (FE) analysis is nowadays alternative technique to the experimental work to capture the behaviour of fibre metal laminates closely and efficiently, due to time and cost involved in fabricating and testing the fibre metal laminates in laboratory is a tedious process [14]. VCCT, CZM are frequently used numerical techniques to simulate the delamination behaviour in composite materials [15]. CZM, unlike VCCT, is independent of mesh size, time step and it is the efficient method to predict the delamination behaviour in the materials [16][17]. One more advantage of CZM compared to VCCT is that it can forecast the new crack initiation and propagation of existing crack. All these advantages make CZM, the mostly using method to understand the delamination behaviour in the interface region[18].

It can be observed from the literature that numerical modelling of delamination in magnesium alloy (Mg AZ31) based fibre metal laminates has not been discussed in detail. Therefore, in the present study, numerical validation of mode I delamination in AZ31 magnesium alloy-based fibre metal lamination is discussed using cohesive zone method (CZM) and influence of delamination on vibration characteristics of these laminates also studied under free vibration analysis.

2. Methodology

2.1 Modelling of cohesive layer and its strength parameters

Cohesive zone method (CZM) is implemented using either cohesive element or cohesive surface methods. The applicability of both the methods is different but they are mostly similar and follow the same rules. The cohesive surface method is suggested in the case of insignificant thickness of cohesive layer compared to metal or fibre layers, whereas cohesive elements are recommended when there is a failure observed in the adhesion region. Therefore, the modelling of delamination using CZM is nowadays getting popular because propagation of delamination can be easily defined in the front portion of the crack tip. In the numerical modelling the cohesive elements are defined in the delamination path and these elements are controlled by the traction-separation law principle [19].

The traction-separation law is shown in the Figure 1 and it can be noticed from the law that before damage initiation the cohesive elements are obeying the linear behaviour and this behaviour is controlled by the stiffness parameters K_{nn}, K_{ss}, K_{tt}. Once the law reaches to strength parameters N_{max},S_{max},T_{max} then the curve starts to decrease gradually to zero. After damage initiation the behaviour of the curve rely on the softening laws (linear or exponential). The point at which the curve reaches to zero stiffness value of elements then the area under the curve is called fracture energy (G_c) and δ_{fail} is the point at which the elements fail completely. Therefore, all these stiffness, strength parameters can be calculated using following equations.

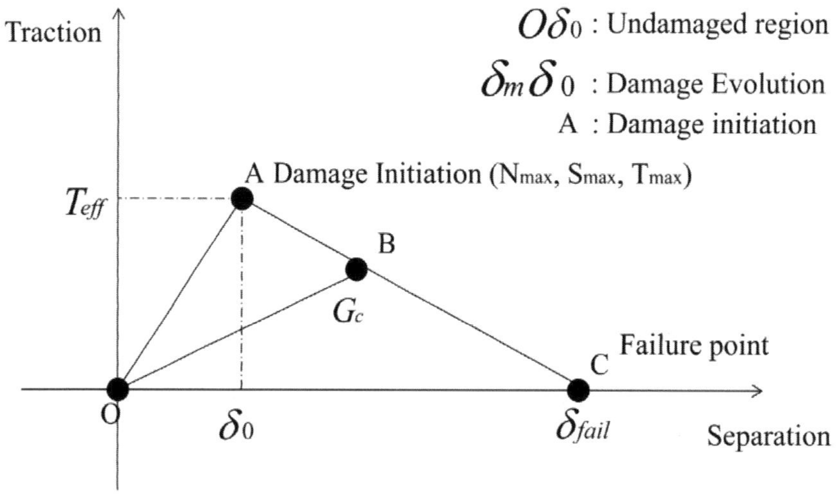

Figure 1. Traction-Separation law

$$K_{nn}=\frac{2G_{IC}}{\delta_{fail}X\delta_{fail}{}^{2}} \quad (1)$$

$$K_{tt}=K_{ss}=\frac{2G_{IIC}}{\delta_{fail}X\delta_{fail}{}^{2}} \quad (2)$$

$$N_{max}=\frac{2G_{IC}}{\delta_{fail}} \quad (3)$$

$$T_{max}=S_{max}=\frac{2G_{IC}}{\delta_{fail}} \quad (4)$$

Where,δ_{fail} = Penalty factor xΔL_C and $\delta_{ratio} = \frac{\delta_{ini}}{\delta_{fail}}$

$$\text{Over meshing factor (OMF)} =\frac{Structural\ mesh}{Cohesive\ region\ mesh} \quad (5)$$

G_{IC},G_{IIC} are the mode I and mode II fracture energy values respectively and over meshing factor (OMF) plays a crucial role in calculatingΔL_C. Generally the structural mesh size is five times greater than cohesive mesh size [20]. In stiffness and strength parameters calculation initially the penalty factor can have assumed to be 0.05. This value can be assumed high for non-convergence and very low for convergence problems.δ_{ratio}value of 0.5 is recommended by Diehl et al [21]. In general, δ_{ratio}will not affect the parameters during simulations. Cohesive elements are generally controlled by stress-strain methods for damage initiation and energy laws for evolution. In the present study Benzeggagh and Kenane (B-K) energy law is used with μ=1.4 to model the damage evolution and for damage initiation, the quadratic normal stress criteria is used shown in Equation 6 and 7 [22].

$$G_{IC} + (G_{IIC} - G_{IC})\left(\frac{G_{II}}{G_T}\right)^{\mu} = G_T \quad (6)$$

$$\left(\frac{N}{N_{max}}\right)^2 + \left(\frac{S}{S_{max}}\right)^2 + \left(\frac{T}{T_{max}}\right)^2 = 1 \quad (7)$$

2.2 Natural frequency (Eigen value) extraction method

Free vibration analysis is performed numerically to observe the influence of delamination on natural frequency response of the AZ31 magnesium alloy-based fibre metal laminates. Cantilever beam is modelled for analysis and linear perturbation with Lanczos Eigen solver is used to extract the natural frequency (Eigen values) during analysis. Therefore, ABAQUS/CAE provides a method to extract the natural frequencies during analysis as follows.

$$(-\omega^2 M^{MN} + K^{MN})\emptyset^N = 0,$$

Where

$M^{MN}$is mass matrix (symmetric and positive);

K^{MN} is stiffness matrix

$\emptyset^N$ is eigen vector

M, N degree of freedom values.

When K^{MN} is positive and definite then all eigenvalues are positive. Due to instabilities and rigid body modes, $K^{MN}$is become indefinite. The instabilities will produce negative eigenvalues when initial stress values are included and zero eigenvalues due to rigid modes.

3. Geometrical modelling of AZ31magnesium alloy-based fibre metal laminates

Double cantilever beam (DCB) and rectangle beam embedded with cohesive layer are modelled with six AZ31 magnesium alloy layers and seven carbon fibre prepreg (T800/5228) layers as shown in Figure 2. The dimension of the beam is taken as per the reference standards i.e., 130x15x2.36 mm with the pre-crack length of 45mm as shown in Figure 3. In both the models cohesive layer is modeled in the mid plane as shown in the Figure 3 and Figure 4 with the mode I, mode II fracture energies 0.23 KJ/m^2 and 5.81KJ/m^2 respectively [23]. The properties of constituent layers are presented in Table.1. During numerical analysis 3D stress (C3D8R), continuum elements (SC8R) are selected to model the AZ31magnesium alloy and T800/5228 carbon fibre prepreg layers respectively. 3D cohesive elements (COH3D8) are used to model the cohesive layer in the interface region. Recent literature suggests that cohesive layer thickness may be taken around 0.01 and 0.02 mm [24][25][26]. So, 0.02mm cohesive thickness has been used in the present study. Mesh sensitivity analysis has been done to improve the accuracy and saving time during analysis. During analysis one side of the beam is entirely fixed and displacement load is applied at another end of the beam as shown in the Figure 5.

Table 1. AZ31magnesium/Carbon fibre prepreg layer mechanical properties[23].

Materials	E (GPa)	ρ(g/cm^3)	$\sigma_{tensile}$ (MPa)	$t_{thickness}$(mm)
AZ31Magnesium	45	1.78	230	0.2
T800C/5228 epoxy resin	125	1.53	1600	0.2

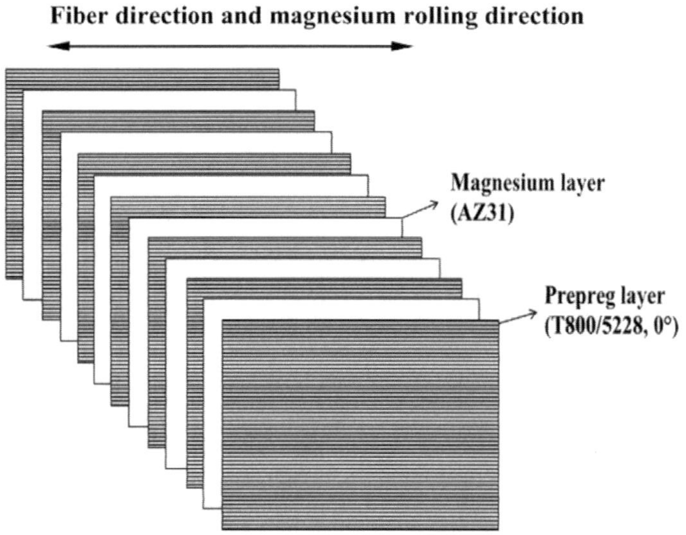

Figure 2. AZ31magnesium alloy and T800/5228 carbon prepreg ply lay-up[23]

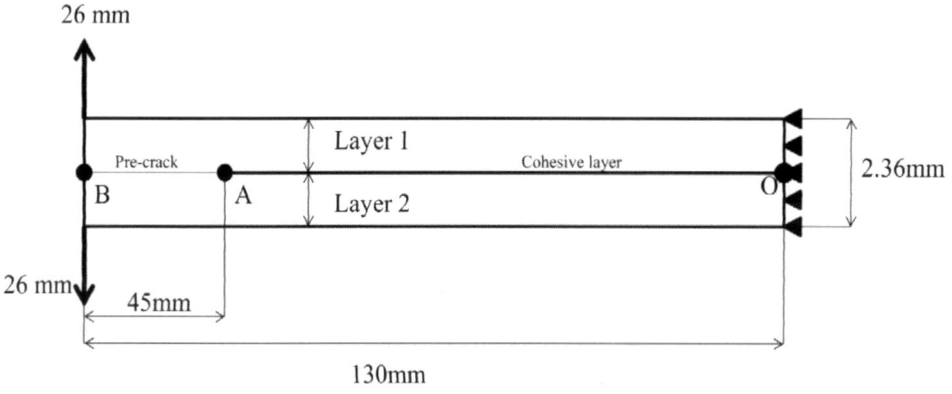

Figure 3. Schematic view of DCB model

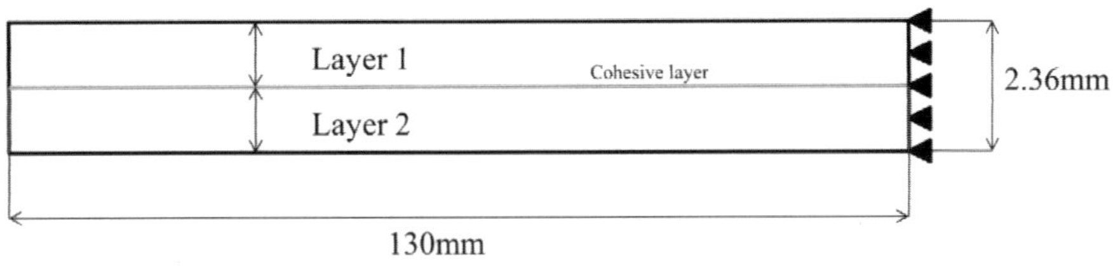

Figure 4. Schematic view of delamination model

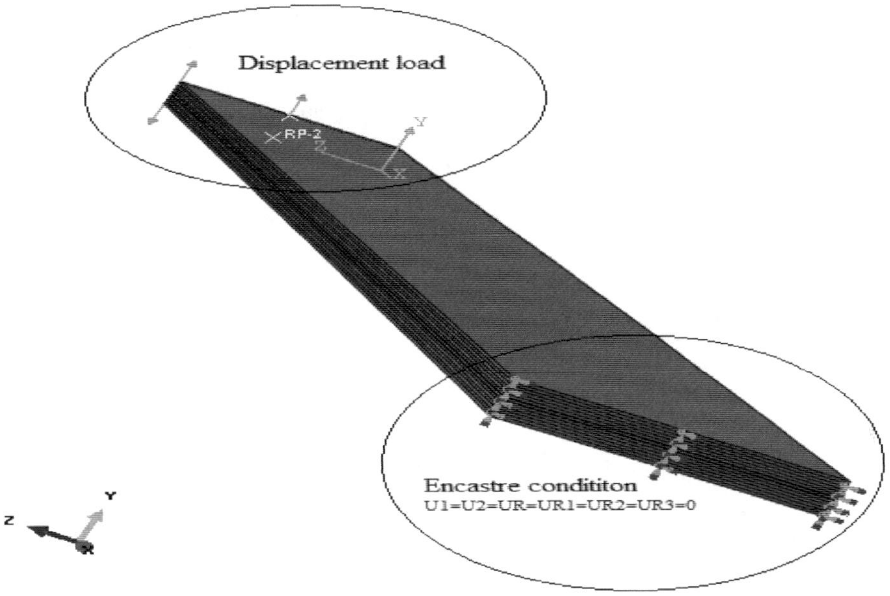

Figure 5. Boundary conditions during analysis

4. Results and Discussion

In the present study, numerical analysis of DCB test for AZ31 magnesium alloy and T800/5228 carbon fibre prepreg based fibre metal laminates is performed to validate the experimental load-displacement curve. Cohesive zone method (CZM) is used to analyse the DCB test results numerically. The numerical results are in the order of experimental results presented by Y. Pan *et al* [23] and are as shown in Figure 6 and further, these cohesive layer properties are used in the free vibration analysis. From the curve, it may be depicted that linear behaviour till the damage initiation point (peak load in the curve) and after that there is a gradual decrease in the curve during propagation of delamination. The delamination propagation is also shown in the in Figure 7.

It is observed from the Figure 8 that there is a significant reduction in the natural frequency due to delamination in the FMLs. It may be due to the high reduction in the stiffness compared to the mass of the fibre metal laminates. Approximately 51%,84%,55%,65%,61% are the percentage reductions observed in the natural frequency shifts of respective modes from mode 1 to mode 5. The lowest and highest percentage reductions corresponding to mode1 and mode 2 respectively. Therefore, it is observed that reduction in natural frequency shifts may be depends on the mode shape and number.

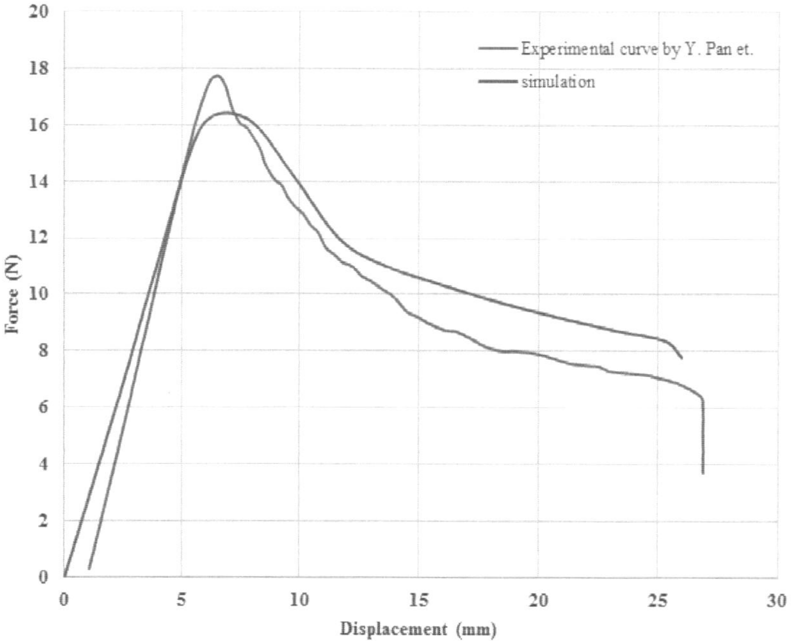

Figure 6. load vs displacement curve

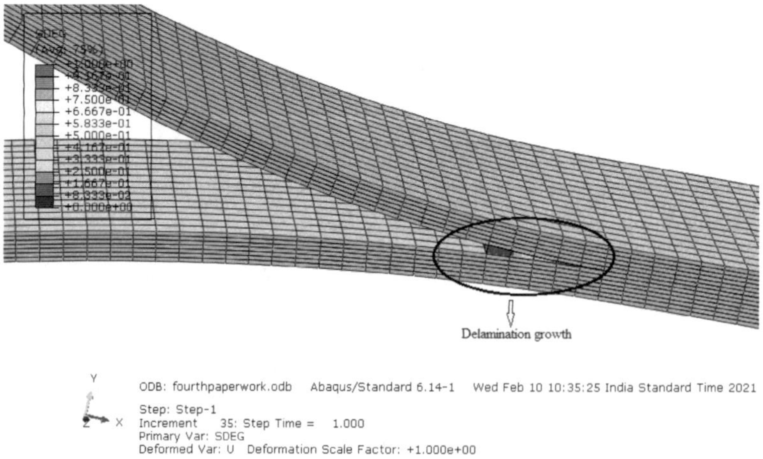

Figure 7. Delamination propagation during DCB test

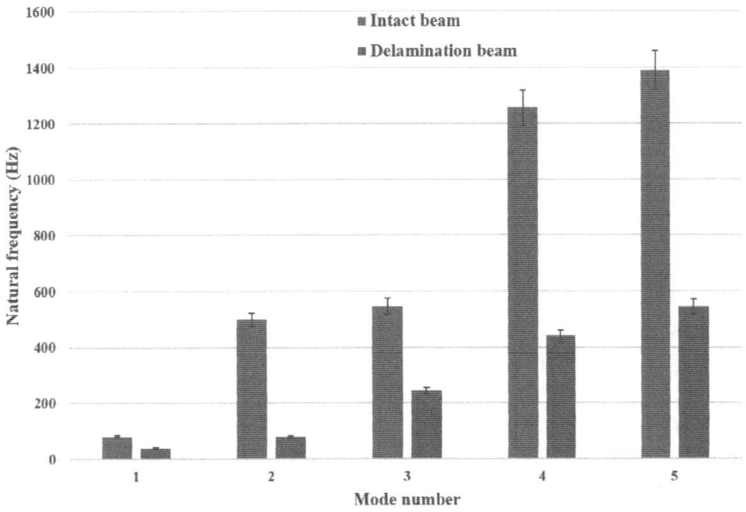

Figure 8. Effect of delamination on natural frequency

5. Conclusions

This research paper investigated the numerical validation of mode I delamination using DCB test and studies the influence of delamination under free vibration analysis of Magnesium (Mg AZ31) alloy-based fibre metal laminates. After summarizing all the results obtained from this study, it can be depicted that for the presented model, the numerical values showed good acceptance with the experimental results and the natural frequency is sensitive to the delamination damages present in the fibre metal laminates. The lowest and highest percentage reductions corresponding to mode1(51%) and mode2 (84%) respectively. Therefore, further it may be observed that reduction in natural frequency shifts are depends on the mode shape and its number.

References

[1] Zhang X, Ma Q, Dai Y, Hu F, Liu G, Xu Z, Wei G, Xu T, Zeng Q and Xie W 2018 Effects of surface treatments and bonding types on the interfacial behavior of fiber metal laminate based on magnesium alloy *Appl. Surf. Sci.* **427** 897–906

[2] Abouhamzeh M, Nardi D, Leonard R and Sinke J 2018 Effect of prepreg gaps and overlaps on mechanical properties of fibre metal laminates *Compos. Part A Appl. Sci. Manuf.* **114** 258–68

[3] Kali N, Pathak S and Korla S 2019 Effect on vibration characteristics of fiber metal laminates sandwiched with natural fibers *Mater. Today Proc.* **28** 1092–6

[4] Sinmazçelik T, Avcu E, Bora M Ö and Çoban O 2011 A review: Fibre metal laminates, background, bonding types and applied test methods *Mater. Des.* **32** 3671–85

[5] Alderliesten R, Rans C and Benedictus R 2008 The applicability of magnesium based Fibre Metal Laminates in aerospace structures *Compos. Sci. Technol.* **68** 2983–93

[6] Pärnänen T, Alderliesten R, Rans C, Brander T and Saarela O 2012 Applicability of AZ31B-H24 magnesium in Fibre Metal Laminates - An experimental impact research *Compos. Part A Appl. Sci. Manuf.* **43** 1578–86

[7] Zhang X, Zhang Y, Ma Q Y, Dai Y, Hu F P, Wei G B, Xu T C, Zeng Q W, Wang S Z and Xie W D 2017 Effect of surface treatment on the corrosion properties of magnesium-based fibre metal laminate *Appl. Surf. Sci.* **396** 1264–72

[8] Ostrovsky L and Henn Y 2007 Present state and future of magnesium application in aerospace industry *Int. Conf. New Challenges Aeronaut. ASTEC 07* 1–5

[9] Moslemi M and Khoshravan M 2015 Cohesive zone parameters selection for mode-i prediction of interfacial delamination *Stroj. Vestnik/Journal Mech. Eng.* **61** 507–16

[10] Kaveh A 2014 Determining the crack propagation in DCB composite specimens by fiber bridging modeling *Rev. Tec. la Fac. Ing. Univ. del Zulia* **37** 1–9

[11] Moheimani R, Sarayloo R and Dalir H 2020 Failure study of fiber/epoxy composite laminate interface using cohesive multiscale model *Adv. Compos. Lett.* **29** 2633366X2091015

[12] Heidari-Rarani M and Sayedain M 2019 Finite element modeling strategies for 2D and 3D delamination propagation in composite DCB specimens using VCCT, CZM and XFEM approaches *Theor. Appl. Fract. Mech.* **103**

[13] Botelho E C, Campos A N, De Barros E, Pardini L C and Rezende M C 2005 Damping behavior of continuous fiber/metal composite materials by the free vibration method *Compos. Part B Eng.* **37** 255–63

[14] Khan S H, Sharma A P, Kitey R and Parameswaran V 2018 Effect of metal layer placement on the damage and energy absorption mechanisms in aluminium/glass fibre laminates *Int. J. Impact Eng.* **119** 14–25

[15] Heidari-Rarani M, Shokrieh M M and Camanho P P 2013 Finite element modeling of mode i delamination growth in laminated DCB specimens with R-curve effects *Compos. Part B Eng.* **45** 897–903

[16] Di Caprio F, Saputo S and Sellitto A 2019 Numerical-Experimental Correlation of Interlaminar Damage Growth in Composite Structures: Setting Cohesive Zone Model Parameters *Adv. Mater. Sci. Eng.* **2019**

[17] Sitnikova E, Li D, Wei J, Yi X and Li S 2019 On the Representativeness of the Cohesive Zone Model in the Simulation of the Delamination Problem *J. Compos. Sci.* **3** 22

[18] Masood S N, Singh A K and Viswamurthy S R 2016 Simulation & validation of delamination growth in CFRP specimens under mixed mode loading using cohesive elements *Proc. of 14th ISAMPE National Conference on Composites (INCCOM-14) Hyderabad* **14** 1-9

[19] Maleki S, Rafiee R, Hasannia A and Habibagahi M R 2019 Investigating the influence of delamination on the stiffness of composite pipes under compressive transverse loading using cohesive zone method *Front. Struct. Civ. Eng.* **13** 1316–23

[20] Diehl T 2006 Using ABAQUS Cohesive Elements to Model Peeling of an Epoxy-bonded Aluminum Strip : A Benchmark Study for Inelastic Peel Arms *2006 ABAQUS Users' Conf.* 1–20

[21] Diehl T 2008 On using a penalty-based cohesive-zone finite element approach, Part I: Elastic solution benchmarks *Int. J. Adhes. Adhes.* **28** 237–55

[22] Kenane M and Benzeggagh M L 1997 Fracture Toughness of Unidirectional Glass / Epoxy Composites Under Fatigue Loading *Compos. Sci. Technol.* **3538** 597–605

[23] Pan Y, Wu G, Cheng X, Zhang Z, Li M and Ji S 2016 Mode I and Mode II interlaminar fracture toughness of CFRP / magnesium alloys hybrid laminates *Compos. Interfaces* **23** 453-65

[24] Yamanaka T, Heidari-Rarani M, Lessard L, Feret V and Hubert P 2016 A new finite element method for modeling delamination propagation without additional degrees of freedom *Compos. Struct.* **147** 82–98

[25] Turon A, Dávila C G, Camanho P P and Costa J 2007 An engineering solution for mesh size effects in the simulation of delamination using cohesive zone models *Eng. Fract. Mech.* **74** 1665–82

[26] Curiel-Sosa J L, Tafazzolimoghaddam B and Zhang C 2018 Modelling fracture and delamination in composite laminates: Energy release rate and interfacial stress *Compos. Struct.* **189** 641–7

Investigation of Jump Resonance of a Horizontal Axis Washing Machine for Nonlinear Vibration using the Harmonic Balance Method

Mohit Dhanda [a*], **Arpan Gupta** [a]

[a] School of Engineering, Indian Institute of Technology, Mandi, Himachal Pradesh, India
* *Corresponding author e-mail*: s18006@students.iitmandi.ac.in

Abstract. In this work, a drum-type horizontal axis washing machine with two springs and two dampers has been considered to study the jump resonance phenomenon. Here two springs and two dampers are used and effects of nonlinear spring stiffness on vibration level in both horizontal and vertical directions have been investigated. In this study, a mathematical model of a horizontal axis washing machine is formed. A separate analysis is carried out in both vertical and horizontal directions by considering the whole system as a two degree of freedom system. As vibration magnitude is large due to unbalance mass of wet clothes, hence nonlinear stiffness of spring is considered. The analytical solution for the mathematical model is found using the harmonic balance method. One term expansion is used in this method to study the response of the system. To study the vibration response of the system, one term expansion is used. From this study, it has been found that the spring with cubic nonlinear stiffness increases the amplitude of vibration. Also, with the increase in cubic non-linear stiffness, the system tends to have multiple equilibrium positions. These equilibrium positions may be stable or unstable. It has been detected that the vibration amplitude has three different values for a few excitation frequencies due to cubic stiffness. So, the response of the system suddenly increases and then decreases due to cubic non-linearity which indicates the jump resonance.

Keywords: Resonance, Stiffness, Vibration

1. Introduction

The recent advances in technology have continuously eased the life of a human being. We are surrounded by several such examples in our daily life. Among such, the notable example is the washing [1]. Washing machines are categorized as horizontal axis and vertical axis washing machines based upon the viewpoint of the rotational axis of the drum during washing and drying operation [2].

A washing machine generally consists of two springs, two dampers or three springs, three dampers attached to the drum. But in this study, two spring and two dampers are used to minimize the manufacturing cost and assess the vibration level due to unbalancing of wet clothes during washing and drying operation. Throughout the washing operation, damp clothes cling to the inner side of the drum and create an unbalanced mass. This unbalance mass makes the vibration level severe and hence owing to the large deflection of spring, non-linear cubic stiffness comes into the picture.

Resonance is an unwanted phenomenon associated with vibration. A very large amplitude of vibration is observed during the resonance conditions. Due to unbalance mass of wet clothes, the washing machine sometimes shakes violently, which may cause injury to the person and a breakdown of its part may take place [3]. Beatriz et al. studied the outcome of the padding in estimating the fleeting movement of the tub [4]. They provide a D-optimal design approach to anticipate the movement of the tub during resonance and used it in the design phase of the washing machine to avoid tub collision. Vibration amplitude in a washing machine can be minimized by employing a hydraulic balancer. This hydraulic balancer contains saltwater and acts as a counter mass to unbalance mass[5]. Cristiano et al. used the control technique for magnetorheological dampers for noise and tremor lessening in the washing machine [6]. They have proposed to swap the standard passive dampers with electronically controlled one and used 3-axis MEMS sensors instrumented to the washing machine body to study the noise and vibration level. Baris and Haluk came up with a method to regulate the oscillation to deal with the dynamic constancy of a horizontal axis washing machine. They have achieved it using feedback control of step motors. The oscillation data was received by the step motors attached to the friction dampers and accordingly expand or narrow the bracelets radius positioned on dampers in a washing machine. Friction damper non-linearity was studied by Kang et al. by using Bouc-wen model. They have assumed the hysteresis behaviour of the damper for the nonlinear model [8]. Yasuaki et al. have predicted the resonant frequency range in both horizontal and vertical directions for the outer drum of the washing machine. They have proposed the G fall balancer to reduce the unbalance mass due to wet clothes and reduces the vibration level by 70 % at resonance [9]. Pinar and Mutlu optimized the oscillation features of the washing machine throughout the spinning cycle using a genetic algorithm [10]. The suspension block of the washing machine was modelled dynamically in 2D to study the genetic algorithm for vibration characteristics optimizations. The input was given as the spin speed and translations were obtained as the output in the x-z plane. Mingyu et al. predicted the rebalancing method using a robot balancer system to reduce the unbalance mass for the front-loading machine. Here rotor balancer detects the unbalance mass and moves the zigzag layer in reverse direction[11]. Walking instability and its control during the spinning cycle was predicted by Evangelos et al. They have proposed a design-based and active method to reduce the vibration amplitude [12].

So far, negligible work has been done for investing jump resonance and instability of a horizontal washing machine considering cubic stiffness. So present work focuses on predicting the jump resonance frequency of a washing machine with two springs and two dampers using the harmonic balance method. Here the effect of cubic stiffness on vibration amplitude, jump resonance frequency and the possibility of multiple equilibrium states in both horizontal and vertical directions have been investigated.

2. Mathematical Prototype

The mathematical prototype of the horizontal axis washing machine is obtained considering the whole system as two degrees of freedom system and it is displayed in Figure 1. Here two spring and two viscous dampers are used. Large deflection of spring and dampers from equilibrium position is assumed and equation of motion in both horizontal and vertical direction is derived using D' Alembert's principle. The spring stiffness is assumed as cubic. The mathematical prototype of the washing machine with cubic stiffness is represented by equation (1) and (4). As equation (1) and (4) does not obey the superposition principle, hence these are non-homogenous, nonlinear ordinary differential equation. To get the analytical solution for equations (1) and (4), method of harmonic balance is used. Here the solution of the system is assumed in the form of Fourier series expansion.

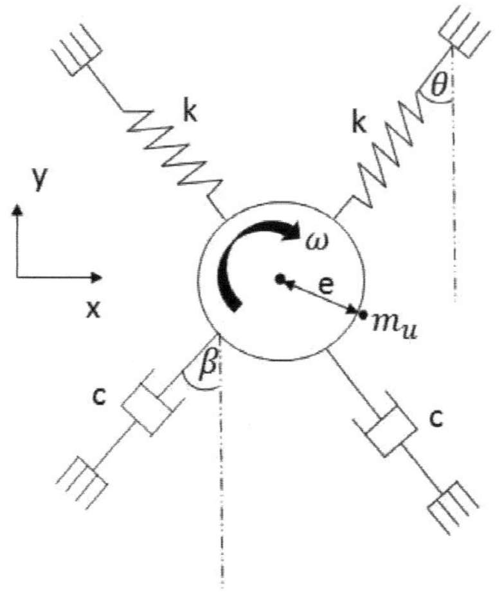

Figure. 1. Mathematical prototype of the washing machine with two springs and two dampers.

$$(m_u + m)\ddot{x} + 2(c_e)_x\dot{x} + 2(k_e)_x x + k_3x^3 = m_u e\omega^2 \sin\omega t \quad (1)$$

$$x = x_1 \cos(\omega t + \phi) \quad (2)$$

$$x_1^2\left[\left(\frac{3k_1x_1^2}{4} + (k - m\omega^2)\right)^2 + \omega^2c^2\right] = F^2 \quad (3)$$

Here equation (2) represents the assumed solution in horizontal direction using harmonic balance method. One term expansion is used to get frequency amplitude relation. Equation (3) represents the final solution after putting equation (2) in (1) and comparing the coefficient of $'\omega'$ and $'\phi'$ terms.

$$(m_u + m)\ddot{y} + 2(c_e)_y\dot{y} + 2(k_e)_y y + k_3y^3 = m_u e\omega^2 \cos\omega t \quad (4)$$

$$y = y_1 \cos(\omega t + \phi) \quad (5)$$

$$y_1^2\left[\left(\frac{3k_1y_1^2}{4} + (k - m\omega^2)\right)^2 + \omega^2c^2\right] = F^2 \quad (6)$$

Here equation (5) represents the assumed solution in the vertical direction using the harmonic balance method. One term expansion is used to get frequency amplitude relation. Equation (6) represents the final solution after putting equation (5) in (4) and comparing the coefficient of $'\omega'$ and $'\phi'$ terms.

Here m is mass of drum, k is linear spring stiffness, k_3 is spring stiffness due to cubic nonlinearity, c is damping coefficient, ω is the speed of rotation of the drum, e is the eccentricity of unbalanced mass, m_u is unbalanced mass due to wet clothes, θ is the inclination angle of spring with vertical and β is the inclination angle of the damper with vertical. $(k_e)_x$ and $(k_e)_y$ are the effective stiffness in x and y-direction respectively. The values of $(k_e)_x$ and $(k_e)_y$ are $k\sin^2\theta$ and $k\cos^2\theta$ respectively. Similarly, $(c_e)_x$ and $(c_e)_y$ are the effective damping coefficient in x and y-direction respectively. The values of

$(c_e)_x$ and $(c_e)_y$ are $c\,sin^2\,\beta$ and $c\,cos^2\,\beta$ respectively. The value of F is $m_u e\omega^2$. The value of linear spring constant k along the spring direction is taken as 1500 N/m.

Table 1. Values of the parameters used for analysis.

m_u (kg)	m (kg)	e (m)	θ (degree)	β (degree)
7	10	0.066	30	30

3. Results and Discussion

The governing nonlinear differential equation obtained from the mathematical model is solved analytically. To get the amplitude-frequency relation, the harmonic balance method is used. In the harmonic balance method, the solution is assumed in the form of the Fourier series. Here one term expansion in the Fourier series is used. From the frequency versus amplitude relation, it has been observed that for a small value of cubic spring stiffness, the system has only one equilibrium position. With the increase in cubic stiffness, multiple equilibrium positions exist at a particular frequency. These equilibrium positions may be stable or unstable. The amplitude of vibration also increases with cubic stiffness. The results are shown below in Figures 2 and 3 respectively for horizontal and vertical directions.

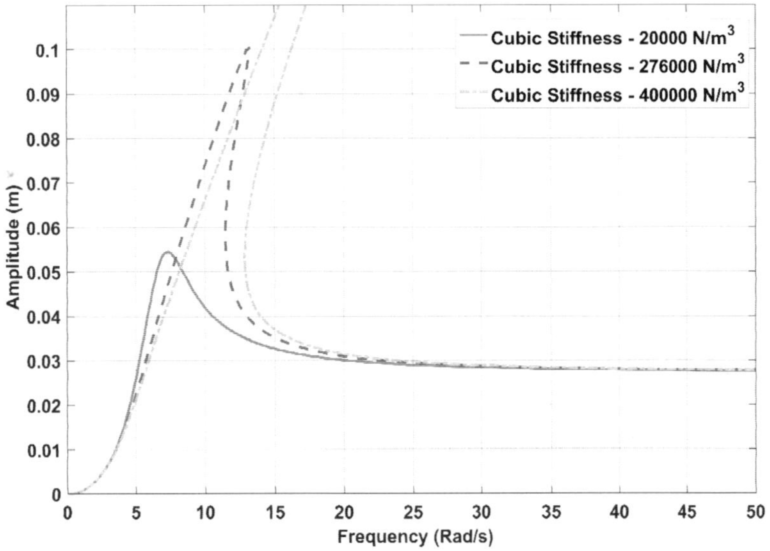

Figure 2. Frequency versus amplitude relation showing the jump resonance for cubic stiffness in the horizontal direction.

From the results shown in Figures 2 and 3, it has been observed that the jump resonance frequency value is more for vertical direction response. Also, vibration amplitude first increases with excitation frequency become maximum at the resonant frequency and then become constant for any value of cubic stiffness. The pattern is the same for both horizontal and vertical direction deflection.

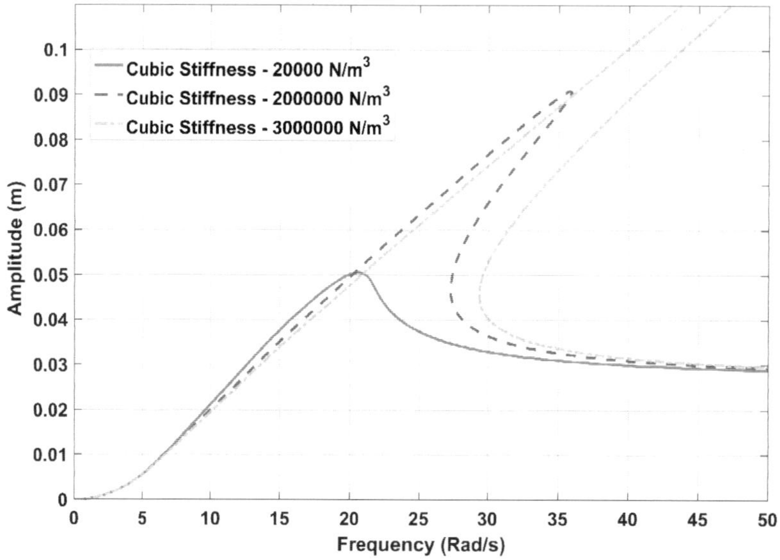

Figure 3. Frequency versus amplitude relation showing the jump resonance for cubic stiffness in the vertical direction.

4. Conclusion

In this study, the jump resonance of a horizontal axis washing machine with two springs and two dampers have been investigated. Cubic non-linear stiffness of both the spring is considered. The non-linear differential equation of the mathematical model is solved by the harmonic balance method to get frequency amplitude relation. Separate analysis is carried out in both vertical and horizontal directions by considering the whole system as a two degree of freedom system. From this study, it has been found that, by increasing the cubic stiffness, oscillation amplitude increases in both horizontal and vertical directions. Also, the system (washing machine) starts having multiple equilibrium positions with an increase in cubic stiffness at particular excitation frequency. These multiple equilibrium positions may be stable or unstable. As the number of solution changes with cubic stiffness, so the cubic stiffness act as a bifurcation parameter. This study also predicts the jump resonance frequency range in both horizontal and vertical directions. This study can be extended to design a new control technique to minimize the system vibration amplitude in this jump frequency range. Also, this amplitude can be used by piezoelectric energy harvesting MEMS sensor to generate power for the internet of things sensors of washing machine.

Informed consent: Informed consent was obtained from all individual participants included in the study.

Acknowledgement: Authors would like to acknowledge the support received from DST for the project DST/INT/CAN/P-04/2020. This was helpful in carrying out the present work.

References

[1] Somheil T 1992 *New Ways of Washing* Appliance pp.74-7

[2] Waverly W 2021 3 Types of Washing Machines Retrieved from https://www.duerdensappliance.com/blog/3-types-of-washing-machines

[3] Homepage 2021 *Why does washing machine sometimes shake violently* Retrieved from https://youaskweanswer.net/why-does-washing-machine-sometimes-shake-violently

[4] Tabuenca B, Galé C, Lladó J, Albero C and Latre R 2020 Washing machine dynamic model to prevent tub collision during transient state *Sensors* **20** 1–17

[5] Bae S, Lee M, Kang J, Kang S and Yun R 2002 Dynamic analysis of an automatic washing machine with a hydraulic balancer *J. Sound Vib.* **257** 3–18.

[6] Spelta C, Previdi F, Savaresi S, Fraternale G and Gaudiano N 2009 Control of magnetorheological dampers for vibration reduction in a washing machine *Mechatron.* **19** 410–21

[7] Yalçin B C and Erol H 2015 Semiactive vibration control for horizontal axis washing machine. *Shock Vib.* 1-10

[8] Kang D, Jung S, Nho G, Ok J, Yoo W 2010 Application of Bouc-Wen model to frequency-dependent nonlinear hysteretic friction damper *J. Mech. Sci. Technol* **24** 1311–17

[9] Sonoda Y, Yamamoto H and Yokoi Y 2003 Development of the vibration control system g-fall balancer for a drum type washer/dryer *Proc. IEEE/ASME Inter. Conf. on Advanced Intelligent Mechatronics* **2** 1140–44

[10] Boyraz P and Gündüz M 2013 Dynamic modeling of a horizontal washing machine and optimization of vibration characteristics using Genetic Algorithms. *Mechatron.* **23** 581–93

[11] Jo M, Kim J and Choi J 2020 Rebalancing Method for a Front-loading Washing Machine Using a Robot Balancer System. *Int J Control Autom Syst.* **18** 1053–60

[12] Papadopoulos E and Papadimitriou I 2001 Modeling, design and control of a portable washing machine during the spinning cycle. Proc. *IEEE/ASME Inter. Conf. on Advanced Intelligent. Mechatronics* **2** 899–04

Static and Dynamic Mechanical Characterization of Polydimethylsiloxane (PDMS) under Uniaxial Tensile Loading

David Kumar[1,*] and Sarthak S. Singh[2]

[1]Department of Aeronautical Engineering and Center for NDT, Chaoyang University of Technology, Taichung 413310, Taiwan
[2]Department of Mechanical Engineering, Indian Institute of Technology (ISM) Dhanbad, Dhanbad 826004, India

[*]Corresponding author: david.kumar1988@gmail.com

Abstract. The present study is based on performing mechanical characterization of polydimethylsiloxane (PDMS) under tensile loading. PDMS samples are developed by mixing prepolymer and curing agent with 10:1 ratio. Mechanical characterization of the PDMS samples is carried out using static and dynamic mechanical testing methods. In particular, the material is characterized under monotonic loading, different strain rates, and force-controlled cyclic loading. The monotonic loading tests are performed at different grip tightening levels to optimize gripping that avoids slipping and reaches failure point. The strain rate dependency and cyclic loading tests are performed to study the viscoelastic behaviour of the material. Strain rate dependency tests are carried out at the strain rates of 6.7×10^{-3} s^{-1}, 3.2×10^{-2} s^{-1}, and 6.2×10^{-2} s^{-1}. The hysteresis tests are performed at the force rates of 0.1 – 2 N/s. The obtained results show that fixing the testing specimen made of soft hyperviscoelastic material such as PDMS is challenging and can significantly affect its stress-strain behaviour and mechanical properties. A higher strain rate increases the stiffness of the material due to the viscoelastic nature of the material. Observing the stress-strain curves, with respect to the lowest strain rate case, the curve for 6.2×10^{-2} s^{-1} strain rate diverges at 40% strain compared to 3.2×10^{-2} s^{-1} strain rate curve that diverges at 80% strain. The material shows a nonlinear hysteresis behaviour and similar energy dissipation for the considered force-controlled cyclic loading cases.

1. Introduction

Polydimethylsiloxane (PDMS) is a high-performance most widely used silicone polymer due to its unique mechanical, physical and chemical properties. As an elastomer, PDMS is used in various applications including microfluidics, biomedical, cosmetics and soft robotics. Recently, PDMS has gained huge attention in the development of soft and flexible sensors that are useful to many aerospace and mechanical applications. Kamat et al. [1] developed a PDMS-graphene cantilever-form airflow sensor with excellent sensitivity and repeatability. Montazerian et al. [2] mixed chopped carbon fibers with PDMS to develop piezoresistive composite yarns. Chen et al. [3] developed stretchable strain sensor using PDMS with CNTs (carbon nanotubes) in a sandwich-like form as PDMS/CNTs/PDMS. Kou et al. [4] fabricated a pressure sensor using PDMS and graphene. Development of such sensing soft materials could be useful to fabricate smart and flexible flapping wings for small robotic flyers known as micro aerial vehicles (MAVs) [5,6]. Hence, it is important to study the mechanical behavior of PDMS carefully to develop accurate material models for designing purposes. However, it is a challenging task due to soft and hyperviscoelastic nature of the material. Some studies have considered the mechanical characterization aspect of PDMS but only a few did it systematically, under tensile mode and characterized the material for static and dynamic mechanical properties.

Liu et al. [7] investigated tensile mechanical properties of PDMS membranes of different thicknesses. Authors reported that Young's modulus and tensile strength depend on the thickness of PDMS films due to shear stress developed with the cross-linked chains during the spin coating process. Liu et al. [8] studied the effect of heating temperature and time on mechanical properties. This study showed that the tensile strength starts to decrease at a temperature higher than 200°C. It was also observed that the heating time does not affect the tensile strength for low-temperature heating. Mata et al. [9] prepared the PDMS samples with different weight percentages (wt%) of cross-linker with prepolymer. They found that the tensile strength increases up to 14.3 wt% and decreases for higher ones, i.e., 21.4 wt% and 42.9 wt%. Khanafer et al. [10] studied the effect of mixing ratio and strain rate on the elastic modulus. The authors reported that the elastic modulus increases with mixing ratio up to 9:1. It was also observed that the strain rate increases the modulus. This effect is more significant for the lower mixing ratio of 6:1 as compared to 9:1. Nunes [11] characterized PDMS in shear mode and measured large deformations using digital image correlation technique. The authors discovered the nonlinear behaviour of shear strain, while other studies reported linear shear stress-strain. Kim et al. [12] also studied the effect of the mixing ratio on the stress-strain behaviour under tensile mode. It was reported that the stress softening occurs during the first unloading cycle and decreases with the mixing ratio (5:1 to 15:1). There are a few studies that carried out dynamic mechanical characterization at different frequencies. Placet and Delobelle [13] reported that the storage and loss moduli increase with the frequency (10^{-2} - 10^{5} Hz). However, Lin et al. [14] reported an increase in storage modulus for 0.1 to 100 Hz, but the loss modulus starts to decrease after 1 Hz.

It has been understood that the proper mechanical characterization and properties of PDMS material are not yet established and still in exploring state. The material has to be characterized systematically by considering proper testing methods and precautions. Therefore, the present study was aimed at performing systematic mechanical characterization of PDMS under tensile loading using static and dynamic mechanical testing methods.

2. Materials and Methods

Given the soft and hyperviscoelastic nature of PDMS, the present study explored a systematic mechanical characterization process. The study employs appropriate testing equipment including low-capacity high-accuracy load cell and screw-controlled rubber-coated lightweight tensile grips. The present study also paid considerable attention to the level of grip tightening of the soft material samples. The material is characterized under monotonic loading, different strain rates and cyclic loading to investigate the hyperelastic as well as viscoelastic characteristics of the material. This section describes the preparation of the PDMS material and the characterization process used for investigating its mechanical properties.

2.1. Material preparation

The PDMS prepolymer and the curing agent (Sylgard 184 manufactured by Dow Corning Corporation) were procured from Sigma-Aldrich. Figure 1 shows the procedure used for making PDMS samples for characterization. The base polymer (prepolymer) and the curing agent are mixed with the ratio of 10:1, as the same proportion is considered in many other studies. The curing is done in an oven at 100°C for 35 min based on the instructions from the maker. The circular shape samples are cured in the petri dish after placing them at a horizontal surface inside the oven. The testing samples are cut using a surgical cutter based on the testing standard used for a particular characterization.

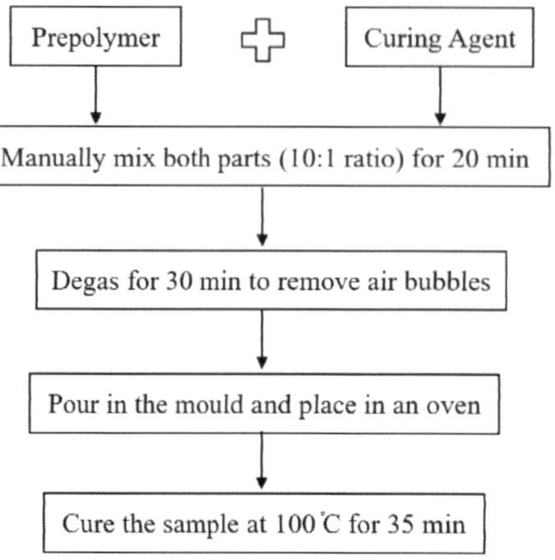

Figure 1. Procedure for making PDMS samples.

2.2. Static mechanical characterization

It is a well-known fact that testing a soft material such as PDMS is a complicated task, especially in tensile mode. Gripping the specimen at both ends should be done carefully to avoid unwanted initial stress due to compression from the grip surfaces. This stress would cause deformations that would affect the actual initial stress-strain behaviour of the material. At the same time, the gripping should be enough to make the sample fail during testing. Therefore, selecting the grip type and the gripping force is critical to perform tensile testing of PDMS material.

The static mechanical characterization is carried out to perform two studies; one is simple monotonic loading to observe the stress-strain behaviour, and another is done with different testing speeds to observe the effect of strain rate. These tests are performed using Tinius Olsen H10-KT universal testing machine (UTM). Due to the soft nature of the material, 250 N load cell is used to measure force values, and the samples are clamped at both ends using manual screw-controlled rubber-coated light-weight grips to appropriately fix the specimen and control the specimen tightening (see Figure 2). The dimension of the testing specimen and testing parameters are based on ASTM D-412. The specimen dimensions are: gauge length 25 mm, breadth 6 mm, and thickness 1 mm.

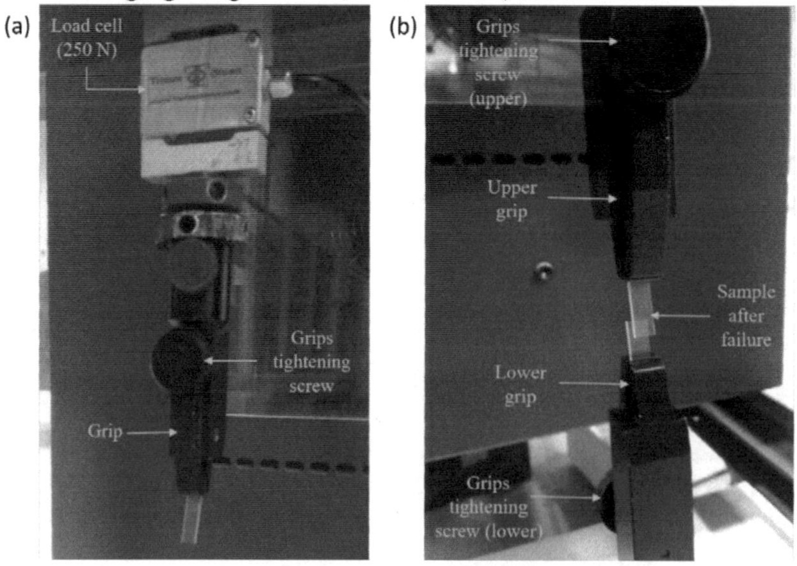

Figure 2. Illustration of components of the setup for testing PDMS using UTM, (a) load cell and grips, (b) grips and the failed sample.

2.3. Dynamic mechanical characterization
The dynamic mechanical characterization is performed using a dynamic mechanical analyzer (DMA (MetraVib^{+}100) shown in Figure 3). The tests are performed under force-controlled cyclic mode to study the viscoelastic behaviour of the PDMS material. Testing specimen and parameters are decided using the instruction manual provided by the machine maker. The specimen are fixed at both ends using suitable grips for the tensile testing. The relation which is used to find dimensions of the testing sample is

$$b > \frac{2tl}{t+l}$$

where, l is the length, b is the breadth, and t is the thickness.

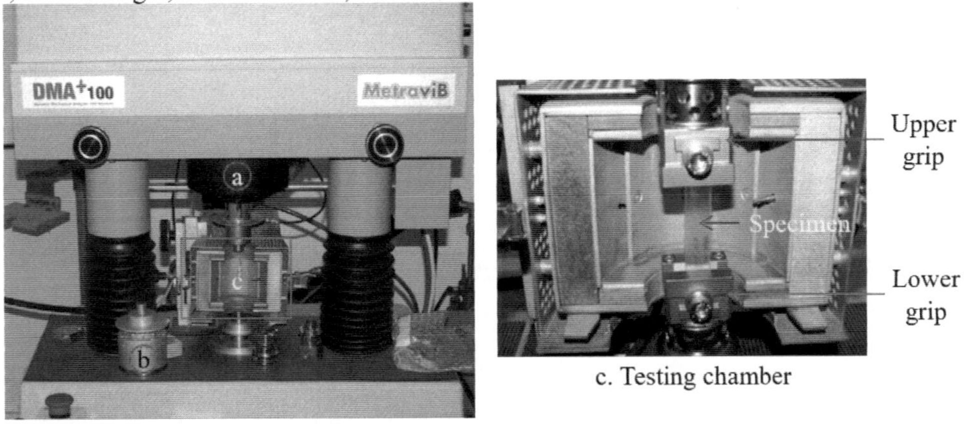

c. Testing chamber

a. Electrodynamic shaker, b. Load cell

Figure 3. Dynamic mechanical analyzer with the illustration of the testing setup.

3. Results and Discussion

This section presents the results obtained using UTM and DMA. The obtained results are discussed to study the stress-strain behaviour, strain rate dependency, and energy dissipation capability of the PDMS material.

3.1. Stress-strain behaviour under monotonic loading
We started the monotonic tensile testing of the material with less tightening and increased it until the sample stopped slipping from the grips and reached its failure point. The initial tightening level where the sample did not slip and failed is considered as the medium or less tightening. The full tightening level is considered when the sample is fully tightened by rotating the screw-controlled tightening valve to the maximum. Figure 4 shows two stress-strain tests conducted at the strain rate of 6.7×10-3 s^{-1} (crosshead speed of 10 mm/min), one is for medium tightening, and another is for full tightening. Note that the sample failed in both of the cases. It can be seen that tightening a sample at different forces changes the stress-strain behaviour of the material. The material shows less initial stiffness for full tightening in comparison to the one with less tightening level. This can be attributed to the fact that due to higher tightening level, the material will be under deformation before applying tensile force during testing. Basically, the compression due to grips leads to a large desired axial stretch, which is constrained at the surface (no-slip). Hence, local compressive axial stress develops in the grip which would lead to an initial compression in the gauge region. Now as the load is applied, additional stress is experienced by the material, which sees an apparent increase in strain for the same stress. This leads to an apparent loss of stiffness and reduction of strength. In terms of the mechanical properties, the

respective values of Young's modulus, tensile strength, and failure strain for full tighten condition are 0.89 MPa, 3.79 MPa, 177.6%, and for medium tighten condition are 1.8 MPa, 7.73 MPa, 141.4%. These are significant differences as far as consistency in mechanical properties is concerned. Hence, the selection or rather designing of grips as well as fixing of samples must be done very carefully to achieve actual stress-strain behaviour and mechanical properties of soft materials such as PDMS.

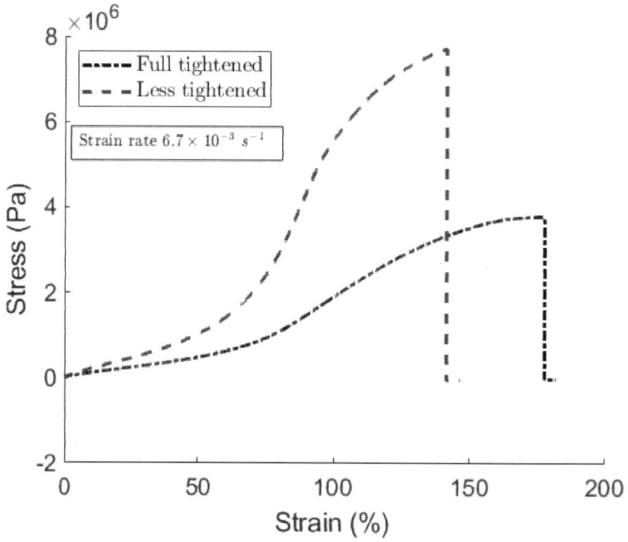

Figure 4. Stress-strain behaviour under monotonic loading at two different tightening levels, sample failed in both cases.

3.2. *Effect of strain rate*

The strain rate tests are performed to observe the strain rate dependency of PDMS. The selection of the range of strain rates or crosshead speeds is done on the basis of ASTM D-412. The standard recommends a crosshead speed of 5 mm/min to 500 mm/min for various rubbers and elastomers. In the present study, to study the effect of varying strain rate, three testing speeds are considered, i.e., 10 mm/min, 50 mm/min and 100 mm/min to comply with the ASTM standard and testing speed capability of the UTM machine. The corresponding strain rates based on the gauge length of 25 mm are 6.7×10^{-3} s^{-1}, 3.2×10^{-2} s^{-1}, and 6.2×10^{-2} s^{-1}. Figure 5 shows the results for the considered strain rate cases. The comparison is shown only up to 80% strain level considering that the materials generally work under 80% of strain. Under this strain limit, it can be seen that the material shows negligible strain rate dependency until 40% strain. The strain rate of 6.2×10^{-2} s^{-1} diverges at around 40% strain, whereas the other two low strain rate cases are overlapping till the considered 80% strain limit. From these results, it can be concluded that strain rate dependency or viscoelastic nature of the material depends on the strain rate and the strain limit. The increase in strain rate increases the stiffness of the material as it increases the slope of the stress-strain curve. The similar observation has been observed in literature at the testing speed of 5 mm/min and 500 mm/min for PDMS samples made using 9:1 and 6:1 mixing ratio [10]. Another observation is that the stress-strain curve for the higher strain rate case diverges earlier than that with the lower strain rate curve.

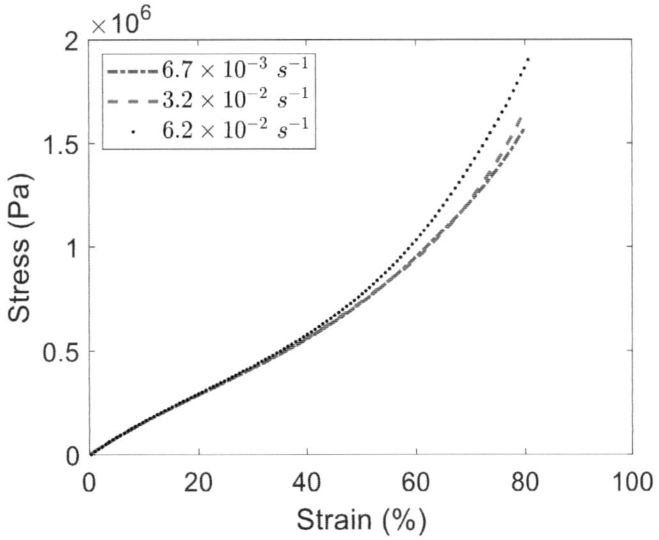

Figure 5. Stress-strain curves at different strain rates.

3.3. Viscoelastic behaviour via cyclic loading

The cyclic loading tests are performed to estimate the energy dissipation capacity of the PDMS material. The tests are performed in force-controlled mode at 0.1 N/s, 0.5 N/s, 1 N/s and 2 N/s. The length, breadth and thickness of the samples are 4 mm, 4 mm, 2.32 mm, respectively. These dimensions are calculated using the dimension selection equation given in Section 2.3. The results from the cyclic tests are shown in Figure 6. It can be seen that the material shows hysteresis behaviour as the unloading does not follow the loading path. It indicates that the material is viscoelastic. However, the energy dissipation capacity is low due to less area of the hysteresis loop, for all the force rate cases. It can also be observed that the stress-strain curves are nonlinear that indicate the nonlinear viscoelastic behaviour of the PDMS material. Comparing the loading rates, for high force rate loading cycles, the hysteresis loop approaches consistency as the material follow the same path for all the cycles of loading and unloading. Another observation is that the energy dissipation capacity decreases with the increase in force rate but with a minimal difference.

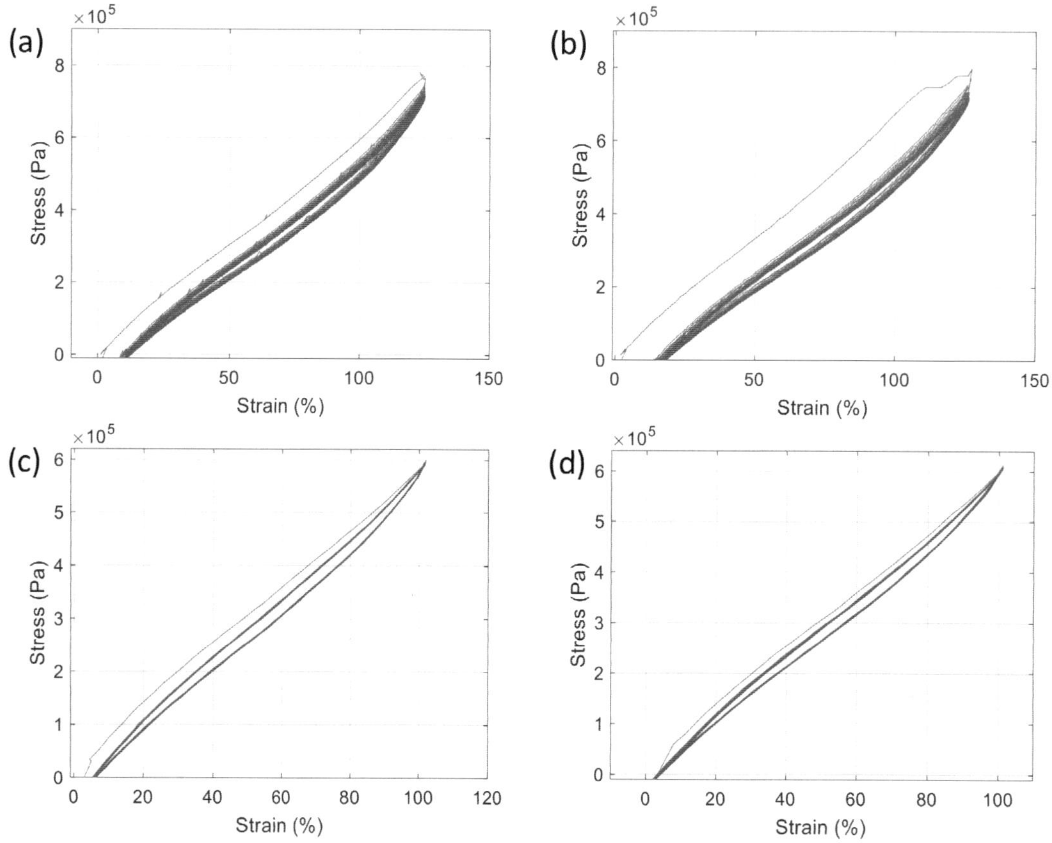

Figure 6. Results from the cyclic loading tests performed using dynamic mechanical analyser, (a) 0.1 N/s, (b) 0.5 N/s, (c) 1 N/s, (d) 2 N/s.

4. Conclusions

The present study provides interesting insights on the hyperviscoelastic nature of soft material PDMS subjected to static and dynamic mechanical loadings in the tensile mode. It is observed that the grip tightening can alter the actual stress-strain behaviour and mechanical properties of the material. Increasing strain rate increases the stiffness of the material owing to its viscoelastic nature. The strain rate tests also showed that the stress-strain curve subjected to a higher strain rate diverges significantly earlier than those with the lower strain rate. Cyclic tests showed the nonlinear hysteresis nature of the material. The hysteresis behaviour happens to be more consistent at a high force rate cycle than the lower ones. However, there was no significant difference in the energy dissipation capacity of the material.

The present study employed appropriate testing equipment and processes to characterize the material with a systematic approach as it is important for a better characterization and modelling of PDMS. The study provides new observations and results that could be helpful to establish systematic mechanical characterization methods for soft materials.

Acknowledgements
The authors thank the members of Aerospace Materials and Structures Laboratory, Indian Institute of Technology Kanpur for their help with the characterization of the material. The authors also thank Dr G M Kamath, Dr C S Upadhyay, and Dr Rajesh Kitey for their valuable suggestions.

IOP Conf. Series: Materials Science and Engineering 1225 (2022) 012041 doi:10.1088/1757-899X/1225/1/012041

References

[1] Kamat A M, Jayawardhana B and Kottapalli A G P 2020 *2020 IEEE Sensors* Rotterdam
[2] Montazerian H, Dalili A, Milani A S and Hoorfar M 2019 *Compos. Part B Eng.* **164** 648–58
[3] Chen J, Zhu Y and Jiang W 2020 *Compos. Sci. Technol.* **186** 107938
[4] Kou H, Zhang L, Tan Q, Liu G, Lv W, Lu F, Dong H and Xiong J 2018 *Sensors Actuators A Phys.* **277** 150–6
[5] Kumar D, Mohite P M and Kamle S 2019 *J. Bionic Eng.* **16** 894–903
[6] Kumar D, Kamath G M, Mohite P M and Kamle S 2019 *Mechanics of Biological Systems & Micro and Nanomechanics* Grady M E Springer International Publishing **4** 35–7
[7] Liu M, Sun J, Sun Y, Bock C and Chen Q 2009 *J. Micromechanics Microengineering* **19** 35028
[8] Liu M, Sun J and Chen Q 2009 *Sensors Actuators A Phys.* **151** 42–5
[9] Mata A, Fleischman A J and Roy S 2005 *Biomed. Microdevices* **7** 281–93
[10] Khanafer K, Duprey A, Schlicht M and Berguer R 2009 *Biomed. Microdevices* **11** 503–8
[11] Nunes L C S 2011 *Mater. Sci. Eng. A* **528** 1799–804
[12] Kim T K, Kim J K and Jeong O C 2011 *Microelectron. Eng.* **88** 1982–5
[13] Placet V and Delobelle P 2015 *J. Micromechanics Microengineering* **25** 35009
[14] Lin I-K, Ou K-S, Liao Y-M, Liu Y, Chen K-S and Zhang X 2009 *J. Microelectromechanical Syst.* **18** 1087–99

Optimization of Reinforced Concrete Cantilever Retaining Wall using Particle Swarm Optimization

Shubham Srivastava[a], Saurabh Pandey[b], Rajesh Kumar[c]

[a] *Research Scholar, Department of Civil Engineering, IIT BHU Varanasi, India,* *shubhamsrivastava.rs.civ19@itbhu.ac.in*
[b] *M.Tech, Department of Civil Engineering, IIT BHU Varanasi, India,* *jazzysaurabh@gmail.com*
[c] *Professor, Department of Civil Engineering, IIT BHU Varanasi, India,* *rkumar.civ@iitbhu.ac.in*.

ABSTRACT The design of structures depends on designer's experience and generally, the designer proceeds with trial and error until he arrives at a design which satisfies prescribed limit states. This is especially true for reinforced concrete structures in which the structural configuration is decided first and then reinforcement requirements are determined, resulting in higher cost. Retaining walls involve a large number of variables and therefore have been far from optimization. However, since retaining walls comprise of 20-30 percent of the cost of highways in hilly regions, their optimization is critical to economy of the project. Particle Swarm Optimization (PSO) does not require the objective function to be linear or differentiable and hence is ideally suited for optimization of retaining walls. This study uses PSO as a tool for the optimizing a Cantilever RC Retaining Wall. The objective function consists of two parts - cost function and penalty term. A program and GUI was developed to implement PSO. A reduction in cost was achieved from 8% -17% depending on the height of retaining wall, while the optimization in weight varied from 9%-34%.
Keywords: Particle Swarm Optimization, Retaining Wall, MATLAB, Cantilever, Optimization

1. INTRODUCTION

Retaining walls are structures used to support soils or other such material behind them. Earth retaining walls have several application such as in bridge abutments and for getting the desired carriageway in case of roads in hilly regions. According to an estimate the cost of retaining wall comprises of 20-30% of the cost of a highway project in a hilly region. The design of structures is largely dependent on the experience of the designer and rule of thumb. The designer proceeds with the practice of trial and error and arrives at a design which satisfies the limit states prescribed in the design codes. This is especially true for concrete structures in which the structure configuration is decided by the designer who then proceeds to find out the reinforcement requirements. This leads to a design, the cost of which is highly dependent on the experience and expertise of the design engineer.

Particle swarm optimization (PSO) is an optimization method developed by Eberhart and Kennedy [1]. They explored its efficacy on several benchmark functions and training of neural networks. They also described the relationships between particle swarm optimization and artificial life and genetic algorithms. The particle swarm optimization tested by them achieved 92% correct training of Neural Networks as compared to the conventional back-propagation method which achieved 89%. However, the origin of particle swarm optimization may be traced to the work of Reeves (1983), who introduced particle systems to simulate dynamic objects which could not easily be represented by particles like smoke, fire, clouds etc. in the field of computer graphics. Later, particle systems were used to simulate the behaviour of a flock of birds. PSO caught the attention of several researchers which lead to many

IOP Conf. Series: Materials Science and Engineering 1225 (2022) 012042 doi:10.1088/1757-899X/1225/1/012042

improvements and variants [2, 3]. It has also been used with other evolutionary and non-evolutionary algorithms like Genetic Algorithms, Ant-Colony Optimization, and Simulated Annealing etc. to derive what are known as hybrid swarm algorithms. Recently, Plasma Generation Optimization technique was used to optimize retaining structures based on ACI 318-19 [4]. Studies were carried out to access the performance of other optimization methods example differential evolution (DE), evolutionary strategy (ES) and biogeography based optimization algorithm (BBO), for nonlinear constrained optimum design of cantilever retaining wall [5]. The application of optimization under earthquake loads was also studied [6,7].

A modified version of Particle Swarm Optimization known as passive congregation for cost-effective design of Retaining Wall [8]. PSO was also applied for solving the problem of Beam-Slab layout design [9]. Genetic Algorithms was used to the optimization of cantilever reinforced concrete retaining wall [10] and to optimize reinforced concrete retaining walls giving heuristic rules for the proportioning of retaining walls [11]. The hybrid metaheuristic algorithm has also been applied for optimization for retaining walls [12]. Genetic Algorithm (GA), Simulated Annealing (SA) and Particle Swarm Optimization (PSO) - the three heuristic optimization algorithms for the optimization of retaining walls were compared for efficiency and PSO was recommended to be used as it was found to be superior in terms of effectiveness and efficiency [13]. Further, the Optimtool in MATLAB was used for optimized design of cantilever wall [14]. The application of evolutionary algorithms applied to the optimization of structures was studied the structural optimization was identified to be of three types: topology, shape and size [15]. Most recently, flower pollination algorithm [16] and reliability based design optimization [17] was also utilised for optimum design of reinforced concrete retaining walls.

This study aims at testing PSO as a tool for the optimization of structures, particularly reinforced concrete cantilevered retaining walls. The study uses inbuilt MATLAB PSO toolbox to implement the PSO algorithm. A program was written to calculated constraints, bounds and to implement the PSO. A GUI was also made to make the changing of design and Particle Swarm Parameters easier. Observations were taken at height of 3-12 m for both non-cohesive and cohesive soil to find the reduction in cost and weight. The effect of swarm size and number of iterations was also observed on the cost of computations.

2. PARTICLE SWARM OPTIMIZATION (PSO)

PSO is primarily inspired by swarms of living organisms like swarm of bees, flocks of birds etc. It is a stochastic computational method that optimizes a problem over many iterations. It was originally introduced to simulate the social behaviour of living organisms. PSO is quite similar to other evolutionary methods like Genetic Algorithms (GA) in the sense that it is also a population based optimization method and the optimization takes places due to the updating of generations. It however differs from GA as it does not employ mutation, cross-over or any other such evolutionary operators. The objective solution is represented as food source (as shown in Figure 1). The particles try to find the path to the food source using their own intelligence and the collective intelligence of the swarm.

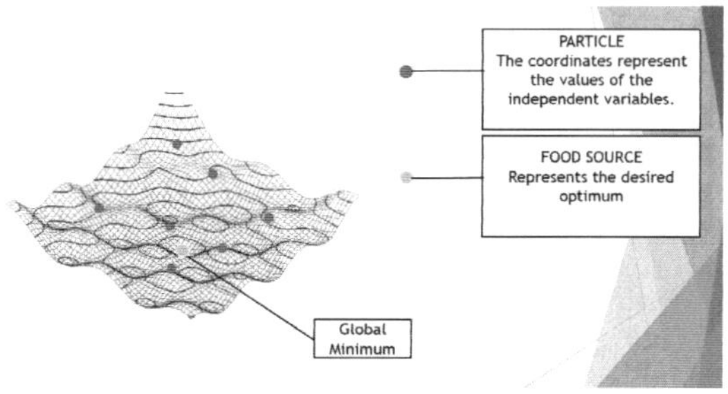

Figure 1. Illustration of PSO

The PSO algorithm starts by randomly allocating initial velocities to the initial particles created. Objective function at each location is estimated and the best (lowest) function value and the best location are determined. Based on current velocity, particles' individual best locations and the best locations of their neighbours the new velocities were selected. The particle locations, velocities, and neighbours were iteratively updated. The iterations continued until the algorithm reached a threshold criterion.

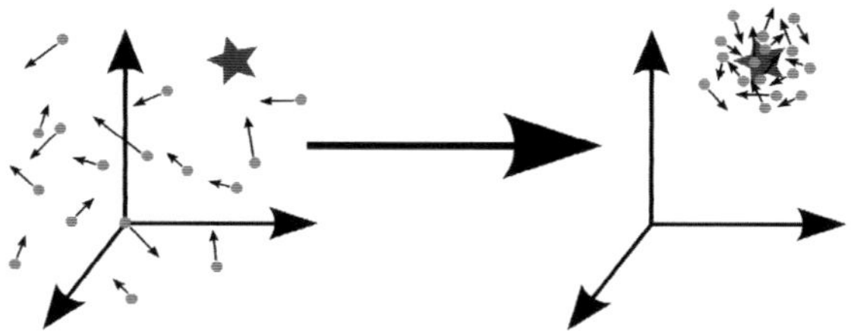

Figure 2. Convergence of swarm particles in PSO

The reiteration leads to convergence of swarm particles in PSO as shown in Figure 2. The cost function represents the actual function to be optimized while the penalty term is a means for the implementation of the constraints on the design. The algorithm flowchart for PSO is shown in Figure 3.

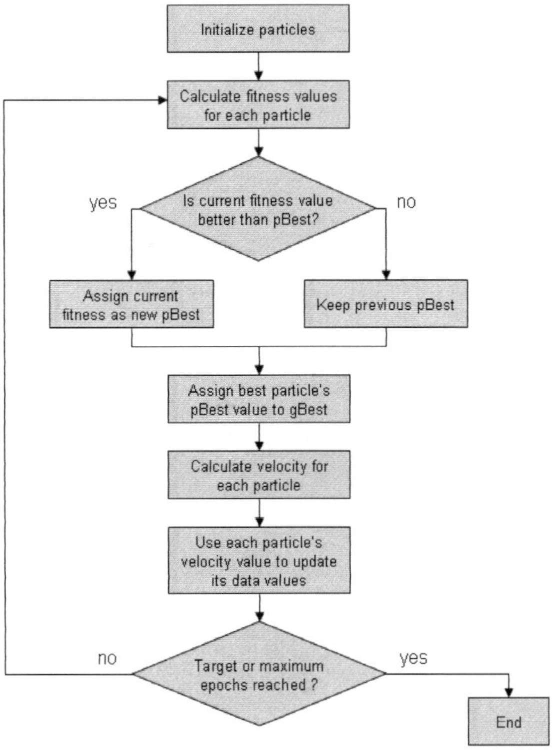

Figure 3. PSO Algorithm Flowchart

3. DESIGN OF RETAINING WALL

For design of retaining wall, the various forces acting on the retaining wall as shown in Figure 4 are needed to be determined. The coefficients of active (k_a) and passive (k_p) earth pressure were calculated using Rankine's Earth Pressure theory, where i is the angle of inclination of backfill and ∅ is the friction angle of the soil. For the stability of the retaining wall, the resultant of the forces must lie in the middle one third of the foundation base width. For safety against bearing failure, the maximum base pressure must not exceed the safe bearing capacity (SBC) of soil. The horizontal component of the earth pressure tends to overturn the retaining wall which is resisted by the combination of weight of the retaining wall and the earth over the heel portion. Further, it tends to slide the retaining wall which is resisted by the cohesive forces (in case of cohesive soils), the frictional forces between the base of the retaining wall and the soil and the shear key (if provided).

The design of the retaining wall consists of determining the parameters from (i) to (vi).

(i) Estimation of earth pressure:

$$k_a = \cos i . \frac{\cos i - \sqrt{\cos i^2 - \cos \emptyset^2}}{\cos i + \sqrt{\cos i^2 - \cos \emptyset^2}}$$

$$k_p = \cos i . \frac{\cos i + \sqrt{\cos i^2 - \cos \emptyset^2}}{\cos i - \sqrt{\cos i^2 - \cos \emptyset^2}}$$

(ii) Eccentricity: $e \geq B/6$

(iii) Bearing Pressure:

$$q_{min} = \frac{R}{L}\left(1 - \frac{6e}{L}\right) \geq 0$$

$$q_{max} = \frac{R}{L}\left(1 + \frac{6e}{L}\right) \leq SBC$$

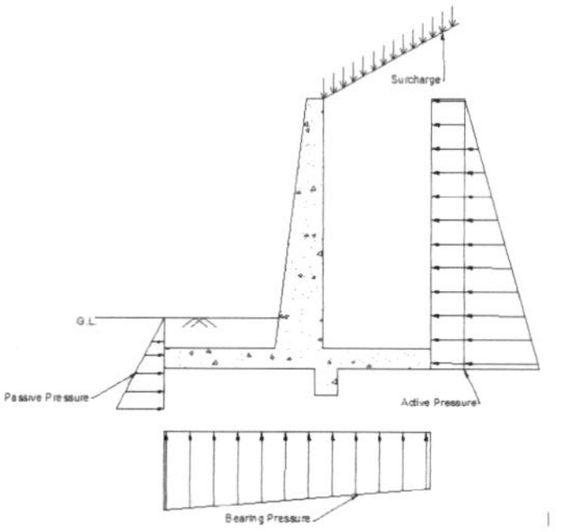

Figure 4. Forces acting on the retaining wall.

(iv) Moments and Shear Forces at Toe and heel sections

(v) Stability against overturning:

$$FOS_{overturning} = \frac{0.9 \times M_{correcting}}{M_{overturning}} \geq 1.4$$

(vi) Stability against sliding:

$$FOS_{sliding} = \frac{0.9 \times (F_{friction} + F_{cohesion} + F_{key})}{F_{sliding}} \geq 1.4$$

The minimum and maximum tension reinforcement areas were taken as 0.12 and 4 percent respectively according to the directives of IS: 456 – 2000.

4. FORMULATION OF DESIGN OPTIMIZATION PROBLEM

The Optimization of both weight and cost was undertaken and the results compared with conventional design results for heights 3 to 12 metres based on the guidelines provided in IS 456:2000. Also, the variation of cost with grade of steel and concrete was studied.

4.1 Design Variables

As shown in the Figure 5 and Table 1, a total of nine design variables were considered for optimization. They consisted of six geometric variables and three variables giving areas of steel in various sections, namely: the heel, the toe and the stem.

Table 1. Design Variables

Sr. No.	Design Variable	Description
1.	Lh	Length of the heel
2.	Lt	Length of the toe
3.	wb	Width of the base slab
4.	Ass	Area of steel in the stem
5.	Ash	Area of steel in the heel
6.	Ast	Area of steel in the toe
7.	Lb	Width of the stem slab at the base
8.	ws	Width of the shear key
9.	Ds	Depth of the shear key

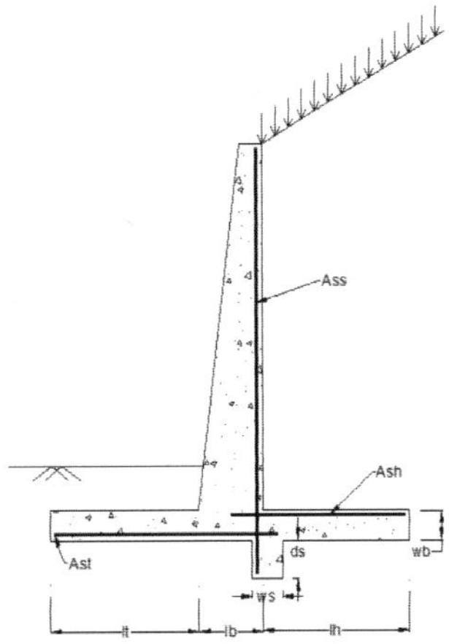

Figure 5. Design variables

4.2 Design Parameters

Design parameters (as listed in Table 2 and illustrated in Figure 6) were used to define site conditions and were taken as input from the Graphic User interface (GUI) in MATLAB. The various parameters were soil properties, loadings, material properties and material cost rates.

Table 2. Design Parameters

Design Parameter	Description	Input values
H	Height Above GL	
hf	Depth of Foundation	
i	Inclination of backfill	10°
q	Surcharge	20kN/m^2
γ_b	Unit weight of the backfill	16 kN/m^3
φ_b	Friction Angle of backfill	25° (cohesive) 35° (Non-cohesive)
c_b	Cohesion of backfill	20 kN/m^2
γ	Unit weight of the soil	18 kN/m^3
φ	Friction Angle of Soil	30°
c	Cohesion of Soil	10 kN/m^2
SBC	Safe Bearing capacity	250 kN/m^2
μ	Friction Coefficient	0.4
fck	Concrete Grade	25 MPa
fy	Steel Grade	500 MPa
γc	Concrete Density	25 kN/m^3
γs	Steel Density	78.5 kN/m^3
r_{con}	Concrete Price Rate	Rs. 7000/m^3
r_{steel}	Steel Price Rate	68.1 per kg

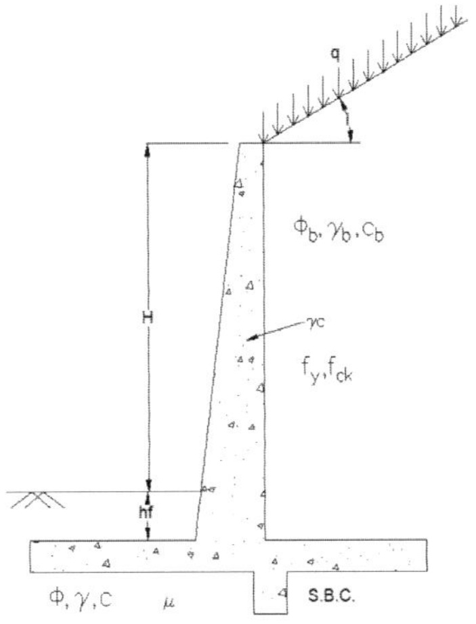

Figure 6. Design parameters

4.3 Design Variables bounds

The design variable bounds as suggested by Saribus and Erabatur [18] and based on IS: 456-2000, were taken. The design variable bounds (as listed in Table 3) are necessary to reduce the computation time and prevent impractical results. The variable bounds are related to the total height of the retaining wall.

Table 3. Design variable bounds

S.No.	Design Variable	Description	Lower Bound	Upper Bound
1.	Lh	Length of the heel	2.3*h/11	0.7*h
2.	Lt	Length of the toe	(0.4*12/11)*h/3	0.7*h
3.	wb	Width of the base slab	h/15	max{lh, lt}
4.	Ass	Area of steel in the stem	0.0012*0.2	0.04*(0.1*h+0.2)
5.	Ash	Area of steel in the heel	0.0012*h/15	0.04*max{lh, lt}
6.	Ast	Area of steel in the toe	0.0012*h/15	0.04*max{lh, lt}
7.	Lb	Width of the stem slab at the base	0.2	0.1*h+0.2
8.	ws	Width of the shear key	0.3	0.1*h
9.	ds	Depth of the shear key	0.3	0.1*h

4.4 Design Constraints

Design constraints (as given Table 4) were utilized to implements various constraints on design such factors of safeties in stability and limit states of strength. Additional design constraints were also utilized for implementing minimum and maximum steel areas and preventing the design of over-reinforced sections. Constraints were converted into dimensionless quantities for their weights to be considered equally while calculating penalty term.

Table 4. Design Constraints

Sr. No.	Design Constraint	Description
1.	$g_1(x)$	Stability against overturning
2.	$g_2(x)$	Stability against sliding
3.	$g_3(x)$	Eccentricity
4.	$g_4(x)$	Maximum base pressure
5.	$g_5(x)$	No tension condition at base
6.	$g_6(x)$	Shear capacity at toe slab
7.	$g_7(x)$	Moment capacity at toe slab
8.	$g_8(x)$	Shear capacity at heel slab
9.	$g_9(x)$	Moment capacity at heel slab
10.	$g_{10}(x)$	Shear capacity of the stem
11.	$g_{11}(x)$	Moment capacity of the stem
12.	$g_{12}(x)$	Shear capacity of the shear key

4.5 Objective Function

The objective function consisted of two parts namely the cost function and the penalty term. The cost function represents the actual function to be optimized while the penalty term is a means for the implementation of the constraints on the design. As the iteration proceeds the penalty term approaches zero and the objective function equals the cost function. The cost function is the total amount of concrete and steel used multiplied by their respective rates. Other costs like shuttering and formwork, temperature and nominal steels were neglected to make the computations simpler. The objective function used for cost optimization was:

$$f(x) = Rconc * Vol_{conc} + Rsteel * Wt_{steel} + h(k) * H(x) \quad (1)$$

Where the first two terms represent the cost of the structure and the third term is the penalty factor based on the violation of the constraints. The combination of continuous assignment function along with linear penalty parameter function gave the best results hence was adapted for the problem. Similarily the objective function used for the optimization of weight was:

$$f(x) = dcon * Vol_{conc} + dsteel * Vol_{steel} + h(k) * H(x) \quad (2)$$

5. ANALYSIS AND RESULT

Reinforced retaining walls of height 3 to 12 m were analysed for the savings in cost and reduction in weight. The optimization was done for both the cases of cohesive and non-cohesive backfill. Also, the effect of swarm size and the iterations was studied. The effect of grade of concrete and the friction angle on optimal cost was also studied.

5.1 Non-cohesive backfill

The comparison of optimized cost and weight with conventional design was carried out for cohesionless backfill. Figure 7 shows graph obtained height optimization, Figure 8 shows graph obtained for weight optimization and Figure 9 shows the graph for various dimensions versus height for cost optimization.

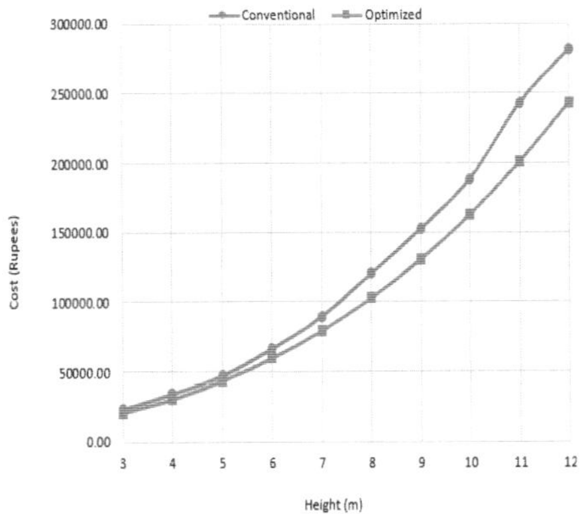

Figure 7. Cost vs Height

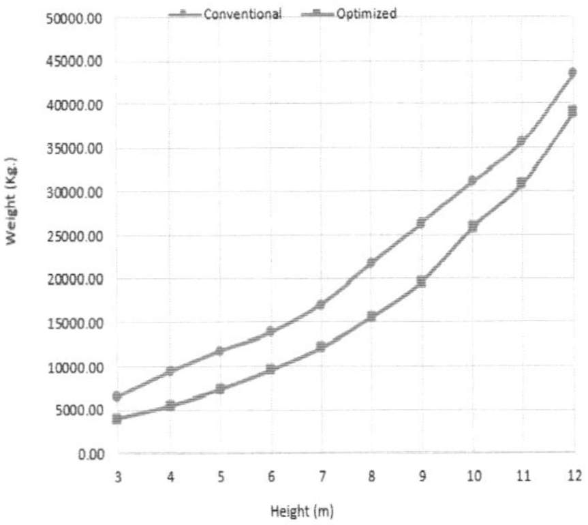

Figure 8. Weight vs Height

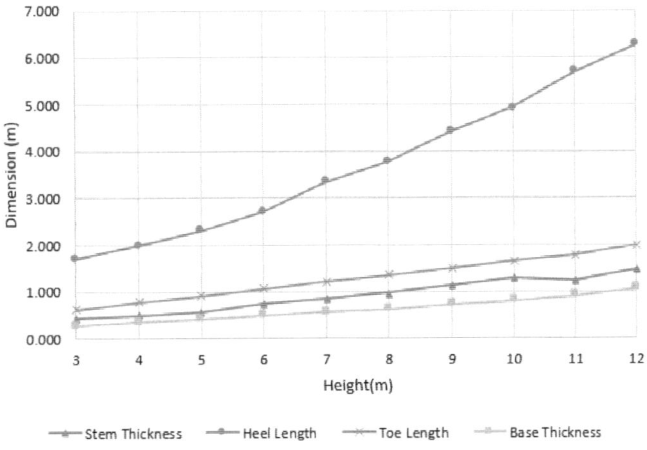

Figure 9. Dimensions vs Height for Cost Optimization

5.2 Cohesive backfill

The comparison of optimized cost and weight with conventional design was carried out for cohesive backfill. Figure 10 shows graph obtained height optimization, Figure 11 shows graph obtained for weight optimization for cohesive backfill soil.

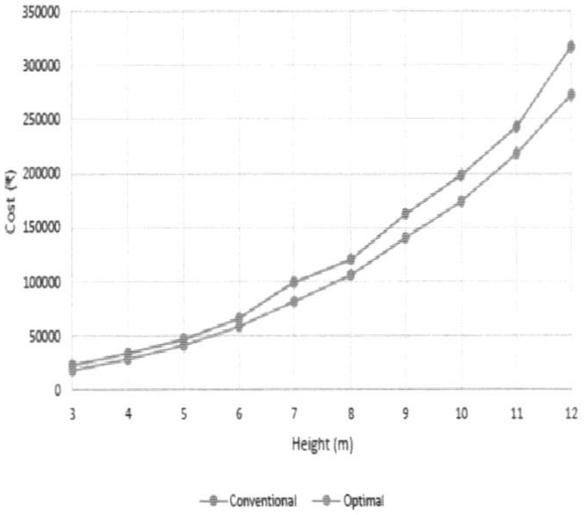

Figure 10. Cost vs Height

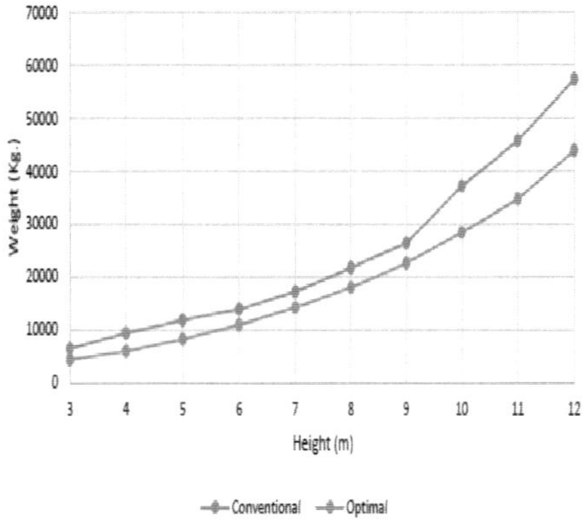

Figure 11. Weight vs Height

5.3 Variation of cost with friction angle of the soil for 6m wall

Figure 12 shows the variation of the optimal cost with the friction angle of soil for 6m wall.

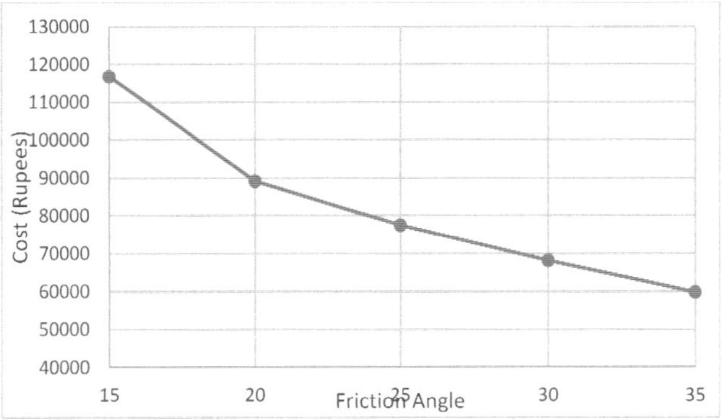

Figure 12. Variation of the optimal cost with the friction angle of soil

5.4 Variation of optimal cost with the grade of concrete

Figure 13 shows the variation of the optimal cost with grade of concrete.

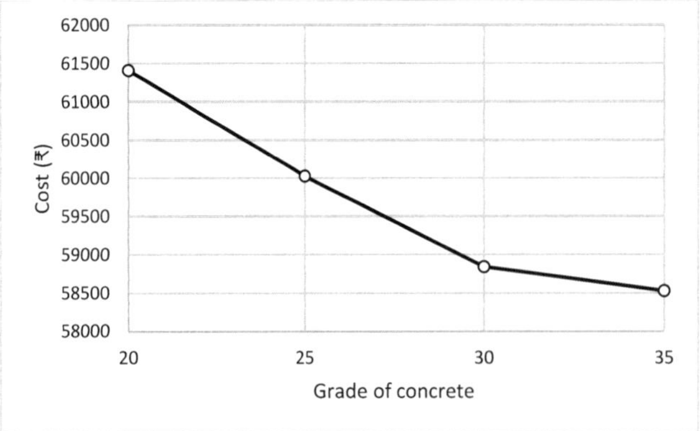

Figure 13. Cost of retaining wall for different concrete grades

The program was able to obtain an 8-17 % reduction in the cost depending on the height of the retaining wall. Also, a reduction of about 27 % was observed on an average in the case of weight optimization. The difference in conventional and optimized weight was however, lower for higher retaining wall.

5.5 Effect of Number of Iterations and Swarm Size

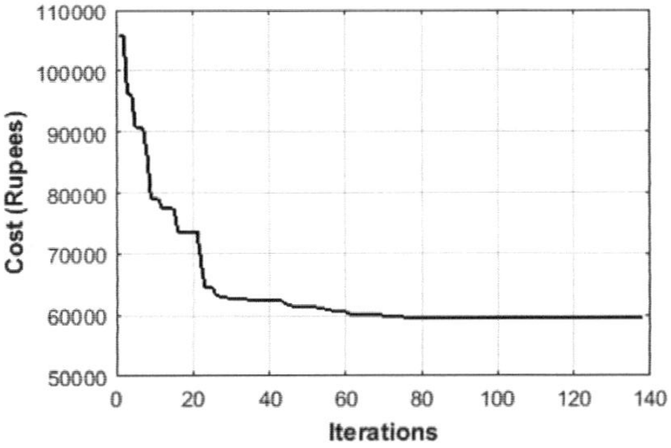

Figure 14. Cost vs Iterations

Figure 14 shows the variation of cost with iterations for the retaining wall of height 6m. It may be noted that the programs started with a high initial cost and slowly converges to the final cost. The cost shown in the plot is of the best particle in the swarm. It is also evident from the plot that the improvement in the solution after about 100 iterations is negligible. Therefore, the number of iterations for a run can be limited to about 150 without affecting the accuracy by a significant amount. Figure 15 shows the relationship between the swarm size and the time taken for computation.

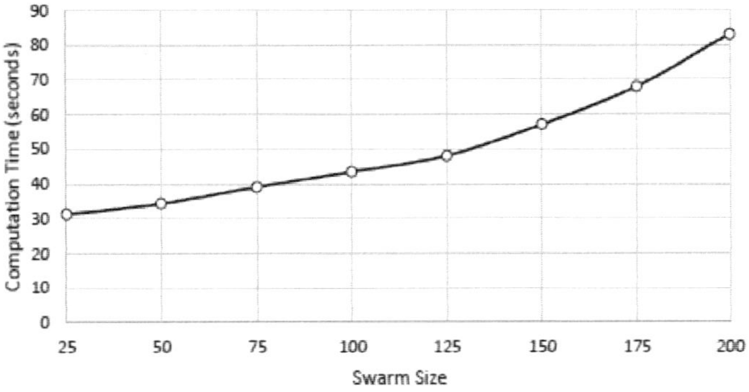

Figure 15. Computation time vs size of the swarm

It was found that the computation time increase drastically with the increase in the number of particles in the swarm, but there is not much significance in the quality of the solution. Therefore, a compromise has to be made between the computation time and the optimization. It is found that a population size of 25 is adequate for a satisfactory solution but the chances of the algorithm getting stuck in a local minimum is high. Therefore, a swarm size of 75-100 is suggested. Figure 16 shows the variation of the shape of the retaining wall as the iterations progress. The initial candidate solution in represented in a lighter shade of grey and the darker lines progressively show later iterations. It can be seen that the program starts with a comparatively heavier section which is slowly modified to a more effiecient section.

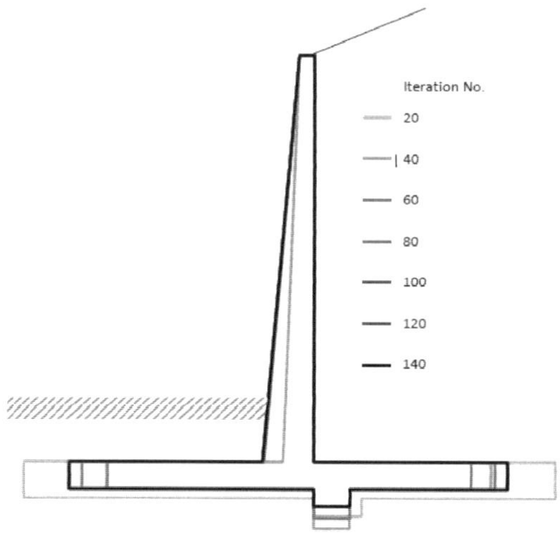

Figure 16. Shape of the retaining wall at different iterations

6. CONCLUSION

The present study proposed to optimize reinforced concrete cantilever retaining walls using PSO (Particle Swarm Optimization). The optimization both in terms of cost and weight was performed using a program developed in MATLAB. The program featured the capabilty to vary various site and material parameters.

- The reduction in cost achieved was from 8 -17 percent depending on the height of the retaining wall, while the optimization in weight varied from 9-34 percent.
- The percentage saving for cost showed a slight increasing trend with the height of the retaining wall while the percentage saving in terms of weight showed a decrease with the height of the retaining wall.
- The design solution in each case satisfied all the factors of safety as prescribed in the cost.
- The time taken for the optimized design was 50-70s on an average for a swarm of 100 particles over 150 iterations.
- The program was successful in optimizing the cost and weight and also reducing the design times dramatically.
- The effect on computation times due to the number of iterations and the number of particles in the swarm was also observed and it was concluded that a swarm size of 75-100 with the number of iterations at 150 provided satisfactory balance between computation times and quality of solutions.

Thus the study was successful in establishing PSO as a robust method for the optimization of RC cantilevered retaining walls.

REFERENCES

[1] Kennedy J and Eberhart R C 1995 Particle Swarm Optimization *Proc. of the IEEE Inter. Conf. on Neural Networks* **4** 1942–48

[2] Reynolds C W 1987 Flocks, Herds and Schools: A distributed behavioural model *Proc. of the Inter. Conf. on Computer Graphics and interactive techniques* **21** 25-34

[3] Heppner F and Grenander U 1990 A Stochastic Non-Linear Model for Coordinated Bird Flocks, The *Ubiquity of Chaos* 233-38.

[4] Kaveh A, Akbari H and Hosseini S M 2021 Plasma Generation Optimization for Optimal Design of Reinforced Concrete Cantilever Retaining Wall Structures *IJST-T CIV ENG* 1-24

[5] Gandomi A H, Kashani A R, Roke D A and Mousavi M 2017 Optimization of retaining wall design using evolutionary algorithms *Struct Multidiscipl Optim.* **55** 809–25

[6] Ravichandran N, Wang L, Rahbari P and Juang C H 2021 Robust design optimization of retaining wall backfilled with shredded tire in the face of earthquake hazards *Bull. Eng. Geol. Environ.* **80** 1351–6,

[7] Aydogdu I. Cost optimization of reinforced concrete cantilever retaining walls under seismic loading using a biogeography-based optimization algorithm with Levy flights *Eng. Optim.* **49** 384-400

[8] Khajehzadeh M, Taha M R, El-Shafie A, and Eslami M 2017 Economic Design of Retaining Wall Using Particle Swarm Optimization with Passive Congregation *Aust. J. Basic & Appl. Sci* **4** 5500–07

[9] Nimtawat A and Nanakorn P 2011 Simple particle swarm optimization for solving beam-slab layout design problems *Procedia Eng.* **14** 1392–1398.

[10] Bazaz J.B. and Fahmideh A. 2012 Optimum Cost Design of Reinforced Concrete Cantilever Retaining Wall using Genetic Algorithm *In Proc. of 9th International Congress on Civil Engineering*

[11] Donkada S and Menon D 2012 Optimal design of reinforced concrete retaining avails *Indian Concr. J.* 9–18

[12] Sharma, S, Saha A K and Lohar G 2021 Optimization of weight and cost of cantilever retaining wall by a hybrid metaheuristic algorithm *Eng. Comput.*1-27

[13] Pei Y and Xia Y 2012 Design of Reinforced Cantilever Retaining Walls using Heuristic Optimization Algorithms *Procedia Earth Planet.* **5** 32–36

[14] Sable K S and Patil A A Comparison between Optimization and Conventional Cantilever Retaining Wall by Using Optimtool in Matlab 2012 *Int. J. Eng. & Adv.Tech.***1** 255–261

[15] Cazacu R and Grama L 2013 Structural Optimization with Genetic Algorithms and Particle Swarm Optimization *FMTE* **1** 45–48

[16] Mergos P E and Mantoglou F 2020 Optimum design of reinforced concrete retaining walls with the flower pollination algorithm *Struct Multidiscipl Optim* **2** 575-85

[17] Al Sebai H M, Barakat S and Arab M A 2021 Reliability-Based Design Optimization of Cantilever Reinforced Concrete Retaining Walls *Int. Foundations Congress and Equipment Expo* 267-77

[18] Saribus A and Erbatur F 1996 Optimization and Sensitivity of Retaining Structures *J. Geotech. Eng.* **122** 649-56

Biogenic Metal and Metal Oxides Nanoparticles as Anticancer Agent: A Review

N Rani[1], K Saini[1]*

[1]Department of Chemistry, Miranda House, University of Delhi, New Delhi-110007

* Corresponding Author E-mail : kalawati.saini@mirandahouse.ac.in

Abstract. Herein this review we have summarized a number of cytotoxic studies which have been done using various biosynthesized metal nanoparticles (MNPs) and metal oxide nanoparticles (MONPs) on different cancer cell lines. Plants can serve as a good candidate to provide natural surfactants for the green approach in the preparation of nanoparticles. Numerous biomolecules are present in the plants. Also, numerous plant extract-based MNPs and MONPs have been synthesized and used in several fields of applications particularly in biomedicine. This property can be attributed because of their low cost, biocompatibility and favourable to the environment. In the past few years, the utilisation of these biogenic nanoparticles has increased tremendously particularly in cancer therapy. These biogenic nanoparticles considered as an excellent tool for cancer diagnosis and drug delivery at the tumour site preferentially. By utilising the unique properties of nanoparticles and antioxidant and antitumor nature of plants, these biosynthesised nanoparticles selectively destroy the tumour cells and do not harm the normal healthy cells. In this review, we have compiled the most significant results obtained by the biosynthesized MNPs and MONPs like silver (Ag), gold (Au), Fe_2O_3/Fe_3O_4, ZnO, and CuO respectively.

Keywords: Biogenic synthesis, Apoptosis, Reactive Oxygen Species, Cytotoxicity, Antitumor

1. Introduction

1.1. Importance of metal /metal oxide nanoparticles

Nanoparticles are the particles whose size lie between 1nm to 100 nm. Owing to very small size, their chemical and physical properties are size dependent. This feature is different from properties associated with bulk material [1]. The utilization of MNPs and MONPs in various areas is increasing day by day. Nanoparticles have immense possibilities to be used in diverse area of applications like electronics, biotechnology, biomedical and environmental sciences. The use of these nanoparticles has increased tremendously in several everyday care products like soaps, cosmetics, shampoos, detergent, medical and pharmaceutical products [2,3]. Nature has the ingenious beauty of developing most efficient small sized functional materials. These materials can be synthesized by utilising either green approach or the conventional physical and chemical approaches. However, the conventional methods of synthesis are suffered from various disadvantages like high energy requirements, use of toxic chemicals, time consuming and expensive respectively. However, the biosynthesis is the promising way of synthesizing variety of nanoparticles with unique size and shape. Green synthesis avails non-toxic, sustainable and economical way to produce nanoparticles. The biosynthesis of nanoparticles is a bottomup technique where biological entities like as the extracts of various parts of plants (leaf, flower, fruit peel, bark, root and stem), microorganisms such as algae, fungus and bacteria are taken in use for the preparation of nanoparticles. Herein, **Figure 1** displays the use of different sources of the green approach for preparation of nanoparticles.

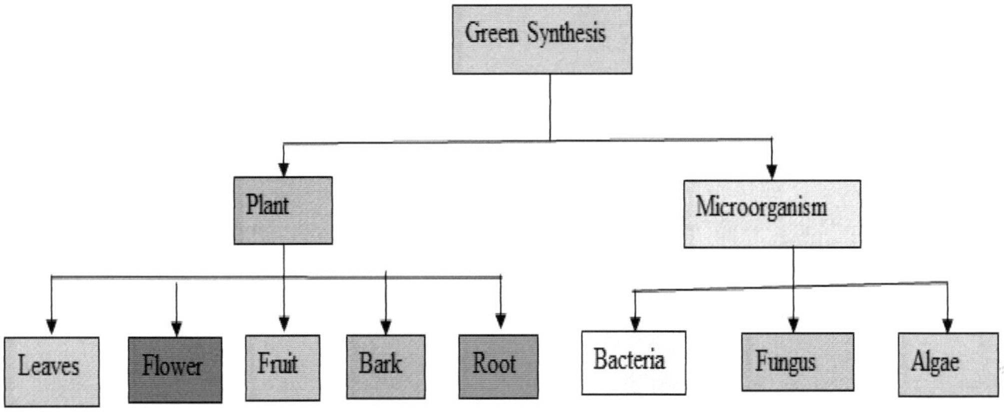

Figure 1. Schematic representation showing utilization of various sources for green synthesis

1.2. Importance of plants for nanoparticle synthesis

Very significant phytochemicals such as flavonoids, derivatives of phenols, polyphenolic acids, monoterpene, quercetin and many more have been investigated in the extracts of various parts of different plants. These phytochemicals serve the role as capping and reducing agents in the synthesis of variety of MNPs and MONPs which are non-toxic. Since the last few years, the extensive study has been performed on utilisation of extract of different plants parts for the preparation of nanoparticles and their anticancer studies. Metal oxide nanoparticles are prepared from metal precursors. These nanoparticles can be formed by several ways like electrochemical, sol-gel, co-precipitation, hydrothermal, solvothermal, microemulsion technique, laser ablation technique, thermal decomposition, high energy ball milling technique etc. The biogenic synthesis of nanoparticles with fungi, bacteria and plant extracts provides several benefits over the abovementioned methods. Similarly, if we compare the biosynthesis with fungi, bacteria and extract of plants, then synthesis with extract of plants is easier in terms of decreased number of synthesis steps like elaborate culturing procedures of bacteria and other things which are required for maintenance. In green approach for preparation of nanoparticles with extract of plants, unique size and shape of nanoparticles can be obtained for cancer targeted properties.

Because the extract of various plants has specific biomolecules which have their own biological significance namely anticancer, antimicrobial, antioxidant, drug delivery and antiflammetry respectively. The plant extracts have specific biomolecules such as polyphenols, polysaccharides, terpenoids, flavonoids, amide, aldehydes, carboxylic acids and heterocyclic compounds respectively [4-8]. Very recently, Nutan et al also reported in vitro investigation of ZnO nanoparticles prepared by green approach on (A549) human lung cancer cell lines [92]. Gupta et al have reported the synthesis of MNPs and MONPs (inorganic nanoparticles) and their implimentation in cancer therapy and as antibacterial agent. Gupta et al have prepared gold nanoparticle with green tea extract and utilized in catalytic reduction of methylene blue [93]. Singh et al have synthesized metal nanoparticles namely platinum, silver and gold using green approach [94]. Herein, these nanoparticles have been used for degradation of methyl orange. Herein, they have studied an electron relay effect. Byrne et al have synthesized biogenic magnetite nanoparticle (Fe_3O_4) using extract of *Geobacter sulfurreducens* at industrial level [95]. Herein, *Geobacter sulfurreducens* is a gram-negative metal and sulphur-reducing roteobacterium. **Figure 2** shows the general procedure which is followed for the preparation of plant extracts based MNPs and MONPs. Herein, first step demonstrates the preparation of fresh extract with various parts of plants and second step demonstrates preparation of aqueous metal salts precursor solution after that by adding definite amount of extract in the metal salt precursor solution, specific morphology of MNPs and MONPs are obtained.

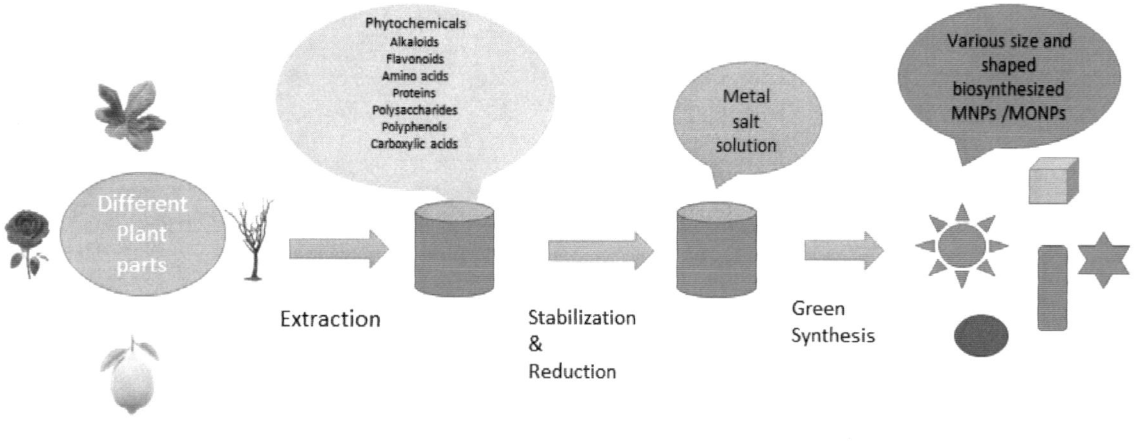

Figure 2. Showing the common method for green synthesis of MNPs and MONPs using plant parts.

Literature survey reveals that numerous MNPs and MONPs have been synthesized using extract of different plants parts such as stem, bark, seed coat, fruit etc. **Table 1** summarizes the various biosynthesized MNPs and MONPs having different shape and size. These biosynthesized nanoparticles have also been studied for their application as anticancer agent on different cancer cell lines.

Table 1. Showing various metal/metal oxide nanoparticles synthesized by utilizing green methods and their study on various cancer cell lines.

Sr. no.	Source	Plant part utilised	Metal/ metal oxide prepared	Size/Shape	Cancer cell line used	Ref.
1	*Costus pictus D. Don*	leaf	Zinc oxide	20-80 nm/rod shaped	DLA	75
2	*Tabernaemontana divaricata*	leaf	Zinc oxide	~ 36 nm/ spherical	MCF-7	9
3	*Capparis zeylanica*	leaf	Zinc oxide	28-30 nm	A549	10
4	*Deverra tortusa*	Aerial parts	Zinc oxide	9.26-31.18 nm/ spherical	Caco-2 A549	84
5	*Albizia lebbeck*	Stem bark	Zinc oxide	66.25 nm/ spherical	MDA-MB 231, MCF-7	65
6	*Borassus flabellifer*	Seed coat	Iron oxide	35 nm/ hexagonal	NIH3T3	57
7	*Rosemary extract*	leaf	Iron oxide	20-80 nm/ spherical	4T1, C26	58
8	*Juglan regia*	Green husk	Iron oxide	~5.77 nm/ cubical	NIH-3T3/ HT-29	11
9	*Rhamnella gilgitica*	leaf	Iron oxide	~21 nm/ spherical	Hep G2	12
10	*Punica Granatum*	Fruit peel	Iron oxide	11 nm/ spherical, cubical	HCT-116, MCF-7, HeLa, A549, HONE1	13
11	*Acalypha indica*	leaf	Copper oxide	26-30 nm/ spherical	MCF-7	88
12	*Ficus religiosa*	leaf	Copper oxide	577 nm/ spherical	A549	89
13	*Ormocarpum cochinchinese*	leaf	Copper oxide	1-2 μm/agglomerated	HCT-116	85
14	*Black bean*	Seed extract	Copper oxide	~ 26 nm/ spherical	HeLa cells	90
15	*Syzygium alternifolium*	Bark	Copper oxide	~ 17 nm/ spherical	MDA-MB 231	14
16	*Cotton plant*	Saffron stigma	Gold	4-10 nm/ spherical	MCF-7	54

17	*Carica papaya and Catharanthus*	leaf	Gold	15-28 nm/ spherical	MCF-7, Hep G2	53
18	*Curcuma wenyujin*	Rhizome	Gold	200 nm/ spherical	A498, Sw-156	52
19	*Rabdosia rubescens*	leaves	Gold	130 nm/ spherical	A549	47
20	Scutellaria barbata	Whole plant	Gold	0.4-1 μm/ spherical	PANC-1	46
21	*Dolichos lablab*	leaf	Silver	9 nm/ spherical	Hep G2	15
22	*Plumeria alba*	flower	Silver	36.19 nm/ spherical	COLO 205	16
23	*Syzygium aromaticum*	Fruit	Silver	5-20 nm/ spherical	MCF-7, A549	17
24	*Punica granatum*	peel	silver	15.6 nm/ spherical	MCF-7, BT-20	32
25	*Nepta deflersiana*	Aerial parts of the plant	silver	33 nm/ spherical	HeLa	33

2. Conditions required by MNPs and MONPs for anticancer application

2.1. Size

The size of MNPs and MONPs is a foremost criterion to be considered for biomedical application [21-23]. Since, the size of nanoparticles controls their movement inside the blood stream and tumour vasculature. In cancer treatment, the nanoparticles follow either active or passive pathway to treat tumour cells. The utilisation of nanoparticle has been observed as an important tool for the discovery of new age cancer drug [18]. For the application of nanoparticles in medicine and in vivo studies, the biocompatibility, biodegradability, long retention time and non-toxic properties are required to minimize the aftereffects [19]. The anticancer activity of various metal and metal oxide nanoparticles might be either due to intrinsic effect or by external stimuli such as hyperthermia. The intrinsic antitumor effect such as antioxidant behaviour of nanoparticles is due to specific physicochemical properties of nanoparticles. When external applied magnetic field or infrared radiations are applied, the external stimuli work by producing reactive oxygen species (ROS) which kills the tumour cells [20]. Enhanced permeability and retention (EPR) effect is applied in passive drug delivery system. The leaky vascular nature of cancerous cells permits metal and metal oxide nanoparticles to enter easily and kill the tumour cells [24]. In active pathway the nanoparticles are functionalized with different biomolecules or ligand to directly target at the cancerous cell site.

2.2. *Shape*

The shape of nanoparticles is another important criterion. Nanoparticles with spherical shape is most preferred for drug delivery application [25], because these have higher surface volume ratio on which sufficient drug amount can be loaded for better efficacy. Finally, by conjugating different targeting ligand the efficacy of the delivery system can be further increased. Li et al has observed the increased cytotoxicity into different tumour cells by N-(2-hydroxylpropyl) methacryl amide copolymer loaded with doxorubicin through the folate receptor [26]. Herein, **Figure 3** illustrates the graphical representation of biosynthesized MNPs and MONPs in the treatment of cancer as anticancer agent. Where three important anticancer applications of various biosynthesized MNPs and MONPs have been demonstrated. Those are cytotoxic study in mouse model, DNA damage of cancerous cell (apoptosis) and targeted drug delivery in the human beings due to biocompatibility.

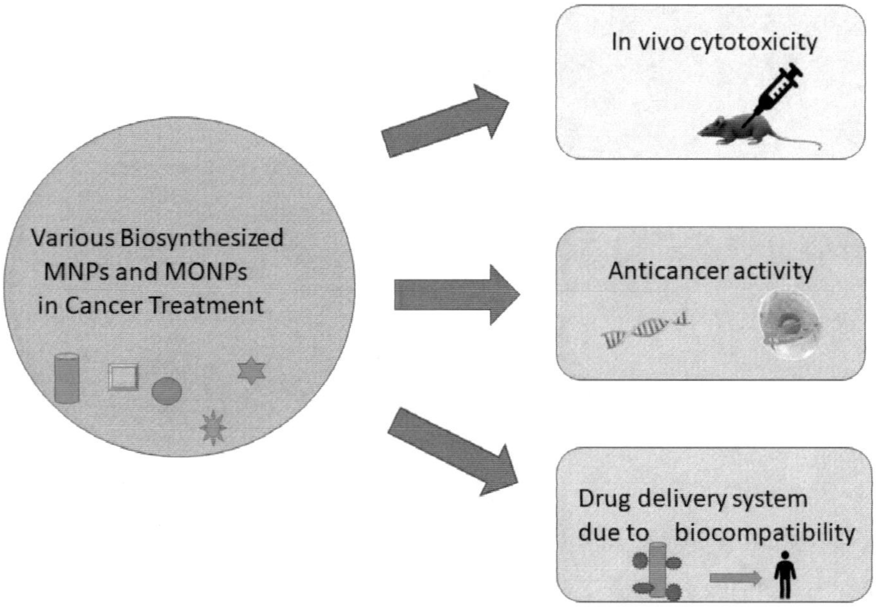

Figure 3. Graphical Illustration of utilisation of biosynthesized MNPs and MONPs as anticancer agents in cancer treatment.

3. Cancer

Cancer, considered as a deadly disease where an uncontrolled growth of abnormal cells occurs in any organ or tissue of the body. The most common types of cancer are carcinoma, melanoma (mainly skin cancer), sarcoma, lymphoma (cancer of lymphocytes) and leukemia (cancer of bone marrow where blood cells generate) respectively. The sarcoma is a cancer of connective tissues like as blood vessels muscles, cartilage and bones. We know that the carcinoma is the most common type of cancer. As we also know that the carcinoma starts in the cells or the tissue that covers the surface of internal organs, system and glands respectively. Some carcinoma are like as pancreatic cancer (in pancreas), prostatic cancer (in prostate), colorectal cancer (in colon) and so on. GLOBOCAN 2020 survey estimates that almost 10 million cancer death and 19.3 million new cancer cases occurred worldwide [27]. The common approaches utilised for the diagnosis and treatment of cancer involve chemotherapy, surgery, radiotherapy and use of alkylating agents. But the major problem of these approaches are the side effects which occurred due to problems in identifying the normal and cancerous cells which leads to systemic toxicity [28]. Hence there is an utmost requirement to discover alternate methods and drug development in the field of cancer treatment.

4. General mechanism of cytotoxic activity

The exact mechanism is not yet clear about the cytotoxic effect of various MNPs and MONPs towards various cancer cell lines. The vital mechanism of cell death has been considered as apoptosis till yet [29-31]. The MTT (3-(4, 5-dimethylthiazolyl-2)-2, 5-diphenyltetrazolium bromide) assay is commonly used to measure the cytotoxic effect of various MNPs and MONPs in tumour cells. The MTT assay is generally used to access the metabolic activities of cells. This assay involves the conversion of soluble MTT yellow salt to formazan crystals which are insoluble in water by the living cells. **Figure 4** demonstrates the various steps involved in the cytotoxicity caused by biosynthesized MNPs and MONPs.

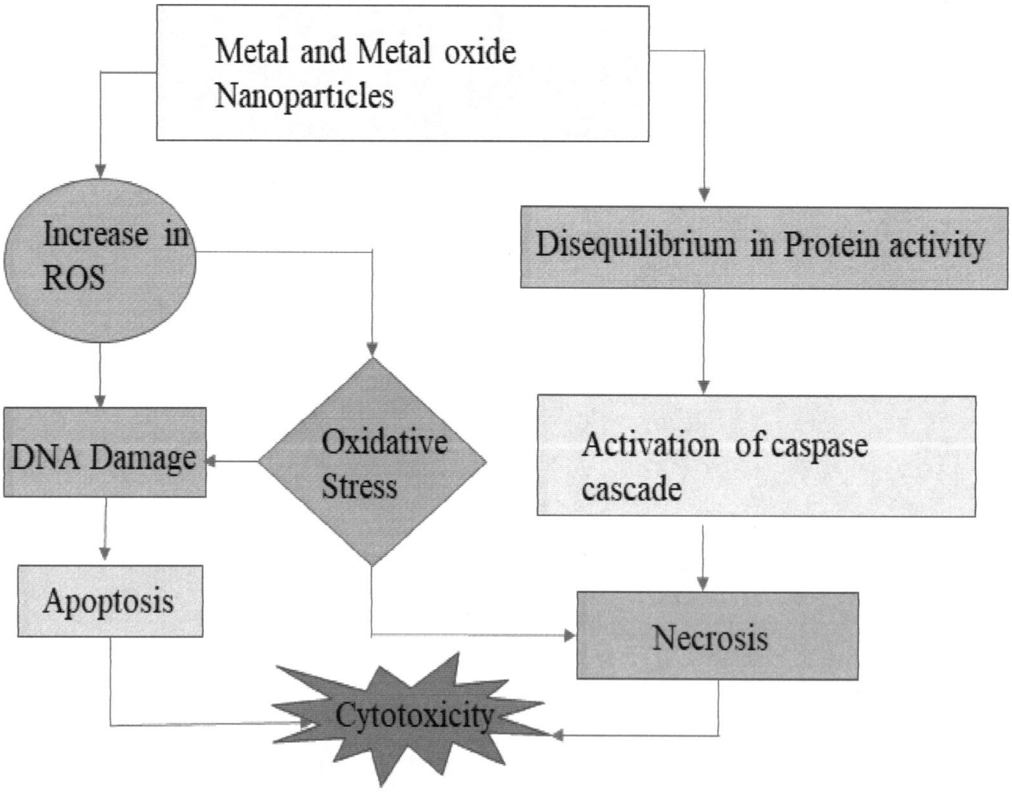

Figure 4. Schematic representation showing general mechanism of cytotoxicity

5. Silver Nanoparticles

During the past few years, many researchers have proved and explained the cytotoxic effects of silver nanoparticles which are synthesized using green methods on various cancer cell lines. The researchers have used the water mediated leaf extract of pomegranate (*Punica granatum*) for synthesizing the silver nanoparticles [32]. The size obtained in the range from 41.69 nm to 69.1 nm of the as formed silver nanoparticles. The cell viability of human cervical cancer cells has been tested by employing MTT assay. Herein, the cell viability decreases dose dependently with increase in concentration of silver nanoparticles. Similarly, the lactate dehydrogenase (LDH) assay has been performed to know the cytotoxic effect of synthesized silver nanoparticles. All the normal mammalian cells contain lactate dehydrogenase (LDH) but only damaged and necrotic cells release LDH. Therefore, by using the obtained nanoparticles on the above mentioned cancer cell lines the cytotoxic effect has been explored by knowing the released amount of LDH. The experimental results show that the released amount of LDH increases as the concentration of silver nanoparticles increased from 50 - 250 μg/mL. The DNA fragmentation assay of human cervical cancer cells also demonstrates that ladder pattern has been observed for the cleaved DNA. Which indicates that apoptosis has been induced by silver nanoparticles. No fragmentation of DNA has been observed in the untreated cells. These findings show that biosynthesised silver nanoparticles serve the role as therapeutic agent for cancer treatment. Similarly, *Nepeta Deflersiana* aqueous extract has been taken for synthesizing the silver nanoparticles [33]. The cell viability of HeLa cells was assessed by both MTT and neutral red uptake (NRU) assay. Herein, the cell viability of human cervical cancer cells affects very significantly on variation of concentration of silver nanoparticles. Where, very low value of IC_{50} (5 μg/mL) has been reported. These biosynthesised *Nepeta Deflersiana* mediated silver nanoparticles (ND-Ag NPs) show cytotoxicity on increasing the lipid peroxidation level and ROS level and simultaneously decreasing the glutathione (GSH) level. Herein, the flow cytometry analysis has been done through the arrest of SubG1 cell cycle. These results show that cell death occurs due to the induced apoptosis and necrosis in the human cervical cancer cells. Khatami et al have researched on MCF-7 cancer cell lines, herein they have investigated the inhibitory effect of phytomediated silver nanoparticles with extract of waste grass [34]. Furthermore, Western blot assay has been performed to extract the Cyclin D1 protein and to study the effect of the synthesised silver nanoparticles on the cell division and growth of MCF-7 cell line. It has been found that the expression of Cyclin D1 protein has reduced to 64.42% at 20 μg/mL of concentration of silver nanoparticles.

Khorrami et al have prepared silver nanoparticles with black peel pomegranate which have been found highly selective and show enhanced cytotoxicity against two tumour cell lines namely MCF-7 and BT-20 [35]. The maximum death of MCF-7 cell has been found to be 89.6% after 48 h incubation. Similarly, 81% cell death occurs in BT-20 cancer cell lines. Herein, same cytotoxicity study has been performed against non-tumour L-929 cell lines which indicates that these silver nanoparticle as well as only the extract of black peel pomegranate do not exhibit any cytotoxic effect.

Sankar et al have synthesised silver nanoparticles using *Origanum vulgare* aqueous extract [36]. In vitro cytotoxic study has been done A549 cell lines by performing MTT assay. The experimental findings show the 50% cell death occurs in A549 cell line at 100 μg/mL concentration. The researchers have reported the presence of bioactive compounds like thymol, linalool, quercetin, apigenin, sabinene, carvacol, terpinolene, terpinene in *Origanum vulgare* helps in increasing the cytotoxic activity of silver nanoparticles by acting as capping agent. Kathiravan et al have investigated the potent anticancer activity of phytomediated silver nanoparticles against MCF-7 cell line [37]. The aqueous leaf extract of *Melia dubia* has been taken for the preparation of silver nanoparticles. Herein, the cell death of MCF-7 cell line enhanced with increasing the concentration of biosynthesised silver nanoparticles. The IC_{50} value (31.2 μg/mL) has been determined with crude leaf extract of *Melia dubia*.

Subramanian *et al* also have formed silver nanoparticles with *Sargassum polycystum* and used them in vitro cytotoxic study of human colon cancer (HT-29) [38]. The authors have found that the cell viability decreases as the concentration of silver nanoparticles increases from 5 to 50 μg/mL. Herein, the IC_{50} value has been reported 20 μg/mL. The lower value of IC_{50} indicates the higher cytotoxic effect of the biosynthesised silver nanoparticles on cancer cell lines. Herein, a study on nuclear condensation has been done using the propidium iodide method. A large number of propidium iodide-stained cells have been observed when the HT-29 cells are treated with silver nanoparticles at 20 μg/mL after 24 h.

Lakshmanan et al have similarly synthesized silver nanoparticles by taking the fruit extract of *Cleome viscosa* [39]. Herein, the biosynthesized silver nanoparticles have been assessed for checking the in vitro anticancer activity via MTT assay against human lung (A549) and ovarian (PA1) cancer cell lines. Where, DAPI (4,6-diamidino-2-phenylindole) nuclear staining method has been used for finding the morphological changes happened in lung (A549) and ovarian (PA1) cancer cell lines. The IC_{50} values have been observed to be 28 μg/mL and 30 μg/mL respectively for lung and ovarian cancer cell lines. In another study, Afurayadi et al have reported the biosynthesis of silver nanoparticles by utilising the sesame oil cake [40]. Herein, the antitumor activity has been investigated at two distinct concentrations (2.5 μg/mL and 7.5 μg/mL) of silver nanoparticles on MCF-7 cell lines.

Similarly, the aqueous extract has prepared by taking aerial parts of *Lippia nodiflora* [41]. Silver nanoparticles have been synthesised using this water mediated extract which act as a reducing agent. The MCF-7 cancer cell line has been utilised for determining the cytotoxic activity of biosynthesised silver nanoparticles. The observed cytotoxicity is time and dose dependently, where the value of IC_{50} has been obtained 40 μg/mL for incubation time of 24 h. The morphological changes have been seen in the MCF-7 cancer cell lines using phase contrast microscopy. The Retraction, detachment and suspended cells are seen, when the MCF-7 cells are treated with silver nanoparticles. Whereas untreated cells show flattened and smooth morphology.

From the literature survey, it is clear that *Alternanthera sessils* mediated synthesis of silver nanoparticles show better cytotoxic effect as compared to standard drug cisplatin on breast cancer cells [42]. Herein, the cytotoxic effect has been studied on (A431) cell line using various shaped silver nanoparticles such as cuboidal and spherical respectively. Nayak et al have reported IC_{50} values from 78.58 to 83.57 μg/mL for epidermoid (A431) carcinoma [43]. Asharani et al have reported an anti-Proliferative activity of as prepared silver nanoparticles [44]. The specific participation of silver nanoparticles in the interference of mitochondrial respiratory chain which leads to generation of ROS and disturbance in the synthesis of ATP cycle that is cause of nuclear damage. Abovementioned these activities of silver nanoparticles show that it is a promising anticancer agent.

6. Gold Nanoparticles

Albutilon indicum leaf extract mediated gold nanoparticles (AIGNPs) have been synthesized by Mata et al [45]. Herein, the cytotoxic effect of these biosynthesized nanoparticles has been studied on HT-29 colon cancer cells. The cytotoxicity has been observed in term of the damage in cell membrane as well as nuclear morphological changes. This observed cytotoxicity has been explained on the basis of increase in ROS and concurrent decrease in cellular antioxidants like glutathione (GSH), catalase and superoxide dismutase (SOD). Further, there has been seen loss in mitochondrial membrane potential, G1/S phase cell cycle arrest and DNA damage due to reduction of antioxidants. Herein, the results also show that increase in the expression level of particular protease markers like Caspase-3, Caspase-8, Caspase-9, poly (ADP-ribose) polymerase (PARP) and Lamin A/C supports the induction of apoptosis by intrinsic and extrinsic pathways in the cells treated with AIGNPs. With increasing the concentration of AIGNPs, cause simultaneous increase in cytotoxic effect and IC_{50} values have been obtained 210 μg/mL at 24 h and 180 μg/mL at 48 h incubation respectively. The experimental data also reveal the selective cytotoxicity of AIGNPs against HT-29 cells. Herein, in vitro study has been done on normal HaCaT cells. This study demonstrates that these gold nanoparticles do not show cytotoxic effect on these normal cell lines.

The gold nanoparticles have been synthesised using the aqueous extract of *Scutellaria barbata* [46]. The cytotoxic study has been done using MTT assay against pancreatic cancer cells (PANC- 1) at two concentration (25 μg/mL and 50 μg/mL) of gold nanoparticles. The oxidoreductase enzyme presents in the mitochondria of viable cells react with MTT dye and gets converted into formazan crystals. Herein, the cell viability decreases with time in dose dependent manner. Herein, the anticancer activity of *Scutellaria barbata* mediated gold nanoparticles have been reported due to generation of ROS in the cancer cell line. There are also some factors such as upregulation of Caspase 3, Caspase 9 and Bax and down regulation of Bid and Bcl-2 which occur at 50 μg/mL concentration of gold nanoparticles.

Similarly, Zhang et al have also revealed the apoptotic study of gold nanoparticles which are biosynthesised using *Rabdosia rubescens* [47]. Herein, the ROS production, protein expression of apoptotic signalling pathways has been confirmed on the basis of estimation of caspase level which show the marked anticancer activity of biosynthesized gold nanoparticles.

Zheng et al have synthesized gold nanoparticles using extract of *Magnolia officinalis*. These nanoparticles also demonstrate anticancer activity on lung cancer cell line (A549). Where, the apoptosis has been reported through intrinsic pathway [48]. Herein, DAPI (4′,6-diamidino-2-phenylindole) and terminal deoxynucleotidyl transferase (TdT) dUTP nick end labelling (TUNEL) staining have been used for assessing the apoptosis in A549 cancer cell lines. Herein, the gold nanoparticles cause fragmentation of DNA and western blotting results shows the decrease in expression level of Bcl-2, Bid and increase in expression level of Bax, Caspase-3 and Beclin-1 in A549 cancer cells.

The preparation of gold nanoparticles by a green route method using *Cordyceps militaris* extract have reported [49]. The formed gold nanoparticles have been assessed for their antiproliferative study against liver cancer cells (HepG2). The IC_{50} value has been found to be 12.5 μg/mL. Herein, the experimental data show that induction of apoptosis in liver cancer cells occur via the pathway that is caspase dependent. The Caspase are the family of cysteine protease enzymes demonstrates a significant part in apoptosis process. Herein, various parameters have been reported like as up regulation of apoptosis pathway genes (Bax, Bid, Caspase 3 and Caspase 9) expression and downregulation of anti apoptotic genes (Bcl-2) expression of HepG2 cells.

Mishra et al have synthesized gold nanoparticles using the extract of *Hibiscus sabdariffa* and their stability has been observed in different concentration of glucose [50]. The literature survey reveals that for production of energy, most cancer cells make use of glucose. Moreover, it was observed that gold nanoparticles are stable in glucose solution (concentration 1-30mM) for seven days. Under hyperglycaemia condition the cytotoxic study has been performed with the synthesized gold nanoparticles against U87 Glioblastoma cells. The experimental data reveal that 85% and 80% cell death occur under normal and hyperglycaemia conditions respectively. The cytotoxic effect also depends on the amount of gold nanoparticles used in study of U87 Glioblastoma cells.

The utilisation of gold nanoparticles in biomedical applications, mainly in molecular imaging specifically depends on the evaluation of particular biomarkers and the adherence of gold nanoparticles to particular ligands on the tissue of tumour cells [51]. In vitro apoptotic effect of biogenic gold nanoparticles synthesized using *wenyujin* extract has been evaluated by MTT assay against human renal cell carcinoma A498 cells [52]. The study shows that induction of apoptosis in the A498 was due to enhancement of caspase3, caspase 9, Bax, Bid and decrease in the expression of antiapoptotic protein Bcl-2 in a dose dependent manner which has been analyzed by qPCR and immunoblotting assay. Herein, the Rhodamine 123 staining results show that there is an increase in mitochondrial membrane potential that leads to induction of apoptosis. This observation has been further confirmed by flow cytometry assay.

Similarly, Kumar et al have prepared the gold nanoparticles using the leaf extracts of two plants namely *Carica papaya* (CP), *Catharanthus roseus* (CR) and mixture of these two extracts (CPCRM) [53]. The gold nanoparticles have been tested for their antibacterial and antitumor activity. MTT assay has been used for testing the cytotoxic effect against two cell lines, MCF-7 and HepG2 cell. The as formed gold nanoparticles have been found biocompatible against 3T3 fibroblast cell line at a concentration of 250 μg/mL of gold nanoparticles. Herein, gold nanoparticles have been synthesized using combined (mixture) extract (CPCRM) showed the enhanced cytotoxic effect. Where, a comparative study has been performed on the gold nanoparticles prepared using individual extract and combined (mixture) extract to two plants respectively. The experimental data reveal that the live cells percentage decreases as the concentration of the gold nanoparticles increases. Crocin, an important component of saffron stigma has been taken for the phytosynthesis of gold nanoparticles and investigated for their in-vitro anticancer effect on MCF-7 cells using MTT and LDH assay [54]. It has been found that crocin mediated gold nanoparticles prevents the growth of MCF-7 cell lines in both manner i.e., dose and time dependent. The cytotoxic effect of crocin-gold nanoparticles was less on MCF-7 cell lines than with only crocin.

7. Iron oxide Nanoparticles

The iron oxide nanoparticles have also been applied in magnetic resonance targeted drug delivery, theranostics, imaging (MRI) and CT-scan for detection of several types of cancer. Kudr and Vallabani et al have reported various medical applications for cancer treatment and diagnosis like targeted drug delivery, thermoablation, magnetic fluid hyperthermia, contrast agent in MRI etc [55,56]. An environmentally benign green synthesis pathway has been suggested as an efficient method for the production of iron oxide nanoparticles because of their potential cytocompatibility for biological system. The iron oxide nanoparticles have been prepared with taking the seed coat extract of *B. flabellifer* which are highly biocompatible [57]. The cytocompatibility of these magnetic iron oxide nanoparticles has been studied in NIH-3T3 fibroblast cell lines using MTT assay. Different concentrations of iron oxide nanoparticles (50-500 μg/mL) have been taken for studying the cytocompatibility at incubation time of 24 h and 48 h respectively. Herein, the cell viability has been reported more than 80% which indicates their high biocompatibility towards the fibroblast cells. The hemolytic study has also been performed for finding the biocompatibility by taking different concentration of iron oxide nanoparticles 25, 50, 75 and 100 μg/mL respectively. These iron oxide nanoparticles do not show the hemolytic activity at abovementioned different concentrations.

Farshchi et al have prepared iron (Fe) nanoparticles with Rosemary aqueous extract [58]. In vitro, also cytotoxic activity of biosynthesised Fe nanoparticles has been done on two different cancer cell lines i.e., breast cancer 4T1 cell line and colon carcinoma C26 cell line using MTT assay. The cytotoxicity of biosynthesized Rosemary-Fe nanoparticles and rosemary extract has been determined at concentration of 6.25-200 μg/mL against 4T1 cell line at 48 h incubation time. The cell viability was investigated 80% and 45% with Rosemary-Fe nanoparticles and only rosemary extract respectively. These values of cell viabilities have been reported that at the same concentration of Rosemary-iron nanoparticles and only extract. Similarly, Rosemary-iron nanoparticles shows 20-90% cytotoxicity in the colorectal cancer C26 cell line at concentration of 3.125- 200 μg/mL. The IC_{50} values have been obtained 20.98 μg/mL for Rosemary-Fe nanoparticles and 47.87 μg/mL for rosemary aqueous extract respectively. Which indicates the more cytotoxicity of biosynthesized Rosemary-iron nanoparticles against C26 colon cancer cell line as compared to 4T1 cell line.

The functionalized iron oxide nanoparticles with conventional drugs have also used in cancer treatment and diagnosis. The biosynthesized iron oxide nanoparticles could be conjugated with enzymes, drugs or proteins and can be targeted to a particular organ, tissue or tumour site with the aid of external applied magnetic field or for hyperthermia application could be heated up in alternating magnetic field [59]. The core shell systems of iron oxide nanoparticles show marked activity for cancer treatment and diagnosis when they are functionalized with natural compounds [60]. Iron oxide nanoparticles have been biosynthesized by co-precipitation method using glucose. The average size of iron oxide nanoparticles formed has been reported 20 nm. These nanoparticles have also utilised in targeted hyperthermia process for cancer cells [61].

Similarly, in the hyperthermia treatment of breast cancer, iron oxide nanoparticles have been conjugated with dextran-spermine. Avazzadeh et al have found that 63% of cancer cells are died within 20 minutes when these cells are treated with iron oxide nanoparticles [62]. The stability of iron oxide nanoparticles can be improved by coating with metal oxides like ZnO, SnO, TiO_2, ZrO_2 and WO_3 and it also helps in increasing the generation of heat by iron oxide nanoparticles in hyperthermia treatment. Kandasamy et al have reported in vitro study on liver cancer cell lines (HepG2) at temperature (42 °C) [96]. They have reported the heating profiles of the synthesized surface functionalized superparamagnetic iron oxides (SPIOs). Herein, these SPIOs have been used as nanomedicine for treating liver cancer. Herein, these SPIOs have been used in the form of alternating magnetic fields (AFFs). Herein, the calorimetric magnetic fluid hyperthermia (C-MFH)-based thermotherapy has been demonstrated using a magnetic hyperthermia instrument (magneTherm-nanoTherics). The researchers have reported that by using superparamagnetic Fe_3O_4 nanoparticles heat dissipation occurred due to spin resonance and rise in temperature was seen [97].

The characteristic small size and advantage of having more surface area of the iron oxide nanoparticles permits their use as in targeted drug delivery application. The researchers synthesized iron oxide nanoparticles by coprecipitation method. These iron oxide nanoparticles were conjugated with bovine serum albumin and curcumin exhibiting more than 90% cytocompatibility in HFF-2 cells. Moreover, the IC_{50} value of conjugated iron oxide nanoparticles have been found to be 915 μM and 275μM at 72 h and 96 h incubation respectively. Whereas the IC_{50} value of curcumin only has been found 730 μM and 300 μM at 72 h and 96 h incubation. An enhanced cytotoxic activity has been investigated by conjugating these nanoparticles with bovine serum albumin and curcumin towards MCF 7 cell lines [63]. Pirayesh et al have coated SPIO nanoparticle with mesoporous hydroxyapatite nanocomposites (mHANs) for the treatment of breast cancer cells [64]. These nanocomposites have been loaded with doxorubicin and 2-Deoxy-D-glucose by diffusion process. T47D and SKBR3 human cancer cell lines have been used for studying the effect of these synthesised binary nanocomposites. The researchers have investigated that combined chemoradiotherapy is best method to treat the cancerous cells. Where, the cytotoxicity effect can be increased towards the tumour cells and side effect can be minimized on healthy cells due to single mode of cancer treatment. The authors have explained that increased cytotoxicity may be due to more localization of chemodrug at the tumour site. Since the loading efficiency of iron oxide nanocomposites has been reported experimentally 93% and 45% for doxorubicin and 2-Deoxy-D-glucose respectively.

8. Zinc Oxide Nanoparticles

Umar et al, in their research work reported the significant cytotoxic activity of the Zinc oxide on two breast cancer cell lines namely MDA-MB 231 and MCF-7 in a concentration dependent manner [65]. Herein, these nanoparticles have been biosynthesized using *Albizia lebbek* stem bark extract. They have also observed the formation of membrane blebs in the cells of as taken breast cancer cell lines. When these cells have been treated with zinc oxide nanoparticles in a concentration dependent manner at 48 h incubation. Herein, the formation of plasma membranes blebs indicates the induced apoptosis in the cells. These plasma membrane blebs have not seen in the control cells. Prashant et al have demonstrated cytotoxic effect of zinc oxide nanopowder against MCF-T Bca cell lines in a concentration dependent manner [66]. These zinc oxide nano powders have been prepared using *Punica granatum* and *Tamarindus indica L extracts*.

Hanley et al have investigated that the increase in Zn^{2+} ions and other molecules which are released from the zinc oxide in the intracellular level causes the increase in generation of ROS and apoptosis [67]. For improved targeted therapy, in-vitro and in-vivo tumour model have been developed. Where quercetin loaded PBA functionalised zinc oxide nanoparticles (PBA-ZnO-Q) have been utilised for monitoring the anti-tumour efficacy. Herein, for in-vitro study human breast cancer MCF-7 cells were taken. In vivo study has been performed on Swiss albino mice having ehrlich ascites carcinoma (EAC) solid. These synthesised quercetin loaded PBA functionalised nanohybrids show a strong green and blue fluorescence under a fluorescent microscope. Hence due to fluorescent nature of these nanohybrids, they can also be used as bioimaging agent. The cancer cells are characterised by the expression of sialic acid on its cell membrane [68]. Also, it has been reported that presence of sialic acid on the membrane of cancer cells are more than the normal cells. The PBA-ZnO-Q have showed the cytotoxicity in a dose dependent manner. Herein, the amount of dose has been taken from 3.5 to 35 μg/mL. These nanocarriers have shown a notable cytotoxic effect in breast cancer cells. The particular cytotoxic effect has not seen in non-tumorigenic (normal) human epithelial cell lines. These nanohybrids induce mitochondrial dysfunction according to the dose amount taken by decreasing the mitochondrial membrane potential (MMP). Saha et al have found that the oxidative stress occurred with increase in the level of intracellular ROS [69]. They have also reported that downregulation of Bcl2 and proteins which are present on the mitochondrial membrane and upregulation of Bax are associated with the oxidative stress. From the literature survey it has been suggested that functionalization of nanoparticles with PBA increases their stability and more absorption in the tissues [70]. It has been found that these nanohybrids reduces toxicity associated with tumour in liver, kidney and spleen respectively. Due to tumour targeting ability of PBA-ZnO-Q nanoparticles, the accumulation of quercetin has been found to be very less in kidney and liver. The biocompatible nature of zinc oxide nanoparticles makes the significant loading of quercetin at the tumour site. The prominent increase of anticancer effect of quercetin based zinc oxide nanoparticles has been observed due to increase in their bioavailability. These naturally occurring anticancer bioactive compounds show distinctive cytotoxicity in cancer cells. These are seemed to be non- toxic to the other body organs [71, 72]. Sadhukhan et al have developed quercetin loaded PBA functionalised zinc oxide nanoparticles (PBA-ZnO-Q) in monitoring the anti-tumor efficacy [72]. Using this type of nanohybrids, the antitumor efficacy of quercetin at the tumor site has increased due to high bioavailability of zinc oxide nanoparticles. The authors have demonstrated that the PBA-ZnO-Q nanohybrids do not show the systemic toxicity in the mice having tumour. They also concluded that zinc oxide nanoparticles can behave like a remarkable system for delivering and loading of anticancer drugs which are hydrophobic in nature.

There is a need of a lot of research in therapeutic technique to improve cancer therapy in order to minimise the tumour associated side effects. Kandasamy et al synthesized Trichoderma strains mediated zinc oxide nanoparticles which have been conjugated with β-D glucan from barley (T-β-D-glu-ZnO) nanoparticles [73]. These synthesised nanoformulation have been found non-toxic against NIH3T3 cells. Also, they show cytotoxic effect in (A549) human lung carcinoma cells with IC_{50} 56.25 μg/mL. Similarly, the authors have found that dose dependent cytotoxicity of zinc oxide nanoparticles in-vitro conditions. These nanoparticles have been synthesized using pure bioflavonoid rutin [74]. The cytotoxicity has been reported against MCF-7 cell lines. Herein, it is seen that when the concentration of synthesised zinc oxide nanoparticles increases from 10-50 μg/mL, viability of (MCF-7) breast cancer cells increase. Similar observations have been investigated by Suresh et al where the researchers synthesized zinc oxide nanoparticles with the extract of leaves of *Costus pictus*. In-vitro cytotoxicity study has been done on mice having DLA (Dalton's ascites cells) cell lines [75]. Herein, the investigation show that cell viability and cell growth inhibition increase as the concentration of zinc oxide nanoparticles increases from 10-50 μg/mL. The maximum cell growth inhibition has been observed at 50 μg. Also, it has been revealed that zinc oxide nanoparticles show the cytotoxic effect towards highly proliferating cells because of increased ROS generation [76]. The increase in lipid peroxidation level in the liposomal membrane was due to production of ROS by zinc oxide nanoparticles [77]. Many reported studies have demonstrated that the zinc oxide nanoparticles are very selective and highly cytotoxic towards various cell lines but do not show toxic effect towards the normal cells. The selective anticancer activity of zinc oxide nanoparticles has been shown against (HepG2) i.e., human hepatocellular carcinoma, (BEAS-2B) i.e., human bronchial epithelial and (A549) i.e., human lung adenocarcinoma as a result of apoptosis. These nanoparticles do not show the cytotoxicity towards the normal cells such as astrocytes and hepatocytes respectively [78].

It has also been suggested that the apoptosis is considered the vital mechanism of cell death due to cytotoxicity caused by zinc oxide nanoparticles [79]. Similar results have been reported about the apoptosis which occurs in HT29 colon cancer cells on exposing with zinc oxide nanoparticles [80]. The increase in early stage apoptotic cell number with increase in concentration of zinc oxide nanoparticles have seen. The zinc oxide nanoparticles react with the chemical species present inside the cells and produce larger amount of ROS, that is due to output of oxidative stress. Owing to this stress inside the cell, there occurs damage to proteins, nucleic acid and lipids through protein denaturation and lipid peroxidation, followed by DNA damage and necrosis, finally resulting in cell death by apoptosis [81,82]. Literature survey reveals that surface modifications of zinc oxide quantum dots with silver and polyvinylpyrrolidone (PVP) show excellent anticancer activity in (MCF-7) breast cancer cell lines [83]. Similarly, high cytotoxicity effect has been seen in HT29 colon cancer cells using zinc oxide nanoparticles conjugated with hydrophobic peptides. The authors have reported that the expense of cytotoxicity depends on the type of cells and also on the modification of nanoparticles by using various formulations [84]. The researchers have prepared the zinc oxide nanoparticles with the water mediated plant extract of *Deverra tortuosa* and studied the cytotoxic activity of these nanoparticles against two cancer cell lines i.e., (Caco-2) human colon adenocarcinoma and (A549) human lung adenocarcinoma and compared their activities using MTT assay on human lung fibroblast cell line(W138). It has been found that both zinc oxide nanoparticles and plant extract displayed a marked particular cytotoxic effect on (A549) and (Caco-2) cancer cell lines and remarkable lower cytotoxic effect on the normal W138 cells.

The ZnO nanoparticles synthesized with the leaves of *Hyssopus officinalis* have been tested in a mouse model to control tumour growth in liver and spleen cancers [98]. These nanoparticles also have been used in vitro study of and then in HUH7 and HepG2 human liver cancer cell lines respectively. Although *in vivo* application of ZnO NPs has been rare, this study is remarkable in revealing a synergistic mechanism for cytotoxicity wherein apart from increased ROS concentration, the NPs also activate expression of pro-apoptic genes Bax and p53.

9. Copper Oxide Nanoparticles

Gnanavel et al have also prepared copper oxide nanoparticles with the leaves of *Ormocarpum cochinchinense* [85]. These biosynthesized copper oxide nanoparticles have been assessed for their anticancer activity against (HCT-116) human colon cancer cell lines by performing the MTT assay. These results show that biosynthesized copper oxide nanoparticles have strong cytotoxic efficiency against HCT-116 with IC_{50} value observed at 40 μg/mL. The cell viability also decreases to 22% at 100 μg/mL concentration of copper oxide nanoparticles.

Rehana et al have reported the in vitro investigation to show the cytotoxic effect of biosynthesized copper oxide nanoparticles against the cancer cell lines namely (A549, HeLa, Hep-2, MCF-7) and normal human dermal fibroblast cell line (NHDF) respectively [86]. Herein, the cisplatin has been used as a standard. The Hoechst 33258 staining method has been used to see the morphological changes in the cell lines. Herein, the utilised Copper oxide nanoparticles have been phytofabricated with the leaves extract of five different plants namely *Azadirachta, indica, Hibiscus rosa-sinensis, Moringa oleifera, Murraya Koenigi, Tamarindus indica*. The experimental results show that the cell viability decreases as the concentration of CuO nanoparticles increases. The cytotoxic effect of copper oxide nanoparticles on normal cell line (NHDF) has been reported less than 20%. The copper oxide nanoparticles show enhanced cytotoxic effect against human breast cancer cell line (MDA-MB-231). Herein, these copper oxide nanoparticles have been prepared with the extract of *Pterolobium hexapetalum* leaves [87]. The authors have synthesized the copper oxide nanoparticles by utilizing aqueous extract of *Acalypha indica* [88]. The cytotoxic effect has been studied on MCF-7 cancer cell line at various concentrations of copper oxide nanoparticles (from 6.5 to 100 μg/mL). The IC_{50} value has been obtained 56.16 μg/mL.

Sankar et al have reported *Ficus religiosa* arbitrated preparation of copper oxide nanoparticles and their cytotoxicity effect towards A549 cell lines [89]. Herein, highly crystalline copper oxide nanoparticles have been synthesized. These synthesized nanoparticles exhibit cytotoxicity in vitro in the concentration range 50-500 μg/mL. Herein, strong cytotoxic effect has been studied against A549 lung cancer cells by apoptosis induction with more generation of ROS and altering the potential of mitochondrial membrane. The phase contrast microscope has been used for visualising the change in cell morphology. The results prove that the cytotoxic effect of biogenic copper oxide nanoparticles has observed dose dependently where IC_{50} has been determined 200 μg/mL.

In another research paper, Nagayjyothi et al have described the induction of apoptosis and suppression of proliferation of human cervical carcinoma cells using copper oxide nanoparticles [90]. The water mediated black bean extract has been taken for the preparation of copper oxide nanoparticles. The sulphordamine -B assay has been performed for studying the cytotoxic effect of as prepared copper oxide nanoparticles. These copper oxide nanoparticles have been found very efficient to reduce colonies of cervical carcinoma as depicted by clonogenic assay. In another study, Kalaiarasi et a have concluded that copper oxide nanoparticles are responsible for the regulation of several classes of enzymes called histone deacetylases (HDACs) [91]. Herein, the induction of apoptosis might be due to inhibition of HDACs which control the transcription process. Herein, the copper oxide nanoparticles have been prepared with the leaf extract of *Ficus religiosa*. It has been observed that copper oxide nanoparticles inhibited the HDAC level in lung cancer cells (A549). When, lung cancer cells are treated with copper oxide nanoparticle then the expression of inflammatory responsible genes (TNF-, COX-2) are downregulated. The results also show that the significant activation of caspase i.e., (mRNA expression of caspase 3, caspase 8 and caspase 9) pathway and downregulation of oncogenes (MMP-2, MMP-9) occurred in the lung cancer cells (A549).

10. Conclusion and future

The unique physicochemical properties of various biogenic MNPs and MONPs makes the enhanced application in cancer research and therapy. Nanoparticles can be regarded as modern structures medicines for the cancer treatment. The application of nanosized metal and metal oxide nanoparticles leads to the better efficacy and minimal toxicity of novel anticancer drugs. Herein, this review specifically focussed on the cytotoxic effect of several metals and metal oxides on different cancer cell lines in vitro and in vivo respectively. Presently the use of biosynthesized nanoparticles in medical field is a new environmentally benign emerging field of research. Various cell lines of cancer have been taken for in vitro study for showing the cytotoxicity of biogenic nanoparticles. For in-vivo study and clinical trials, the nanoparticles should have higher selectivity and efficacy towards cancer cells and less toxicity on normal cells. So, there is need of thorough understanding of cancer biology and metal and metal oxide nanoparticles for clinical purpose. Cancer is an epidemic disease, affecting people worldwide at a large scale. Inorganic nanoparticle has been classified into superparamagnetic (SPIO), nanoshells, nanocages silica nanoparticles, gold nanoparticles and quantum rods, quantum dots and some antibody-conjugated nanoparticles (ACNPs) have been developed as diagnostic biomarkers which can be used as screening tool for anticancer and other biomedical applications [99]. Anselmo et al have performed the clinical trials on few drugs based on nanoparticles [100]. Various modification in synthesis of nanoparticles and their formulations with cancer drug can be best way to cure various cancers in the future. Hence, in future detailed investigations and efforts are needed for controlled morphology, biocompatibility and pharmacokinetic studies for making MNPs and MONPs as a major tool for treatment of cancer.

Acknowledgment

Authors acknowledge The Principal, Miranda House, University of Delhi for providing lab space to conduct this research work. Kalawati Saini would like to thank Dr. Amrita Tripathi Sheikh, Associate Professor, Miranda House, University of Delhi for her immense help, encouragement and assistance.

Disclosure

The authors hereby declare no conflict of interest.

References

[1] Baker S, Rakshith D, Kavitha K S, Santosh P, Kavitha U, Rao Y and Satish S 2013 *Bioimpacts* **3**(3) 111-117.
[2] Bhattacharya R and Mukherjee P 2008 *Adv. Drug Deliv. Rev.* **60** 1289e1306.
[3] Bhumkar D R, Joshi H M, Sastry M and Pokharkar V B 2007 *Pharm. Res.* **24** 1415e1426.
[4] Nasrollahzadeh M, Sajadi S and Mohammad M 2015 *RSC Adv.* **5** 40628-40635.
[5] Nadagouda M N and Varma R S 2008 *Green Chem.* **10**(8) 859-862.
[6] Duan H, Wand D and Li Y 2015 *Chem. Soc. Rev.* **44** 5778-5792.
[7] Mittal A K, Chisti Y and Banerjee U C 2013 *Biotechnol. Adv.* **31** 346-356.
[8] Vijayaraghavan K, Nalini S K, Prakash N U and Madhankumar D 2012 *Mater. Lett.* **75** 33-35.
[9] Sivaraj R, Rahman PKSM, Rajiv P and Venckatesh R 2014 International conference on advances in agricultural, Biological & Environmental Sciences.
[10] Nilavukkarasi M, Vijayakumar S and Parthipkumar S 2020 *Material Science for Energy Technologies* **3** 335-343.
[11] Izadiyan Z, Shameli K, Miyake M, Hara H, Mohamad S E B, Kalantari K, Taib S H M and Rasouli E 2020 *Arab. J. chem.* **13**(1) 2011-2023.
[12] Iqbal J, Abbasi B A, Ahmad R, Shahbaz A, Zahra S A, Kanwal S, Munir A, Rabbani A and Mahmood T 2020 *J. Mol. Struct.* **1199** 126979.
[13] Yusefi M, Shameli K, Ali R R, Pang S W and Teow S Y 2020 *J. Mol. Struct.* **1204** 127539.
[14] Yugandhar P, Vasavi T, Devi P U M and Savithramma N 2017 *Appl. Nanosci.*, **7** 417-427.
[15] Kahsay M H, Devi D R, Kumar Y P, Mohan B S, Tadesse A, Battu G, Basavaiah K 2018 *OpenNano*, **3** 28-37.
[16] Mata R, Nakkala J R and Sardas S R 2015 *Mater. Sci. Eng. C*, **51** 216-225.
[17] Venugopal K, Rather H A, Rajagopal K, Shanthi M P, Sheriff K, Illiyas M, Rather R A, Manikandan E, Uvarajan S and Bhaskar M 2017 *J. Photochem. Photobiol.B Biol* **167** 282-289.
[18] Vinardell M P and Mitijans M 2015 *Nanomaterials* **5**(2) 209-212.
[19] Wu W, Wu Z, Yu T, Jiang C and Kim W S 2015 *Sci. technol. Adv. Mater.* **16** 1-43.
[20] Laurent S, Dutz S, Hafeli U O and Mahmoudi M 2011 *Adv. Colloid Interface Sci.* **166** 8-23.
[21] Tang L, Yang X, Yin Q, Cai K, Wang H, Chaudhary I and Yao C 2014 *Proc. Natl. Acad. Sci. U.S.A.* **111**(43) 15344-15349.
[22] Ramkumar V S, Pugazhendhi A, Gopalakrishanan K, Sivagurunathan P, Saratale G D, Dung T N B and Kannapiran E 2017 *Biotechnol. Rep. (Amst)* **14** 1-7.
[23] Karmous I, Pandey A, Haj K B and Chaoui A 2020 *Biol. Trace Elem. Res.* **196** 330-342.

[24] Danhier F, Feron O and Preat V 2010 *J. Control Release* **148** 135-146.
[25] Yew Y P, Shameli K, Miyake M and Khairudin N B B A 2020 *Arab. J. Chem.* **13**(1) 2287-2308.
[26] Li L, Yang Q, Zhou Z, Zhong J and Huang Y 2014 *Biomaterials* **35** 5171-5187.
[27] Sung H, Ferlay J, Siegel R L, Laversanne M, Soerjomataram I, Jemal A and Bray F 2021 *CA CANCER J CLIN* **71** 209-249.
[28] Rasmussen J W, Martinez E, Louka P and Wingett D G 2010 *Exp. Opin. Drug Deliv.* **7** 1063-1077.
[29] Elmore S 2007 *Toxicol Pathol.* **35**(4) 495-516.
[30] Pistritto G, Trisciuoglio D, Ceci C, Garufi A and D'orazi G 2016 *Aging (Albany NY)* **8**(4) 603-619.
[31] Yakop F, Ghafar S A A, Yong Y K, Yazan L S, Hanafiah R M, Lim V and Eshak Z 2018 *Artificial cells, Nanomedicine & Biotechnology* **46** 131-139.
[32] Sarkar S and Kotteeswaran V 2018 *Advances in Natural Sciences: Nanoscience and Nanotechnology* **9** 025014.
[33] Al-Sheddi E S, Farshori N N, Al-Oquail M M, Al-Massarani S M, Saquib Q, Wahab R, Musarrat J, Al-Khedhairy A A and Siddiqui M A, 2018 *Bioinorg. Chem. Appl.* 9390784.
[34] Khatami M, Sharifi I, Nobre M A L, Zafarnia N and Aflatoonian M R 2018 *Green. Chem. Lett. Rev.* 125-134.
[35] Khorrami S, Zarepour A and Zarrabi A 2019 *Biotechnol. Rep.* **24** e00393.
[36] Sankar R, Karthik A, Prabhu A, Karthik S, Shivashangari K S and Ravikumar V 2013 *Colloid Surfaces B* **108** 80-84.
[37] Kathiravan V, Ravi S and Kumar S A 2014 *Spectrochim. Acta A Mol. Biomol. Spectrosc.* **130** 116-121.
[38] Subramanian P, Periyannan R, Gandhi V, Ganesan R, Ramar M and Marimuthu P N 2016 *Mater. lett.*
[39] Lakshmanan G, Sathiyaseelan A, Kalaichelvan P T and Murugesan K K 2018 *Karbala International Journal of Modern Science*, **4** 61-68.
[40] Afurayadi A A, Devanesan S, Al-Ansari M, Alsalhi M S and Ranjitsingh A J 2019 *J. Photochem. Photobiol. B* **192** 83-89.
[41] Sudha A, Jeyakanthan J and Srinivasan P 2017 *Resource-Efficient Technologies* **3**(4) 506-515.
[42] Jannatbul F M and Lalitha P 2015 *Prog. Biomater.* **4** 113-121.
[43] Nayak D, Pradhan D S, Ashe S, Rauta P R and Nayak B 2015 *J.Colloid Interface Sci.* **457** 329-338.
[44] Asharani P V, Hande M P and Valiyaveettil S 2009 *BMC Cell Biol.* **10** 65.
[45] Mata R, Nakkala J R and Rani S 2016 *Colloid Surface B* **1** 81-86.
[46] Wang L, Xu J, Yan Y, Liu H, Karunakaran T and Li F 2019 *Artificial Cells, Nanomedicine and Biotechnology.* **47**(1) 1617-1627.
[47] Zhang X, Tan Z, Jia K, Zhang W and Dang M 2019 *Artificial Cells, Nanomedicine and Biotechnology* **47** 2171-2178.
[48] Zheng Y, Zhang J, Zhang R, Luo Z, Wang C and Shi S 2019 *Artificial Cells, Nanomedicine and Biotechnology* **47** 3101-3109.
[49] Ji Y, Cao Y and Song Y 2019 *Artificial Cells, Nanomedicine and Biotechnology* **47** 2737-2745.
[50] Mishra P, Ray S, Sinha S, Das B, Khan Md I, Behera S K, Yun S, Tripathy S K and Mishra A 2016 *Biochemical Engineering Journal* **105** 264-272.
[51] Jain P K and El-Sayed I H 2007 *Nano today* **2(1)** 18-29.
[52] Liu R, Pei Q, Shou T, Zhang W, Hu J and Li W 2019 *Int. J. Nanomed.* **14** 4091-4103.
[53] Kumar T M, Kumari S, Sambandam B, Aravinthan A, Sastry T P and Kim J H 2016 *Process Biochemistry* **51(3)** 384-391.

[54] Hoshyar R, Khayati G R, Poorgholami M and Kaykhaii M 2016 *Photochem. photobiol.* B **159** 237-242.
[55] Kudr J, Haddad Y, Richtera L, Hager Z, Cernak M, Adam V and Zitka O 2017 *Nanomaterials* **7(9)** 243.
[56] Vallabani N V S and Singh S 2018 *3 biotech* **8(6)** 2-23.
[57] Sandhya J and Kalaiselvam S 2020 *Mater. Res. Express* **7** 015045.
[58] Farshchi H K, Azizi M, Jaafari M R, Nemati S H and Fotovat A 2018 *Biocatal. Agric. Biotechnol.* **16** 54-62.
[59] Yew Y P, Shameli K, Miyake M, Khairudin N B B, Mohamad S E B, Naiki T and Lee K X 2020 *Arab J. Chem.* **13(1)** 2287-2308.
[60] Zhou L, Yuan J and Wei Y 2011 *J. Mater. Chem.* **21** 2823-2840.
[61] Wu C Y, Lin C H and Chen Y C 2015 *J. Nanomed. Nanotechnol.* **6(1)** 1-7.
[62] Avazzadeh R, Farahani E V, Soleimani M, Amanpour S and Sadeghi M 2017 *Prog. Biomater.* **6(3)** 75-84.
[63] Nosrati H, Sefidi N, Sharafi A, Danafar H and Manjili H K 2018 *Bioorg. Chem.* **76** 501-509.
[64] Islamian J P, Hatamian M, Rashidi M R, Aval N A, Mesbahi A, Mohammadzadeh M and Jafarabadi M A 2017 *The Breast* **33** 97-103.
[65] Umar H, Kavaz D and Rizaner N 2019 *Int. J. Nanomed.* **14** 87-100.
[66] Prashanth G K, Prashanth P A and Utpal B 2015 Karbala Int. J. Mod. Sci. **1(2)** 67-77.
[67] Hanley C, Layne J, Punnoose A, Reddy K M, Coombs I, Coombs A, Feris K and Wingett D 2008 *Nanotechnology* **19(29)** 295103.
[68] Cui H, Lin Y, Yue L, Zhao X and Liu J 2011 *Oncol. Rep.***2** 1365-1371.
[69] Saha S, Khan P S, Sinha K, Agarwal N and Sil P C 2016 *Biochem. Biophys. Rep.* **5** 313-327.
[70] Ma R and Shi L 2014 *Polym. Chem.* **5** 1503-1518.
[71] Harvey A L, Edrada-Ebel R and Quinn R J 2015 *Nat. Rev. Drug Discov.* **14** 111-129.
[72] Sadhukhan P, Kundu M, Chatterjee S, Ghosh N, Manna P, Das J and Sil P C 2019 *Mater Sci Eng C Mater Biol Appl.* **100** 129-140.
[73] Kumar K S, Jeevithan E, Hu X, Chelliah R, Oh D H and Wang M H 2020 *J. Photochem. Photobiol. B* **203** 111728.
[74] Bharathi D and Bhuvneshwari V 2019 *Res. Chem. Intermed.* **45** 2065-2078.
[75] Suresh J, Pradheesh G, Alexramani V, Sundrarajan M and Hong S I 2018 *Adv. Nat. Sci.: Nanosci. Nanotechnol.* **9** 015008.
[76] Taccola L, Raffa V and Riggio C 2011 *Int J. Nanomedicine* **6** 1129-1140.
[77] Premanathan M, Karthikeyan K and Jeyasubramanian K 2011 *Nanomedicine Nanomed-Nanotechnol.* **7** 184-192.
[78] Akhtar M J, Ahamed M, Kumar S, Khan M and Ahmad J 2012 *Int J. Nanomedicine* **7** 845-857.
[79] Bist G and Rayamajhi S 2016 Nanobiomedicine **3** 264-271.
[80] Aswathanarayan J B, Vittal R R and Muddegowda U 2018 *Artificial cells, nanomedicine and Biotechnology.* **46** 1444-1451.
[81] Manke A, Wang L and Rojanasakul Y 2013 *Biomed. Res Int.* **2013** 942916.
[82] Valko M, Rhodes C J, Monocol J, Izakovic M and Mazur M 2006 *Chem. Biol. Interact.* **160** 1-40.
[83] Abdolmohammadi M H, Fallanhian F, Fakhroueian Z, Kamalian M, Keyhanvar P and Harsini F M 2017 *Artificial cells, nanomedicine and biotechnology,* **45(8)** 1769-1777.
[84] Selim Y A, Azb M A, Ragab I and Abd El-Azim M H M 2020 *Scientific reports nature research* **10** 3445.
[85] Gnanavel V, Palanichamy V and Roopan S M 2017 *Photochem. Photobiol. B,* **171** 133-138.
[86] Rehana D, Mahendira D N, Kumar R S and Rahiman A K 2017 *Biomed. Pharmacother.* **89** 1067-1077.
[87] Elavarasan N, Kokila K, Prakash S and Sujatha V 2019 *J. Cluster Sci.* **30** 1157-1168.

[88] Sivaraj R, Rahman P K S M, Rajiv P, Narendhran S and Venckatesh R 2014 *Spectrochemica Acta Part A: Molecular and Biomolecular Spectroscopy.* **129** 255-258.
[89] Sankar R, Maheswari R, Karthik S, Shivashangari K S and Ravikumar V 2014 *Mater. Sci. Engin. C* **44** 234-239.
[90] Nagayjyothi P C, Muthuraman P, Sreekanth T V M, Kim D H and Shim J 2017 *Arab. J. Chem.* **10(2)** 215-225.
[91] Kalaiarasi A, Sankar R, Anusha C, Saravanan K, Aarthy K, Karthic S, Mathuram T L and Ravikumar V 2018 *Biotechnol. Lett.* **40** 249-256.
[92] Rani N, Rawat K, Shrivastava A, Ekta, Yadav S, Gupta K, Saini K, In Vitro study of green synthesized ZnO nanoparticles on human lung cancer cell lines. Materials Today: Proceedings xxx (2021) xxx. Source link: https://doi.org/10.1016/j.matpr.2021.07.203
[93] Gupta N, Singh H P, Rakesh Kumar Sharma R K 2010 *Colloids and Surfaces A: Physicochem. Eng. Aspects* **367** 102–107.
[94] Singh H P, Gupta N, Sharma S K, Sharma R K 2013 *Colloids and Surfaces A: Physicochem. Eng. Aspects* **416** 43-50.
[95] Byrne J M, Muhamadali H, Coker V S, Cooper J and Lloyd J R 2015 J R Soc Interface 12 20150240. Published:06 June 2015https://doi.org/10.1098/rsif.2015.0240
[96] Kandasamy G, Sudame A, Luthra T, Saini K, and Maity D 2018 *ACS Omega* 2018, **3**, 3991−4005. doi: 10.1021/acsomega.8b00207
[97] Lee J H, Kim B, Kim Y and Kim S K, 2021 *Scientific Report* 11(1). https://doi.org/10.1038/s41598-021-84424-1
[98] Rahimi Kalateh Shah Mohammad G, Seyedi S M R, Karimi E and Homayouni-Tabrizi M 2019 J Biochem Mol Toxicol 33 (7) e22324.
[99] Kolahkaj F F, Derakhshandeh K, Khaleseh F, Azandaryani A H, Mansouri K and Khazaei M 2019 J Drug Deliv Sci Technol 53 101136
[100] Anselmo AC, Mitragotri S. Nanoparticles in the clinic: An update. Bioeng Transl Med. 2019;4:e10143 10.1002/btm2.10143

Probe into DNA interaction, cell toxicity and antifungal activities of phyto-synthesizedyttrium oxide (Y2O3) nanoparticles

Rugmani Meenambal[a*], Hema S K[b], Vedika Tomar[c], Anita Puyam[d]

[a]*Department of Clinical Psychopharmacology and Neurotoxicology, National Institute of Mental Health and Neuro Sciences, Bangalore, India*
[b]*Department of Bio-Sciences, Vellore Institute of Technology, Vellore, India*
[c]*Amity Institute of Nanotechnology, Amity University, Noida, India*
[d]*Department of Plant Pathology, Rani Lakshmi Bai Central Agricultural University, Jhansi, India*
[*]Corresponding Author: mrugmani@yahoo.com

Abstract Green synthesis is considered to be eco-friendly approach in comparison to chemical mediated synthesis and hence the study demonstrates facile *Nyctanthes arbortristis* mediated phyto-synthesis of biocompatible yttrium oxide (Y2O3) nanoparticles. X-ray diffraction (XRD) and Fourier Transform Infrared Spectroscopy (FTIR) were employed to confirm the formation of cubic Y2O3 nanoparticles. Further, the interaction of Y2O3 with calf thymus-DNA were scrutinized by UV-visible, fluorescence spectroscopy and agarose gel electrophoresis techniques and the results suggested the stabilization of CT- DNA by the biosynthesized Y2O3. Also, the synthesized Y2O3 does not exhibit any toxicity in SK-MEL, human melanoma cell lines. Further, the studies on antifungal activities of Y2O3 nanoparticles revealed significant inhibitory effectson the growth of plant pathogen *Exserohilum rostratum* causing Exserohilum leaf blight in pearl millet. Hence, the findings deliver the foundation for furtherbiological research investigations of Y2O3 and its application in phytopathogen control.

Keywords: Phyto-synthesis, yttrium oxide, DNA interaction, antifungal activity

1. Introduction

The development of metal oxide nanoparticles has received attention in biomedical applicationsowing to larger surface area, magnetic and optical characteristics, antioxidant and antimicrobial activities, mechanical stability and biocompatibility. Yttrium oxide (Y2O3) is of substantial interest owing to their antioxidant and radical scavenging activity under physiological conditions. A recent study demonstrated that Y2O3 nanoparticles encumbered with a biomaterial scaffold can induce angiogenesis and vascularization with its potential application in tissue engineering [1]. Y2O3 nanoparticles have been prepared using different synthesis techniques but most of these techniques suffers from shortcomings and therefore, bio or green synthesis stands an advantage by being less toxic, economic, environmentally friendly and can be processed for large-scale production [2]. Although few studies report phytosynthesis of yttrium oxide nanoparticles, the potential plants extract as biological materials for the synthesisof nanoparticles is yet to be explored [3]. In the present study, the leaf extract of the plant, *Nyctanthes arbor-tristis* is used as stabilizing and capping agents in the synthesis of Y2O3 nanoparticles. The extracts of the plant are reported to act as immune enhancer and hepatoprotective agent along with antimicrobial and anti-parasitological properties *in vitro*[4].

To the best of our knowledge, there are no other reports on *N. arbortristis* mediated phytosynthesis of yttrium oxide nanoparticles.

The studies on the interaction of nanoparticles with biomolecules including DNA is well explored owing to the likely effects that the nanoparticles bring about in various molecular mechanisms such as synthesis, replication and structural integrity of nucleic acids [5,6]. It is significant to study the molecular interactions as the interacted nanoparticles may alter the properties of DNA and also results in the conformational change of DNA, which in turn affects gene expression, replication, repair, transcription and signalling mechanism [7]. Hence, nanoparticle-DNA interaction is considered as one of the most pomising field of investigation for molecular and targeted therapies.

The present study demonstrates *N. arbortristis* mediated phytosynthesis of yttrium oxide nanoparticles along with its material characterization. The interaction of Y2O3 nanoparticles with calf thymus DNA were also assessed with spectroscopic techniques and agarose gel electrophoresis. The cytotoxic effect of the nanoparticles on SK-MEL viability was assessed using MTT assay. Further, their antifungal properties against *Exserohilum rostratum* isolated from pearl millet were examined.

2. Materials and Methods

2.1 *Phyto-synthesis and characterization of Y2O3 nanoparticles*

The leaves of *Nyctanthes arbor-tristis* were collected, processed and the crude extract was prepared in deionized water. The stock solution containing 0.1 M and 0.5M yttrium nitrate hexahydrate was added dropwise to 10ml of leaf extract under continuous stirring conditions at room temperature. The homogenous suspension was then kept in a hot air oven at 120 °C overnight and the dried samples were calcined at 400 °C in a muffle furnace and used for further characterization. The phase purity and composition of Y2O3 nanoparticles were analysed usingpowder X-ray diffractometer (Rigaku, Ultima IV, Japan) with scan range of 2θ between 10° and 60°. For determining the functional groups, the transmittance of the prepared samples wasrecorded using a FTIR spectrophotometer (Perkin–Elmer, USA) in the scan range of 4000- 400cm^{-1} in the infrared region.

2.2 *Effects of DNA interaction with nanoparticles*

A stock solution of CT- DNA (10 µg/ml) were prepared in 10 mM Tris HCl buffer with mild quaking until the formation of homogenous solution mixture which was maintained at a pH of 7.4 and the resultant mixture was stored at 4°C. DNA interaction with nanoparticles was performed by treating nanoparticles of different concentration (20 and 40 µg/ ml) with 10µg ofCT-DNA [6]. All the samples including the control without nanoparticles, and suitable blank were utilized to check the interaction of Y2O3 nanoparticles with DNA through UV spectrophotometry within the wavelength region of 200-400 nm with constant DNA concentration and steady state fluorescence assay with emission spectrum scanned at 350 to 650 nm.

2.3 *Agarose gel electrophoresis studies*

The cleavage activity of samples was tested through agarose gel electrophoresis technique [8]. In a reaction mixture of 20 µl, 3200 µM of calf thymus DNA was treated with different concentrations of Y2O3 nanoparticles. Both the control DNA and Y2O3 conjugated DNA were exposed to agarose gel electrophoresis embedded in 1% gel in 0.5 % TAE buffer for 1h. The resultant gel was imaged using the GELSTAN 4X Advanced gel-documentation system.

2.4 Cytotoxicity test

All the samples and controls were seeded with fibroblast line SK-MEL, human melanoma cell lines during 24 hours, in Dulbecco's modified eagle's medium (D-MEM) accompanied with fetal bovine serum (FBS) and penicillin-streptomycin. The viability of the Y2O3 treated cells in comparison with control cells was evaluated using an Inverted phase contrast microscope (Labomed TCM-400 with MICAPS™ HD camera) and was further quantified by MTT assay by measuring absorbance at 570 nm using an ELISA plate reader [9].

2.5 Antifungal test

Two concentrations of 0.1M and 0.5M were used for the study and potato dextrose agar plates were divided into two parts, A (control) and B (Y2O3 nanoparticles) which were inoculated 1mm away from the edge of the plate. The growth rate was recorded on alternate days for both the parts and the test was performed in triplicates. Antagonistic activity was checked after incubation by measuring the radial growth rate of *Exerholium rostratum* in Y2O3 treated plates(R2) and in the control plate (R1). Further, the percentage inhibition of radial growth (PIRG) was determined using Skidmore and Dickinson formula [10]:

$$PIRG = \frac{(R1 - R2)}{R1} * 100 \qquad (1)$$

Further, to see the impact of the particles on fully grown fungal mycelium, 0.16 mg of the Y2O3 nanoparticles were sprinkled over the culture plate that have been inoculated three days before and another plate of same incubation period was kept as control without any treatment.

3. Results and Discussion

Yttrium oxide (Y2O3) nanoparticles were successfully synthesized through *Nyctanthes arbortristis* mediated phyto-synthesis route. The powder X-Ray diffraction (XRD) patterns as shown in Figure 1a corresponds to crystalline yttrium oxide prepared by phytosynthesis route which was calcinated at 400 °C and all resultant peaks were in good ordinance for a cubic structure of Y2O3 (ICDD: 01-071-0049). FTIR peak were recorded for stretching of Y-O bondand bending of C-O at 566 cm^{-1} and 674, 854 and 1381, 1525 cm^{-1} respectively [11]. FTIR spectra of the calcinated samples also intense peak at 566 cm^{-1} due to Y-O stretching [12]. Both the results from XRD and FTIR spectra are in good agreement with earlier literature reports.

The absorption spectra of various concentrations of Y2O3-DNA with respect to constant concentration of CT-DNA (10 µg/ ml) is depicted in the Figure 2a. A maximum absorption at 206nm is witnessed in the samples owing to the electronic transitions from components of DNA molecules such as purine and pyrimidine [13]. With increasing concentration of Y2O3, the samples exhibited a significant red shift, offering that there exists a strong interaction between the Y2O3 nanoparticles with CT- DNA. In steady state fluorescence, the emission spectrum of Y2O3 nanoparticles exhibited a sharp emission peak at 510 nm at an excitation wavelength of 260 nm. A hypsochromic shift of photoluminescence spectrum is observed with enhanced luminescence intensity with increasing concentration of Y2O3 that indicates interaction of Y2O3 with DNA.

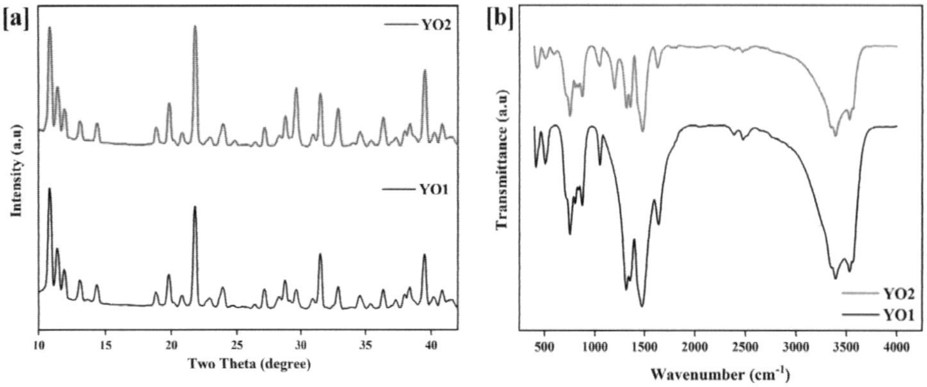

Figure 1. (a) XRD patterns (b) FTIR spectra of Y2O3 samples heat treated at 400 °C

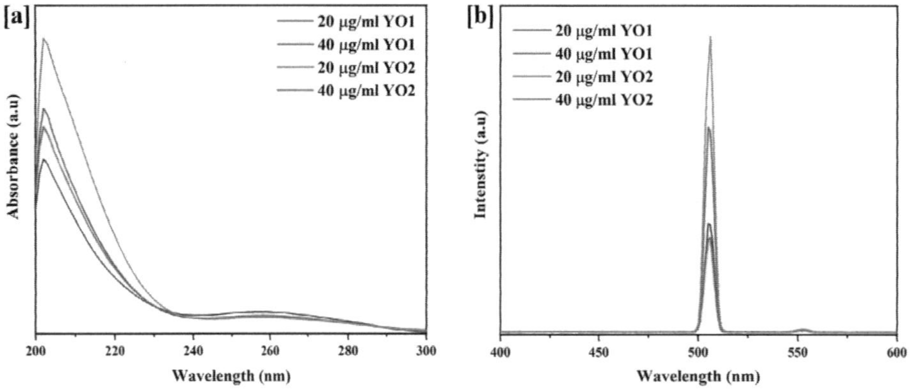

Figure 2. (a) UV-Visible absorption spectra and (b) Steady state fluorescence spectra of CT-DNA along with different concentration of Y2O3 nanoparticles

Figure 3 shows the gel electrophoretogram of CT-DNA with different concentration of Y2O3 nanoparticles which was subjected to a run time of 60 min at100V. 3200μM of CT-DNA was maintained as control as shown in the first lane of Figure 3. Further, significant mobility of DNAwas observed in lanes 1 to 4 on addition of varying concentrations of Y2O3 in CT-DNA. The mobility might be an indication of binding between CT-DNA and Y2O3 nanoparticles [14].

It is important to check the biosafety of developed nanoparticle prior to its biological applications and here MTT assays were performed to validate cell viability. Dose dependent decrease in cell viability was detected in SKMEL cancer cells with the administration of different concentrations of the sample labelled as Y01. The IC 50 value was observed with 24.95 μg concentration of this test sample.

The IC 50 value of YO2 was observed with 57.75 µg concentration of this test sample. After sample addition, the treated as well as the control wells were observed at regular intervals up to 24hrs in an inverted phase contrast tissue culture microscope and the observations were photographed. Although no cytotoxic effects were observed at lower concentration of Y2O3 nanoparticles, higher concentrations exhibited toxic behaviour which is attributed to over-accumulation of particles inside the cell. The observed minor changes in the structure of the cells, including girdle, shrinkage, agglomeration, and vacuolization in the cytoplasm were treated as pointers of toxicity [15].

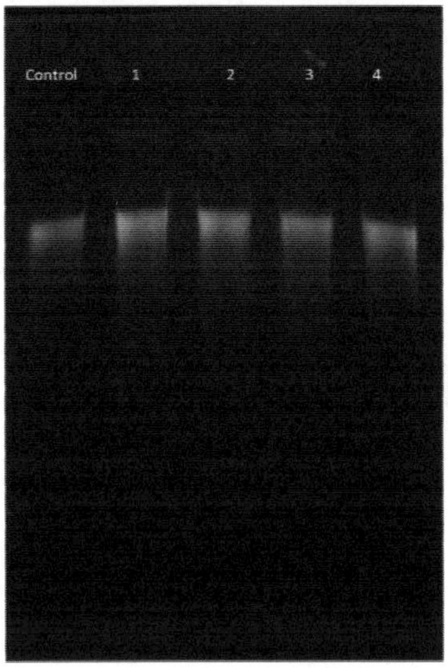

Figure 3. Agarose gel of control and CT-DNA at different concentrations of Y2O3 nanoparticles from lane 1 to 4 respectively

Further, in the anitfungal test there was variation in growth rate in the treated plates as compare to untreated control. It was found that YO2 was found more effective as compared to YO1. Although incase of YO1, the inoculum could initiate its growth. 0.16 mg of YO2 was used for conducting the experiment further and it was found effective. It completely restricted the growthof the pathogen (Figure 5). The radial colony on part B of the treated plate completely lie in the origin when plotted on the graph as indicated in Figure 5 and the growth on part A of the same plate was greatly reduced as compared to part A and B of the control plate. This clearly indicatesthe nanoparticle in used has anti-fungal property. Percent inhibition zone ranges from 51.50 to60.53% in three replicates with an average of 56.39%. It could be stated that Y2O3 nanoparticles might inhibit the growth of *Exserohilum rostratum* by pervading the fungal cell and destructingthe cell wall and cellular mechanisms, thereby restricting the mycelial growth [16].

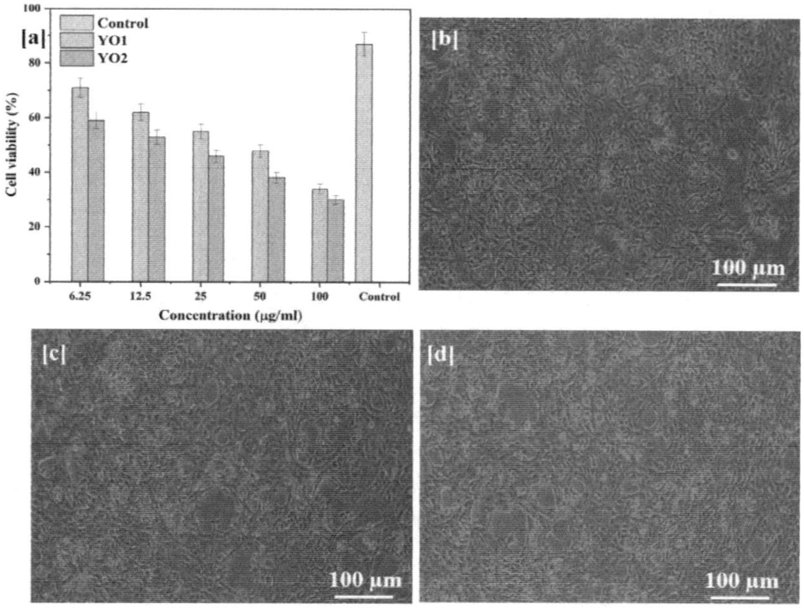

Figure 4. (a) Cell viability of different concentration of Y2O3 nanoparticles in SK-MEL, human melanoma cell lines. Microscopic images of (b) control, (c) 6.25μg/ml YO1 and (d)6.25μg/ml YO2 nanoparticles incubated for 24 hours

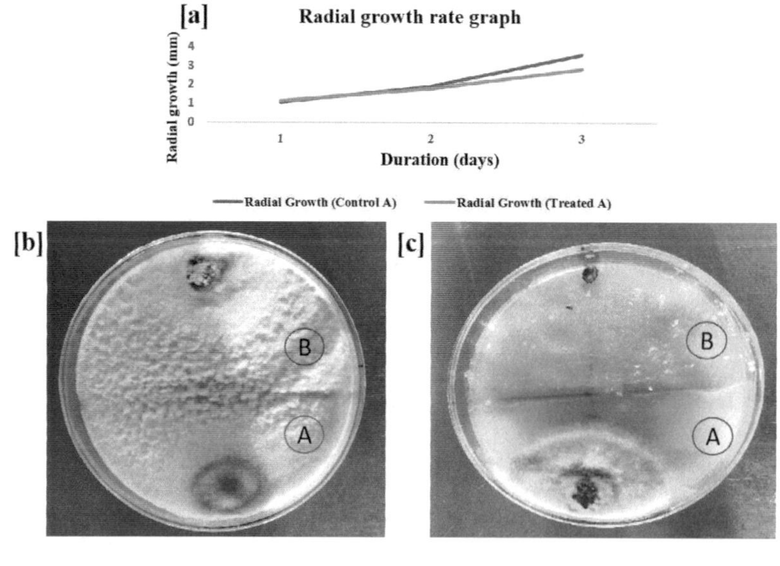

Figure 5. (a) Graph showing the difference in Radial growth of *Exserohilum rostratum* on potato dextrose agar. Antagonistic effect of Y2O3 nanoparticles on *Exserohilum rostratum* culture plates marked with part A and B (a) control and (b) treated

4. Conclusion

Yttrium oxide (Y2O3) nanoparticles have successfully synthesized using aqueous leaf extract of *Nyctanthes arbortristis* and the results from X-ray diffraction (XRD) and Fourier Transform Infrared Spectroscopy (FTIR) confirm the formation of cubic Y2O3. The UV–Visible absorption and steady state fluorescence studies suggested the interaction of CT-DNA with Y2O3 nanoparticles. The mobility of gel detected trough electrophoresis might be an indication of binding between CT-DNA and Y2O3 nanoparticles. Further, the toxicity tests showed that Y2O3 nanoparticles does not show cytotoxic effects at 24 hours of exposure in SK-MEL, human melanoma cell lines at lower concentrations. The results also demonstrate the potential antifungal activity of Y2O3 nanoparticles have good antifungal activity against *Exserohilum rostratum*. Therefore, the results obtained from this study can be used for studying various mechanisms underlying biological applications.

Acknowledgments
The financial assistance received from Department of Science and Technology (DST) [Reference: [DST/INSPIRE/04/2018/000925], India is acknowledged. The authors declare thatthere are no conflicts of interest.

References

[1] Augustine R, Dalvi Y B, Nath V Y, Varghese R, Raghuveeran V, Hasan A, Thomas S and Sandhyarani N 2019 Yttrium oxide nanoparticle loaded scaffolds with enhanced cell adhesion and vascularization for tissue engineering applications *Mater. Sci. Eng. C* . **103** 109801.

[2] Das R K, Pachapur V L, Lonappan L, Naghdi M, Pulicharla R, Maiti S, Cledon M, Dalila L M A, Sarma S J and Brar S K 2017 Biological synthesis of metallic nanoparticles: plants, animals and microbial aspects *Nanotechnol. Environ. Eng.* **2** 1-21

[3] Kannan S K and Sundrarajan M 2015 Biosynthesis of Yttrium oxide nanoparticles using Acalyphaindica leaf extract *Bull. Mater. Sci.* **38** 945–50

[4] Ilyas U, Katare D P, Aeri V and Naseef P P 2016 A review on hepatoprotective and immunomodulatory herbal plants *Pharmacogn Rev.* **10** 66–70

[5] Ribeiro A P C, Anbu S, Alegria E C B A, Fernandes A R, Baptista P V, Mendes R, Matias, A S, Mendes M, da Silva M G and Pombeiro A J L 2018. Evaluation of cell toxicity and DNA and protein binding of green synthesized silver nanoparticles *Biomed. Pharmacother.* **101** 137-144

[6] Babu E P, Subastri A, Suyavaran A, Rao P L, Kumar M S, Jeevaratnam K and Thirunavukkarasu C 2015. Extracellularly synthesized ZnO nanoparticles interact with DNA and augment gamma radiation induced DNA damage through reactive oxygen species *RSC Adv.* **5** 62067-77

[7] An H and Jin B 2012. Prospects of nanoparticle–DNA binding and its implications in medical biotechnology *Biotechnol. Adv.* **30** 1721-32

[8] Singh A and Kaushik M 2019. Physicochemical investigations of zinc oxide nanoparticles synthesized from Azadirachta Indica (Neem) leaf extract and their interaction with Calf-Thymus DNA *Results Phys.* **13** 102168

[9] Meenambal R and Kannan S 2018. Design and structural investigations of Yb3+ substituted β-Ca3 (PO4) 2 contrast agents for bimodal NIR luminescence and X-ray CT imaging *Mater. Sci. Eng. C* . 91 817-23

[10] Schoeman M W, Webber J F and Dickinson D J 1996 The effect of diffusible metabolites of Trichoderma harzianum on in vitro interactions between basidiomycete isolates at two different temperature regimes *Mycol. Res.* **100** 1454-58

[11] Krishna R H, Nagabhushana B M, Nagabhushana H, Chakradhar R P S, Sivaramakrishna R, Shivakumara C and Thomas T 2014 Auto-ignition based synthesis of Y2O3 for photo- and thermo-luminescent applications *J. Alloys Compd.* **585** 129–37

[12] Shivaramu N J, Lakshminarasappa B N, Nagabhushana K R and Singh F 2016 Synthesis characterization and luminescence studies of gamma irradiated nanocrystalline yttrium oxide *Spectrochim. Acta A Mol. Biomol. Spectrosc.* **154** 220–31

[13] Nagababu P, Kumar D A, Reddy K L, Kumar K A, Mustafa M B, Shilpa M and Satyanarayana S2008 DNA binding and photocleavage studies of cobalt(III) ethylenediamine pyridine complexes: [Co(en)2(py)2] 3+ and [Co(en) 2(mepy)2]3+ *Metal-Based Drugs* **100** 275084

[14]Das S, Chatterjee S, Pramanik S, Devi P S and Kumar G S 2018 A new insight into the interaction ofZnO with calf thymus DNA through surface defects *J. Photochem. Photobiol. B: Biol.* **178** 339–47

[15]Miller M A and Zachary J F 2017 *Mechanisms and Morphology of Cellular Injury, Adaptation, and Death In Book Pathologic Basis of Veterinary Disease.*, ed Miller M A and Zachary J F (Elsevier) Chapter 1 p.2

[16]Kumari M, Giri V P, Pandey S, Kumar M, Katiyar R, Nautiyal C S and Mishra A 2019 An insight into the mechanism of antifungal activity of biogenic nanoparticles than their chemical counterparts*PESTIC BIOCHEM PHYS* **157** 45–52

Comparative Study of Chemo- Bio Synthesized Mgo Nanoparticle on Maize Seed Germination

Karanpreet Kaur[1], N Jayarambabu[2] and K Venkateswara Rao[2]

[1]The department of Nanotechnology, Sri Guru Granth Sahib World University, Fatehgarh Sahib, Punjab, India.
[2]Department of Physics, National Institute Of Technology, Warangal-500604, India.
[2]The centre for Nano Science and Technology, Institute of Science and Technology, Jawaharlal Nehru Technological University, Hyderabad, India.
*E-mail: karanpreetbajwa4@gmail.com

Abstract A nanotechnology is expanding area of research is speedily gaining significance and a prominence interest among the farmers. In the present study, the chemical and green synthesize MgO nanoparticles are used for the treatment of maize seeds germination. The green method is a simple, reducing toxic chemical's concentration, an eco-friendly and economically viable. Hence, a locally available plant cissus quadrangularis was selected for MgO nanoparticle's synthesis. The result of XRD, TG/DTA, SEM, TEM, and PSA analysis confirmed formation of MgO nanoparticles. XRD showed the crystalline size of chemical and green synthesized nanoparticles are 25 nm and 24 nm, and the morphology is spherical analysed by SEM. This study is carried out to determine the nanoparticles penetration inside the seed coat, root growth, shoot growth and stimulate the growth hormones. The MgO nanoparticles capability to amplify the dry weight by delivering water to the whole seedling which help to enhance the water nutrient transport and the biomass. The experiment was administered under greenhouse conditions. From the results obtained it is suggested that a green MgO could be used effectively in agricultural fields.

Keywords: A chemo-bio synthesis method, MgO nanoparticle, Maize seed, Seed germination.

1. Introduction

Nanotechnology could come up as a very profound tool for the future. This technology changed the concept of agriculture itself to all the way. Nanomaterials have been increasing attention due to various applications in industries (e.g. electronics, cosmetics, fuel cells, textiles, biosensors), health care, energy.[1] The engineered nanoparticles have unique physio chemical property so, we can expect there are multiplication of applications in global market.[2] In agriculture food sector, nanotechnology has been used to enhance the food quality and food ingredients.[3] Some metal nanoparticles have huge applications in industry such as water treatment and antibacterial properties.[4]

This study reported the novel synthesis of MgO nanoparticles by using magnesium acetate and cissus quadrangularis plant extracted solution in biological method. For a comparative study of maize yield a chemical method has been used to synthesized MgO nanoparticles by using sodium hydroxide as reducing agent. MgO nanoparticles is a non –toxic, odorless, and tasteless white powder. In the biological method for production of MgO nanoparticles we used cissus quadrangularis plant extract as reducing agent instead of chemical compounds. Cissus quadrangularis has many applications in ayurvedic medicines. cissus quadrangularis can be found in rangareedy district of Telangana, India. Cissus quadrangularis plant has shown strong anti-inflammatory, analgesic and antioxidant properties. It has been found to contain carotenoids, triterpenoids and ascorbic acid. The plant contains β-sitosterol, calcium oxalate, ascorbic acid and 3-ketosteriods.[5] The extract of cissus quadrangularis plant has been used successfully to the synthesized single crystalline triangular MgO nanoparticles and spherical MgO nanoparticles.

Over the past decades, Many metal oxides nanoparticles researched as an antibacterial agent. Zhen-Xing Tang et al. (2014) has been used three methods (sol-gel method, hydrothermal method and micro-emulsion method) to synthesized MgO nanoparticles which shown antibacterial activity and this activity of MgO nanoparticles has been depended upon the concentration and size.[6] Mei-Rong Song et al. (2007) has been used same nanoparticles by using tetrahydrofuran to provide significance to investigate the senstivity, catalysis and other properties.[7] Anagno stakos et al. (2008) described Inorganic nanoparticles has play an important role because its antimicrobial activity has been very productive at high surface area with low concentration which permit a wide range of reactions on this bacterial surface.[8] Haung et al. (2005)[9] and Makhulf et al. (2005)[10] investigate the MgO which has been showing activity against gram positive and negative and are safe to use for environment and mammal's cells. The lichens of algae and fungi has been used to synthesized magnesium nanoparticles for many biomedical applications T Devasena et al.(2014).[11] Siavash Iravani (2011)[12] shown that the presence of polyphenols and enzymes in plants help to reduce metal ions faster than microorganisms (funi or bacteria) which play an important role to provide more stable metal nanoparticles. CH. Ashok et al.(2015)[13] prepared MgO nanoparticles by green synthesis technique using neem extract for the germination of Ricinus seeds at different heating temperature of MgO nanoparticles. Environmental friendly technique by using extracts of Phyllanthus recticulatus for preparing silver nanoparticles that showed antibacterial activity against the tested microorganisms and can be further analysed for the bacterial growth inhibition (Karunakar Rao Kudle et al. 2013).[14] Anamika Mubayi et al. (2012)[15] the study represents antimicrobial activity of nanoparticles in both gram positive and gram negative bacteria by using extract of Moringa Oleifera and silver metal ion. If we embraced the power of nanotechnology to increase our yield, increase our food production, preservation of food grains as well as pest control for the food grains we can make an enormous difference for many masses.

2. Materials and methods

2.1. Experimental details

Green Method: In this experiment Cissus quadrangularis plant extract was prepared by boiling the 25gm of plant leaves in 100ml water. The solution was filtered and 30ml of filtered extract was used as reducing agent. 0.1M of magnesium acetate in 200ml distilled water was added separately to plant extract and a vigorous stirring was done at 60^0C. The resulted sample was calcinated at 400^0C in a muffle furnace for removing the any by products and organic compounds in the sample.

*Chemical Method :*In a chemical method adds 0.1M of magnesium nitrate in 100ml of distilled water and kept under stirring. 100ml of water of 0.2M NaOH was added a drop wisely into above solution to allow continue reaction for four hours. The supernatant was discarded and remaining the solution was filtered and washed several times with water to remove impurities. Then sample was placed in oven at 80^0C for 12 hours and white powder was collected in ceramic crucible. The resulting powder was calcinated for 4 hours in the muffle furnace at 400^0C.

3. Characterization techniques

3.1. XRD analysis : As per given XRD pattern in figure 1 showing that the structure of synthesized MgO nanoparticles was cubic. The XRD characterization is employed for the structure and impurities within the sample. According to the reported JCPDS data, all the crystal structure matched to it. The broad peaks indicate synthesized materials are in nanometer range. The peaks were absorbed with 37°, 42°, 62°, 74° and 78° in conjunction with intensities of (111), (210), (220), (311), (222) are comparable JCPDS (01-0420) values.

Scherrer formula (D= 0.9λ/βcosθ) was used to calculate the average grain size.

In Scherrer formula, λ, θ, β was represented x-ray wavelength, angle of the bragg diffraction peak And line width at half maximum respectively.

The crystalline size of chemical and green synthesized MgO was calculated around 25 nm and 24 nm by implementation of this formula to major peaks. The XRD patterns were recorded within the range of 30^0-80^0 with a scan speed of 2/min[16].

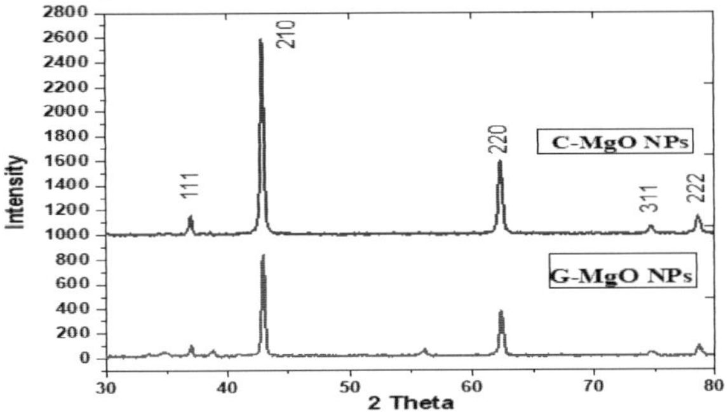

Figure 1. XRD pattern of chemical and green synthesized MgO Nanoparticles.

3.2. Thermo gravimetric analysis : The synthesized MgO nanoparticles are analysed for the thermo gravimetric properties was carried out by the raising temperature gradually and plotting weight against temperature.The temperature of $Mg(OH)_2$ was an important component at which $Mg(OH)_2$ has been converted into MgO from its hydroxide precursor. With the help of alumina crucible the precursor was heated at the temperature of 10^0C/min in nitrogen atmosphere and then it started loses its weight in a first step of the process. This process helped to obtain a end product by losing water from $Mg(OH)_2$ as shown in Fig.2. TGA analysis was observed from room temperature to 800^0C in both cases of chemical synthesized MgO nanoparticles and Green synthesized MgO nanoparticles[17].

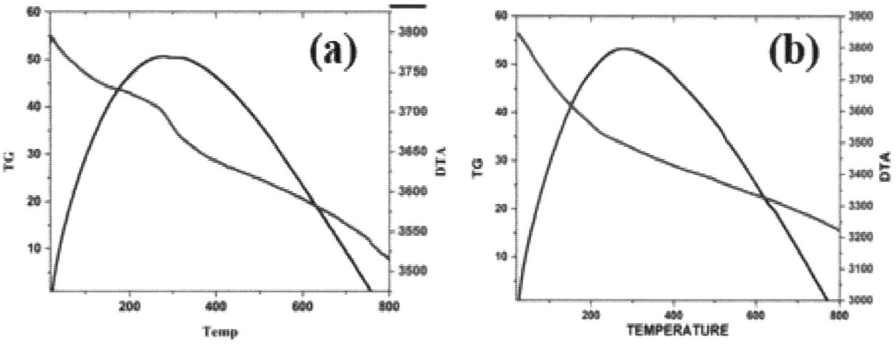

Figure 2. TGA curves of (a) Green and (b) chemical synthesized MgO Nanoparticles.

3.3.Particle size analyser : The prepared Magnesium Oxide nanoparticles were ultra-sonicate by using the distilled water. The sizes of the MgO nanoparticles in the suspensions were determined using a particle size analyser. A fig.3 shows the particle sizes verse undersize percentages in the dispersed distilled water, and the average particle size of chemical synthesized MgO Nanoparticles is 28 nm and Green synthesized MgO Nanoparticles is 26 nm [18].

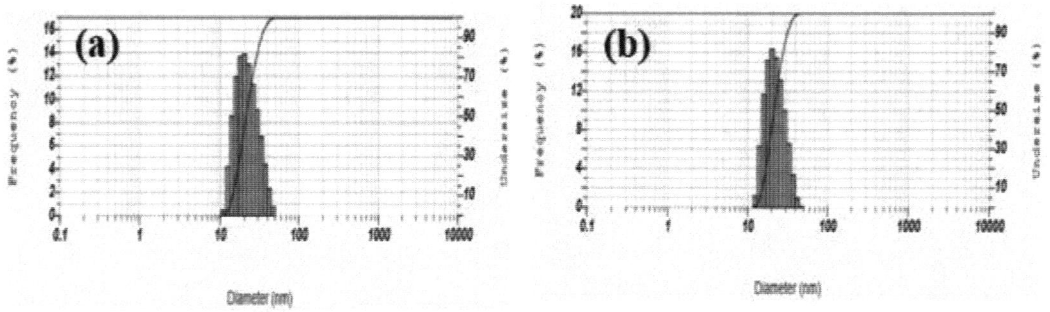

Figure 3. Particle size Analysis of (a) Green and (b) chemical synthesized MgO Nanoparticles.

3.4. SEM analysis and EDS : The shape of chemical and green nanoparticles is spherical and the average size of chemical and green synthesized MgO nanoparticles are 200 nm and 1 µm[19] which is shown as Fig.4. EDS spectrum shows that the sample primarily contain Mg, O and their respective atomic content of chemical MgO are 55.93%, 44.07% and Green MgO are 62.96%, 37.04%. Magnesium oxides species & residue solvent on the surface of the particles might represent O peak.

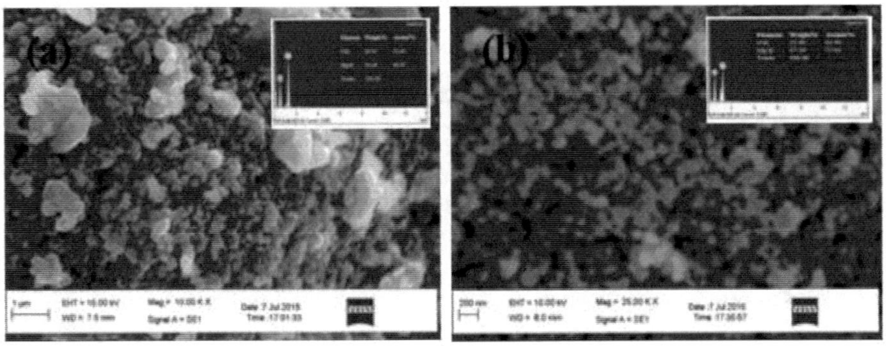

Figure 4. SEM and EDS images of (a) Green and (b) Chemical synthesized MgO Nanoparticles.

3.5.TEM Analysis: TEM is playing important roles for a characterization of grain boundaries, a morphology and crystalline materials. In a Fig.5 the shape of green MgO nanoparticles is spherical and there size measurement dimension is 26.7 nm, 39.3 nm, 49.6 nm where as the chemical MgO nanoparticles has different size measurement which is 17.2 nm, 18.4 nm, 20.5 nm[20].

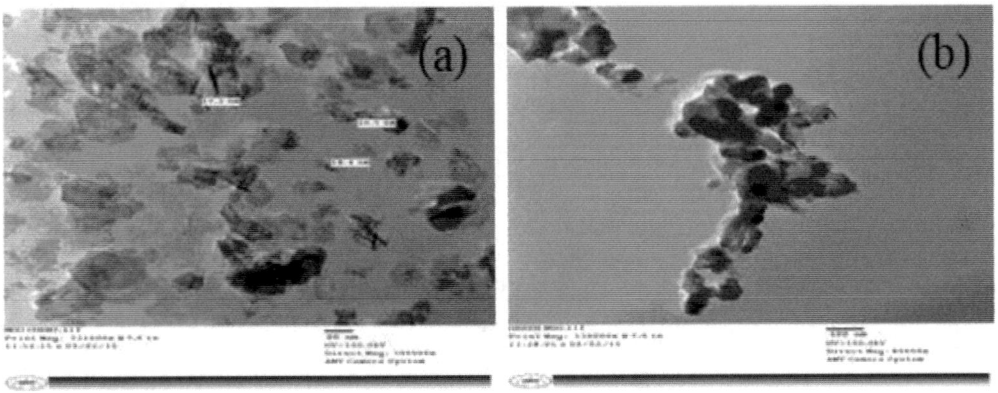

Figure 5. TEM images of (a) Green and (b) Chemical MgO Nanoparticles.

4. Preparation of particle Suspensions and Seed Treatment

The maize seeds were purchased from market kept in the dark and dry place under room temperature before using. For the Green and chemical synthesized MgO Nanoparticles suspensions, maize seeds were surface-sterilized in sodium hypo chloride for 10 minutes, then thoroughly rinsed with distilled water. By using a sonicator instrument the maize seeds were sonicated in prepared nanoparticles suspensions for 2 hours. For the maize seeds growth characterization, the soaked seeds were put in soil pots.

4.1. Seed germination test : After 10 days, we observed the maize seed germination in the presence of the chemical and green synthesized MgO nanoparticles. The green synthesized MgO nanoparticles shown the higher seed germination percentage as compared to the chemical synthesized MgO nanoparticles as shown in Fig.6. By using thread and scale we determined the fresh biomass and the dry biomass of the plant where as digital weighing machine was used to determined the fresh weight of root and shoot of seedlings . After measuring fresh weight, the seedlings was kept in incubator at 80^0C. The fresh and dry biomass of the control plant were 0.1g and 0.09g. The fresh and dry biomass of the chemical Nanoparticle treated plant were 0.2g and 0.18g. The fresh and dry biomass of green Nanoparticles treated plant were 0.22g and 0.19g. The green synthesized Nanoparticles treated plant has more biomass than chemical and the control as shown in Fig.6. The MgO nanoparticles mechanism on maize seed germination is helpful for water and minerals absorption at same time increase the biomass of the plant.

5. Effect of MgO nanoparticles Suspension on maize seed Germination

The interaction of MgO nanoparticles significantly affected the percentage of seed germination. The green method synthesized MgO Nanoparticles suspension shown the higher root growth of the maize seed as compare to the control germination and chemical synthesized MgO Nanoparticles suspension. This treatment also increase the root length and shoot length notably.

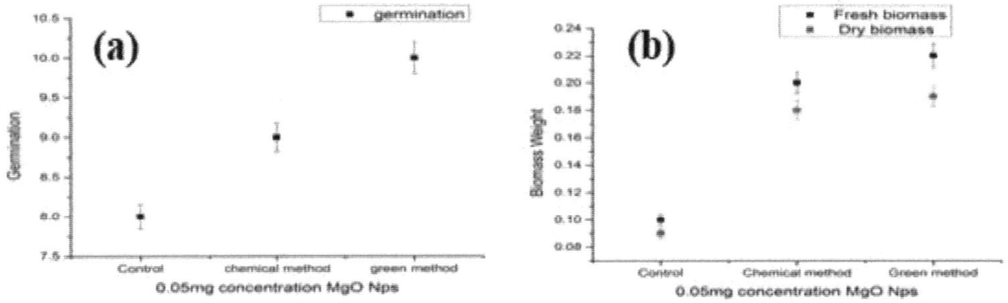

Figure 6. (a) Seed germination of control, chemical, green method nanoparticles and (b) Biomass weight of Control, Chemical, and Green Synthesized maize seed.

5.1. Scanning electron microscopic analysis : We separated the roots of the grown plant of control, chemical and the green nanopartcles and washed with the ethanol solvent. Cut the small pieces of roots and added the glutaraldehyde solution along with roots in test tubes. The surface structure near the roots of the germinated maize seeds in the presence/ absence of the MgO nanoparticles was inspected by the SEM as shown in Fig. 8. In this experiment the MgO nanoparticles capability to amplify the dry weight by delivering water to the whole seedling to enhance the water nutrient transport and the biomass. MgO nanoparticles has faster ability to transported through the plant (including) within the metabolic processes. Based on SEM images of plant root surface analysis the MgO nanoparticles were helpful for absorption of minerals and water at the same time support for the photosynthesis plants.

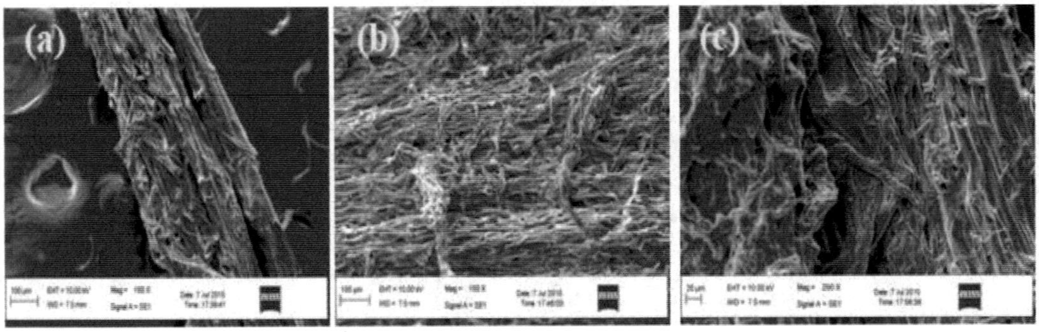

Figure 7. (a), (b) and (c) SEM images of control, Chemical, Green plant germination.

6. Conclusions

The increasing demand for nanoparticles because their wide applications in various areas like seed germination, biosensors, cosmetics etc. Magnesium oxide nanoparticle was synthesized by chemical method from magnesium nitrate and green method from magnesium acetate using plant extract (cissus quadrangularis). Various parameter's viz. stirring temperature, concentration of plant extract, calcinations temperature were noteworthy effect on the result. The nanoparticles by using chemical techniques is toxic and flammable in nature while the green method or biological method is an environmental friendly technique. The structure and morphology of particles were investigated by XRD, TG/DTA, PSA, SEM, EDS and TEM. XRD showed the structure of MgO nanoparticles was cubic and crystalline in nature. The crystalline sizes of chemical and green synthesized nanoparticles are 25 nm and 24 nm. The particle size of chemical and green synthesized nanoparticle are 28 nm and 26 nm. SEM was used for characterized the morphology of the MgO nano crystal. The average size of the MgO nanoparticles of chemical and green synthesized are 200 nm and 1 μm. EDS pattern infers the presence of element Mg and O. The chemical and green synthesize MgO nanoparticles are utilized on the maize seed to review its growth characterization and the effect on a bioavailability of seed germination. We were observed in maize seeds germination, green synthesize MgO nanoparticles shown the good shoot, root growth results compare to the chemical and the control during this experiment. The study results also revealed that the germination process was influenced by MgO nanoparticles in maize seed. The nanoparticles exposure involves nutrient transport to support photosynthesis of plant. In agriculture, the antimicrobial properties of nanoparticles make it more suitable functional to use it as pesticide or herbicide.

Acknowledgement

Our special thanks to Dr. Ashok Chinthakuntla, Senior Scientist, Nanospan India Pvt. Limited and Centre for nanoscience and technology JNTU-Hyderabad, Telangana, India for supporting our work.

References

[1] Boonyanitipong P, Kumar P, Kositsup B, Baruah S and Dutta J 2011. Effects of zinc oxide nanoparticles on roots of rice Oryza sativa L. *In Int. Conf. Environ. Bio. Sci.* **21**172-6

[2] Zhang R, Zhang H, Tu C, Hu X, Li L, Luo Y and Christie P 2015. Phytotoxicity of ZnO nanoparticles and the released Zn (II) ion to corn (Zea mays L.) and cucumber (Cucumis sativus L.) during germination *Environ. Sci. Pollut. Res.* **22** 11109-17

[3] Jayarambabu N, Siva Kumari B, Rao K V and Prabhu Y T 2014 Germination and growth characteristics of mungbean seeds (Vigna radiata L.) affected by synthesized zinc oxide nanoparticles *Int. J. Curr. Eng. Technol.* 5 2347-5161

[4] Naseem T and Durrani T 2021 The role of some important metal oxide nanoparticles for wastewater and antibacterial application *Environ. Chem. Ecotoxicol.* **3** 59-75

[5] Mishra G, Srivastava S and Nagori B P 2010 Pharmacological and therapeutic activity of Cissus quadrangularis: an overview *Int. J. Pharmtech Res.* **2** 1298-01.

[6] Zhen-Xing T and Bin-Feng L 2014 MgO nanoparticles as antibacterial agent: preparation and activity *Braz. J. Chem. Eng.*596-597.

[7] Song M R, Chen M and Zhang Z J 2008. Preparation and characterization of Mg nanoparticles *Mater. Charact.* **59** 514-18

[8] Anagnostakes K, Hitzler P, Pape D, Kohn D and Kelm J 2008 Persistence of bacterial growth on antibiotic-loaded beads *Acta Orthop.* **79** 302-07

[9] Haung L, Li D Q, Lin Y J, Wei M, Evans D G and Duan X 2005, Controllable preparation of Nano-MgO and investigation of its bactericidal properties *J Inorg Biochem* **5** 986-93

[10]Makhulf S, Dror R, Nitzan Y, Abramovich Y, Jelinek R and Ge-danken A 2005 Microwave-assisted synthesis of nano-crystalline MgO and its use as a bactericide *Adv Funct Mater* **15** 1708-15

[11]Devasena, T, Ashok V, Dey N and Francis A 2014. Phytosynthesis of magnesium nanoparticles using lichens *World J. Pharmaceut. Res* **3** 4625-32

[12]Iravani, S 2011 Green synthesis of metal nanoparticles using plants *Green Chem.* **13** 2638-50.

[13]Ashok C H, Rao K V, Chakra C H and Rao K G 2015 Temperature Effects on MgO Nanoparticles: Prepared by Green Route Method and Application of Seed Germination *Mater. Focus* **4** 290-294

[14]Manisha D R, Alwala J, Kudle K R and Rudra M P 2014 Biosynthesis of silver nanoparticles using flower extracts of Catharanthus roseus and evaluation of its antibacterial efficacy *World J. Pharm. Pharm. Sci* **3** 877-885.

[15]Anamika M, Sanjukta C, Rai P M and Watal G 2012 Evidence based green synthesis of nanoparicles, *Adv. Mat. Lett.* **3** 519-25

[16]Ashok C H, Rao K V, Chakra C H and Rao K G 2015 Temperature Effects on MgO Nanoparticles: Prepared by Green Route Method and Application of Seed Germination *Mater. Focus* **4** 290-294

[17]Zhu P, Wang L Y, Hong D and Zhou M 2012. A study of cordierite ceramics synthesis from serpentine tailing and kaolin tailing *Sci. Sinter.* **44** 129-134

[18]Ashok C H , Rao K V, Chakra C S and Rajendar V 2014 Structural properties of CdS nanoparticles for Solar cell applications *Int. J. Pure Appl. Sci. Technol* **23** 8-12

[19]Park J Y, Lee Y J, Jun K W, Baeg J O and Yim D J 2006 Chemical synthesis and characterization of highly oil dispersed MgO nanoparticles *J. Ind. Eng. Chem* **12** 882-887

[20]Ashok C H, Rao K V and Chakra C S 2014 Structural properties of CdS nano particles prepared in the presence of organic solvent *Appl. Sci. Res* **5** 99-105

A Study on Milk Adulteration and methods of detection of various Chemical Adulterants qualitatively

Riya Chugh[1], Gurmeet Kaur*[1]

[1]Chemistry Department (UIS), Chandigarh University, Gharuan, Mohali - 140413

Email ID- riya.r62@cumail.in, gurmeet.uis@cumail.in

ABSTRACT: The food Adulteration now becomes a worldwide issue as the emergent nations are at very high risk associated with this Adulteration. The milk is an important part of our daily life and is consumed on the daily basis but, now a days milk is being adulterated by several chemicals due to lack of monitoring, unhygienic conditions and for financial gain. Unfortunately, various chemical Adulterants can pose danger health risks that lead to many lethal diseases including heart problems, diarrhoea, CNS disorders, irritation and gastrointestinal disorders. Various common Adulterants like benzoic acid, urea, water, skimmed milk powder, sugar, detergent were detected in the pure milk samples which concludes that the milk samples doesn't meet the expected quality accepted as standard. Many studies have been carried out on milk Adulteration and detection techniques. This Present work gives a clear view of the quality of milk. This article gives a comprehensive review of the common Adulterants added in milk along with the various methods of detection qualitatively.

1. Introduction

Milk is a pale white nutrient rich liquid food produced by mammals. It is a major source of nutrition and provides essential nutrients to maintain the growth of body. Since ancient time in India, It is a significant and valuable part of our diet as it is readily digested and absorbed. Humans have been preserving the milk for centuries. Milk is made of water, fat, protein, lactose and some minerals. This milk is used as a source of food for the infant humans and animals who are unable to eat at that time. It provides various essential nutrients like protein, vitamins, carbohydrates, fat and minerals. Milk alone contains these all-important nutrients than any other single food; this is the main reason for milk to be called as complete food. The lactation milk contains many antibodies like IgA which contains many antibodies which helps in fighting from disease in the infants [1]. It is easily obtained by various milk-producing animals like cow, buffalo and goat etc.

India's share in world milk production stands at 17 % as it produces 90% of the total volume of buffalo milk in world and is on second rank for worldwide production of cow's milk i.e. 54 million metric tons followed by USA.

Being the largest milk exporter of milk and skimmed milk yet it is needed for country like India to import some milk varieties [2,3].

The process of production and exportation of milk is done via a proper supply chain that creates network between different companies, which helps in the distribution of the sample. More than six billion of the world consumes milk and its products [4].

The milk constitutes many particles containing lactose, water, fat, protein, milk sugar and salt mineral. Trace amount of various phospholipids, vitamins, enzymes are there in the milk.

Rajasthan, Maharashtra, Uttar Pradesh, Madhya Pradesh, Punjab, Haryana, Karnataka and Andhra Pradesh are some of the states having highest production rate. The following map shows the three major states having highest production of milk (figure.1). The survey shows the contribution of different states of India to the growth of milk production [5].

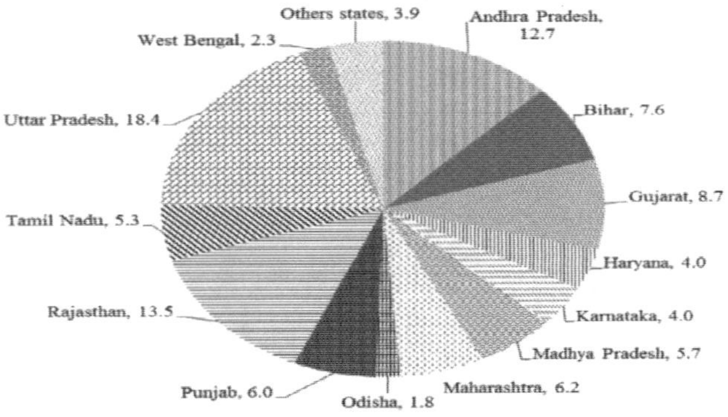

Figure 1. Contribution of States in the production of milk

However, the quality of milk is being exploited continuously since era due to its Adulteration in the market. Adulteration is done by adding various cheaper Adulterants that increases the quantity of the milk and to make more profit. The major and common adulterant of milk is water, which consequently decreases its nutritive value and also become a source of water- borne diseases if added water is contaminated.

In order to know more about the milk Adulterants and its effects on human health, a research has been done determine the quality of raw milk in Delhi-NCR.

1.1 Adulteration

The Deterioration of Milk is a socio economic issue especially in the developing nations the act of addition of various substances to the food products in order to increase its quantity results in the loos of its actual quality of the food items and to makes it unfit for consumption. The adulterated food product fails to meet the legal standards [6].

The most common chemical Adulterant which are added in the milk includes starch, Chlorine, Sodium carbonate, Formalin, Ammonium Sulphate. These chemicals like urea, starch, flour, detergent etc. are added in the milk to increase profit margin [7].

To meet the paucity of milk the artificial milk is then processed by adding urea, refined oil, Sodium hydroxide, and some ordinary detergent which has adverse and toxic effects [8–11,11,12]. The consumption of adulterated milk causes various critical health regarding problems like heat diseases, cardiovascular diseases, skin diseases etc. Hence, it has become a great concern to the food safety. The major food Adulteration occurs with widely publicized Adulteration of milk and milk products. Adulterated milk is the one in which the milk quality is intentionally degraded by mixing substandard toxic substances or chemicals to increase the profit.

Generally, the major adulterant in the milk is water in high percentage and it is a very serious concern. When milk is adulterated with water there is a decrease in the quantity of ions present in the milk and electrical conductivity also get changed [11,13–18].

The contaminated water if added in the milk, it will cause severe health problems especially to infants [11,19–22].

To reduce the increased demand, the Contamination of milk has progressively prevalent in India. Most of the times, the Adulteration is intentionally done to make high profits, but sometimes may be due to lack of proper detections. The chemicals and various substituents added to the milk to meet the rapid demand of milk. Constantly the natural milk is contaminated with low value ingredient like water and detergent. Several other reasons may also be the major issues for milk Adulteration.

Milk used by the common people is such adulterated that it has become very less nutritive and may cause several serious diseases like gastrointestinal illness, heart problems, cancer, irritation in eyes, kidney problems, etc. The common Chemical which are intentionally or unintentionally added to milk are glycogen, urea, chlorine, soda ash, formaldehyde, detergent, caustic soda, skimmed milk powder, flour, vegetable oils etc.

In order to know more about the milk Adulterants and its effects on human health, a research has been done determine the quality of raw milk in Delhi-NCR. FSSAI in 2011 conducted a survey on milk Adulteration in some selected urban and rural areas, found that the most common adulterant of milk is water.

2. SOME COMMON FOOD ACTS

The Act ensure that the milk used is safe for drinking and consumption purpose .The data contains all the information about various offensive content present in the sample. Various food acts which are applicable to preserve and package milk and milk products in India are Vegetable and Oil Products Order (1947), Prevention of food Adulteration act (1954), The Fruit Products order (1955), Meat Food Products Order (1973), Edible Oil Packaging Order (1988), Milk and Milk Products Order (1992), Food Safety and Standards Authority of India (2006).

3. LITERATURE REVIEW

In the Department of Public Health and Epidemiology, College of veterinary science, Proddatur town, Kadapa, Andhra Pradesh a study was reported, Fifty samples of raw milk were found to be adulterated with highest percentage of salt and sugar that is 54% and 40% respectively and did not confirmed to the legal standards of FSSAI [23].

Nirwal et al. (2013) collected 100 samples from different localities of Dehradun. Out of which only two milk samples were found to be pure and rest was adulterated with high amount of glucose i.e. 80%, of the skimmed contains 58% of milk powder, 51% of salt and 35% of urea. 32% of the result of milk samples was found to be safe for consumption and follow the act and the rest of the 68% failed to follow the law. Study concluded that the most common adulterant added in milk is water that not only reduces the nutritional value but also spread various water-borne illnesses.

Testing of milk Adulteration was done on 50 samples obtained in sterilized screw capped glass bottles to avoid external contamination were from Godaulia and Pandeypur of Uttar Pradesh (Varanasi) India. The extent of Adulteration in these milk samples with NaCl, neutralizers and NH_2CONH_2 were 80%, 28% and 60% respectively. The preservative used increases the shelf life of the milk. Same testing was done in Tandojam, Hyderabad with 20 samples among which 40% of the milk sample were found to be adulterated followed by 30% with starch and 25% with urea. The Adulterants had strong potential to degrade the qualities and physico-chemical characteristics of milk [19].

A 2007 report in the journal of American Heart Association found that consumption of vanaspati elevates cholesterol level in the body causing diabetes and coronary ailments. A national survey conducted shows that nearly 70% of nations the milk is adulterated with neutralizers and detergent, but water again is the major contaminant found in milk [24].

According to study of Adulteration in Hyderabad, total 50 samples were collected for determination and extent of Adulteration. It was found that sucrose and SMP were present in 22% and 88%

respectively. The amount of Urea, neutralizers and salt (NaCl) were found to be 26%, 80%, and 82% of the milk samples. Some other chemical Adulterants like Formalin, detergent and hydrogen peroxide were present in 32%,32%, and 44% is obtained (figure.2). All the mentioned Adulterants found in the milk samples proved about the milk that it did not conform any legal standards and might cause critical health hazards [19].

Figure 2. Percentage of chemical Adulterants in milk.

According to the national survey in India by FSSAI in 2011 in concluded that after detergent water is found to be the most commonly used adulterant in the milk. Further in 2012, it concluded in a survey that 68% were found to be adulterated and 88% was there in Uttrakhand [19].

Research on milk Adulteration done in Koti, Hyderabad concluded that out of ten samples of milk, six samples were highly adulterated with water and starch and one was adulterated with detergent (figure. 3) [25].

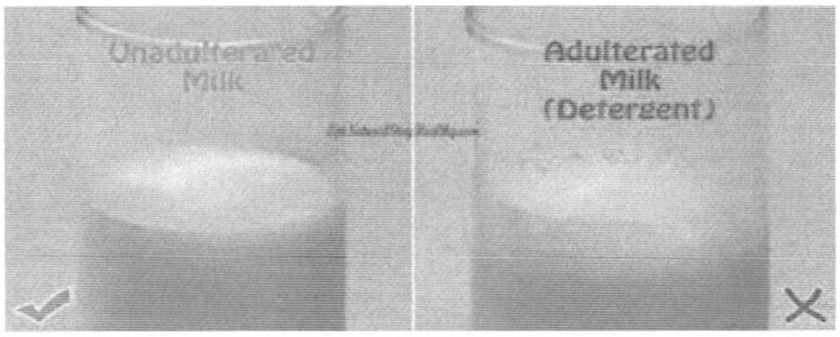

Figure 3. Adulterated milk with detergent

India's milk production is 18.5% of the world population. According to FAO the increase in worlds milk production is 3.1% while as that of India is 18.5% in the year 2014-15. Though India ranks 1st in milk production still it is importing milk products and on the other hand exporting less quantity of milk and milk products to the neighbouring countries [26]. From China 300 samples were collected from 3 localities in Sudan. Ninety five percent of the samples were found to contain water and 35.5% contained starch.

As per the studies carried out in Hyderabad, Pakistan and its vicinity it was found that most of the milk simples contained most common Adulterants practiced out by market milkman. Out of 100 unprocessed milk samples, no sample was found to be wholesome/unadulterated. Beniwal and Akhetar paul (1999) reported the Adulteration of water to be 70% while almost 94% samples of milk adulterated with water was recorded by Lateef et al. (2009). Other Adulterants like formalin, starch, skimmed milk powder and NaCl were 34%, 13%, 8%, and 5% respectively [27].

Karabi et al. have conducted a study that shows constant phase element sensor for the detection of microorganisms and bacteria in milk. The growth of bacteria and microbes changes the ionic properties of the medium, which results in changing the conductivity value of the milk. [28].

In 2011, it was found that 260 million dairy produced 730 million tonnes of milk. But in today's world, 600 million tonnes of milk was produces by 264 million dairy cow every year and goat milk is the number one consumed milk in the world. Six billion of consumer of milk and its product are there worldwide.

Dr. Sitaram dixit, chairman, CGSI Maharashtra , reported that milk Adulteration has increased by 5% in the year 2018 as per the research conducted, the 340 low-graded or loose milk packets were tested, out of which 76 samples, which made up for 22% were standardised, while the remaining 264 packets that were made for 78% were found to be adulterated.

Pawan Agarwal, CEO, FSSAI conducted a detailed survey on the presence of Aflatoxin M1 a kind of fungus, gets mixed in the milk through feed and fodder , majorly found in packaged milk products in top three states including Delhi, Tamil Nadu, Kerala, total 368 samples (out of 6,432) that is 5.7% in the country were found to be adulterated by the Aflatoxin M1 [29]The (table 1) shows the common Adulterants present in the milk

Table 1. Some Places where Major Adulterants were detected in Milk Samples.

S No.	Area	No. of Samples	Major Adulterants	References
1.	Kadapa, Andhra Pradesh	50	Salt, sugar	[23]
2.	Dehradun	100	Glucose, skimmed milk powder	[22]
3.	Pandeypur, Varanasi	50	Sodium chloride, urea	[22]
4.	Tandojam, Hyderabad	20	Starch, urea	[11]
5.	Hyderabad, Andhra Pradesh	50	Urea, salt	[19]
6.	Koti, Hyderabad	10	Water, Starch	[19]
7.	China	300	Water, starch	[27]
8.	Hyderabad, Pakistan	100	Water, formalin	[28,29]

4. METHODOLOGY

4.1 Procurement of milk samples

The present study was conducted in Manav Rachna University, Faridabad, Haryana. In total twenty-five samples were collected from local vendors, dairies and farms from different areas and brands that come under Delhi-NCR. The raw samples were collected into clean dry sterilized bottles to prevent any type of contamination. The testing was done for given eight Adulterants i.e. Starch, Pulverised soap, Vanaspati, SMP (Skimmed Milk Powder), sugar, synthetic milk, Benzoic acid and Salicylic acid.

Table 2. Number of milk samples collected from different areas.

S No.	Milk Samples	Area	Milk Brand
1.	A	N.I.T - 5	Dairy
2.	B	Jawahar Colony	Dairy
3.	C	Dabua Colony	Amul Toned
4.	D	Okhla	Dairy
5.	E	Dabua Colony	Dairy
6.	F	N.I.T – 5	Mother Dairy Toned
7.	G	S.G.M Nagar	Dairy
8.	H	N.I.T - 5 , Block C	Dairy
9.	I	N.I.T – 4	Dairy
10.	J	Badarpur Border	Dairy
11.	K	Manav Rachna	Milk Powder
12.	L	N.I.T -2	Dairy
13.	M	Manav Rachna	Mother Dairy Toned
14.	N	Sec-20b	Dairy
15.	O	Okhla	Amul
16.	P	Sec-20b	Dairy
17.	Q	Sec-15	Amul
18.	R	N.I.T-3	Dairy
19.	S	Okhla	Vita
20.	T	Sec-21c	Paras
21.	U	Sec-16	Mother Dairy
22.	V	N.I.T-2	Ananda
23.	W	S.G.M Nagar	Dairy
24.	X	Gandhi Colony	Country Delight
25.	Y	Sec-16	Parag

5. ANALYSIS OF MILK SAMPLES

5.1 Sugar -Take 3ml of milk sample in the test tubes. Add 2ml of concentrated HCL and 50 mg of resorcinol, then heat the test tubes and solution turns red, which shows the existence of sugar in milk sample (figure.4).

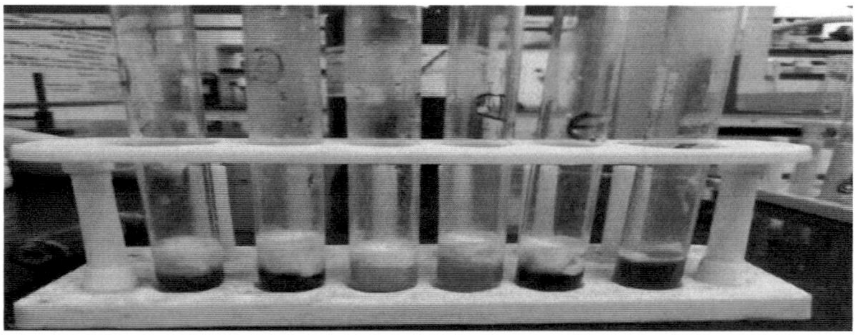

Figure 4. Detection of sugar in milk

5.2. Pulverized soap -In the given milk samples add 10ml of hot water with 2-3drops of phenolphthalein as an indicator. The presence of soap in the milk is analysed by the appearance of pink colour (figure.5).

Figure 5. Detection of soap in milk

5.3 *Skim milk powder* -Take some milk in test tubes, add few drops of Nitric acid drop wise, the yellow colour if appears the milk is unadulterated (figure.6).

Figure 6: Detection of skim milk powder in milk

5.4. Benzoic acid- 5ml milk was taken in test tubes, and to it sulphuric acid was added drop wise and 0.5% ferric chloride in it. The buff colour of the milk shows the presence of benzoic acid (figure.7)

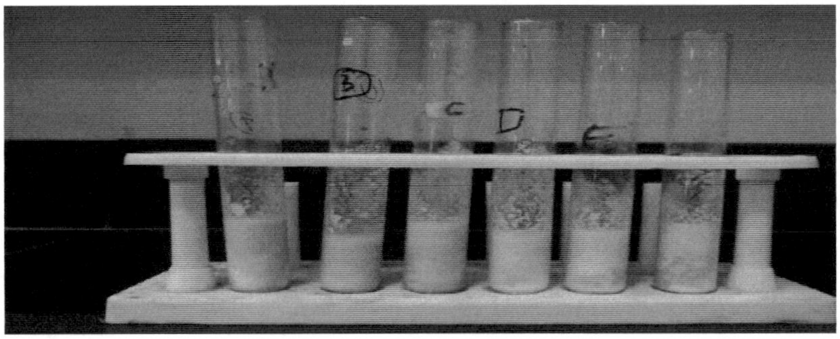

Figure 7. Detection of benzoic acid in milk.

5.5. Salicylic acid -Take some amount of milk sample in test tubes, add few drops of H_2SO_4 and 0.5% $FeCl_3$ in it. The colour changes to violet that indicates that salicylic acid is present in the given milk sample.

5.6. Vanaspati - In a test tube, Take 5 ml of milk sample and add 10 drops of HCL with 1 teaspoon of sugar is added. If the colour of the milk changes to red, it shows the presence of ghee (figure.8).

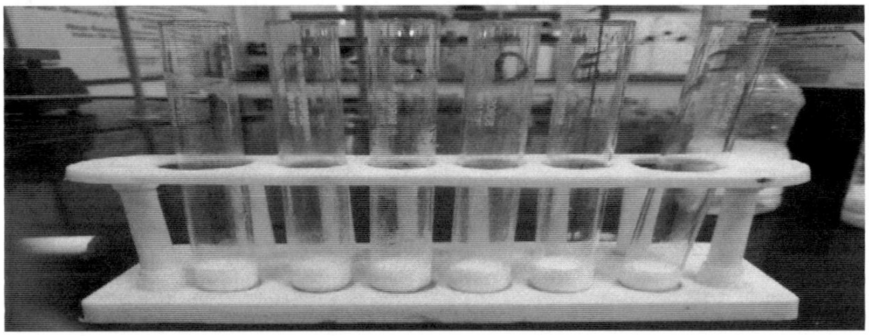

Figure 8. Detection of Vanaspati ghee in milk

5.7. Synthetic milk -Take some milk in test tube and heat the test tube. The yellow colour indicates that the milk sample is adulterated (figure.9).

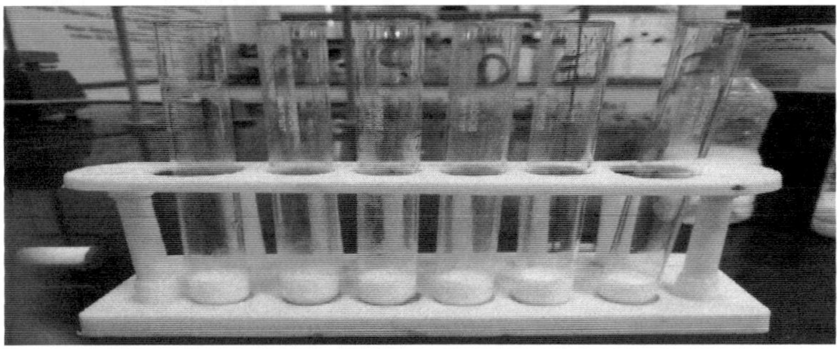

Figure 9. Detection of synthetic milk.

IOP Conf. Series: Materials Science and Engineering 1225 (2022) 012046 doi:10.1088/1757-899X/1225/1/012046

5.8. Starch -Take 3ml milk in test tubes and was kept for incubation for 5 minutes in boiling water. Then, cool it and add (1%) iodine solution, mix well. Black colour appeared which confirms the presence of starch.

6. RESULTS AND DISCUSSIONS

The most harmful Adulterants found in the milk are Water, Benzoic acid and Sugar.

The presence of sugar in the milk is observed generally, as it is a common source of natural sweetness in the milk. Some Adulterants like Hydrogen peroxide, urea and various other neutralizers also used as cheap preservatives that assist to enhance the shelf life of milk. Among all the Adulterants, water and benzoic acid both were found to be the common and major Adulterants present in the milk that degrades the nutritional value of milk and may cause various critical water-borne diseases.

All the 25 samples were tested for common Adulterants like sugar, vanaspati, starch etc. as depicted in the (table3). After the determination of milk samples benzoic acid, synthetic milk, sugar and skimmed milk powder were found to be major Adulterants. A, B, G, I are milk samples from different localities were highly adulterated with SMP, sugar and benzoic acid whereas the milk samples named as E, F and L were commonly adulterated with sugar, synthetic milk and benzoic acid.

Table 3: Twenty-five different brands of Milk were samples tested with presence of different Adulterants.

Sno.	Adulterants	A	B	C	D	E	F	G	H	I	J	K	L	M	N	O	P	Q	R	S	T	U	V	W	X	Y
1.	**Starch**	-	-	-	-	-	-	-	-	-	-	-	-	-	-	-	-	-	-	-	-	-	-	-	-	-
2.	**Pulverized Soap**	-	-	-	-	-	-	-	-	-	-	-	-	-	-	-	-	-	-	-	-	-	-	-	-	-
3.	**Skim Milk Powder**	-	-	-	-	-	-	-	-	-	-	-	+	-	-	-	-	-	-	-	-	-	+	-	-	-
4.	**Sugar**	+	+	+	+	+	+	+	+	-	+	+	+	+	+	+	-	+	+	+	+	+	+	+	+	+
5.	**Synthetic Milk**	-	-	-	-	-	-	+	+	-	-	-	-	+	+	-	-	-	-	-	+	-	-	+	-	-
6.	**Benzoic Acid**	+	+	+	+	+	+	+	+	-	-	+	+	+	+	+	+	-	+	+	+	+	+	-	+	+
7.	**Salicylic Acid**	-	-	-	-	-	-	-	-	-	-	-	-	-	-	-	-	-	-	-	-	-	-	-	-	-
8.	**Vanaspati**	-	-	-	-	-	-	-	-	-	-	-	-	-	-	-	-	-	-	-	-	-	-	-	-	-

7. CONCLUSION

The various chemical substances that are added in milk due to financial gain. But, it is mostly done due to carelessness and lack of hygienic condition of processing and storing and is mainly occurs in both developing and backward countries both. The consumers are often become victim of diseases consuming the Adulterated milk. This present work shows that Benzoic acid, water, synthetic milk and sugar were the most common Adulterants present in all the milk samples taken. Salicylic acid and vanaspati ghee found to be Non-existent whereas, benzoic acid and sugar both found in highest amount as the adulterant. The purpose of adding benzoic acid is to preserve the milk and milk products for long time, which causes various adverse effects on human health. According to the research conducted, most of the collected milk samples are Contaminated with different chemical Adulterants which were added during the production or processing of milk. Whether intentionally and unintentionally in order to deal with maximum production of milk and to generate more money these harmful and poor quality chemicals added in milk. In India, as milk is the common and complete food for the human beings, this analysis is carried out to bring awareness to public about the malpractices in milk production and their health hazards.

ACKNOWLEDGEMENT

The authors are grateful to the Chandigarh University for giving the appropriate environment and necessary resources to carry out this study.

REFERENCES:

[1] Pehrsson P R, Haytowitz D B, Holden J M, Perry C R and Beckler D G 2000 USDA's National Food and Nutrient Analysis Program: Food Sampling *J. Food Compos. Anal.* **13** 379–89

[2] Kumar A 2013 *India emerging as a leading milk product exporter* Retreived from dawn.com

[3] PTI 2014 Government scraps incentive on milk powder exports to check prices Retreived from economictimes.indiatimes.com

[4] Hemme T and Otte J 2010 *A Living from Livestock Pro-Poor Livestock Policy Initiative Status and Prospects for Smallholder Milk Production A Global Perspective* Food and agriculture organization of the united nations Rome

[5] National Diary Development Board 2017 Contribution of different states to the growth of milk production in India

[6] J S and DC M 2011 Defining the public health threat of food fraud *J. Food Sci.* **76**

[7] M P, CM B, B R, A S and J O 1999 On-farm quality assurance programs: a survey of producer and industry leader opinions *J. Dairy Sci.* **82** 2224–30

[8] Mudgil D and Barak S 2013 Synthetic milk: a threat to Indian dairy industry. Carpathian J *Food Sci Technol* **5** 64-8.

[9] Renny E F, Daniel D K, Krastanov A I, Zachariah C A and Elizabeth R 2014 Enzyme Based Sensor for Detection of Urea in Milk *http://mc.manuscriptcentral.com/tbeq* **19** 198–201

[10] Singh M, Verma N, Garg A K and Redhu N 2008 Urea biosensors *Sensors Actuators, B Chem.* **134** 345–51

[11] Das S, Goswami B and Biswas K 2016 Milk Adulteration and Detection: A Review *Sens. Lett.* **14** 4–18

[12] Paradkar M M, Singhal R S and Kulkarni P R 2000 An approach to the detection of synthetic milk in dairy milk: 2. Detection of detergents *Int. J. Dairy Technol.* **53** 92–5

[13] S K, W T and A K 2007 Feasibility of near-infrared spectroscopy to detect and to quantify adulterants in cow milk *Anal. Sci.* **23** 907–10

[14] Yu H, Wang J and Xu Y 2007 Identification of Adulterated Milk Using Electronic Nose *Sensors Mater.* **19** 275–85

[15] Mabrook M F and Petty M C 2003 A novel technique for the detection of added water to full fat milk using single frequency admittance measurements *Sensors Actuators, B Chem.* **96** 215–8

[16] Mabrook M F and Petty M C 2002 Application of electrical admittance measurements to the quality control of milk *Sensors Actuators B Chem.* **84** 136–41

[17] Büttel B, Fuchs M and Holz B 2008 Freezing point osmometry of milk to determine the additional water content - An issue in general quality control and German food regulation *Chem. Cent. J.* **2**

[18] Shabir Barham G 2014 Detection and Extent of Extraneous Water and Adulteration in Milk Consumed at Hyderabad, Pakistan *J. Food Nutr. Sci.* **2** 47

[19] Singuluri H 2014 Milk Adulteration in Hyderabad, India – A Comparative Study on the Levels of Different Adulterants Present in Milk *J. Chromatogr. Sep. Tech.* **05**

[20] Shaikh N, Soomro A H, Sheikh S A and Khaskheli M 2013 Extent of water adulteration and its influence on physical characteristics of market milk *Pakistan J. Nutr.* **12** 178–81

[21] Indumathi J and Obula Reddy B 2015. Quality evaluation of milk samples collected from different intermediaries at the vicinity of Chittoor district, Andhrapradesh, India *Int. J. Curr. Adv. Res.* **4** 436-40.

[22] Azad T and Ahmed S 2016 Common milk adulteration and their detection techniques *Int. J. Food Contam.* **3**

[23] A S, P M and IA K 2006 Determining the adulteration of natural milk with synthetic milk using ac conductance measurement *J. Food Eng.* **77** 472–7

[24] Ganapathy A, Pandey N, Srisailapathy CS, Jalvi R, Malhotra V, Venkatappa M, Chatterjee A, Sharma, M, Santhanam R, Chadha S and Ramesh A **2014** Non-syndromic hearing impairment in India: high allelic heterogeneity among mutations in TMPRSS3, TMC1, USHIC, CDH23 and TMIE. *PloS one* **9** 84773

[25] Palsra T, Khurana S K and Mane B G Milk Adulteration and Its Impact on Consumer Health: A Review *J. Food Res. Technol.*

[26] SM H, R R, SD D, A S, NS G and SS H 2017 In vivo Comparative Evaluation of Mineral Trioxide Aggregate and Formocresol Pulpotomy in Primary Molars: A 60-month Follow-up Study *Contemp. Clin. Dent.* **8** 122–7

[27] Jones J M B 1957 The Chronology of the "Maghāzī"-- A Textual Survey *Bull. Sch. Orient. African Stud. Univ. London* **19** 245–6

[28] Biswas K, Sen S and Dutta PK 2006. A constant phase element sensor for monitoring microbial growth. *Sens. Actuators B Chem.* **119** 186-91.

[29] 2018 *National Milk Safety and Quality Survey Executive Summary* Standards Authority of India (FSSAI)

A Mini Review on Application of Boron Neutron Capture Therapy in Cancer Treatment

Shruti Namdev[1], Gurmeet Kaur*[1]

[1]Chemistry Department (UIS), Chandigarh University, Gharuan, Mohali - 140413

Email ID- shrutinamdev4@gmail.com, gurmeet.uis@cumail.in

Abstract. Nowadays, Cancer is undoubtedly a consequential and potentially life-threatening illness. In the U.S., Around 1.7 million people discerned with Cancer last year. The application of boron in cancer treatment is applicable as therapy known as *Boron Neutron Capture Therapy (BNCT)*. It is an analeptic technique that depends on the nuclear capture and fission reaction that results in various particles such as Li, He nuclei with their kinetic energy and ɤ-radiation. These radiations kill the malignant cells in our body without damaging the normal tissue. Boron particles delivered selectively to only malignant cells by minimizing their concentration in normal cells.
In this article, The clinical trials, clinical investigation in Different countries with a different type of Cancer as Lung, Brain, Head/Neck, Hepatic and gastrointestinal. Various delivery methods of Boron agents in the tumor cells recently developed boron delivery agents, and different techniques of their dose distribution explain. In recent years, boron compounds were applied with porphyrin, copolymers, nanoparticles, other peptides, EGRF's, and Liposomes to intensify their killer properties toward target cancer cells. BNCT is also effectively used for the medication of various kinds of Cancers explained. The purpose of this article is to indicate the intelligible way of BNCT(Reactor-Based) for the treatment of malignant cells.
Keywords: *Malignant cell, Analeptic, Neutron capture therapy, Boron in cancer medication*

1. INTROODUCTION

Boron Neutron Capture Therapy (BNCT) has an inventive form of radiotherapy that has the potential to treat various types of cancers. It depends upon the nuclear capture and fission reaction in which the boron's stable isotope B-10 irradiated with the thermal neutron of low energy (0.002eV) [1] or epithermal neutron of high energy (10,000eV) Figure 1. The reaction results in the formation of boron B-11 isotope, which is unstable to give a high value of linear transfer energy (LET) of α-particles ($^4He_2$) with recoiling lithium nuclei (Li^+) and γ-radiation. These radiations destruct the malignant cells selectively in our body that contains energetic and cytotoxic ions [2].
It liberates a high value of LET energy and the timid path length of α-radiation with a range of 4-10μm. The extensive amount of energy released in this reaction dispatch in a limited volume, approximately one cell diameter (0.01mm) [3]. So, these confine radiations only damaged the cells from which they arise without affecting those normal cells. The two main difficulties are recognized to carry out this therapy, the sufficient amount of boron atom (thousand) deposited in the cell and the balanced amount of neutron release per second (tens of millions) to target the tumor cell.

$$^{10}B + {}^{1}n \longrightarrow [^{11}B] \longrightarrow \begin{cases} ^{4}He + {}^{7}Li + 2.79\text{ MeV } (6\%) \\ ^{4}He + {}^{7}Li + 0.48\text{ MeV } + 2.31\text{ MeV } (94\%) \text{ (gamma rays)} \end{cases}$$

Figure 1. Reaction of B-10 neutron to form Li and He nuclei [5]

Specifically, Boron utilizes as an anticancer agent in neutron capture therapy (NCT) due to its high nuclear cross-section value with a epithermal or thermal neutron, i.e., 3837 B (1B=10-24cm^2) [4]. The term nuclear cross-section defines as the ability of atomic nuclei to interact with neutrons that depends upon the actual structure of the Nuclei. The following table shows the Nuclear cross-section value of different atoms concerning their abundance. The Boron atom proves the highest Barn value with sufficient presence in the earth's crust.

Table 1. Different Elements' cross-section values concerning their abundance in our body[1][4]

Serial No.	Element	Abundance (%)	Nuclear cross section value (Barn)
1.	$^{14}N_7$	99.6	1.8
2.	$^{12}C_6$	98.9	3.4x10^{-3}
3.	$^{16}O_8$	99.8	1.8x10^{-4}
4.	$^{1}H_1$	99.8	0.33
5.	$^{10}B_5$	99.8	3.8x10^{3}
6.	$^{32}S_{16}$	95.0	0.53
7.	$^{4}He_2$	1.4x10^{-4}	5.3x10^{3}
8.	$^{157}Gd_{64}$	14.8	6.1x10^{3}

Table 1 also represents that Gd-157 is also suitable for nuclear capture reactions. But the application of Boron is wider than the Gd-157 isotope due to its less abundance. The main challenge is to achieve the desired effect of BNCT is that the delivery of the boron compound would be selectively within doomed cells and its concentration also adjust in our body. The study reported that the acceptance concentration of Boron (B-10) in malignant cells is 10μ to 30μ per gram tumor as it depends upon the position of B-10 mass in tumor cell structure [6]. It also depends on the location of boron nuclei for its drug dosage. If the isotope of B-10 Nuclei is bound to the cells' external membrane, a 30ppm concentration is required [5]. Otherwise, If B-10 nuclei located within the cells' cytoplasm having 10-12 ppm concentration requires only [7].

Around 60 years, BNCT was performed with the help of two delivery agents such as (BPA) boronophe-nylalanine or a boron-containing amino acid (L)- 4-dihydroxy-borylphenylalanine and BSH sodium mercaptoundecahydro-closo-dodecaborate ($Na_2B_{12}H_{11}SH$) or sodium borocap- tate. Here, BPA is generally used to treat high-grade gliomas patients in Figure 2, and BSH is applicable for a wide range of cancer such as cutaneous melanomas, 9L gliosarcoma, etc. In recent years, to enhance the versatility of BNCT, several delivery agents were introduced as boron derivatives, boron compound fusion with nanoparticles, peptide, amino acid, polymers, etc [8].

Besides this, several new plans are introduced for the future development of BNCT. The researcher investigated that phenylboronic acid (PBA) performed the dual function (i.e., targeting effect and also contain the capacity to capture neutron) and formed with the fusion of nanoparticle, polymer. Further, this supramolecular structure act as a sialic acid-targeting agent to treat the B16 melanoma [9].

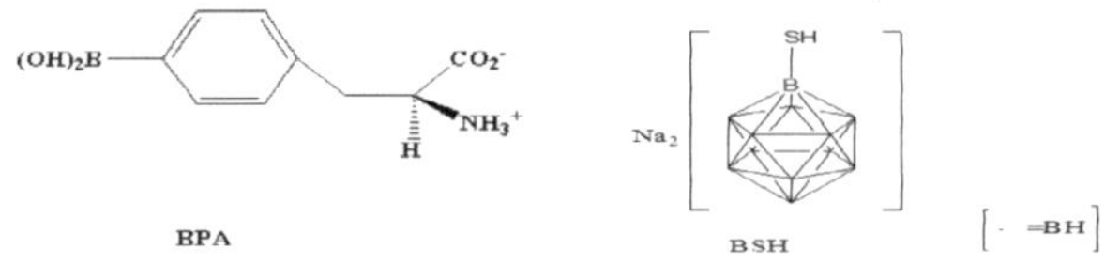

(A) BPA

(B) BSH (polyhedran edges was occupied by BH except upper central edge)

Figure 2. Main Boron delivery Agents [5]

Transition metal carboranes that are based on photosensitizer were internalized in the SKBR-3 cells. Mono-carborane-phenanthroline complexes show low toxicity under photodynamic conditions. Thus, these complexes can be used for BNCT as anti-cancer agents due to the high content of the boron atom. In this paper, recent development in the boron anti-cancer agents for BNCT are explained with their clinical trials and clinical investigation [10].

2. HISTORY

In 1954, This therapy was introduced to 40 patients by using a simple inorganic boron compound. But their result gives severe side effects to the treated patient, such as scalp's radio dermatoses, cerebral oedema. The two boron compounds are introduced as BPA (p-boronophenylalanine) and BSH (di-sodium mercaptoundecahydro-closo-dodecaborate) to remove these side effects. These two delivery agents are used to treat severe diseases by targeting the damaged cell with the nuclear reactor at BNL and MIT [10]. Malignant melanoma was treated by using BPA in Japan in 1987.

There are a lot of clinical trials and clinical investigations performed for BNCT in Finland [11], Sweden [12], Argentin [13], Japan [14]. This technique attracted the Researchers as the epithermal neutron is targeted the dead cell effectively. This therapy comes to an end due to a lack of facilities and equipment in hospitals and research centers. The presence of only two delivery agents for clinical trials also led to a drawback for BNCT. Remarkable progress was seen in this field in recent years that gives rise to the BNCT's future. A short detail represents in the following (Figure 2) throughout the history of BNCT [4], [7]

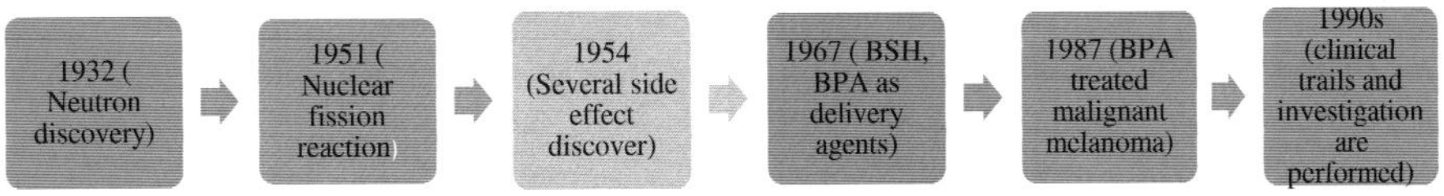

Figure 3. Historical Presentation of BNCT [8], [4]

3. CLINICAL TRAILS

The capability of BNCT enhances with Several clinical trials contains new drugs. Further, these drugs were approved by the authority of national regularity for treatment purposes after a three-phase clinical trial. The first one is too concerned with the toxicity of the drug combination. The next one is to evaluate the effectiveness of the BNCT treatment, its dosage rate, its anti-tumor effect [12], and the third one leads to evaluate the therapeutic effect of the BNCT treatment on a large number of patients. These clinical trials would be held in various areas such as Finland, Japan, Taiwan. Here, Boron therapy proves by treating the maximum number of patients in glioblastoma [15], brain[16] and Head/Neck Cancer [17], Recurrent glioblastoma multiforme (GBM)[18], Malignant meningioma[15]. The most important use is that the damaged capability of normal cells and its side effect on the other parts of the cell are less. It is explained in the given (Table 2.) the clinical trial for different diseases in various areas.

Table 2. Clinical Trails of BNCT in Different Countries

S.No.	Place	Disease (number of patients)	Delivery Agent	Result	Reference
1.	Finland	**Head and neck tumor :** (inoperable's 30 patients, SCC'S 24 patient, T4 's 14 patient)	^{18}F BPA	Median progressive-free survival (PFS) was 7.5 months in 27% patient had no clue of HN cancer within 2 years	[11], [17]
		Recurrent malignant glioma: (GBM): 22 patients	BPA	Median survival time (MST) is 7 months	[11]

2.	Japan	**Head and neck malignancies (HNM)** : (26 patients)	BPA, BSH	Survival periods increase as MST (33.36 months), Overall srvival (2 years)	[14],[19]
		Newly diagnosed glioblastoma : (15 patients)	BPA, BSH	BNCT show postive result with mean and median survival rate of 20.7 and 15.6 months respectively	[18]
		Malignant meningioma (19 patients)	BPA-PET	18/19 pateients shows postive result with survival period of 14.1 months.	[14]
3.	Taiwan	**Head and neck region : (10 patients)**	BPA-fructose	MST of 11.3 months , it cure 3 patient fully, 3 partially anf other are in stable disease.	[20]

4. CLINICAL INVESTIGATIONOF BNCT:

During medical research, Scientists were concern whether BNCT is promising for a different disease or not. Several investigations were performed to evaluate the versatility of BNCT. For this purpose, the Japanese medical framework performs BNCT for the treatment of different diseases. With their recent work, it has been found that BNCT exploits the medication of many types of cancer such as hepatic cancer [21], lung cancer [22], Recurrent gastrointestinal cancer [23].
In the clinical investigation of BNCT, it concludes that the treatment of a particular disease in BNCT required a specific technique and doses as described in (Figure 3)

4.1. Treatment for Hepatic Cancer:

For the recurrent hepatic cancer investigation, Yanagie synthesized a technique in which BSH (B-10 isotope) contains WOW (water-in-oil-in-water) emulsion. It is applicable to treat recurrent hepatic cancer as previously Epitubic that containing WOW emulsion in hepatic arteries chemotherapy for patient treatment[24]. Here, SERA is using as Treatment planning has an equivalent dose of 12.044 Gy for mean tumor, 19.0 Gy for maximum, and 5.044 Gy Doses for normal mucosa uses in 56 minutes irradiation. The result shows that BSH -WOW emulsion is applicable for an intraarterial boron carrier in recurrent hepatic cancer with no side effects [21].

4.2. Treatment for Gastrointestinal Cancer:

In gastrointestinal cancer, BPA can decrease the tumor size when a severe lymph node is observed. It gets a high accumulation of 10B with left metastasize. After, the surgery it regrew again. But, in this case, we use the 18F-BPA as a delivery agent, and SERA is used as treatment planning and represents a comparable dose of 20 Gy in more than 4:5 ratios of the tumor. Its highest Dose is delivered 5.6 Gy in the ordinary case. The result shows a high suppression in tumor growth after two months of BNCT treatment [15].

4.3. Treatment for Lung Cancer:

In 2006, the Research Team of the Kyoto University investigated all the viability of BNCT for doomed pleural mesothelioma. The results are in favor of BNCT without any side effects. The Neutron source energy proposed for the treatment of lung cancer in BNCT. The oncology Institute Roff (University of Buenos Aires) and the University of Pavia Collaborated with R. Farias from the commission Nacional de Energia Atomica. During BNCT, the Dose distribution in lung cancer treatment effected by the respiratory system [25]. The Change in tumor location affects the neutron flux distribution where the tumor locates. It may also affect the patient dose distribution in BNCT. In 2018, Lee et al. Verified the dose distribution result from the different amplitude of motion with tumor movement. The difference between the end-inhaled and ex-haled phases represents the mean tumor dose distribution difference of the virtual respiratory patient model. The radial direction of the neutron beam causes the dose change by tumor movement [26]. To solve this, researchers evaluated the technique, and LDL is also used as a delivery agent. Throughout the respiratory cycle, the anatomy of a patient represents by six phases at different moments. The Monte-Carlo tool kit exploits to calculate the tumor-related dose-volume histogram (DVH) indices at different Phases. Under the treatment planning Configuration, the movement in the anterior-posterior (AP) direction caused the Dose to change intensively with the increase of respiratory motion amplitude. the dose difference ranges from -13.8% to15.8% increased [27].

An innovative approach for lung cancer is that a low-Density Lipoprotein (LDL) as a carrier used to increase the selective update of tumor cells, and magnetic resonance imaging (MRI) could quantify the boron distribution.

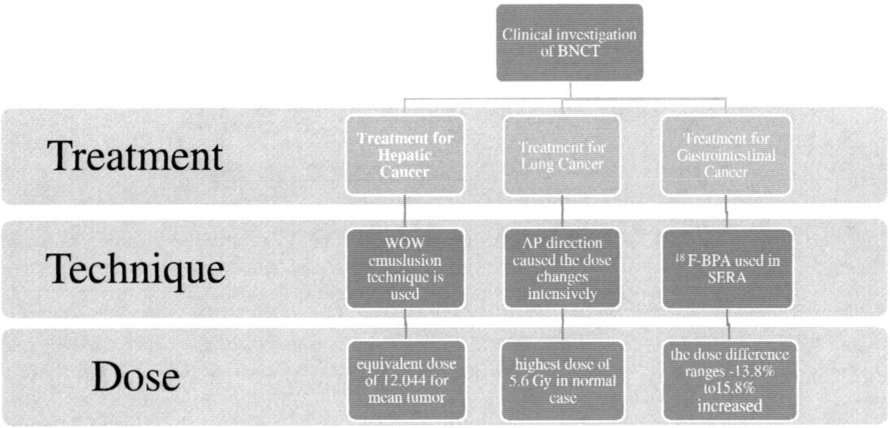

Figure 4. Clinical Investigation on Treatment of Cancer with Technique and Doses [21],[25] ,[12]

5. OTHER BORON DELIVERY AGENTS

Several new boron delivery agents have come to enhance this synthetic technique and raised several necessary biochemical properties. The main challenge to synthesize a new delivery agent is the specification of tumor cell killing and the process of delivering the drug with a minimum concentration in undoomed tissue [8]. For the synthesizes of the delivery agents, the boron compound undergoes fusion with nanoparticles, biomaterials (lipids, peptides, carborane, etc.). It contains three main parameters such as boron presence in the tumor should be 20-35 mg, the ratio of tumor to normal tissue should be more than 3 with less toxicity. The molecular weight should be at a particular range for the synthesis of new delivery agents because Glioma cells have a high infiltration capability. Thus, the blood-brain barrier (BBB) and their genomic heterogeneity are responsible for excluding the molecular weight agents greater than 200Da [28][29]. Tumor targeting moieties use to enhance the selectivity of boron delivery agents.

Many boron anti-cancer agents are currently under evaluation with lower and higher molecular weight. Boric acid, Boron containing immuno-liposomes, Boron containing Lipiodol, Boronated Co-polymers, Boron nitride nanotubes Boronated cyclic peptides, Boronated DNA intercalators, Boronated EGF/anti- EGFR monoclonal antibodies, Boronated polyamine, Boronated porphyries, Boronated sugar, Boronated unnatural amino acid, Boronated containing thymidine, analogs Decacaborane (GB10), Dodecahydro-closo-dodecaborate clusters, linear and cyclic peptides, Transferrin-polyethylene glycol liposomes, and Polyanionic polymers are some examples of boron delivery agents [29].

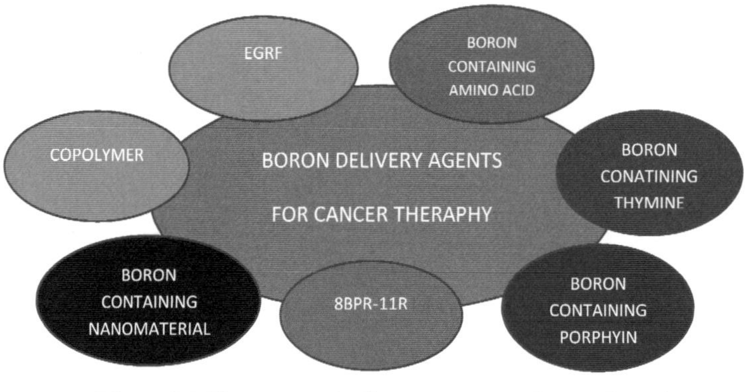

Figure 5. Different Boron Delivery Agents For Cancer Theraphy

5.1. Boron containing amino acid
Boron-containing unnatural amino acids have high metabolic stability as compared to natural ones. For example, (ABCHC)1-aminocyclobutane-1-carboxylic acid, the 1-amino-3-boronocyclo-pentane carboxylic acid (ABCPC) [30]. Dodecaboratethio-l-amino acid has been synthesis by S- alkylation from Bromo-l-α-amino acids and (*S*-cyanoethyl-10BSH, [10B12H11]2−SCH2CH2CN) *S*-cyanoethylthioundecahydro-*closo*-dodecaborate [31]. Boron-containing amino acids contain non-immunogenic properties, are easy to synthesize, are low Noxious, and have high tissue pervasive properties. The boron-containing peptides and linear conjugated bonds with boron captate have been studied. Phenylboronic acid-enhanced its functionalities, i.e., containing a capacity to capture a neutron toward itself and highly erect the activated BNCT drugs and kill the malignant cells within a particular range [32]

5.2. Boron containing Thymidine
3-Carboranyl thymidine analogs (3CTAs) are compounds of Boron-containing purines, Pyrimidine, thymidine. It uses for targeted thymidine kinase-1(TK1) especially. For example, N5-2OH represents the selective tumor uptake due to its low toxicity and high phosphorylation. (CED) Convection enhanced delivery which delivered the therapeutic agents to the brain directly and bypassing the Blood-brain barrier (BBB) completely [33].
Rats bearing RG2 gliomas treated by CED of N5-20H show an effective result. In F98 glioma, N5-20H is not promising as a boron delivery agent by producing a modest increase in MST [34].

5.3. *Boron containing porphyrin*
BOPP, Cu TCPH, and H2OCP are boron-containing porphyrins that investigate due to their low Noxious, natural empathy for tumors. Octa-anionic 5,10,15,20-tetra[3,5-(Nido-carboranylmethyl)phenyl] porphyrin (H2OCP) accelerated within the cells to a greater extent than BSH/BPA, and the median survival time with H2OCP treated patients is greater than the untreated patients. They show similar characteristics as BPA and localize in macrophages rather than the tumor cells. β-carboranyl porphyrins show great potential in the treatment of BNCT because it contains (32-43%) boron content. It concluded that boron-containing porphyrin use as boron drug delivery agents by specifically localized the compound within the tumor cell rather than macrophages [35].

5.4. Cell membrane penetrating peptide
Amino acid Transporters are also proposed to kill the malignant cells as the cell membrane permits it to pass through the tissue. But the cell membrane doesn't permit BSH to pass through the tissue. At the distinct location of Boron, the energy transfer simulated and showed a different mark between outside the cell and the inner nucleus [28]. This drawback of BSH in BNCT is deduced by using a cell

pervasive peptide system for transduction. Here, CPP (Cell membrane penetrating peptide) is a simple peptide contains various biological properties that transduces Biological active substance into cells or tissue by fusing with BSH. For example-8BSH fuse with 11R contains a dendritic lysine structure localized in the cell and shows as a killing agent to the doomed cell with neutron irradiation. Its effect is 100 times better than the high concentration dose of BSH. These results show its promise value for another stage trial of BNCT [7].

5.5. Boron containing Nanomaterials

In Biomedical sciences, Nanomaterials play a notable role in the medication of severe disease because the fusion of the nanoparticles with the organic or inorganic compound gives rise to explore innovative techniques. Here, Boron compounds get composed with the nanoparticles to form several delivery agents such as nanotubes, nanofibers, nanoparticles, Nanocapsules, etc. Several biological processes get influenced by the morphology of nanomaterials such as Cellular blood uptake, Blood circulation time, and the Biodistribution of drugs [36].

5.5.1. Boron nitride nanotubes: For the diagnostic purpose and drug delivery, Boron nitride (BN) shows a remarkable property such as a high value of thermal neutron capture cross-section of BN. It can be sufficient for interaction and absorption of B10 enriched atom with the thermal neutron. At Kyoto University, the Research reactor of Boron nitride was synthesized using an effective solvothermal method with the highest purity distorted electron cloud [37]. It forms by the distorted Boron nitride that forms the structural material becomes water dispersal and partial polar. Because of these properties, the formed nanosheets are less Noxious for the aggregation of Boron with cells in proper quantity for neutron irradiation and can easily replace the L-BPA. Effective anti-tumor neutron irradiation of 10B in 10BN is possible due to the high-water dispensability. 10BN can be used promisingly as a delivery agent for the medication of the doomed cell in BNCT [38]. The boron nitride nanotubes in which DSPE-PEG2000 are used to prepare BNNT's in a suspended aqueous solution. It shows an anti-tumor effect against B16 melanoma cells, and the treated rate is three times higher than the BSH [39].

5.5.2. Liposomes: BNCT could make more effective with a greater concentration of Boron atoms in cancer cells. Boron-containing liposomes that have low bioavailability and poor tumor retention are used to enhance the efficiency of BNCT which also leads to the law of Therapeutic efficiency.

Polyborane has been synthesized from decaborane that is implanted in bare and pegylated liposomes. These liposomes contain a large amount of boron atom with a hydrophobic property and highly accumulated delivery agent. In the diameter range of 35-40 nm, these liposomes get Biologically distribution in the body after ejecting into tumor-bearing mice. A Certain condition is marked for the whole process as the tumor/boron ratio should be exceeding 5, and the concentration of boron in the tumor should be 30 μg/g of tissue. It has been found PEGylated liposomes having a diameter of 40nm were capable to achieve sufficient results without exchanging the 11B with 10B [40].

Incorporating active drugs of polymeric nanoparticles and PS offered the way to deliver drugs to the tumor selectivity due to its high permeability and retention effect. But during the blood circulation, the leakage of incorporative Drugs from the nanoparticles takes place to deduce this hydrophilic drug that loads in a large amount into phospholipid-based liposomes [29]. It improves bioavailability in the body and blood circulation based on selective tumor uptake. Folic acid enhances the functionalization of NPS. Water in oil strategy uses to create liposomes to increase the bioavailability of Boron. Folic acid is conjugated equally on two amino-functionalized PEG to produce the folate functionalized PEG polyethylene glycol [40],[41]. The water in oil emulsion-based liposomes containing poly (maleic anhydride-alt-1-octadecane) (PMAO) and PEG on the upper face. The Boron nanoparticles and the fluorescent dine cy5 near-IR region in the Core. The average diameter of these liposomes is about 100 to 120 nm, and the Zeta potential of liposomes -38.0 +-1.5 mV are formed. Fluorescence microscopy monitored the cellular uptake for folic acid that is conjugated with the liposome's capability. It also is noted that the aggregation of FA-conjugated liposomes was much more effective than non-effect conjugated liposomes in C6 brain tumor cells under the same conditions [42]. This boron delivery agent represents less toxicity and high stability under physiological conditions and the blood-brain barrier. A significant concentration of intracellular Boron gets aggregated with the cancer cells. It contains four therapeutic benefits from BNCT that are confirmed by inductively coupled plasma mass spectroscopy (ICP-MS) [3].

5.5.3. *EGFR:* Some Special antibodies are used as a molecular target of EGFR/EGFRviii and EGF ligands due to their high specificity for the medication of F98 rat glioma. Monoclonal antibodies (MoABs) and L8A4 can identify their type in which it presents a class of tumor-targeting agents. In clinical research of Barth, Wu, Yang, and their coworkers propose that five dendritic generations use a heavily Boronated precision dendrimer. In this way, these are associates with using heterobifunctional agents to targeting the vehicles [43]. For example, EGFR chooses to target MoAbs, cetuximab, and EGFRviii targets MoAbs LSA4. These MoAbs give a promising result in F98EGFR glioma when these bioconjugates combine with BBA administration. But, the F98 wild-type tumor is ineffective as they can't express the amplified EGRF [44]. Here the use of a combination of two drugs gives better results for the treatment of BNCT.

5.5.4. *Carbon nanotubes: Carbon* nanotubes are used in drug delivery due to their many pharmaceutical applications, highly developed cellular uptake, and large surface area, and they can also functionalize with other bioactive compounds. It shows effective results as a anti-cancer agent in BNCT. Due to its low toxicity and biocompatibility, Carbon nanotubes can also be used for treatment purposes in BNCT [45].

The specification for the delivery of CNT's in targeted tumors is that Within 48 hours after administration, a hyper concentration of CNT's compound 21.5 UG per gram tumor approached with high selectivity of greater than 3:1 of the tumor to blood ratio in patient body [45]. The result shows a better quality of water solubility and High retention in tumor tissue than normal tissue. By making a little associate of (SWCNT-C(CO)-[O(CH2)2]2-O-metallaborane], a single world carbon nanotube functionalized with Cobalt polaroid and graphene-oxide-close-do-deca- carbonate non-hydrides [30]. When these single-walled carbon nanotubes (SWCNT's) hybrid with dendrimer (SB12)4 contain the fluorescence of NIR-I/NIR-II [46]. That will prove boron clusters can be extensively usable in BNCT.

5.6. Copolymers

With the occurrence of Nanomaterials as a drug carrier, this technique faces a problem in penetrating the tissue within the cell membrane. Thus, to solve this problem Polymers, copolymers are used for the treatment process in BNCT. These copolymers help to penetrate the tissue deeply with homogeneous delivery of 10-B atom specifically. Based on the clinical uses of sodium Boro-captate (BSH), block copolymers conjugated with Boron cluster such as poly(ethylene -glycol)-b-poly(glutamine acid) copolymers[47]. The Boron cluster was synthesized to deliver into the tumor cell selectively without affecting the normal surrounding cells that also penetrate the tumor tissue. Higher cellular uptake and tumor accretion are biological functions. They were demonstrated by these PEGylated block copolymers Boron cluster (BSH) conjugate. These conjugates reached the intracellular space in the tumor due to the selective delivery with their superior intratumorally penetration. Block copolymers contain the Boron cluster has the potential to damage the tumor cell [48].

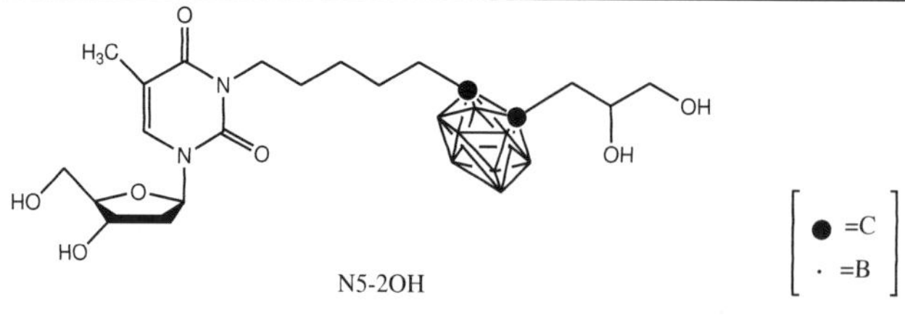

N5-2OH

A) N5-2OH (3-[5-{2-(2,3-dihydroxyprop-1-yl)-o-carboran-1-yl} pentan-1-yl] thymidine) is a (CAT) carboranyl thymidine analogue that showed promising results in the RG2 [33]. (All the atom in the polyhedron are occcupied by boron atom exceprt the two dark dots)

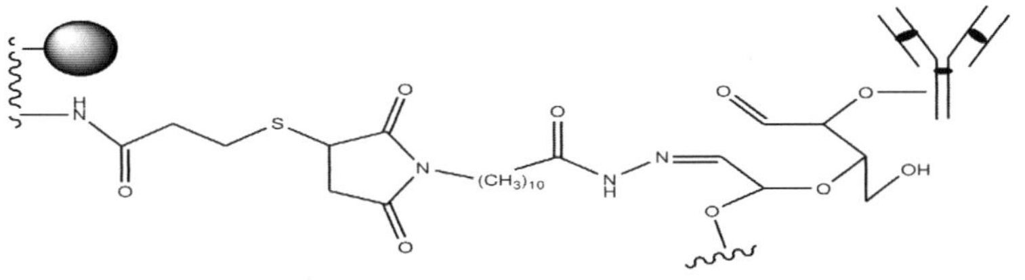

EGFR- targeting MoAb, C225-G5-B_{1100}

B) C225-G5-B1100 is a monoclonal antibody cetuximab that contain heavily boronated atoms (Further attached by Linker) have minor groove binder and targets the EGFR (epidermal growth factor receptor), which has been used for the F98EGFR rat glioma treatment in BNCT [45].

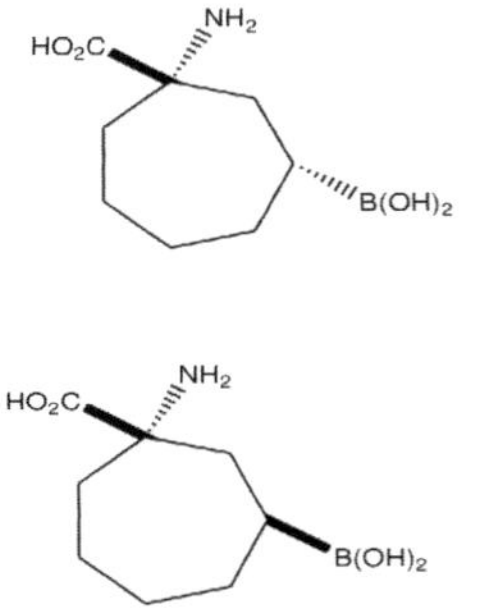

Trans-ABCHC and cis-ABCHC

C) cis-ABCHC and trans-ABCHC (1-amino -3-borono-cycloheptanecarboxylic acid) is a racemic mixture of amino acids that is Compatible to BPA in the B16 melanoma [30]

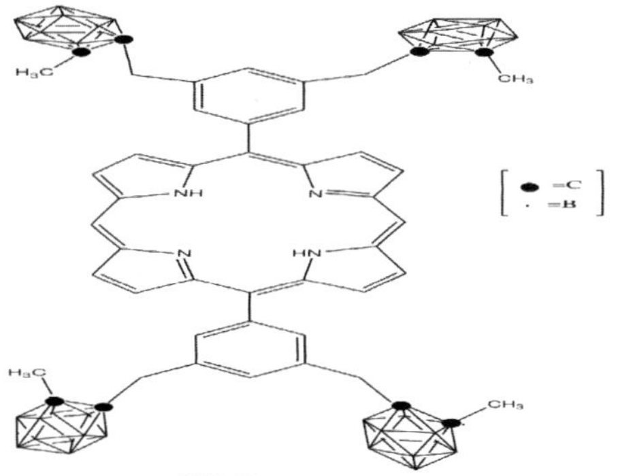

H_2DCP

D) H2-DCP (di [3,5-(nido-carboranylphenyl) tetra-benzoporphyrin]) is carboranyl porphyrins boron conjugate, which gives large in vitro cellular uptake [35]. (All the atom in the polyhedron are occcupied by boron atom exceprt the two dark dots)

Figure 6. Different Boron agent Conjugates Used in BNCT

6. Conclusion

In this article, we have discussed the BNCT of Clinical trials for the treatment of therapeutic malignant cells in the human body. These Trails indicates positive results for Head/Neck and Brain tumor. The Clinical investigation shows promising outcomes for hepatic, Head/Neck, and Gastrointestinal Cancer through different techniques such as WOW emulsion, extra. That indicates that BNCT is applicable to attract the people of the Twenty Century for a reactor-based treatment. In this discussion, new types of Boron delivery agents give a clinical approach for the Evaluation. Other than simple boron compounds, Liposomes, Caron nanotubes are also introducing the best aspects for BNCT. In last, BNCT proves the best treatment therapy for various types of Cancers. But, the complication links with this process due to lack of equipment, the unreliability of neutron source accelerator. The future for the therapeutic agents of BNCT clear with a variety of boron compounds with sources of neutron-based accelerators.

References

[1] Allen D A and Beynon T D 1995. A design study for an accelerator-based epithermal neutron beam for BNCT *Phys. Med. Biol.* **40** 807

[2] Moss R L 2014 Critical review, with an optimistic outlook, on Boron Neutron Capture Therapy (BNCT) *Appl. Radiat. Isot.* **88** 2–11

[3] Perl J, Shin J, Schümann J, Faddegon B and Paganetti H 2012 TOPAS: An innovative proton Monte Carlo platform for research and clinical applications *Med. Phys.* **39** 6818–37

[4] Soloway A H, Tjarks W, Barnum B A, Rong F G, Barth R F, Codogni I M and Wilson J G 1998 The Chemistry of Neutron Capture Therapy *Chem. Rev.* **98** 2389-2390.

[5] Wittig A, Michel J, Moss R L, Stecher-Rasmussen F, Arlinghaus H F, Bendel P, Mauri P L, Altieri S, Hilger R, Salvadori P A, Menichetti L, Zamenhof R and Sauerwein W A G 2008 Boron analysis and boron imaging in biological materials for Boron Neutron Capture Therapy (BNCT) *Crit. Rev. Oncol. Hematol.* **68** 66–90

[6] Barth R F, Mi P and Yang W 2018 Boron delivery agents for neutron capture therapy of cancer *Cancer Commun.* 1–15

[7] Roesch S, Fermi V, Rominger F, Herold-Mende C and Romero-Nieto C 2020 Correction: Gold (I) complexes based on six-membered phosphorus heterocycles as bio-active molecules against brain cancer *Chem. Comm.* **56**15088-15088.

[8] Hawthorne M F and Lee M W 2003 A critical assessment of boron target compounds for boron neutron capture therapy *J. Neurooncol.* **62** 33–45

[9] Kaur M, Singh P, Singh K, Gaharwar U S, Meena R, Kumar M, Nakagawa F, Wu S, Suzuki M, Nakamura H and Kumar A 2020 Boron nitride (10BN) a prospective material for treatment of cancer by boron neutron capture therapy (BNCT) *Mater. Lett.* **259** 126832

[10] Conway-Kenny R, Ferrer-Ugalde A, Careta O, Cui X, Zhao J, Nogués C, Núñez R, Cabrera-González J and Draper S M 2021 Ru(ii) and Ir(iii) phenanthroline-based photosensitisers bearing o -carborane: PDT agents with boron carriers for potential BNCT *Biomater. Sci.* **9** 5691–702

[11] Savolainen S, Kortesniemi M, Timonen M, Reijonen V, Kuusela L, Uusi-Simola J, Salli E, Koivunoro H, Seppälä T, Lönnroth N, Välimäki P, Hyvönen H, Kotiluoto P, Serén T, Kuronen A, Heikkinen S, Kosunen A and Auterinen I 2013 Boron neutron capture therapy (BNCT) in Finland: Technological and physical prospects after 20 years of experiences *Phys. Medica* **29** 233–48

[12] Capala J, Stenstam B H, Sköld K, Rosenschöld P M, Giusti V, Persson C, Wallin E, Brun A, Franzen L, Carlsson J, Salford L, Ceberg C, Persson B, Pellettieri L and Henriksson R 2003 Boron neutron capture therapy for glioblastoma multiforme: Clinical studies in Sweden *J. Neurooncol.* **62** 135–44

[13] González S J, Bonomi M R, Santa Cruz G A S, Blaumann H R, Larrieu O A C, Menéndez P, Rebagliati R J, Longhino J, Feld D B, Dagrosa M A, Argerich C, Castiglia S G, Batistoni D A, Liberman S J and Roth B M C 2004 First BNCT treatment of a skin melanoma in Argentina: Dosimetric analysis and clinical outcome *Appl. Radiat. Isot.* **61** 1101–5

[14] Nakagawa Y, I K P, Kobayashi T, Kageji T, Uyama S, Matsumura A and Kumada H 2003 Clinical review of the Japanese experience with boron neutron capture therapy and a proposed strategy using epithermal neutron beams *J. Neurooncol.* **62** 87–99

[15] Kawabata S, Miyatake S, Kuroiwa T, Yokoyama K, Doi A, Iida K, Miyata S, Nonoguchi N, Michiue H, Takahashi M, Inomata T, Imahori Y, Kirihata M, Sakurai Y, Maruhashi A, Kumada H and Ono K 2009 Boron Neutron Capture Therapy for Newly Diagnosed Glioblastoma *J. Radiat. Res.* **50** 51–60

[16] Diaz A Z 2003 Assessment of the results from the phase I/II boron neutron capture therapy trials at the Brookhaven National Laboratory from a clinician's point of view *J. Neurooncol.* **62** 101–9

[17] Kankaanranta L, Seppälä T, Koivunoro H, Saarilahti K, Atula T, Collan J, Salli E, Kortesniemi, M, Uusi-Simola J, Välimäki P and Mäkitie A 2012. Boron neutron capture therapy in the treatment of locally recurred head-and-neck cancer: final analysis of a phase I/II trial *Int. J.*

Radiat. Oncol. Biol. Phys. **82** 67-75.
[18] Vos M J, Turowski B, Zanella F E, Paquis P, Siefert A, Hideghéty K, Haselsberger K, Grochulla F, Postma T J, Wittig A, Heimans J J, Slotman B J, Vandertop W P and Sauerwein W 2005 Radiologic findings in patients treated with boron neutron capture therapy for glioblastoma multiforme within EORTC trial 11961 *Int. J. Radiat. Oncol. Biol. Phys.* **61** 392–9
[19] Leoncini E, Ricciardi W, Cadoni G, Arzani D, Petrelli L, Paludetti G, Brennan P, Luce D, Stucker I, Matsuo K, Talamini R, La Vecchia C, Olshan A F, Winn D M, Herrero R, Franceschi S, Castellsague X, Muscat J, Morgenstern H, Zhang Z F, Levi F, Dal Maso L, Kelsey K, McClean M, Vaughan T L, Lazarus P, Purdue M P, Hayes R B, Chen C, Schwartz S M, Shangina O, Koifman S, Ahrens W, Matos E, Lagiou P, Lissowska J, Szeszenia-Dabrowska N, Fernandez L, Menezes A, Agudo A, Daudt A W, Richiardi L, Kjaerheim K, Mates D, Betka J, Yu G P, Schantz S, Simonato L, Brenner H, Conway D I, Macfarlane T V., Thomson P, Fabianova E, Znaor A, Rudnai P, Healy C, Boffetta P, Chuang S C, Lee Y C, Hashibe M and Boccia S 2014 Adult height and head and neck cancer: A pooled analysis within the INHANCE Consortium *Head Neck* **36** 1391
[20] Barth R F, Vicente M G H, Harling O K, Iii W S K, Riley K J, Binns P J and Wagner F M 2012 Current status of boron neutron capture therapy of high grade gliomas and recurrent head and neck cancer *Radiat. Oncol.* **7** 1–21
[21] Yanagie H, Higashi S, Seguchi K, Ikushima I, Fujihara M, Nonaka Y, Oyama K, Maruyama S, Hatae R, Suzuki M, Masunaga S ichiro, Kinashi T, Sakurai Y, Tanaka H, Kondo N, Narabayashi M, Kajiyama T, Maruhashi A, Ono K, Nakajima J, Ono M, Takahashi H and Eriguchi M 2014 Pilot clinical study of boron neutron capture therapy for recurrent hepatic cancer involving the intra-arterial injection of a 10BSH-containing WOW emulsion *Appl. Radiat. Isot.* **88** 32–7
[22] Suzuki M, Suzuki O, Sakurai Y, Tanaka H, Kondo N, Kinashi Y, Masunaga S, Maruhashi A and Ono K 2012 Reirradiation for locally recurrent lung cancer in the chest wall with boron neutron capture therapy (BNCT) *Int. Cancer Conf. J.* **1** 235–8
[23] Farhood B, Samadian H, Ghorbani M, Zakariaee S S and Knaup C 2018 Physical, dosimetric and clinical aspects and delivery systems in neutron capture therapy *Reports Pract. Oncol. Radiother.* **23** 462–73
[24] Ikushima I, Higashi S, Ishii A, Seguchi K, Iryo Y and Yamashita Y 2012 Ultraselective transcatheter infusion of epirubicin in water-in-oil- in-water emulsion for small hepatocellular carcinoma *Brit. J. Radiol.* **85** 584-89.
[25] Erridge S C, Seppenwoolde Y, Muller S H, Van Herk M, De Jaeger K, Belderbos J S A, Boersma L J and Lebesque J V. 2003 Portal imaging to assess set-up errors, tumor motion and tumor shrinkage during conformal radiotherapy of non-small cell lung cancer *Radiother. Oncol.* **66** 75–85
[26] George R, Vedam S S, Chung T D, Ramakrishnan V and Keall P J 2005 The application of the sinusoidal model to lung cancer patient respiratory motion *Med. Phys.* **32** 2850–61
[27] Wu S, Geng C, Tang X, Bortolussi S, Han Y, Shu D, Gong C, Zhang X and Tian F 2020 Dosimetric impact of respiratory motion during boron neutron capture therapy for lung cancer *Radiat. Phys. Chem.* **168** 108527
[28] Miyabe J, Ohgaki R, Saito K, Wei L, Quan L, Jin C, Liu X, Okuda S, Nagamori S, Ohki H, Yoshino K, Inohara H and Kanai Y 2019 Boron delivery for boron neutron capture therapy targeting a cancer-upregulated oligopeptide transporter *J. Pharmacol. Sci.* **139** 215–22
[29] Thirumamagal B T S, Zhao X B, Bandyopadhyaya A K, Narayanasamy S, Johnsamuel J, Tiwari R, Golightly D W, Patel V, Jehning B T, Backer M V., Barth R F, Lee R J, Backer J M and Tjarks W 2006 Receptor-targeted liposomal delivery of boron-containing cholesterol mimics for Boron Neutron Capture Therapy (BNCT) *Bioconjug. Chem.* **17** 1141–50
[30] Futamura G, Kawabata S, Nonoguchi N, Hiramatsu R, Toho T, Tanaka H, Masunaga S I, Hattori Y, Kirihata M, Ono K, Kuroiwa T and Miyatake S I 2017 Evaluation of a novel sodium borocaptate-containing unnatural amino acid as a boron delivery agent for neutron capture therapy of the F98 rat glioma *Radiat. Oncol.* **12** 1–11
[31] Kusaka S, Hattori Y, Uehara K, Asano T, Tanimori S and Kirihata M 2011 Synthesis of optically active dodecaborate-containing l-amino acids for BNCT *Appl. Radiat. Isot.* **69** 1768–70

[32] Kim A, Suzuki M, Matsumoto Y, Fukumitsu N and Nagasaki Y 2021 Non-isotope enriched phenylboronic acid-decorated dual-functional nano-assembles for an actively targeting BNCT drug *Biomaterials* **268** 120551

[33] Al-Madhoun A S, Johnsamuel J, Barth R F, Tjarks W and Eriksson S 2004 Evaluation of human thymidine kinase 1 substrates as new candidates for boron neutron capture therapy *Cancer Res.* **64** 6280–6

[34] Barth R F, Yang W, Wu G, Swindall M, Byun Y, Narayanasamy S, Tjarks W, Tordoff K, Moeschberger M L, Eriksson S, Binns P J and Riley K J 2008 Thymidine kinase 1 as a molecular target for boron neutron capture therapy of brain tumors *Proc. Natl. Acad. Sci. U. S. A.* **105** 17493–7

[35] Hiramatsu R, Kawabata S, Miyatake S I, Kuroiwa T, Easson M W and Vicente M G H 2011 Application of a novel boronated porphyrin (H2OCP) as a dual sensitizer for both PDT and BNCT *Lasers Surg. Med.* **43** 52–8

[36] Zhu Y and Hosmane N S 2018 Nanostructured boron compounds for cancer therapy *Pure Appl. Chem.* **90** 653–63

[37] Datcu A and Ciobanu D 2020 Boron Nanoparticles *Characterization* **23** 257–64

[38] Li L, Dai K, Li J, Shi Y, Zhang Z, Liu T, Jun Xie, Ruiping Zhang and Liu Z 2021 A Boron-10 nitride nanosheet for combinational boron neutron capture therapy and chemotherapy of tumor *Biomaterials* **268** 120587

[39] Ferreira T H, Miranda M C, Rocha Z, Leal A S, Gomes D A and Sousa E M B 2017 An assessment of the potential use of BNNTs for boron neutron capture therapy *Nanomaterials* **7** 1–11

[40] Takeuchi I, Ishizuka Y, Uchiro H and Makino K 2017 Detailed biodistribution of liposomes prepared with polyborane instead of cholesterol for BNCT: effects of PEGylation *Colloid Polym. Sci.* **295** 1455–61

[41] Ueno M, Ban H S, Nakai K, Inomata R, Kaneda Y, Matsumura A and Nakamura H 2010 Dodecaborate lipid liposomes as new vehicles for boron delivery system of neutron capture therapy *Bioorganic Med. Chem.* **18** 3059–65

[42] Singh A, Kim B K, Mackeyev Y, Rohani P, Mahajan S D, Swihart M T, Krishnan S and Prasad P N 2019 Boron-nanoparticle-loaded folic-acid-functionalized liposomes to achieve optimum boron concentration for boron neutron capture therapy of cancer *J. Biomed. Nanotechnol.* **16** 1714–23

[43] Yang W, Wu G, Barth R F, Swindall M R, Bandyopadhyaya A K, Tjarks W, Tordoff K, Moeschberger M, Sferra T J, Binns P J, Riley K J, Ciesielski M J, Fenstermaker R A and Wikstrand C J 2008 Molecular targeting and treatment of composite EGFR and EGFRvIII-positive gliomas using boronated monoclonal antibodies *Clin. Cancer Res.* **14** 883–91

[44] Yang W, Barth R F, Wu G, Tjarks W, Binns P and Riley K 2009 Boron neutron capture therapy of EGFR or EGFRvIII positive gliomas using either boronated monoclonal antibodies or epidermal growth factor as molecular targeting agents *Appl. Radiat. Isot.* **67** 328–31

[45] Bianco A, Kostarelos K and Prato M 2008 Opportunities and challenges of carbon-based nanomaterials *Expert Opin Drug Deliv.* **5** 331–42

[46] Yamagami M, Tajima T, Ishimoto K, Miyake H, Michiue H and Takaguchi Y 2018 Physical modification of carbon nanotubes with a dendrimer bearing terminal mercaptoundecahydrododecaborates (Na2B12H11S) *Heteroat. Chem.* **29** 1–7

[47] Mi P, Yanagie H, Dewi N, Yen H C, Liu X, Suzuki M, Sakurai Y, Ono K, Takahashi H, Cabral H, Kataoka K and Nishiyama N 2017 Block copolymer-boron cluster conjugate for effective boron neutron capture therapy of solid tumors *J. Control. Release* **254** 1–9

[48] Yoneoka S, Park K C, Nakagawa Y, Ebara M and Tsukahara T 2018 Synthesis and evaluation of thermoresponsive boron-containing poly(N-isopropylacrylamide) diblock copolymers for self-assembling nanomicellar boron carriers *Polymers (Basel).* **11**

Green Synthesis of Rod Shaped ZnO using Extract of *Origanum majorana* Leaf and Investigation for Antibacterial Applications

M Saini[1], A Mushtaq[2], S Yadav[1], S Rawat[3], Nutan Rani[1], K Gupta[4], K Saini[1*]

[1]Department of Chemistry, Miranda House, University of Delhi, Patel Chest Marg, New Delhi-110007, India
[2]Department of Botany and Microbiology, H.N.B. Garhwal University, Srinagar-241674, Uttarakhand, India
[3]School of Life Sciences, Central University of Gujarat, Sector- 30, Gandhinagar-382030, Gujarat, India
[4]Department of Chemistry, Raj Rishi Government College, Alwar-301001, Rajasthan, India

* Corresponding author email address: kalawati.saini@mirandahouse.ac.in

Abstract. In this research article, we have fabricated the zinc oxide (ZnO) nanoparticles (NPs) with help of aqueous extract of leaf of *Origanum majorana* and studied their antimicrobial activity. *Origanum majorana* is one of medicinally important plant of *Lamiaceae/Labiatae* family. Green-approach of NPs has been favoured over traditional synthesis methods, as plant-based extracts have phytochemicals, which are non-toxic and biologically safe. The phytochemicals such as flavonoids, derivatives of phenol and polyphenolic biomolecules are found to be in aqueous leaf extract of *Origanum majorana* which used as capping and reducing agents. They include many functional groups such as -OH, -C=O and more which improve the physicochemical properties of NPs and consequently affect their targeting towards specific molecules. Plant mediated synthesis of ZnO NPs with fresh leave extract of Origanun majorana is also simple, quick and provides a vast array of functionalised NPs of particular size and morphology. Herein, ZnO NPs have been prepared with 3 mL of 25% (w/v) of leaf extract of *Origanum majorana* (Maruva). These synthesized NPs have been characterized using PXRD (Powder X-ray diffraction), FT-IR (Fourier transform-infra-red spectroscopy), UV-Visible spectroscopy, SEM (Scanning electron microscopy) with EDS (Energy dispersive spectroscopy). UV-Visible spectrum shows maximum absorbance at 379.75 nm and energy band gap have been evaluated 2.84 eV using tauc plot. Obtained PXRD pattern shows hexagonal wurtzite crystalline structure which is similar as reported in the literature (JCPDS No-36-1451). The morphology of synthesized NPs has been obtained with SEM images. These NPs are rod shaped with width calculated approximately 90 nm-125 nm and length 0.5μm-1.2μm respectively. The EDS analysis shows the 51.41% of zinc (Zn) and 48.49% of oxygen (O)elemental composition of fabricated nanoparticles. Antimicrobial activity has been performed on gram negative and gram positive microbes with the help of broth dilution method. These synthesized NPs shows very great bactericidal activity against *Staphylococcus* *aureus, Pseudomonas aeruginosa, Escherichia coli* and *Streptococcus pneumoniae* respectively. Minimum inhibitory concentrations (MIC) have been investigated 175μg/mL for *Pseudomonas aeruginosa*, 125 μg/mL for *Escherichia coli* and 100 μg/mL for both the *Staphylococcus aureus* and *Streptococcus pneumoniae* respectively.

Keywords: Green approach; ZnO NPs; Extract of leaf ; *Origanum majorana*; Antimicrobial activity; MIC

1. Introduction

ZnO NPs are very inexpensive, stable and biocompatible. These NPs have been investigated for various biomedical uses like as antibacterial, antifungal, anticancer, drug delivery bio-imaging and wound healing due to their physic-chemical properties [1-5]. ZnO NPs also have been used in many miscellaneous applications like as for supercapacitors, UV-shielding materials, nano-generators, functionalized air filters, gas sensing, optoelectronic as well as photocatalytic applications [6-14]. It has been investigated that the green approach of synthesis of NPs are more beneficial over the conventional chemical wet method, where the hazardous chemicals are used in their synthesis. These hazardous chemicals absorbed on the surface of the NPs that can cause toxic impact in biomedical applications. Recently, ZnO NPs have been prepared with help of extract of different parts of plants (leaf, seeds, bark, stem, fruits peels) and used in microbial activities [15-18]. Asif et al have also synthesized ZnO NPs using cyanobacteria [19]. Herein, a comparative study has been performed with commercial supplied ZnO NPs and green synthesized ZnO NPs regarding biocompatibility. Ann et al has done antibacterial study on *Staphylococcus aureus*, *Psuedomonas aeruginosa* and *Streptococcus pyogenes* using ZnO structures [20].

Antibacterial efficiency has been improved using coral-like ZnO/C-$ZnFe_2O_4$ hierarchical structure by Gu et al [21]. Herein, antibacterial study has been done on *Escherichia coli* and *Staphylococcus aureus* respectively. The optimized sterilization efficiencies have been investigated 98.0 % for *Staphylococcus aureus* and 99.4 % for *Escherichia coli* respectively. The outcomes of ZnO NPs on prohibition rate of *Staphylococcus aureus* and *Escherichia coli* have been studied by Pranjali et al [22]. Where, abovementioned ZnO NPs have been modified with help of polyethylene glycol (PEG). This study has been done by dispersing these modified ZnO NPs in peritoneal dialysis (PD) fluid at two concentrations such as 0.5mg/mL and 1 mg/mL respectively. Nano–bio interface interactions also have been studied using PD fluid by Guleria et al [23]. A Comparative study has been demonstrated on antibacterial and photocatalytic activities using ZnO NPs, also ZnO NPs based nanocomposite with therapeutic drug 9-aminoacridine hydrochloride hydrate (9AA-HCl) for anticancer activity [24]. Herein, Mitra et al has investigated that these ZnO nanoconjugates with this therapeutic drug shows synergistic catalytic effect than the bare ZnO. Gil et al has investigated the physical and chemical interactions between microorganism (gram-negative and gram-positive) and nanoporous ZnO spheres [25].

Ray et al has reported excellent bactericidal efficacy with good cytocompatibility using Ag-NPs decorated on ZnO Nanoflower impregnated eggshell membrane [26]. Shape-dependent antibacterial activity of ZnO nanostructure has been determined against several pathogens by Visinescu et al [27]. Herein, the surface reactivity of ZnO nanostructures and interactions with microbial and mammalian cells have been investigated. Synergistic effect of ZnO NPs synthesized with *Cassia auriculata* Leaf extract has been investigated on microbes and cancer cell lines [28]. Toxicity of the ZnO nanopowders has been studied on gram negative bacteria such as *Escherichia coli* and *Pseudomonas aeruginosa* through disk diffusion procedure [29], which is similar to well diffusion method. Where, micro-broth dilution procedure has been employed for investigation of MIC (minimum inhibitory concentration) value. Antimicrobial activity of the ZnO NPs synthesized with *Trifolium pratense* flower extract have been demonstrated on *Staphylococcus aureus*, *Pseudomonas aeruginosa* and *Escherichia coli* respectively [30]. Herein, 100 μL of the ZnO NPs have been used at different concentrations of 1028 μg/mL, 516 μg/mL and 256 μg/mL and 125 μg/mL. This study has been performed with reference to standard antibiotic Gentamicin. Divya et al has investigated antibacterial application of ZnO NPs synthesized with Hibiscus rosa-sinensis leaf extract [31]. Divyapriya et al has done antimicrobial study on *Murraya Koenigii* [32]. Gunalan et al has studied ZnO NPs which was synthesized by green method for antibacterial and antifungal applications [33]. A novel microbial route has been investigated by Jayaseelan et al for the fabrication of ZnO NPs [34]. Herein, these synthesized NPs have been used to study for the antibacterial and antifungal applications. The maximum zone of inhibition has been reported for these synthesized ZnO NPs at concentration of 25 μg/mL against *Aspergillus flavus* (19±1.0 mm) and *Pseudomonas aeruginosa* (22±1.8 mm) respectively.

The antibiotic effects and cellular internalization of superfine zinc oxide nanoparticles have been demonstrated on *Escherichia coli* bacteria by Brayner et al [35]. Herein, ZnO nanoparticles have been fabricated by forced hydrolysis of ionic Zn^{2+} salts in diethylene glycol (DEG) medium. Where, *Escherichia coli* cells have been exposed with DEG and ZnO NPs, which leads to damage as well as disorganization of cell membrane of these bacteria. Herein, the disorganization in cell membrane of tested bacteria leads the increment in permeability of membrane, which causes the accession of ZnO NPs. in the membrane of bacterial cells. Memarzadeh et al have investigated 100% ZnO NPs and its composite 25% nanohydroxyapatite (nHA) coated 75% ZnO NPs for dental and orthopedic implants [55]. Herein, ZnO NPs have been reported for inhibition of bacterial adhesion and help osteoblast growth. Herein, nano-coated material has been studied for antibacterial activity. *Staphylococcus aureus* stains are used to expose to nano-coated glass substrate in saline buffer or bovine serum and their antibacterial activity has been determined. Herein, in vitro study has also been done on MG-63 cells, UMR-106 cells and human mesenchymal cell (hMSC) respectively. Above literature survey reveals that the green synthesized ZnO NPs and chemically synthesized ZnO NPs are highly investigated for their various applications.

In this paper, our research goal is to synthesized ZnO NPs with leaf extract of *Origanum majorana* (Maruva) and their study for antibacterial applications. These green synthesized ZnO nanoparticles are characterized with the help of FT-IR, PXRD, UV-Visible absorption spectroscopy, SEM and EDS respectively. Herein, we have done the antimicrobial study with these synthesized NPs on *Escherichia coli*, *Staphylococus Aureus* and *Streptococus Pneumoniae* and *Pseudomonase Aeruginosa* using broth dilution method.

Though, few research work has been reported on antibacterial activities with *Origanum majorana* leaf extract and its essential oil. But, according to our best knowledge, this is first time, we have examine antibacterial activity of ZnO NPs fabricated with leaf extract of *Origanum majorana*, In future, we would like to use these ZnO NPs as antibacterial agent in place of some selective antibiotic which are resistant for some pathogens. Literature survey reveals that green approach of preparation of ZnO NPs is beneficial over the conventional chemical methods and these NPs are non-toxic and biocompatible. These green synthesized ZnO NPs may also be utilized in food packing industry to inhibit the pathogen growth in the packed food.

So, this study opens a new gate for the researchers about the synthesis of NPs with leaf extracts of *Origanum Majorana and to explore them for various biomedical applications.* Specifically, our reported results in this paper indicate that ZnO NPs synthesized with leaf extract can be an optimum coating material in future for orthopedic and dental implants which is stable, biocompatible and antimicrobial. Our, obtained MIC values for bacteria like *Escherichia Coli, Pseudomonas aeruginosa, Streptococcus pneumonia* and *Staphylococcus aureus* respectively, demonstrate that these synthesized ZnO NPs can also be used selectively for gram positive bacterial infection in the human beings in the future.

2. Materials and Methods

$ZnSO_4.7H_2O$ (Zinc sulphate heptahydrate) and NaOH (Sodium hydroxide) have been bought from Indian pharmaceutical company (Central Drug House (P) Ltd – CDH, India). All reagents and chemicals are utilized analytical reagent (A.R) grade.

2.1. Origanum majorana Leaf Extract Preparation

Origanum majorana leaves were taken from home garden, AG-Block-476, Shalimar Bagh, New Delhi-110088, India. These leaves were washed by using double deionized water and dried in the oven. These leaves were dried for two 2 hours at temperature of sixty degree centigrade. In next step, these leaves were grounded. Then, the 25 % solution was prepared in deionized H_2O. This 25 % solution of grounded leaves was heated at forty degrees centigrade with continuous stirring. Then, the prepared aqueous leaves extract was filtered utilizing the Whatman filter paper no.1 after that this freshly prepared extract was utilized in fabrication of zinc oxide nanoparticles.

2.2. Fabrication of ZnO NPs.

ZnO NPs were fabricated by green approach. Herein, 50 mL (0.2 M) of zinc sulphate heptahydrate ($ZnSO_4.7H_2O$) aqueous solution was prepared in standard volummetric flask. In next step, 50 mL (1.0 M) of sodium hydroxide (NaOH) aqueous solution and abovementioned 3 mL leaves extract of *Origanum majorana* and 50 mL (0.2 M) zinc sulphate solution were transferred into the standard conical flask of 250 mL. After that reaction was run on hot plate with magnetic stirrer. The reaction solution was stirred at 400 rpm at temperature of forty degree centigrade (313 K) for two hours. Later on, the synthesized ZnO NPs were separated through centrifugation. These synthesized zinc oxide nanoparticles were washed with 90 % alcohol. In last step, these washed zinc oxide nanoparticles were dried for two hours at temperature of sixty degree centigrade (333 K). All solutions were prepared in double deionized water.

2.3. Advanced techniques used for characterization of zinc oxide nanoparticles.

The optical properties were obtained using advanced UV-Visible spectrophotometer. The spectrum was recorded using the model (Spectromax M2e). The presence of functional group in the synthesized ZnO NPs were identified through Fourier transform infrared (FT-IR). Herein, Potassium bromide was employed as the standard substance. The crystalline nature and phase purity of the synthesized ZnO NPs (dried powder) were investigated by model D8 Discover diffractometer (λ = 0.154056 nm). Herein, the 2θ ranges from 10° - 80° with a step per second (0.02^o). The morphological determination of zinc oxide (ZnO) nanoparticles were investigated using scanning electron microscope (JEOL Japan Mode: JSM 6610LV) at 30 kV accelerating voltage and Magnification: X5 to X 3,00,000. Herein, the chemical composition was acquired using software of energy-dispersive X-ray spectroscopy (EDS) that is coupled with SEM.

2.4. Antibacterial activities

The antimicrobial study of fabricated zinc oxide nanoparticles against the pathogens viz., *Streptococcus pneumonia, Staphylococcus aureus, Pseudomonas aeruginosa* and *Escherichia coli* was evaluated through broth dilution process (Garcia, 2010) [36]. A stock solution of 2mg/mL of nanoparticles was prepared by dispersing them in pre-sterilized deionized water by ultra-sonication. The varying concentration ranging from 0.2 mg/mL to 200 mg/mL were prepared from stock solution and were added to the test tubes containing varying amount of sterile Luria Bertani broth (Muniyan et al., 2017) [37]. 0.5 mL of overnight culture (0.5 McFarland turbidity standards) of test pathogens was added to these tubes aseptically. Then, the tubes were incubated for 24 h at 37 °C . The bacterial cultures without any test solution and the tubes with only sterile media were kept as positive and negative controls respectively. The results were obtained by recording the optical density at 600 nm of the inoculated broth. MIC was recorded as the lowest concentration of the test sample prohibiting the growth of the inoculated test pathogen.

3. Results and Discussions

3.1. UV-Visible Spectroscopic and FT-IR Analysis

Figure 1(a) shows the UV-Visible absorption spectrum of synthesized ZnO NPs. The range of UV-Visible spectrum has been taken from 200 nm to 1000 nm. Herein, a sharp absorption band has been obtained at 379.75 nm and also a broad absorption band at near infra-red region 969.36 nm for synthesized ZnO NPs. This obtained value of λ maximum has been found in good agreements with the reported peak at 380 nm by Khara et al [38]. Alike outcomes have also been reported by Jayaseelan et al and Padalia et al respectively [34, 39]. Herein, they have prepared ZnO NPs by extract of *Aeromonas hydrophila* and *Ziziphus nummularia respectively*. The tauc's plot was made with the help of abovementioned spectrum data of fabricated ZnO NPs with *Orignum majorana* leaves, which is given in **Figure 1(b).** The energy band gap is obtained using the equation investigated by Tauc. This is given as below.

$$(\alpha h\nu)^2 = A(h\nu - E_g)^{1/2} \quad (1)$$

Herein, the constant A is not affected by magnitude of hν. But the absorption coefficient (α) depends on nature of the synthesized material. Herein, band gap (E_g) has been determined via drawing intercept in $(\alpha h\nu)^2$ vs hν curve as given in **Figure 1(b)**. From this curve, the energy band gap (E_g) value have been obtained 2.84 eV for as-synthesized ZnO. Approximately similar, energy band gap value has been found by Rani et al., where, the ZnO NPs have been prepared by leaf extract of *Tabernaemontana divericata* [6]. In similar way the value of (E_g) has been evaluated by Arvanag et al [40].

The FTIR spectrum for synthesized ZnO NPs is given in **Figure 1(c)**. Herein, the -Zn-O- bands have been obtained from 640.87 cm^{-1} to 835.56 cm^{-1} because of zinc oxide (-Zn-O-) stretching. There are other bands present in FT-IR spectrum like as 939.50 cm^{-1}-1044.27 cm^{-1}, 1384.96-cm^{-1} because of presence of aromatic -C-H bond, 1498.80 cm^{-1} because of presence of -C-O bond and 3299.63 cm^{-1} a strong broad in H_2O or -O-H stretching. These bands show the functional groups of phytochemicals. These phytochemicals are derivative of phenol (Thymol), monoterpene, trans-hydrate Sabineno, polyphenolic acids and flavonoids. These functional compounds are found due to presence of these phytochemicals in extract of *Origanum majorana leaves which are involved in surface coating of zinc nanoparticles*. Such type of data are investigated by Ramesh et al, Soto-Robles et al and Guerra-Boone et al respectively [47-48, 54]. These phytochemicals also governs the size and shape of green synthesized ZnO NPs.

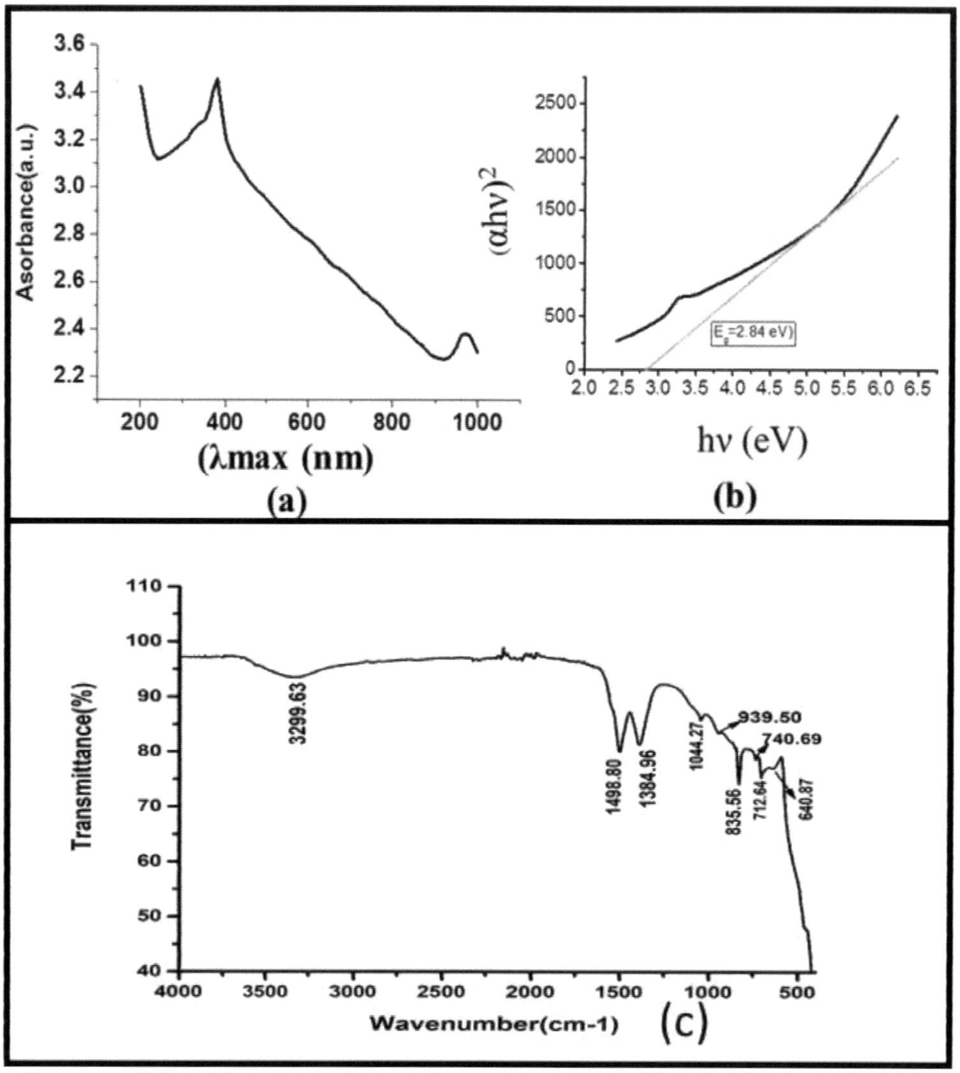

Figure 1. (a) UV-Visible absorption spectrum of ZnO NPs prepared by green approach and (b) Tauc plot $(\alpha h\nu)^2$ vs hν (eV) of synthesized ZnO NPs, (c) FT-IR spectrum of ZnO NPs prepared by green approach.

3.2. Powder X-Ray Diffraction (PXRD) Analysis

The PXRD pattern of fabricated ZnO NPs has been depicted in **Figure 2.** The PXRD pattern of these NPs has been obtained using X-ray diffraction. The PXRD pattern (A) shows for fresh prepared ZnO NPs using using 3mL of aqueous leaf extract of *Origanum majorana* where as PXRD pattern (B) for same ZnO NPs after six months. These is no decrease in peak intensity of synthesized ZnO NPs after six months and obtained peak positions are same, it means synthesized ZnO NPs are very stable. The peaks intensity have been obtained at 2θ = 31.67°, 34.31°, 36.14°, 47.40°, 56.52°, 62.73°, 66.28°,

67.91°, 69.03° and 72.48° respectively. The corresponding (hkl) values have been assigned to (100), (002), (101), (102), (110), (103), (200), (112), (201) and (004) respectively. These values indicate the successful synthesis of hexagonal wurtzite crystalline structure of ZnO NPs. These 2θ values are well indexed with the JCPDS No. 36-1451 [41]. This crystalline structure also matches with the reported structure JCPDS No.5-0664 [42].

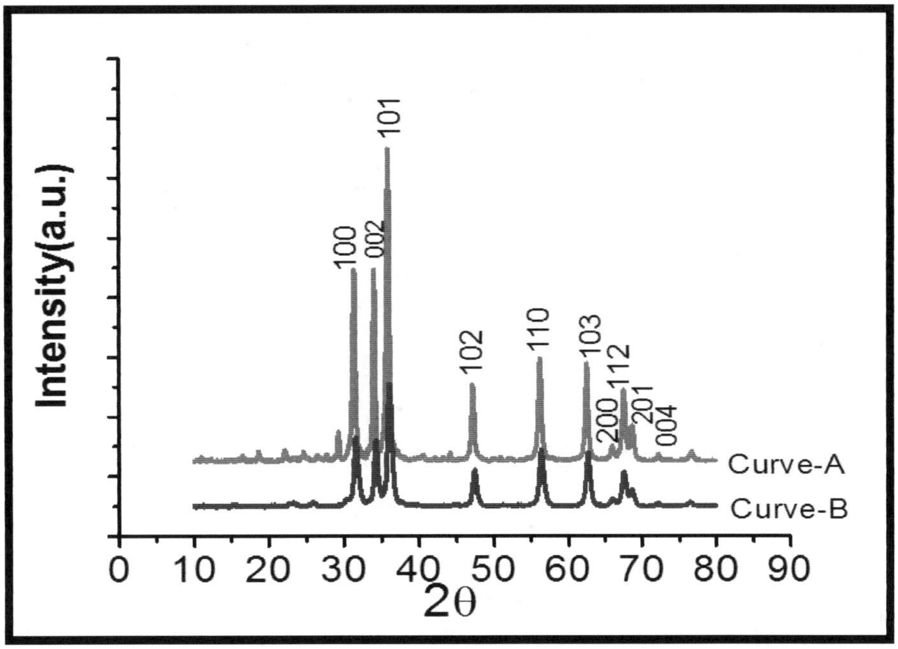

Figure 2. Powder X-ray Diffraction Pattern for ZnO NPs prepared by green approach(A) For fresh prepared Sample, (B) After Six months

3.3. SEM and EDS Analysis

The scanning electron micrograph of green fabricated zinc oxide nanoparticles with help of leaf extract of *Origanum majorana* is given in **Figure 3.** It is clear from these **Figures 3 (a, b, c)** that all reported ZnO NPs are of rod shaped. The width of rod has been obtained approximately from 90 nm to 125 nm and from 0.5 µm to 1.2 µm in length. Our results are good agreement with reported dimension of ZnO NPs by Mahdizadeh et al [43]. Herein, ZnO NPs have been fabricated by green method using rough shell extract of *Cucumis melo* inodorus. These rod shaped NPs have been found agglomerated. The width and length of synthesized ZnO NPs have been found approximately 75-125 nm and 0.5 µm -1µm respectively. Basri et al has reported the ZnO NPs having diameters is in the scale of 73–123 nm with flower rod shapes [44]. Nutan et al and Zhong et al have obtained rod shaped morphology of synthesized ZnO NPs [6, 9]. The EDS spectrum of prepared ZnO NPs have been demonstrated in **Figure 4.** The EDS results (in set of **Figure 4**) reveal the elemental composition of this fabricated ZnO NPs and also reveal about the purity of fabricated ZnO NPs. The percentage atomic ratio has been obtained for Zinc (Zn) 51.41% and for Oxygen (O) 48.59 % respectively. Approximately similar percentage ratio of zinc (Zn) and oxygen (O) have been obtained by Rani et al [6].

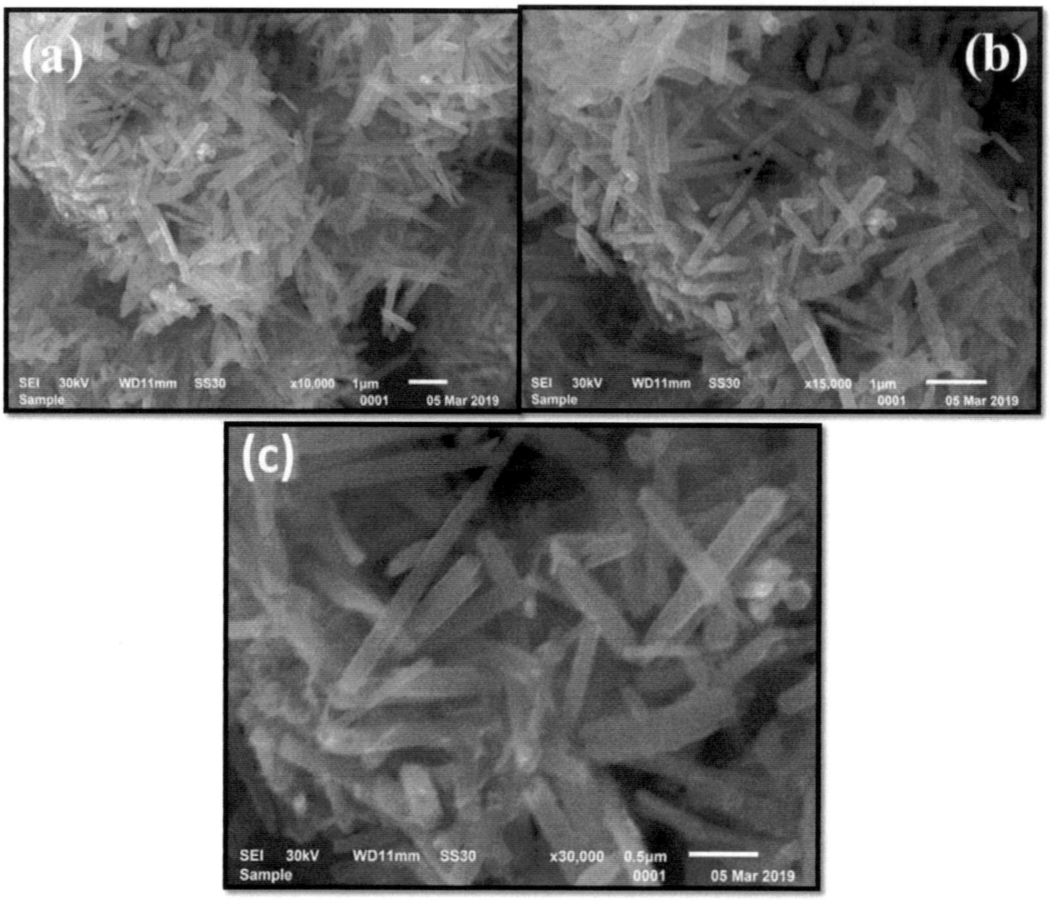

Figure 3. SEM images of ZnO NPs prepared by green approach (a) at 10,000 Magnification, (b) at 15,000 Magnification (c) at 30,000 Magnification

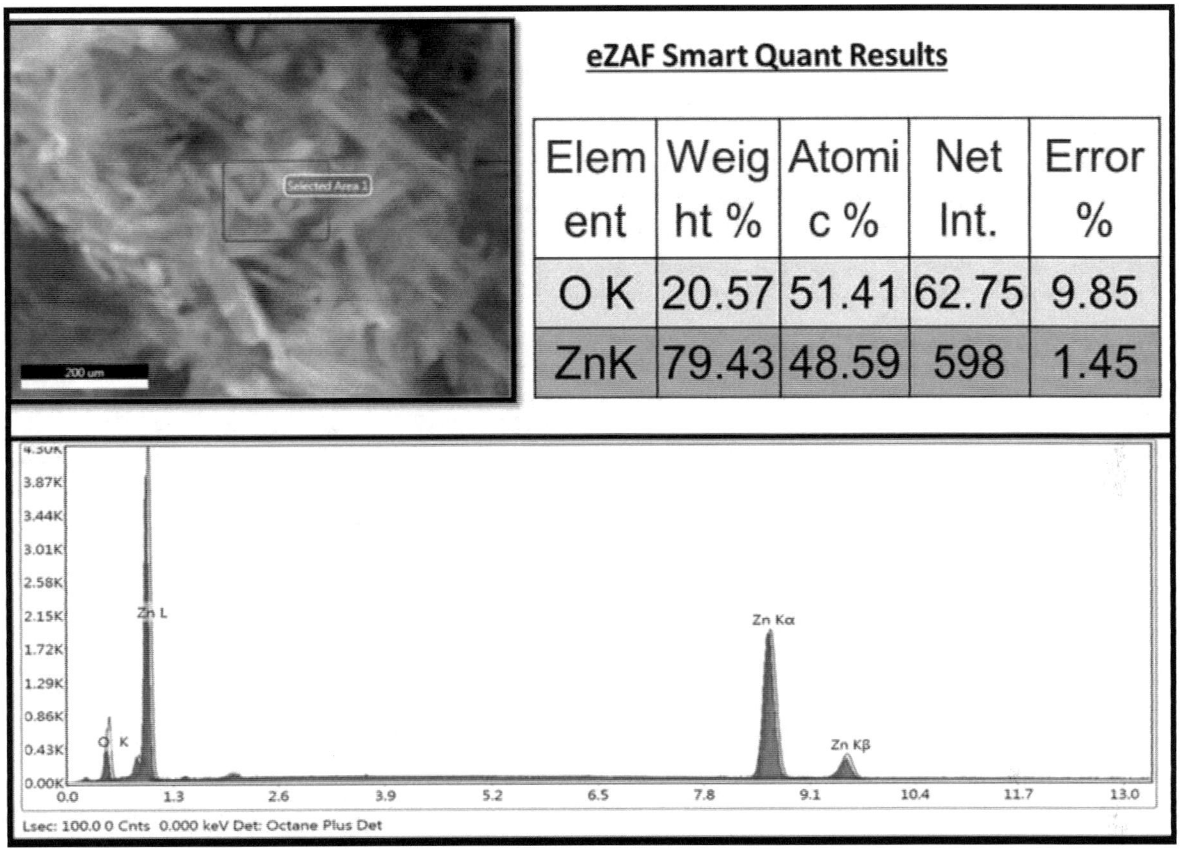

Figure 4. EDS Spectra of green fabricated zinc oxide nanoparticles.

3.4. Antibacterial activities

The antibacterial results have been given in **Table 1.** Herein, we have studied the antimicrobial activity of nanoparticles on *Staphylococcus aureus*, *Streptococcus pneumonia*, *Escherichia coli* and *Pseudomonas aeruginosa*. Herein, gram-negative microbes are *Escherichia coli* and *Pseudomonas aeruginosa* while gram-positive microbes are *Staphylococcus aureus* and *Streptococcus pneumoniae*. This study has been performed using the broth dilution method. The MIC has been found 150-175 µg/mL for *Pseudomonas aeruginosa,* 100-125 µg/mL for *Escherichia coli*, and 76-100 µg/mL for *Staphylococcus aureus* as well as for *Streptococcus pneumoniae.* More MIC value means that a higher concentration is needed to inhibit the growth of microbes. **Table 1 shows** that these synthesized ZnO NPs are more effective against all clinical isolates except *Pseudomonas aeruginosa*. It means *Pseudomonas aeruginosa* shows more resistance against the synthesized ZnO NPs. Our results are in good agreement with reported antimicrobial study by Sana *et al* [17]. Herein, Sana et al has studied the antibacterial study at the concentrations of 25 µg/mL, 50 µg/mL and 100 µg/mL respectively. Similar study has been reported with ZnO NPs using *Peltophorum pterocarpum* flower-mediated synthesis in the literature [38]. Herein, these NPs are tested against four types of gram-positive microbes and four types of gram-negative microbes. This study has been done at two concentrations of ZnO NPs 100 mg/mL and 150 mg/mL respectively. Hazra et al has also evaluated the antimicrobial activity of ZnO NPs synthesized using extract of Aloe vera and MIC values have been determined in the range of 3-5 mg/mL [45]. It means, our results are best than the previously reported results. Similar types of results have also been reported only with essential oil, whole plant extract and extract of leaves of *Origanum majorana* and *Origanum compactum Benth* respectively [46,54]. Herein, reported very high values of MIC with only extract indicates that the impact of only extract has been found very less on antimicrobial activity as compared with our synthesized ZnO NPs. Herein, an elaborative **Table 2** has been given which indicates the type of system, synthesis methodology, application and advantage of the respective system as compared to the reported system [17, 29-30, 33, 35, 38, 45, 49-53]. Herein, Barani et al. have reported same value of MIC (125 µg/mL) for *Escherichia coli* but they have reported very high MIC value (250 µg/mL) for *Staphylococcus aureus.*

Table 1. Minimum Inhibitory Concentration (**MIC in µg/mL**) of ZnO nanoparticles for various bacterial pathogens.

Isolates	MIC Range
Escherichia coli (E. Coli)	100-125
Pseudomonase Aeruginosa	150-175
Staphylococcus Aureus	76-100
Streptococus Pneumoniae	76-100

Table 2. Comparison of earlier reported ZnO nanoparticles with synthesized ZnO nanoparticles for antibacterial activities.

S.N.	Nanoparticles	Plant extract/Bacteria used for biosynthesis	Shape/Size	Organism/MIC value	References
1	ZnO	*Origanum majorana*	Rod shape/90-125nm	*Pseudomonas aeruginosa/175µg/mL Escherichia coli/125µg/mL Staphylococcus aureus/100µg/mL Streptococcus pneumonia/100µg/mL*	Present work
2	ZnO	*Punica granatum L and Tamarindus indica*	Spherical shape/10nm and 14nm	*Escherichia coli/512µg/mL Pseudomonas aeruginosa/1024µg/mL*	29

3	ZnO	*Aloe Leaf broth*	-/40nm and 25nm	*Staphylococcus aureus/0.40mM* *Serratia marcescens/1.40mM* *Proteus mirabilis/1.0mM*	33
4	ZnO	*Aeromonashy drophila*	Spherical/57-72nm	*Pseudomonas aeruginosa/2.9µg/mL* *Staphylococcus aureus/2.4µg/mL*	35
5	ZnO	*Aloe vera*	Rod shape/lesser than 100nm	*Bacillus subtilis/3.0mg/mL* *Escherichia coli/3.0mg/mL* *Klebsiella pneumonia/4.0mg/mL* *Pseudomonas aeruginosa/5.0mg/mL* *Staphylococcus aureus/4.0mg/mL*	45
6	ZnO	*Cissus quadrangularis*	Spherical shape/75-90nm	*Mycobacterium smegmatis/50µg/mL*	49
7	ZnO	*Marinobacter sp. And Vibrio sp.*	Spherical shape/~10nm and ~20nm	*Escherichia coli/125µg/mL* *Pseudomonas aeruginosa/62.5µg/mL* *Staphylococcus aureus/250µg/mL* *Bacillus subtilis/500µg/mL* *Listeria innocua/1000µg/mL*	50
8	ZnO	*Livistona jekinsiana*	Not uniform spherical, oval and hexagonal/3.6nm	*Staphylococcus aureus/15µg/mL* *Escherichia coli/25µg/mL* *Bacillus subtilis/10µg/mL* *Klebsiella pneumonia/25µg/mL*	51
9	ZnO	*Gum Arabic*	-/180nm	*Staphylococcus aureus/31.25µg/mL* *Escherichia coli/62.5µg/mL*	52
10	ZnO	*Brassica rapa*	Irregular shape/27.48nm	*Escherichia coli/25µg/mL* *Bacillus*	53

				subtilis/12.5μg/mL	
11	Ni-doped ZnO	*Brassica rapa*	Irregular shape/27.48nm	*Escherichia coli/25μg/mL* *Bacillus subtilis/50μg/mL*	53

4. Conclusions

Rod shaped ZnO NPs was prepared with the help of 3 mL leaves extract of *Origanum majorana*. These green synthesized ZnO NPs are very stable, energy saving, low cost, biocompatible and more environmental friendly. These ZnO NPs was also characterized with the help of different advanced techniques like as UV-Visible absorption spectroscopy, FT-IR, EDS, PXRD, SEM and EDS respectively. These rod shaped ZnO NPs have also been utilized for antibacterial activity. The four microbes namely (Pseudomonas aeruginosa, *Escherichia coli*, Staphylococcus aureus and *Streptococcus pneumonia*) were tested for the antimicrobial study. The obtained value of MIC with broth dilution method indicates that our prepared ZnO NPs are more effective towards the gram-positive microbes as compare to gram-negative microbes. This, in vitro study reveals that these ZnO NPs synthesized with leaves extract of *Origanum majorana* can be developed as a new nano-medicine for bacterial infection in future.

Acknowledgment

Kalawati Saini wants to present a sincere thanks to Principal, Miranda House, University of Delhi for providing laboratory for research work. The authors are appreciating to USIC, University of Delhi, India for getting SEM, EDS and PXRD of green fabricated zinc oxide nanoparticles. Kalawati Saini and Mona Saini also very grateful to DBT, New Delhi, India for providing fund (Ref. No. BT/IN/Indo/Foldscope/39/2015 dated: 20.03.2018). One of the author, Sapna Yadav thanks to Council of Scientific & Industrial Research (CSIR), India for providing fellowship (File Reference No. 08/700/(0004)/2019-EMR-1). Kalawati Saini would like to thank Dr. Amrita Tripathi Sheikh, Associate Professor, MH, DU for her constant encouragement and support.

Disclosure

The authors claims no conflicts of interest in publishing of this research article.

References

[1] Taşdemir A, Aydin R, Akkaya A, Akman N, Altinay Y, Çetin H, Şahin B, Uzun A and Ayyildiz E 2021 *Ceram. Int.* **47(14)** 19362-19373.

[2] Mishra P K, Mishra H, Ekielski A, Talegaonkar S and Vaidya B 2017 *Drug Discov. Today* **22** 1825–1834.

[3] Martínez-Carmona M, Gun'ko Y and Vallet-Regí M 2018 ZnO *Nanomaterials* **8** 268.

[4] Xiong H M 2013 *Adv. Mater* **25** 5329–5335.

[5] Bayrami M, Bayrami A, Habibi-Yangjeh A, Shafeeyan M S, Feizpoor S, Arvanagh F M, Nourani M R Taheri R A 2020 *IET Nanobiotechnol.* **14(7)** 548-554

[6] Rani N, Saini M, Yadav S, Gupta K, Saini K and Khanuja M 2020 *AIP Conf. Proc.* **2276** 020042

[7] Li R, Yabe S, Yamashita M, Momose S, Yoshida S, Yin S and Sato T 2002 *Solid State Ion.* **151** 235-241.

[8] Lu M P, Song J, Lu M Y, Chen M T, Gao Y, Chen L J and Wang Z L 2009 *Nano Lett.* **9** 1223-1227.

[9] Zhong Z, Xu Z, Sheng T, Yao J, Xing W and Wang Y 2015 ACS Appl. Mater. Interfaces **7(38)** 21538-21544

[10] Yu A, Qian J, Pan H, Cui Y, Xu M, Tu L, Chai Q and Zhou X 2011 *Sen. Actuators B Chem.* **158** 9-16.

[11] Elilarassi R and Chandrasekaran G 2010 *Optoelectron. Lett.* **6** 6-10.

[12] Lv T, Pan L K, Liu X J and Sun Z 2012 *Catal. Sci. Technol.* **2** 2297-2301.

[13] Gao P, Ng K and Sun D D 2013 *J. Hazard. Mater.* **262** 826-835.

[14] Salehi R, Arami M, Mahmoodi N M, Bahrami H and Khorramfar S 2010 *Colloid Surface B* **80** 86-93.

[15] Li J, Li Y, Wu H, Naraginti S and Wu Y 2021 *Environ. Res.* **200** 111433.

[16] Soltanian S, Sheikhbahaei M, Mohamadi N, Pabarja A, Abadi M F S and Tahroudi M H M 2021 *BioNanoScience* **11 (2)** 245-255.

[17] Sana S S, Kumbhakar D V, Pasha A, Pawar S C, Grace A N, Singh R P, Nguyen V H and Le Q V 2020 *Molecules* **25(21)** 4896.

[18] Arvanag F M, Bayrami A, Yangjeh A H and Pouran S R 2019 *Mater Sci Eng C.* **97** 397-405.

[19] Asif N, Fatima S, Md Aziz M A, Shehzadi, Zaki A and Fatma T 2021 *Bioorg. Chem.* **113** 104999.

[20] Ann L C, Mahmud S, Bakhori S K M, Sirelkhatim A, Mohamad D, Hasan H, Seeni A and Rahman R A 2014 *Ceram. Int.* **40** 2993-3001.

[21] Gu Y, Teng G, Jin X, Wang L, Qiang Z, Ma W and Zhang C 2020 *Indust. Eng. Chem. Res.* **59(24)** 11219-11231.

[22] Pranjali P, Meher M K, Raj R, Prasad N, Poluri K M, Kumar D and Guleria A 2019 *ACS Omega* **4(21)** 19255-19264.

[23] Guleria A, Meher M K, Prasad N, Poluri K M and Kumar D 2017 *J. Phys. Chem. C* **121(34)** 18598-18607.

[24] Mitra P, Dutta D, Das S, Basu T, Pramanik A and Patra A 2018 *ACS Omega,* **3(7)** 7962-7970.

[25] Lucas-Gil E, Leret P, Monte-Serrano M, Reinosa J J, Enríquez E, Campo A D, Cañete M, Menéndez J, Fernández J F and Marcos F R 2018 *ACS Appl. Nano Mater.* **1(7)** 3214-3225.

[26] Ray P G, Biswas S, Roy T, Ghosh S, Majumder D, Basak P, Roy S and Dhara S 2019 *ACS Sustainable Chem. Eng.* **7(16)** 13717-13733.

[27] Visinescu D, Hussien M D, Moreno J C, Negrea R, Birjega R, Somacescu S, Ene C D, Chifiriuc M C, Popa M, Stan M S and Carp O 2018 *Langmuir* **34(45)** 13638-13651.

[28] Padalia H, Moteriya P and Chanda S 2018 *BioNanoSci.* **8(1)** 196-206.

[29] Prashanth G K, Prashanth P A, Utpal Bora U, Manoj Gadewar M, Nagabhushana B M, Ananda S, Krishnaiah G M and Sathyananda H M 2015 *Karbala Int. J. Mod. Sci.* **1** 67-77.

[30] Dobrucka R and Dugaszewsk J 2016 *Saudi J. Biol. Sci.* **23** 517–523.

[31] Divya M J, Sowmia C, Joona K and Dhanya K P 2013 *Res. J. Pharm. Biol. Chem.* **4(2)** 1137–1142.

[32] Divyapriya S, Sowmia C and Sasikala S 2014 *World J. Pharm. Pharm. Sci.* **3(12)** 1635–1645.

[33] Gunalan S, Sivaraj R and Rajendran V 2012 *Prog. Nat. Sci. Mater. Int.* **22(6)** 693–700.

[34] Jayaseelan C, Rahuman A A, Kirthi A V, Marimuthu S, Kumar T S, Bagavana A, Gaurav K, Karthik L and Bhaskara Rao K V 2012 *Spectrochimica Acta Part A.* **90** 78– 84.

[35] Brayner R, Ferrari-Iliou R, Brivois N, Djediat S, Benedetti M F and Fiévet F 2006 *Nano Lett.* **6(4)** 866–870.

[36] Garcia L 2010 Broth Microdilution MIC Test, In Clinical Microbiology Procedures Handbook, 3rd Edition. ASM Press, Washington, DC. 25-41.

[37] Muniyan A, Ravi K, Mohan U and Panchamoorthy R 2017 *World J Microbiol Biotechnol.* **33(7)** 1-12.

[38] Khara G, Padalia H, Moteriya P and Chanda S 2018 *Arab. J. Sci. Eng.* **43** 3393-3401.

[39] Padalia H and Chanda S 2017 *Nanomed. Biotechnol.* **45(8)** 1751-1761. DOI: 10.1080/21691401.2017.1282868

[40] Arvanag F M, Bayrami A, Habibi-Yangjeh A and Pouran S R 2019 *Mater Sci Eng C* **97** 397-405.

[41] Seo H K and Shin H S 2015 *Mater. Lett.* **159** 265-268.

[42] Akhtar M J, Ahamed M, Kumar S, Khan M M, Ahmad J and Alkokayan S A 2012 *Int. J. Nanomed.* **7** 845-851.

[43] Mahdizadeh R, Homayouni-Tabrizi M, Neamati A, Seyedi S M R and Afshar H S T 2019 J. Cell. Biochem. **10** 1-10.

[44] Basri H H, Talib R A, Sukor R, Othman S H and Ariffin H 2020 *Nanomaterials* **10(6)** 1061.

[45] Lakshmi V J, Sharath R, Chandraprabha M N, Neelufar E, Hazra A and Patra M 2012 *J. Biochem. Technol.* **3 (5)** S151–S154.

[46] Charai M, Mosaddak M and Faid M 1996 *J. Essent. Oil Res.* **8(6)** 657-664. (Online published in 2011).

[47] Ramesh M, Anbuvannan M and Viruthagiri G 2015 *Spectrochim. Acta A. Mol. Biomol. Spectrosc.* **136** 864-870.

[48] Soto-Robles C A, Luque P A, Gomez-Gutierrez C M, Nava O, Vilchis-Nestor A R, Lugo-Medina E, Ranjithkumar R and Castro-Beltran A 2019 *Results Phys.* **15** 102807.

[49] Sathappan S, Kirubakaran N, Gunasekaran D, Gupta P K, Verma R S and Sundaram J 2021 *Proc. Natl. Acad. Sci. India, Sect. B Biol. Sci.* **91(2)** 289–296.

[50] Barani M, Masoudi M, Mashreghi M, Makhdoumi A and Eshghi H 2021 *Int. J. Pharm.* **606** 120878.

[51] Baruah R, Yadav A and Das A M 2021 *Spectrochim. Acta A Mol. Biomol. Spectrosc.* **251** 119459.

[52] Pauzi N, Zain N M, Kutty R V and Ramli H 2021 *Mater.Today: Proc.* **41** 1-8.

[53] Khan M I, Fatima N, Shakil M, Tahir M B, Riaz K N, Rafique M, Iqbal T and Mahmood K 2021 *Physica B Condens. Matter* **601** 412563.

[54] Guerra-Boone L, Alvarez-Románz R, Salazar-Aranda R, Torres-Cirio A, Rivas-Galindo V M, Waksman de Torres N, González G and Pérez-López L A 2015 *Pak. J. Pharm. Sci.* **28(1)** 363-369.

[55] Memarzadeh K, Sharili A S, Huang J, Rawlinson S C F and Allaker R P 2015 *J. Biomed. Mater. Res. A* **103(3)** 981-989.

Role of nanofluids as an anti-freezing agent

S Soni[1,*], J Singh[2] and B K Gill[2]
[1]ME Chemical (2019-21) Dr S.S.B.UICET, Punjab University Chandigarh-160014.
[2]Assistant Professor, Dr S.S.B. UICET, Punjab University, Chandigarh-160014.
[*]Email address: shivamsoni896@gmail.com

Abstract In the era of the advancement, finding better and efficient solution for the problems is the real deal. Nanofluids discovery and related research work are giving efficient results in coping up the problems. Usage of nanofluids in different spheres and now finding application as an anti-freeze agent really helping normal people and industrialist. The contribution of nanofluids towards the high heat transfer rate is because of the improved properties. This paper reviews the features which make nanofluids as an anti-freeze agent and also deals with the published research papers to see how efficient the nanofluids are and what made them that efficient and the chances of the improvement to get better outcome.

Keywords: Nanofluids, anti-freezing agent, heat transfer, properties

Nomenclature

C_p	Specific heat J(kg.K)$^{-1}$
φ	volume concentration in %
μ	dynamic viscosity (cP)
ρ	Density (kg.m^{-3})
k	Thermal conductivity (W(m.K)$^{-1}$)nf nanofluid.
Bf	base fluid.
EGW	ethylene glycol water.
V_B	Brownian movement.

1. INTRODUCTION

The technological development leads to more use of unique features of nanofluids in various fields has been increasing day by day. Nanofluids have been used more frequently in heat transfer industries. The term Nano fluid was brought in to picture by **Eastman and Choi** [1]. They have given the term "nanofluids". When compared to the qualities of base fluids such as engine oil, water, transformer oil, ethylene glycol and so on, nanofluids have better viscosity (μ) , thermal conductivity (k), and heat transfer coefficients (h). According to **Rashmi et al., 2014**[2]; **Mahmoudi et al., 2016** [3] the nanofluids shows the enhanced properties in thermal performance than the conventional heat transfer fluids. The role of nanoparticles and then nanofluids was investigated by **Ravisankar and Chand [4]** and **Azmi et al. [5]**. Different materials along with respective thermal conductivity are included in figure1 which shows the importance of a particular material.

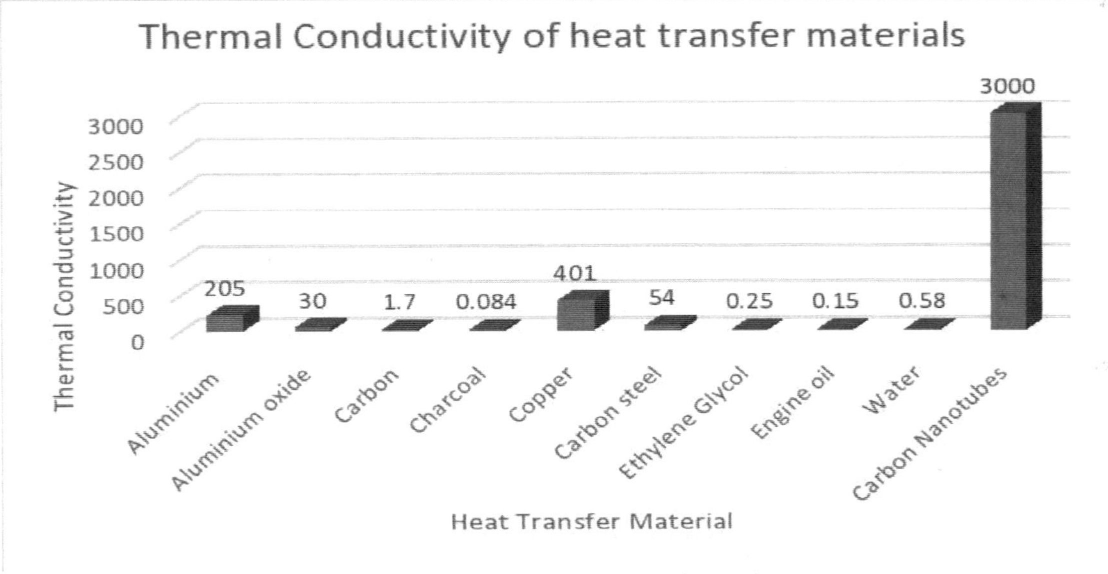

Figure 1. Thermal conductivity of materials

This review paper deals with the properties of nanofluids to act as an anti-freeze agent and how that affects the behavior, what changes comes in the thermo physical properties and use in industry.

2. Nanofluids Preparation and Nanofluid Stability

Creating a stable Nano fluid is a difficult task. To the greatest extent possible, the stability of nanofluids is determined by the manner of preparation. Nanofluids preparation is basically from two methods. They are 2 step method and 1 step method.

2.1 Two-Step preparation Method- Material is made in dry powder form and then small sized powder is immersed into a fluid. This immersing is done by agitation, ball milling techniques. **Eastman et al. (2001)** [6] prepared Nano fluid from two step method and thus dispersed Cu Nanomaterial in ethylene glycol and result showed EG shows less thermal conductivity that the mixture has higher effective thermal conductivity than pure EG. Figure 2 shows the schematic view of the preparation.

2.2 One-Step Method- Nanoparticles are first created and then dispersed immediately into the base fluid. This approach does not use the processes i.e., drying, storage, transportation, and distribution of nanoparticles. **Jwo et al. (2007**) [7] dealt with CuO–water Nano fluid by the use of the temperature arc method and results in enhancement of the thermal conductivity by 9.6%. **Kurihara et al. (1995)** [8] prepared Cu nanoparticles by suing polypol solvent.

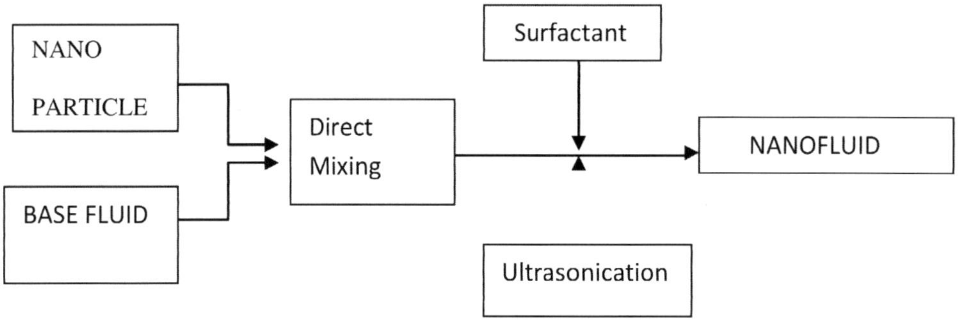

Figure 2. Preparation Method of Nano fluid (2 step approach)

2.3 Nanofluid Stability - Nanofluid stability is one of the most important characteristics of nanofluids. If agglomeration occurs it ultimately decreases the thermal conductivity property. Thus stability of nanofluid is termed as the important property because it plays an important role in influencing the properties of nanofluids for application purposes. There are many methods to measure or check the stability but generally following two methods have been used by the researchers.

2.3.1 Sedimentation method- Sedimentation is the tendency that allows nanoparticles to settle down in the base fluid. Sedimentation is the most basic approach.The nanoparticles sediment weight or volume is set under an external applied force, which shows the stability of the characterized Nano fluid. The variation happens in the properties like the size of the particle and the concentration with time is measured and on seeing the visual we depict the stability of the nanoparticles.

2.3.2 Zeta potential method –Stability of nanofluid can also be checked from zeta potential value. Zeta potential magnitude helps in showing the stability, more the value high stable is the prepared nanofluid. If colloids are having high zeta potential no matter if it is negative or positive, they are considered to be electrically stabilized. On the other hand lower value of the zeta potential of colloids shows its tendency to coagulate or flocculate.

2.4Methods for Improving Nanofluid Stability - Sometimes the nanofluids formed are not stable and therefore for the enhancement different methods are applied.

2.4.1 Surfactants-Surfactant or dispersants affect the surface property. These are made of a hydrophobic tail, a hydrophilic polar head, which leads to continuous phase between the nanoparticles and fluids. It increases the dispersibility of nanofluids but sometimes causes problems too like heat transfer media decay. Sometimes they produce foams when they are heated. Surfactant sometimes get attached to nanoparticles surface and thus increase the thermal resistance which means less increase in the thermal conductivity [9].

3. Thermo physical Properties

Three major thermo physical properties can change their values due to the addition of nanoparticles into the base fluid i.e. thermal conductivity, viscosity and density. These properties are checked and analysed. These are helpful in convection studies.KD2 pro analyzer is used to find the value of conductivity. The working principle includes a transient hot wire, a heating circulator that maintains the temperature constant. Viscosity is measured through Rheometer, Ostwald viscometer. Rheocalc software can be used to find the value or in viscometer, calculation can be done from the formula. The calculated readings of the properties is compared with **ASHRAE Handbook** [10] for EG-DW mixture. **Pak and Cho** [26] equation were used to check various thermo physical properties and those are then estimated by using these relations;

$$C_{p\,nf}=\left\{C_{p\,bf}(1-\varphi)\left(\frac{\rho_{bf}}{\rho_{nf}}\right)\right\}+\varphi C_{p\,p}\left(\frac{\rho_p}{\rho_{nf}}\right) \quad (1)$$

$$\rho_{nf}=(1-\varphi)\rho_{bf}+\rho_p\varphi \quad (2)$$

$$\mu_{nf}=\mu_{bf}(1+2.5\varphi) \quad (3)$$

$$\mathrm{k}=\left\{\mathrm{k}_p+(n-1)\mathrm{k}_{bf}-\varphi(n-1)\left(\mathrm{k}_{bf}-\mathrm{k}_p\right)\right\}\left\{\mathrm{k}_p+(n-1)\mathrm{k}_{bf}+\varphi(n-1)\left(\mathrm{k}_{bf}-\mathrm{k}_p\right)\right\}^{-1} \quad (4)$$

Equation (1), (2), (3) and (4) are taken from Pak and Cho model [26] from their study of hydrodynamic and heat transfer study of dispersed fluids with submicron metallic oxide particles.

For EG based and mixtures based Nano fluids,**Masoumi et al** [27] proposed one formula as

$$\mu_{nf}=\mu_{bf}+\rho_p \mathrm{V_B} \mathrm{d_p^2}(72\mathrm{C}\delta)^{-1} \quad (5)$$

Where δ is the distance between the centre of nanoparticles and C is the correction factor.

$$\Delta=\left(\pi \mathrm{d_p}/6\varphi\right)^{\frac{1}{3}} \quad (6)$$

$$\mathrm{C}=\mu_{bf}^{-1}(a\varphi+b), \quad (7)$$

Where a and b are experimental parameters, which were estimated to be 0.00004 and $7.1274*10^{-7}$ for engine coolant–Al_2O_3 nanofluids, respectively [28].

Ethylene-glycol and propylene-glycol are usually blended with water to make an anti-freezing agent. Table 1 shows the comparison of their different properties with that of water at 20°C. Researchers usually check these properties at 20°C and then compared with each other to know the variations [25].

Table 1. Thermo Physical Properties Data

	ρ in $kg.m^{-3}$	Cp in $J(kg.K)^{-1}$	k in $(W(m.K)^{-1})$	μ in (Pa.s)
EG	1126	2354	0.256	$21*10^{-3}$
PG	1035.3	2479	0.1962	$57.571*10^{-3}$
water	999	4185	0.6	$1.002*10^{-3}$

4. Hybrid Base Fluid – Antifreeze

The solution obtained by mixing liquid alcohols like ethylene glycol, methanol etc. with water is known as antifreeze as it lowers down the freezing point of the base water. Nanofluids are thus helpful in cryogenic conditions. Nanofluids are formed by spreading nanoparticle in a blend of base fluids. Such nanofluids prove to be best suitable for heat transfer applications in cryogenic conditions.

In refrigeration systems, nanoparticles are actually mixed with liquid water and ethylene- or propylene-glycol thus forms EG-W, PG-W and nanoparticle mixture which thus helps in lowering the freezing point which ultimately leads to prevention from ice. In cold regions due to less temperature, the amount of energy required for heating purposes is very high, the temperature extends up to -40°C. As a result, differing quantities of EG-W and PG-W are used as heat transfer fluids. They actually decrease the freezing point of water. Heat exchangers or car radiators, where the high temperature is required the EG-W and PG-W used there in order to raise the aqueous boiling points.

4.1Antifreeze material and the need for antifreeze material

EG-DW and PG-DW act as antifreeze. The thing is to know the need for such antifreeze agents and their synthesis. Ethylene glycol (EG) [11]. EG possess certain characteristics which lead to its use as an antifreeze material. EG has a high boiling point (B.P.), lower freezing point and possess good stability when seen at different ranges of temperatures. The specific heat, thermal conductivity are also high and has a low viscosity. These features of EG reduces the pumping requirements. The melting and boiling points of a base fluid made up of ethylene glycol and water at various volume percentages are shown in Table [2] to highlight the variance and pattern.

Table 2. EG-water solution freezing and boiling points [12]

EG Solution(%by vol)	0	10	20	30	40	50	60
Temp(°C) MP	0	-3.4	-7.9	-13.7	-23.5	-36.8	-52.8
BP	100	101.1	102.2	104.4	105.7	107.2	111.1

Though, EG-water mixture shows low properties when compared with distilled water because water have high value of the thermal conductivity property than the other base fluid constituent. To increase this property addition of additives is done like EG is mixed with water which enhance the thermal property. The addition of nanoparticles to base fluids also aids in heat transfer augmentation, resulting in an increase in thermal performance [13-16].

4.2 Ethylene glycol as an antifreeze

EG helps in lowering the freezing point of water and when added to water it easily gets miscible. EG when seen in chemical bonding shows that it contains a polar O-H group; higher electronegativity of oxygen as compared to hydrogen makes them behave as a polar group. Being polar it starts polarizing the electron pair in the O-H bond towards itself, When this occurs, oxygen acquires a fractional negative (-ve) charge while hydrogen acquires a slightly positive (+ve) charge. Now being the opposite charges they start attracting each other and making the bond stronger as shown in Figure 3 thus it's very hard to separate the EG molecules. Thus this leads to having a higher boiling point of EG.

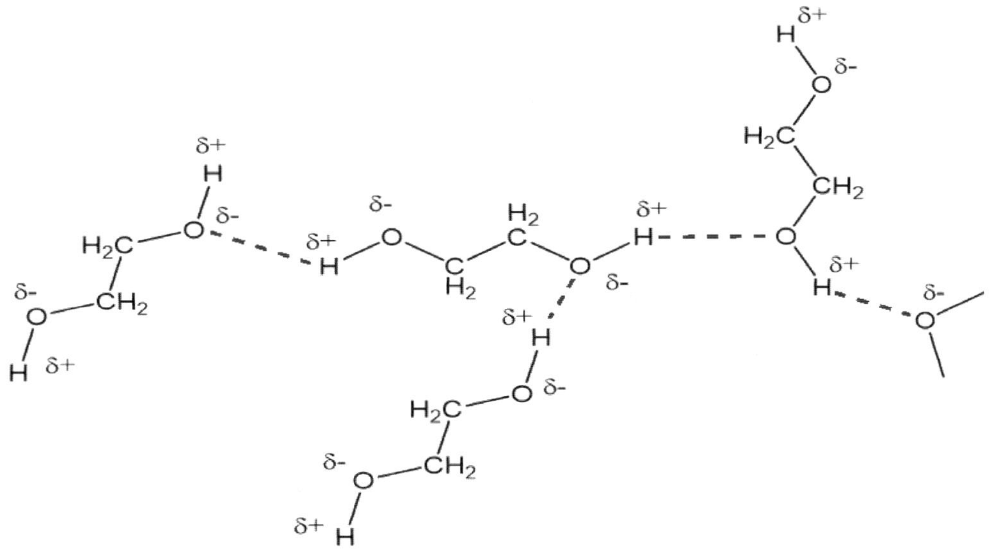

Figure 3. Hydrogen bonding in ethylene glycol [17]

Water also has O-H groups, the same things happen here. Bonds get stronger and thus leads to a higher boiling point of water. Now when both EG and water mix together they both form the hydrogen bond with each other as shown in Figure 4.

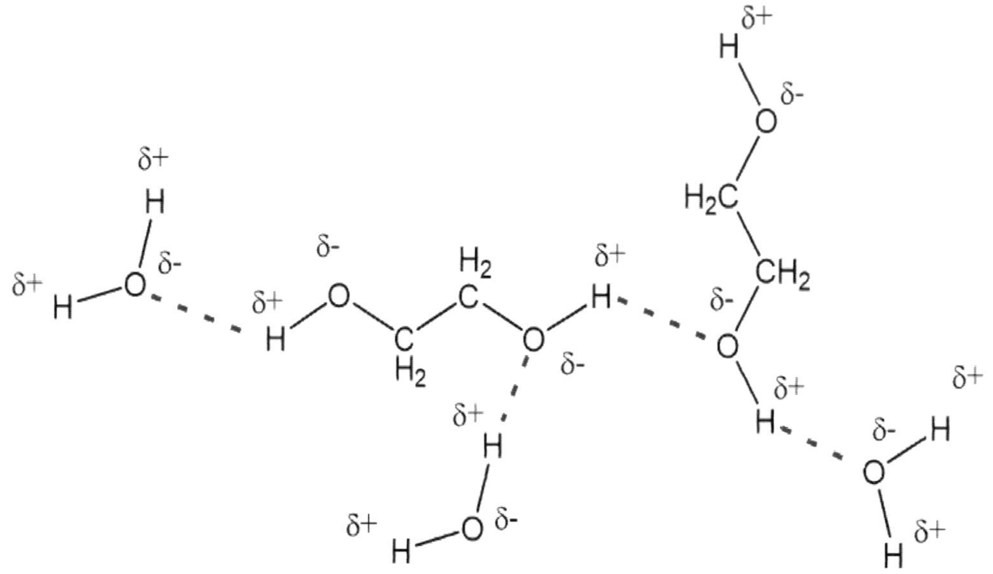

Figure 4. EG and water hydrogen bonding [17]

Ethylene glycol and water having hydrogen bonding with each other. As we know that water freezes at 0°C and the pure EG freezing point is at -12°C. When these two mixes with each other (30:70, water: EG) and forms the mixture its melting points reaches up to -55°C. This decrease in the melting point actually helps when this is used in the cooling system of a vehicle as even in adverse temperature condition it will be in liquid form and don't get solidified. Thus it also finds application in de-icing in aircraft [17]

4.3 Challenges in Anti freezing role of nanofluids- In the heat transmission process, nanofluids play a crucial role. But researchers while doing the research faces challenges too, which actually make it tough for them to carry out the research. Few of the challenges are as follows.

4.3.1 Sedimentation issue – Without using the surfactant, the prepared Nano fluid settles downfast and thus not that good for the research purpose. Optimum quantity of the surfactant is to be choose, otherwise the impact of the surfactant changes a lot of thermo physical properties and also introduce defect at the molecular interface as studied by **Xien H et al** [29]

4.3.2 Size- The effect of nanoparticle size on thermal conductivity was investigated by Vajjha and Das [30]. The efficient heat transfer mechanism necessitates the selection of an optimal nanoparticle size.

4.3.3 Rheological behavior -For application in the industries where the heat transfer fluids is to be transferred, it is an important concern because the dynamic viscosity impacts the pressure drop directly and so this ultimately impacts the required power to transport it.

4.3.4 Thermal conductivity – It is a well-known thing that conductivity increase with volume Concentration of nanoparticle. Finding an optimum value is necessary as more volume concentration is not efficient all the time and can cause decrease in the efficiency of the diesel engine.

4.4 Summary of Nano fluid studies using anti-freezing agents

The various researchers performed experiments by taking hybrid base fluids generally consisting of water and ethylene glycol. They reported a significant increase in heat transfer.

In TiO2 nanoparticles, **Bhanvase et al. (2014)** [18] investigated the distilled water, EG, and EG-water combination in the ratio 40:60. Heat transfer coefficient is improved, with a 105 percent improvement for a 0.5 percent volume fraction**. Zafar Said et al** [19] showed TiO_2 Nanofluid a better for a long run coolant, the results of this study can be used in automotive car radiators applications.

Devireddy et al [20] studies show when TiO_2 nanoparticle are used with 40:60% EG-W, heat transfer rate in the automobile radiator got increased. At 0.5% vol conc, 35% increase was observed. **Sanjeevawitharana et al** [21] PG and 1 wt. % ZnO, Al_2O_3 and TiO_2 nanofluids studied and they showed good stability even after 2 months of preparation. Al_2O_3 are more stable than others, whereas TiO_2, ZnO were found stable only in PG. **Yiamsawas et al.** [22] studied EG/water base fluid in Al_2O_3, TiO_2 nanofluid for 15 to 60°C.The results yielded the increase in dynamic viscosity with increase of size whereas there is decrease in this property with increasing temperature.

S.M. Peyghambarzadeh et al [23] investigated the characteristics of water and pure EG in binary mixes. When compared to their own base fluid, nanofluids improved heat transmission capabilities, according to the results of the trials. The maximum improvement was discovered to be 40%.

Hamid et al. [24] dealt with TiO_2 Nanofluid under different concentration 0.5, 1 and 1.5 % volume concentration, they found there is exponential decrease of the dynamic viscosity of the Nano fluid with temperature for a mixture of EG: DW by 40:60 volume ratio.

Ali Taghizadeh et al [31] obtained results of antifreeze with cerium oxide nanoparticle in the concentration range 0.015-0.035 vol%, temperature range 20-50°C found the maximum enhancement up to 36.13% for 0.025 vol% at 50° C, showing optimum value of the concentration is the key parameter for the enhancement.

Sara Rostam et al [32] found out that the value of thermal conductivity is found to be a direct function of temperature. As temperature increase the slope also increase in a good rate and thus thermal efficiency also increased.

Enhancement in the properties is the common feature seen after studying all the research papers. Different authors put forward the extent up to which improvement can be seen, it is generally found out that increasing the temperature leads to high thermal conductivity and generalized fluid flow characteristic can be seen. Different parameter leads to a generalized result and it can be seen the up to which extent a particular hybrid is efficient.

5. CONCLUSION

- Nanofluids and anti-freezing agent are very helpful in automobile, thermal industries for cooling or heat transfer, especially at low temperature. Nanofluids have a good potential in the heat transfer enhancement and this property actually help the working fluid to be cooler which is used in car radiators.
- Thermodynamic conductivity increases as the volume concentration and temperature rise.
- The viscosity increases as the volume fraction increases, and the viscosity reduces as the temperature rises, indicating an inverse relationship between temperature and viscosity.

REFERENCES

[1] Choi S U S and Eastman J A 1995 Enhancing thermal conductivity of fluids withNanoparticles *ASME Int. Mech. Eng. Congr. Expo.* **66** 99–105

[2] Rashmi W, Khalid M, Ong S S and Saidur R 2014 Preparation, thermophysicalproperties and heat transfer enhancement of nanofluids *Mater.Res. Express* **10159** 163–32001

[3] Mahmoudi M, Tavakoli M R, Mirsoleimani M A, Gholami A and Salimpour M R 2016 Experimental and numerical investigation on forced convection heatTransfer and pressure drop in helically coiled pipes using TiO_2 /water nanofluid *Int. J. Refrigeration* **74** 627–643

[4] Ravisankar B and Chand V T Influence of nanoparticle volume fraction, particle Size and temperature on thermal conductivity and viscosity of nanofluids- A review 2013*Int. J. Automot. Mech. Eng.* **8** 1316-38

[5] Azmi W H, Sharma K V, Mamat R and Anuar S Nanofluid properties for forced Convection heat transfer: An overview 2013 *J. Mech. Eng. Sci.* **4** 397-08

[6] Eastman J A, Choi S U S , Li S, Yu W and Thompson L J 2001 Anomalously Increased effective thermal conductivities of ethylene glycol-based nanofluidsContaining copper nanoparticles Appl. Phys. Lett. **78** 718-20

[7] Jwo C S, Teng T P and Chang J 2007 A simple model to estimate the thermal Conductivity of fluid with acicular nanoparticles *J Alloys Compd* **434** 569–71

[8] Kurihara L K, Chow G M, Schoen P E 1995 Nanocrystalline metallic powders andFilms produced by the polyol method *Nanostruct Mater* **6** 607–613

[9] Chen L, Xie H, Li Y, and Yu W 2008 Nanofluids containing carbon nanotubes treated by mechanochemical reaction *ThermochimicaActa* **477** 21–24

[10] Handbook A F 2009 *American Society of Heating, Refrigerating and Air-Conditioning Engineers* ASHRAE Inc. (Atlanta, GA, USA)

[11]Sidik N A, Yazid M N and Mamat R 2017 Recent advancement of nanofluids in engine cooling system *Renew Sustain Energy Rev* **75** 137–44

[12]Engineering toolbox. Ethylene Glycol Heat-Transfer Fluid 2018 Available online: *https://www. engineeringtoolbox.com/ethylene-glycol-d_146.html* (accessed on 15 November 2018)

[13]Kahani M, Zeinali Heris S and Mousavi S M 2013 Multiwalled carbon nanotube/water nanofluid or helical coiling technique, which of them is more effective? *IndEngChem Res* **52** 13183–91

[14]Kumar A and Subudhi S 2019 Preparation, characterization and heat transfer analysis ofNanofluids used for engine cooling *Appl Therm Eng* **160** 114092.

[15]Naddaf A, Heris S Z and Pouladi B 2019 An experimental study on heat transfer Performance and pressure drop of nanofluids using graphene and multi-walledCarbon nanotubes based on diesel oil *Powder Technol* **352** 369–80

[16]Gupta A, Roy A, Gupta S and Gupta M 2020 Numerical investigation towards Implementation of punched winglet as vortex generator for performance Improvement of a fin-and-tube heat exchanger *Int J Heat Mass Tran* **149** 119171

[17]Cotton Simon 2018 *Ethylene glycol - Molecule of the Month June* 2018 [Archived version] Available online: *http://www.chm.bris.ac.uk/motm/ethylene-glycol/glycolh.htm*

[18] Bhanvase B A, Sarode M R, Putterwar L A, Abdullah K A, Deosarkar M P and Sonawane S H 2014 Chemical Engineering and Processing: Process Intensification Intensification of convective heat transfer in water /ethylene glycol- Based nanofluids containing TiO_2 nanoparticles *Chem. Eng. Process. Process Intensif.* **82** 123–131.

[19]Said Z, Assad M E H, Hachicha A A, Bellos E, Abdelkareem M A, Alazaizeh D Z and Yousef B A 2019 Enhancing the performance of automotive radiatorsUsing nanofluids. *Renew. Sust. Energ. Rev.* **112** 183-194

[20] Devireddy S, Mekala C S R and Veeredhi V R 2016 Improving the cooling Performance of automobile radiators with ethylene glycol water-based TiO_2 Nanofluids *Int. Commun. Heat Mass Transf*.**78** 121-126

[21] Witharana S, Palabiyik I, Musina Z and Ding Y 2013. Stability of glycol nanofluids—the theory and experiment *Powder Technol.* **239** 72-77
[22]Yiamsawas T, Mahian O, Dalkilic A S, Kaewnai S and Wongwises S 2013 Experimental studies on the viscosity of TiO2 and Al2O3 nanoparticles suspended in a mixture of ethylene glycol and water for high temperature applications *Appl. Energy* **111** 40-45
[23]Peyghambarzadeh S M, Hashemabadi S H, Hoseini S M and Jamnani M S 2011 Experimental study of heat transfer enhancement using water/ethylene glycol based nanofluids as a new coolant for car radiators *Int. Commun. Heat Mass Transf.* **38** 1283-90
[24]Hamid K A, Azmi W H Mamat R and Sharma K V 2016. Experimental Investigation on Heat transfer performance of TiO_2 nanofluids in water ethylene Glycol mixture *Int. Commn. Heat Mass Transf.* **73** 16–24
[25]Patil M S, Seo J H, Kang S J and Lee M Y 2016 Review on Synthesis, Thermo- Physical Property, and Heat Transfer Mechanism of Nanofluids *Energies* **9** 840
[26]Pak B C and Cho Y I 1998 Hydrodynamic and heat transfer study of dispersed fluids withsubmicron metallic oxide particles *Exp. Heat Transf.* **11** 151–170.
[27]Masoumi N, Sohrabi N and Behzadmehr A 2009 A new model for calculating the Effective viscosity of nanofluids *J. Phys. D: Appl. Phys.* **42** 055501
[28]Kole M and Dey T K 2010 Viscosity of alumina nanoparticles dispersed in car enginecoolant *Exp. Therm. Fluid Sci.* **34** 677–83
[29]Xie H and Chen L 2011 Review on the preparation and thermal performances of carbonnanotube contained nanofluids *J. Chem. Eng. Data* **56** 1030-41
[30]Vajjha R S and Das D K 2009 Experimental determination of thermal conductivity of threenanofluids and development of new correlations *Int. J. Heat Mass Transf.* **52** 4675–82
[31]Taghizadeh A, Taghizadeh M, Azimi M, Alsagri A S, Alrobaian A A and Afrand M 2020 Influence of cerium oxide nanoparticles on thermal conductivity of antifreeze. Journal of Thermal Analysis and Calorimetry **139** 225-36
[32]Rostami S, Nadooshan A A and Raisi A 2019 An experimental study on the thermal conductivity of new antifreeze containing copper oxide and graphene oxide nano-additives *Powder Technol.* **345** 658-67.

IOP Conf. Series: Materials Science and Engineering 1225 (2022) 012050 doi:10.1088/1757-899X/1225/1/012050

Synthesis and characterization of Graphene Oxide and its reduction with different reducing agents

Karan Bansal*, Jagdeep Singh and A. S. Dhaliwal

Department of Physics, Sant Longowal Institute of Engineering and Technology, Longowal-148106, Punjab, India
**corresponding author e-mail*: karban083@gmail.com

Abstract: - The present work reports the simplistic formation of graphene oxide (GO) and reduced graphene oxide (rGO) by oxidizing graphite powder with the 9:1 volume combination of concentrated sulfuric acid/ and ortho-phosphoric acid (H_2SO_4/H_3PO_4) in the company of $KMnO_4$ as an oxidant. The acid combination solution of H_2SO_4/H_3PO_4 produces the improved oxidized hydrophilic GO without any evolution of hazardous gases instead of only H_2SO_4, which differentiates this procedure from the Hummers' method. Then, the synthesized GO is diminished to rGO by employing various reducers namely hydrazine hydrate, sodium borohydride and, ascorbic acid through the chemical route. The properties of synthesized materials have been investigated by diverse investigative systems that are X-ray diffraction (XRD), Field Emission Scanning Electron Microscope (FESEM), Raman spectroscopy and, Energy Dispersive X-ray spectroscopy (EDS) which approved its high crystallinity and reduction of GO into rGO. EDS analysis confirms that the carbon to oxygen ratio is higher in rGO prepared by using hydrazine hydrate in comparison to rGO made by using sodium borohydride and ascorbic acid which reveals the good reducing capability of hydrazine hydrate. Hence, it may be concluded that the rGO synthesized by the reduction of GO in the presence of hydrazine hydrate is better as compared to others and it is expected that the synthesized material may be used for various applications.

1. Introduction

Graphene is the thinnest and strongest material having a two-dimensional (2D) structure with sp^2-hybridized carbons [1-3]. The material based on graphene has been widely utilized in support of several advanced purposes namely Li-ion batteries, Supercapacitors, etc[4-15]. For the synthesis of GO, various synthesis methods in particular thermal reduction [16], chemical reduction [17], electrochemical reduction [18] have been used. In the chemical reduction technique for the oxidation of graphite, the Tour method is widely used because it is cost-effective and environmentally friendly [19]. Moreover, some modification in this process has been employed to develop the efficiency of the oxidation of graphite [20-21]. The properties and performance of synthesized material in the above-stated applications are dependent on the oxygen content [22-24]. The complete reduction of GO into reduced graphene oxide (rGO) and complete exclusion of oxygen from it is still a contest. To achieve this goal, some modification in the well-known reported method [25] has been carried out in the current design for the formation of rGO by using various reducing agents. To the best of our knowledge, the comparison of effects of various reducers to produce rGO from GO has been not available in the literature. Hence, the present studies related to the synthesis of graphene oxide by using graphite powder by Tour method and then reduced into rGO by various reductants such as ascorbic acid, sodium borohydride, Hydrazine

Hydrate. Furthermore, the properties of synthesized samples have been investigated by several diagnostic systems such as X-ray diffraction (XRD), Raman spectroscopy, Field Emission Scanning Electron Microscope (FESEM), and Energy Dispersive X-ray spectroscopy (EDS).

2. Synthesis of GO and rGO

2.1 Chemicals and materials

Graphite flake (natural, 325 mesh) along with a purity of 99.8% is purchased from Sigma Aldrich. Sulphuric acid (H_2SO_4), potassium permanganate ($KMnO_4$), barium chloride ($BaCl_2$), hydrogen peroxide (H_2O_2), Phosphoric acid (H_3PO_4), Hydrochloric acid (HCl) are obtained from Merck, India. Distilled water has been utilized during all the experiments.

2.2 Preparation of GO

GO has been synthesized by the Tour method [25-26]. Firstly, an acid solution having 9:1 (volume/volume ratio) of concentrated H_2SO_4 and H_3PO_4 has been prepared. Then, the mixture of 9 g of $KMnO_4$ (potassium permanganate) and 1 g of graphite powder (in 9:1 ratio) is added to the prepared acid mixture. After that, the reaction mixture is heated and stirred for 12 hours at 50°C. After this time interval, the thick paste having greenish-black colour is formed and then cool down the reaction mixture at room temperature. Finally, the distilled water of quantity 500 millilitres is added into the reaction mixture for stopping this reaction. In order to decrease the ions of manganese to soluble manganese sulphate ($MnSO_4$) and manganese oxides, 20 millilitres of 30 wt.% hydrogen peroxide (H_2O_2) is added, as defined by the reaction given below [27]:

$$2KMnO_4 + 5H_2O_2 + 3H_2SO_4 \longrightarrow 2MnSO_4 + K_2SO_4 + 8H_2O \text{ (steam)} + 5O_2 \quad (1)$$

With the addition of 30 weight percent of H_2O_2, bubbling occurs, and a clear pale colour appears that signifying a high level of oxidation. The reaction product is centrifuged by centrifugation at 4000 rpm and cleaned with distilled water and 5% HCl mixture again and again till the sulphate ions are completely removed. The verification of elimination of the metal sulphate ions has been done by applying barium chloride ($BaCl_2$) mixture. Finally, the product is dried up at 60 °C.

Figure 1. (a) Shows the bright yellow colour when 30 wt.% H_2O_2 is added to the mixture.

(b) the obtained solution of graphene oxide (GO) after centrifugation at 4000 rpm and rinsed with distilled water and 5% HCl mixture continually for removal of the metal sulphate ions.

2.3 Reduction of Go with Ascorbic acid

Added 25 mg GO in 50 ml of Deionized water and sonicated it for 1 hour till a homogenous solution would be obtained. Maintained the PH value to 10-11 (by using 0.1 M of NaOH solution). Then added 100 mg of Ascorbic acid and stirred for 1 hour at 95 °C. Filtered out the solid product by using Whatman filter paper and cleaned with ethanol and distilled water. Then dried up in the oven at 60 °C.

2.4 Reduction of GO with sodium borohydride

Added 25 mg GO in 50 ml of Deionized water and sonicated it for 1 hour till a homogenous solution would be obtained. Maintained the PH value to 10-11 (by using 0.1 M of NaOH solution). Then added 200 mg of sodium borohydride ($NaBH_4$) and stirred for 1 hour at 95 °C. Filtered out the solid product by using Whatman filter paper and cleaned with ethanol and distilled water. Then dried up in the oven at 60 °C.

2.5 Reduction of GO with Hydrazine Hydrate

Added 25 mg GO in 50 ml of Deionized water and sonicated it for 1 hour till a homogenous solution would be obtained. Maintained the PH value to 10-11 (by using 0.1 M of NaOH solution). Added 4 ml Hydrazine Hydrate and stirred at 90 °C for 1 hour. Filtered out the solid product by using Whatman filter paper and cleaned with ethanol and distilled water. Then dried up in the oven at 60 °C.

3. Characterization

Bruker D8 advance X-ray diffractometer having Cu Kα radiation (λ = 1.54060 Å) has been used for noting X-ray diffraction (XRD) patterns to explore the crystallinity and structure of formed samples. The surface morphology of synthesized samples has been recorded by using a field emission scanning electron microscope (JSM-7610FPlus). To enhance the surface conductivity, the samples have been coated with gold by using a DII-29030SCTR smart coater. Furthermore, the elemental analysis has been carried out using Energy-dispersive X-ray spectroscopy (EDS). Renishaw In-Via confocal micro-Raman spectrometer with 488 nm argon-ion laser at room temperature has been used for noting the Raman spectra of synthesized materials.

3.1 X-ray diffraction

The XRD pattern of Graphite, GO and rGO synthesized by using different reducing agents have been presented in Figure 2. In the case of graphite, the diffraction peak is available at 26.7° with an interlayer spacing of 3.7 Å. Moreover, in the case of GO, the diffraction peak appears at 10.9° with an interlayer distance of 8.5 Å. The other small peaks around 43° in the case of GO are not observed which confirms that synthesized GO is not in limited order in mounded sheets [28]. The shifting in the position of the peaks in the case of rGO_AA, rGO_HH, rGO_SB is due to the decline in the interlayer gap owing to the expulsion of groups comprising oxygen in the reduction procedure. It can be too observed that the XRD spectrum of different reduced graphene oxide is not smooth and small peaks are noticed around 43° because there are some residuary functional units on the rGO that affects the rGO assembly [29].

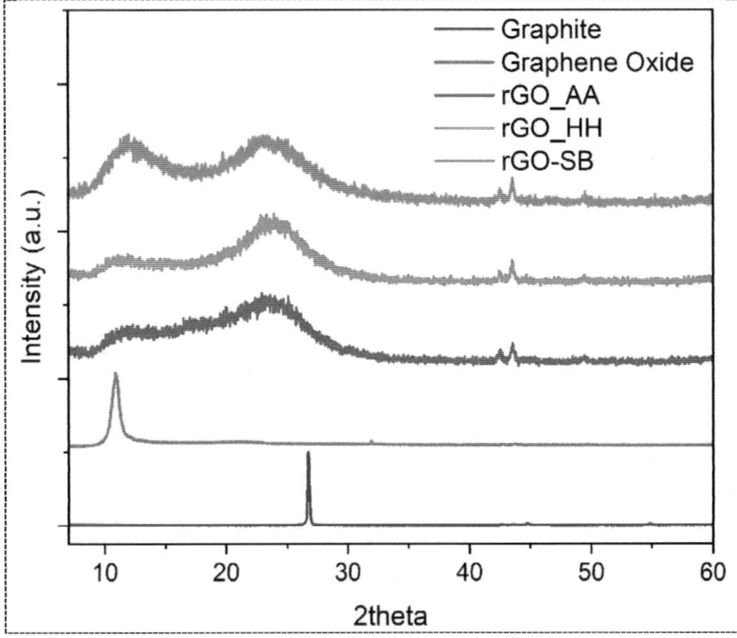

Figure 2.
XRD spectrum of Graphite, GO, and reduced graphene oxide (rGO) synthesized by using hydrazine hydrate (rGO_HH), sodium Borohydride (rGO_SB), and ascorbic acid (rGO_AA).

3.2 Raman Spectra

The Raman spectra of Graphite, GO and rGO produced by utilizing distinct reducers, have been presented in Figure 3. In the case of graphite, the graphitic pattern (G band) occurs at 1575 cm^{-1} due to highly ordered graphite. Unlike peaks are noted at 1365 cm^{-1} and 1591 cm^{-1} in the case of graphene oxide which correspond to the D band and G band of the carbonic matter. The fraction of the intensity of D height and that of G height (I_D/I_G) are about 0.9 and 1.0 for GO and rGO synthesized by using various reducing agents, respectively. This (I_D/I_G) ratio confirms the mean dimension contraction of sp^2 realms [30].

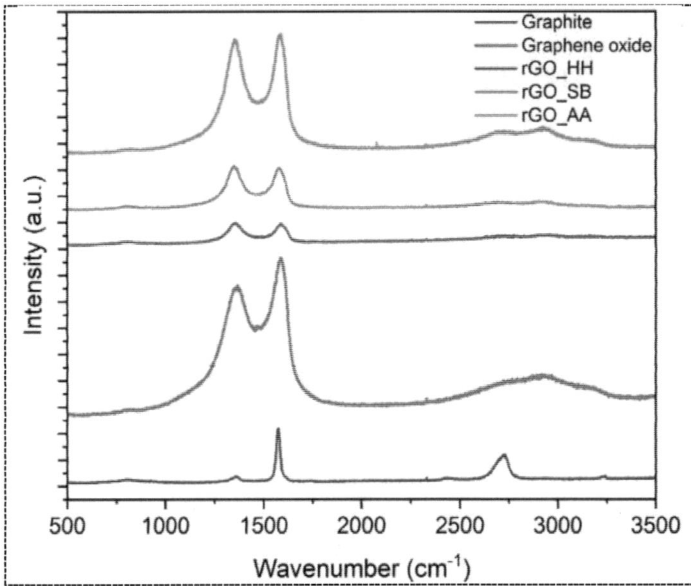

Figure 3.
Raman spectra of Graphite, graphite, GO, and reduced graphene oxide (rGO) synthesized using hydrazine hydrate (rGO_HH), sodium Borohydride (rGO_SB), and ascorbic acid (rGO_AA)

IOP Conf. Series: Materials Science and Engineering 1225 (2022) 012050 doi:10.1088/1757-899X/1225/1/012050

3.3 FESEM

In Figure. 4, the micrograph of GO confirms its sheet-type structure with a magnification of 25,000x. The surface morphologies of synthesized rGO_AA, rGO_HH and rGO_SB confirm its wrinkled stacked layers structure [31]. In rGO, several edges are visible on it, and it is very essential for the interaction of synthesized material with other materials. It is noticeable, the complete change in the surface morphologies has been brought by the conversion of GO into rGO. The more wrinkled type of structure has been noted in the case of sodium borohydride as correlated to hydrazine hydrate and ascorbic acid.

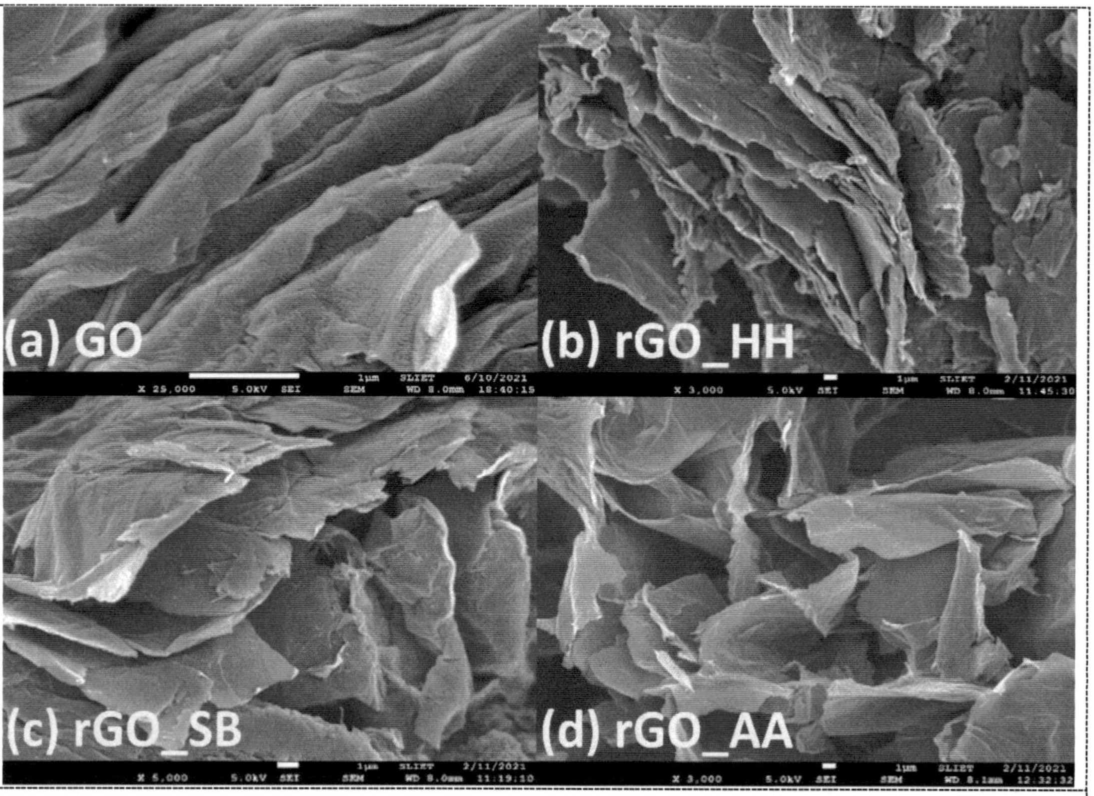

Figure 4 (a) FESEM image of GO at a magnification of 25,000x (b) FESEM image of rGO prepared by hydrazine hydrate. (c) FESEM image of rGO prepared by sodium borohydride. (d) FESEM image of rGO prepared by ascorbic acid.

3.4 Energy-dispersive X-ray spectroscopy (EDS).

The EDS spectra of synthesized rGO_AA, rGO_HH and rGO_SB is displayed in Figure 5. The carbon to oxygen ratio is a significant parameter that determines the oxygen content present in the sample. In the case of GO, the C:O fraction is about 2.53. The C:O fraction after the reduction of GO into rGO is about 9.99, 7.62 and 6.69 respectively, by employing a variety of reducers such as hydrazine hydrate, ascorbic acid, sodium borohydride. The higher C:O fraction in the case of hydrazine hydrate confirms its effectiveness in the reduction procedure.

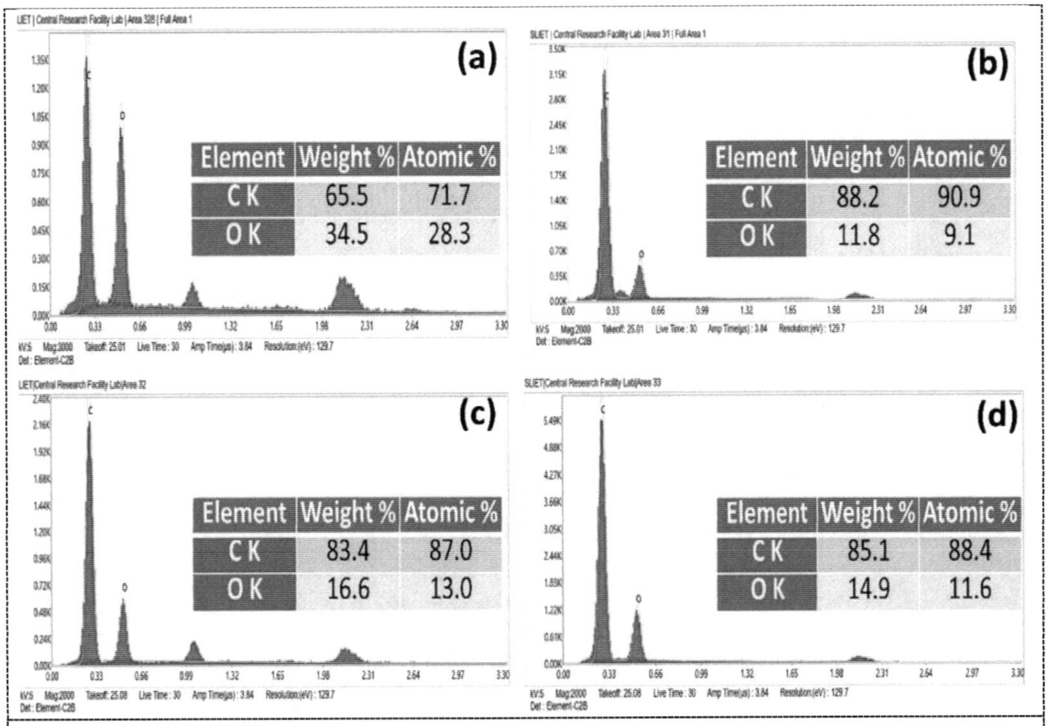

Figure. 5 EDS of synthesized (a) graphene oxide (b) rGO_HH (c) rGO_SB (d) rGO_AA

4. Conclusion and Summary

Due to the simple synthesis and astonishing properties of graphene oxide, it is an auspicious material. Although the process of formation decides the Graphene oxide assembly. Here, the synthesis of graphene oxide by the Tour method is less hazardous, more efficient, and eco-friendlier than Modified Hummer's method without the release of toxic gases. Further, reduction of graphene oxide was done using a different reducing agent and characterized by XRD, FESEM, Raman Spectra, and EDS. The qualities of graphene oxide are affected using different chemicals and reducers because they produce different types of impurities. The reduction GO into rGO is effective in the case of hydrazine hydrate as a reducing agent as indicated by the higher carbon to oxygen ratio. Hence, it may conclude that the rGO synthesized by the reduction of GO in the presence of hydrazine hydrate is better as compared to other reducing agents and it is expected that the synthesized material has improved properties and may be used for various advanced applications.

References

[1] Rao C N R, Sood A K, Subrahmanyam K S and Govindaraj A 2009 Graphene: The new two-dimensional nanomaterial *Angew. Chemie - Int. Ed.* **48** 7752–77

[2] Geim A K 2009 Graphene: Status and prospects *Science (80-.).* **324** 1530–4

[3] Morozov S V., Novoselov K S, Katsnelson M I, Schedin F, Elias D C, Jaszczak J A and Geim A K 2008 Giant intrinsic carrier mobilities in graphene and its bilayer *Phys. Rev. Lett.* **100**

[4] Singh J and Dhaliwal A S 2022 Electrochemical and photocatalytic degradation of methylene blue by using rGO/AgNWs nanocomposite synthesized by electroplating on stainless steel *J. Phys. Chem. Solids* **160** 110358

[5] Choi J H, Lee J, Byeon M, Hong T E, Park H and Lee C Y 2020 Graphene-Based Gas Sensors with High Sensitivity and Minimal Sensor-to-Sensor Variation *ACS Appl. Nano Mater.* **3** 2257-65

[6] Chen X and Tian Y 2021 Review of Graphene in Cathode Materials for Lithium-Ion Batteries *Energy and Fuels* **35** 3572–80
[7] Gao J, Wu H, Li A, Yue Y, Xie D and Zhang X 2019 Graphene Nanofluids as Thermal Management Materials: Molecular Dynamics Study on Orientation and Temperature Effects *ACS Appl. Nano Mater.* **2** 6828–35
[8] Singh J and Dhaliwal A S 2020 Plasmon-induced photocatalytic degradation of methylene blue dye using biosynthesized silver nanoparticles as photocatalyst *Environ. Technol. (United Kingdom)* **41** 1520–34
[9] Liu J, Cui L and Losic D 2013 Graphene and graphene oxide as new nanocarriers for drug delivery applications *Acta Biomater.* **9** 9243–57
[10] Zhou T, Zhou X and Xing D 2014 Controlled release of doxorubicin from graphene oxide based charge-reversal nanocarrier *Biomaterials* **35** 4185–94
[11] Han T H, Kim H, Kwon S J and Lee T W 2017 Graphene-based flexible electronic devices *Mater. Sci. Eng. R Reports* **118** 1–43
[12] Velasco A, Ryu Y K, Boscá A, Ladrón-De-Guevara A, Hunt E, Zuo J, Pedrós J, Calle F and Martinez J 2021 Recent trends in graphene supercapacitors: From large area to microsupercapacitors *Sustain. Energy Fuels* **5** 1235–54
[13] El-Kady M F, Shao Y and Kaner R B 2016 Graphene for batteries, supercapacitors and beyond *Nat. Rev. Mater.* **1**
[14] Su H and Hu Y H 2021 Recent advances in graphene-based materials for fuel cell applications *Energy Sci. Eng.* **9** 958–83
[15] Czerniak-Reczulska M, Niedzielska A and Jędrzejczak A 2015 Graphene as a Material for Solar Cells Applications *Adv. Mater. Sci.* **15** 67–81
[16] Chen W, Yan L and Bangal P R 2010 Preparation of graphene by the rapid and mild thermal reduction of graphene oxide induced by microwaves *Carbon N. Y.* **48** 1146–52
[17] De Silva K K H, Huang H H, Joshi R K and Yoshimura M 2017 Chemical reduction of graphene oxide using green reductants *Carbon N. Y.* **119** 190–9
[18] Shao Y, Wang J, Engelhard M, Wang C and Lin Y 2010 Facile and controllable electrochemical reduction of graphene oxide and its applications *J. Mater. Chem.* **20** 743–8
[19] Singh J and Dhaliwal A S 2018 Synthesis, characterization and swelling behavior of silver nanoparticles containing superabsorbent based on grafted copolymer of polyacrylic acid/ Guar gum *Vacuum* **157** 51–60
[20] HIRATA M, GOTOU T, HORIUCHI S, FUJIWARA M and OHBA M 2004 Thin-film particles of graphite oxide 1:High-yield synthesis and flexibility of the particles *Carbon N. Y.* **42** 2929–37
[21] Nakajima T, Mabuchi A and Hagiwara R 1988 A new structure model of graphite oxide *Carbon N. Y.* **26** 357–61
[22] Bansal S A, Singh S, Srivastava A, Singh A P and Kumar S 2021 Covalent attachment of 2D graphene oxide (GO) sheets with poly allylamine (PAA) for enhanced mechanical performance: Theoretical and experimental study *Polymer (Guildf).* **213**
[23] Singh J, Kumar S and Dhaliwal A S 2020 Controlled release of amoxicillin and antioxidant potential of gold nanoparticles-xanthan gum/poly (Acrylic acid) biodegradable nanocomposite *J. Drug Deliv. Sci. Technol.* **55** 101384
[24] Singh J and Dhaliwal A S 2019 Novel Green Synthesis and Characterization of the Antioxidant Activity of Silver Nanoparticles Prepared from Nepeta leucophylla Root Extract *Anal. Lett.* **52** 213–30
[25] Higginbotham A L, Kosynkin D V., Sinitskii A, Sun Z and Tour J M 2010 Lower-defect graphene oxide nanoribbons from multiwalled carbon nanotubes *ACS Nano* **4** 2059–69
[26] Twinkle, Singh K, Bansal S A and Kumar S 2019 Graphene oxide (GO)/Copper doped Hematite (α-Fe2O3) nanoparticles for organic pollutants degradation applications at room temperature and neutral pH *Mater. Res. Express* **6**
[27] Shahriary L, Ghourchian H and Athawale A A 2014 Graphene-multiwalled carbon nanotube hybrids synthesized by gamma radiations: Application as a glucose sensor *J. Nanotechnol.*

[28] Stobinski L, Lesiak B, Malolepszy A, Mazurkiewicz M, Mierzwa B, Zemek J, Jiricek P and Bieloshapka I 2014 Graphene oxide and reduced graphene oxide studied by the XRD, TEM and electron spectroscopy methods *J. Electron Spectros. Relat. Phenomena* **195** 145–54
[29] Krishnamoorthy K, Veerapandian M, Yun K and Kim S J 2013 The chemical and structural analysis of graphene oxide with different degrees of oxidation *Carbon N. Y.* **53** 38–49
[30] Xue Y, Zhu L, Chen H, Qu J and Dai L 2015 Multiscale patterning of graphene oxide and reduced graphene oxide for flexible supercapacitors *Carbon N. Y.* **92** 305–10
[31] Marcano D C, Kosynkin D V., Berlin J M, Sinitskii A, Sun Z, Slesarev A, Alemany L B, Lu W and Tour J M 2010 Improved synthesis of graphene oxide *ACS Nano* **4** 4806–14

Electro-Magnetic switching in NiO-Graphene film

S. Goswami[1], M. Chakraborty[1#], D. De[1,2*]
[1]Material Science Research Lab, The Neotia University, Sarisa, D.H. Road, 24 Pgs (South) West Bengal 743368, India
[2]Sukumar Sengupta Mahavidyalaya, State Highway 7, Keshpur, Paschim Medinipur 721150, West Bengal, India.
E-mail: #manashi.chakraborty@tnu.in, *debajyoti.de@tnu.in

Abstract. Nickel oxide (NiO) thin film is grown via pulse laser deposition (PLD) technique and it is trapped in between conducting graphene films deposited through the same technique. Epitaxial crystalline growth of both NiO and graphene films are confirmed from X-ray diffraction studies. Raman studies propose creation of pure graphene film with acceptable defects. Electrical transport of the NiO film reveal resistance switching properties for an wide range of temperature which is useful for resistive random access memory (RRAM) and *electric-switch*. Besides electrical switching, the transport properties of the NiO film depict a systematic response in influence of magnetic field. Resistance of the NiO film changed significantly with external magnetic field which makes the system useful as a *magnetic-switch*.

1. Introduction

Nickel oxide (NiO) is widely studied since decades because of its rich applications in gas sensors, spin valves, supercapacitors besides nonvolatile resistive random access memory (RRAM) in 2D geometry (film) because of its simple structure, low power consumption, high-speed performance etc. NiO thin films formed via chemical/physical vapor deposition, drop/dip casting, pulse laser deposition (PLD) technique etc. occasionally form metal-insulator-metal (MIM) structure and reveal voltage-induced switching useful for potential applications in resistive random access memory (RRAM). In 1964 Gibbons *et al* [1] observed electrical switching in NiO thin pallet and since then this has been extensively studied by the scientific community. Formation of localized metallic conducting filaments (as observed from transmission electron microscopy (TEM) and atomic force microscopy (AFM)), phonon-associated tunneling mechanism, Schottky barriers with trapped charges in the interface states, space-charge-limited current are proposed to be responsible in different studies for this switching of resistive values between two applied voltages of same polarities [2, 3, 4, 5, 6, 7]. However, the exact reason behind switching mechanism in RRAM is yet to be understood completely. Emphasis has been given to study the nonvolatile performance, tuning the switching voltage and explore applications at/near room temperature after preparing films of NiO on different substrates [8, 9]. In this regards introduction of suitable doping and variation film thickness has been extensively studied whatsoever [3]. Semiconducting metal oxide NiO shows promising resistive switching memories (RSM) characterized via hysteresis in the $I-V$ loop. Because of widespread application on nanostructure RSM, controlling the same by externally applied parameters has been a key issue of research. Controlling RSM by externally applied magnetic field using the strong influence of the same on NiO film, is one important probable solution [10].

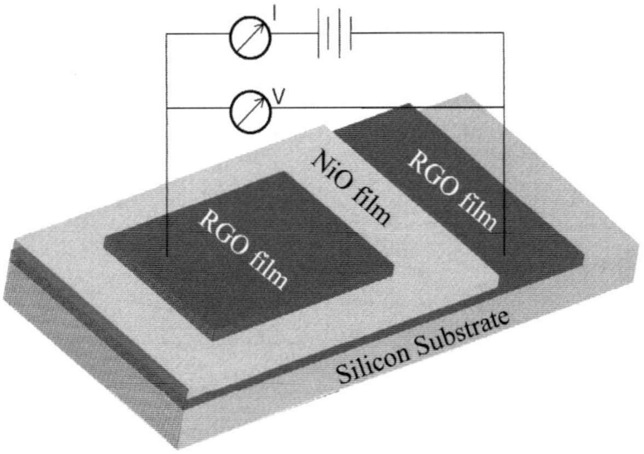

Figure 1. Schematic representation of the graphene-NiO-graphene film.

Nowadays, exfoliated Graphite and Graphene films are extensively studied because of its rich electrical and thermal conductivity, mechanical stability, flexibility, transparency and unique physico-chemical properties. Graphene film is also a suitable alternate of indium tin oxide (ITO) for using as conducting electrode [11]. Multilayer films using graphene and metal oxide may explore the combined physical properties and thus may be used as multifunctional smart materials, useful for potential device fabrications. Pulse laser deposition (PLD) is one simplified technique which provides the platform to grow graphene films on metal/glass/Si substrates directly from Graphite [12, 13]. At the same time NiO film deposition is possible via the same technique in next step without hampering the morphology or unwanted air contamination in one advanced PLD system with multiple targets. However, at the same time PLD extends the flexibility of tuning the characteristics of the film by controlling substrate temperature, laser power, wavelength, gas pressure etc. Herein, we have prepared multilayer thin film of graphene-NiO-graphene for exploring resistive switching properties under influence of electrical and magnetic fields [10]. The fabricated film demonstrates a switching from high resistance to low resistance state at a particular threshold voltage. At a small applied voltage, the resistance of NiO film is generally very high (known as OFF state), by increasing the voltage gradually, at a critical threshold, the resistance of the system reduces significantly – which is known as the ON state.

2. Fabrication

Tri-layer graphene-NiO-graphene is deposited on Si(100) substrate using a pulsed laser deposition (PLD) system (Model No: CompexPro 205 F, KrF) [14, 15]. Thick pallets of Graphite and NiO are prepared and calcined well before using as source. Target Si substrate is carefully cleaned several times with acetone and water in ultrasonic bath before being placed into the vacuum chamber as target. At first, for deposition of graphene film, the operating energy of the laser and target-to-substrate distance set to 450 mJ (λ=248 nm) and 10 cm respectively. Substance temperature is maintained at 650° C, number of laser pulse applied 10,000 with frequency 10 Hz and the whole procedure is done in $\sim 10^{-6}$ Torr pressure with flow of Ar gas in the deposition chamber. Ar is used to prevent the possibility of oxidation leading to formation of graphene oxide (GO). After completion of deposition, the substrate temperature is brought to room temperature slowly (4 K/min) to allow crystallization and formation of epitaxial film [16, 17].

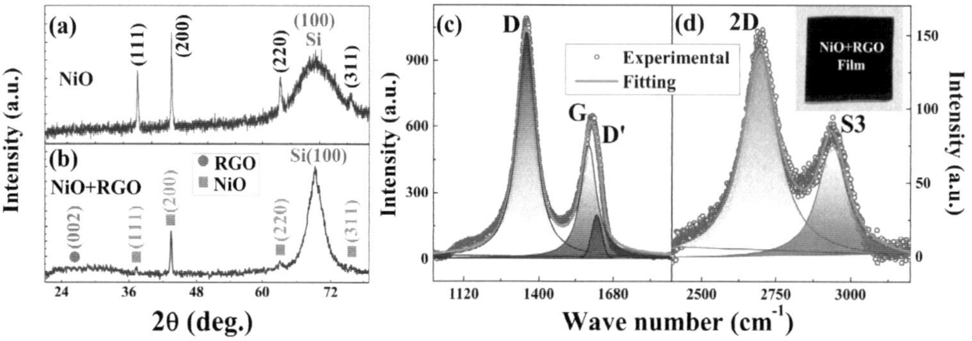

Figure 2. (a) and (b) X-ray diffraction patterns of NiO-graphene film. Well crystallinity and absence of impurity phases are confirmed from the pattern. (c) and (d) Raman spectra of NiO-Graphene film at room temperature revealing *D*, *G*, *2D* and *S3* peaks. After deconvolution of the experimental curve, individual components of corresponding peaks are generated. Shaded regions represent intensities (area) of individual curves.

After the substrate reaching room temperature, second step of the deposition is started with suitable masking, so not expose the whole film. Now, for the deposition of NiO film, 60,000 laser pulse is applied on the NiO source in $\sim 10^{-3}$ Torr oxygen ambient atmosphere and other conditions are kept unaltered. Substrate temperature is maintained at 400° C [12, 13]. After the laser exposure is over, the substrate is again brought to room temperature slowly (2 K/min). The final layer of graphene is fabricated on NiO similarly like the initial layer after desired masking.

Characterization of the as-synthesized film is performed by powder X-ray diffractometer (PXRD) (Bruker D8 Advanced Diffractometer) using Cu K_α (λ=1.54184 Å) radiation source; PXRD pattern of the samples are recorded with a scan speed of 0.02 °/2 s. Dc electrical transport measurements are performed via commercial PPMS using a source meter (Keithley, 2400) coupled with a computer by the GPIB network and magnetization measurements are performed in commercial SQUID magnetometer (MPMS-3).

3. Experimental Results

3.1. X-Ray and Raman Studies

X-ray diffraction pattern of the thin film recorded at room temperature is depicted in Figs. 2(a) and (b). Figure 2(a) focuses on the layer of only NiO deposition whereas Fig. 2(b) reveals the diffraction pattern after deposition of both the layers (graphene and NiO). Most intense peaks in both the figures around 70° appears from the contribution of Si(100) substrate. Significantly, simultaneous presence of honeycomb ($P63/mmc$) plane (022) for graphene and ($Fm3m$) planes (111,200,220,311) for NiO are noted in Fig. 2(b) [18, 19, 20]. All the planes are well indexed which are in accordance with standard JCPDS. Lattice constants may be roughly estimated from the peak positions (for graphene a=2.45 Å, for NiO a=4.18 Å). Significant mismatch between the lattice constants of the samples and substrate (a=5.43 Å) provides moderate tensile strain at the interface [15, 14].

NiO-graphene film has undergone Raman spectroscopy studies (measured at 514.5 nm excitation) to characterize vibrational, rotational and other low-frequency modes present in the system [21, 22, 23, 24]. As expected for graphitic structure, characteristic D ($\sim$1350 cm^{-1}), G ($\sim$1600 cm^{-1}), $2D$ ($\sim$2700 cm^{-1}) and $S3$ ($\sim$2940 cm^{-1}) peaks are noticed with increase in

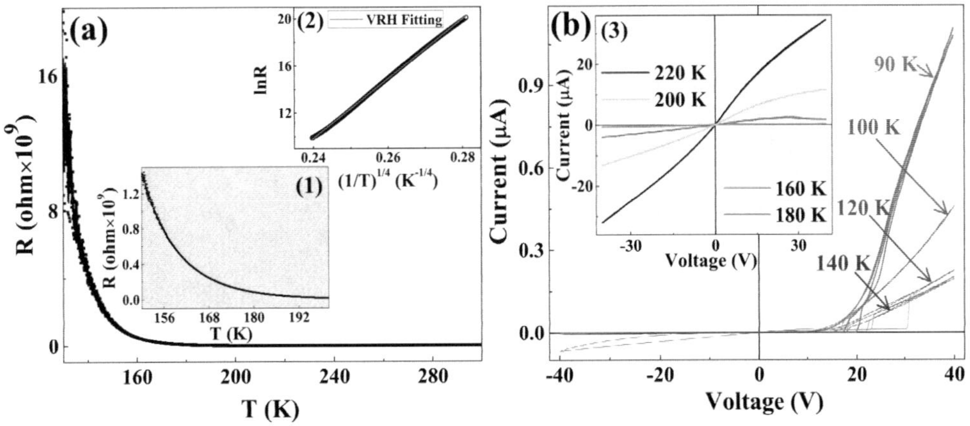

Figure 3. (a) Variation of resistance with temperature reveling semiconducting behavior of NiO film. Inset [1] magnifies the semiconducting nature. Inset [2] depicts the fitting of $R-T$ curve with VRH model. (b) $I-V$ curves measured at different temperatures exhibit sharp switching between low and high conducting states above a threshold voltage (V_{Th}). The inset [3] depicts $I-V$ curves with no switching effect at higher temperature.

wave number as depicted in Figs. 2(c) and (d). As per earlier reports, characteristic feature of the Raman spectra of graphite-like structure i. e. G and D bands are well indexed. This not only points formation of well crystalline graphene films, but also confirms graphitic structure. Careful observation reveal notable asymmetric G peak which is confirmed from deconvolution. Existence of small satellite D' peak appears indicating defects in RGO sheets; I_D/I'_D ~14, primarily indicates sp^3 type defects in the graphene film [24, 25]. Deconvolution further indicates the creation of single layered RGO structure because of appearance of symmetric D and $2D$ peaks (depicted in Figs. 2(c) and (d)) [26, 27, 18].

3.2. Electric Transport Properties

Variation of electrical resistance with temperature is measured using two probe technique on the NiO film considering the graphene films as contact electrodes (see Fig. 3(a)). In inset [1] of Fig. 3(a) magnified region of $R-T$ curve is plotted. Thermal variation of resistance exhibits typical manifestation of a semiconductor [18, 19]; although $\rho(T)$ could not be calculated, $R-T$ curve may be fitted according to Mott's variable-range hopping (VRH) model [$ln(\rho_{dc}/\rho_0)=(T_{\rho0}/T_0)^{1/4}$, where, ρ_0 and T_0 are constants; instead of ρ_{dc} a fitting with R is proposed]. With VRH formula, curve is fitted well between 160 K to 300 K (see inset [2] of Fig. 3(a)). Current-Voltage ($I-V$) curves measured in between ± 40 V at different temperatures are shown in Fig. 3(b). $I-V$ curves at 90 K, 100 K, 120 K and 140 K reveal that I increases abruptly after a certain threshold voltage (V_{Th}) as a result of certain change in conductance [28, 10]. The low and high conductance states are generally denoted as OFF and ON states and thus this NiO film may be used as an electrical switch. At a fixed temperature and fixed voltage sweep rate, $I-V$ curves are highly reproducible and they repeats exactly same V_{Th}. However, lots of reports are there where V_{Th} shows a tunable property with variation of voltage sweep rate [10]. It may be noted that after switching, the current does not follow the same path with decreasing voltage. Rather current follows the identical path during the next cycle with increasing voltage. We further note that the resistive switching has a strong dependence on measurement temperature and amount of

IOP Conf. Series: Materials Science and Engineering 1225 (2022) 012051 doi:10.1088/1757-899X/1225/1/012051

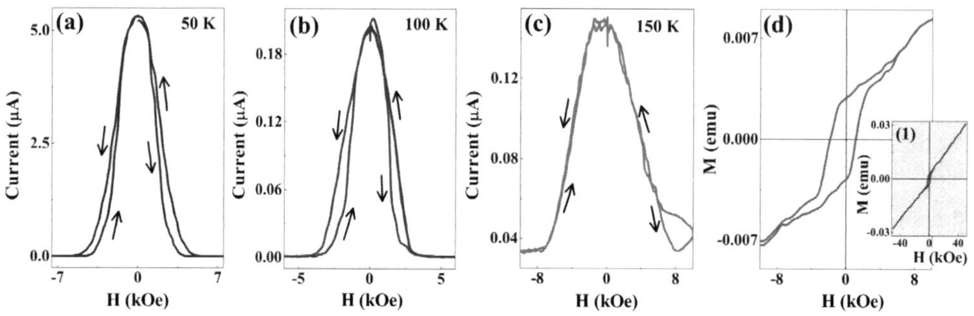

Figure 4. (a)-(c) Plot of current (I) as a function of magnetic field (H) at different temperatures exploring magnetic switching. (d) $M - H$ loop of the NiO film revealing antiferromagnetic characteristics at 150 K.

change in resistance is inversely proportional to T. Inset [3] of Fig. 3(b) reveals the absence of switching tendency and almost semiconducting $I - V$ characteristics of the film on and above 160 K.

It is observed that the prepared film has a strong dependence on magnetic field. To explore the probability of magnetic switch, where the resistance can be controlled by external influence of magnetic field, a fixed voltage is applied (well above V_{Th}) to the film and variation of current (I) is recorded with change in applied magnetic field (H). Figures 4(a)-(c) show that I eventually falls to almost zero value on application of less than 1 T field. This experiment is good example of exploring the probability of using the prepared film as a magnetic switch. To understand the magnetic state of NiO at low temperature, the film undergone a magnetic hysteresis study at 150 K and that reveals antiferromagnetic characteristics primarily, as evident from Fig. 4(d).

4. Conclusion

Switching mechanism in NiO films due to local filamentary conduction accompanied with Joule heating is one common phenomenon. Here we have prepared tri-layer graphene-NiO-graphene film in a single step in PLD with a view to explore the provision of using graphene films as contact electrodes with the NiO film. At the same time, in future we aim to analyze the superiority of this type of films to enhance the the magnetoresistance, explore resistance switching near room temperature, magnetic switching possibilities etc. The prepared film is well characterized via XRD and Raman studies, where the purity of the film and crystalline growth is established. However, analysis via cross-sectional electron microscopy may be done after a variation of thickness of different layers. Variation of film thickness may introduce unique characteristics and thus physical properties of the film may show suitable tunable dependence on thickness. Though electrical resistance switching property is observed for an wide range of temperature (90 K- 140 K), the film preparation demands further modification to explore similar properties near room temperature. For any kind of switching properties, order of magnitude of switch in resistance/conductance is important for device fabrication. This is not very high here (one order of magnitude) despite systematic and reproducible change in resistance on application of a threshold voltage. Most importantly, this film demonstrates a resistive switching property which may be controlled via application of external field. On application of magnetic field over a threshold value at a particular temperature it becomes possible to switch the current to almost zero. This film requires further exercise to understand completely and have a control over the observed electro-magnetic switching effect.

Acknowledgement D. De, M. Chakraborty and S. Goswami thank SERB Project EMR/2017/001195 and TNU Project R&D/2020/F1 for financial support and The Neotia University authority & Sukumar Sengupta Mahavidyalaya for their cooperation and encouragement. All authors thank Prof. N. Sarkar, The Neotia University for research support.

5. References
[1] Gibbons J F and Beadle W E *Solid State Electron.* **7** 785
[2] Liu C, Chae S C, Lee J S, Chang S H, Lee S B, Kim D W, Jung C U, Seo S, Ahn S E, Kahng B and Noh T W *J. Phys. D: Appl. Phys.* **42** 015506
[3] You Y H, So B S, Hwang J H, Cho W, Lee S S, Chung T M, Kim C G and An K S *Appl. Phys. Lett.* **89** 222105
[4] Ye J Y, Li Y Q, Gao J, Peng H Y, Wu S X and Wu T *Appl. Phys. Lett.* **97** 132108
[5] Son J Y and Shin Y H *Appl. Phys. Lett.* **92** 222106
[6] Seo S, Lee M J, Seo D H, Jeoung E J, Suh D S, Joung Y S, Yoo I K, Hwang I R, Kim S H, Byun I S, Kim J S, Choi J S and Park B H *Appl. Phys. Lett.* **85** 5655
[7] Lu W, Xiao J, Wong L M, Wang S and Zeng K *ACS Appl. Mater. Interfaces* **10** 8092
[8] Choi J S, Kim J S, Hwang I R, Hong S H, Jeon S H, Kang S O, Park B H, Kim D C, Lee M J and Seo S *Appl. Phys. Lett.* **95** 022109
[9] Kawai M, Ito K and Shimakawa Y *Appl. Phys. Lett.* **95** 012109
[10] Das S, Majumdar S and Giri S *J. Phys. Chem. C* **114** 6671
[11] Bae S Y, Jeon I Y, Yang J, Park N, Shin H S, Park S, Ruoff R S, Dai L and Baek J B *ACS Nano* **5** 4974
[12] Cappelli E, Scilletta C, Mattei G, Valentini V, Orlando S and Servidori M *Appl. Phys. A* **93** 751
[13] Wang K, Tai G, Wong K H, Lau S P and Guo W *AIP Adv.* **1** 022141
[14] Ghosh A, Pal A, Dey K, Majumdar S and Giri S *J. Mater. Chem. C* **3** 4162
[15] Dey K, Ghosh A, Modak P, Indra A, Majumdar S and Giri S *Appl. Phys. Lett.* **105** 142905
[16] Franta D, Negulescu B, Thomas L, Dahoo P R, Guyot M, Ohlídal I, Mistrík J and Yamaguchic T *Appl. Surf. Sci.* **244** 426
[17] Fasaki I, Koutoulaki A, Kompitsas M and Charitidis C *Appl. Surf. Sci.* **257** 429
[18] De D, Chakraborty M, Majumdara S and Giri S *Phys. Chem. Chem. Phys.* **16** 19661
[19] De D, Majumdara S and Giri S *Adv. Mat. Res.* **856** 299
[20] Goswami S, Chakraborty M and De D 2020 *AIP Conf. Pro.* **2265** 030579
[21] Mao A, Zhang D, Jin X, Gu X, Wei X, Yang G and Liu X *J. Phys. Chem. Solids* **73** 982
[22] Jeong H Y, Kim J Y, Kim J W, Hwang J O, Kim J E, Lee J Y, Yoon T H, Cho B J, Kim S O, Ruoff R S and Choi S Y *Nano Lett.* **10** 4381
[23] Moon I K, Lee J, Ruoff R S and Lee H *Nat. Commun.* **1** 73
[24] Eckmann A, Felten A, Mishchenko A, Britnell L, Krupke R, Novoselov K S and Casiraghi C *Nano Lett.* **12** 3925
[25] Fan W, Lai Q, Zhang Q and Wang Y *J. Phys. Chem. C* **115** 10694
[26] Kalita G, Wakita K and Umeno M *RSC Adv.* **2** 2815
[27] Ferrari A *Solid State Commun.* **143** 47
[28] Panda D, Dhar A and Ray S K *J. Appl. Phys.* **108** 104513

Experimental study of Carbon Nanotubes to enhance Tribological Characteristics of Lubricating Engine Oil SAE10W40

Nitesh Kumar[1], Parveen Goyal[2]

[1]UIET, Panjab University, Chandigarh-160014, INDIA
E-mail:Nitesharora297@gmail.com
[2] Assistant Professor,UIET, Panjab University, Chandigarh-160014, INDIA
E-mail: pgoyal@pu.ac.in

Abstract. Wear and friction had always been key factors in the functioning of advanced machines. Machine components containing high-performance oil cannot withstand increasing temperatures and intense pressures; therefore, additives improve different lubricants' parameters and performances. Lubricating oil is the only element that can minimize wear and friction while simultaneously protecting materials from wear. The relative movement between two surfaces is impossible without a lubricant. As a result, for a more complex machine, stern lubrication is necessary. Nanomaterials play a vital role and have tendency to improve tribological properties. Carbon nanotubes (CNTs) are most popular due to numerous applications in different fields on account of their excellent chemical, electrical, thermal, mechanical, and optical properties. This study aims to research the tribological activity of carbon nanotubes (CNTs) with a four-ball tribotester. Furthermore, SWCNTs and MWCNTs as an additives in SAE10W40 motor oil were also tested to evaluate its tribological characteristics. In compared to industrial engine oil without SWCNT and MWCNT, using MWCNTs as additives in the engine oil reduced the worn scar diameter by 67 percent, while utilising SWCNTs as an addition reduced the worn scar diameter by 38 percent. The average friction coefficient with MWCNTs is decreased by 56%, and with SWCNTs is decreased by 48%; this may be due to the effect of viscosity. The provided experimental result conveys that the lubricating oil with multiwall carbon nanotubes exhibit higher performance in anti-wear and decrease friction reduction compared with single-walled carbon nanotubes.

Keywords: Lubricant, Tribological properties, Carbon Nanotubes, Friction, Additive

1. Introduction

In the manufacturing process, machines play a vital role in different fields of the production industry. Wear and friction have limited the life of mechanical parts of devices, which may result in mechanical energy loss and, as a result, more fuel is needed to accomplish the same task. Also, it consumes more power and quickly overheats its working parts, resulting in equipment failure [1]. Therefore, critical challenges to be resolved effectively for an advanced production industry are friction and wear [2]. To avoid equipment failures, efficient lubrication is crucial. Oil is the most ancient method of reducing friction and wear. However, with the advancement in manufacturing technology, high pressure, and temperature occurred due to friction. Also, high speed makes the oil unfit, which needs the oil's lubrication characteristics to be enhanced. The high-quality lubricants are continuously formulated with improved additives to overcome these problems [3].

Lubrication additives are an essential feature of all lubricant systems because they add properties to the formulated lube - oil that the base oil lacks. Thus, additives are those chemical matter which are mixed with lube oil to transmit certain characteristics to the finished oil [4]. Because of the limitations of basic oils' characteristics, they are unable to match the needs of a high-performance lubricant. Additives improve the qualities of the existing base oil, reduce unfavourable properties, and introduce new features to the base oil in a number of ways [5]. By improving the properties of lubricating oils, additives have contributed in the development of better prime movers and manufacturing equipment[6]. Anti-friction, anti-wear, extreme pressure, anti-corrosion, and anti-oxidant properties of lubricants can all be improved by adding additives to them. Few examples of these additives are organo-zinc phosphate compounds and organo-molybdenum compounds [7].Traditional additives such as sulphides [8, chlorides [9], and phosphates [10] are used to protect materials from producing excessive wear and tear. However, these may have at least one of the following disadvantages like copper and/or lead corrosion, less soluble, finished lubricant grow dark, and increased levels of sulphur and/or phosphorus in the finished lubricant. Hence, there is a demand for additives in lubricant which will alter the overall qualities, performance of lubricants and also act as more eco - friendly and dealing with pollution control systems used in automotive and diesel engines[11].

Automobile engine manufacturers and machines are increasingly focusing on features such as low fuel consumption, high speed, and maximum kinetic energy [2]. They are constantly experimenting with the fabrication of automotive engine designs that contribute to higher friction and quick component wear. A high-performance lubricant [12] is one of the most important options for meeting all of these criteria. Engine oil is the most widely used lubricant. As a result, one of the most critical problems in advanced manufacturing processes is to boost properties like fuel consumption and friction. The search for improved lubricant additives to enhance tribological qualities including lubrication, friction, and wear continues [13].

The name tribology comes from the Greek word tribos, which means "study of things that rub" or simply "rubbing" [14]. Friction, lubrication, and wear in all contacting pairs are dealt with in this branch of engineering. As a result, tribology is known as the science and technology of interacting surfaces moving relative to one another, as well as the problems that come with it. Tribological awareness increases the service life, protection, maintenance, repair, and wear of technical equipment, as well as the efficiency of interacting machine components ranging from spacecraft to household items, resulting in major cost savings and economical benefits [15]. Industrial lubricants are mainly used to increase component life and quality by reducing material wear and surface friction [16]. This quest for energy efficiency has led to the use of nanoparticles as lubricant additives in current research. Nanoparticles are the newest class of lubricant additives that gained popularity among other additives utilized in the past [17]. Nanotechnology, by the use of nano additives, has the potential to enhance lubricant efficiency. Nanomaterials can thus be used as lubricant oil modifiers to improve tribological properties and reduce wear and friction [18].

Nanotechnology is today one of the world's most rapidly expanding topics of research, with applications in many fields of technology and science [19]. Nanomaterials, due to its unique features, such as the quantum and small size effect, surface and interface effects, have emerged as a new additive alternative in recent years. The use of nanoparticles in Tribology has attained much attention recently [20]. The addition of nanoparticles to oil has also been confirmed to improve extreme-pressure properties and load-carrying capability, as well as lowering friction coefficients [18]. Small scale effects, temperature insensitivity, and strong tribological properties are all significant advantages of nanoparticles employed as lubricant additives to improve conventional lubricants [21]. They are excellent as lubricant additives because their size, which is generally less than 100 nm, permits them to enter the contact zone and is thermally stable at high temperatures [22].

Nanomaterials are frequently employed in a variety of human activity domains, such as cancer therapy, modified textiles, energy storage, rocket explosives and propellants, as illustrated in figure 1. Nano lubricants, according to several studies, are efficacious in reducing friction and wear. They useda range of nanoparticles obtained from polymer, metal, organic, polymer, and inorganic components to synthesize nanolubricants. Carbon Nanotubes, graphite [24], molybdenum disulfide [25], Silica Oxide [26], Aluminum Oxide [27], Copper Oxide [28], Titanium Oxide [29], and Zinc Oxide [30] are nanoparticles that have been studied as pure oil additives. Carbon nanotubes (CNTs) have aroused a lot of attention due to their exceptional mechanical, thermal, electrical, chemical, and optical capabilities. Superior mechanical strengths, CNTs in various forms have been investigated for tribological applications[31]. This paper looked at the tribological characteristics of Single-Walled Carbon Nanotubes (SWCNTs) and Multi-Walled Carbon Nanotubes (MWCNTs) as an additive in Engine Oil to improve friction-reduction, load-carrying capacity of pure oil, and wear-reduction capabilities.

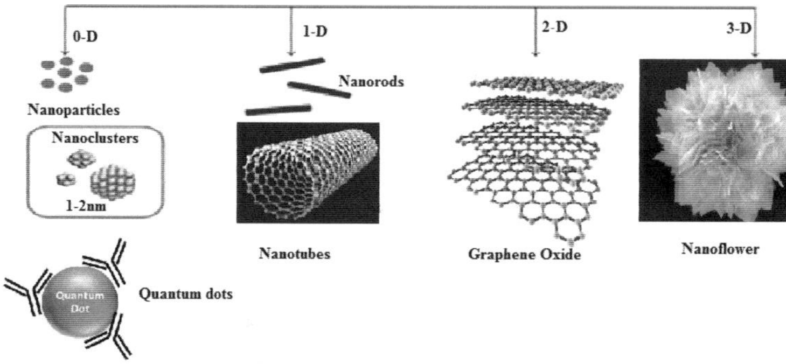

Figure 1. Examples of Nanomaterial with different structures [32]

1.1 Overview of lubrication and lubricants

The oldest method of reducing wear and friction in sliding contact bodies was liquid lubrication. Lubricant isolates adjoining parts by creating a fluid film between them. Metal-to-metal interaction is reduced by the fluid film [33]. Therefore, Lubricants are substances that are used to reduce the force of friction between the moving parts of a system in contact. The process of minimizing friction between the moving parts of a machine that comes into contact is known as lubrication.

Lubricants play a major role in machinery element safety, as they:

- Separate moving parts from one another
- Remove heat from contact by passing through.
- Ensure a clean surface.
- Transport functional additives toward the surface
- Power transfer in applications such as hydraulics, automatic transmission, and brakes [34].

1.2 Lubrication Types

When it comes to forms, there are several types to choose from. The three types of lubrication are represented in Figure 2: boundary, mixed, and complete film. Despite the fact that each one is different, they all rely on additives or lubricants contained in oils to preserve them from wear and tear. Aside from this, there are two types of full-film lubrication: hydrodynamic and elastohydrodynamic.

(a) Hydrodynamic lubrication occurs when a fluid layer effectively separates two sliding surfaces (relative to each other).

(b) In Elastohydrodynamic lubrication, the film is thinner, the friction is higher, and the effect is more noticeable when the surfaces are rolling. Elastohydrodynamic describes how the elasticity of the film deforms the rolling surface. [35].

(c) In conditions with frequent starts and stops, as well as shock loading, boundary lubrication occurs. If complete film cannot be created owing to speed, load, or other circumstances, certain lubricants contain extreme-pressure (EP) or anti-wear (AW) compounds to protect surfaces [36].

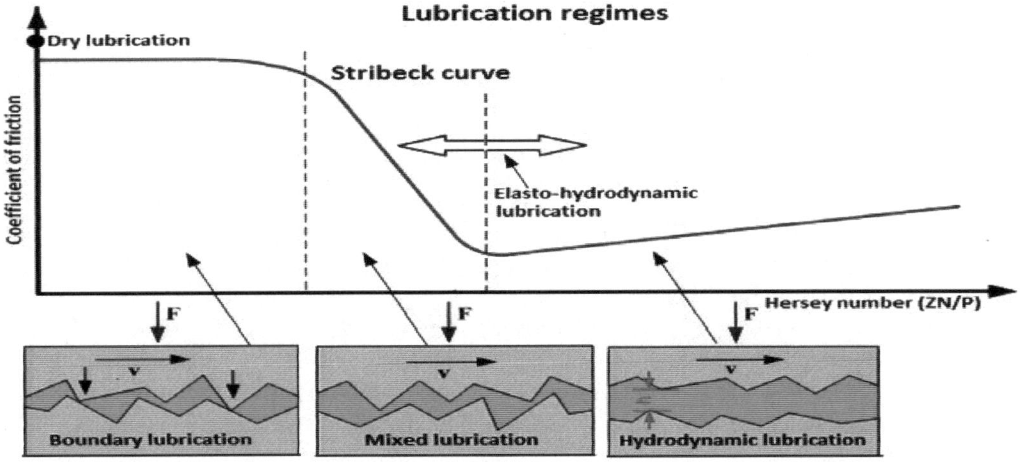

Figure 2. Stribeck curve and Regimes of Lubrication [37]

1.3 Role of Nanomaterials in lubricants

Since most existing lubricants have reached their maximum performance capacity, new ones must be formulated that are capable of achieving energy efficiency in a number of fields and under increasingly difficult conditions [42].The most significant attribute of nanomaterial is their small size, which allows them to enter microscopic spaces between rough surfaces in contact and modify the contact's tribological performance. As a result, nanoparticles provide an alternative to lubrication by directly introducing third-body entities into the interface. To improve tribological properties, a minimal concentration of nanoparticles (up to 1%) is enough.

The following are some of the benefits of lubricant having nanoparticles as additives:

a) Insolubility in non-polar base oils.
b) Their reactivity with other lubricant additives is low.
c) A huge film formation on a variety of surfaces.
d) Longer lasting.
e) Non-volatility is high, allowing it to withstand high temperatures.

1.4 Nanomaterial parameters responsible for enhancing the performance of lubricant oils

Nanomaterial size and morphology: The nanoparticles' size has a great impact on the tribologicalbehavior of nano lubricants in a number of different ways. The nanoparticles' small size allows them to reach the rubbing surface, reducing friction as well as wear mechanisms [44]. The size of nanoparticles can also affect lubricant homogeneity. It was noticed that decreasing the particle size increased the dispersion of nanoparticles[45]. The shape and structure of nanoparticles are extremely important. Nanoparticles come in five different shapes: spherical, onion-shaped, granular, tube-shaped, and sheet-shaped [46].

Dispersion stability of nanoparticles: Because of their large surface area to volume ratio, nanoparticles possess high surface energy. The particles in suspension can bind together to form agglomerates in order to achieve the equilibrium state. This phenomenon is known as Agglomeration/flocculation [47]. The sedimentation rate is affected by the agglomeration of nanoparticles, as well as the ability to protect against wear and friction. As a result, the nanoparticles' dispersion stability is essential [48].

1.5 Carbon nanotubes (CNT) – lubricants

Carbon nanotubes (CNTs) are streamlined, tubular structures made by rolling c-c covalent bonds in a hexagonal plane. CNTs have the highest yield strength (~ 1 TPA) of any substance [49]. Nanometer-sized carbon nanotubes can self-assemble at sliding asperity contacts, reducing the sliding contact area and shear stresses, leading to reduced friction and wear. Recent progress in nanoparticles have caught the eye of researchers for their critical number of applications. The nanoparticles have been identified to have superior mechanical and chemical properties. Carbon Nanotubes are an example of a breakthrough that has fascinated the attention of researchers due to their superior mechanical and chemical properties [50].

Carbon nanotubes are of two types: single-walled carbon nanotubes (SWCNT) and multiple-walled carbon nanotubes (MWCNT). Single-walled carbon nanotubes (SWCNT) are cylindrical in shape and can be formed by rolling a graphene sheet. As graphene is derived from honeycomb lattice, and sheets are seamless cylinders that reflect a single atomic layer of crystalline graphite. An MWCNT is a graphene sheet stack that has been folded up into concentric cylinders. Millions of atoms make up each of these nanotubes. This molecule is tens of micrometers long and has a diameter of 0.7 nanometers [51]. SWCNTs are typically just ten atoms around the circumference and have a one-atom thickness. Nanotubes can be thought of as a one-dimensional structure with a length-to-diameter ratio (aspect ratio) of around 1000 [52].

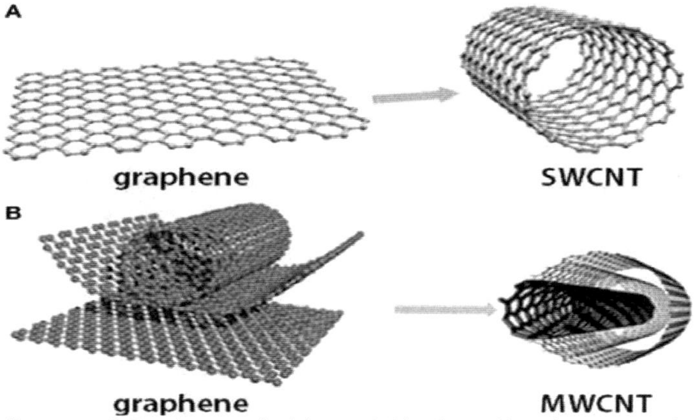

Figure 3. Schematic diagram of Graphene rolled into A) Single-wall carbon nanotubes (SWCNT), B) Multiwall carbon nanotubes (MWCNT)[53].

2. Experimental

In this study, the tribological effects of single-wall carbon nanotubes and multiwall carbon nanotubes are investigated by mixing it into the engine oil at definite proportions.

2.1 Materials

Commercial Engine oil SAE10W40 with a viscosity of 83.5 cSt at 40 degrees Celsius was used as the base lubricant. The lubricants' properties are described in Table 1. Single wall carbon nanotubes (SWCNTs) and multiwall carbon nanotubes (MWCNTs) bought from Nanoshel LLC were employed in varied ratios as solid additions to the base oil (0.1, 0.5, 1, and 2 wt%). SWCNTs had an average length and size of 8 um and 2 nm, respectively, while MWCNTs had an average length and size of 8 um and 20 nm, respectively.

Table 1. Specifications of Engine oil SAE10W40 used

Property or Test	SAE Grade10W40
Viscosity (at 40°C), cSt	83.5
Viscosity (at 100°C), cSt	13.0
Viscosity index	155
Density, g/ml at 15.6°C	0.86
Flash Point, °C	212
Pour Point,°C	45

2.2 Synthesis

Oil blends were formulated by adding SWCNTs and MWCNTs at different concentrations of 0.1%, 0.5%, 1%, and 2%, respectively, which were then mechanically stirred for 30 minutes and ultrasonically agitated for 60 minutes. Ultrasonication was employed to achieve uniform dispersion of SWCNTs and MWCNTs in the lubricantto minimise aggregation. The nano lubricants were collected after ultrasonication for 2 hours. Figure 4 depicted the illustration of nano lubricants

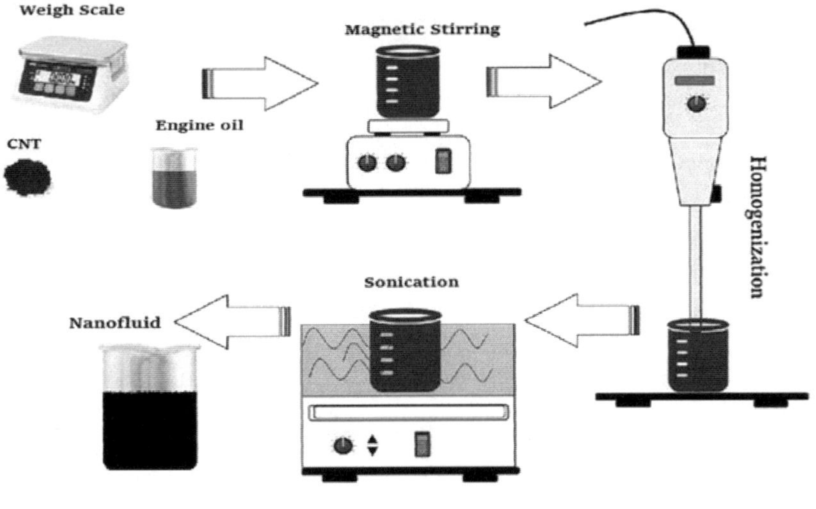

Figure 4. Schematic illustration of the synthesis of CNTs (SWCNT and MWCNT) added in Engine Oil

3. Results and discussion

3.1 Fourier transforms infrared (FTIR) spectroscopy

FTIR analysis uses infrared light to scan samples in order to distinguish biological, inorganic, and polymeric components. Fourier transform infrared (FTIR) spectral tests of SWCNT and MWCNT were conducted at intelligent materials Pvt.Ltd.to confirm potential structural variations between the two series of samples. FTIR spectra of SWCNT and MWCNT ground in KBr pellets recorded using a spectrometer with 200 scans averaged at a resolution of 1 cm^{-1}.The FTIR spectra of the SWCNT and MWCNT are shown in Figure 5. The characteristic response of FTIR spectra for SWCNT and MWCNT shows peaks C=C, which are considered as a result of carbon groups. Carbon nanotubes' C=C stretching mode is more likely to be responsible for the absorption band at 1584 cm^{-1}, 1678cm-^{1}.Based on this result and the wave numbers it can be concluded that MWCNTs a strong and broad stretch peak at 3480 cm^{-1} and a bending peak at 1584 cm^{-1} are assigned the C=C bond[23].

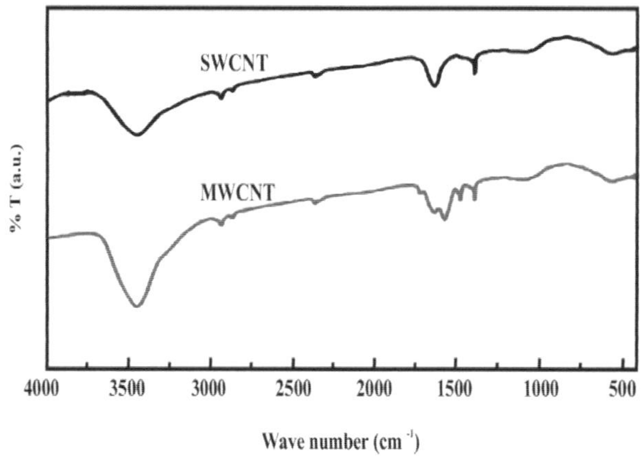

Figure 5. FTIR spectra of SWCNT and MWCNT

3.2. Kinematic viscosity with the temperature at different concentrations

The lubricating mechanism relies heavily on kinematic viscosity. At temperatures of 40 °C and 100 °C, the effect of various concentrations on the viscosity of nano lubricant is calculated using SWCNT and MWCNT(shown in Tables 2 and 3). The result showed that enhancing the concentration of carbon nanotubes increased the kinematic viscosity of nano lubricants. At less concentration, though, the rate of viscosity shifts is much lower than at higher concentrations.

Table 2. Viscosity change of engine oil with additive SWCNT at different concentration with temperature

Property	Engine Oil (SAE10W40)	SAE10W40+ SWCNT (0.1%)	SAE10W40+ SWCNT (0.5%)	SAE10W40+ SWCNT (1%)	SAE10W40+ SWCNT (2%)
Viscosity at 40°C, cSt	83.5	98.2	105.9	139.2	180.6
Viscosity at 100°C, cSt	13.0	18.4	23.6	67.3	73.0
Viscosity index	155	164	179	186	198

Table 3. Viscosity changes of engine oil with additive MWCNT at different concentration with temperature

Property	Engine Oil (SAE10W40)	SAE10W40+M WCNT (0.1%)	SAE10W40+ MWCNT (0.5%)	SAE10W40+ MWCNT (1%)	SAE10W40+ MWCNT (2%)
Viscosity at 40°C, cSt	83.5	101.1	112.9	147.2	195.6
Viscosity at 100°C, cSt	13.0	22.4	34.1	73.3	87.0
Viscosity index	155	168	182	190	112

3.3. Scanning electron microscope (SEM)

SEM (scanning electron microscope) is used to examine surfaces of materials by scanning a beam of directed low-energy electrons over the material on a regular basis. The behaviour of the electron beam allows high-energy backscattered electrons to be released, and low-energy secondary electrons from the specimen's surface are captured with a detector. Figure 6 shows images of SWCNT and MWCNT captured with a SEM at a magnification scale of 500 nm. It shows that the SWCNT has a smooth surface, whereas MWCNT has bundles of tangled tubes due to the multiwall structure.

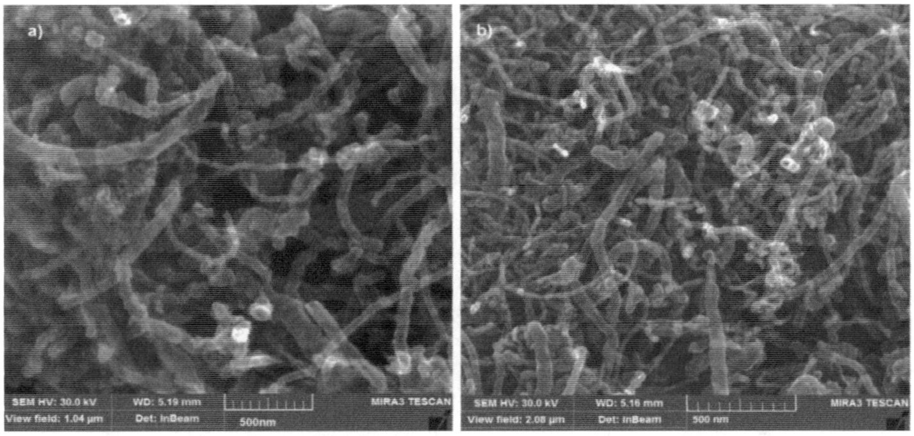

Figure 6. SEM image of (a) SWCNT and(b) MWCNT

3.4. Four ball tribotester

Figure 7 shows the four ball tribotester that were utilised in this study to analyse the tribological characteristics of nano lubricants. There are four balls in this device: one on top and three on the bottom. The lubricant sample to be analysed is held in a ball pot with the three bottom balls. The one ball on top, which is fastened on a collet, rubs against the below balls. Each ball's surface is cleansed with acetone. The upper ball spinws at appropriate rpm. As the three lower balls touched the fourth upper ball, the load was divided over three spots. The temperature of the lubricant-filled pot (oil cup) rises to 75 degrees Celsius. The drive motor was then switched on and set at 1200 rpm to drive the top ball. The rotating spindle creates frictional torque, resulting in a scar on the bottom pair of balls. The heater was then turned off for an hour, and the oil cup assembly was taken out from the cup. The test balls had a diameter of 12.8 mm and a roughness of Ra = 0.017 mm and were constructed of AISI E 52100 steel. The variation in temperature of a liquid sample and the friction torque between the balls were both measured. The wear scar diameter was used to measure the wear on the steel ball surfaces (WSD). An optical microscope is used to measure the sizes of the wear scars on the lower stationary balls.

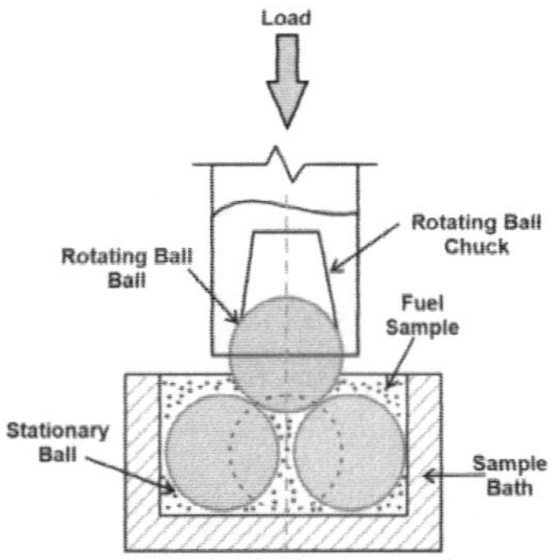

Figure 7. Fourballtribotester

3.5 Testing of tribolological properties and surface analysis

The diameters of worn scars are used to evaluate the anti-wear capabilities. Recalculating the worn scar diameter of the three spheres yielded the average WSD for each lubrication test. After adding SWCT and MWCNTs to SAE10W40 engine oil, Figure 8 illustrates the worn surfaces. The worn scar diameters of SWCNT and MWCNT engine oil-lubricated specimens are smaller than those of engine oil without CNTs. With the addition of SWCNTs and MWCNTs in the engine oil as additives, the worn scar diameter is decreased by 38 percent and 67 percent, respectively, when compared to commercial engine oil without SWCNT and MWCNT. The inability of the tribochemical reaction film to form on the rubbing surfaces may be due to the higher shear strength of the adsorbed oil on the balls' surface. As a result, the rolling effect of the SWCNT and MWCNT nanoparticles may be to contribute the smaller worn scar diameter.

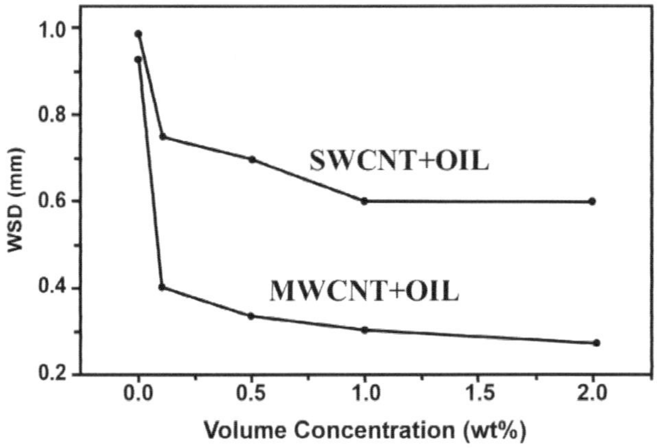

Figure 8. SWCNT and MWCNT in Engine oil SAE10W40 at various volume concentrations as a function of wear scar diameter.

During tribological experiments, the friction coefficient was calculated based on friction torque measurements. The effect of SWCNT and MWCNT concentrations on the friction coefficient of SAE10W40 engine oil is shown in Figure 6. As can be seen, adding SWCNT and MWCNT to engine oil SAE10W40 changes the friction coefficient when applied to commercial engine oil SAE10W40. The friction coefficient of CNTs is smaller than the friction coefficient of engine oil without additive. Furthermore, for MWCNTs, the average friction coefficient is reduced by 56 %, while with SWCNTs, it is reduced by 48 %; this may be due to the effect of viscosity. Higher viscosity fuel oils, such as SAE10W40, have a lower friction coefficient.

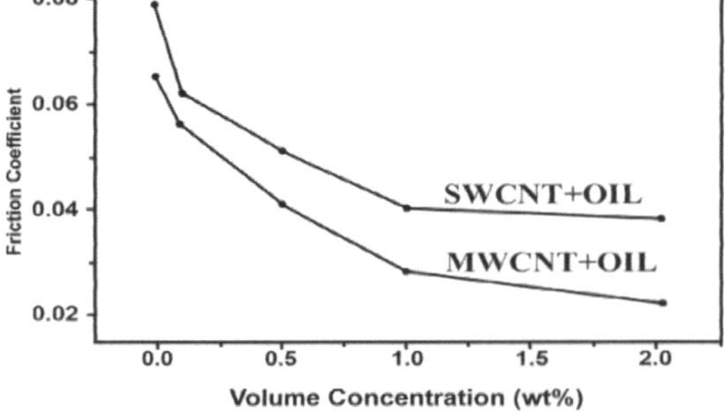

Figure 9. The Coefficient of Friction variations with SWCNT and MWCNT in Engine oil SAE10W40 at different volume concentrations

Figures 10 and 11 illustrates the coefficient of friction and wear scar diameter for nano lubricants and engine oil at optimal concentrations as a result of applied loads. With the addition of SWCNT and MWCNT to the engine oil, the friction coefficients and wear scar diameter were significantly decreased. It has been observed that adding MWCNT to engine oil produces better results than using SWCNT at the same concentration.

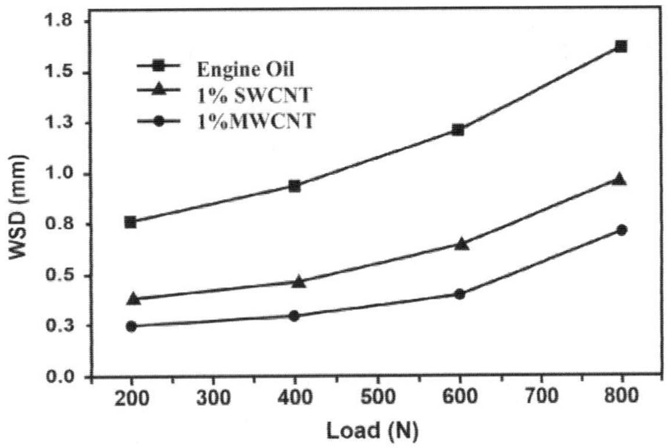

Figure 10. The wear scar diameter variations with load applied containing Engine Oil SAE10W40 , 1 wt. % SWCNT and 1wt. %MWCNT.

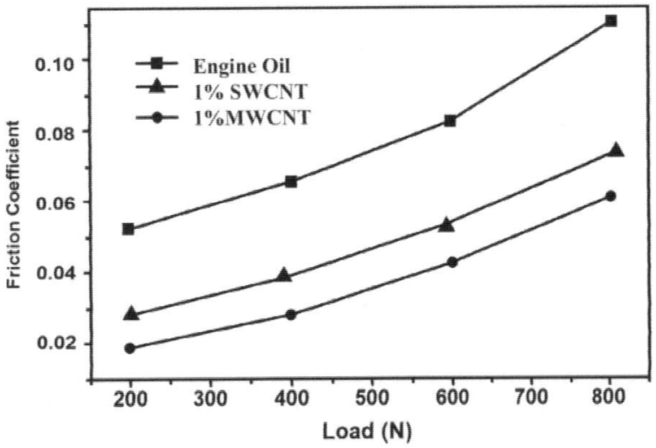

Figure 11. TheCoefficient of Friction variations with the load applied containing Engine Oil SAE10W40, 1 wt. % SWCNT and MWCNT.

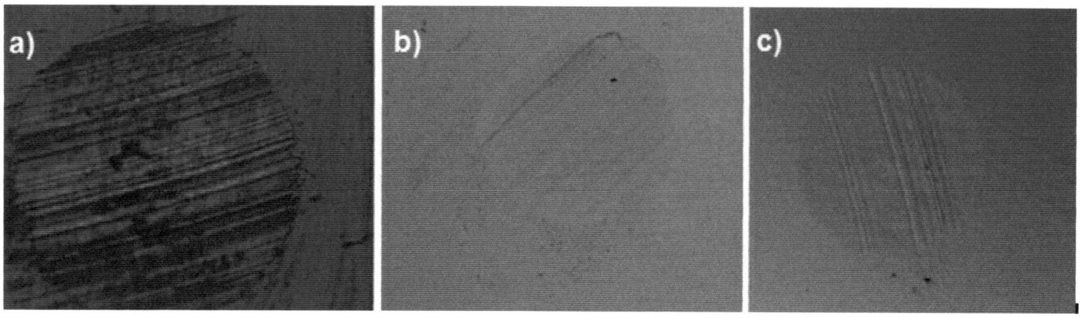

Figure 12. SEM analysis of the wear scar of steel balls lubricated by (a) Engine Oil SAE10W40 (b) 1wt. % SWCNTs Engine Oil SAE10W40 (c) 1wt. % MWCNTs Engine Oil SAE10W40.

The specimens' worn surfaces were lubricated with base oils comprising SWCNTs and MWCNTs nanoparticles were analyzed using SEM analysis, as seen in Fig 9. The MWCNT Engine oil SAE10W40-lubricated wear scar track on the ball has a larger, smoother surface than the SWCNT Engine oil SAE10W40-lubricated wear scar track. Engine oil SAE10W40 without any additive, on the other hand, indicates roughening on the wear surface of the ball, with narrow and deeper grooves, as shown in Fig 9. A lubricating film may have formed due to the existence of carbon nanotubes, avoiding or reducing direct steel-to-steel contact that results in a smoother surface.

4. Conclusion

The four-ball tribotester was used to find the nano lubricant properties of Engine oil SAE10W40 containing SWCNTs and MWCNTs. The tests show that the addition of MWCNTs and SWCNTs to the Engine oil SAE10W40 improved the lubrication properties of engine oil without additives.

- The kinematic viscosity of nano lubricants had increased by increasing the concentration of CNTs.
- These results showed that using MWCNTs as additives in the engine oil decreased the worn scar diameter by 67%. With SWCNTs as an additive, the worn scar diameter is reduced by 38% in comparison to the industrial engine oil without SWCNT and MWCNT.
- The average friction coefficient with MWCNTs is decreased by 56%, and with SWCNTs is decreased by 48%; this may be due to the effect of viscosity.
- The observations prove that by adding MWCNTs has enhanced the anti-wear friction coefficient of Engine oil SAE10W40 compared to SWCNTs when both are taken at the same concentration, i.e., 1 %w/v. These experimental results revealed that on adding MWCNT nanoparticles, wear has been reduced and resulted in a smoother surface with less scars, implying that the MWCNTs greatly decreased metal contact.
- The anti-wear mechanism is because of accumulation of MWCNTs and SWCNTs on the surface, that reduces shearing stress and improves tribological properties.

References

[1] Iliuc I 1980 Tribology of thin layers*Elsevier* **4**
[2] Dowson D 1979 *Men of Tribology* (William Bate Hardy 1864–1934 and Arnold Sommerfeld 1868–1951)
[3] Su Y, Gong L, Chen D 2015 An investigation on tribological properties and lubrication mechanism of graphite nanoparticles as vegetabl based oil additive *J. Nanomater.* **16** 203
[4] Srivyas P D, Charoo M S 2018 A Review on Tribological Characterization of Lubricants with Nano Additives for Automotive Applications *Tribol. Ind.* **40** 594–623

[5] Rizvi S Q 2009 *Lubricant chemistry, technology, selection, and design* ASTM Int., Conshohocken
[6] Leslie R R 2003 *Lubricant Additives Chemistry and Applications* Marcel Dekker Inc, New Yorkpp 254-293
[7] Ludema KC and Ajayi OO 2018 *Friction, wear, lubrication: a textbook in Tribolog y*CRC press
[8] Saidur R, Lai YK 2010 Parasitic energy savings in engines using nanolubricants *Energy Education Science and Technology Part a-Energy Science and Research* **266** 1–74
[9] Socoliuc A, Gnecco E, Maier S, Pfeiffer O, Baratoff A, Bennewitz R, Meyer E 2006 Atomic-scale control of friction by actuation of nanometer-sized contacts *Science* **313** 207–10
[10] Gunda RK, Narala SK 2016 Tribological studies to analyze the effect of solid lubricant particle size on friction and wear behavior of Ti-6Al-4V alloy *Surf. Coat. Technol.* **308** 203–12
[11] Dorinson A 1973 The additive action of some organic chlorides and sulfides in the four-ball lubricant test *ASLE Trans.* **16** 22–31
[12] Johnson D W 2016 *The tribology and chemistry of phosphorus-containing lubricant additives* IntechOpen
[13] Lin J, Wang L and Chen 2011 Modification of graphene platelets and their tribological properties as a lubricant additive*Tribol. Lett.* **41** 209–15
[14] Salah N, Abdel-Wahab MS, Alshahrie A, Alharbi ND and Khan ZH 2017 Carbon nanotubes of oil fly ash as lubricant additives for different base oils and their tribology performance *RSC Adv.* **7** 40295–302
[15] Booser E R 1993 *CRC Handbook of Lubrication and Tribology, Volume III: Monitoring, materials, synthetic lubricants, and applications* **3**(CRC Press)
[16] Bhushan B 2013 *Introduction to Tribology* (John Wiley & Sons)
[17] MalleswaraRao K N D, Sankararao V and Somasekha K 2017A review on tribological properties of lubricating oil with nanoparticles additive *Int. J. Adv. Eng. Res. Dev.* **4** 197-99
[18] Siniawski MT, Saniei N, Pfaendtner J 2007 Tribological degradation of two vegetable-based lubricants at elevated temperatures *J. Synth. Lubr.* **24** 167–79
[19] Zin V, Agresti F, Barison S, Colla L, Gondolini A and Fabrizio M 2013 The synthesis and effect of copper nanoparticles on the tribological properties of lubricant oils *IEEE Trans Nanotechnol* **12** 751–59
[20] Tao X, Jiazheng Z, Kang X 1996 The ball-bearing effect of diamond nanoparticles as an oil additive*J. Phys. D: Appl. Phys.* **29** 2932
[21] Hamid N, Thakur A 2018 Assessing the capability of CNTs in improving the properties of Engine Oil*J. Univ. Shanghai Sci.* **23** 1321-27
[22] Lee CG, Hwang YJ, Choi YM, Lee JK, Choi C and Oh JM 2009 A study on the tribological characteristics of graphite nano lubricants *Int. J. Precis. Eng. Manuf.* **10** 85–90
[23] Wu YY, Tsui WC, Liu TC 2007 Experimental analysis of tribological properties of lubricating oils with nanoparticle additives *Wear* **262** 819–25
[24] Dai W, Kheireddin B, Gao H, Liang H 2016 Roles of nanoparticles in oil lubrication *Tribology Int.* **102** 88–98
[25] Nunn N, Mahbooba Z, Ivanov MG, Ivanov DM, Brenner DW, Shenderova O 2015 Tribological properties of polyalphaolefin oil modified with nanocarbon additives *Diam. Relat. Mater.* **54** 97–102.

[26] Kalin M, Kogovšek J, Remskar M 2012 Mechanisms and improvements in the friction and wear behavior using MoS2 nanotubes as potential oil additives *Wear* **280** 36–45

[27] Li X, Cao Z, Zhang Z, Dang H 2006 Surface-modification in situ of nano-SiO2 and its structure and tribological properties *Appl. Surf. Sci.* **252** 7856–61

[28] Luo T, Wei X, Huang X, Huang L and Yang F 2014 Tribological properties of Al2O3 nanoparticles as lubricating oil additives *Ceram. Int.* **40** 7143–49

[29] Kedzierski MA and Gong M 2009 Effect of CuOnanolubricant on R134a pool boiling heat transfer *Int. J.Refrig.* **32** 791–9

[30] Sabareesh RK, Gobinath N, Sajith V, Das S, Sobhan CB 2012 Application of TiO2 nanoparticles as a lubricant-additive for vapor compression refrigeration systems–An experimental investigation *Int. J.Refrig.* **35** 1989–96

[31] Battez AH, González R, Viesca JL, Fernández JE, Fernández JD, Machado A, Chou R, Riba J 2008 CuO, ZrO2 and ZnO nanoparticles as antiwear additive in oil lubricants *Wear* **265** 422–8

[32] Farzin L, Shamsipur M, Samandari L, Sheibani S 2020 HIV biosensors for early diagnosis of infection: The intertwine of nanotechnology with sensing strategies *Talanta* **206** 120201

[33] Chen CS, Chen XH, Xu LS, Yang Z, Li WH 2005 Modification of multi-walled carbon nanotubes with fatty acid and their tribological properties as lubricant additive *Carbon* **43** 1660–66

[34] Arai S, Fujimori A, Murai M, Endo M 2008 Excellent solid lubrication of electrodeposited nickel-multiwalled carbon nanotube composite films. *Mater. Lett.* **62** 3545–48

[35] Mobarak HM, Mohamad EN, Masjuki HH, Kalam MA, Al Mahmud KA, Habibullah M, Ashraful AM 2014 The prospects of biolubricants as alternatives in automotive applications *Renew. Sust. Energ. Rev.* **33** 34–43

[36] Hsu S M, Gates RS 2005. Boundary lubricating films: formation and lubrication mechanism *Tribol Int.* **38** 305–312

[37] Robinson JW, Zhou Y, Bhattacharya P, Erck R, Qu J, Bays JT, Cosimbescu L 2016 Probing the molecular design of hyper-branched aryl polyesters towards lubricant applications *Sci. Rep.* **6** 1–10

[38] Hori Y 2006 *Hydrodynamic lubrication*Springer Science & Business Media

[39] Höglund E 1999 Influence of lubricant properties on elastohydrodynamic lubrication *Wear* **232** 176–184

[40] Stachowiak G and Batchelor AW 2004 *Experimental methods in tribology*Elsevier

[41] Stachowiak GW and Batchelor AW 2013 *Engineering tribology* (Butterworth-heinemann)

[42] Nagendramma P and Kaul S 2012 Development of ecofriendly/biodegradable lubricants: An overview *Renew. Sust. Energ. Rev.* **16** 764–74

[43] Fox NJ and Stachowiak GW 2007 Vegetable oil-based lubricants—a review of oxidation *Tribology Int.* **40** 1035–46

[44] Akbulu M 2012 Nanoparticle-based lubrication systems *J Powder Metall Min* **1** 1–3

[45] Saidur R, Kazi SN, Hossain MS, Rahman MM and Mohammed HA 2011 A review on the performance of nanoparticles suspended with refrigerants and lubricating oils in refrigeration systems *Renew. Sust. Energ. Rev.* **15** 310–323

[46] Liu G, Li X, Qin B, Xing D, Guo Y and Fan R 2004 Investigation of the mending effect and mechanism of copper nano-particles on a tribologically stressed surface *Tribol. Lett.* **17** 961–6

[47] Thottackkad MV, Perikinalil RK andKumarapillai PN 2012 Experimental evaluation on the tribological properties of coconut oil by the addition of CuO nanoparticles *Int. J. Precis. Eng. Manuf.* **13** 111–6

[48] Xie H, Jiang B, He J, Xia X and Pan F 2016 Lubrication performance of MoS2 and SiO2 nanoparticles as lubricant additives in magnesium alloy-steel contacts *Tribol. Int.* **93** 63–70

[49] Yu W, Xie H 2012 A review on nanofluids: preparation, stability mechanisms, and application *JJ. Nanomater.* **2012**1-17

[50] Muzakkir SM, Lijesh KP and Hirani H 2015 Effect of base oil on the anti-wear performance of multi-walled carbon nano-tubes (MWCNT) *Int. J. Curr. Eng. Technol.* **5** 681–4

[51] Sengupta J 2018 *Carbon nanotube fabrication at industrial scale: Opportunities and challenges* Handbook of Nanomaterials for Industrial Applications 172–194

[52] Meyyappan M, Delzeit L, Cassell A and Hash D2003 Carbon nanotube growth by PECVD: a review. *Plasma Sources Sci. Technol.* **12** p 205

[53] Vidu R, Rahman M, Mahmoudi M, Enachescu M, Poteca TD, Opris I 2014 Nanostructures: a platform for brain repair and augmentation*Front. Syst. Neurosci.* **8**p 91

Effect of Surface Modification of Silica Nanoparticles with Thiol group on the Shear Thickening Behaviors of the Suspensions of Silica Nanoparticles in polyethylene glycol (PEG)

Shuchi Sharma[a], Yogesh Kumar Walia[a*], Gunjan Grover[b], Sanjeev K. Verma[b]*,.

[a]*Career Point University, Hamirpur, 176041,India*

[b]*Terminal Ballistics Research Laboratory, DRDO, Chandigarh,160030, India*
skv002002@gmail.com,

ABSTRACT. The Fine-tuning of Shear Thickening Fluids (STFs) by surface modification of silica particles has fascinated scientist's interest worldwide as it results in performance enhancement of STF based on armor systems. In the current study, surface modified Silica nanoparticles (average diameter of 600 nm) possess thiol functional groups which were attained through a reaction with 3-mercaptopropyl-trimethoxysilane in absolute ethanol at 90 °C. Shear thickening fluid of Thiol functionalized Silica nanoparticles were prepared by sonochemical method in polyethylene glycol (PEG-200). The rheological parameters of STFs (modified and unmodified silica Nano particles) were measured using Rheometer MCR 52, Anton Par, Germany. The shear thickening behavior of thiol-based STF shows shear thickening at a higher shear rate compared to only silica-based STF with the decrease in viscosity maximum.

Keywords: Bullet Impact; Shear Thickening Fluid; Strain-Thickening; Particle Dispersion; Rheology.

1. INTRODUCTION:

The guardians of our country are provided with body armors, which are designed to weaken slashing or penetrating attacks by resisting the low-velocity bullet penetration. Multiple layers of high modulus, high strength, and low-density fabrics eg. Kevlar & Dyneema have been reported to be used for making soft armor panel (SAP) for protection against 9 x 19mm ammunition[1–4]. However, increases the weight of the armor observed because of large number of layers of ballistic materials, thereby posing restrictions to the comfort and mobility of the user. A lot of work has been reported in lliterature cites for the reduction in the areal density of the armor. One such potential method is application of liquid armor or shear thickening fluid (STF) on high strength fabric like Kevlar. [5. 6].

A large number of reports have been published on shear thickening phenomena of colloidal suspensions [7-9]. Under sudden impact the liquid armor or the colloidal suspension of hard particles in polymer matrix converts to a solid-alike state, however, at low shear rates STF flows easily because of lower viscosity. A number of methods have been put forward in literature to elaborate the Non-Newtonian behavior of concentrated suspensions. In one of the early studies, a rise in viscosity is proposed by Hoffman because of order-disorder transition of particles [10, 11].

Nonetheless, shear-induced hydro clusters formation is the most widely accepted mechanism, in which inter particle repulsive forces was completely overcome by hydrodynamic lubricating forces that results to increase in viscosity. This simulation studies, rheological and optic-SANS supported this hypothesis. [12-15]. The possibility to control rheological property of STFs for desired applications makes STF a suitable candidate for a lot of applications in different fields such as damping devices, body armors, medical devices smart structures etc [16- 21].

The shear thickening behaviour of STF depends on different parameters such as particle shape, particle size, particle concentration, and polydispersity of dispersed particles, the method of preparation of STF and the molecular weight of polymeric matrix. The ST behavior of STF is considerably influenced by the surface properties of dispersed particles [22-29]. The surface modification of silica nanoparticles significantly alters particle–media and the particle-particle interactions.. Thus, shear thickening fluids consisting of surface-modified silica nanoparticleshas bee in interest worldwide specially for its energy absorption behavior with potential application in base protective system.

Mahfuz et al. prepared STF using surface-modified silica particles with aminopropyl triethoxy silane and its application to Kevlar fabric was analyzed [30,31]. The performance of the resulting armor composites was observed to improve significantly in comparison to that of the neat fabric. Butola et al reported that a reduction in surface hydroxyl group increases the energy absorption behavior of shear thickening fluids [32,33]. This paper includes the synthesis of thiol-modified Silica Nanoparticles based shear thickening fluid and its rheological behaviors in continuation of our earlier research works related to fine-tuning of shear thickening fluids. To the best of our knowledge, the STF of Thiol modified Silica particles has not been reported in literature. This study also believes in the development of STF based armor systems [1, 2, 15, 34-36].

2. EXPERIMENTAL:

2.1 Materials

The spherical silica nanoparticles of particle size 600nm and PEG 200 were used for the synthesis of the Shear Thickening Fluids. The average particle size of unmodified and surface- modified silica nanoparticles was measured through DLS (Fig. 1a), and also through SEM (Fig. 1b). 3-Mercaptopropyltrimethoxysilane (MPTMS) was used for the modification of dispersed particles (silica particles 600nm). The absolute ethanol 99.9 % was used in synthesis of STFs.

2.2 Synthesis of Thiol-modified Silica Nanopartilcles.

Thiol-modified Silica nanoparticles with different concentration of surface thiol groups were synthesized by dispersion of the nanoparticles (10g) in a solution of 3- Mercaptopropyltrimethoxysilane (MPTMS) (25-100 mmol) in toluene (400.0 mL) as shown in table 1. The mixture was reflux for 12 hours. The final centrifuged product was left for another 2 hours and than washed with toluene, ethanol, and dried under vacuum. Ellman's reagent and Emmett–Teller (BET) surface area is used to determine the quantity of thiol group (□SH) per particle of functionalized silica nanoparticles.

Table 1. Mercaptopropyltrimethoxysilane (MPTMS) mmols added to silica nanoparticles of average particle size 600nm.

S. No.	Sample Code.	Silica Nanoparticles -600nm (grams)	MPTMS (mmol)
1	Silica-SH-25 mmol	10	25
2	Silica-SH-50 mmol	10	50
3	Silica-SH-75 mmol	10	70
4	Silica-SH-100 mmol	10	100

IOP Conf. Series: Materials Science and Engineering 1225 (2022) 012053 doi:10.1088/1757-899X/1225/1/012053

2.3 Synthesis of Shear thickening fluid

In this study, shear thickening fluid with 70 wt% of modified and unmodified silica nanoparticles were synthesized using sonochemical & homogenization method with ethanol solvent. The treated and untreated silica nanoparticles (70% w/W) were dispersed in ethanol (excess) using homogenisation at 5000 rpm for 10 minutes followed by sonication at a frequencyof 24 kHz, 75 W/cm^2 for 10 min. After it the polyethylene glycol (PEG-200) (liquid, 30 % w/W) was added and the mixture is homogenised for 10 minutes at 5000 rpm followed by sonication. The reaction mixture was maintained below 40°C. Ethanol was evaporated under vacuum and product is collected as pale yellow or white coloured liquid.

2.4 Rheology of Shear thickening fluid

ST behavior of the synthesized STF samples was analyzed by rheological studies with MCR-52 Rheometer, Anton Paar, Germany, with cone angle of 1 ° in cone–plate geometry plate diameter of 40 mm and a gap of 0.08 mm. All the rheology experiments were performed at 25 °C with shear rate in the range of 0 - 2000/s .All the STF samples were subjected to 6 seconds pre-shearing at 0.1 to 1 S-1 to set up an equilibrium structure of the sample and to remove any air bubble in the system.

3. RESULTS AND DISCUSSION

The dispersed silica nanoparticles of average size 600 nm were treated with different concentration of Mercaptopropyltrimethoxysilane (MPTMS) (25-100 mmol) to achieve different concentration of surface thiol groups (Table 2). Mercaptopropyltrimethoxysilane is usually used for surface modification of nanoparticles with thiol groups as shown in figure 1. Huang and co- workers functionalized mesoporous silica materials with MPTMS [34].

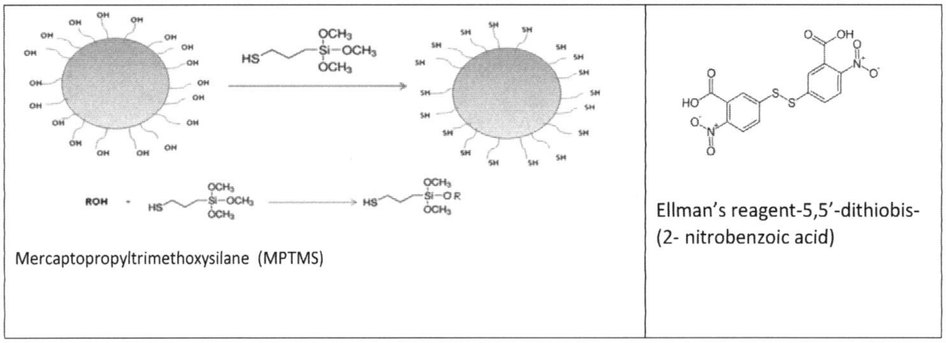

Figure 1. Schematic representation of surface modification of silica nanoparticles with thiolfunctional group.

The morphology and average diameter of unmodified and modified silica nanoparticles (Silica- SH-100mmol) was studied by SEM (Scanning Electron Microscope) analysis as shown in Figure 2.It was observed that functionalization of silica nanoparticles doesn't result to significant change in particles size.

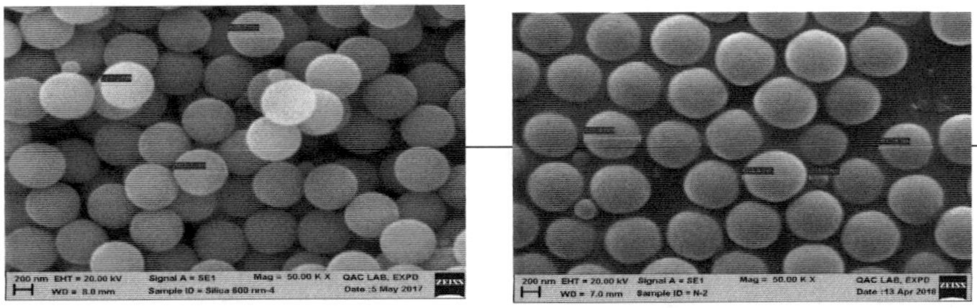

Unmodified Silica Nanoparticle **Modified Silica Nanoparticle (Silica-SH-100mmol)**

Figure 2.SEM analysis on unmodified and modified Silica Particles with thiol group

The BET (Brunauer–Emmett–Teller) surface area of unmodified silica particles and thiol functionalized silica nanoparticles were evaluated. Based upon the results it was further utilized along with Ellman's reagent 5,5'-dithiobis-(2-nitrobenzoiac acid)] (DTNB) (figure 1) to calculate the number of thiol groups per silica nanoparticles (Table 2) [35].

The surface charges of unmodified and modified silica particles dispersed in ethanol were evaluated using the zeta potential at pH of 6.7. The unmodified silica shows zeta potential of -7.1 ± 2 mV due to the deprotonation of surface hydroxyl groups (Table 2). Whereas, after modification with M3-Mercaptopropyltrimethoxysilane (MPTMS), the zeta potential of the thiol functionalised silica particles was observed at -11.1 to -17.4 mV.

Table 2. Zeta potential/surface area and Thiol group per particles of modified and unmodified shear thickening fluids.

S. No.	Sample	Zeta Potential(mV)	Surface area per particle (nm^2)	Thiol (-SH) group per Particle
1	Unmodified Silica	-7.1 ± 2	5.20×10^5	0
2	Silica-SH-25mmol	-11.1± 2	4.50×10^4	1.20×10^6
3	Silica-SH-50 mmol	-14.2± 2	4.35×10^4	1.90×10^6
4	Silica-SH-75mmol	-17.1± 1	4.20×10^4	2.21×10^6
5	Silica-SH-100mmol	-17.4 ± 1	4.15×10^4	2.23×10^6

Table 3. Rheological parameters of surface modified shear thickening fluids of different Thiolconcentration

S. No.	Sample Name	Critical Shear Rate (s^{-1})	Critical Viscosity (Pa.s)	Shear Max (s^{-1})	Viscosity Max (Pa.s)
1	Unmodified Silica	2.23	7.7089	81.7	238
2	Silica-SH-25mmol-70%	4.82	4.5176	37.7	187.39
3	Silica-SH-50mmol-70%	13.5	2.007	48.8	143.84
4	Silica-SH-75mmol-70%	16.1	3.4073	48.8	128.47
5	Silica-SH-100mmol-70%	29.2	2.9346	106	67.433

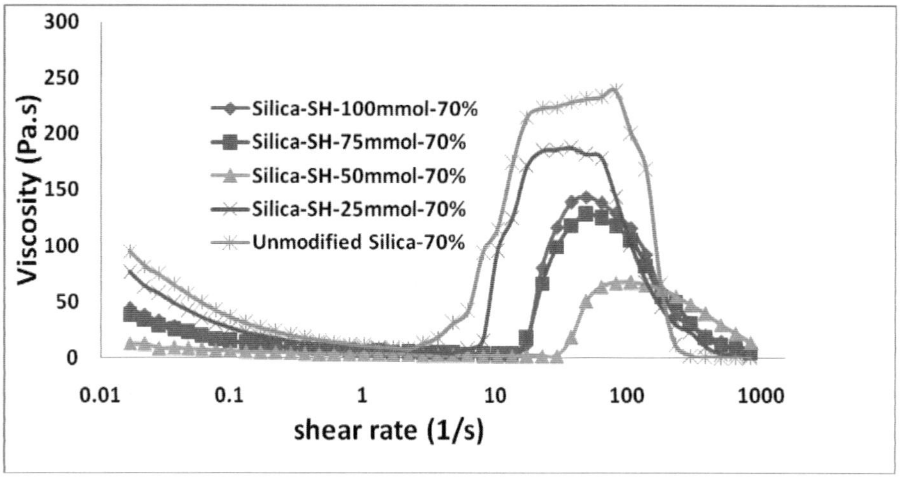

Figure 3. Steady state viscosity as a function of shear rate of STF of unmodified and surfacefuntionalised silica particles.

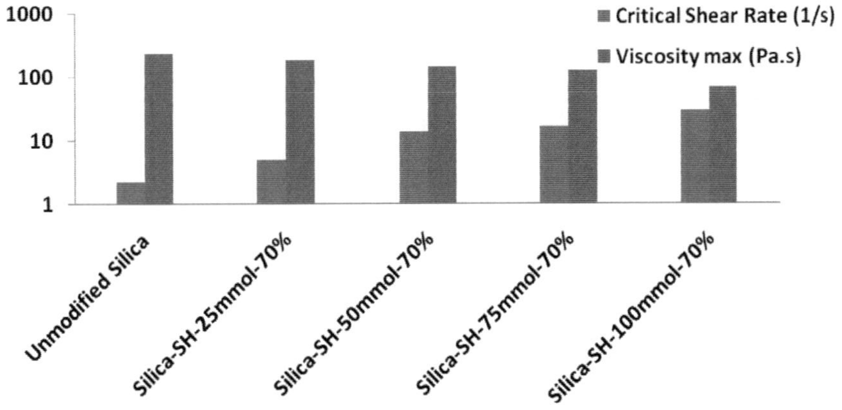

Figure 4. Viscosity max profile and Critical shear rate of the STFs with different SH concentration.

The steady viscosity as a function of shear rate from 0 to 1000 s^{-1} of STF of unmodified and surface functionalised silica particles was depicted in Figure 3. It can be seen that all the STFs of both kind of particles show a similar rheological behaviour to the applied shear stress. At lower shear rates the viscosity of STFs decreases. It is followed by a spontaneous rise in viscosity aftera critical value of shear rate. It exhibits shear thickening response of both kind of shearthickening fluids. This is an important aspect for the impact energy absorption of shear thickening fluids . Steady-shear rheological analysis (Figure 3) also shows the effect of □SH onthe steady shear response of the synthesized STFs. It was observed from Figure 3 & 4 and Table3 that with increasing –SH concentration of modified silica nanoparticles, a fall in the maximum viscosity and rise in shear rate was observed. The maximum viscosity value decreases from 238 to 67.43 Pa.s as the thiol concentration on the surface of silica particles was increased to 2.23 x 10^6 per particle. It was also observed that the critical shear rate value increases from 2.23 to 29.2 sec^{-1} with increase in thiol concentration on silica nanoparticles.

In case of the thiol modified silica nanoparticles, the STF formed shows lesser attraction between hydrophilic polyethylene glycol and hydrophobic thiol modified silica particles This results to increase in critical shear rate (29.2 s^{-1}) with the lowering of the peak viscosity (67.43 Pa.s) for the onset of shear thickening of STFs.

Previous studies have reported that the hydroxyl density on silica influences the inter-particle interactions of particles [30-32]. It is understandably obvious from Fig. 3 that modified silica based STF shows significant increase in the critical shear rate profile compared to that of silica- based STFs. In our study, it was found that the STF of -SH functionalized silica nanoparticles have shown a lesser inter-particle attraction as compared to unmodified silica particles due to lower surface hydroxyl density. It will result to increase in value of critical shear rate. Butola etal [36] also found increase in performance of shear thickening fluid with decrease in surface hydroxyl group. It was also observed that with the decrease in surface OH group, the viscosity max of the STF decreases with increase in critical shear rate. The critical shear rate is one of the important parameter for ballistic performance of armor panels. The STF with higher shear rate maintain the flexibility of armor panels and results in shear hardening at higher shear rate. The developed STF of thiol modified silica particles with higher critical shear rate and lower surface hydroxyl group is therefore a plausible candidate for STF based armor systems.

4. CONCLUSION

This study opine a method in which silica nanoparticles of different concentration with surface thiol groups were synthesized by the reaction of silica nanoparticles with average particle size of 600nm and MPTMS. Shear thickening fluid of thiol functionalized silica nanoparticles were characterized using steady state rheology. It was observed that thiol functionalized silica nanoparticle based STFs shows shear thickening at higher shear rate compared to unmodified silica nanoparticles. Therefore, the STFs of modified silica NP is well-advised as more suitable fluid for STF based armor system.

Conflict of interest

The authors declare that they have no conflict of interest.

REFERENCES

[1] Bajyan M, Majumdar A, Butola B S, Verma S K, Bhattacharjee D 2020 *Composites Part B: Engineering* **83** 107721.

[2] Grover G, Verma S K, Thakur A, Biswas I, Bhatacharjee D 2020 *Mater. Today: Proc.* **28** 1472-76

[3] Hassan T A, Rangar V K and Jeelani S 2010 *Mater. Sci. Eng.* **527** 2892

[4] Luz F S, Junior E P L, Louro L H L and Monteiro S N 2015 *Mater. Res.* **18** 170-77

[5] Majumdar A, Laha A, Bhatacharjee D, Biswas I and Verma S K 2019 *J. Text. Inst* **10** 1515-18

[6] Manukonda B H, Chatterjee V A, Verma S K, Bhattacharjee D, Biswas I and Neogi S 2019 *Period. Polytech. Chem. Eng.* **64** 75-84

[7] Barnes H A, 2000 *A handbook of elementary rheology* Aberystwyth: University of Wales, Institute of Non-Newtonian Fluid Mechanics

[8] Bender J and Wagner N J 1996 *J. Rheol.* **40** 899-16

[9] Brady J and Bossis G 1985 *J Fluid Mech* **155** 105-29

[10]Hoffman R L 1972 *Discontinuous and Dilatant Viscosity* **161** 155–73

[11]Catherall A A, Melrose J R and Ball RC 2000 *J Rheol* **44** 1-25

[12]Bossis G and Brady J F 1989 *J. Chem. Phys* **91** 1866-74

[13]Bender J W, Wagner N J 1995 *J Colloid Interface sci* **172** 171–84

[14]Farr R S, Melrose J R and Ball R C 1997 *Phys Rev E* **55** 7203–11

[15]Maranzano B J and Wagner N J 2002 **10** 1063

[16]Helber R, Doncker F and Bung R 1990 *J Sound Vib* **138** 47–57

[17]Laun H M 1991 *J Rheol* **35** 999

[18]Fischer C, Braun S A and Bourban P E 2006 *Smart Mater Struct* **15** 1467–75

[19]Zhang X Z, Li W H, Gong X L 2008 *Smart Mater Struct* **17** 35027.

[20]Jiang W, Gong X, Xuan S, Jiang W, Ye F, Li X and Liu T 2013 *Appl. Phys. Lett.* **102** 101901

[21]Afeshejani S H A, Sabet S A R, Zeynali M E and Atai M 2014 *J Mater Eng Perform* **23** 4289–97

[22]Hasanzadeh M and Mottaghitalab V 2014 *J. of Mater Eng and Perform* **23** 1182-96

[23]Ye Y, Xiao H, Reaves K, McCulloch B, Mike J F and Lutkenhaus J L 2018 *ACS Applied NanoMaterials* **1** 2774-84

[24]Konijn B J, Sanderink O B J and Kruyt N P 2014 *Powder Technology* **266** 61-69

[25]Genovese D B 2012 *Adv colloid interface Sci* **171** 1–16

[26]Singh M, Verma S K, Biswas I and Mehta R 2018 *J Mater .Res. Express* **05** 05570

[27]Asija N, Chouhan H and Bhatnagar N 2017 *J Chemistry Sci* **06** 09-17

[28]Kang T J, Hong K H and Yoo M R 2010 *J. Fibers and polymers.* **11** 719-24

[29]Chu B, Brady A T, Mannhalter B D and Salem D R 2014 *J.Phys D Appl Phys* **47** 1–7

[30]Islam E, Kaur G, Bhattacharjee D Singh S, Biswas I and Verma S K 2018 *Colloid Polym Sci.* **296** 883-893

[31]Singh M, Verma S K Biswas I and Mehta R 2018 *Mater. Res. Express* **5** 55704

[32]Singh M, Mehta R, Verma S K and Biswas I 2018 *Mater. Res. Express* **5** 14001

[33]Sabzi M, Mirabedini S M, Zohuriaan M S and Atai J M 2009 *Prog. Org. Coat* **65** 222–22

[34]Liberman A, Mendez N, Trogler W C, and Kummel A C 2014 *Surf Sci Rep.* **69** 132–58.

[35]Ellison P A, Chen F, Goel S, Barnhart T E, Nickles R J, DeJesus O T, and Cai W **2017** *ACS ApplMater Interfaces* **01** 6772–81

[36]Talreja K, Chauhan I, Ghosh A, Majumdar A, and Butola B S 2017 *RSC Adv.* **70** 49787

Synthesis and Characterization of $La_{0.70}Ca_{0.30}MnO_3$ System for Solid Oxide Fuel Cell Applications

Surinder Paul[1]*, Manokamna[2], Arun Kumar[3], Arvind Kumar[4]

[1]Department of Physics, Arni University, Kathagarh Indora-176401, (H.P) India
[2]Department of Applied Sciences, I.K.G.P.T.U, Jalandhar-144001, (Punjab) India
[3]Department of Physics, Swami Premanand Mahavidyalaya, Mukerian-144211 (Punjab) India
[4]Department of Applied Sciences, B.C.E.T, Gurdaspur- 143521 (Punjab) India

E. Mail: surinderpaul79@gmail.com

Abstract. $La_{0.70}Ca_{0.30}MnO_3$ system has been prepared by solid state reaction method for cathode of solid oxide fuel cell applications. The structural studies have been investigated by X-Ray diffraction method and the experimental results have confirmed that the prepared system has been well crystallized into single phase, crystal structure is orthorhombic and size of the particle has been calculated to be 0.35 μm. The surface morphology has been studied by scanning electron microscopy which confirmed that grain sizes are irregular, non uniform and randomly oriented. Archimedes principle has been used to investigate the density of the prepared material. Thermal properties of the system confirmed its stability at high temperature and thermal expansion coefficient well matched with other cell components. The dielectric as well as impedance properties have been investigated at different ranges of temperature and frequency. Electrical conductivity of the prepared system has been found to be more than 100 S/cm as well as obtained value of activation energy signifies that the synthesized system is suitable for cathode of solid oxide fuel cell applications.

Keywords: solid oxide fuel cell, X-Ray diffraction, density, thermal expansion coefficient, activation energy

1. Introduction

Solid oxide fuel cell (SOFC) represent one of the sustainable, clean alternate non conventional energy sources and versatile way of power production by converting the chemical energy from fuel directly into electric energy. SOFCs work at higher value of temperatures in comparison to other fuel cells [1-3]. Because of clean, portable and renewable character of the hydrogen, it can be considered as the next generation energy carriers [4, 5]. SOFCs are one of the chief sources of generation of hydrogen. In addition to this its efficiency arrives at sixty percentages and reaches at eighty percentages on heat recovery [6, 7]. Cathode is one of its components which is a conducting electrode and play a vital role in increasing the efficiency of the SOFCs. The main requirement of the cathode of SOFCs includes its adequate conductivity preferably more than 100 S/cm and its compatibility of thermal expansion coefficient (TEC) along with neighbouring components like electrolyte and anode [8]. High temperature working condition of SOFCs limits the properties of cathode and hence affects the overall efficiency of the cell [9-11]. In recent years a variety of mixed ionic and electronic cathode has been developed which prove their potential when SOFCs worked at intermediate temperature range [12, 13]. Metal doped $LaCO_3$, $LaMnO_3$ and $LaFeO_3$ perovskite oxide are commonly used inexpensive cathode material for SOFC. However at low temperature their electrochemical performance being affected [14]. Calcium is another reported pronouncing element for doping result in to elevated

IOP Conf. Series: Materials Science and Engineering 1225 (2022) 012054 doi:10.1088/1757-899X/1225/1/012054

electrical conductivity in addition to thermal expansion compatibility among YSZ electrolyte [15- 17]. Therefore in this work calcium is substituted in $LaMnO_3$ system to study the structural, thermal, dielectric and conductive properties in order to use $LaMnO_3$ as cathode component of solid oxide fuel cell.

2. Experimental

$La_{0.70}Ca_{0.30}MnO_3$ prepared by solid state reaction method. Calcium carbonates ($CaCO_3$), Lanthanum oxide (La_2O_3), Manganese oxide (MnO_2) from sigma Aldrich and 99.9% pure has been taken in stoichiometric ratio. Acetone has been used as solvent and this mixture has been further milled with zirconium oxide balls for 6 hours. Dried precursor powder further has been calcinated conventionally at 950^{0}C for twelve hours and pellet of 1mm diameter has been prepared. Pellets have been conventionally sintered 1400°C for 2 hours.

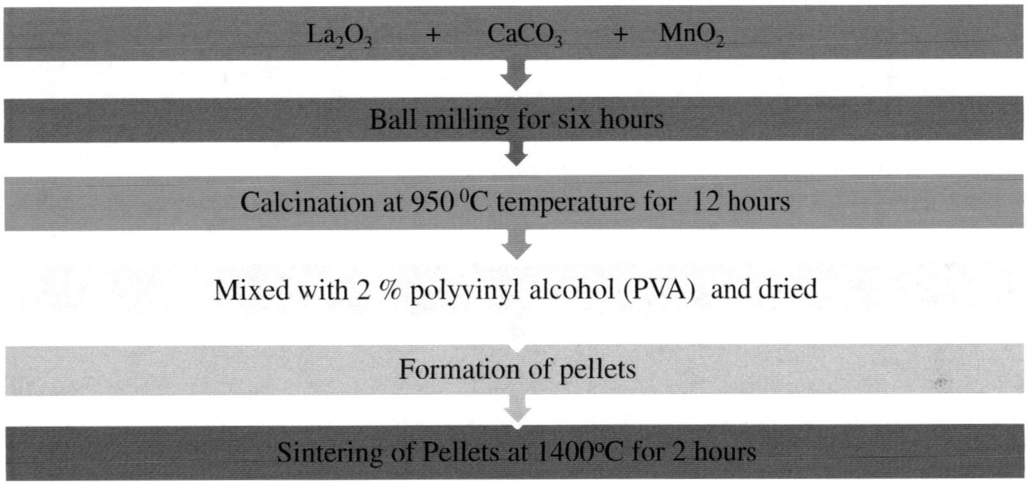

3. Results and Discussion

3.1 X-ray diffraction and Density

The as-prepared $La_{0.70}Ca_{0.30}MnO_3$ solid solution has been characterized by X-ray diffraction and the obtained patterns at 1400°C has been shown in figure 1. X-rays of wavelength equal to 1.54 Å, range 20°–80°, size of step 0.02 degree as well as speed limit of 2 degree per minutes have been used to obtain the XRD diffraction patterns. The crystal structure of the as prepared perovskite is orthorhombic and has been well crystallized, moreover recognized by sharp crystalline peaks. None of the peak remain unassigned which confirm that sample is synthesized in single phase. Crystalline size has been determined by scherrer formula and found to be 0.35 μm.

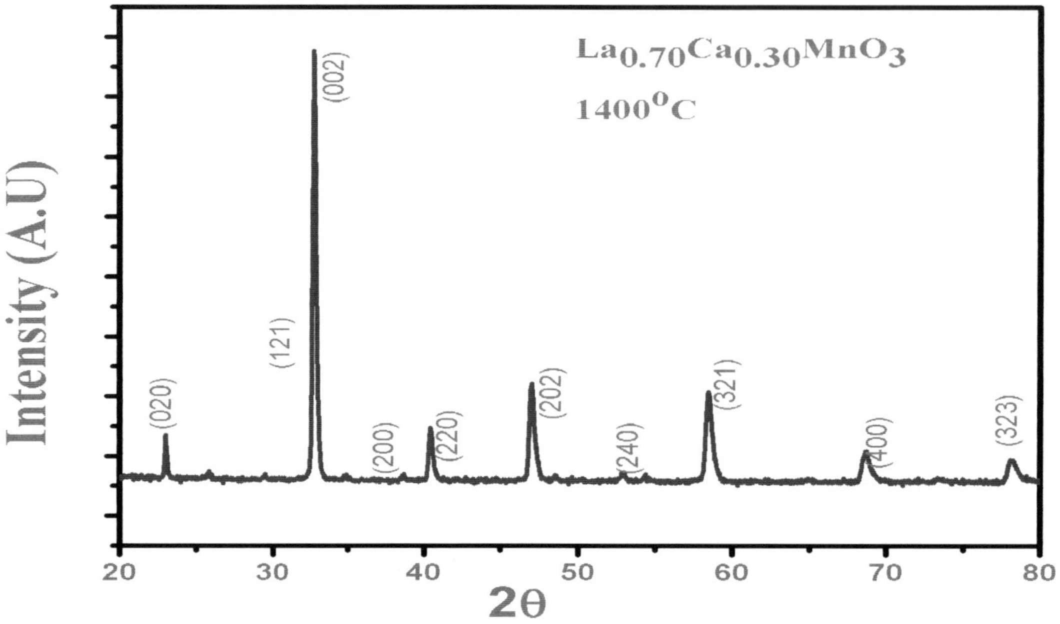

Figure 1. X-ray diffraction patterns of $La_{0.7}Ca_{0.3}MnO_3$ system sintered at 1400°C.

Archimedes principle has been used to measure the density of the as-prepared $La_{0.70}Ca_{0.30}MnO_3$ system and tabulated in table 1. Calculated value of specific free volume as well as tolerance factor is also given in the table 1.

Table 1. Density, specific free volume and Tolerance factor of sintered $La_{0.7}Ca_{0.3}MnO_3$ perovskite material.

Sample	Density			Specific free volume	Tolerance factor
	Theoretical density d_{th} (g/cm^3)	Experimental density d (g/cm^3)	(d/ d_{th}) %		
$La_{0.70}Ca_{0.30}MnO_3$	5.12	4.16	81.25	0.821	0.79

3.2 Scanning Electron Micrograph

The SEM micrograph of prepared $La_{0.70}Ca_{0.30}MnO_3$ has been shown in figure 2. The grains have been formed of non-regular in size and shape which are evidently observable from the image. Average grain size is found to be in well agreement with the XRD results. Average grain size is observed to be near 0.35 μm which is in well agreement with size calculated from XRD analysis.

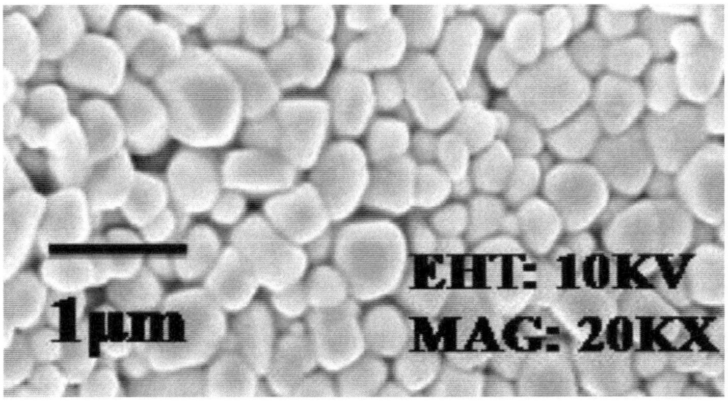

Figure 2. SEM micrograph of $La_{0.7}Ca_{0.3}MnO_3$ perovskite material.

3.3 Thermal Properties

Thermo-gravimetric analysis (TGA) and Thermal expansion coefficient (TEC) has been used to study the thermal properties if the as prepared material. Material used as cathode of SOFCs must be thermally stable in order to avoid electrolyte cracking or delaminations at interface of electrodes. Thermo-gravimetric analysis (TGA) and thermal expansion coefficient (TEC) graphs has been shown in figure 3 (a) and 3 (b). Consequence of heat on weight change is analyzed by thermo-gravimetric analysis (TGA) technique. At low temperature due to loss of moisture there is a sharp weight drop and above 350°C temperature material start gaining stability. This is because of oxidation at high temperature of Mn present at B site of ABO_3 perovskite [18, 19]. Dilatometer is used to investigate the TEC of the $La_{0.7}Ca_{0.3}MnO_3$ sample and therefore obtained value is found to be 10.9×10^{-6} °C^{-1}.

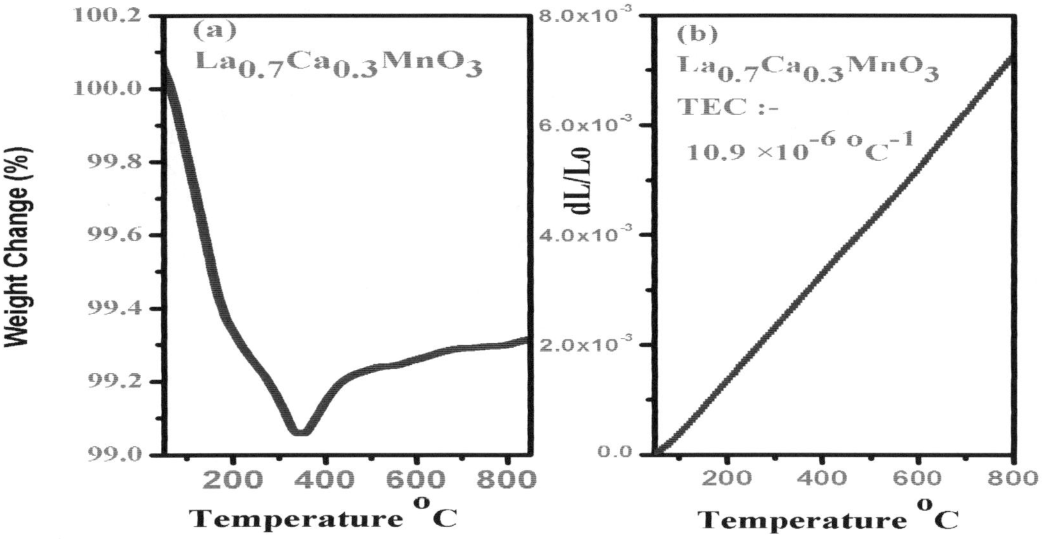

Figure 3. (a) Thermo-gravimetric analysis (TGA) and (b) Thermal expansion coefficient (TEC) curves of $La_{0.7}Ca_{0.3}MnO_3$ perovskite material.

3.4 Dielectric properties

The dielectric properties in the $La_{0.70}Ca_{0.30}MnO_3$ has been studied by impedance spectroscopy. At different temperature, ε′ (real) and ε′′ (imaginary) dielectric constant with respect to variation in frequency has been given in figure 4 (a) as well as 4 (b) respectively. The graphs undoubtedly expose that with rising frequency (low frequency region), both ε′ (real) and ε′′ (imaginary) continually decreases while in higher frequency together shows approximately linear behavior. At low frequency maximum polarizability arise due to dipolar, ionic, electronic and space charge while at high frequency dielectric constant ε′ (real) and ε′′ (imaginary) lags behinds the switching signal of dipolar orientation result into linear variation. Figure 4 (c) and 4 (d) has been show variation of part Z′ of impedance and part Z′′ of impedance with respect to frequency respectively. Continuous decrease in Z′ with temperature and frequency give rise to negative temperature coefficient of resistance (NTCR) in the sample. The merging of impedance Z′′ vs. frequency values curves bear out the disappearing of polarization (space charge) [20].

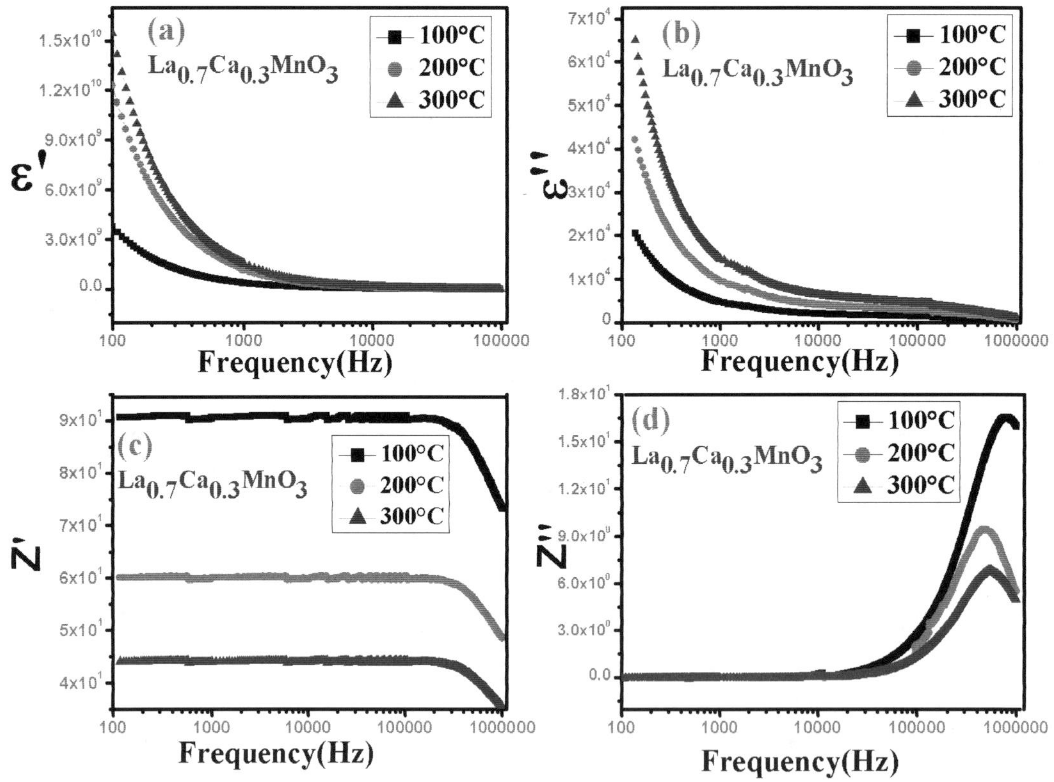

Figure 4. Variation of (a) dielectric constant ε′ (real), (b) ε′′ (imaginary) , (c) real part (Z′)of impedance and (d) imaginary part (Z′′) with respect to frequency.

3.5 Electrical Properties

Electrical conductivity Value at diverse frequencies range for $La_{0.70}Ca_{0.30}MnO_3$ up to 600°C has been shown in figure 5. Graph moreover explains that conductivity improved with temperature as well as frequency. The maximum calculated conductivity value has been found to be 103 S/cm at 600 ^{0}C and activation energy calculated by using Arrhenius fitting is 0.19 eV.

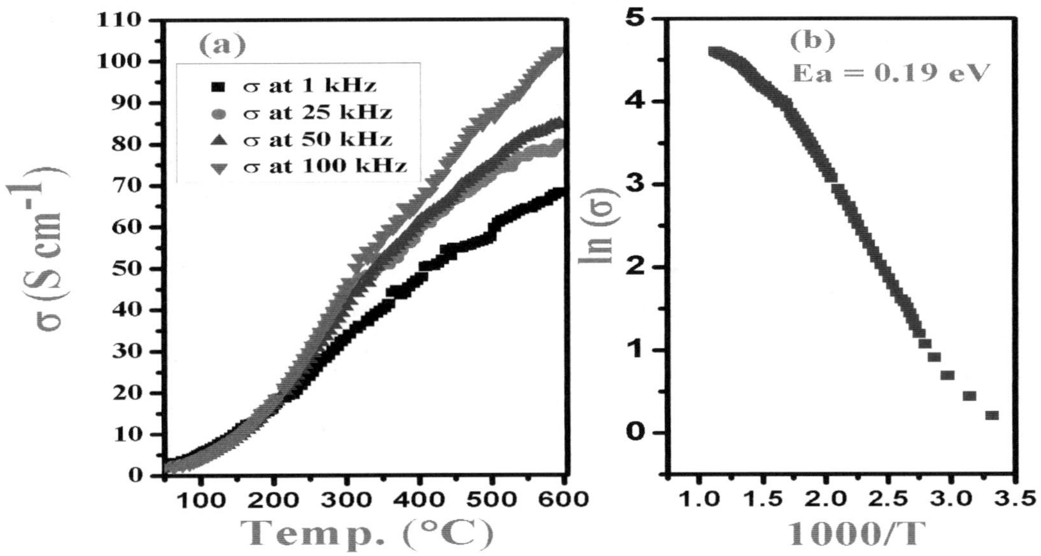

Figure 5. Varition of (a) conductivity with temperature at different frequencies and (b) activation with respect to temperature

4. Conclusions

$La_{0.70}Ca_{0.30}MnO_3$ perovskite material has been prepared by solid state reaction method. X-Ray diffraction confirmed that the prepared system has been well crystallized into single phase, crystal structure is orthorhombic and size of the particle is calculated to be 0.35 µm. The surface morphology has been studied by scanning electron microscopy which confirmed that grains are non uniform, randomly oriented and density of the prepared material is found to be 4.16 g/cm^3. TEC of the material is found to be 10.9 ×10^{-6} °C^{-1} in addition to this material show weight gain above 350°C. The dielectric constant lowered with temperature as well as frequency and also material showing negligible dielectric loss. Electrical conductivity has been found to be 103 S/cm at 600 ^{0}C and activation energy 0.19 eV signifies that the synthesized system is suitable to be used as cathode electrode of solid oxide fuel cell.

References

[1] Carter S, Selcuk A, Chater R J, Kajda J, Kilner J A, Steele B C H 1992 Oxygen transport in selected nonstoichiometric perovskite-structure oxides *Solid State Ion.* **53–56** 597–05

[2] Jiang S P 2003 Issues on development of (La,Sr)MnO3 cathode for solid oxide fuel cells *J. Power Sources* **124** 390–02

[3] Choudhury A, Chandra H, Arora A 2013 Application of solid oxide fuel cell technology for power generatio a review *Renew. Sustain. Energy Rev.* **20** 430–42

[4] Hauch A, Jensen S H, Ramousse S, Mogensen M 2006 Performance and durability of solid oxide electrolysis cells *J. Electrochem. Soc.* **153** A1741–A1747.

[5] O'Brein J E, Stoots C M, Herring J S, Hartvigsen J J 2006 Hydrogen production performance of a 10-cell planar solid-oxide electrolysis stack *J Fuel Cell Sci Technol.* **3** 213–219.

[6] Singhal, S C 2000. Advances in solid oxide fuel cell technology *Solid State Ion.* **135** 305-13

[7] Tinoco R R, Mansilla C, Bouallou C 2010 Competitiveness of hydrogen production by High Temperature Electrolysis: Impact of the heat source and identification of key parameters to achieve low production costs *J. Energy Convers. Manag.* **51** 2623–34.

[8] Sun C, Hui R and Roller J 2010 Cathode materials for solid oxide fuel cells: a review *J Solid State Electrochem* ***14*** 1125-44

[9] Kumar S A, Kuppusami P and Vengatesh P 2018. Auto-combustion synthesis and electrochemical studies of La0. 6Sr0. 4Co0. 2Fe0. 8O3-δ–Ce0. 8Sm0. 1Gd0. 1O1. 90 nanocomposite cathode for intermediate temperature solid oxide fuel cells *Ceram. Int.* **44** 21188-96.

[10] Letilly M, Joubert O. and Salle A L G 2012 Characterisation and optimisation of the cathode/electrolyte couple for SOFC LSCF/BIT07 *J. Power Sources* **212** 161-68

[11] Carter S, Selcuk A, Chater R J, Kajda J, Kilner J A, Steele B C H 1992 Oxygen transport in selected nonstoichiometric perovskite-structure oxides *Solid State Ion.* **53** 597–05

[12] SA M A, Anwar M, Raduwan N F, Muchtar A and Somalu M R 2018. Optical, mechanical and electrical properties of LSCF–SDC composite cathode prepared by sol–gel assisted rotary evaporation technique J. Sol-Gel Sci. Technol. **86** 493-04

[13] Baharuddin N A, Muchtar A and Somalu M R 2017 Short review on cobalt-free cathodes for solid oxide fuel cells *Int. J. Hydrog. Energy* **42** 9149-55

[14] De Souza R A, Kilner J A and Walker J F 2000 A SIMS study of oxygen tracer diffusion and surface exchange in La0. 8Sr0. 2MnO3+ δ *Mater. Lett.* **43** 43-52.

[15] Takeda Y, Sakaki Y, Ichikawa T, Imanishi N, Yamamoto O 1994 Stability of La1− xAxMnO3− z (A= Ca, Sr) as cathode materials for solid oxide fuel cells *Solid State Ion.* **72** 257–64

[16] Jie Z H A O, Chen L I, Linglong, K O N G, Xuewei W U and Yongchang M A 2011 Synthesis and characterization of calcium and manganese-doped rare earth oxide La1-xCaxFe0. 9Mn0. 1O3-δ for cathode material in IT-SOFC *J. Rare Earths* **29**1066-69

[17] Sahoo K K, Mishra P C, Kumar M 2020 *PhD journey made easy* Orange publisher edn 1,p.120.

[18] Richter J, Holtappels P, Graule T and Gauckler L J 2008. Non-stoichiometry, thermal expansion and electrical properties of Pr1− xSrxMn1− yInyO3− δ perovskites *Solid State Ion.* **179** 2284-89

[19] Chen F, Sorensen O T, Meng G, Peng D 1997 Chemical stability study of BaCe 0.9 Nd 0.1 O 3-α high-temperature proton-conducting ceramic *Journal of Mater. Chem.* **7** 481–85.

[20] Badapanda T, Sarangi S, Behera B and Anwar S 2014 Structural and impedance spectroscopy study of Samarium modified Barium Zirconium Titanate ceramic prepared by mechanochemical route *Curr. Appl. Phys.* **14** 1192-1200.

Electrochemical sensing of malathion using doped MOFs

S Nikhar[1] and P Kumar[1]*

[1]Materials Application Research Lab, Department of Nano Sciences and Materials,

Central University of Jammu,Jammu - 181143 (India)

*Email- pawan.nsm@cujammu.ac.in, pawannano10@gmail.com

Tel./Fax No: +91-1923-249658

Abstract Coordination polymerization produces a variety of metal–organic frameworks (MOFs), each with its own set of physical & chemical properties. Making MOFs suitably conductive in nature, is one of the hurdles faced to be utilized its potential practical applications. To explore the electrical conductivity's features of MOFs, NH_2-Al-MOF, ZnQ@ZIF-8 and Al-MIL-53 MOF have been treated with NaOH, H_2SO_4, and HCl. Acid/Alkaline doped MOFs have been utilized to prepare conducting thin films on indium tin oxide (ITO) slides using drop-casted method. Electrical feature analysis has indicated reduction in overall resistance of NH_2-Al-MOF < ZnQ@ZIF-8 < Al-MIL-53. The prepared MOFs thin films have been used for malathion detection. We have found limit of detection (LOD) i.e., 1.668 mg/L 2.386 mg/L and 2.397 mg/L for malathion using doped MOFs, respectively.

Keywords: NH_2-Al-MOF, ZnQ@ZIF-8, Al-MOF electrochemical sensors, Organophosphate Pesticides

1. Introduction

In present literature, electrochemical sensors arewidely popular in diverse application including environmental nursing, food production process regulator systems, and biotic investigation. It is due to their disposing quick response output, better sensitivity with cheaper price [1-4]. As like, pesticide electrochemical sensors has shown greater accessibility as a sensitive sensor for pesticides especially organophosphate pesticide (OPPs) determination. Basically, OPPs are widely employed in farming, profitable, and suburban applications. Importantly, they have terrible health and environmental adversely affect. As a result, the detection of OPPs hasfind out wide attention [5-7]. In present literature, insufficient/limited stability, sensitivity character loss, improper transduction methods and many more are matter of concern for OPPs sensing in real world application [8]. To counter these concern use of novel and advanced materials for pesticides sensors have shown great results [9]. For examples, AuNPs tinted graphene nanosheets (GNs) on glassy carbon electrode for enzymeless OPPs sensing.Au nanoparticle-decorated graphene hybrid nanosheet was demonstrated for rapid and, unpretentious of methyl parathion. The better sensitivity with linearity range 0.2-1.0

g mL^{-1} and limit of detection (LOD) 0.6 ng mL^{-1} was reported -via stripping voltammetric performances to capture the MP [3].

Here, we utilized doped metal organic framework (MOFs) for malathion sensing. Basically, MOFs are intriguing nano-structured materials made up of metallic ions with organic linker at certain experimental conditions [10,11]. Basically, they have comprising tunability behaviour, high surface area, excellent chemical and thermal features [12,13]. Surpassing interior MOFs surface area; they are greatly explored for gas adsorbent, electrochemical sensor with an establishment of H_2, CO_2, N_2 and methane [14-17]. We have synthesized ZnQ@ZIF-8, Al-MIL (53), and NH_2-Al-MIL as per reported in literature [18,19].

Afterward, the drop casting method was used for deposition of above said MOFs on ITO glass slides. Alkaline/acidic doping of NaOH, H_2SO_4, and HCl on synthesised MOFs was performed. Morphological and conductive analysiswas performed at each stage confirm analytical changes. In order to sense the selected OPPs i,e., Malathion, linear sweep voltammetry (LSV) was performed using current range (100mA – 10nA).

2. Experimental Details

2.1 Materials Used

Aluminium nitrate, Dimethyl formamide (DMF), terephthalic acid, 2-methyl imidazole, zinc acetate, triethylamine (TEA), 8-HQ (hydroquinidine), NH_2H_2BDC (2-aminoterepthalic acid), $AlCl_3 \cdot 6H_2O$ (aluminium chloride hexahydrate), sodium hydroxide (NaOH), Indium Tin oxide (ITO) slides, Ethanol (EtOH), H_2SO_4 (sulfuric acid), HCl (hydrochloride), malathion were of analytical reagent grade and used without further purification. All solution was prepared and diluted in deionized water.

2.2 Synthesis of MOFs

To synthesis of NH_2-Al-MOF, ZnQ@ZIF-8 and Al-MIL-53 MOF have been performed as per literature (Sánchez et al., 2015; Chaudhari et al., 2016).

2.3 MOF@Thin Film Preparation

After analysis, thin film of lab synthesized NH_2-Al-MIL ZnQ@ZIF-8, and Al-MIL (53) MOF was prepared on ITO using drop casting technique. A 10 mg/ml denser conjugated stock solution oflab synthesised MOFs in ethanol was prepared. The homogenously stock solution waswell mixed before the drop casting. Drop casting of the prepared stock solution was poured using a pipette drop wise over the ITO, simultaneously. This was then allowed to air dry for 10-15 minutes for the drop casting and analysed.

2.4 Alkaline/Acidic Doping @ MOF

0.1 M solution of sodium hydroxide (NaOH), sulphuric acid (H_2SO_4) and hydrochloric acid (HCl) was prepared in distilled water. Afterward, NaOH, H_2SO_4, and HCl are doped to each NH_2-Al-MIL ZnQ@ZIF-8, and Al-MIL (53) sample independently on ITO slide. We have used doped NH_2-Al-MIL ZnQ@ZIF-8, and Al-MIL (53) for pesticides sensing. Importantly; the electrical features were measured using electrochemical workstation.

2.5 Linear sweep voltammetry

Linear sweep voltammetry (LSV) protocol has been introduced to measure the current (I) versus voltage (V) linearity (I-V) graph. LSV was used to measure the resistance of blank ITO in the current range (100mA – 10nA).

2.6 Malathion sensing

The different dilution ranges (5ppm – 0.001ppm) solutions have been prepared in ethanol for malathion sensing. All dilutions were used for IV analysis using doped MOFs@ITO. The resistance was analysed over doped surfaces to measure the all-investigative alteration of analyte.

3. Results and Discussion

3.1 Morphological and spectroscopic analysis

NH_2-Al-MIL ZnQ@ZIF-8, and Al-MIL (53) were synthesized as per reported literature. Selection of above said MOFs was made due to and Metal based reversible and multi-electron transfer characters. Basically, the drop casting technique was used for the thin film deposition on ITO slides. Afterwards, 0.1 M solution of NaOH, H_2SO_4, and HCl were used for doping of ZnQ@ZIF-8, Al-MIL (53), and NH_2-Al-MIL. The morphological and spectroscopic techniques was utilized for undoped and doped MOFs analysis.

The SEM analysisshows nine images of Blank ITO lab synthesized undoped and doped NH_2-Al-MIL ZnQ@ZIF-8, and Al-MIL (53) thin film, respectively (**Figure 1**A, D & G). The morphological analysis specifies the pure crystalline pristine MOFs with rhombic dodecahedral (ZnQ@ZIF-8), irregular shape (Al-MIL (53)), microneedles and nanorods (NH_2Al-MIL (53)) with estimated crystal size, respectively (**Figure 1.** B, E & H). Afterward, significant changes in morphological of doped MOFs were find out and confirmed through SEM analysis (**Figure 1.** C, F& I).

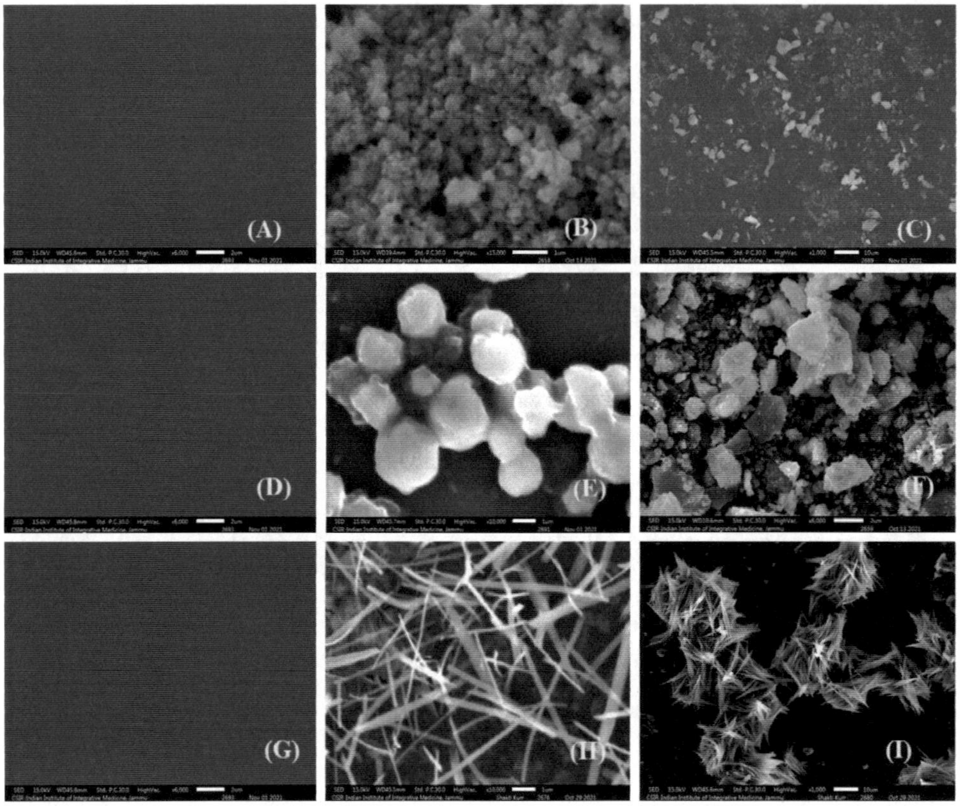

Figure 1. SEM Analysis on undoped and doped MOFs on ITO: Blank ITO (A, D & G), lab synthesized undoped (B, E & H) and doped ZnQ@ZIF-8, Al-MIL (53), and NH_2-Al-MIL thin film, respectively (C,F & I).

The voltametric analysis was performed using LSV technique. Swept linearity was measured with respect to time when in absence of potential between the reference electrode (RE) and working electrode (WE). Basically, the redox behaviour was observed in the current range (100mA – 10nA).

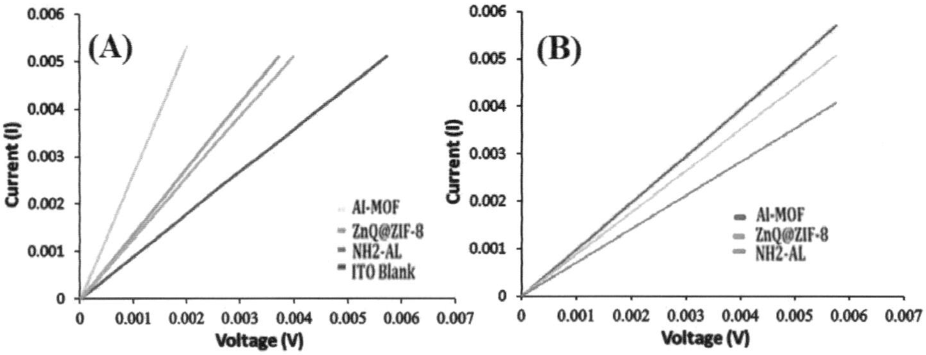

Figure 2 (A & B). Current–voltage (I-V) curve to resistance analysisof undoped and doped NH_2-Al MOF, AL-MOF and ZnQ@ZiF-8 at constant temperature using two probe techniques.

Figure 2 present the current–voltage (I-V) features of MOFs film developed on ITO. The data analysis with the linear visuality of I-V peaks achieved by performing LSV with scan rate of 21 sec^{-1}. All analysis was considered to following power law ($I \propto V$) dependency at constant temperature. This I-V peak allows measuring the outflow of resistance over the doped and undoped materials. A power scaling mechanism can be predicted with the effect of SCL (space-based charge limited) to show conduction modal and power-law $V^2 \propto I$ efficacy [20,22]. The conduction nature was shown by the doped MOFs surface which was Ohmic resistivity (V_{Ohm} = aI) expression accompanied in series, i.e., $V=aI+bI_{0.5}$. Finally, we find out thatNH_2Al-MOFwashave excellent conductive due to conjugated π orbital and bridging the dimeric aluminium subunits in MOF. It allows first-rate continuous conducting electrons path to support electron transfer [21-24]. The resistance values recorded was measured for undoped and doped MOFs (**Table 1).**

Table 1. Resistance (Ω) variations analysis undoped and doped MOF on ITO surfaces

Description	Undoped MOF Resistance (Ω)	Doped MOF Resistance (Ω)		
		H_2SO_4	*HCl*	*NaOH*
ITO – Blank	140.231	-	-	-
NH_2-Al-MOF	184.759	128.046	136.908	158.746
ZnQ@ZiF-8	242.814	134.059	150.104	166.069
Al-MOF	288.908	149.945	174.532	196.822

Our results indicate that the resistance level i.e., NH2-Al -MOF< ZnQ@ZIF8 < Al-MIL (53), respectively. On comparing norms, the conductivity of undoped MOF has given a evidential exposure to show the efficacy of the doped MOFs. It is clearly observed from the given date in Table 1.1, treatment with alkaline and acidic medium, i.e., NaOH, H2SO4, and HCl has caused increase in surface conductivity on applying potential range of -0.6 V to +0.6 V. The conductivity scale implies as. H2SO4>HCl> NaOH, means the hydrogen sulphide drags a more conductive behaviour than other doping solvents. These outcomes have proved that the doping of H2SO4, HCl and NaOH implies a conductive surface.

3.2 Malathion Sensing

The different concentration levels of malathion i.e., 1 ppb -5 ppm was made to explore the sensing application of doped MOFs. The current–voltage (I-V) data for malathion (with concentrations;1 ppb, 10 ppb, 100 ppb, 500 ppb, 1 ppm, 2ppm, 5 ppm), with different doped MOFs have been shown in **Figure 3, 4 and 5,** Basically, interaction between doped MOFs and malathion sensing was noted through decreases in the resistance with respect to increasing concentration of malathion. Basically, doped MOF have p-moieties which allowing them to interact with analyte and provide a transport channel. The delocalization of charge caused due to donor/acceptor interactions between analyte and MOFs[25].

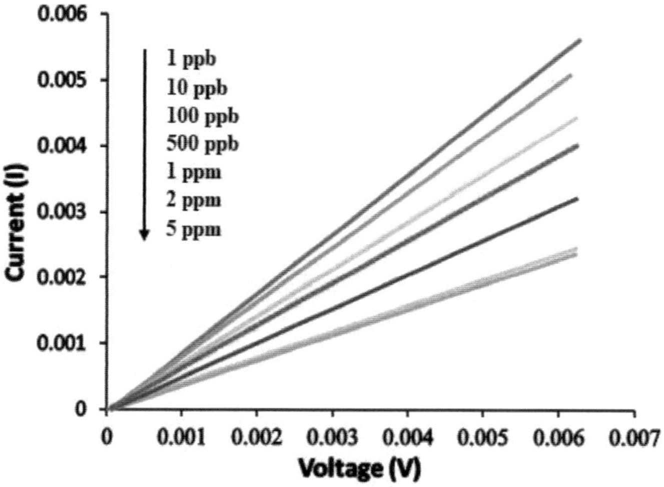

Figure 3. Current–voltage (I-V) curve to resistance analysis between H_2SO_4doped ZnQ@ZIF-8 and different concentration of malathion.

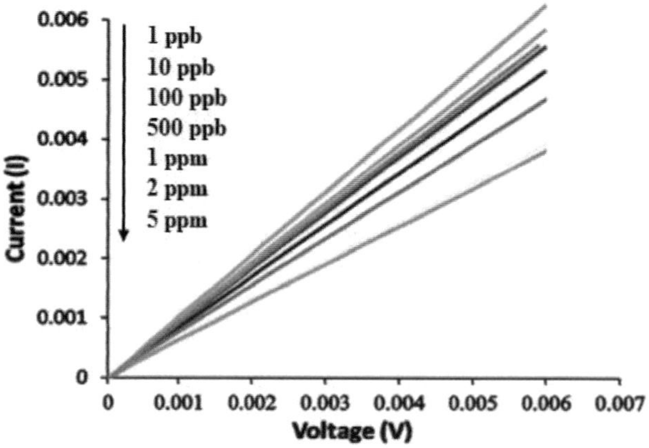

Figure 4. Current–voltage (I-V) curve to resistance analysis between H_2SO_4 doped Al-MIL (53) and different concentration of malathion.

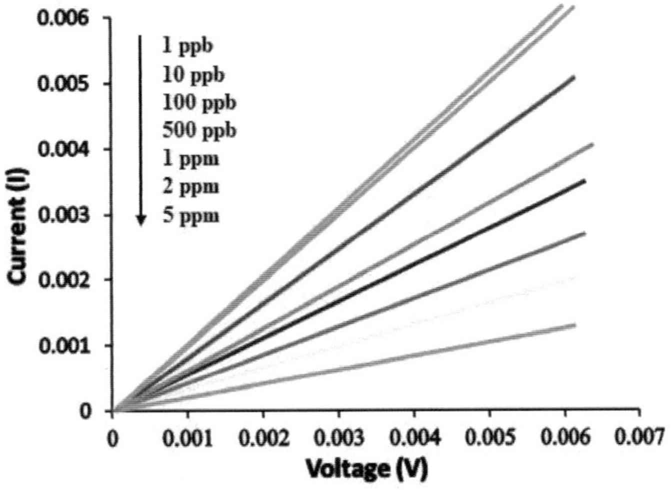

Figure 5. Current–voltage (I-V) curve to resistance analysis between H_2SO_4doped NH_2-Al-MOF and different concentration of malathion.

Interestingly, poor analyte interaction was found with Al-MIL-53 than ZnQ@ZIF-8 and NH_2-Al-MOF, respectively. The change in resistance during the interaction with all different pesticides concentration is shown in **Table 2**. Interaction comparison of analyte and doped MOFs clearly revealed that NH_2-Al-MOF performs better over ZnQ@ZIF-8 and NH_2-Al-MOF for malathion sensing. Finally, our results indicate limit of detection (LOD) i.e., 1.668 mg/L 2.386 mg/L and 2.397 mg/L, for malathion using doped NH_2-Al-MOF < ZnQ@ZIF-8 < Al-MIL-53, respectively.

Table 2. Interaction comparison on resistance (Ω) variations between malathion and doped MOFs

Sr. No.		**NH_2-Al-MOF**			**ZnQ@ZIF-8**			**Al -MIL (53)**		
	Pesticides Conc'n	***H_2SO_4***	***HCl***	***NaOH***	***H_2SO_4***	***HCl***	***NaOH***	***H_2SO_4***	***HCl***	***NaOH***
1	**1 ppb**	120.32	134.011	157.345	127.541	135.637	173.732	143.318	177.947	193.284
2	**10 ppb**	131.480	141.407	164.290	135.429	144.533	182.480	161.658	185.394	161.171
3	**100 ppb**	139.333	145.197	169.086	143.456	151.634	190.133	191.238	194.055	191.238
4	**500 ppb**	151.287	153.015	180.094	149.167	159.285	200.877	198.712	201.654	198.267
5	**1 ppm**	160.913	162.341	188.428	157.88	163.914	206.193	220.726	213.035	214.854
6	**2 ppm**	168.648	167.106	197.201	166.635	179.864	218.468	235.255	218.022	225.689
7	**3 ppm**	178.294	177.734	205.712	174.707	188.354	226.294	243.263	232.298	238.122
8	**5 ppm**	218.03	186.932	213.316	183.451	218.563	238.030	256.602	239.120	245.256

4. Conclusion

In this model study, we have performed the acid and alkaline doping of lab synthesized NH_2-Al-MOF, ZnQ@ZIF-8, and Al-MIL-53. Our results indicated that electrical features of MOFs have sustainably improved after doping of NaOH, H_2SO_4, and HCl. More importantly, we explore doped MOFs for pesticide sensing and find out suitable for the same. Interestingly, our comparison study indicated that H_2SO_4 doped NH_2-Al-MOF is most suitable doped MOFs for malathion sensing with limit of detection (LOD) i.e., 1.668 mg/L. We are exploring more possibility on electrochemical sensing based doped MOFsfor organophosphate pesticides (OPPs) detection. Hopefully, we will present more good results in near future.

Acknowledgments
We thank Science and Engineering Research Board (SERB), New Delhi, for funding under "Empowerment and Equity Opportunities for Excellence in Science" (Project's reference no. **EEQ/2016/000484**).

References

[1] İpek Y, Dinçer H and Koca A 2014 Selective electrochemical pesticide sensor modified with "click electrochemistry" between cobaltphthalocyanine and 4-azidoaniline *J. Electrochem. Soc.* **161** B183.

[2] İpek Y, Dinçer H and Koca A 2014 Electrode modification based on "click electrochemistry" between terminal-alkynyl substituted cobalt phthalocyanine and 4-azidoaniline *Sens. Actuators B: Chem.* **193** 830-37

[3] Gong J, Miao X, Zhou T and Zhang L 2011. An enzymeless organophosphate pesticide sensor using Au nanoparticle-decorated graphene hybrid nanosheet as solid-phase extraction *Talanta* **85** 1344-49

[4] Wang F and Hu S 2009 Electrochemical sensors based on metal and semiconductor nanoparticles *Microchim. Acta* **165** 1-22.

[5] Tjeerdema R S, d Voogt P, Giesy J, Stevens J T and Whitacre D M 2012 *Reviews of Environmental Contamination and Toxicology* Springer

[6] Wang M, Huang J, Wang M, Zhang D and Chen J 2014 Electrochemical nonenzymatic sensor based on CoO decorated reduced graphene oxide for the simultaneous determination of carbofuran and carbaryl in fruits and vegetables *Food Chemistry* **151** 191-97

[7] Mehrvar M and Mustafe A 2004 Recent developments, characteristics, and potential applications of electrochemical biosensors *Anal Sci* **20** 1113-26

[8] Kimmel D W, LeBlanc G, Meschievitz M E and Cliffel D E 2012 Electrochemical sensors and biosensors *Anal. Chem.* **84** 685-707.

[9] Hayat A and Marty J L 2014 Disposable screen printed electrochemical sensors: Tools for environmental monitoring *Sensors* **14** 10432-53.

[10] Yaghi O M, O'Keeffe M, Ockwig N W, Chae H K, Eddaoudi M and Kim J 2003 Reticular synthesis and the design of new materials *Nature* **423** 705-14

[11] Kitagawa S, Kitaura R and Noro S I 2004 Functional porous coordination polymers *Angew. Chem. Int. Ed. 43* 2334-75

[12] Allendorf M D, Foster M E, Leonard F, Stavila V, Feng P L, Doty F P, Leong K, Ma E Y, Johnston S R and Talin A A 2015. Guest-induced emergent properties in metal–organic frameworks *J. Phys. Chem.* **6** 1182-95

[13] Deria P, Mondloch J E, Karagiaridi O, Bury W, Hupp J T and Farha O K 2014 Beyond post-synthesis modification: evolution of metal–organic frameworks via building block replacement *Chem. Soc. Rev.* **43** 5896-912

[14] Kreno L E, Leong K, Farha O K, Allendorf M, Van Duyne R P and Hupp J T 2012. Metal–organic framework materials as chemical sensors *Chem. Rev.* **112** 1105-25.

[15] Rosi N L, Eckert J, Eddaoudi M, Vodak D T, Kim J, O'Keeffe M and Yaghi O M 2003 Hydrogen storage in microporous metal-organic frameworks *Sci.* **300** 1127-29

[16] Sumida K, Rogow D L, Mason J A, McDonald T M, Bloch E D, Herm Z R, Bae T H and Long J R 2012 Carbon dioxide capture in metal–organic frameworks *Chem. Rev.* **112** 724-81

[17] Furukawa H, Ko N, Go Y B, Aratani N, Choi S B, Choi E, Yazaydin A Ö, Snurr R Q, O'Keeffe M, Kim J and Yaghi O M 2010. Ultrahigh porosity in metal-organic frameworks *Sci. 329* 424-28

[18] Chaudhari A K, Ryder M R and Tan J C 2016. Photonic hybrid crystals constructed from in situ host–guest nanoconfinement of a light-emitting complex in metal–organic framework pores *Nanohl* **8** 6851-59

[19] Sanchez-Sanchez M, Getachew N, Díaz K, Díaz-García M, Chebude Y and Díaz I 2015 Synthesis of metal–organic frameworks in water at room temperature: salts as linker sources *Green Chem.* **17** 1500-09

[20] Rose A 1955 Space-charge-limited currents in solids *Phys. Rev.* **97** 1538.

[21] Talin A A, Centrone A, Ford A C, Foster M E, Stavila V, Haney P and Allendorf M D 2014 Tunable electrical conductivity in metal-organic framework thin-film devices *Sci.* **343** 66-69.

[22] Chin J M, Chen E Y, Menon, A G, Tan H Y, Hor A T S, Schreyer M K and Xu J (2013). Tuning the aspect ratio of NH 2-MIL-53 (Al) microneedles and nanorods via coordination modulation *CrystEngComm* **15** 654-57

[23] Campbell M G, Sheberla D, Liu S F, Swager T M and Dincă M 2015 Cu3 (hexaiminotriphenylene) 2: an electrically conductive 2D metal–organic framework for chemiresistive sensing. *AngewandteChemie Int. Ed.* **54** 4349-52

[24] Shinde S S, Lee C H, Jung J Y, Wagh N K, Kim S H, Kim D H and Lee J H 2019 Unveiling dual-linkage 3D hexaiminobenzene metal–organic frameworks towards long-lasting advanced reversible Zn–air batteries *Energy Environ. Sci.* **12** 727-38

[25] Day R W, Bediako D K, Rezaee M, Parent L R, Skorupskii G, Arguilla M Q and Dincă M 2019 Single crystals of electrically conductive two-dimensional metal–organic frameworks: Structural and electrical transport properties *ACS Cent. Sci.* **5** 1959-64

Piezoelectric sensor voltage variation with different adhesives at elevated temperature

Sukesha Sharma

UIET, Panjab University, Chandigarh, India

sukeshauiet@gmail.com

Abstract. Piezoelectric sensors and actuators are used in various applications. They have electromechanical properties. Electric voltage is generated when strain is applied on them and vice versa. These patches can be used at cryogenic temperatures as well as elevated temperatures according to different applications. Piezoelectric patches are pasted on the surface of the structure to sense the vibrations. Different adhesives can be used to paste piezoelectric patches on the structure. In this paper, two different Adhesives, Araldite and Cyanoacrylate are used to paste piezoelectric patches. Two test set up are prepared using two different adhesives. Sensor voltage is obtained from piezoelectric patches at reference temperature and at elevated temperatures. Results are obtained, compared and analyzed. Performance of adhesives is compared at reference temperature and at elevated temperature. Where, Cyanoacrylate is found to be better adhesive to paste piezoelectric patches on the surface of structure as compared to Araldite adhesive.

Keywords: Piezoelectric sensor, Araldite adhesive, Cyanoacrylate adhesive.

1. Introduction

Piezoelectric patches can be used as sensor and actuator for various applications, such as: Active vibration control (AVC)[1]–[3], structural health monitoring (SHM)[4] etc. In AVC, piezoelectric sensor is used to sense the vibrations and actuator is used to control the vibrations [5]. In AVC, piezoelectric sensors and actuators are pasted on the structure at specific locations with the help of adhesives. Various types of adhesives are available, such as epoxy paste adhesive [6], [7], araldite epoxy [8], metallic epoxy[4], Hysol EA 9309.3NA [9], Cyanoacrylate etc. Quality assessment of adhesive layer between piezoelectric patches and the structure is very important [10]. Wandowski et al has used electromechanical impedance method for quality assessment of adhesive bonds of piezoelectric transducers. Bonding layer may lead to change in electromechanical impedance. Xuerong et al has worked on effect of adhesive thickness, accelerated aging, and monitoring signal [11]. It is observed that thickness of adhesive and shear strength effects the signal of piezoelectric. Performance ofadhesives is also affected with change in temperature from the room temperature. Various types of adhesives are available in the market with different working temperature ranges [8], [9], [12].

In this paper, effect of temperature on piezoelectric sensor output with change in adhesive is observed. Experiments are done in the range from 15 °C to 75 °C. Results are obtained and compared for two different types of adhesives, i.e. Araldite and Cyanoacrylate. Organization of paper is as: In section 2, experimental set up is discussed. Comparison of sensor voltages at reference temperature and at elevated

IOP Conf. Series: Materials Science and Engineering 1225 (2022) 012056 doi:10.1088/1757-899X/1225/1/012056

temperature using two different adhesives is done in section 3. Finally, conclusions are drawn in section 4.

2. Experimental Setup

Two square cantilevered smart structures are used as test set up. Cantilevered smart structures are placed in the thermal unit (Fig (1) a & b). Complete test set up consists of:

I. Two square plates instrumented with piezoelectric patches using two different adhesives.
II. Piezoelectric sensing and amplifying unit
III. Thermal chamber
IV. Digital storage oscilloscope.

Two square plates of size 16 cm × 16 cm × 0.5 mm are used and each one is divided into 64 elements (Fig (2) a& b). In one square plate, piezoelectric patch is pasted using Araldite adhesive. In the second plate, Cyanoacrylate is used as adhesive to paste piezoelectric patch. Araldite adhesive used here has working temperature range from -30 °C to 80 °C. Whereas, working temperature range of cyanoacrylate adhesive is -30 °C to 90 °C Two separate test set up are prepared to be placed in thermal chamber one by one. Wires are soldered on both the piezoelectric patches, that serve the purpose of electrodes. These wires are connected to the piezoelectric sensing and amplifying unit. Output from piezoelectric sensing unit is fed to the PXI unit. Program to capture sensor voltage is written using Labview software.

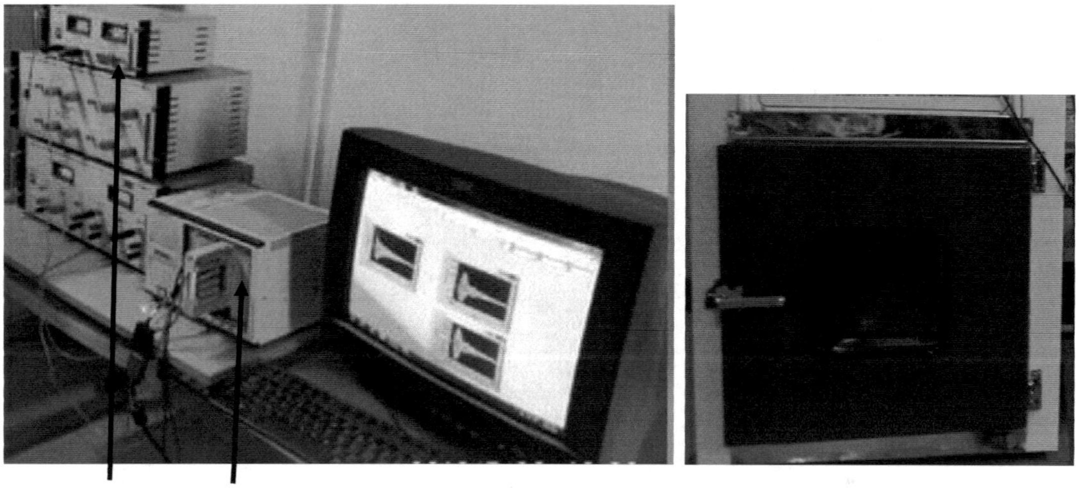

(a) (b)

Figure 1. a: Experimental set up and b) Thermal chamber

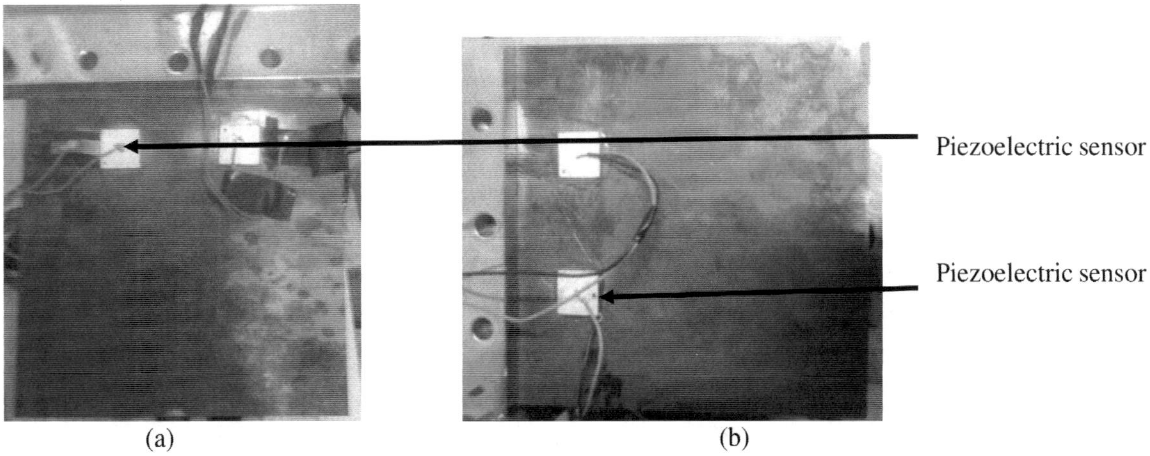

Figure 2. (a) Piezoelectric instrumented smart structure (using Cyanoacrylate as adhesive), (b) Piezoelectricsmart structure (using Araldite® as adhesive)

3. Comparison of Piezoelectric sensor voltages at room temperature and at elevated temperature

Displacement in the square plate structure leads to the vibrations. These vibrations are sensed by the piezoelectric sensor. In this work, tip displacement of 1 mm is given to the cantilevered plate and output in form of sensor voltage is sent to piezo sensing unit. Amplified piezoelectric output voltage is fed to PXI unit for analysis. Experiments are performed at 15 °C, 30 °C, 55 °C, 65 °C and 75 °C. Sample is kept at the particular temperature for at least 15 minutes for uniform distribution of temperature in whole test set up. According to available literature, sensor voltage increases with increase in temperature [3], [13]. Reference temperature is considered to be 15 °C. All other observations at elevated temperatures are compared with the observation at 15 °C. First observation is obtained for test set up where Cyanoacrylate adhesive is used to paste piezoelectric sensor on the square plate. Sensor voltage is captured at both 15 °C and 55 °C. As shown in Fig (3), piezoelectric sensor voltage increases with increase in temperature.

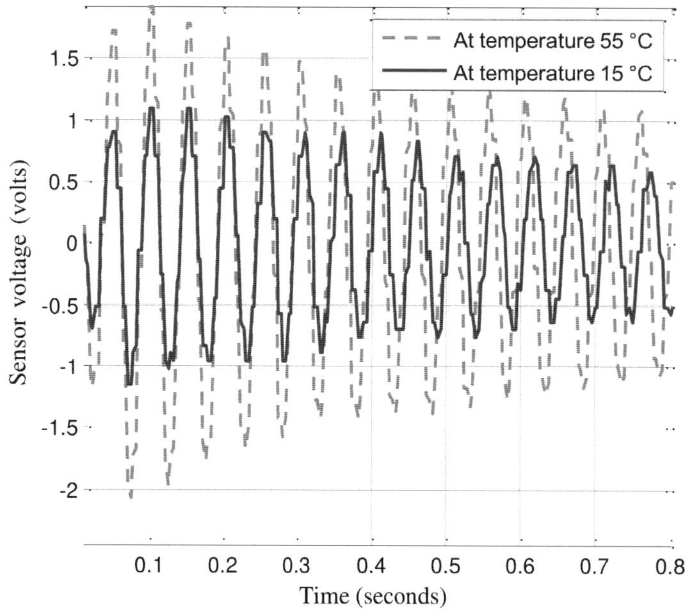

Figure 3. Comparison of sensor voltage obtained using Cyanoacrylate adhesive at room temperature (15 °C) and elevated temperature (55 °C).

Thereafter, sensor voltage is obtained for test set up where Araldite adhesive is used to paste piezoelectric sensor on the square plate. Sensor voltage is captured at both 15 °C and 30 °C. As shown in Fig (4) piezoelectric sensor voltage decreases with increase in temperature. Similar results are obtained at 55 °C, 65 °C and 75 °C. As shown in Fig (5), Fig (6) and Fig (7), sensor voltage is decreasing with increase in temperature. At 75 °C, decline in sensor voltage is almost more than 50 %. Effect of elevated temperature is visible starting from 30 °C while using Araldite adhesive. Whereas, working range given in specification sheet is -30 °C to 80 °C. It is visible from these results that Araldite is melting at micro level.

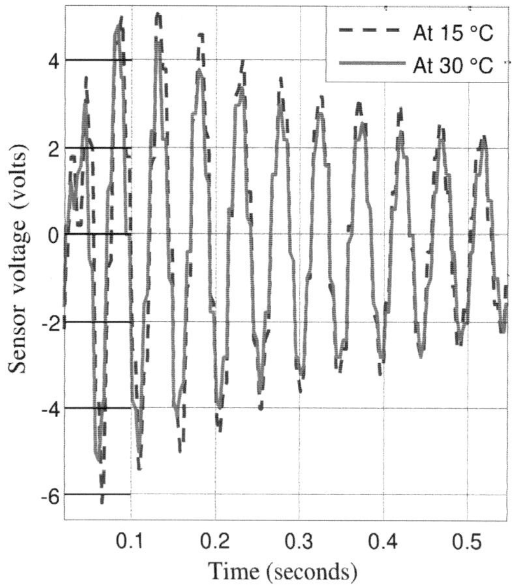

Figure 4. Comparison of sensor voltage obtained using araldite® adhesive at room temperature (15 °C) and at elevated temperature (30 °C).

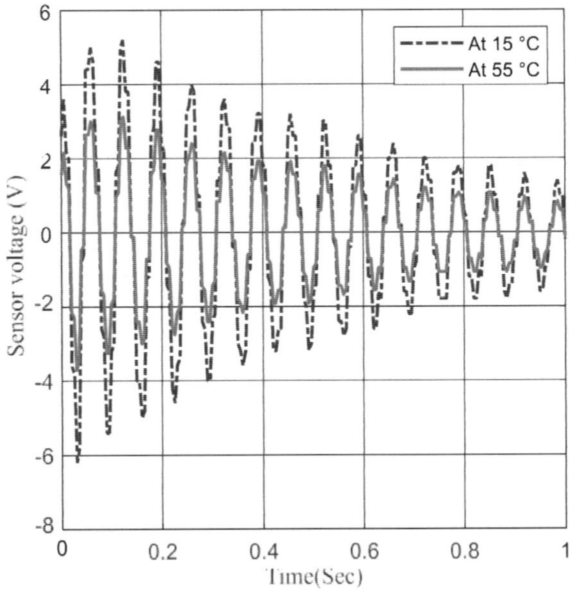

Figure 5. Comparison of sensor voltage obtained using araldite® adhesive at room temperature (15 °C) and at elevated temperature (55 °C).

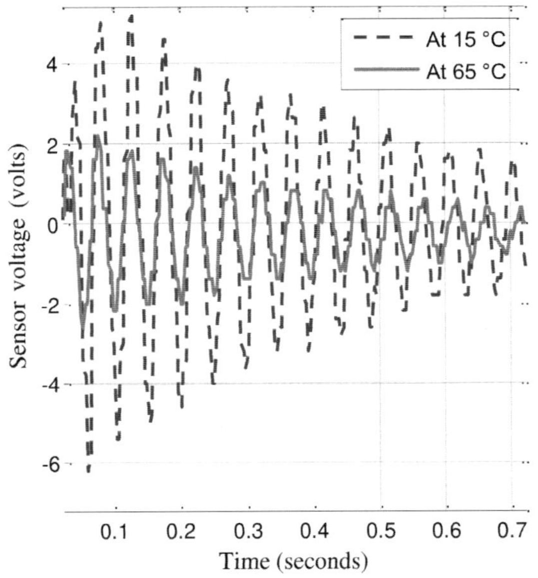

Figure 6. Comparison of sensor voltage obtained using araldite® adhesive at room temperature (15 °C) and elevated temperature (65 °C).

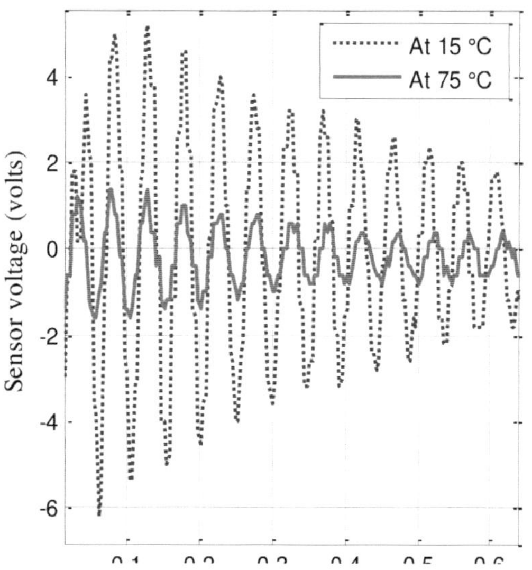

Figure 7. Comparison of sensor voltage obtained using araldite® adhesive at room temperature (15 °C) and elevated temperature (75 °C).

Results are also tabulated in Table 1. Comparison of peak sensor voltage at different elevated temperature is given. These results are also compared with the data available in the literature. As tabulated in Table 1 (Column 2 and 4), peak sensor voltage is increasing with increase in temperature for Cyanoacrylate adhesive. Whereas (column 1 and 3), sensor voltage is decreasing with increase in temperature in case of Araldite/epoxy adhesive.

Table 1. Comparison of peak sensor voltage at elevated and room temperature.

Temperature	Adhesive: Araldite/ epoxy (In this paper)	Cyanoacrylate (In this paper & [3])	Araldite/ epoxy (A Sharma et al) [14]	Cyanoacrylate (V Gupta et al) [13]
--	Plate structure 1mm tip displacement	Plate structure 1mm tip displacement	Beam structure 2.5 mm tip displacement	Plate structure 2mm tip displacement
--	Peak sensor voltage (V)	Peak sensor voltage (V)	Peak sensor voltage (V)	Peak sensor voltage (V)
15 °C	3	1.1	4	12
30 °C	2.9	1.3	3.75	-
55 °C	2	2	3.55	16
65 °C	2.1	2.6	3.5	17
75 °C	1.5	-	3.4	18

In active vibration control sensor voltage is used to generate control voltage. Change in sensor voltage leads to change in control voltage. If sensor voltage observation is incorrect then control voltage calculation is also incorrect. Therefore, it is not advisable to use Araldite adhesive at elevated temperatures. Another option is to include adhesive layer in the finite element modelling of smart structure.

4. Conclusions

Two test set up are prepared for experiments. One set up is prepared using Araldite adhesive to paste piezoelectric patches. Second test set up is prepared using Cyanoacrylate adhesive to patch piezoelectric patches. Experiments are performed at reference temperature and elevated temperatures for both test set up separately. Sensor Voltage is obtained for both test set up and results are compared with available literature. It is concluded here that:

i. Sensor voltage of piezoelectric patches increases with increase in temperature.
ii. Sensor voltage decreases with increase in temperature if Araldite/epoxy adhesive is used to paste the piezoelectric patches. Decrease in sensor voltage (using Araldite adhesive) means that adhesive is melting at micro level.

Therefore, at elevated temperatures adhesive like Cyanoacrylate with high working temperature range should be used instead of Araldite/epoxy.

References

[1] S Sharma, Vig R, and Kumar N 2016 Variation of piezoelectric coefficient and permittivity of piezoactuator with electric field in AVC experiments *Ferroelectr.* 502 170–186
[2] S Sharma, Vig R, and Kumar N 2016 Finite element modeling of smart piezo structure:considering dependence of piezoelectric coefficients on electric field *Mech. Based Des.Struct. Mach.* **44** 372–83
[3] S Sharma, Vig R, and Kumar N 2016 Temperature compensation in a smart structure by application of DC bias on piezoelectric patches *J. Intell. Mater. Syst. Struct.* **27** 2524–35
[4] Duan W H, Wang Q and Quek S T 2010 Applications of piezoelectric materials in structural health monitoring and repair: Selected research examples *Materials (Basel).* **3** 5169–94
[5] S Sharma, Vig R, and Kumar N 2015 Active vibration control: considering effect of electric field on coefficients of PZT patches *Smart Struct. Syst.* **16** 1091–1105
[6] Hazimeh R, Challita G, Khalil K, and Othman R 2015 Finite element analysis of adhesively bonded composite joints subjected to impact loadings *Int. J. Adhes. Adhes.* **56** 24– 31
[7] Yokoyama T and Nakai K 2015 Determination of the impact tensile strength of structural adhesive butt joints with a modified split Hopkinson pressure bar *Int. J. Adhes. Adhes.* **56** 13–23
[8] Sankar H R, Adamvalli M, Kulkarni P P and Parameswaran V 2015 Dynamic strength of single lap joints with similar and dissimilar adherends *Int. J. Adhes. Adhes.* **56** 46– 52, 2015
[9] E. P. Adhesive *Hysol ® EA 9309.3NA* 2–5.
[10] Wandowski T, Moll J, Malinowski P, Opoka S, and Ostachowicz W 2015 Assessment of piezoelectric sensor adhesive bonding *J. Phys. Conf. Ser.* **628**
[11] Liu X, Xu Y, Wang X, Ran Y, and Zhang W 2020 Effect of adhesive and its aging on the performance of piezoelectric sensors in structural health monitoring systems *Metals (Basel).* **10** 1–14

[12] Rodewald R, Korfe S, Marnitz K, Yezerska O, Plagemann P, and Gesang T *Adhesives for piezo bonding.*

[13] Gupta V, Sharma M, Thakur N, and Singh S P 2011 Active vibration control of a smart plate using a piezoelectric sensor-actuator pair at elevated temperatures *Smart Mater. Struct.* **20** 105023

[14] Sharma A, Kumar R, Vaish R, and Chauhan V S 2016 Experimental and numerical investigation of active vibration control over wide range of operating temperature *J. Intell. Mater. Syst. Struct.* **27** 1846–60

IoT based Resuscitation and Hypoxemia Diagnosis System using Embedded Microcontroller Architecture and IFTTT Implementation

N Mohanty[1], A Ghosh[1] and S K Jindal[1]

[1]School of Electronics Engineering, Vellore Institute of Technology, Vellore, Tamil Nadu, India.

E-mail: sumitjindal08@gmail.com

Abstract. In recent times, the World has taken a critical blow from the novel Coronavirus. Moreover, with its research ongoing and multiple variants emerging, uncertainty is on the rise. Lung damage and cardiac issues occur during and post recovery which is why there is a requirement to constantly monitor the patient's vitals. This also applies to patients suffering from other lung diseases and heart conditions. They need immediate attention to avoid complications and further deterioration in their health. The proposed system aims to do so by observing the health parameters such as oxygen saturation in blood and heartbeat constantly and making the data available for the user by displaying it. In case of abnormality, it alerts the person related to the patient about the developments and relays the required information to help like symptom details and the patient's location, this ensures that actions are taken quickly. Simultaneously, it provides assistance in breathing during the interlude till professional help arrives and the patient receives the necessary treatment. This is achieved with a combination of sensors, modules and other components integrated with a microcontroller as well as the implementation of Internet of Things (IoT) using a web service.

Keywords. Internet of Things (IoT), Remote Patient Monitoring (RPM), Pulse Oximeter, Resuscitators, Global Positioning System (GPS).

1. Introduction

For the normal functioning of a human being, it is suggested that the oxygen saturation in blood, otherwise commonly called the SpO2 levels lie in the 92-96% range [1]. This is often compromised in patients suffering from lung diseases, conditions like sleep apnoea and underlying heart illnesses. Reduced SpO2 levels is medically termed as hypoxemia. Although respiratory disorders have always been a cause of concern, lung damage in patients has increased owing to the virus during and post recovery also known as the long covid effect [2]. The World is facing a huge scramble in the oxygen supply due to the exponential increase in cases. With such load on the hospital infrastructures, many patients are being left unattended and don't receive the prescribed treatment, which is why the usage of pulse oximeters, that recognize the measure of oxyhaemoglobin immersion (SpO2) which is a roundabout assessment of blood vessel oxygen immersion (SaO2), is advised to detect symptoms earlier and avoid mishap [3].

The demand for a patient-centric and commercially accessible health care setup has increased as technology advances and industries intertwine with each other, It is also partially due to the independent lifestyle approach in the unprecedented times of Covid-19 [4,5]. In the health industry, the development of IoT combined systems allow professionals to identify the best treatment processes for patients and has significantly led to reduction in cost, easier access, and effective long-term monitoring [6]. Remote Patient Monitoring (RPM) also enhances health services and makes it affordable which in turn makes it widely available [7]. Such systems are recommended for those who do not need heavy medical care until unavoidable hospitalization because there is a risk of misdiagnosis which may be fatal [8].

In this work, a resuscitation and hypoxemia diagnosis system is proposed. It is integrated with other facilities such as a text and email alerts, as well as location tracking using IoT and various modules. It centers around a manual resuscitator that operates on the application of pressure. This is automated to simulate a replica of inhaling and exhaling actions to assist a patient. Various existing literature and works related to the concept were studied to have a better understanding in section 2. Section 3 elaborates on the proposed system with subsections 3.1, 3.2 and 3.3 dealing with the hardware explanation, the system architecture and working of the model, respectively. The results obtained are discussed in section 4 while section 5 includes the conclusion and future aspects of the system.

2. Related Works

An IoT based health monitoring system is implemented in [9] where data is collected by the sensors attached to an Arduino and Raspberry Pi. The vital parameters are constantly monitored as well as stored in real time using an application on a smartphone or any device that can connect to the internet. This is attained using Bluetooth technology, Message Queue Telemetry Transport (MQTT) protocol and cloud computing for client-based approach.

In [10], the proposed system works heavily around monitoring health parameters of patients suffering and isolating due to the Coronavirus. It keeps track of the patient's location to ensure that they are following the norms while visualizing their temperature, SpO2 levels and pulse in real time with the help of an IoT gateway to index and store the data of multiple such patients to be surveilled by the medical facility.

The system in [11] uses various body sensors to collect data. The data is correlated with various activities to calculate mean values that are unique to the user and Exploratory Data Analysis (EDA) approach helps identify the activity being performed by the user. It also makes it possible for real time monitoring of the parameters such as temperature and pulse, collected and recorded on a cloud for further analysis.

The work mentioned in [12] aims at reducing hospitalization of a patient by early detection of conditions by embedding a toilet seat with various modules like electrocardiogram, ballistocardiogram, and photoplethysmogram. This model is capable of clinical grade measurement and estimates blood pressure, stroke volume and peripheral blood oxygenation for analysis. The values collected were compared with those collected in a hospital centric system to validate the results.

Commercial off the Shelf (COS) sensors are used to obtain figures and evaluate one's sleep cycle using random forest model in [13]. It is competent to analyze sleep and snoring patterns through the user's bio status that includes physical movement, pulse, and oxygen saturation. The system accounts for the analysis in a statistical method and provides 95% accuracy when tested and compared thus making it an effective system for monitoring disturbed sleep at minimum expenses.

In [14], the system proposed is an automated ventilator that is implemented using a programmable logic controller that runs the mechanical ventilation process of a bag-valve-mask-based emergency ventilator. It uses a human-machine interface to check on the vital parameters, keep a record of the values and identify when the patient needs oxygen supply and carries out the process accordingly.

The system in [15] is an automated bag valve mask (BVM) compression system that is based on the Arduino Controller and 3-D printable parametric component-based structure. It provides flexible adjustments in breathing modes by possessing the ability to alter breathing rates, tidal volumes and inspiratory to expiratory ratio thus making it an efficacious system.

Various testcases in [16] helped infer that Personal Emergency Response System (PERS) spread a sense of security within the users and make them feel independent. However, the systems need to be compact to have a greater impact in the market which was backed by [17] in which a study was conducted in order to compare the thoughts of PERS users and non-users. Likewise, [18] suggests that research and approaches must be expanded in the field of wireless technology to be incorporated in the health industry as technology continues to strive.

These works provided the inspiration to work and extend on the existing literature thus making a system that can monitor the health of a patient, alert a health personnel or a relative in case of an emergency to avoid negligence, contribute by providing the patient's location and keep their condition stable till they receive proper care. This system is extremely helpful since it helps cope with the shortage of ventilators and supply chain disruptions as well as heavy load on the hospitals in general.

3. Proposed System

In this work, a resuscitation system is combined with IoT technology and implemented to monitor a patient's health condition with reference to their SPO2 levels and pulse while simultaneously cautioning the concerned person about the situation in case of an emergency since low SPO2 levels imply that the patient needs to be hospitalized. As mentioned, the system can be segregated into its offline working and online working where the hardware model observes vitals and contributes in keeping the patient stable until they have access to better medical equipment in case of deteriorating health. It also establishes a connection with the internet in order to alarm the person related, about the crisis and providing them with essential details such as the patient's location and health, so that they can take necessary actions to help the patient.

3.1. Experimental Setup

Shown in figure 1 is the implementation of the model. With an uncomplicated configuration, the system is user friendly and easy to work with. The Arduino Uno which is a microcontroller-based development board, is the heart of the model with all the processing done within it. MAX30102 is a pulse oximetry and heart rate sensor which is connected to sense the pulse and SPO2 levels. The Direct Current (DC) motor and BTS7690 43A motor driver combination is employed to enable the resuscitator with the help of Infra-Red (IR) sensors. The ESP8266 Wireless-Fidelity (Wi-Fi) module is attached to establish a connection with the internet while the GPS module is used to retrieve location coordinates. The Liquid Crystal Display (LCD) helps visualise the values and situation.

IOP Conf. Series: Materials Science and Engineering 1225 (2022) 012057 doi:10.1088/1757-899X/1225/1/012057

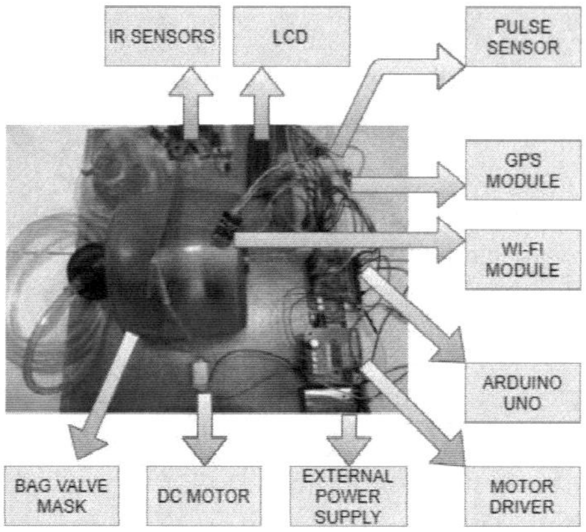

Figure 1. Hardware setup of the proposed model.

3.2. System Architecture

The ATmega328P within the Arduino processes the code for the entire model by being interfaced with the components to receive input and provide output while the board also acts as a power supplier to most of the components attached to it.

The pulse oximeter reads the patient's vitals by using the method 'photoplethysmography'. The patient places their finger on the sensor where a Light Emitting Diode (LED) along with a photodiode measures the perfusion of blood that helps deduce comprehensible data as required.

The resuscitation system is built using a Bag Valve Mask (BVM) that works on the application of pressure. This is achieved with the help of a DC motor and IR sensors. A rod is linked to the motor by the means of a nylon thread to govern the inflation and deflation of the bag. This is made possible by altering the motor's rotation between clockwise and anticlockwise directions to push and pull the rod thus exerting force on the resuscitator. The motion is switched according to the readings collected by the IR sensor that are placed in appropriate positions to detect the state of the BVM. It is administered if the SPO2 and heart rate don't lie in the normal range thus causing hypoxemia, to assist the patient with breathing. Since this is effective for a short time, the system essentially works on forcing air supply to the patient, but an oxygen reservoir can be fixed for high-risk patients.

It also connects to the internet so as to trigger a predefined Short Message System (SMS) and an email to alert about the patient's condition using web services. At the same time, the GPS module acquires the patient's location in terms of latitude, longitude and altitude in real time using satellites to help navigate to them. Figure 2 showcases the design of the model proposed.

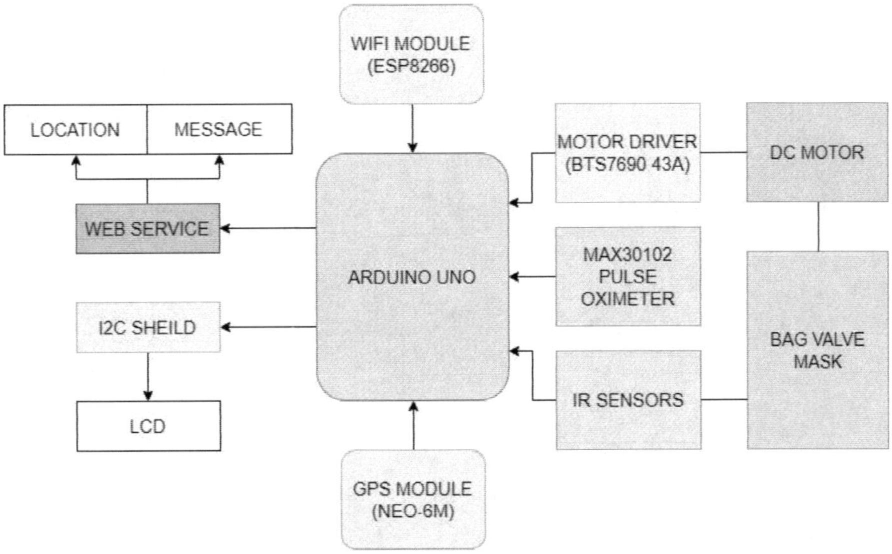

Figure 2. Block diagram of the proposed model.

3.3. *Workflow*

Patients often lose their ability to comprehend the state that they are in, which makes it difficult for them to take control of the situation. They become incapable of helping themselves and need support immediately. Hence, the main goal of the model is to avoid the requirement of applying pressure manually on the bag valve mask in order to operate it as a resuscitator till the patient receives proper medical attention which is made possible by automatically letting people related to the patient know about the situation through messages to ensure prompt response. Figure 3 elaborates on the working of the proposed model.

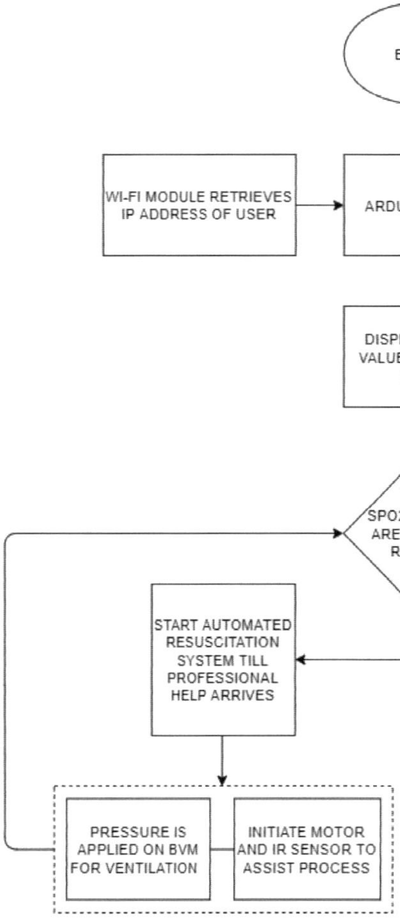

Figure 3. Flowchart of the proposed model

The ATmega328P is fed with the values collected by the sensor and displays the same on the LCD while comparing the values obtained with a set condition to check if the patient's health is stable. If it is not, then the dc motor runs to create pressure on the bag valve mask to assist the patient in breathing. The bag is inflated and deflated at regular intervals to mimic respiration using IR sensors to control the rotor of the DC motor.

On the other hand, the Wi-Fi module attains the IP address of the patient by connecting to the internet. This helps in implementing the IF This Then That (IFTTT) web service where the 'this' condition is to connect and trigger an event through a Unique Resource Locator (URL) using a unique Application Programming Interface (API) key and the 'that' condition is to send a text to inform about the patient's condition and an email that expatiates on the same respectively as two different events. The GPS also gains and sends the live coordinates of the patient to ensure that they receive proper medical attention within crucial time.

4. Results and Discussions

The working and results of the system can be measured at the patient's end where the resuscitation as well as hypoxemia detection system is implemented and at the receiver's end where the alert system is accomplished. The working of the system at the patient's end mainly involves data retrieval and hardware functionality while most of the work on the receiver's side is processing of the data and implementing online features.

4.1. *Patient's end*

Figure 4 displays the values recorded by the MAX30102 pulse sensor. It senses the amount of light reflected back and calculates the SPO2 and heart rate level accordingly. The validity of the acquired values are also taken into consideration while observing them by returning Boolean values to indicate the same. It is displayed on the serial monitor which is set at a fixed baud rate.

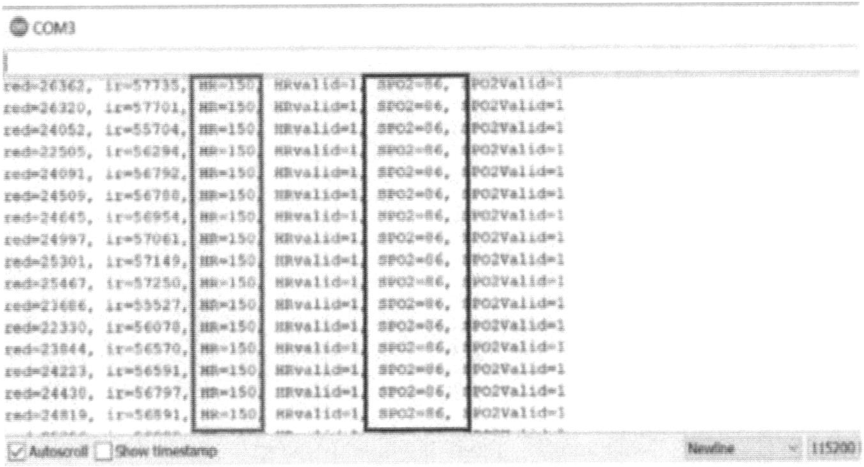

Figure 4. Serial monitor display when Arduino runs the program.

The system primarily concentrates on the monitoring hypoxemia and maintaining the dropping values of oxygen saturation in blood. If the values are valid, only then are the figures compared with the conditions set and while being shown on the LCD persistently. Figure 5 exhibits that the system constantly monitors the SPO2 levels and displays the same thus making it available at the user end.

Figure 5. SPO2 levels being monitored using the LCD.

Figure 6 provides insights on how the resuscitation system operates in the unfortunate event of alarmingly low levels of SPO2 detected to assist the patient until emergency service arrives. It indicates that air is forcefully being supplied to the patient but with enough gaps to exhale. This is done in order to restore SPO2 levels. As a test case, the value conditioned to be compared with the fed values of SPO2 was set high enough just to operate the system when assessing it with people whose SPO2 levels were normal.

Figure 6. Deflation and inflation of the BVM.

4.2. Receiver's end

Figure 7 exhibits the coordinates received from the GPS module that is then displayed in real time on a developed webpage that refreshes the coordinates continuously. It also links google maps in order to map the position hence helping in comprehending the exact location of the patient to commute to them as soon as possible and treat them as required.

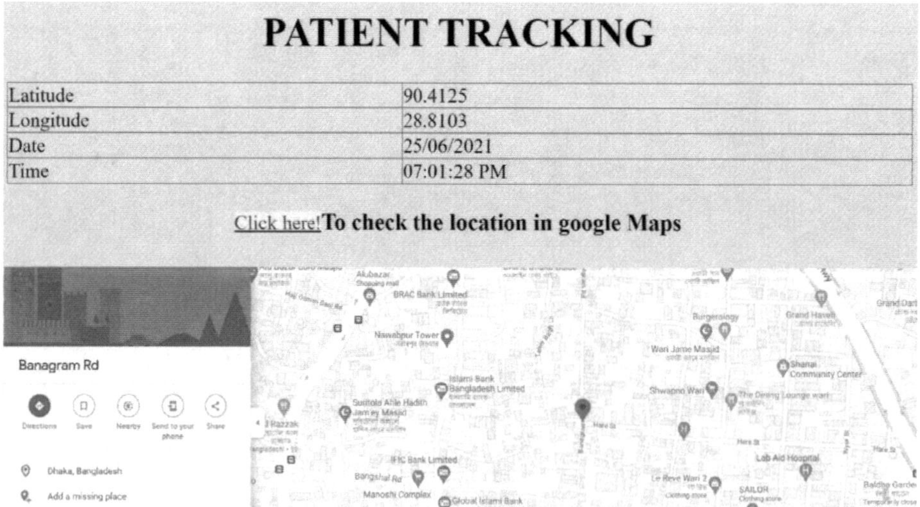

Figure 7. Coordinates and mapping of location received.

Figure 8 describes the working of the IFTTT web service where an event in defined and so are the 'this' and 'that' conditions. For this system, two such events are established. One to send an SMS to notify about the sudden drop in SPO2 values and another to email a detailed description of the situation and possible deductions of the same. On establishing a connection with the web, the webhooks condition is satisfied which triggers the latter part of the service - text and email. Both of them have a predefined format along with values picked up from the program such as the time the event got triggered and the location obtained by the GPS.

Figure 8. IFTTT implementation of an event.

Figure 9 presents the outcome of the text event once triggered. An SMS is sent from the registered mobile number to the provided mobile number. SMS is the most effective way to relay a message since it is the most accessed form of communication in present day. Although, it only mentions the occurrence of the worrisome event with values, the time of the event and the measures to be taken since there is a limit on character count.

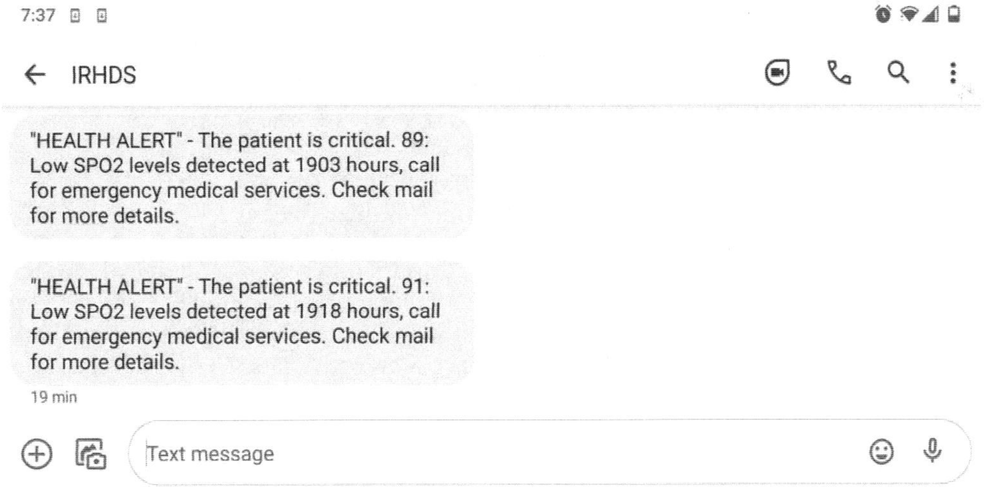

Figure 9. Resultant SMS sent to provided mobile number.

An email, however, allows us to pass on more information thus making it a better carrier. Similar to the text event, the email is sent from the registered email address to the email address opted for. It provides all the details to be sent for reference including the location of the patient as well as an inference of the circumstances such as other symptoms that the patient might experience and the likely conditions that could have caused it. Figure 10 presents the email sent once the event is set off.

IOP Conf. Series: Materials Science and Engineering 1225 (2022) 012057 doi:10.1088/1757-899X/1225/1/012057

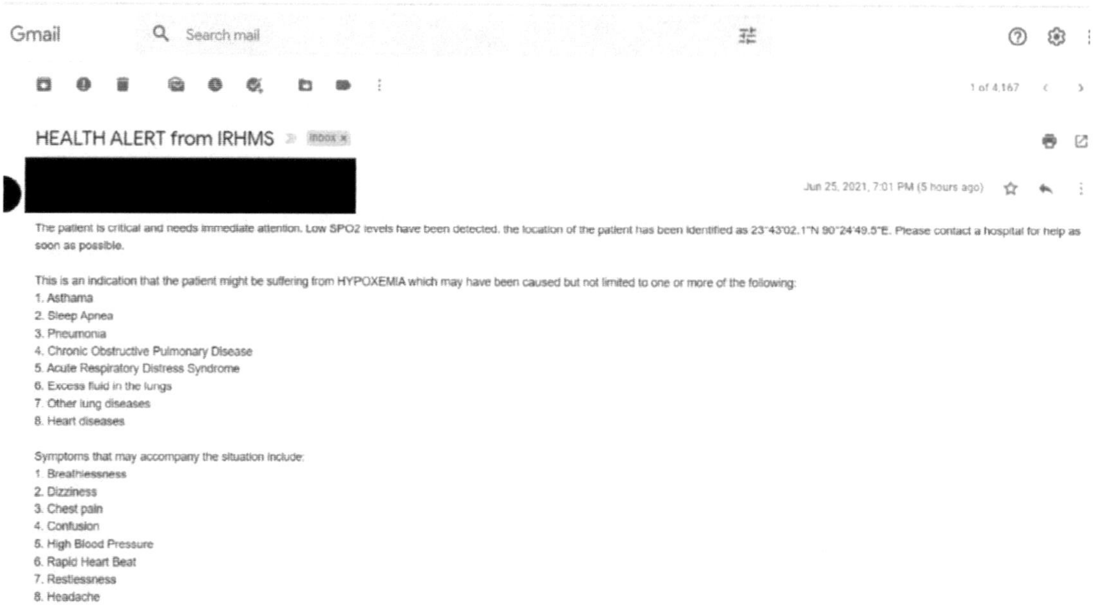

Figure 10. Resultant email sent to provided email address.

5. Conclusion and Future Aspects

With heavy load on hospitals due to the increasing cases of Coronavirus, it is important to adapt to Remote Patient Monitoring (RPM) in the long run. Contributing to RPM, a resuscitation, hypoxemia detection and alert system is implemented and analysed. The main aim of the proposed system is to keep a check on a patient's levels of oxygen saturation and heart rate since hypoxemia makes a patient highly vulnerable. Conceding that the SPO2 levels drop more than the level that is considered safe, the system activates the automated BVM that functions to provide assistance in respiring thus avoiding further collapse of health till they receive professional medical support. It also alerts health workers or people associated with the patient about their condition and plausible reasonings of the same so that timely measures are taken effectively. This is achieved using a microcontroller-based development board along with few models and components attached. The location identified by the GPS ensures that the patient receives timely attention. To implement IoT, services from IFTTT were utilised since the application is easy to grasp and available to all as it is an open platform. Overall, the system is reliable, user friendly and economical. Both, the software, and hardware components were tested and debugged to achieve the results.

The current system focuses on oxygen saturation whose readings fluctuate due to multiple reasons, it is difficult to narrow down the cause of concern and provide targeted treatment without investigation. To overcome this, other biomedical sensors can be integrated within the system to detect other symptoms that can be coupled to identify probable conditions. To provide better user experience, an application can be developed to monitor the vitals of the patient which also implies that the data is stored and can be revisited for analysis.

References

[1] Beasley R, Chien J, Douglas J, Eastlake L, Farah C, King G, Moore R, Pilcher J, Richards, M, Smith S and Walters H 2017 Target oxygen saturation range: 92–96% Versus 94–98% *Respirol.* **22** 200-02

[2] Callard F and Perego E 2021 How and why patients made Long Covid *Soc. Sci. Med.* **268** 113426.

[3] Seifi S, Khatony A, Moradi G, Abdi A and Najafi F 2018 Accuracy of pulse oximetry in detection of oxygen saturation in patients admitted to the intensive care unit of heart surgery: comparison of finger, toe, forehead and earlobe probes *BMC Nurs.* **17** 1-7

[4] Saranya M, Preethi R, Rupasri M and Veena S 2018. A survey on health monitoring system by using IOT *Int. J. Res. App. Sci. Eng.Tech.* **6** 778-82

[5] Shin W and Rho J 2020 A Study on the Design of Real-Time Monitoring System Using IoT Sensor in Respirator *Int. J.Adv. Smart Convergence* **9** 169-75

[6] Patil P, Waichal R, Kumbhar U and Gadkari V 2017. Patient health monitoring system using IOT *Int. Res. J. Eng. Technol.* **4** 638-41

[7] Parihar V R, Tonge AY and Ganorkar P D 2017. Heartbeat and temperature monitoring system for remote patients using Arduino *Int. J. Adv. Eng. Res. Sci.* **4** 237161.

[8] Arora S 2016 IoMT (Internet of Medical Things): Reducing Cost While Improving Patient Care *IEEE pulse* **11** 24-27

[9] C S Krishna and N Sampath 20017 Healthcare Monitoring System Based on IoT *2nd Int. Conf. on Computational Systems and Information Technology for Sustainable Solution (CSITSS)* 1-5

[10] R. Priambodo and T. M. Kadarina 2020 Monitoring Self-isolation Patient of COVID-19 with Internet of Things *IEEE Int. Conf. on Communication, Networks and Satellite (Comnetsat)* 87-91

[11] Thaung S M, Tun H M, Win K K K and Than M M 2020 Exploratory Data Analysis Based on Remote Health Care Monitoring System by Using IoT *Comm..* **8** 1-8.

[12] Conn N, Schwarz K and Borkholder D 2019 In-Home Cardiovascular Monitoring System for Heart Failure: Comparative Study *JMIR Mhealth Uhealth* **7** e12419

[13] K Saleem, I S Bajwa, N Sarwar, W Anwar and A Ashraf 2020 IoT Healthcare: Design of Smart and Cost-Effective Sleep Quality Monitoring System *J. Sens.* **2020** 1-17

[14] Vicerra R R, Calilung E, Española J, Dadios E, Culaba A, Sybingco E, Bandala A, Madrazo A B, Lim L G, Billones R K and Lopez S 2020 Implementation of a Programmable Logic Controller (PLC)-Based Control System for a Bag-Valve-Mask-Based Emergency Ventilator. *In 2020 IEEE 12th Int. Conf. on Humanoid, Nanotechnology, Information Technology, Communication and Control, Environment, and Management (HNICEM)* 1-5

[15] Petsiuk A, Tanikella N G, Dertinger S, Pringle A, Oberloier S and Pearce J M 2020. Partially RepRapable automated open source bag valve mask-based ventilator HardwareX **8** p.e00131

[16] Stokke R 2016 The Personal Emergency Response System as a Technology Innovation in Primary Health Care Services: An Integrative Review *J. Med. Internet Res.* **18** e187

[17] De San Miguel K, Lewin G, Burton E L, Howat P, Boldy D and Toye C 2017. Personal emergency alarms: do health outcomes differ for purchasers and nonpurchasers?. *Home Health Care Serv. Q.* **36** 164-77

[18] Shetty H and Shetty A 2018 A review on health monitoring system using IoT *Int. J. Eng, Res. Tech.* **6** 1-3

Design of low cost IoT enabled embedded control system for covid free smart home

Khushi Singh[1,2] and Sumit Kumar Jindal[1,3]

[1] School of Electronics Engineering, VIT, Vellore, Tamil Nadu

Email Ids: [2] khushi35b@gmail.com ; [3] sumitjindal08@gmail.com

Abstract. In the era of automation, the Internet of Things (IoT) has come up as a solution and innovation for many applications to improve the basic facilities required in the day-to-day routine. IoT, itself is the interconnectivity of different physical devices with the help of wireless networks e.g., Bluetooth, Wi-Fi without human intervention. The low-cost embedded control unit for Covid free Smart Home is based on the idea of home automation by keeping the current situation of the Covid pandemic in mind. Irradiation of Ultraviolet C (UVC) spectrum is used to sanitize household goods, where convectional sanitization process is not suitable. Simple home automation means controlling or monitoring different electronic/electrical home appliances, devices, and other such things with smartphones or by doing sensor-based operations. All the required devices are connected with one embedded control unit which is managed through a mobile app. This functionality can be accessed either by manually operating a mobile app or with voice assistants e.g., Alexa or Google Assistant, etc. The basic point of communication among devices is based on real-time sensor inputs, which trigger the devices for action and react as per requirements. Ambient light sensors are used to control the on/off the functionality of lights based on the intensity of light present naturally. In this scenario, lights will be switched off in day mode and switched on in night mode. Advancement in wireless technologies enables the researchers to apply various modes of communications among the sensors, gateway, and applications. Bluetooth, Wireless sensor network (WSN), Wi-Fi, Global system for mobile communication (GSM) are a few of them, which are being used either in sensing nodes or in communication gateway to make home automation free of the wired network.

Keywords: Smart house, Home automation, Covid, Internet of Things (IoT), UVC Sanitizer, Access control

1. Introduction

Bluetooth enabled embedded devices are getting wide applications in home automation. 2.4 GHz ISM band gives secure short-range communication. It exhibits a secure communication range from 10m to 100m with a bandwidth of up to 3Mb. Various home appliances are controlled using smartphone and Bluetooth gateway [1]. Home automation makes home functioning more energy efficient with reduced efforts. Sensors based actuator control, save lots of electricity bill by reducing unwanted uses of lighting, cooling or heating systems [2].

Various IoT technologies like Bluetooth, Wi-Fi, GSM are being used widely in smart home applications. Sensors are used to detect environmental parameters like ambient light, room temperature, etc. to automate the lighting, heating, and cooling systems [3]. Machine to Machine (M2M) communication, IoT protocol reduces the need for human interventions in data communication and related computing work for decision making. Intelligent embedded devices are programmed to operate on predefined threshold values of the sensor's inputs [4]. IoT enables home monitoring from remote locations at any time. Real-time sensor data are uploaded to the cloud and can be accessed by the end-user mobile application. One can monitor his\her home without physically being there [5]. IoT-based home automation is trending these days. It allows users to securely control and monitor home appliances with ease [6]. Appliances can be controlled remotely in autonomous mode or through user inputs. In autonomous mode, action is taken based on real-time sensor data [7]. The Bluetooth enabled embedded control unit for covid free smart home is developed which is paired with Mobile App to access functionality. The smart home control unit has inbuilt sensors which can monitor different parameters of devices e.g., motion, temperature, and ambient light to gather basic information. [8]. This unit also provides contactless access control by controlling the switching of power supply of the electromagnetic locking system for safe, almirah, or any restricted area at the home [9]. During Covid, it is need of the hour to have a covid free smart home to avoid contact with the virus and to stop further spread of the pandemic. In a normal routine, one must sanitize all the goods, items brought in-house on daily a basis before keeping and using them at home [10]. To simplify and verify the process of sanitization of products purchased on a daily basis, a smart cabinet is developed, in which sanitization can be done with the help of ultraviolet C (UVC) irradiation in a controlled manner. Arrays of UVC LED are fitted in cabinets and its operations are controlled by mobile app.

Smart cabinet is a simple cabinet made of metallic, plastic, or wooden materials in which UVC lights are fitted. Items to be sanitized are placed inside the cabinet and UVC lights are switched on for one or two minutes. Received UVC radiation inside the cabinet\ chamber is measured and displayed on an OLED display positioned outside the chamber to ensure accurate and efficient operation. After switching off the UVC supply, sanitized items can be taken out for use. A Smart home control unit will definitely ease the life challenges and secure from diseases too. Such a simple smart home is itself a money saver in means of heating, cooling, and saving electricity bills by optimized control of the appliances. This also provides greater safety with security modules, locks, and other systems. Automations of the electrical devices make them intelligent devices as human interventions to operate them are reduced to a minimum.

Existing home automation systems do not offer such a comprehensive solution in one unit, whereas the developed unit exhibits various functions like appliance control by Mobile App, smart lighting, activity-based lighting, day\night mode-based lighting, smart acccss control, and Covid19 free sanitization chamber in a single unit. Developed low-cost IoT enabled embedded control system for Covid free Smart Home is a comprehensive solution, it is controlled and monitored by Mobile App, using Bluetooth wireless interface, so it is not dependent on the availability of the mobile network. Bluetooth enabled Smart home control unit is developed around a low cost robust 8051 microcontroller which can operate in Indian conditions very effectively. An Android Mobile App is developed to communicate with developed Embedded Device over Bluetooth Interface. The device has IoT (Bluetooth-SPP) capability and can receive instructions on UART. Electrical loads are connected to the control unit using suitable relays through the ULN relay driver. Home Appliances can be switched ON/OFF using Mobile App, without physically touching the switch. Integration of Google voice assistant to the developed unit, makes it more user-friendly. Convectional sanitization process which is based on isopropyl alcohol is not suitable for sanitization of household goods. The Smart cabinet is fitted with UVC LED arrays for sanitization against Covid19 viruses or any other microbes and bacteria. Sanitization is controlled by mobile app to avoid any accidental exposure of

UVC to human skin. Experiments are performed by switching on UVC led arrays for pre-defined duration and the received UVC radiation index is measured to ensure the effectiveness of the sanitization process. Cascading of 14W UVC panels is done to meet desired 10 mW of UVC irradiation inside the chamber, which is essential to destroy viruses, bacteria, and microbes.

2. Literature Review

Home automation systems are mostly controlled by embedded devices. It is built around a microcontroller interfaced with various sensors for monitoring environmental parameters. Mobile application is used to send switching instructions to embedded devices for controlling loads through actuators [11]. There are few approaches, where control is done through web pages or cloud-based client applications. Sensors data from home is fed to the internet cloud and computing is done at the server end for decision making before sending the control signal back to the control unit [12]. The smart home is an intelligent home these days. IoT makes it very convenient to use various wireless protocols like Bluetooth, Wi-Fi to connect with home and user applications. [13].

Home security is the key concern when a person is out from his\her home. Security sensors like the PIR motion sensor alert when any motion is detected in the house. It works seamlessly in daylight and in night mode. A Security camera can be used to click and send images if any motion or activities are detected. Motion or activities-based function of camera reduces the need for high bandwidth or more cloud storage spaces [14]. Things in IoT, i.e., sensors are for real-time inputs of what they sense in the environment, it depends on what critical parameters are to be included in the design of smart home systems. Sensors like gas sensor, air quality sensor, CO2 sensor, fire sensor, or sensors for ambient parameters are selected as per need of the information which is required to be collected for decision making. Information collected from the sensor terminal is transmitted to a server in real-time or periodically or in polling mode depending on application requirements [15].

Few smart home solutions use wireless sensor network (WSN) deployment for real-time sensing of parameters and remote monitoring. WSN allows using 255 sensor nodes in a single network, although cascading of such networks is possible. Such a solution is feasible in big-size deployments, where sensing locations are more and wireless networking is most suited. WSN technology allows sensing motes to operate in sleep mode at an ideal time between two consecutive samplings of input data. This saves battery life and power consumption. Such deployment can run for months on a single battery. The Self-healing feature of WSN allows to add or drop any node as per requirement and the routing table is updated accordingly without rebooting the network. WSN based smart home system is best suited for remote monitoring, where data payload is less. Monitoring of room temperature, humidity, gas leakage, ambient light, human activities one few use cases of such solutions [16].

3. System Architecture

Embedded system is built around an 8-bit 8051 microcontroller NXP89V51RD2, Bluetooth Module (SPP), ULN Relay driver, and Relays. Home appliances are connected to Embedded Devices through Relays. UVC LED arrays are fitted in the Smart Cabinet and it is controlled by the mobile app.

Bluetooth module is connected to the serial port of the microcontroller, which is paired with mobile application to receive instructions for controlling electrical loads. The use of an 8051 microcontroller makes the design suitable to operate in Indian conditions. The system block diagram is depicted in figure 1.

IOP Conf. Series: Materials Science and Engineering 1225 (2022) 012058 doi:10.1088/1757-899X/1225/1/012058

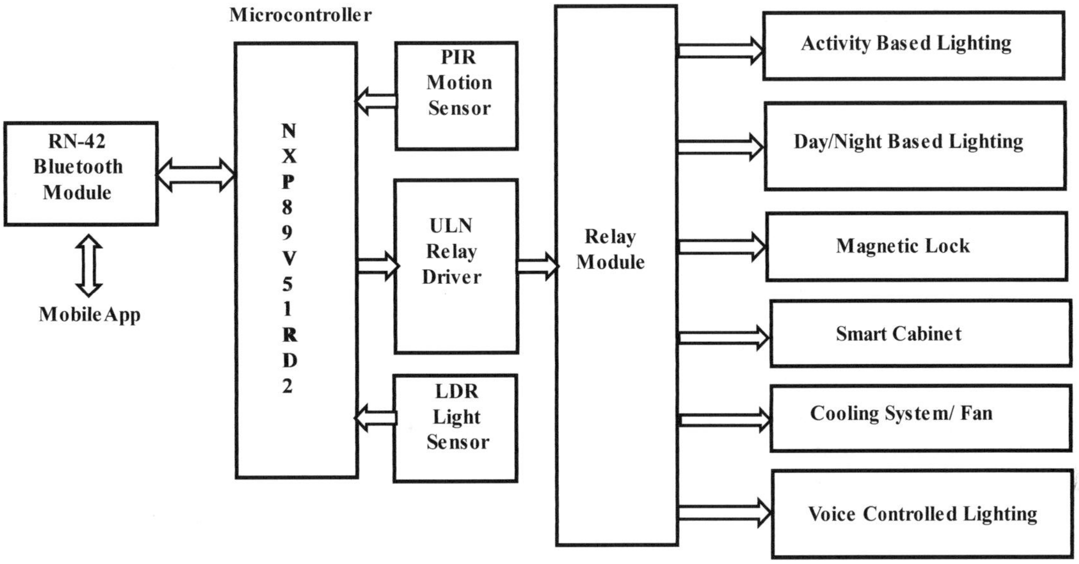

Figure 1. System Architecture

System receives instructions from Mobile App using wireless Bluetooth protocol and based on predefined instructions, Embedded System will be switching ON/OFF a particulate home appliance.

Microcontroller NXP89V51RD2 is used in the design. It is an 8051 based microcontroller, having more code memory and ease of flashing options than traditional 8051 microcontroller. It is having an on-chip bootloader with 32Kb of flash memory. Its port 1 pins are interfaced with the ULN relay driver to have the required current support. Output pins of ULN are connected to relay ports. All the electrical home appliances, which are to be controlled, are connected to relays.

RN42 Bluetooth module is connected to the UART of the microcontroller. RN42 Bluetooth module works on Serial Port Profile (SPP) protocol. It is used for serial communication between user applications and the microcontroller`s UART. It is configured for required baud rate, authentication mode, and user name using AT commands. The transmission and reception range of this Bluetooth modem is approximately 100 meters. Relay`s specifications are decided according to the requirement of electrical loads which are to be attached to the control units. Relays of 220V AC/16A rating are widely used in home automation applications. It can handle loads of most of the electrical appliances which are used in the home environment. The required number of relays depends on the number of electrical loads to be connected.

4. Hardware Specification and Features

4.1. Activity-Based Lighting

Activity based lighting is used to reduce the load of lighting as per the requirement of lights. It is achieved by sensing the human activities in the target area to either increase/decrease or turn on/off the light automatically. This is done with the help of PIR sensors which are commonly used in these day's basic appliances and gadgets. PIR motion sensors are used to detect the motion of humans in the targeted area under the range of sensors. Based on human activity, these sensors react in the system. PIR motion sensor is available in 3 PIN configuration, beside VCC and Ground PIN, it has output PIN, which is interfaced to the microcontroller.

4.2. Day/Night Mode Based Lighting

In development of Day/Night mode lighting, Light Dependent Resistor (LDR) based light sensors are used. Most of the light sensors are developed using light dependent register (LDR) and LM393 voltage comparator unit. It is available in two versions, first one produces continuous analog voltages corresponding to ambient light falling on it, which is called an analog light sensor. Another one, the digital light sensor, produces logic Low/High signal corresponding to day\night mode. LDR works as a variable resistor and changes its values depending upon the intensity of light falling on it. When there is no light surrounding or falling on it, its resistance becomes very large and vice versa in absence of light. Our circuit makes use of this feature to automatically switch on or off the light. The main advantages of such type of lighting are low energy consumption, longer life of electric bulbs/tube lights and very economical in long term.

4.3. Ultraviolet C Spectrum (UVC) Irradiations Based Sterilizer\Sanitizer Cabinet\Chamber

To overcome the fear and protect the family from covid-19 virus, Ultraviolet C Spectrum (UVC) Irradiations Based Sterilizer\Sanitizer Cabinet\Chamber has been developed. The cabinet can be switched on using the Smart Home Controller unit and Mobile App for a predefined duration of two minutes to successfully carry out sanitization. The Irradiation of UVC has the ability to destroy RNA and DNA of micro-organism which disables them to reproduce and survive. This ability of UVC enables it to sanitize items and reduce the infections due to them. It uses a pure physical sterilization method. As the UVC spectrum of light can destroy the viruses and bacteria from items that are exposed to UVC irradiations, so irradiation of UVC rays is used in smart cabinet in which items are placed for sanitization. The sanitization process can be carried out whenever any items are brought home, to reduce the risk of infection. Irradiation in UVC chamber can be switched on/off through mobile app. It irradiates UVC light on the target area for a predefined duration of 2 minutes at 20mW which is enough to destroy Viruses, microbes, and bacteria. In comparison with recommendations from WHO and US-FDA to use 10 mW of UVC exposure for a minimum of 10 seconds, it is quite effective. The Block diagram of the Smart cabinet is shown in figure 2.

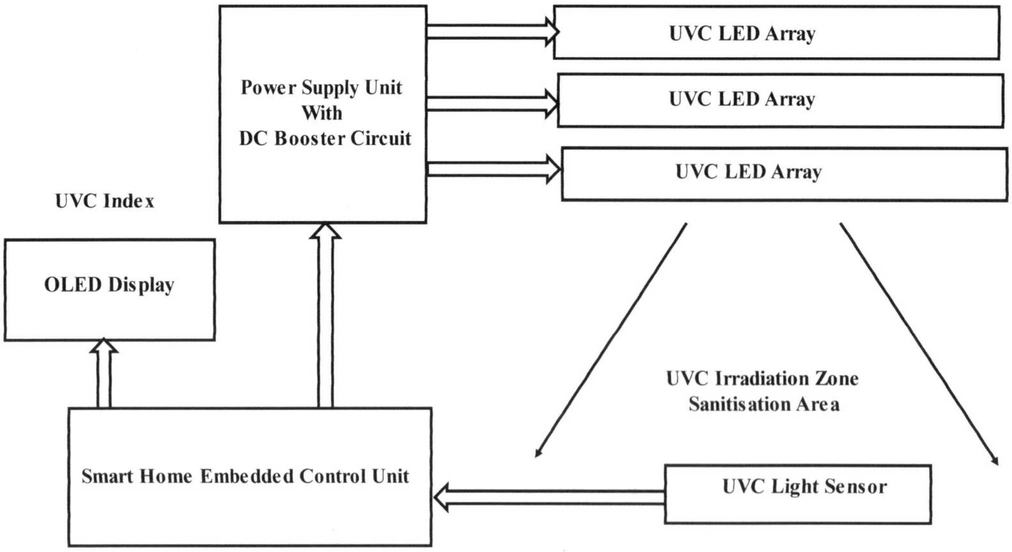

Figure 2. Block diagram of UVC Sanitizing cabinet

IOP Conf. Series: Materials Science and Engineering 1225 (2022) 012058 doi:10.1088/1757-899X/1225/1/012058

5. Experimental Setup

Working Prototype is assembled around 8051 microcontroller NXP89V51RD2. Six relays are interfaced through the ULN relay driver for isolation of the microcontroller from Relays. PIR motion sensor is interfaced with the microcontroller for implementation of activity-based lighting. RN42 Bluetooth module is configured before interfacing it to the UART of the microcontroller. Picture of the developed prototype is shown in figure 3. RN42 Bluetooth module is configured using USB to Serial converter and some terminal communication software like Realterm. Terminal software is opened at the default baud rate of the Bluetooth Module and AT commands are used to configure the required baud rate, authentication mode, switching timer, and other required parameters. The configured module is ready to be used for serial communication between the embedded control unit and mobile phone.

PIR motion sensor HC-Sr501 is used in the design to detect human activities in the target area of the home. It works on detecting IR signals which depend on the body temperature of living things. It sends a high signal to the microcontroller when motion is detected and lights are switched on by the control unit. In case of no activity in the target area, lights are switched off to save electricity. Light Dependent Resistor (LDR) based Digital light sensor with LM393 is interfaced for detection of ambient light. Resistance of LDR changes with changes in ambient light. Sensors send logic high to the microcontroller when dark mode is detected. Sensor`s night mode detection can be adjusted using onboard variable resistors. When night mode is detected by the microcontroller, it switches on the lights and the same is switched off when day mode is detected. Positioning of LDR is done in such a way that it detects ambient light only, lights installed in the home should not illuminate it, otherwise false day mode detection may disturb the switching operation.

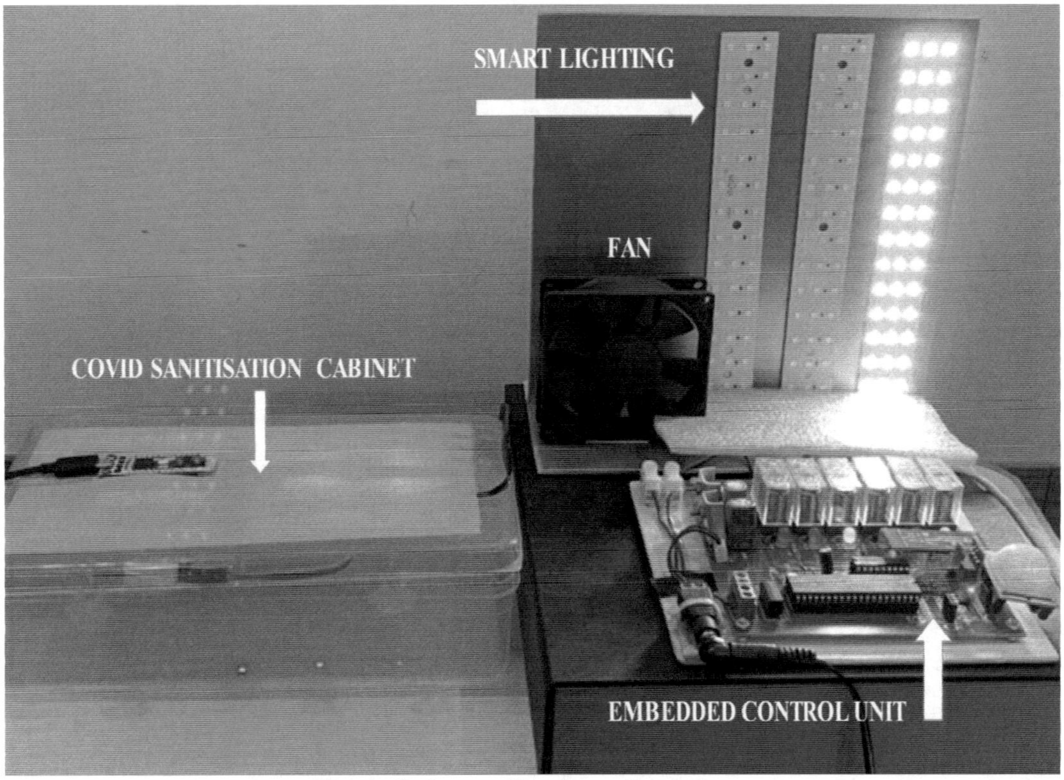

Figure 3. Setup of hardware Prototype

Ultraviolet C Spectrum rays use 270-280nM irradiation-based sanitization. Its Irradiation can instantly destroy the molecules of Covid19 viruses and prevent pandemic to spread further. It can be used for UVC disinfection of any object which is kept in the smart cabinet very effectively. IoT-enabled systems are developed, to control and monitor UVC irradiation systems. UVC-LED arrays are deployed inside the cabinet and it is switched on for a predefined duration when disinfection is to be carried out. UVC light destroys the viruses and bacteria from inside surfaces of Public Transport. The sanitization process can be carried out whenever any suspected item is bought home. The embedded control unit is paired with mobile app using Bluetooth SPP protocol.

6. Results

Prototype is assembled using a Microcontroller, ULN relay driver, Relays, Light sensor, PIR sensor, and RN42 Bluetooth module. An android mobile app is developed to interact with the control unit. Mobile app Bluetooth connection manager API is used for creating a wireless connection with devices. Google voice assistance APIs are used to control devices with voice commands. The Prototype has achieved 100% Switching Control without any noticeable lag time.
Mobile app is developed using Bluetooth connection manager and google voice assistant APIs. Buttons are placed in App GUI corresponding to each electrical load and some command instructions are defined and programmed for switching ON/OFF of every appliance. The Mobile app sends these predefined commands to the microcontroller through Bluetooth for switching operations. Mobile app interfaces are shown in figure 4.

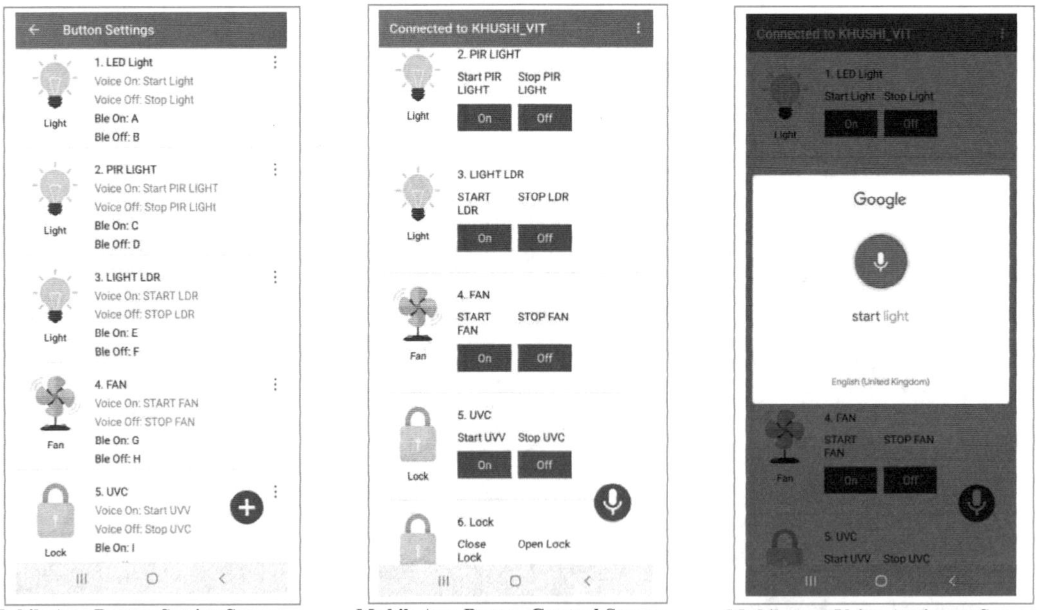

Mobile App Button Setting Screen **Mobile App Button Control Screen** **Mobile App Voice Assistant Screen**

Figure 4. Mobile app for smart home

Google voice assistance`s speech-to-text conversion module is used to generate predefined text commands against predefined voice commands for switching operation of every port. When user inputs voice command like "Switch on Light", corresponding ASCII character of text command is transmitted to microcontroller and the microcontroller calls pre-programmed subroutine and light is switched on. Switching instructions are received from mobile phone and the corresponding electrical device is switched ON/OFF. PIR Motion sensors are used to exhibit activities based on lighting. When human activities are detected in the target area then lights are switched on. In absence of activities, lights are switched OFF. Day\Night modes are detected using LDR light Sensors and Lights are switched ON in night mode and switched OFF in day mode.

For access control, the electromagnetic lock is interfaced with the microcontroller through the relay. The electromagnetic lock consists of two parts, an electromagnet part that needs to be energized to close the lock, and the second part is a metallic plate that is attracted by the magnet to close the door. It works on 12 V DC, which is switched ON/OFF by the control unit based on switching instruction generated by the mobile app. To allow access to Safe or restricted areas, the power supply of the electromagnetic lock is switched off to de-energize the magnet, and hence lock is opened to grant the access. Sanitization process is tested in a smart cabinet with mobile App using Bluetooth wireless interface. Desired UVC switching is achieved with 100% accuracy. Received UVC irradiation inside the cabinet is measured by a self-developed UV index meter, which is more than 11.5 on average in 50 samples (UV index >=5 for more than 10 seconds is desired for sanitization of covid infected items). Received UVC power inside the chamber is constantly more than 20mW, which is much more than desired 10mW UVC irradiation for 10 second for killing for viruses, bacteria and microbes.

When the prototype is powered on, RN42 Bluetooth module turns on in discoverable mode using device discovery protocol. It is paired using mobile app running on Smartphone. Based on user input through button or voice command, the mobile app sends pre-defined switching commands to the embedded device. Signal transmitted by mobile app is received by onboard RN42 Bluetooth module and it further sends the command to Rx pin of Microcontroller. Based on received instruction, Microcontroller calls specific subroutines and specific home appliances are switched ON/OFF. If any human motion/activity is detected in target area then PIR motion sensor sends logic High signal to the microcontroller and corresponding lights are turned ON, else activity-based lights are turned OFF. Day/Night mode is detected by Light sensor, it sends logic Low/High signal to the microcontroller and respective lights are turn OFF/ON.

7. Conclusion

Developed prototype is a low-cost IoT-enabled embedded control unit and worked as per expectation with any noticeable latency and exhibited 100% switching operation with full electrical load. Voice command-based switching makes it more convenient to use. The Bluetooth-based operation makes it free from the need for internet availability for local applications. PIR motion enabled activity-based lighting kept lighting in off mode when no activities are detected, it is a good measure to save electricity. Such lighting is more useful in the wider area, where more lights are fitted and switching off the lighting of the unused area is required. Light sensors-based lightings are found more useful as it automates the lighting and reduces human intervention. IoT enabled access control system enables keyless entry to restricted areas.

Smart cabinet achieved desired ultraviolet index inside the chamber, which is essential to destroy Covid19 like viruses, bacteria, and microbes. Mobile app-based operation makes its monitoring very easy and user-friendly.

Thus, IoT enabled low-cost embedded control unit for covid free smart home, demonstrated home automation and home monitoring application with added advantages of sanitization of household things in no time. Hardware unit is tested successfully for 12 hours without failure. Standardized security modules can be added in the future for digital safety. Further cost reduction may be undertaken in future work to make it more economical.

References

[1] Piyare R and Tazil M 2011 Bluetooth based home automation system using cell phone *IEEE 15th Int. Symp. on Consumer Electronics (ISCE)* 192 95

[2] Asadullah M and Ullah K 2017 Smart home automation system using Bluetooth technology *Int. Conf. on Innovations in Electrical Engineering and Computational Technologies (ICIEECT)* 1-6

[3] Islam A R, Bhowmick K, D Sikder and Zaman H U 2019 A Multifarious Design of a Microcontroller Based Home Security and Automation System *11th Int. Conf. on Computational Intelligence and Communication Networks (CICN)* 1-6

[4] Tseng C, Cheng C, Hsu Y and Yang B 2018 An IoT-Based Home Automation System Using Wi-Fi Wireless Sensor Networks *IEEE Int. Conf. on Systems, Man, and Cybernetics (SMC)* 2430-35

[5] Al-Kuwari M, Ramadan A, Ismael Y, Al-Sughair L, Gastli A and Benammar M 2018 Smart-home automation using IoT-based sensing and monitoring platform *IEEE 12th Int. Conf. on Compatibility, Power Electronics and Power Engineering (CPE-POWERENG 2018)* 1-6

[6] Nitu A M, Hasan M J and Alom M S 2019 Wireless Home Automation System Using IoT and PaaS *1st Int. Conf. on Advances in Science, Engineering and Robotics Technology (ICASERT)* 1-6

[7] Singh H K, Verma S, Pal S and Pandey K 2019 A step towards Home Automation using IOT," 2019 *Twelfth Int. Conf. on Contemporary Computing (IC3)* 1-5

[8] Somani S, Solunke P, Oke S, Medhi P and Laturkar P P 2018 IoT Based Smart Security and Home Automation *Fourth Int. Conf. on Computing Communication Control and Automation (ICCUBEA)* 1-4

[9] Yokogi K, Kitagawa N and Yamai N 2018 Access Control Model for IoT Environment Including Automated Configuration *IEEE 42nd Annual Computer Software and Applications Conference (COMPSAC)* 616-21

[10] Chanprakon P, Sae-Oung T, Treebupachatsakul T, Hannanta-Anan P and Piyawattanametha W 2019 An Ultra-violet sterilization robot for disinfection *5th Int. Conf. on Engineering, Applied Sciences and Technology (ICEAST)* 1-4

[11] Asadullah M and Raza A 2016 An overview of home automation systems *2nd Int. Conf. on Robotics and Artificial Intelligence (ICRAI)* 27-31

[12] Islam M M, Farook M N, Mostafa S M G and Arafat Y 2019 Design and Implementation of an IoT Based Home Automation *1st Int. Conf. on Advances in Science, Engineering and Robotics Technology (ICASERT)* 1-5

[13] Mandula K, Parupalli R, Murty C A S, Magesh E and Lunagariya R 2015 Mobile based home automation using Internet of Things(IoT) *Int. Conference on Control, Instrumentation, Communication and Computational Technologies (ICCICCT)* 340-43

[14] Somani S, Solunke P, Oke S, Medhi P and Laturkar P P 2018 IoT Based Smart Security and Home Automation *Fourth Int. Conf. on Computing Communication Control and Automation (ICCUBEA)* 1-4

[15] Ma L, Li Z and Zheng M 2019 A Research on IoT Based Smart Home *11th Int. Conf. on Measuring Technology and Mechatronics Automation (ICMTMA)* 2019 120-122

[16] Hong X, Yang C and Rong C 2016 Smart Home Security Monitor System *15th Int. Symp. on Parallel and Distributed Computing (ISPDC)* 247-51

MoS_2/CeO_2 Based Composite for Ammonia Sensing

Nitesh Dogra[1], Kajal Dadwal[1], Arun Kumar[2] and Sandeep Sharma[1]

[1]Department of Physics, Guru Nanak Dev University Amritsar, Punjab-143005, India

[2]Swami Premanand Mahavidyalaya, Mukerian, Punjab-144211, India

E-mail: Sandeep.phy@gndu.ac.in

Abstract. Here, we report synthesis of MoS_2/CeO_2 composite using a two-step hydrothermal techniques. A two-terminal sensor device made from the composite was further used for ammonia detection at room-temperature. The device has shown selective behavior toward ammonia gas molecules as compared to acetone, ethanol and N-methyle pyroliddone (NMP). Further, the sensor display delayed recovery and a relative-response of eleven percent for ammonia (30 ppm). The decrease in resistance after interaction with ammonia gas molecules indicate n-type conduction of the composite. Conduction type was further confirmed using Mott-Schottky measurements which gives a positive slope with composite as working electrode. These initial studies indicated that transition-metal dichalcogenides based composite with metal oxide semiconductor could be optimized for selective detection of various volatile organic compounds.

Keywords: ammonia sensor, transition metal dichalogenides (TMDCs), 2D materials, selectivity and CeO_2

1. Introduction

Development of gas sensor that can operate at room-temperature has remained a challenging task. Various factors influence the gas sensing performance of the sensors made from different materials. Past few decades have seen extensive growth in investigating metal oxide semiconductor (MOS) for gas sensing applications[1]. Though, they offer a large sensing response but their operation at elevated temperature and accompanied inferior selectivity limit their use. In addition to MOS, recently, 2D transition metal di-chalcogenides (TMDCs) with a formula MX_2, have emerged as an alternative for gas sensing applications[2]. Here, X (S or Se etc) represents a chalcogen atom whereas M (Mo or W) represents a transition metal atom. These 2D materials possess layer dependent physical properties which make them attractive for various applications ranging from nanoelectronics, optoelectronics and gas sensing also. Their, 2D structure and hence large surface to volume ratio accompanied with semiconducting nature facilitate their potential use for gas sensing applications. Though, gas sensing using MoS_2, WS_2 and their composites has been demonstrated [3, 4] but in a few cases, delayed recovery arising due to thick layered structure has been reported [5, 6]. Therefore, to overcome some of these problems various other methodologies, for instance, grain size control, doping and making composite with other materials have been used. Among these methods, composite formation has proven to be successful as it provides enhanced device performance and better selectivity over the individual host material. For instance, MoS_2/SnO_2 n-n heterojunctions [4, 7], MoS_2/multiwalled carbon nanotubes hybrids [8], MoS_2/WO_3 composite[3] and MoS_2/Co_3O_4 composite[9] were found to have enhanced ammonia sensing performance. Therefore, a composite based approach may offer solution to the problems faced by sensor made from either MOS and TMDCs alone. Here, we report the synthesis of MoS_2/CeO_2 composite via two-step hydrothermal technique. Two-terminal sensing device made from composite has shown n-type conduction with preferential ammonia detection at room-temperature. The large response/recovery times limits device performance and future work is further required to improve the performance of such type of composite based sensor devices.

2. Experimental Section

Synthesis of MoS_2/CeO_2 composite was performed using simple hydrothermal method. Various synthesis steps are shown in Fig.1.

All precursors namely, sodium hydroxide (NaOH), cerium nitrate hexahydrate ($Ce(NO_3)_3.6H_2O$) and ammonium tetra-thio-molybdate ($(NH_4)_2.MoS_4$) were acquired from Sigma Aldrich, India and used as received. Hydrazine was obtained from Loba Chemie India and also used as received. In the first step CeO_2 was synthesized at 130°C. In particular, 0.7 g of $Ce(NO_3)_3.6H_2O$ was added to 40 ml of de-ionized (DI) water and obtained solution was stirred for fifteen minutes. To this solution 20 ml of NaOH was added and obtained mixture was stirred for another two hours until solution changed to pink colour. Obtained mixture was put in an autoclave and kept in a furnace at 130°C for 18 hours. The autoclave was then allowed to cool under ambient conditions and obtained precipitates were washed and filtered. These precipitates were subsequently dried in a vacuum oven and light yellow colored CeO_2. Was obtained. In the next step, 0.3 g of CeO_2 was mixed with DI-water and bath-sonicated for 15 minutes. Thereafter 0.5 g $(NH_4)_2$ MoS_4 and 13 ml of hydrazine were added to the suspension obtained in previous step. Obtained mixture was sonicated for another 15 minutes. This mixture was transferred into an autoclave which was kept at 200°C in a furnace for 12 hours. Autoclave was allowed to cool under ambient conditions and obtained black colored precipitates of MoS_2/CeO_2 composite were washed and dried. A paste made from composite was applied on already deposited electrical contacts (with a separation of 2 mm) on an alumina substrate. A home built 40 L test chamber having a controlled oven and a circulatory fan was used for performing gas sensing measurements. Desired gas environment was achieved by inserting appropriate volume of analyte into the test chamber containing a furnace. The gas concentration is calculated using equation [10]:

$$C = (22.4\rho T\ V_s/273MV\ ) \times 100 \tag{1}$$

here, ρ referes to liquid density (g/mL), T, Vs and M refer to working temperature in K, evaporated liquid volume in μL and molecular weight of liquid in g/mol, respectively. V and C represent the test chamber volume (L) and concentration of different gases in ppm, respectively [3]. Further, Mott-Schottky (MS) plots were acquired with an electrochemical workstation (M204 Autolab, The Netherlands). Three electrode configuration was employed to perform MS measurements in 0.5 M H_2SO_4 solution. The measurement geometry consists of a platinum wire as a counter electrode, Ag/AgCl reference electrode and MoS_2/CeO_2 composite as a working electrode[3].

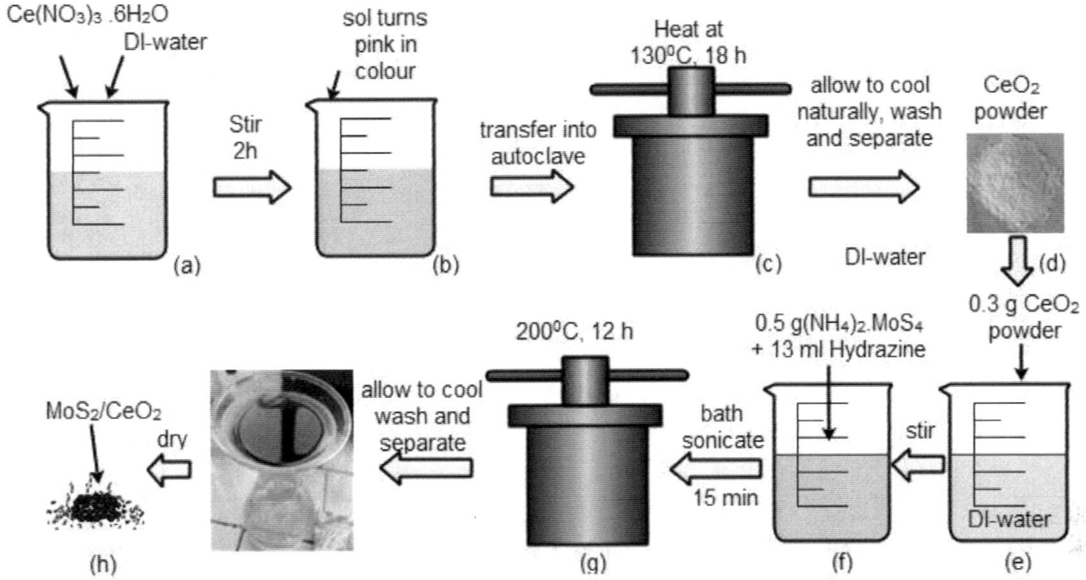

Figure 1. Schematic displaying two-step synthesis procedure for MoS_2/CeO_2 (a) Mixing of cerium nitrate in distilled water (b) solution turns into pink color (c) transfer of solution into an autoclave and keep it at 130 ° C for 18 h (d) after washing light yellow colored precipitates of CeO_2 are obtained (e) and (f) mixing of CeO_2, $(NH_4)_2MoS_4$ and hydrazine followed by (g) transfer into an autoclave (h) at the end black colored precipitates of MoS_2/CeO_2 are obtained.

3. Result and Discussion

3.1 Ammonia sensing using two-terminal device

Fig.2 displays the response-recovery transients for a sensor made from composite. In Fig. (a), as the ammonia gas enters into the measurement chamber, sensor resistance reduces to a certain minimum value and again displays an increase in resistance when measurement chamber is opened to remove the gas molecules. Here, we also define three parameters for a sensor device. The response time '$t_{response}$' is the time taken by the device to attain 90% change w.r.t. base resistance (R_b in air). On the other hand, once sensor has reached minimum resistance value in presence of gas, the time taken to recover to a resistance value lying 10% above the base resistance, is termed as recovery time '$t_{recovery}$'. If ΔR is the maximum resistance change, then $\Delta R/R_b$ defines the relative response for the sensor. As can be seen, the response time in present case is close to 180 s whereas the recovery is quite slow. Various factors may result in delayed recovery of the device. One possible reason could be large adsorption energy for the composite. Larger recovery time may limit the device operation and in a few cases external device heating or ultraviolet light exposure have shown improved results. Fig. (b) displays, sensor response to three different ammonia concentrations given in ppm. The response increases with increase in ammonia concentration. In Fig. (c), the sensor response for successive three cycles with same ammonia concentration is shown. Device displays less than 5% change in the relative response. As our environment consists of various gases, selectivity of sensor device towards a particular gas becomes an important concern. The sensor response was checked for acetone, ethanol and NMP in addition to ammonia and results are shown in Fig (d). The composite based device has shown selective ammonia detection compared with other analytes.

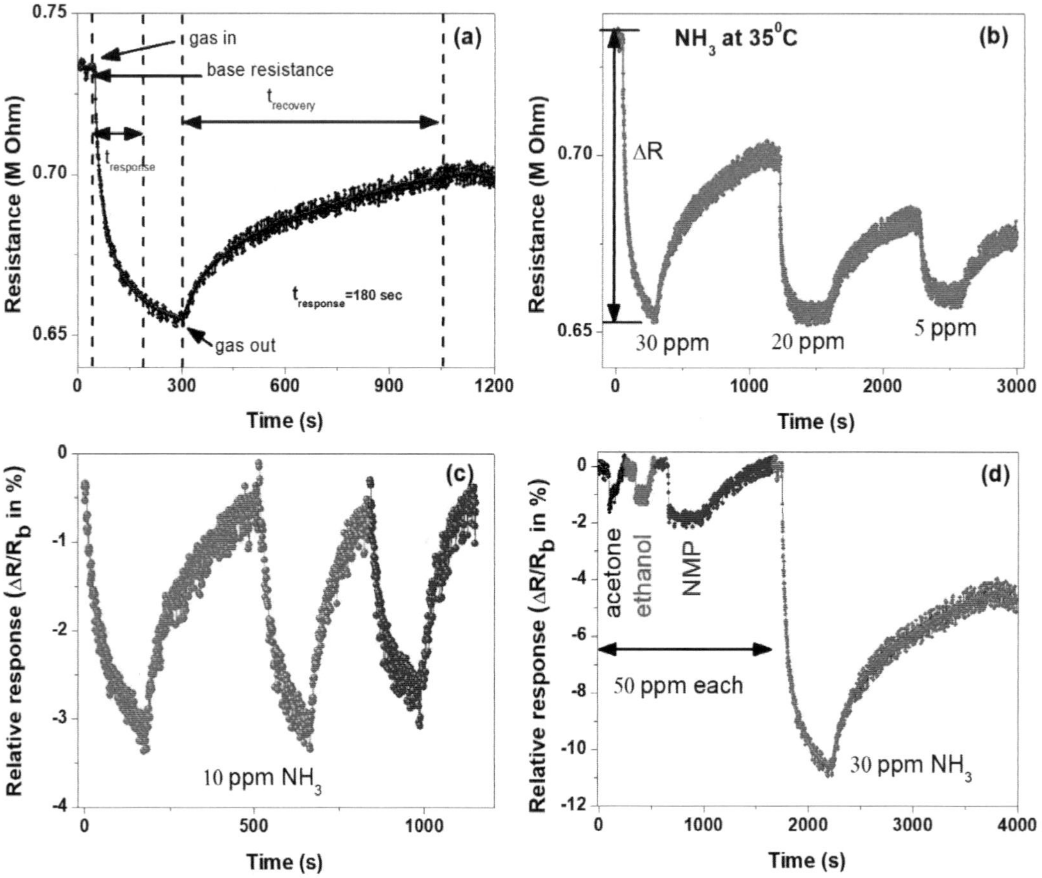

Figure 2. (a) Response-recovery transient when sensor exposed to NH_3. Figure also define response, recovery time and relative response of a device. (b) Response-recovery transient for three different concentrations of NH_3 (c) repeatability and (d) selectivity test of the device. All measurements were performed at room-temperature.

3.2 Mott-Schottky measurements

Further, Mott-Schottky measurements in three electrode configuration were performed to assess the conductivity type of the device. Sensor when exposed to ammonia gas molecules, exhibited a decrease in resistance, implying n-type conduction in the composite. The MS plot shown in Fig.3, displays a positive slope confirming the dominant n-type conduction in sample.

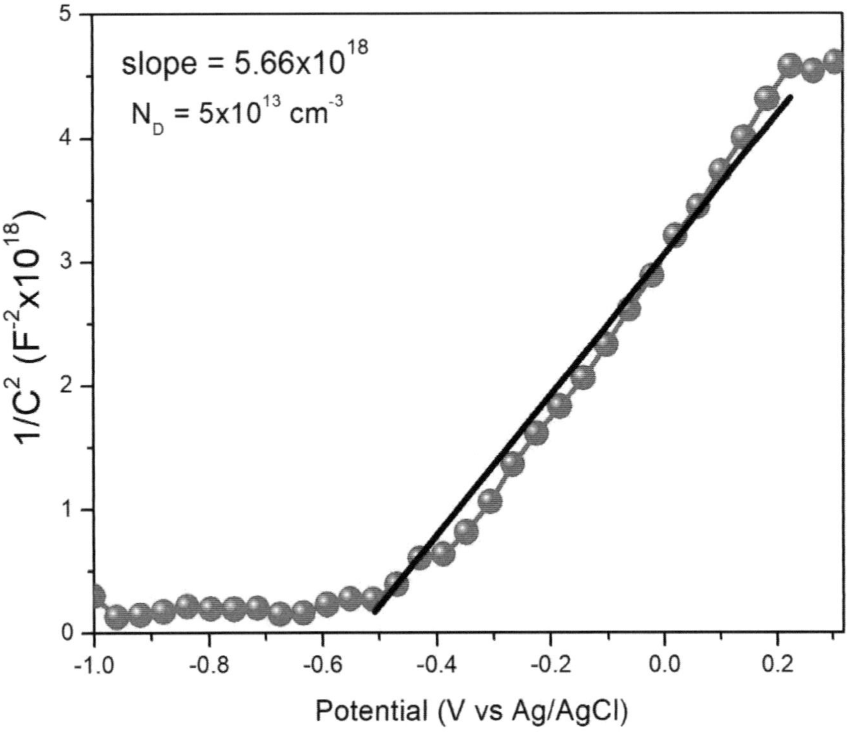

Figure 3. Mott-Schottky plot with positive slope indicating n-type character of the working electrode made from MoS_2/CeO_2 composite.

3.3 Sensing Mechanism

Hydrothermally synthesized MoS_2 behaves like an n-type material [8]. In present study a composite of MoS_2/CeO_2 also displays n-type conduction. A widely acceptable sensing mechanism for MOS assume ambient oxygen adsorbed on MOS surface. However, in case of composite, the mechanism may differ in a sense that both MoS_2 and CeO_2 behave like an n-type material. CeO_2 possesses a band gap of 3.2 eV and work function of 4.69 eV[11]. On the other, the work function for air exposed MoS_2 is 4.47 eV [12]. Lower work function of MoS_2 causes electron transfer from MoS_2 to CeO_2 and give rise to a depletion layer. Depletion layer modulation takes place whenever a charge transfer between composite surface and adsorbed gas molecule occurs. This conductivity modulation is reflected as change in measured resistance of the device. Adsorbed oxygen after interaction (electron capture) with MOS turns in to oxygen ion (O^{2-}). Such a surface when exposed to reducing gas such as ammonia, electron transferred from ammonia to MOS causes a reduction in depletion layer width. Net result will be a reduction in the resistance of the device.

4. Conclusion

Here, we presented a systematic study involving interaction of ammonia gas molecules with MoS_2/CeO_2 composite surface. We also demonstrated that composite based sensor has a linear response with ammonia concentration. Sensor exhibited a decrease in resistance when exposed to ammonia. This implies an n-type conduction process which was cross checked using Mott-Schottky measurements. The sensor possesses a relative response of 3-3.5% and response time of 180 s accompanied with delayed recovery. Sensor exhibited preferential room-temperature ammonia detection as compared to ethanol, acetone and NMP. Though, the obtained response and recovery times are a bit large and further work is required to address these issues.

Acknowledgments
This work was supported financially by DST-New Delhi, India, Under the Grant No: EMR/2016-007483.

References

[1] Ji H, Zeng W and Li Y 2019 *Nanosc.* **11** 22664-84
[2] Tyagi D, Wang H, Huang W, Hu L, Tang Y, Guo Z, Ouyang Z and Zhang H 2020 *Nanosc.* **12** 3535-59
[3] Singh S, Deb J, Sarkar U and Sharma S 2021 *ACS Appl. Nano Mater.* **4** 2594-605
[4] Singh S, Sattigeri R M, Kumar S, Jha P K and Sharma S 2021 *ACS Omega* **6** 11602-13
[5] Li X, Li X, Li Z, Wang J and Zhang J 2017 *Sens. Actuators B Chem.* **240** 273-77
[6] Qin Z, Zeng D, Zhang J, Wu C, Wen Y, Shan B and Xie C 2017 *Appl. Surf. Sci.* **414** 244-50
[7] Wang W, Zhen Y, Zhang J, Li Y, Zhong H, Jia Z, Xiong Y, Xue Q, Yan Y, Alharbi N S and Hayat T 2020 *Sens. Actuators B Chem.* **321** 128471
[8] Singh S, Sharma S, Singh R C and Sharma S 2020 *Appl. Surf. Sci.* **532** 147373
[9] Zhang D, Jiang C, Li P and Sun Y 2017 *ACS Appl. Mater. Interfaces* **9** 6462-71
[10] Singh S, Deb J, Sarkar U and Sharma S 2021 *ACS Sustainable Chem. Eng.* **9** 7328-40
[11] Warule S S, Chaudhari N S, Kale B B, Patil K R, Koinkar P M, More M A and Murakami R I 2012 *J. Mater. Chem.* **22** 8887-95
[12] Lee S Y, Kim U J, Chung J, Nam H, Jeong H Y, Han G H, Kim H, Oh H M, Lee H, Kim H, Roh Y G, Kim J, Hwang S W, Park Y and Lee Y H 2016 *ACS Nano* **10** 6100-07

IOP Conf. Series: Materials Science and Engineering 1225 (2022) 012060 doi:10.1088/1757-899X/1225/1/012060

Semiconductor wafer defect classification using convolution neural network: a binary case

Krishan Kumar Chauhan, Garima Joshi, Manjeet Kaur, and Renu Vig
UIET, Panjab University, Chandigarh, INDIA

joshi_garima5@yahoo.com

Abstract. With the multitude of steps used in semiconductor industry, automation is being practiced extensively in its manufacturing processes to guarantee quality of manufactured chips and improvement in production. At front end of line, wafer are probed and defective chips are segregated. From this data wafer bin map are generated, these show defects on the surface of wafers. If analysis of wafer bin map is done manually, this may result in incorrect categorization of defects due to human error and lack of judgement. Thus, the rationale behind this research study is to determine the scope of vision-based methods for automatic classification of wafer defects. In view of this, a classifier which detects Type A and Type B defects in wafers is proposed. It involves a Convolution Neural Network (CNN) based binary classifier. 250 Images of each class of wafer bin map dataset, is used to conduct the study on five layer CNN architecture. The proposed setup gives the test accuracy of 97.7%.

1. Introduction

In a typical semiconductor processing sequence, there are over hundred process operations performed before the silicon wafers is converted to chips carrying electronic circuits [1]. The semiconductor process steps include processes such as polishing, etching, diffusion, oxidation and metallization. With the progress in semiconductor industry the number of surface defects are escalating due to reduction in size and increasing complexity [2]. These surface defects if inspected manually can consequently lead to incorrect classification of defects. There is a need to identify these defects at an early stage, so as to find the real cause of occurrence of these defects [3]. The pattern of defects can be classified and related to the issues in the particular fabrication process. Timely and proper addressing of these issues can further help the production and improvement in yield [4].

1.1 Computer vision based wafer defect detection

Computer vision is a field that aims to enable computers to interpret, recognize and process images in the same manner as human vision. Defect detection using computer vision techniques can be accomplished in following ways:

- Image processing methods adopt basic techniques, namely, enhancement and image restoration. Comparing the image under test with the standard image help in detecting defects.

- AI-based approach need labelled dataset to develop algorithms for recognizing defect of a particular shape [5]. Furthermore, deep learning is replacing the conventional machine learning and is helping in real-time deployment of such systems.

The frequency of publications from year 2001 to 2020 is shown in Figure 1. Data clearly shows the significant rise in the paper published after 2018 which indicates the rise in research interest in this domain [6]. This graph is extracted from SCOPUS database for keyword "wafer defect classification" and filtered by selecting the articles published in engineering and computer science domain resulting in total 167 articles [7].

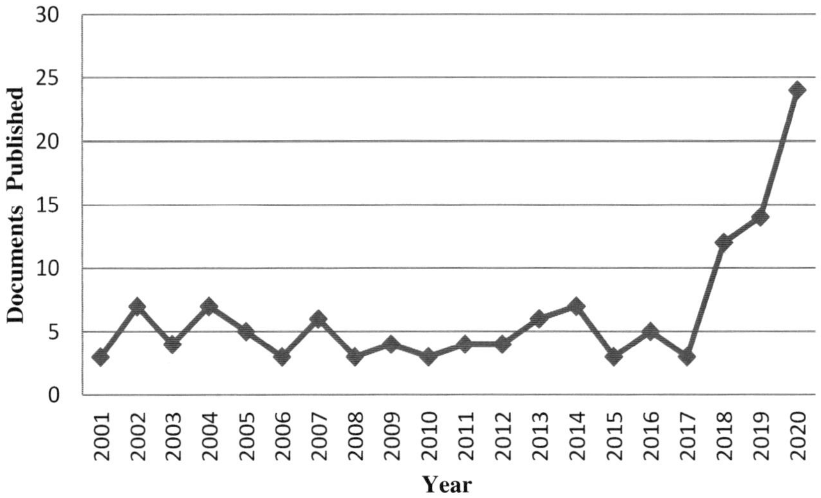

Figure 1. Number of "wafer defect classification" documents published in past 20 years in SCOPUS database

Recently, wafer defects classification using machine learning is being explored. The usual approach is to compute attributes from a wafer map and train a classifier based on these attributes. Contrarily, deep learning method that works directly on a bin map of wafer can be deployed [2]. The idea behind such an application is that it automatically collects defective images of the wafer surface and categorizes the collected images. According to the defect type, it quantifies the tendency of defect generation in the fabrication process [8].

2. Wafer Defects and Dataset Details

Defects on the wafer surface commonly occur due to fault in machinery, stains of chemical, material damages, errors due to manual handling of processes, and ambient settings. The faulty wafer maps illustrate a lot of distinctive patterns such as spot, repetitive shapes and clusters. To locate and visualize these defect patterns wafer map images are observed.

Wafer Bin Map (WBM) represents a particular positions of fault which can be analyzed for locating source of low yield in manufacturing [9]. Publically available real world WBM dataset, WM-811K [10], dataset provided by Multimedia Information Retrieval Lab, Department of Computer Science, National Taiwan University [11], and is created by Wu *et al.* in 2015 [12]. This dataset consists of total 811,459 WBMs among which, 172,950 are labelled and can be used in supervised learning. There are nine labels [13].

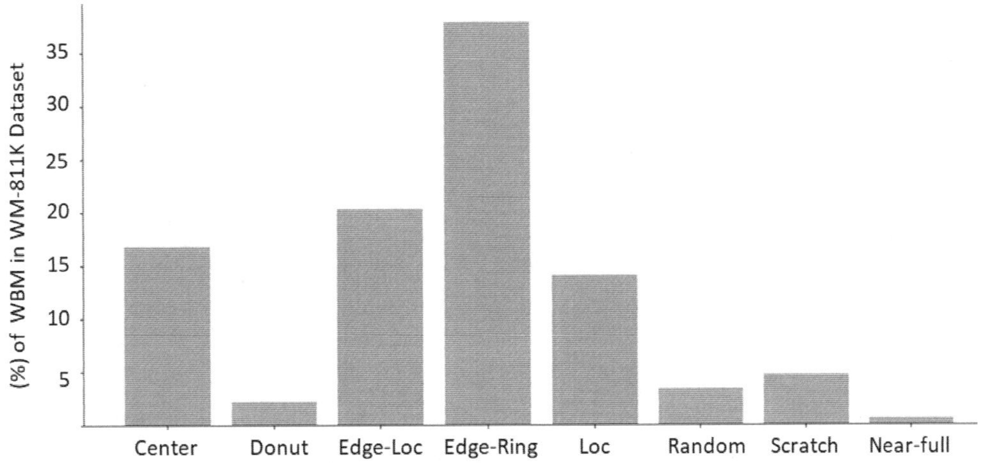

Figure 2. WM-811k dataset analysis

However, it can be noted that the distribution of each type of defect is imbalanced as shown in Figure 2. The most commonly occurring defect is Edge-Ring. In this research work, binary classification case is taken. Total 500 images of Random and Edge-Ring defect are classified as Type A and Type B defect, respectively as shown in Figure 3.

- **Type-A** defects appear haphazardly and no definite pattern is visible. The reason of occurrence of these defects is unknown and it is difficult to point out one specific cause of these defects [2].
- **Type-B** defects are systematic and repeatable. This type of defect has a definite pattern and the reason can usually be established by observing the defect pattern [2].

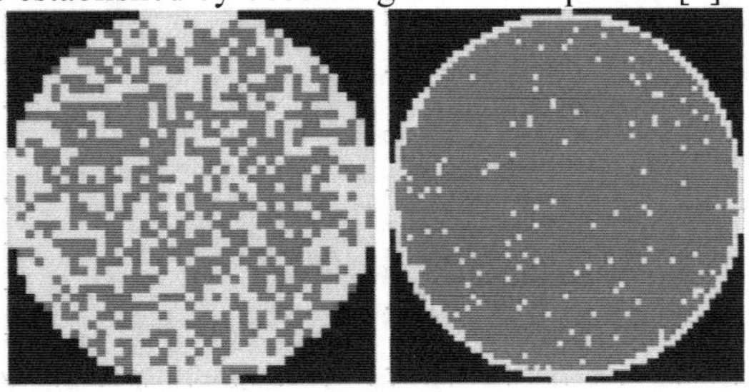

a) Type A - Random b) Type B - Edge-Ring

Figure 3. Defect classes considered for binary case classification

3. Computer Vision Approach Wafer Defect Classification

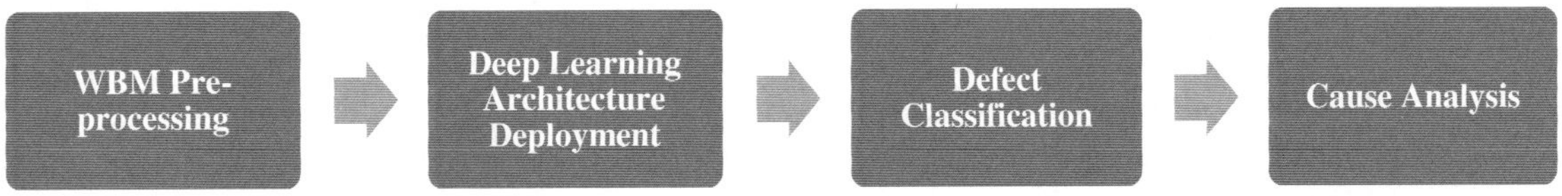

Figure 4. Computer vision approach for wafer defect classification

The methodology to implement computer vision approach for wafer defect classification is shown in Figure 4.

3.1 WBM Pre-processing

It is important to pre-process WBM array in order to apply computer vision techniques for classification. WBM defect pattern is converted to an image. All the bin maps are converted to equal size images of size 26x26.

3.2 Deep Learning Architecture Deployment

In deep learning approach, Convolution Neural Network (CNN) is utilized for image classification. These architectures can be designed from a scratch. To design CNN model for a specific application, early layers learn low-level features while the last layer performs classification on the basis of feature map. Table 1 lists the details of each layer of proposed CNN architecture.

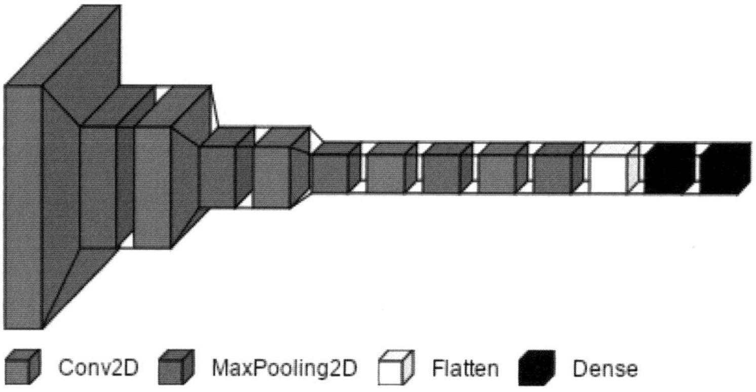

Figure 5. Details of proposed CNN Model

CNN architecture includes:

- Convolution layer performs convolution of image with kernel to extract feature maps. Here five-layer CNN model is designed. It consists of convolution filters of sizes [3*3, 3*3, 5*5, 5*5, 7*7].
- ReLU is used to calculate activations of all convolved feature map.
- Max Pooling layers are utilized to reduce complexity of each CNN layer output.
- Optimizers_1 = "sgdm"; Optimizers_2 = "adam"; loss_function1 = "hinge"; loss_function2 = "squared_hinge" ; classifier = binary support vector machine
- Total Trainable parameters: 1,15,969
- Accuracy is used as the performance metric here. It is the fraction of classes rightly classified to the overall number of input samples fed to the model during testing, training and validation resulting in Test accuracy, Train accuracy and Validation accuracy, respectively. Here, Test accuracy is considered for comparison of results

3.3 Defect Classification

In supervised learning the neural network learns in terms of value of weights and biases from the training dataset. The validation data provides neutral assessment of a model fit on the trained network. The validation dataset is required to fine-tune the hyper parameters. This dataset can also be used for an unbiased evaluation of a trained model. At the output, a fully connected layer and classification layer predict the category of input image. Here binary SVM is used to classify the images on the basis of computed feature map.

3.4 Cause Analysis

The likely causes of each of defect can be related to a process variation. Center defects are caused due to fluctuation in power, liquid flow, or pressure. Local defect can result due to valve leakage, distortion in the pump. Random defect occur because of dirty pipes, showerhead, and scratch are the result of irregularity in robot handoffs. The cause of each defect is mentioned as follows:

- **Type-A** defect does not have a fixed pattern and this can be addressed by improving the constancy and precision of the system involved.
- **Type-B** defect can be caused by microscopic shifting of the mask during photolithography or disproportionate (over or under) etching process. Closed ring shapes usually reflect problem in etching or annealing process.

4. Results and Discussion

250 images of each class is considered and the experiment has been performed for 25 epochs, at 32 batch size, 0.01 learning rate and L2 regularization. For training the proposed network, accuracy is analyzed for the two optimizers selecting hinge loss function and squared hinge loss function at a time. Training is an iterative step until the desired accuracy is achieved. Once the network is trained to a satisfactory loss level, the network is tested on Test dataset. For evaluating the performance of computer vision approach for wafer defect classification the dataset is used as 70%, 15%, and 15% as train, test and validation set, respectively. Training and validation results for different combination of loss function and optimizer are shown in Figure 6 through Figure 9.

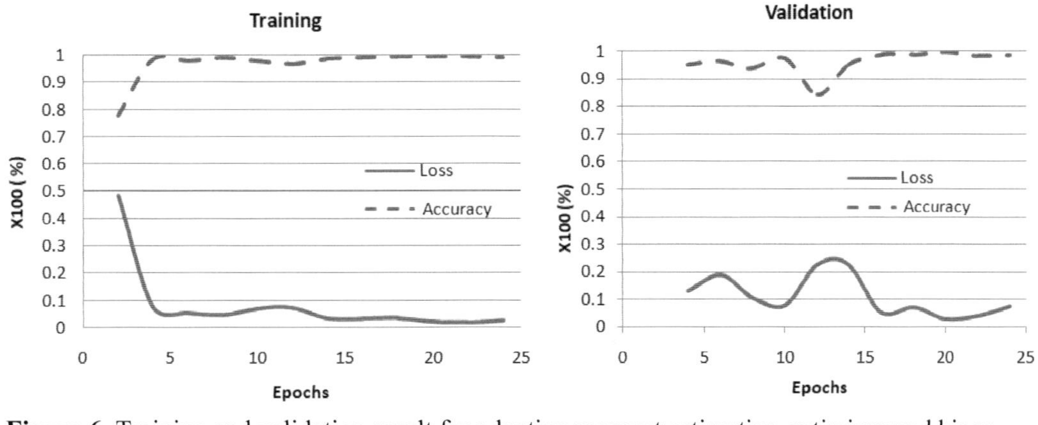

Figure 6. Training and validation result for adaptive moment estimation optimizer and hinge loss function

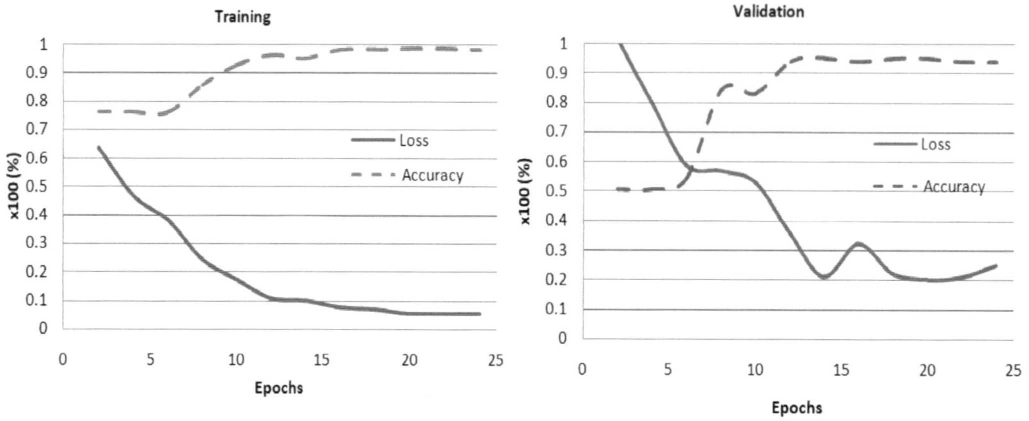

Figure 7. Training and Validation Result for stochastic gradient descent with momentum optimizer and hinge loss function

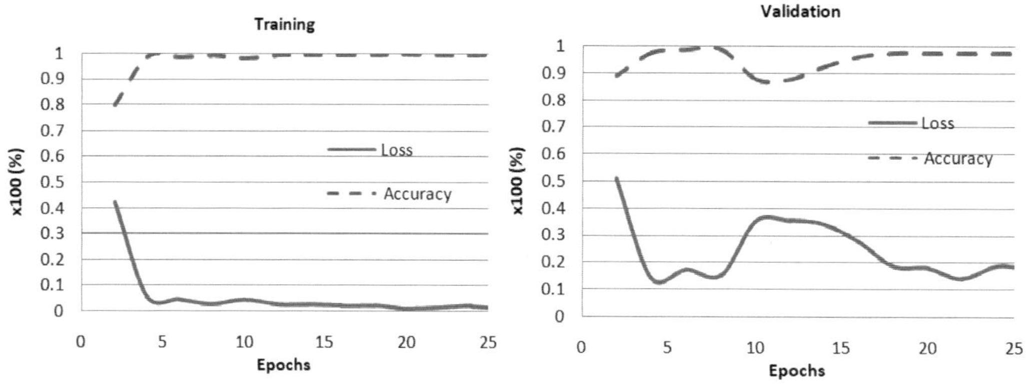

Figure 8. Training and Validation Result for adaptive moment estimation optimizer and squared hinge loss function

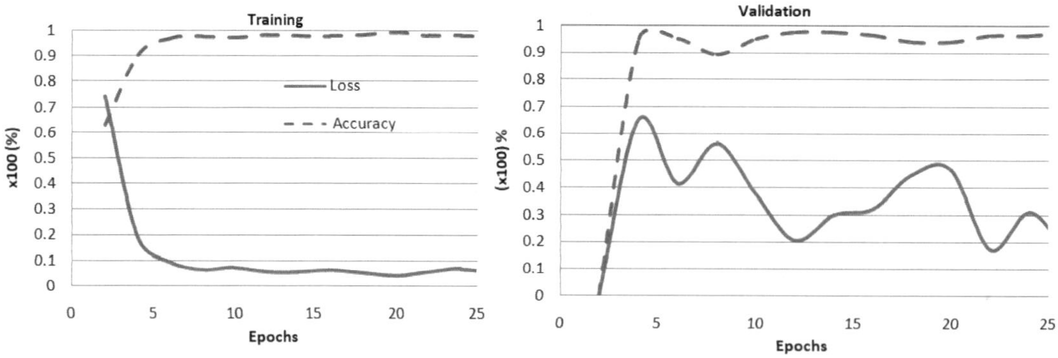

Figure 9. Training and Validation Result for stochastic gradient descent with momentum optimizer and squared hinge loss function

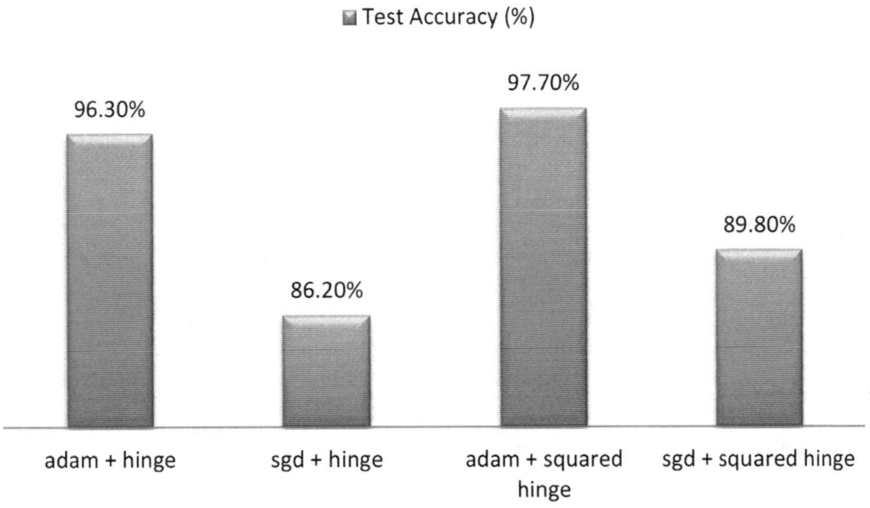

Figure 10. Test accuracy of proposed CNN architecture for wafer defect classification

As shown in Figure 10, the maximum value of test accuracy is 97.7% for proposed five-layer CNN architecture.

5. Conclusion

In semiconductor fabrication steps, analysing the pattern of defect can help in finding the issue in the process. Addressing and correcting the problem may further help in improving the yield. It is critical to identify and classify these bin maps so that the actual cause for these defects can identified and corrected. In general, the engineers examine these wafer maps manually. In order to eliminate human inspection an automatic defect classification mechanism based on computer vision is proposed. In this paper, two classes of WM-811K are used to conduct the study using a five-layer CNN model. For the combination of "adam" optimizer and "squared hinge" loss function, the proposed model attained the maximum test accuracy.

In future, a multiclass CNN classifier will be designed to classify all the classes of wafer dataset. Also, the effect of variation of model hyper parameters for wafer defect classification will be analyzed in detail.

References

[1] Hsu C Y, Chen W J and Chien J C 2020 Similarity matching of wafer bin maps for manufacturing intelligence to empower industry 3.5 for semiconductor manufacturing *Comput. Ind. Eng.* **142** 106358 pp 1–14.

[2] Chien J C, Wu M T and Lee J D 2020 Inspection and classification of semiconductor wafer surface defects using CNN deep learning networks *Appl. Sci.* **10** 5340 pp 1–13.

[3] Batool U, Shapiai M I, Fauzi H and Fong J X 2020 Convolutional neural network for imbalanced data classification of silicon wafer defects *IEEE International Colloquium on Signal Processing and its Applications* pp 230–235.

[4] Ji Y and Lee J H 2020 Using GAN to improve CNN performance of wafer map defect type classification: yield enhancement *SEMI Advanced Semiconductor Manufacturing Conference* pp 1–6.

[5] Kim J, Lee Y and Kim H 2018 Detection and clustering of mixed-type defect patterns in wafer bin maps *IISE Trans.* **50** pp 99–111.

[6] Park S, Jang J and Kim C O 2021 Discriminative feature learning and cluster-based defect label reconstruction for reducing uncertainty in wafer bin map labels *J. Intell. Manuf.* **32** pp 251–263.
[7] Wafer Defect Classification 2020 Available online: http://www.scopus.com
[8] Nakazawa T and Kulkarni D V 2018 Wafer map defect pattern classification and image retrieval using convolutional neural network *IEEE Trans. Semicond. Manuf.* **31** pp 309–314.
[9] Wang C H, Kuo W and Bensmail H 2006 Detection and classification of defect patterns on semiconductor wafers *IIE Trans.* **38** pp 1059–1068.
[10] Wafer Bin Map Data Set Retreived from *https://www.kaggle.com/qingyi/wm811k-wafer-map*
[11] Multimedia Information Retrieval Lab Data Set Department of Computer Science National Taiwan University Retreived From *http://mirlab.org/dataSet/public/*
[12] Wu M J, Jang J S R and Chen J L 2014 Wafer map failure pattern recognition and similarity ranking for large-scale data sets *IEEE Trans. Semicond. Manuf.* **28** pp 1–12.
[13] Jin C H, Kim H J, Piao Y, Li M and Piao M 2020 Wafer map defect pattern classification based on convolutional neural network features and error-correcting output codes *J. Intell. Manuf.* **31** pp 1861–1875.
[14] Vlădăreanu V, Voiculescu V G, Grosu V A, Vlădăreanu L, Travediu A M, Yan H, Wang H and Ruse L 2020 Detection of anomalous behavior in modern smartphones using software sensor-based data *Sensors* **20** 2768 pp 1–22.

Xylene sensing using Dy-doped NiO nanoparticles

Shailja, K J Singh[1], S Sharma

Department of Physics, Guru Nanak Dev University, Amritsar 143005, India

[1]corresponding author: kanwarjitsingh@yahoo.com

Abstract. Present work reports structural and xylene sensing properties of Dy-doped NiO nanoparticles. The samples were prepared using co-precipitation method. XRD pattern reveals incorporation of dysprosium into the host lattice of nickel oxide. SEM images shows the change in morphology from micro-rods to nanoparticles on doping nickel oxide with dysprosium. Raman spectrum reveals the presence of nickel vacancies which serve as adsorption sites for xylene and hence, improves the sensing performance of doped NiO. Doping with dysprosium has not only improved the sensitivity but also enhanced its selectivity towards xylene when tested against ammonia, benzene, ethanol and acetone at 200 ^{0}C. Dysprosium doped device has shown sensor response of 7.4 and response (recovery) time of 50 s (36 s), respectively. Our dysprosium doped sensor show improvement in the response and recovery times of already reported xylene sensing devices employing nickel oxide.

1. Introduction

Sick building syndrome (SBS) is a serious health issue which can cause neuralgia, sore throat, dermatitis and lack of vitality. The chemicals that primarily induce sick-house syndrome include ethanol, formaldehyde, benzene, toluene and xylene. These chemicals emanate from daily usable things like buildings, decorative products, wood accessories, carpets, solvents, paints, coating and fixatives. Xylene is an omnipresent toxic indoor pollutant, that can harm the nervous system and respiratory organs of human body [1, 2]. It is well known that NiO shows better catalytic properties to oxidize methyl radicals of xylene molecule as compared to other metal oxide semiconductors (MOS) gas sensors such as CuO, Cr_2O_3 etc. [3]. However, like other metal oxides, a pristine NiO based sensor offers poor sensing performance. By incorporating an appropriate dopant or catalyst into the host matrix of nickel oxide can improved its sensing properties [4]. Rare earth elements have been successfully employed as dopants in various metal oxide semiconductor gas sensors for reducing their operating temperature and improving the key parameters such as sensitivity and selectivity toward particular analyte [5]. For example, Shaikh et al. has reported selectivity towards ethanol by using dysprosium as a dopant in SnO_2 lattice at 300 ^{0}C [6]. Also, Hastir et al. has shown selective ethanol sensing for Dy-doped ZnO nanorods at 300 ^{0}C [5]. Therefore, doping with rare earth seems to be helpful in improving the sensing characteristics and various other related parameters of sensing devices. Health agencies across the world have fixed the acceptable exposure limit for various volatile compounds (VOCs) [6]. For xylene, it is 100 ppm for continuous eight hours of exposure [7]. Prolonged exposure of such VOCs can result in several health consequences and may cause death of an individual in certain cases. Therefore, a serious attention should be given to improve the indoor air quality all over the world. In order to address some of these issues, we propose the use of dysprosium doped NiO for selective and sensitive detection of xylene at 200 ^{0}C. Dysprosium doping in nickel oxide reduces the operating temperature as well as recovery and response time. We have described the synthesis, structural and xylene sensing parameters of dysprosium doped nickel oxide. The material synthesis was carried out using co-precipitation method.

NiO when doped with Sn and W, gave high response (recovery) time for xylene sensing. Operating temperatures for Sn-doped nickel oxide and W-doped nickel oxide were 225 ^{0}C and 375 ^{0}C , respectively [2, 7]. Although, these dopants have shown better sensor response but they work slower at high temperature. Dysprosium doping in nickel oxide has improved the response time and recovery time, hence it operates at low temperature (~200 ^{0}C). Recent studies have shown that dysprosium as a dopant has improved the sensing properties of various semiconductor metal oxides including ZnO [5] and SnO_2 [6]. So for dysprosium doping was not studied for assessing the gas sensing parameters of NiO. In this work, authors studied structural and xylene sensing properties using dysprosium as a dopant in NiO.

2. Experimental Section

For synthesizing Dy-doped NiO, 0.13 mg (3% by weight) of dysprosium nitrate hydrate and 4.61 mg nickel chloride hydrated were used to prepare 0.2 M aqueous solution and subsequently, ammonium solution was added drop-wise to form precipitates [8]. Obtained precipitates were washed with distilled water during filtration, dried overnight at 80 ^{0}C and calcined for 3 h at 500 ^{0}C to get a black colored nickel oxide powder. Further, 4.75 mg of nickel chloride hexahydrate was taken and similar procedure was followed to obtain pure nickel powder. Thereafter, we carry out structural, electrical and gas sensing characterizations for both samples. The XRD spectra were recorded by using Shimadzu 7000 Diffractometer system (Cu Kα radiation (λ =1.54 Å)) in the 2Θ range. The morphology was investigated with scanning electron microscope (Carl Zeiss SUPRA 55). The Raman data was obtained by using an excitation wavelength of 488 nm on Renishaw Invia Microscope Raman spectrometer. The obtained powders (2 mg) were mixed separately in ethanol to make a slurry which were coated on alumina substrates by using a paint brush. The substrate contains pre-deposited gold electrodes separated by a distance of 3 mm. Xylene sensing parameters were recorded using 40 L home built chamber fitted with a circulating fan, a furnace and a sample holder. The analyte concentration was calculated using equation given below:

$$C= \frac{22.4\,\rho\, T\, V'}{273\, M\, V}\ X\ 100 \quad (1)$$

Here, C, ρ , T , V', M and V correspond to concentration of analyte (in ppm), density of analyte (g/ml), temperature (K), volume of analyte, molecular weight of analyte (g/mol) and volume of testing chamber(L), respectively [9].

3.Results and Discussion

3.1. XRD

XRD patterns of pure NiO and Dy-doped NiO are displayed in Figure 1. These peaks matched well with lattice planes (111), (200), (311) and (222) of NiO (JCPDS card No 78-0423). It is to be noted that after doping with dysprosium, (111) and (200) peaks of nickel oxide shift towards lower angles, indicating incorporation of dysprosium into the host lattice of nickel oxide. This is accompanied with a increase (4.159Å to 4.168Å) in the lattice parameter as obtained from Bragg's law for pure and Dy-doped NiO, respectively. The average crystallite size was evaluated using W-H plot [10] and mentioned as below:

IOP Conf. Series: Materials Science and Engineering 1225 (2022) 012061 doi:10.1088/1757-899X/1225/1/012061

$$\frac{\beta \cos\theta}{\lambda} = \frac{K}{Dw} + \frac{4\,\varepsilon \sin\theta}{\lambda} \quad (2)$$

Where λ, β, θ, D_W, and ε are wavelength of X-ray, full-width at half maxima, Bragg's diffraction angle, crystallite size and micro strain ,respectively. Intercept value from figure 2 gives the average crystallite size and found to be 21.5 nm and 15.7 nm for pure and doped nickel oxide respectively. Tensile strain calculated from the slope of figure 2 increased from $1.3x10^{-3}$ to $2.6x10^{-3}$, respectively, indicating incorporation of dysprosium in host lattice and accompanied lattice expansion resulting from larger size of dysprosium as compared to nickel.

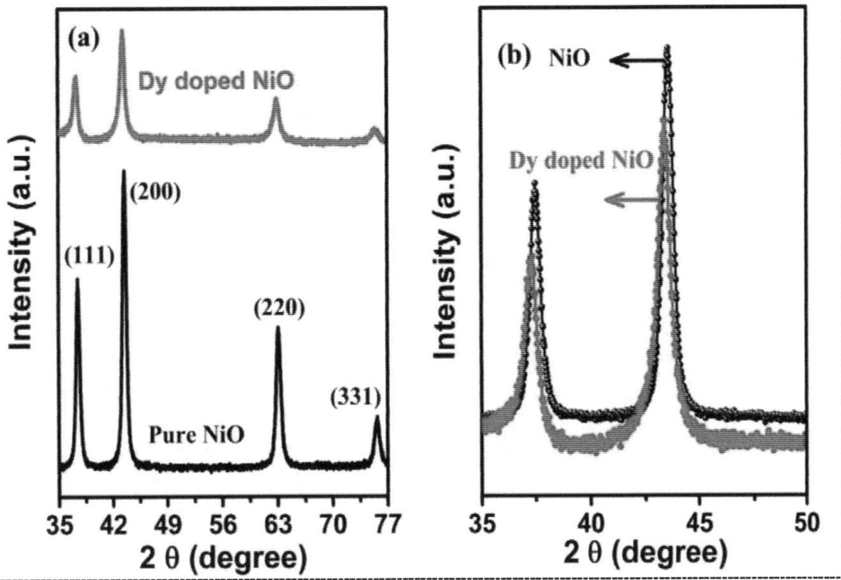

Figure 1. X-ray diffractogram of (a) NiO and Dy-doped NiO (b) Shows the shift of peaks towards lower values

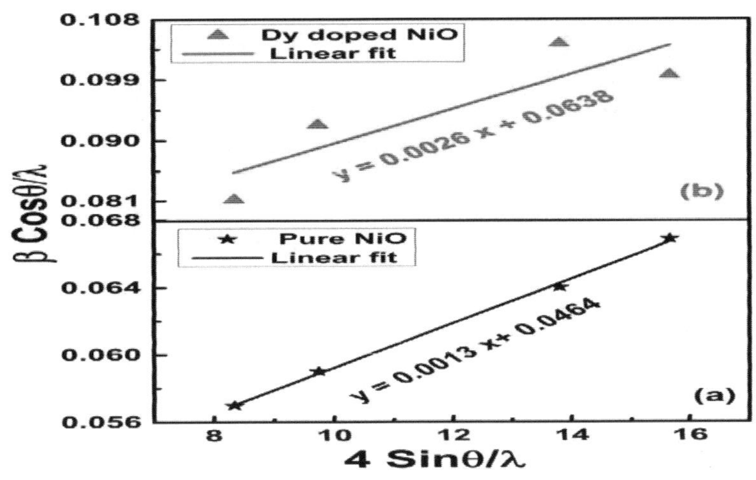

Figure 2. Williamson-Hall plot of pure and Dy-doped NiO.

3.2 Raman

Vibrational spectra of both samples were acquired under ambient conditions and displayed in Figure 3. It can be seen that pure NiO displays first order longitudinal mode (1LO) ascribed to the Ni–O stretching. This mode has been ascribed to nickel vacancies created in the structure. The increased intensity of 1LO mode on dysprosium doping indicates the increase of nickel vacancies in the doped nickel oxide. The second mode at 1003 cm^{-1} corresponds to 2LO mode and describes the vibration of Ni-O bond. After doping, both modes display a red-shift and appear at 484 and 999 cm^{-1}, respectively. The red-shift in Raman modes is assigned to the tensile strain arising from dopant incorporation into host lattice and resulting interaction between them [11,12].

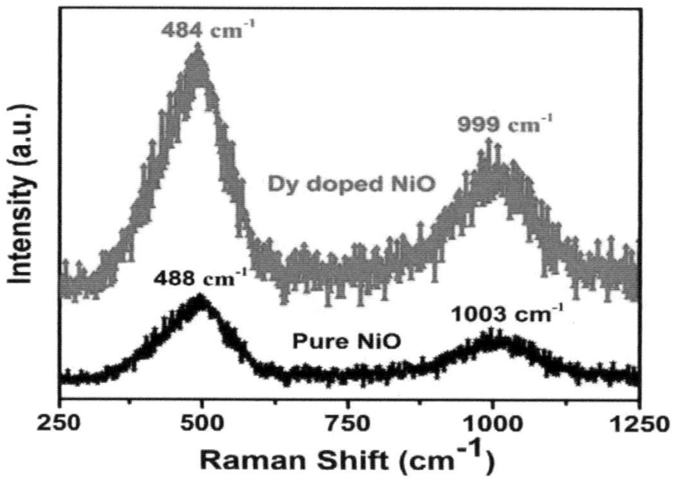

Figure 3. Raman modes of pure and doped NiO particles.

Figure 4. Morphology of pure and Dy-doped NiO.

3.3 SEM

Figure 4 (a) and (b) display the scanning electron micrographs of the samples. Pure nickel-oxide has rod shaped features whereas doped NiO possesses nanoparticle shaped morphology.

3.4 Xylene sensing using two-terminal device

Figure 5 provides the response-recovery transients from two different sensors for 100 ppm of xylene at 200 ^{0}C. Sensors show an enhancement in resistance on exposure to xylene, implying a p-type character of pure and doped NiO. In this figure, we also define response and recovery times as follows: response-time is the time required to reach 90% of maximum resistance change (R_g in gas) for the device from its base value (R_a in air). Similarly, recovery time refers to time required for the device to recover from maximum resistance value to 10% above the base resistance value [13].

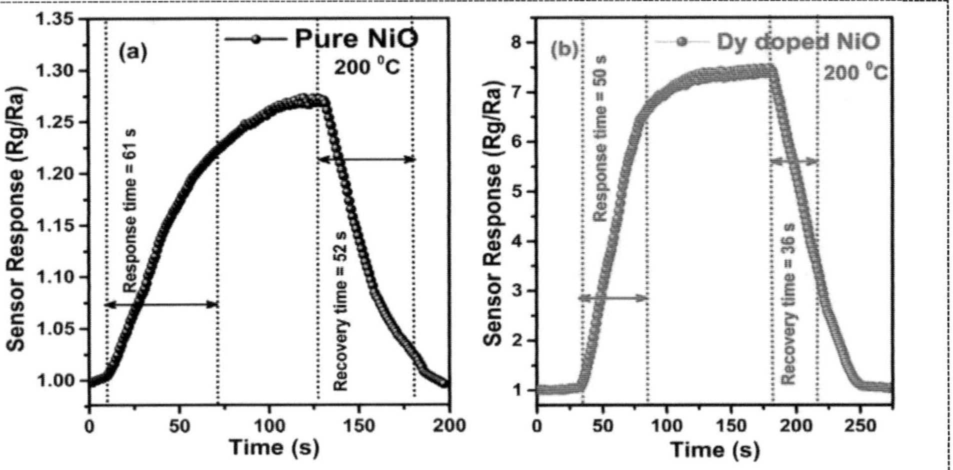

Figure 5. Response-recovery transients of pure and dysprosium doped nickel oxide sensors obtained at 200 ^{0}C for 100 ppm of xylene.

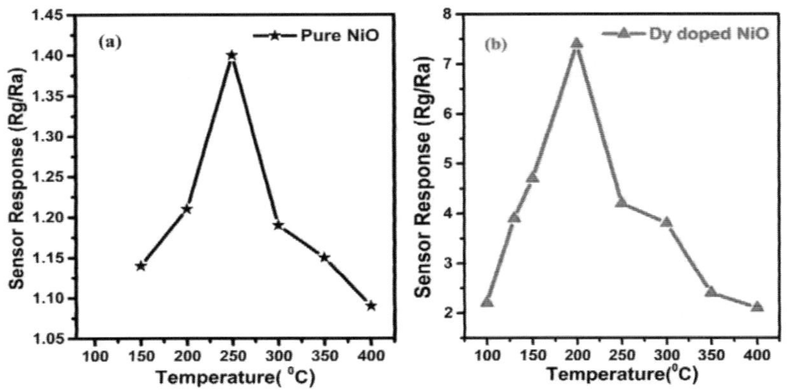

Figure 6. Sensor response of (a) NiO based sensor (b) Dy-doped NiO based sensor at different temperatures toward 100 ppm of xylene.

The response and recovery times for pure (Dy-doped) NiO were found to be 61s (50 s) and 52 s (36 s), respectively. Qiu et. al. reported the response time of 63s and recovery time of 66s towards 100 ppm xylene exposed on Nb-doped NiO nanosheets at 370 ^{0}C[14]. Gao et. al. studied 100 ppm of xylene exposed on NiO/ $NiCr_2O_4$ at 225 ^{0}C and found response time (1217 s) and recovery time (591 s) [2].

Sensor response (S) for device has been defined as S = R_g/R_a . Here, R_g and R_a refer to sensor resistance in the presence of test gas and air, respectively [7]. Further, to evaluate the sensor performance, gas sensing measurements were performed at different temperatures ranging from 100–400 ^{0}C and results are shown in Figure 6 for two samples. Pure NiO provides a maximum value of sensor response at 250 ^{0}C whereas Dy-doped NiO at 200 ^{0}C. Therefore, after doping a decrease in optimum temperature has been observed. Such behaviour shows the temperature dependence of reaction between the chemisorbed oxygen species and xylene molecules. At high temperature, desorption of oxygen species increases over the adsorption process and leads to decrease in sensor response of semiconductor metal-oxide based sensors.

Rare earth oxides as dopants show remarkable catalytic properties and provide a unique pathway to perform reaction with less activation energy even at low temperature [5]. Dysprosium belongs to the rare earth family and its doping in NiO decreases the optimum operating temperature and increases the sensor response. After determining the optimum temperature, rest of the measurements were performed at 200 ^{0}C for sample made from Dy-doped NiO and results are shown in Figure 7. The response-recovery transients at different concentrations ranging from 10-600 ppm of xylene are displayed in Figure 7(a). In (b) sensor response varies linearly with increase in xylene concentration. From the linear fit, one can determine the limit of detection (LOD) as below:

$$\text{LOD} = 3\frac{\sigma}{s} \tag{3}$$

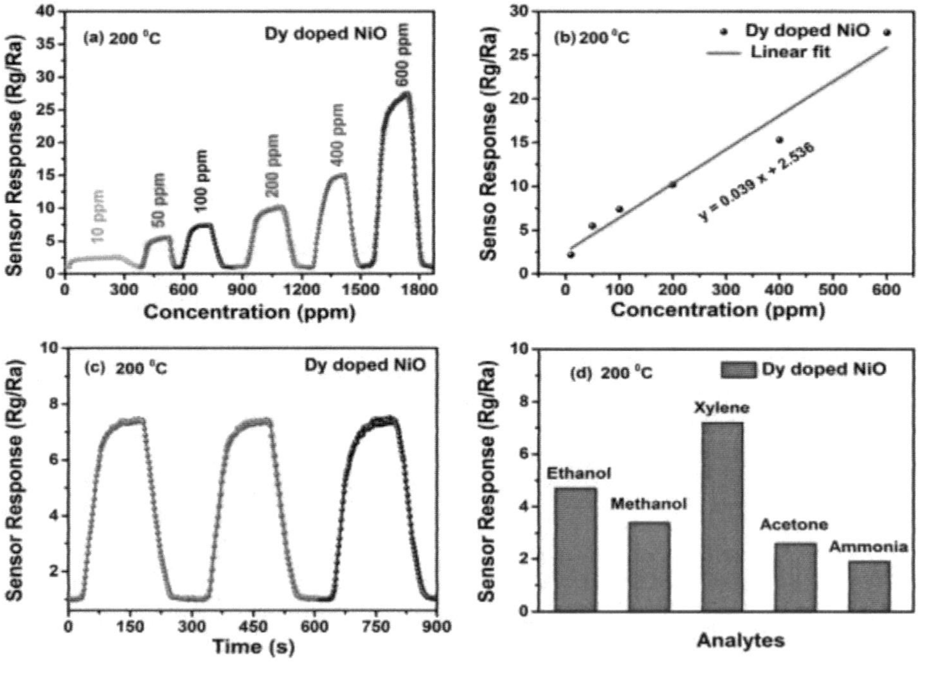

Figure 7. (a) Variation of sensor response with concentrations of xylene in the range 10 to 600 ppm at 200 ^{0}C (b) Linear fitting of concentration (ppm) of xylene vs corresponding sensor response (c) Reproducibility test of Dy-doped sensor when exposed to 100 ppm of xylene at 200 ^{0}C and (d) Selectivity of the sensor device for 100 ppm of various analytes at 200 ^{0}C.

where σ stands for standard deviation and s is the slope of response vs concentration graph [9]. The LOD has been found to be 6 ppm. Repeatability and selectivity are the key parameters of a sensing device and they play the important roles for the use of sensing devices in the practical applications. Figure 7 (c) shows the repeatability test of sensor for three successive cycles of 100 ppm of xylene at 200 ^{0}C. It indicates that the sensor device shows good reversible property. Figure 7 (d) shows selectivity behaviour of Dy-doped NiO sensor when exposed to different analytes. Data reveals preferential detection of xylene over other gas molecules. The basic mechanism of gas sensing for metal oxide based sensors has been understood in terms of oxygen adsorption and hole accumulation layer formation. It involves the interaction between adsorbed oxygen molecules and test gas molecules which leads to an increase in the hole concentration on the surface of metal oxide. Oxygen adsorbs in various ionic forms such as O^{2-}, O^-, O_2^- by taking electron from NiO to form a hole accumulation layer close to surface. When xylene reacts with adsorbed oxygen ions, these oxygen ions returns captured electrons back to NiO which further decreases the hole accumulation layer.

Nickel Oxide works at high operating temperature and it has slow response and recovery time for xylene molecules. For instance, Feng et al. [7] have shown response (~178 s) and recovery times (~152 s) for 200 ppm xylene at 375 ^{0}C using W-doped NiO nanotubes. Gao et al. [2] prepared tin doped nickel oxide nanostructure and have shown response time of ~298 s and recovery time of ~223 s, respectively, for 100 ppm xylene at 225 ^{0}C. Gao et al. [15] fabricated a $NiO/NiCr_2O_4$ composite having response (~1217 s) and recovery times (~591 s) for 100 ppm at 225 ^{0}C. Similarly, Qiu et. al [16] also prepared a composite of NiO/ZrO_2 and have shown response (~240 s) and recovery times (~98 s) for 100 ppm at 330 ^{0}C. A complete description of sensing parameters for doped and composite based xylene sensors are compared with present work and these are displayed in Table 1. As clear from Table 1, Dy-doped NiO nanoparticles show sensor response less than the reported values at 100 ppm but improved response (recovery) time of 50s(36 s) for 100 ppm of xylene at 200 ^{0}C. Our reported values of operating temperature is less than the already reported operating temperatures (225 to 375 ^{o}C) of nickel oxide based xylene sensors. To sum up, our two terminals Dy-doped NiO xylene sensor device has better response and recovery times and it works at low operating temperature as compared to the already reported sensors in the Table 1.

Table 1. Comparative table of our " Dy-doped NiO nanoparticles xylene sensor" with reported NiO based xylene sensors in the literature.

Sample	Sensor Response	Operating Temp. (^{o}C)	Concentration (ppm)	Response Time (s)	Recovery Time (s)	References
W-doped NiO nanotubes	8.5	375	200	178	152	[7]
Sn-doped NiO hierarchical nanostructure	20.2	225	100	298	223	[2]
$NiO-ZrO_2$ nanocomposite	25.6	330	100	240	98	[15]
$NiO/NiCr_2O_4$ nanoparticles	66.2	225	100	1217	591	[16]
Dy-doped NiO nanoparticles	7.4	200	100	50	36	Present Work

4. Conclusions

The Samples were successfully synthesized via co-precipitation method. XRD data reveals that dysprosium as a dopant produces tensile strain which further leads to formation of defects in the host lattice of nickel oxide. Shift of 1LO mode of Raman peaks confirms the presence of large number of nickel vacancies in the doped nickel oxide as compared to nickel oxide. Dysprosium as a dopant provides several favourable factors such as large strain, defect states, small crystallite and grain size, deformed lattice structure etc. to improve the sensor response as well as lowering response and recovery times. The sensor has shown detection down to 10 ppm of xylene with LOD of 6 ppm. The response-recovery transients of Dy-doped NiO were examined at 200^0C. The data revealed linear variation of sensing response with xylene concentration in the range 10-600 ppm.

Acknowledgments
The authors acknowledge Central Instrumental Facility for granting access to characterization facilities such as XRD, SEM and Raman Spectrophotometer of Guru Nanak Dev University, Amritsar.

References

[1] Hempel A J, Kjaergaard K S, Molhave L and Hundnell H K 1999 *Arch. Environ. Health* **54** 416
[2] Gao H, Wei D, Lin P, Xie C, Liu C, Sun P, Shimanoe K, Yamazoe N, and Lu G 2019 *Sens. Actuators B Chem.* **253** 1152
[3] Gao H, Yu Q, Zhang S, Wang T, Sun P, Lu H, Liu F, Yan X, Liu F, Liang X, Gao Y and Lu G 2018 *Sens. Actuators B Chem.* **269** 210
[4] Li X, Tan J F, Hu Y E and Huang X T 2017 *Mater.Res. Express* **4** 045015
[5] Hastir A, Kohli N and Singh R C 2017 *J. Phys. Chem. Solids* **105** 23
[6] Shaikh F I, Chikhale L P, Nadargi D Y, Mulla I S and Suryavanshi S S 2018 *J. Electron. Mater.* **47** 3817
[7] Feng C, Wang C, Zhang H, Li X, Wang C, Cheng P, Ma J, Sun P, Gao Y, Zhang H, Sun Y, Zheng J and Lu G 2015 *Sens. Actuators B Chem.* **221** 1475
[8] Shailja, Singh K J and Singh R C 2021 *J. Mater. Sci.: Mater. Electron.* **32** 11274
[9] Singh S, Sharma S, Singh R C and Sharma S 2020 *Appl. Surf. Sci.* **532** 147373
[10]Goncalves N S , Carvalho J A , Lima Z M and Sasaki J M 2012 *Mater. Lett.***72** 36
[11]Anand K, Kaur J, Singh R C and Thangaraj R 2015 *Mater. Sci. Semicond. Process.* **39** 476
[12]Patel K N, Deshpande M P, Gujarati V P, Pandya S, Sathe V and Chaki S H 2018 *Mater. Res. Bull.* **106** 187
[13] T.T. Wang, Ma S Y, Cheng L, Luo J, Jiang X H, Jin W X 2015 *Sens. Actuators B Chem.* **216** 212
[14] Qiu L, Zhang S, Huang J, Wang C, Zhao R, Qu F, Wang P and Yang M 2020 *Sens. Actuators B Chem.* **308** 127520
[15] Qiu L, Wang P, Zhang S, Qu F, Guo H, Zhao R, Tiju T , Yao D and Yang M 2020 *Sens. Actuators B Chem.* **324** 1286
[16] Gao H, Guo J, Li Y, Xie C, Li X, Liu L, Chen Y, Sun P, Liu F, Yan X, Liu F and Lu G 2019 *Sens. Actuators B Chem.* **284** 305

Green synthesis of vanadium oxide nanoparticles for thin film based sensor application

T.R. Kishan Chand, H.M. Kalpana [1,*], H.S. Lalithamba [2]

[1]Department of Electronics and Instrumentation, Siddaganga Institute of Technology, Tumakuru 572103, Karnataka, India.

[2]Department of Chemistry, Siddaganga Institute of Technology, Tumakuru 572103, Karnataka, India.

*Corresponding author e-mail address: kalpanahm@sit.ac.in

Abstract: A high yield of single phase orthorhombic vanadium oxide (V_2O_5) nanoparticles with solution combustion method is obtained in a very quick succession at 500 °C, by the utilization of Butea monosperma seed extract, to study its suitability towards pressure sensors. The Structural, Transmission spectral, Morphological, Compositional and the Absorption spectral analysis of the deposited vanadium oxide nanoparticle samples are examined using X-Ray diffraction (XRD), Fourier transform infrared spectroscopy (FTIR), Scanning electron microscope (SEM), High resolution transmission electron microscope (HRTEM), Energy dispersive X-Ray analysis (EDX) and Ultraviolet–visible spectroscopy (UV) respectively. The results from the structural analysis denote the nanoparticles with the crystallite size of 4.76 Å, dislocation density of $441.35*10^{16}/m^2$ along with the micro strain of $76.0366*10^{-3}$. Infrared transmission spectral images clearly display three strong peaks at 1633.76 cm^{-1}, 1070.80 cm^{-1} and 537.25 cm^{-1}, stating the presence of a strong and stable vanadyl bonding structure. Images from its morphological analysis depict the nanoparticles are homogeneously porous without any cracks on its surface. The composition of the nanoparticles clearly displays the presence of vanadium and oxygen atoms in a strong bonding structure. Absorption spectral results showed that the obtained samples after synthesis are in nano size with narrow size distribution. The results obtained from the experimental analysis indicate a stable performance, good linearity and stability, indicating its suitability for pressure sensing application.

Key words: Green synthesis; Nanoparticles; V_2O_5; Thin films; Strain gauges.

1. Introduction

In the current trends, pressure is one of those physical quantities which often need to be quantified for many engineering applications such as aerospace industries, bridges in rail monitoring, underground cavities, torque sensors etc. [1]. It is usually defined as the amount of force acting on a surface, per unit area of the substrate or an object [2]. For decades the pressure sensors were usually developed using a strain sensing material attached onto a diaphragm surface with suitable adhesives, but due to their lack of reliability and its detachment from the surface over time, thin films are considered as a remarkable development in the field of strain gauge technology, offering a better alternative [3]. This can be justified from its special characteristics that include low hysteresis loss, good linearity and stability for a longer duration [4]. They also have the ability to maintain the stoichiometry of the target composition and relative simplicity of thickness control, with considerable cost reduction [5, 15].

Vanadium oxide is a promising material and has gained momentum in the field of modern technical and industrial applications because of its ease of synthesis, low cost ingredients with high specific capacity. The material also offers a good safety related properties, high energy density and abundant availability of resources at low price [6]. Out of its many variants amongst the oxide family, the vanadium in its pentoxide form has gained much attention for more than a decade because of its stability and good electrochemical characteristics [7]. They show good metal to semiconductor transition and imply an abrupt change in optical and electrical properties, because of which the material is being utilized for thermal sensing, switching, smart window, architecture, nanomedicine and automotive applications [8].

The nanoparticles find innumerable benefits in day to day life and they are considered as a special material with its miniature form, exclusive physio-chemical characteristics and electrical properties. They also possess desirable shape, size, dispersion and distribution for thin film based applications [9]. There are various techniques for synthesizing the vanadium oxides that include hydrothermal, sol-gel, spray drying, combustion, microwave irradiation and nanostructural assembling. However in this study, the combustion method is preferred compared to its counterparts, because of its less complex procedure and the ease at which the synthesis can be carried out.

The Butea monosperma plant and its seeds being utilized in the current study for green synthesis are depicted in the figure 1. It is normally found in the tropical and subtropical parts of the Indian subcontinent and is known as the flame of the forest or forest fire [10]. In West bengal this plant has lent its name to the town of Palashi. The development of the pressure sensor based on thin films requires the need of a certain type of material which exhibits long term stability, good sensitivity, linearity and a strong resistance against the atmospheric reactions that acts upon it [11, 16]. The material commonly being utilized in the thin film strain gauges include alloys, cermets, metals, polymers and semiconductors [12, 17]. The current research involving an inorganic compound nanomaterial as the strain gauge is a new adventure in meeting the latest trends. From the extensive literature survey, it was noticed that the exploration of the sensor application based on the V_2O_5 nanoparticles are limited [13], this work based on the development of pressure sensors offer a great deal of opportunity in exploring its potential.

Figure 1. (a) Butea Monosperma (BMS) plant. **(b)** Its seeds.

2. Experimentation

2.1 Green synthesis of V_2O_5 nanoparticles using butea monosperma seed extract

The synthesis of vanadium oxide nanoparticles are carried out using combustion, as it offers a simple operating procedure with an excellent yield of product in a very quick succession [14], The materials required for the synthesis are purchased from Vasa scientific company and the Butea monosperma seeds are collected from the Bannerughatta forest, Bengaluru. The seeds are crushed with a mixer grinder to obtain its fine powder form and then transferred onto a beaker containing distilled water. A mixture of 15 g ammonium metavanadate (NH_4VO_3) is dissolved inside the beaker containing 90 ml of monosperma seed extract and sonicated for about half an hour. Then the solution containing the mixture is transferred onto the crucibles for combustion in the muffle furnace for 8 hours at 500 °C to obtain the foam like extract. These samples are crushed with a mortar pestle to bring it down into a small particulate structure, which is subjected for calcinations at 520 °C to obtain the fine vanadium oxide nanoparticles.

3. Results and Discussions

The obtained vanadium oxide nanoparticle samples are subjected to various characterization techniques, to check its suitability for pressure sensing application, which are being discussed in the following section.

3.1 Structural Analysis

The results from the structural analysis of the synthesized nanoparticles obtained from the XRD are displayed in the figure 2. The image indicates strong diffraction peaks at (0 2 0) (0 0 1) (0 1 1) (1 1 0) (0 4 0) (1 0 1) (1 3 0) (0 1 2) (0 6 0) (2 0 0) (2 4 0) (2 3 2) (1 3 3) (2 6 1) and so on, representing the orthorhombic crystalline phase of vanadium oxide nanoparticles. The crystallite size (D) of 4.76 Å, dislocation density (δ) of $441.35*10^{16}/m^2$ and the micro strain (ε) of $76.0366*10^{-3}$ determined from the data, in comparison to that of the existing protocols, presented an optimized process parameters indicating its suitability towards sensor development. They are also tabulated in the Table 1. The observed strong peaks at two location from the results at (0 0 1) and (1 1 0) indicate the absence of any organic impurities [18, 19]. The other small narrow and intense peaks denote the synthesized nanoparticles are highly crystalline, which contributes to the increased sensitivity of the pressure sensor.

Table 1. Crystallite size, Dislocation density and Micro strain of the synthesized V_2O_5 nanoparticles.

Sample	Crystallite Size (D)	Dislocation Density (δ)	Micro Strain (ε)
V_2O_5 (BMS seeds)	4.76 Å	$441.35*10^{16}/m^2$	$76.0366*10^{-3}$
V_2O_5 (Moringa leaf)	3.18 Å	$986.21*10^{16}/m^2$	$113.69*10^{-3}$
V_2O_5 (Sutera cordata)	2.27 Å	$1343.72*10^{16}/m^2$	$159.29*10^{-3}$

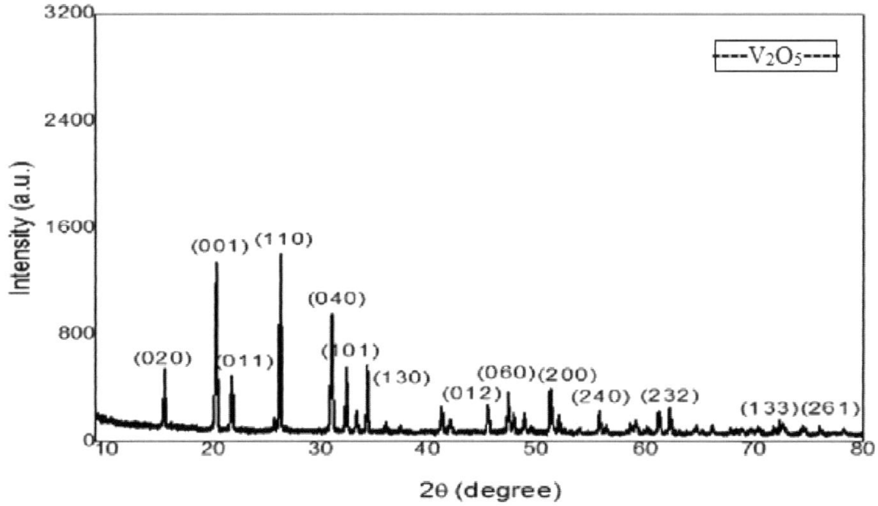

Figure 2. XRD spectrum of the synthesized V_2O_5 nanoparticles.

3.2 Transmission spectral Analysis

The FTIR spectrum is depicted in the Figure 3. It is used to obtain the infrared transmission spectrum, chemical bonds and the functional group of the vanadium oxide nanoparticles, they were analyzed in the range of 400-4000 cm^{-1} wave number (Bruker instrument is used in this analysis). The peaks observed from the results indicate three characteristic vibration modes. The one at 1070.80 denotes V=O vanadyl stretching mode, 537.25 denotes V-O-V symmetric stretching, 1633.76 denotes V-O-V asymmetric stretching around the bonding structure [20]. The results also denote the orthorhombic crystalline structure of the vanadium oxide nanoparticles, which contributes to the better performance of the film if being utilized for the strain gauge application.

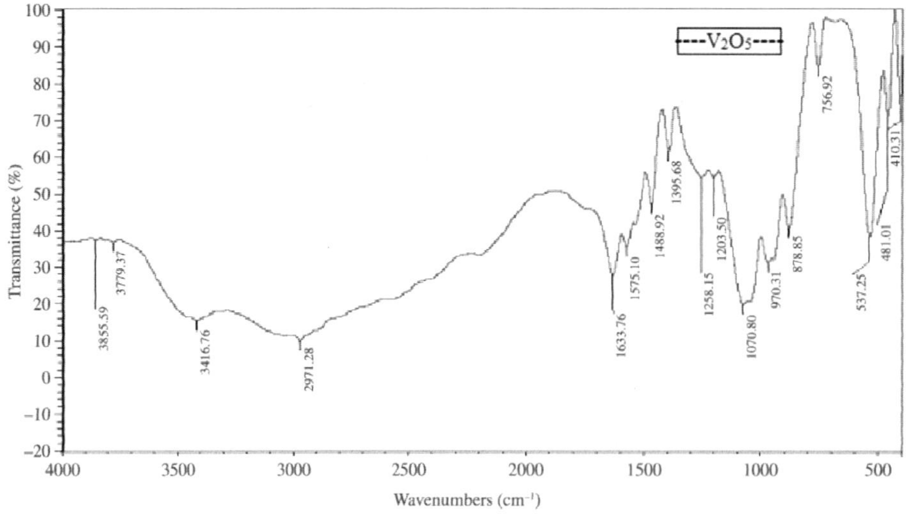

Figure 3. FTIR spectra of the synthesized V_2O_5 nanoparticles.

3.3 Morphological studies

The SEM images are depicted in the Figure 4. They convey the rod like morphological structure of the calcinated samples with a strong vanadium oxide bonds in the size range between 0.2-10 µm, when operated at 10 kV. The structures of the calcinated nanoparticles are homogenously porous without any crack on its surface. The smallest diameter of which was observed to be of 10 nm.

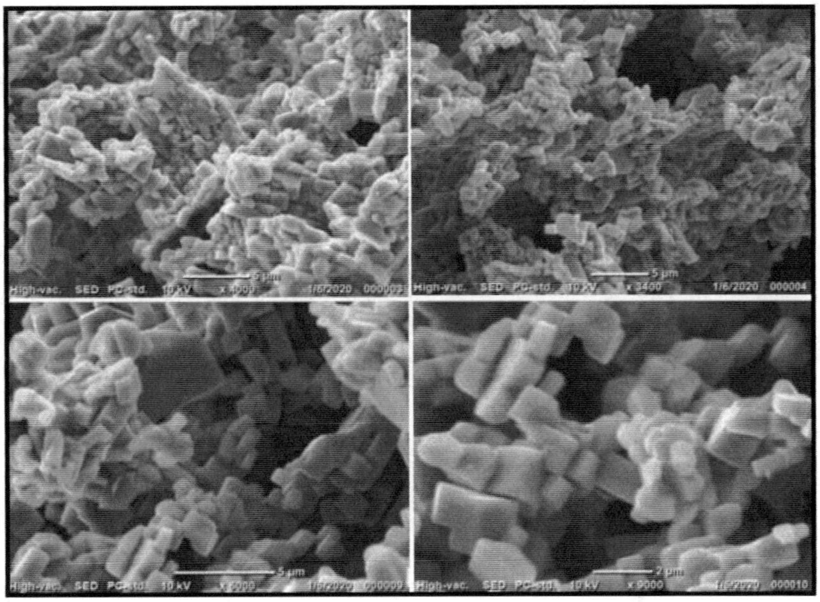

Figure 4. SEM images of the synthesized V_2O_5 nanoparticles.

The HRTEM images of nano V_2O_5 is presented in the Figure 5. The actual size of the particle, their grain growth and the distribution of the crystallites can be determined out of it. The results obtained indicate a sheet like structure of the given vanadium oxide samples. The rod shaped image in the figure denotes the presence of carbon from the precursor and the black spots in it indicates the vanadium oxide bonding structure [21]. The crystalline nature of the samples can be recognized from the given images and the presence of lattice planes in the size ranges between 2-5 nm is clearly visible from it. The distance between the two planes, which are obtained to be of 0.563 nm, show a clear indication of high crystallite size and the lattice constant of the vanadium oxide nanoparticles, which is necessary for pressure sensing application [22]. There is an observed uniformity of the crystallite structure in the size range of 5-10 nm.

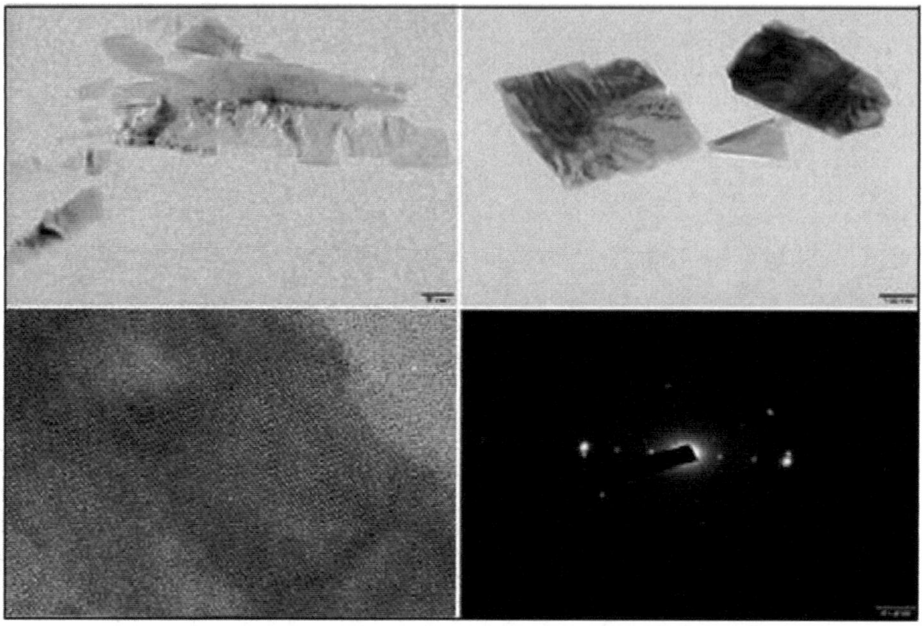

Figure 5. HRTEM images of the synthesized V_2O_5 nanoparticles.

3.4 Compositional analysis

EDX operated at 10 kV is used for the compositional analysis of the given vanadium oxide nanoparticles. The results are displayed in the Figure 6. It confirms the presence of vanadium oxide along with some traces of aluminum and carbon which is from the stub/sample holder, which is not to be quantified. From the results obtained, it was noted that the atomic mass percentage of vanadium and oxygen are 56.01 and 43.98, the weight percentage of the same are noted to be of 50.94 and 15.99 respectively. The results are tabulated in the Table 2.

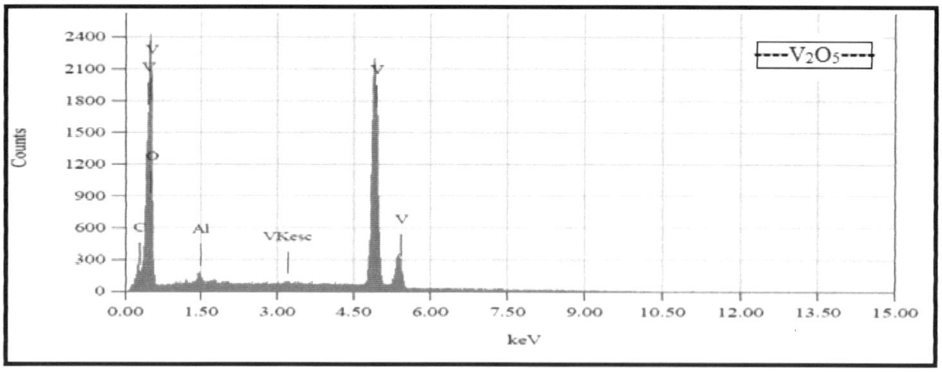

Figure 6. EDX spectrum of the synthesized V_2O_5 nanoparticles.

IOP Conf. Series: Materials Science and Engineering 1225 (2022) 012062 doi:10.1088/1757-899X/1225/1/012062

Table 2. Atomic mass and Atomic Weight Percentages of the synthesized V_2O_5 nanoparticles.

Element	Atomic Mass %	Atomic Weight %
Vanadium (V_2)	56.01	50.94
Oxide (O_5)	43.98	15.99

3.5 Absorption spectral analysis

The absorption spectrum of the V_2O_5 nanoparticles is presented in the Figure 7. It is conducted to find the optical band gap energy level, which is calculated by the relation 1242/λ and their band gap energy was found that to be of 2.5 eV. The spectral width of the absorption spectrum was attributable to the size distribution of the particles with diverse sizes. The absorbance of the samples decreases slightly with increase in annealing. The sharp peaks show that the particles are in nano size with narrow size distribution. The absorption peaks of V_2O_5 shows a blue shift in comparison with that of the bulk V_2O_5, which is attributed to the quantum confinement effect happened during the electronic transition from the occupied 2p bands of oxygen to the unoccupied 3d bands of V_2O_5. A red shift is observed in the characteristic absorption peak due to the decrease in the band gap of the samples [23].

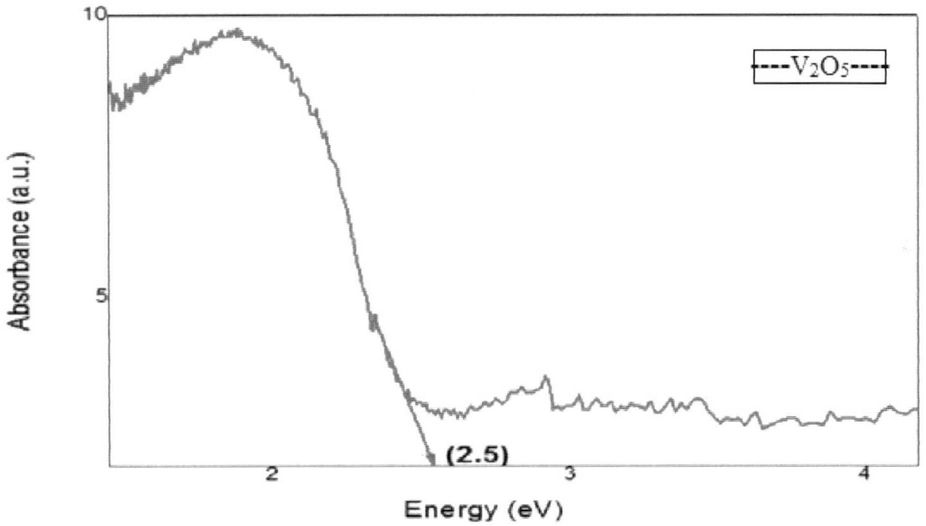

Figure 7. UV spectroscopic data of the synthesized V_2O_5 nanopaticles.

4. Conclusions

Hence by this experimental investigation, it can be concluded that the method of synthesis used in this correspondence gives a better yield of the vanadium oxide samples, whose characterization results clearly indicate a strong vanadium oxide nanoparticles without any organic impurities. The calcinations of the synthesized nanoparticles resulted in good crystalline structure. The adequate crystallite size of 4.76 Å observed from the structural analysis, provides good response to the smallest of change in pressure applied on the surface of the film. The reduced dislocation density $441.35*10^{16}/m^2$ along with the micro strain of $76.0366*10^{-3}$, determined from the diffraction results helps in the increased sensitivity of the pressure sensors. The orthorhombic crystalline structure of the nanoparticles, observed from the infrared transmission spectrum, with the presence of three characteristic stretching modes, contributes to the better performance of the strain gauges. The rod like morphology with the absence of any cracks on its surface, serves to improve the durability of the sensors. Hence from the results obtained, it can be concluded that the V_2O_5 nanomaterial based thin films offer a better alternative for pressure sensor development in overcoming some of the drawbacks associated with the existing pressure sensors.

Acknowledgements

The authors are grateful to, "Siddaganga institute of technology" Tumakuru, for supporting this research work. They thank, "Centre for applied research and nano technology" Tumakuru, and "JNCASR" Bengaluru in providing lab facilities for characterization.

References

[1] Kalpana H M and Prasad V S 2014 Development of the invar36 thin film strain gauge sensor for strain measurement *Meas. Sci. Technol.* **25** 065102

[2] Yuan Y Li M and Xu J 2011 A pressure sensor study and research *IEEE 2nd Int. Conf. on Computing, Control and Industrial Engineering, Wuhan, China* **11** 255-58

[3] Mihara Y and Someya T 2002 A study on the development of a thin-film pressure sensor *IEEE Sensors* **2** 954-59

[4] Kalpana H M, Prasad V S and Nayak M M 2013 Influence of annealing and thickness on the electrical properties of invar36 thin film for strain gauge applications *IJTFST* **2** 155

[5] Witt G R 1974 *Thin Solid Films* **22** 133-56

[6] Liang X, Gao G, Liu Y, Zhang T and Wu G 2017 Synthesis and characterization of Fe-doped vanadium oxide nanorods and their electrochemical performance *J. Alloys Compd.* **715** 374-83

[7] Liu X, Zeng J, Yang H, Zhou K and Pan D 2018 V_2O_5-based nanomaterials: synthesis and their applications *RSC Adv.* **8** 4014-31

[8] Zeng H, Liu D, Zhang Y, See K A, Jun YS, Wu G, Gerbec J A, Ji X and Stucky G D 2015 Nanostructured Mn-doped V2O5 cathode material fabricated from layered vanadium jarosite *Chem. Mater.* **27** 7331-36

[9] Raghavendra M, Yatish K V and Lalithamba H S 2017 Plant-mediated green synthesis of ZnO nanoparticles using Garcinia gummi-gutta seed extract: Photoluminescence, screening of their catalytic activity in antioxidant, formylation and biodiesel production *Eur. Phys. J. Plus* **132** 1-12

[10] Jhade D, Ahirwar D, Sharma N K, Jain R and Gupta S 2009 Butea monosperma (Lam.) taubert: a review *J. Pharm. Res.* **2** 1181-83.

[11] Rajanna K, Muralidhar G K, Nair K G M, Panchapagesan T, Nayak M M, and Mohan S 2014 Effect of strain and temperature on the behavior of oxygen ion implanted manganese films *J. Appl. Phys.* **76** 3573-78

[12] Almassri A M, Wan Hasan W Z, Ahmad S A, Ishak A J, Ghazali A M, Talib D N and Wada C 2015. Pressure sensor: state of the art, design, and application for robotic hand *J. Sens.*

[13] Ramasami A K, Reddy MV, Nithyadharseni P, Chowdari B V R and Balakrishna G R 2017 Gel-combustion synthesized vanadium pentoxide nanowire clusters for rechargeable lithium batteries *J. Alloys Compd.* **695** 850-58

[14] Jagadeesh A, Rattan T M, Muralikrishna M and Venkataramaniah K 2014. Instant one step synthesis of crystalline nano V2O5 by solution combustion method showing enhanced negative temperature coefficient of resistance *Mat. Lett.* **121** 133-36

[15] Kalpana H M, Prasad V S and Satish T N 2019. Effect of Annealing on Hardness and Elastic Modulus of Invar36 Thin Films Deposited by Direct Current Sputtering for Strain Gauge

Applications *IJTFST* **8** 4

[16] Rakshit A, Biswas D and Chakraborty S 2018 Deposition and characterization of vanadium oxide based thin films for MOS device applications *In AIP Conf. Proceed.* **1942** 120024

[17] Jin Y, Basantani H A, Ozcelik A, Jackson T N and Horn M W 2013 High-resistivity and high-TCR vanadium oxide thin films for infrared imaging prepared by bias target ion-beam deposition. *In Infrared Technology and Applications* **8704** 87043C

[18] Chan Y L, Pung S Y and Sreekantan S 2014 Synthesis of V2O5 nanoflakes on PET fiber as visible-light-driven photocatalysts for degradation of RhB dye *J. Catal.*

[19] Mihara Y and Someya T 2002 A study on the development of a thin-film pressure sensor *IEEE Sensors* **2** 954-59

[20] Zhu K, Meng Y, Qiu H, Gao Y, Wang C, Du F, Wei Y and Chen G 2015 Facile synthesis of V2O5 nanoparticles as a capable cathode for high energy lithium-ion batteries *J. Alloys Compd.* **650** 370-73.

[21] Nalini S, Selvakumar B and Periasamy P J I J E M S 2017 Simple synthesis and characterization of V2O5 nanoparticles by microwave assisted wet chemical method *Int. J. Eng. Manuf. Sci* **7** 411-17

[22] Farahmandjou M and Abaeiyan N 2017 Chemical synthesis of vanadium oxide (V2O5) nanoparticles prepared by sodium metavanadate *J. Nanomedicine Res.* **5** 00103

[23] Derkaoui I, Khenfouch M, Elmokri I, Mothudi B M, Dhlamini M S, Moloi S J, Zorkani I, Jorio A and Maaza M 2017 Structural and optical properties of hydrothermally synthesized vanadium oxides nanobelts *IOP Conf. Ser.: Mater. Sci. Eng.* **186** 012007

Electrochemical Corrosion Behaviour Analysis of Mg-Alloys Used for Orthopaedics and Vascular Implants

Rohit Kumar, Puneet Katyal, Munish Gupta and Vijender Singh

Department of Mechanical Engineering
Guru Jambheswar University of Science and Technology, Hisar
Haryana, INDIA – 125001

* *Corresponding author e-mail*: katyalgju@gmail.com

Abstract

Mg-alloys having bone liked mechanical properties are biodegradable, biocompatible and osteoconductive metallic materials are potential candidates used for orthopaedics and vascular implants. Therefore, present experimentation is an effort to analyse the corrosion behaviour of Mg based alloys such as AZ81 and ZM21 for their usage as bio-degradable implant materials. The corrosion behaviour is analysed using the electrochemical workstation. SEM and EDS are used for high-resolution images and composition of magnesium-based alloys. Using Potentiodynamic polarization curves and Nyquist plots, corrosion rates were observed for 3.0 hours studies on ZM21(193.53 mm/year) and AZ81 (24.22 mm/year) for Mg-alloys. AZ81 shows higher corrosion resistance than ZM21. The results of these experimental findings may be helpful for the designers and researchers in selecting and improving the clinical performance of Mg-based implants for biomedical applications.

Keywords: Mg-alloys, Biomaterials, Corrosion rate, SEM, EDS

1. Introduction

Magnesium-alloys; being non-toxic, completely degrades in human body environment, having similar mechanical properties to that of bone have good potential for being used in orthopaedics and vascular implants. These implants should have good biocompatibility, high wear and corrosion resistance and should have degradation rate matched with the bone healing rate [1]. Magnesium, being enzymes cofactor stabilizes the structures of RNA and DNA, is very essential for human metabolism [3-4]. Magnesium and its alloys being lightest as compared with all other metals, having higher specific strength and low elastic modulus (45GPa) nearly same as the natural bone (10-30 GPa) makes them easy to be machined [5] and eliminates the impact of stress shielding effect to a greater extent [6].

For the applications of orthopaedic, the mechanical integrity as well as corrosion rate or degradation rate of biomedical implant plays an important role. An eligible biodegradable implant is one whose rate of corrosion matched with the rate of healing of bone, properties is self enough to supply the expected endorsement during the period of bone healing, completely degrades inside the corpus when the objective bone healed totally [7]. For the implantation of magnesium in the human and animal body, many types of biodegradable Mg based alloys are being developed, but approximately all suffering with fast degradation rate and mechanical support is also relatively insufficient [3,8].

Despite several challenges, Mg-alloy implant clinical performance can be improved by optimising implant design, mechanical functioning, surface treatments, and compositions. Surface changes and their appropriate alloying combinations can be used to achieve regulated deterioration rates. As a result, many alloying options for Mg-based alloys have been considered.

Due to continuous efforts of clinical researchers on Mg based alloys for biomedical applications, significant success has been achieved in vitro and in vivo work, such as Mg–Nd–Zn–Zr [9], Mg–Ca [10,11], Mg–Zn–Ca [12,13] and Mg–Mn–Zn [14–16].

In command to improve the mechanical integrity and controlled degradation rate, various alloying options for magnesium-based alloys have been investigated [17-21]. The selection of alloying elements, should greatly improve the corrosion resistance property, refine the grain structure, forms phase transitions between metals, thus magnify their power and aiding them in their manufacturing [9].

Alloying elements like Zinc (Zn), Manganese (Mn), Calcium (Ca), Strontium (Sr) and Zirconium (Zr), already part of human body showed minimum adverse effects. Therefore, alloying possibilities, which can improve the Mg-alloys performance, are discussed herein.

Zn is a crucial nutrition tenor, which is present in body of all human beings. Zn with Mg is the best alloy combination for biomedical applications. With the adding of Zinc in Mg, the mechanical and corrosion resistance property also improves [20, 21]. One of the most important effects of Zn is in deficiency of H_2 growth [18, 22]. In a binary Mg - alloy, the addition of Zn (maximum 3%) helps in reducing grain size and improving their mechanical strength [17]. It is reported by Zhang et al. [23] that the adding of Zinc content in ternary Mg–Zn–Ca alloy enhances the tensile strength. The elongation, tensile strength and bio corrosion resistance were reported improved by adding of Zinc to Mg [24]. The ultimate tensile strength in Mg-alloy is amplified with 22% with the adding of Zn [25]. With the adding of Zn in Mg alloys, cleavage fracture changed to quasi-cleavage fracture. Moreover, as percentage of Zn content increased up to 3 wt%, Zn particles affects the crack initiation sites [26]. For Mg based alloys, Zn concentration of less than 3% is preferred as beyond 3% Mg-alloys elongation decreases significantly.

Manganese, along with magnesium used as a ternary alloy, is another essential element [27, 28], which improves their corrosion resistive property without disturbing their mechanical properties [14, 29, 30]. Mg–Mn alloy shows good anti-corrosion property with the addition of 1% zinc [20]. Furthermore, its addition in the Mg - alloy controls the impurities harmful effects [18]. For Mg–Zn–Mn based alloys, Rosalbino et al. [31] observed a fourfold growth in resistance of corrosion in the Ringer physiological solution while Zhang et al. [24] concluded about their high mechanical strengths. In addition, Zhang et. al [32] in their in vivo perusal observed that later eighteen weeks, magnesium degradation in Mg–1·2Mn–1·0Zn alloy implant didn't origin any harmful effects on the kidney.

Zr is one of the powerful grain refiners for the alloys of magnesium [33]. In vitro and in vivo study, Zr shows good biocompatibility and is mostly suitable in ternary Mg-alloys [20, 34]. Its addition in Mg based alloys helps in improving their mechanical strength, reduced degradation rate (by 50%) and resulted in damping stresses and vibrations at the implant/bone interface. For biodegradable Mg-alloys applications, its alloying content should be less than 0.8 wt %.

Thus, we can conclude that Mg-based alloys performance can be enhanced through controlled alloying composition. Therefore, for the corrosion analysis, magnesium based alloys such as AZ81 and ZM21 have been selected in the present research work for their use as biodegradable implant materials.

2. Experimental Setup

2.1. *Material preparation*

In the present investigation, Mg-alloys rolled plates are used for the experimental work. The micro structure and chemical composition of work materials are shown in figure 1 & 2. The specimens were prepared on wire-cut EDM (Sprincut-Electronica Ltd., India) and sample size of 20 mm * 20 mm * 4.0 mm of Mg alloys was used in this experimental work. For polishing of WEDM machined surface, grit paper of various grades such as P500, P600, P1000, P1200, P1500, P2000 etc. have been used. These prepared samples are clean-ups ultrasonically with acetone and purified water, and then desiccated in the open air. SEM and EDS analysis are also being used for high-resolution image of the inspected area and to find out the exact chemical composition of the material sample.

IOP Conf. Series: Materials Science and Engineering 1225 (2022) 012063 doi:10.1088/1757-899X/1225/1/012063

2.2. *Electrochemical workstation (Electrochemical corrosion study)*

In the present experimental work, electrochemical work station (Autolab PGSTAT-302N, Metrohm, Switzerland) is used for measurements of electrochemical corrosion. Three electrode configuration: graphite electrode served as a counter and Ag/AgCl (1 M KCL) served as reference electrode and prepared sample with a surface area of $1cm^2$ served as a working electrode. Simulated body fluid (SBF) at 37° C is used an electrolyte in the present investigation. The SBF was prepared according to Kokubo et al.[36]. It was buffered at 7.4 potential of Hydrogen with 0.1 M HCl and tris-hydroxymethyl aminomethane. For stabilizing the OCP, prepared samples were dipped in SBF at 37°C for 30 minutes. After stable OCP, potentiodynamic polarization measurements were taken out within the range of −100 mV to +100 mV for the 3-hour standard testing period.

3. Results and Discussions

3.1. *Elemental analysis*

The high resolution images of the surfaces of Mg-alloys were taken by the SEM technic and the configuration of the Mg-alloys was measured through EDS technic. The SEM micrographics of the polished AZ81 Mg-alloy and EDS spectrum images along with composition table are shown through figures 1(A) & 1 (B) respectivily. Simmilarly, figures 2 (A) & 2 (B) corresponds to ZM21 Mg-alloy.

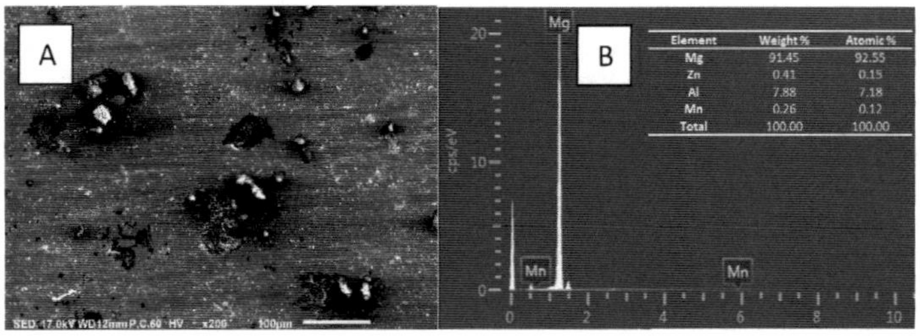

Figure 1. (A) SEM magnificence and (B) EDS spectrums along with composition table of AZ81 Mg-alloy

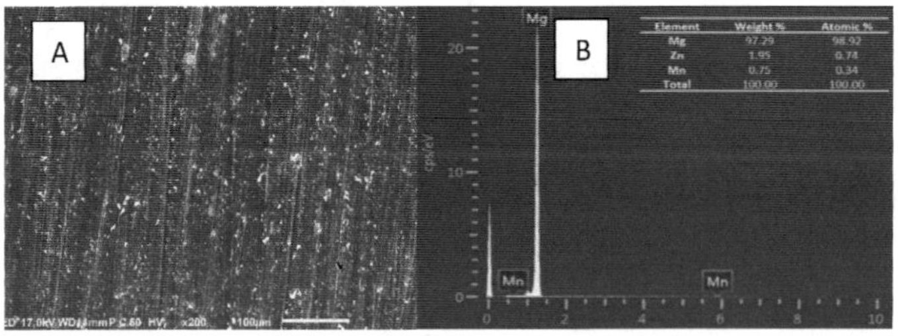

Figure 2. (A) SEM magnificence and (B) EDS spectrums along with composition table of ZM21 Mg-alloy

3.2 Corrosion Analysis

The rate of corrosion of the Mg based alloys was measured by electrochemical workstation. The corrosion potential (E_{corr}) for ZM21 and AZ81 Mg-alloys are -1.6735 V and -1.4988 V respectively tabulated in Table 1. Corrosion potential of ZM21 Mg-alloy is greater than the corrosion potential of AZ81 Mg alloy. In continuation, the current density (I_{corr}) for ZM21 and AZ81 Mg-alloys are 0.00847 A/cm^2 and 0.00106 A/cm^2 respectively. Corrosion current of ZM21 magnesium is higher than the corrosion current of AZ81 Mg-alloy.

Table 1. Electrochemical parameters observed during corrosion studies (test) on electrochemical workstation

Mg –alloys	ZM21 Mg-alloy	AZ81 Mg-alloy
E_{corr} (V)	-1.6735	-1.4988
I_{corr} (A/cm^2)	0.00847	0.00106
Corrosion Rate (mm/year)	193.53	24.22
Polarisation resistance (Ω)	108.64	378.21

Figures 3(A) and 3(B) shows the polished samples before and after the corrosion test. The circular signs observed on the materials surface in figure 3(B) are the cross sectional areas on which solution strikes during a corrosion study on the electrochemical workstation.

The corrosion rates were observed for 3.0 hours studies on ZM21(193.53 mm/year) and AZ81 (24.22 mm/year) for Mg-alloys. Cleraly, AZ81 shows higher corrosion resistance than ZM21. The potentiodynamic polarisation (PDP) curves and Nyquist plots for 3 hours studies on ZM21 and AZ81 are shown through figures 4 and 5 respectively. After 3 hours of immersion, the E_{corr} for ZM21 rise to -1.61 V, whereas this rise is less in AZ81 Mg alloy with value of -1.49 V w.r.t to open circuit potential value. The simultaneous drop in I_{corr} suggest that formation of oxide layers by the presence of aluminium successfully resist the interaction of aggressive ions with Mg matrix. EIS study have been performed to get detailed insight of corrosion phenomena. The Nyquist plot of ZM21 figure 4(b) comprised only two time constants, firstly a capacitive loop of elevated frequency and second one an inductive loop of low frequency. The occurrence of inductive loop signify excessive pitting corrosion due to poor corrosion resistance. In contrast, AZ81 figure 5(b) comprised three time constants, two capacitive loops of elevated frequency and average frequency and a small inductive loop of low frequency which signify the formation of protective oxide layer between corrosive media and substrate due to the presence of aluminium in alloy. Therefore, AZ81 shows higher corrosion resistance than ZM21.

IOP Conf. Series: Materials Science and Engineering 1225 (2022) 012063 doi:10.1088/1757-899X/1225/1/012063

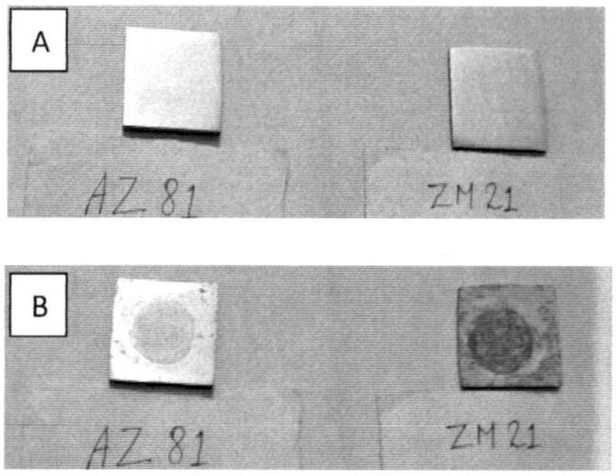

Figure 3: (A) polished samples of Mg-alloys and (B) polished samples of Mg-alloys after corrosion study

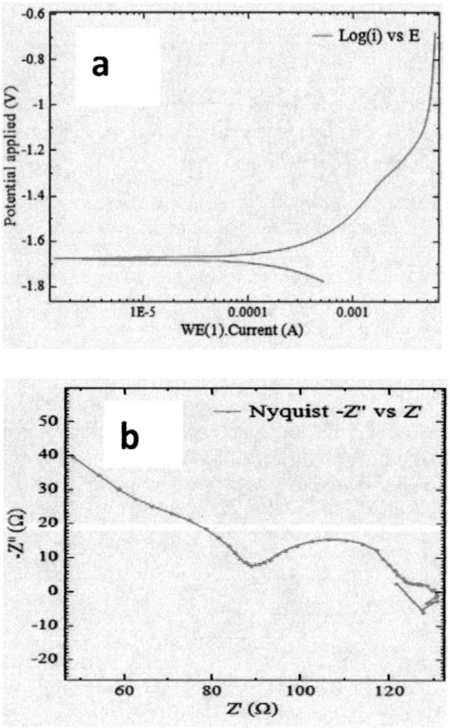

Figure 4. PDP curve and Nyquist plot for ZM21 Mg-alloy

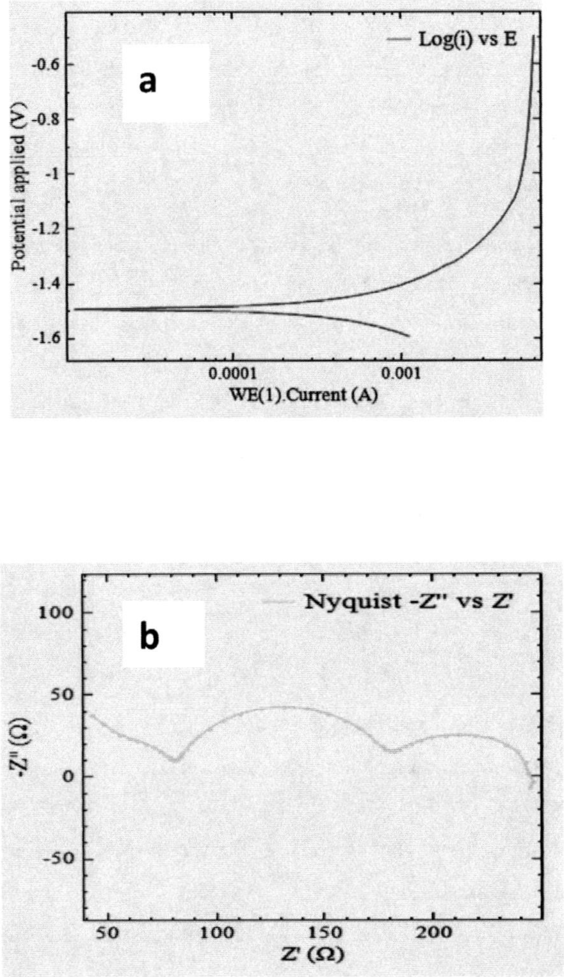

Figure 5. PDP curve and Nyquist plot for AZ81Mg-alloy

4. Conclusions

In the present analysis, the corrosion behaviour of Mg-alloys is analysed using electrochemical workstation. PDP curves and Nyquist plots are used to analyse corrosion rates. SEM and EDS are used for high-resolution images and composition of magnesium-based alloys. The following conclusions drawn from the present research work are

- In the present study, corrosion rates were observed using 3.0 hours studies on ZM21 (193.53 mm/year) and AZ81 (24.22 mm/year) for Mg-alloys.
- In AZ81 Mg-alloy, the simultaneous drop in I_{corr} suggests that formation of oxide layers by the presence of aluminium successfully resist the interaction of aggressive ions with Mg matrix. EIS study have been performed to get detailed insight of corrosion phenomena. Nyquist plots signify the formation of protective oxide layer between corrosive media and substrate due to the presence of aluminium in AZ81 Mg-alloy. Thus, AZ81 shows higher corrosion resistance than ZM21.
- These results may be helpful for the researchers in improving the performance and selection of Mg-alloys implants for biomedical applications.

References

[1]. Hench LL and Polak J M 2002 Third-generation bio materials *Sci.* **295** 1014–17

[2]. Chen Q and Thouas G A 2015, Metallic implant biomaterials *Mater. Sci. Eng.: R: Rep* **87** 1–57.

[3]. Staiger M P, Pietak A M, Huadmai J and Dias G 2006 Magnesium and its alloys as orthopedic biomaterials: A review *Biomaterials* **27** 1728–34.

[4]. Saris N E L, Mervaala E, Karppanen H, Khawaja J A, and Lewenstam A 2000 Magnesium *Clin Chim Acta* **294** 1–26.

[5]. Witte F, Kaese V, Haferkamp H, Switzer E, Meyer-Lindenberg A, Wirth C J and Windhagen H 2005. In vivo corrosion of four magnesium alloys and the associated bone response *Biomaterials* **26** 3557-63

[6]. Gu X N and Zheng Y F 2010 A review on magnesium alloys as biodegradable materials *Front. Mater. Sci. China* **4** 111–15.

[7]. Kumar K, Gill R S, and Batra U 2018 Challenges and opportunities for biodegradable magnesium alloy implants *Mater Technol* **33** 153–72.

[8]. Sunil BR, Ganesh KV 2016 Effect of aluminum content on machining characteristics of AZ31 and AZ91 magnesium alloys during drilling *J Magnes Alloys* **4** 15–21

[9]. Zhang X, Wang Z and Yuan G 2012 Improvement of mechanical properties and corrosion resistance of biodegradable Mg–Nd–Zn–Zr alloys by double extrusion *Mater Sci. Eng B Solid-State Mater Adv. Technol.* **177** 1113–19

[10]. Salahshoor M and Guo Y 2012, Biodegradable orthopedic magnesium-calcium (Mg–Ca) alloys, processing, and corrosion performance *Mat.* **5** 135–55

[11]. Rad H R B, Idris M H, Kadir M R A, 2012 Microstructure analysis and corrosion behavior of biodegradable Mg–Ca implant alloys *Mater. Des.* **33** 88–97

[12]. Sun Y, Zhang B and Wang Y 2012 Preparation and characterization of a new biomedical Mg–Zn–Ca alloy *Mater. Des.* **34** 58–64

[13]. Gao J H, Guan S K and Ren Z W 2011 Homogeneous corrosion of high-pressure torsion treated Mg–Zn–Ca alloy in simulated body fluid *Mater. Lett.* **65** 691–93

[14]. Zhang E L, Yin D S and Xu L P 2009 Microstructure, mechanical and corrosion properties and biocompatibility of Mg–Zn–Mn alloys for biomedical application, *Mater Sci Eng C.* **29** 987–93.

[15]. Yin D S, Zhang E L and Zeng S Y 2008 Effect of Zn on mechanical property and corrosion property of extruded Mg–Zn–Mn alloy *Trans Nonferrous Met Soc China* **18** 763–68

[16]. Yin D S, Zhang E L and Zeng S Y 2009 Study on the influential factors of the content determination of paeoniflorin in Radix *Paeoniae Alba. China Foundry* **6** 43–47

[17]. Chen Q and Thouas G A 2015 Metallic implant biomaterials *Mater Sci. Eng R Reports.* **87** 1–57

[18]. Eddy G, Poinern J and Brundavanam S 2012 Biomedical magnesium alloys: a review of material properties, surface modifications and potential as a biodegradable orthopaedic implant *Am J Biomed Eng.* **2** 218–40.

[19]. Zheng Y F, Gu X N and Witte F 2014 Biodegradable metals *Material Science Eng. R Reports* **77** 1–34.

[20]. Li N and Zheng Y 2013, Novel magnesium alloys developed for biomedical application: a review *J Mater Sci Technol.* **29** 489–502.

[21]. Lotfabadi AFand Bakhsheshi-Rad HR 2015, The role of solution heat treatment on corrosion and mechanical behaviour of Mg–Zn biodegradable alloys, *Mater. Perform.* **55** 53–64.

[22]. Persaud-Sharma D and McGoron A 2012 Biodegradable magnesium alloys: a review of material development and applications *J Biomim Biomater Tissue Eng.* **12** 25–39.

[23]. Zhang B, Hou Y and Wang X 2011, Mechanical properties, degradation performance and cytotoxicity of Mg–Zn–Ca biomedical alloys with different compositions. *Materials Science and Eng. C.* **31** 1667–73.

[24]. Zhang E and Yang L 2010 Microstructure, mechanical properties and bio-corrosion properties of Mg–Si(–Ca, Zn) alloy for biomedical application *Acta Biomater* 1756–62.

[25]. Du H, Wei Z and Liu X 2011 Effects of Zn on the microstructure, mechanical property and bio-corrosion property of Mg–3Ca alloys for biomedical application *Mater Chem Phys.* **125** 568–75.

[26]. Salga MS, Ali HM and Abdulla MA 2012, Gastroprotective activity and mechanism of novel dichlorido-zinc(II)-4-(2-(5-methoxybenzylideneamino)ethyl) piperazin-1-iumphenolate complex on ethanol-induced gastric ulceration *Chem Biol Interact.* **195** 144–53.

[27]. Bailey CS, Weiner JJ and Gibby M 2008 Excessive calcium ingestion leading to milk-alkali syndrome *Ann Clin Biochem.* **45** 527–29

[28]. Nasolodin VV 1985 Biological role of manganese and the prevention of its deficiency in the human body *Vopr Pitan.* **4** 3–6.

[29]. Song S, Song GL and Shen W 2012 Corrosion and electrochemical evaluation of coated magnesium alloys *Corr.* **68** 1-12.

[30]. Avedesian MM and Baker H 1999 *ASM specialty handbook, magnesium and magnesium alloys* Materials Park (OH): ASM International Handbook Committee.

[31]. Rosalbino F, De Negri S and Saccone A 2010 Bio-corrosion characterization of Mg–Zn–X (X = Ca, Mn, Si) alloys for biomedical applications *J Mater Sci Mater Med.* **21** 1091–98.

[32]. Xu LP, Yu GN and Zhang E 2007 In vivo corrosion behavior of Mg–Mn–Zn alloy for bone implant application *J Biomed Mater Res.* **83A** 703–11.

[33]. Wang C and Sun M. 2014 Improvement in grain refinement efficiency of Mg–Zr master alloy for magnesium alloy by friction stir processing *J Magnesium Alloys* 239–44.

[34]. Huan Z G and Leeflang M A 2010 In vitro degradation behaviour and cytocompatibility of Mg–Zn–Zr alloys *J Mater Sci: Mater Med.* **21** 2623–35.

[35]. Gu X, Zheng Y and Cheng Y 2009 In vitro corrosion and biocompatibility of binary magnesium alloys *Biomaterials* **30** 484–98.

[36]. T Kokubo and H Takadama 2006 How useful is SBF in predicting in vivo bone bioactivity *Biomaterials* **27** 2907–15.

[37]. K Chen, J Dai and X Zhang 2015 Improvement of corrosion resistance of magnesium alloys for biomedical applications *Corros Rev* 33 101–17

Evolution of magnetic and structural properties of MgO thin film with vacuum annealing

Harsh Vardhan[1, 2], Gagan Sharma[1,2,*], Kavita Sharma[1,2], R. J. Choudhary[3], D. M. Phase[3], Mukul Gupta[3], V. Raghavendra Reddy[3], Ajay Gupta[4]

[1] Amity Center for Spintronic Materials, Amity University UP, Sector-125, Noida-201313, India
[2] Amity Institute of Applied Sciences, Amity University UP, Sector-125, Noida-201313, India
[3] UGC-DAE Consortium for Scientific Research, University Campus, Khandwa Road, Indore-452001, India
[4] Department of Physics, University of Petroleum and Energy Studies, Dehradun-248007, India

*gsharma2@amity.edu

Abstract: In this work, investigation of MgO thin film, deposited via e-beam evaporation is carried out with increasing annealing temperature. Structural properties, as elucidated using X-ray based techniques viz. grazing incidence XRD and X-ray reflectivity were associated with magnetic properties of film as observed using superconducting quantum interference device (SQUID) magnetometer. As-deposited film is polycrystalline in nature and exhibits ferromagnetism, which is attributed to presence of Mg vacancies. Vacuum annealing results densification of film with concurrent increase in electron density and reduction in film thickness and saturation magnetization. However, at sufficiently high temperature, electron density and saturation magnetization decreases comparative to as-deposited film. The results are significant for the optimization of desired functional properties of MgO particularly in MgO based magnetic tunnel junctions, providing high tunnel magnetoresistance ratio.

Keywords: MgO film, annealing, SQUID, XRR

1. Introduction

The increasing miniaturization of devices induces the requirement of detailed exploration of thin films and multilayers, which exhibit various important and interesting phenomenon like giant magnetoresistance (GMR), exchange bias, tunnel magnetoresistance (TMR), spring effect etc. The magnetoresistance observed in conventional AlO_x barrier based MTJs, was not sufficient for spintronic applications. Moreover, the observation of GMR leads to potential spin transport applications for only two decades. However, relatively larger magnetoresistance and negligible temperature dependence for

MgO based magnetic tunnel junctions triggered investigations of global scientific community in this direction.

Good quality MgO film is important for developing efficient magnetic tunnel junctions (MTJs) having potential applications in magnetic sensors [1], spin logic devices [2], magnetoresistive random access memory (MRAM) [3], read heads for disk drives [4] etc. In fact recent developments by Chen *et al.*, show the applications in wearable medical devices by developing flexible MgO barrier MTJs on ultra-thin Silicon membrane [5]. The high melting temperature, high thermal conductivity, compatibility with reactor materials and small absorption cross-section of thermal neutrons makes MgO potential candidate for wide range of applications [6].

Theoretical predictions and experimental confirmations reveal that MgO film as tunnel barrier in MTJs play an important role in realizing giant tunnel magnetoresistance [7]. In fact, an epitaxial MTJ with crystalline MgO barrier, exhibits cohereni tunneling of Δ_1 majority states resulting in large tunnel magnetoresistance ratio [8]. Perpendicular magnetic anisotropy (PMA) exhibit emerging applications in high-density magnetic recording media [9] was also evidenced at the interface of ferromagnetic (FM) film with MgO [10, 11] and it was observed that MgO capping can increase PMA [10]. Such enhancement in PMA is a result of hybridization of 3d orbitals of FM with O-2p orbitals [12], which depends further on the quality and structure of deposited MgO films. Extensive studies on optical, electrical and structural properties of MgO film prepared by various methods like sol-gel method, ion beam sputtering, pulsed laser deposition etc. have already been carried out in literature.

However, any sort of ferromagnetism in MgO, also referred as d0 magnetism, because of structural defects, may affect the spin transport properties and is an active and unclear topic of current interest. Till date, CoFeB/MgO based MTJs provides a giant TMR ratio at room temperature [13]. Post deposition annealing results in formation of bcc structure of magnetic electrode in TMR multilayer at below 773 K. Therefore, investigation of structural and induced magnetic properties of MgO film with annealing temperature is also important. In this work, structural properties of MgO film is studied using X-ray based techniques viz. X-ray reflectivity; X-ray Diffraction, while magnetic properties are elucidated using superconducting quantum interference device (SQUID) magnetometry.

2. Experimental

MgO powder with 99.99% purity was used to prepare pellets, which were sintered at 1473 K for 12 h for densification. These pellets were loaded in ultra-high vacuum chamber with electron beam evaporation setup to carry out deposition of MgO film on polished Si (100) substrate. Base vacuum prior to deposition was 7 x 10 $^{-8}$ mbar, which reduces to 1 x 10^{-6} mbar during deposition. Deposition rate was maintained 6Å / min. Structural parameters of the samples- like roughness, electron density and film thickness are investigated using X-ray reflectivity (XRR), done using Bruker D8 Advance Diffractometer with a Göbel mirror to generate a monochromatic parallel beam of Cu Kα radiation (wavelength, λ= 1.54 Å). Crystalline quality of deposited film was examined using GIXRD measurements done on the same diffractometer. Post deposition sample annealing at various temperatures were done ex-situ in an evacuated chamber with base vacuum of 1x10^{-6} mbar. Magnetic properties of as-deposited and annealed MgO films are investigated at 300 K using a SQUID magnetometer (Quantum Design).

3. Results and Discussion

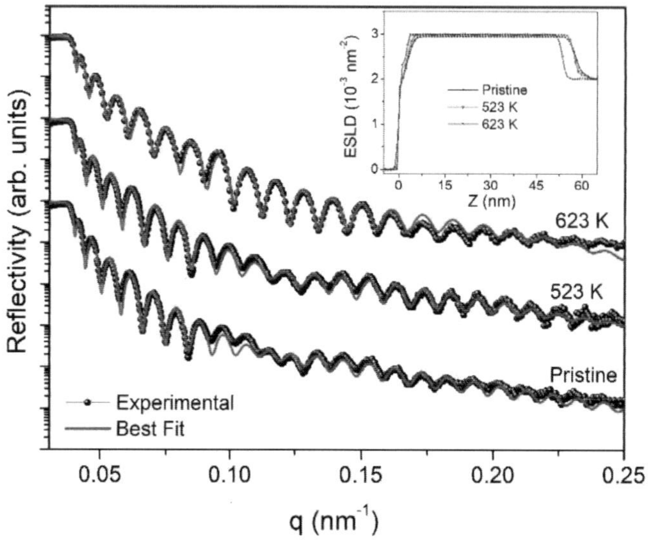

Figure 1. X-ray reflectivity of MgO film at 300 K, 523 K and 623 K. Continuous solid curves characterizes the best possible fit of experimental data. For easy viewing curves are shifted vertically relative to each other. Inset show the electronic scattering length density profiles as function of z (depth)

Figure 1 shows the XRR of MgO film annealed at various temperatures. The data was fitted with Parratt formalism [14]. Electron density profiles is shown in inset of figure as function of increasing z (depth) from air. Figure 2(a) shows the variation of MgO film thickness as calculated from fitting of X-ray reflectivity data, as function of annealing temperature. The as-deposited MgO film exhibits thickness of 62 nm, which reduces to 60 nm and 54 nm for annealing at 523 K and 623 K respectively. With further increase in annealing temperature, film thickness becomes constant. Such reduction in thickness of thin film has also been reported in previous studies for oxide films, which is attributed to film densification [15].

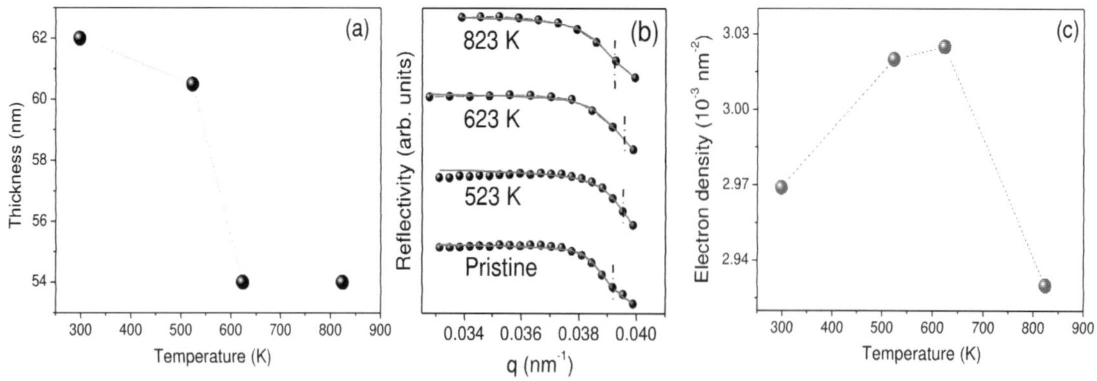

Figure 2. (a) MgO film thickness observed from XRR fitting as function of annealing temperature (b) Fitted XRR in small q-range about critical angle. Dashed line show the qc values, where intensity reduces to half. (c) Variation of electron density as reported from fitting of XRR data in small q-range.

Figure 2b shows the fitted XRR data in small q range about the critical angle region at various temperatures. The dashed line (vertical) intersects the XRR data at point (critical angle), where reflected intensity reduces to half of the initial value. One may note that annealing till 623 K, results in shift of critical angle to higher value relative to as-deposited specimen indicating an increase of electron density and hence mass density. The concurrent increase of mass density and reduction in film thickness with the vacuum annealing has also been reported in previous studies [16, 17]. However, with further increase in annealing temperature at 823 K, critical angle shifts to lower value.

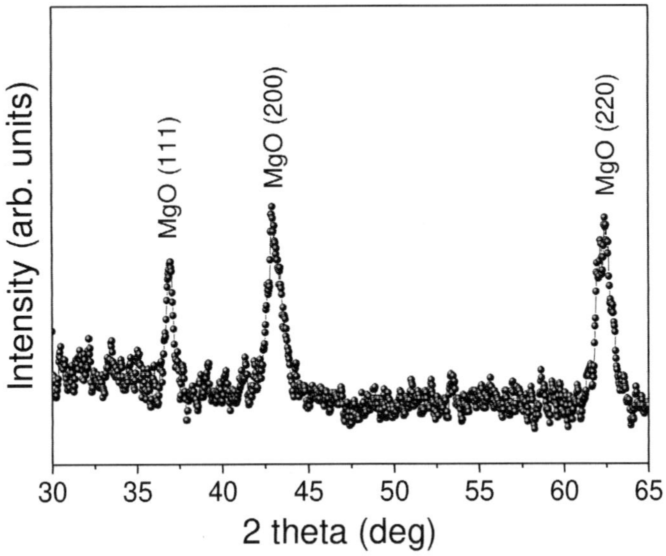

Figure 3. X-ray Diffraction in grazing incidence mode (GIXRD) of as-deposited MgO film

The variation of electron density as calculated from fitting of XRR data, with annealing temperature is shown in Figure 2 (c). Electron density of as-deposited MgO film is observed as 2.97 E-03 $nm^{-2} \pm$ 2.0 E-05 nm^{-2}, which is near to bulk value of 3.04 E-03 nm^{-2}. With thermal annealing till 623 K, electron density and hence mass density increases to theoretical value signalling film densification because of elimination of defects, vacancies and stresses in as-deposited thin film. The results are in agreement with thickness modification of MgO film with annealing temperature.

The crystallinity of MgO film was explored via X-ray diffraction in grazing incidence mode (GIXRD). Figure 3 gives the GIXRD spectrum of as-deposited MgO thin film with incident angle fixed at 0.5 degree. The polycrystalline nature of MgO film was observed with all the peaks attributed to MgO, confirming the absence of any impurity phases.

Figure 4 shows the M-H curves (M = magnetization, H = applied magnetic field) measured at 300 K, for MgO film annealed at various temperatures, after subtracting the signal from Silicon substrate. One may note that as-deposited film exhibits ferromagnetism. Such appearance of hysteresis loop has also been observed in earlier studies [18-21]. Theoretical calculations and experimental confirmation suggest that, magnetism in MgO film is driven by Mg vacancies, which induces spin polarization of 2p electrons of surrounded Oxygen atoms. [18, 19]. However, certain reports suggest that oxygen vacancies and

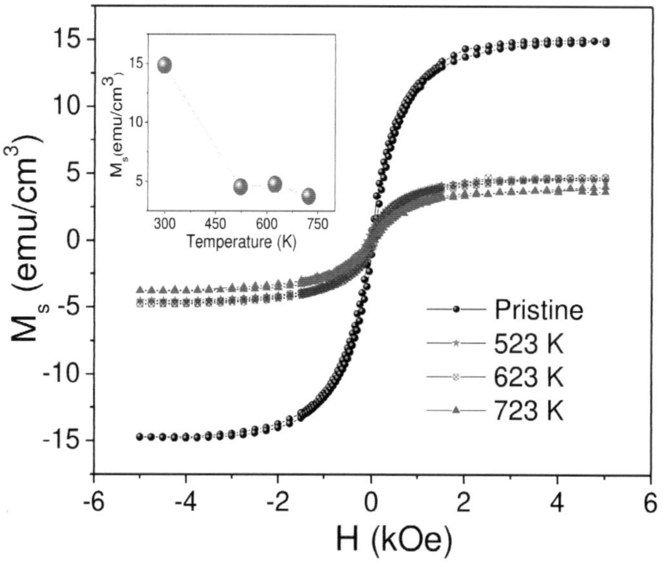

Figure 4. M-H curves of the MgO film annealed at various temperatures. The inset depicts the variation of Ms (saturation magnetization) with annealing temperature

hydrogen, adsorbed at oxygen vacancies may add to room temperature ferromagnetism [22]. For the present case, as deposited MgO film exhibits saturation magnetization of M_s ~ 15 emu / cm^3, which reduces further with vacuum annealing. It is important here to note that MgO being hygroscopic is vulnerable to form diamagnetic $Mg(OH)_2$, which further reduces ferromagnetic ordering [22]. Vacuum annealing is also expected to increase the number of oxygen vacancies, which may degrade ferromagnetism [19, 23]. The variation of saturation magnetization with the annealing temperature is presented in inset of Figure 4. One may note that with increasing annealing temperature, M_s reduces significantly below 5 emu / cm^3 and beyond 523 K decreases slowly. In a study by Li et al., it was also observed that crystallinity of MgO film enhances with annealing in oxygen and vacuum environment, accompanied with reduction of Mg vacancies. However, there is relatively higher reduction in M_s for vacuum annealing relative to oxygen annealing [19]. Therefore, such lowering in saturation magnetization with annealing for MgO can be understood as result of all possible competing contributions discussed here.

4. Conclusions

Structural and magnetic properties of MgO film as function of annealing temperatures have been investigated. As-deposited film was polycrystalline with relatively low electron density, because of presence of defects and vacancies, which results in appearance of ferromagnetism. Vacuum annealing results in film thickness reduction and increase of electron density accompanied with removal of defects and voids, which also lead to increase in crystallinity. The removal of vacancies (Mg) results in lowering of saturation magnetization. However, at sufficiently high temperature, electron density reduces concurrently with reduction of saturation magnetization. The results are important in view of spin-transport applications, where optimization of structural and magnetic properties of MgO are needed for achieving a high value of spin-polarized current across the junction.

Acknowledgments

Mr. Harsh Vardhan is Junior Research Fellow (JRF) in the project funded by Board of Research in Nuclear Sciences (BRNS) (project no. 58/14/07/2019-BRNS/10381). Financial support from the same is thankfully acknowledged. The work was partly funded by UGC-DAE CSR, Indore (project no. CSR-IC-TIMR-04/CRS-271)

References

[1] Chen Y J, Carroll N, Feng J F and Coey J M D 2012 *Appl. Phys. Lett.* **101** 262402.
[2] Behin-Aein B, Datta D, Salahuddin S. and Datta S 2010 *Nat. Nanotechnol.* **5** 266-270.
[3] Gallagher W J and Parkin S S P 2006 *IBM J. Res. Dev.* **50** 5-23.
[4] Tsunekawa K, Djayaprawira D D, Nagai M, Maehara H, Yamagata S, Watanabe N, Yuasa S, Suzuki Y and Ando K 2005 *Appl. Phys. Lett.* **87** 072503.
[5] Chen J-Y, Lau Y-C, Coey J M D, Li Mo, Wang J-P 2017 *Sci Rep.* **7** 42001.
[6] Jiao L, Chen A, Myers M T, General M J, Shao L, Zhang X and Wang H 2013 *J. Nucl. Mater.* **434** 217-222.
[7] Yuasa S and Djayaprawira, DD 2007 *J. Phys D: Appl. Phys.* **40** R337.
[8] Mathon J and Umerski A 2001 *Phys. Rev. B.* **63** 220403(R).
[9] Yoda H, *et. al.*, 2010 *Curr. Appl. Phys.* **10** e87-e89.
[10]Kubota H, Ishibashi S, Saruya T, Nozaki T, Fukushima A, Yakushiji K, Ando K, Suzuki Y and Yuasa S 2012 *J. Appl. Phys.* **111** 07C723.
[11]Yakata S, Kubota H, Suzuki Y, Yakushiji K, Fukushima A, Yuasa S and Ando K 2009 *J. Appl. Phys.* **105** 07D131.
[12]Yang H X, Chshiev M, Dieny B, Lee J H, Manchon A and Shin K H 2011 *Phys. Rev. B.* **84** 054401.
[13]Ikeda S, Hayakawa J, Ashizawa Y, Lee Y M, Miura K, Hasegawa H, Tsunoda M, Matsukura F and Ohno H 2008 *Appl. Phys. Lett.* **93** 082508.
[14]Parratt L G 1954 *Phys. Rev.* **95** 359.
[15]Thomas R, Milanov A, Bhakta, R, Patil U, Winter M, Ehrhart P, Waser R and Devi A 2006 *Chem. Vap. Depos.* **12** 295-300.
[16]Mondal M H and Mukherjee M 2010 *Polymer.* **51** 5550-5555.
[17]Nakamura N, Ogi H, Shagawa T and Hirao M 2008 *Appl. Phys. Lett.* **92** 141901.
[18]Martínez-Boubeta C, Beltrán J I, Balcells L, Konstantinović Z, Valencia, S, Schmitz D, Arbiol J, Estrade S, Cornil J and Martínez B 2010 *Phys. Rev. B.* **82** 024405.
[19]Li J, Jiang Y, Li Y, Yang D, Xu Y and Yan M 2013 *Appl. Phys. Lett.* **102** 072406.
[20]Araujo C M, Kapilashrami M, Jun X, Jayakumar O D, Nagar S, Wu Y, Århammar C, Johansson B, Belova L, Ahuja R, Gehring G A and Rao K V 2010 *Appl. Phys. Lett.* **96** 232505.
[21]Singh J P, Lim W C and Chae K H 2019 *J. Alloy. Compd.* **806** 1348-1356.
[22]Kumar A, Kumar J and Priya S 2012 *Appl. Phys. Lett.*, **100** 192404.
[23]Hu J, Zhang Z, Zhao M, Qin H and Jiang M, 2008 *Appl. Phys. Lett.* **93** 192503.

Study on Defect Reduction in Casting of Ni-Cr alloy (UNS N06003)

Kaustubh Samvatsar[1*] **and Harsh Dave**[2]

[1,2]Faculty of Technology and Engineering, The Maharaja Sayajirao University of Baroda, Vadodara, Gujarat 390001, India

E-mail: *kaustubhsamvatsar@gmail.com

Abstract. In modern times, selection criteria of materials are solely based on its utilization. The characteristics of materials like high temperature sustainability with regard to its strength and corrosion resistance is the primary requirement to satisfy its operations. Industries categorised in sectors like energy, manufacturing, electronics etc. greatly rely on input materials and their purity decides their performance. Amongst many developed alloys, nickel-based alloys find their position in applications having temperatures from cryo to nuclear ranges. One of such unique materials, being used from minute components of semiconductor industry to huge components of marine industry is UNS N06003. It is basically Ni-Cr based alloy having greater than three-fourth fraction of nickel. For major applications wrought material is widely used, but cast material marks their positions in selected applications. This paper deals with casting of UNS N06003 and improving its quality. The reasons associated with cast defects and drawbacks of defects in terms of performance have been discussed. To minimize casting defects and produce a sound casting, this research adopts a novel approach of removal of unwanted gaseous elements. In degassing stage, molten metal is introduced with the combination of calcium-based fluxes which mainly remove the entrapped gases and lighter impurities in the form of slag. It has been found that the addition of quicklime (CaO) in combination with calcium carbide (CaC_2) resulted in reduction of gaseous elements like nitrogen, phosphorus and sulphur. Up to 33% reduction of gaseous elements has been detected.

1. Introduction

Nickel based alloys were developed and introduced to industry for providing magnificent corrosion resistance and strength even under high temperatures or adverse environments.UNS N06003 is one of such alloys is potentially very strong candidate for industrial applications in the field of energy, manufacturing, marine, chemicals/petrochemicals, and electronics[1, 2, 3].It is the combination of hard and ductile nickel, and lustrous and non-oxidizing chromium[4]. Nickel-chromium based UNS N06003 is an alloy with advantage of solid solution strengthening mechanism and precipitation hardening mechanism. This is the primary cause for its high temperature resistance[5, 6]. This alloy is also known as ASTM 80Ni-20Cr, EN NiCr8020 and NICHOME V[7].It forms a preventive layer of chromium oxides on its surface in severe conditions[8, 9].
Components like casings, discs, turbine blades, stator vane segments, vanes, exhaust systems shafts, blades etc. are used in hot gas path. These components are designed using UNS N06003 to achieve serviceability with high temperature strength[5, 6]. Area of applications of UNS N06003 include

marine structures, land-based utilities in power generation, vehicular components[6], heating sources in electronic devices[10, 11, 12],aircraft turbine engine parts (equiaxed or single crystal) etc. These are designed based on applicability of desired creep resistance. Ingots used to make wrought turbine disks can successfully be made by casting[13, 14]. For these applications, various exposure conditions exist in the form of mixture of complex chemical compounds. These chemical compounds impact the base material and degrade it adversely. Such degradations also badly influence the lifecycle of material. This demands the necessity of defect-free material production and assembly stage.

If defects or entrapped residual gases pre-exist, chances of corrosion-based degradation accelerate. Galvanic corrosion happens in the region of complex phases, resulting in microstructural inhomogeneities at microstructural level. These inhomogeneities also translate to generation of localized stresses[15].

The inhomogeneities are minimized with vacuum induction melting technology over the conventional practice. Reduction of the tensile strength occurs due to the such agglomerated elements on grain boundaries at sub-micron levels[16].In practical conditions of casting, proficiency of foundry man, quality of raw material and process parameters play vital role to determine quality[17]. Many studies use crucibles having MgO-Al_2O_3lining,but it forms MgS. Such phenomena shorten the lifespan of lining. Electric arc furnace and Vacuum induction melting are used as alternative methods for casting alloys[18].

In the current study, carbon boil reaction is applicable to enhance melt deoxidation[19]. In real life applications, UNS N06003 shows capabilities of performing at high temperatures (higher than 0.5 times of its melting point) and gives uninterrupted performance. Such capabilities are defined by crystal structure of alloy. For UNS N06003, the equiaxed microstructures are capable of functioning up to 900-950°C whereas single crystal structures can function up to 1100°C[13, 20]. It is essential to remove deleterious phases causing poor mechanical properties, that are formed as a result segregation under non-equilibrium conditions [14, 18]. This research involves experiments in an economical manner for producing cast components from Ni-Cr based alloy.

2. Casting of UNS N06003

Cast alloys are preferred over wrought alloys from hot workability point of view. Major preference of cast alloy as controlling the composition is easier by casting route. Such tailored composition maylead to higher applicability of material owing to its high temperature strength. Cast microstructures generate coarser grains over other manufacturing methods. Hence, even the development of new alternative of UNS N06003 may take a decade or two[21]. Vacuum induction melting is introduced to industries since 1950s[20]. In this study, casting of UNS N06003 has been performed with induction melting furnace. This approach attempts the elimination of impurities and entrapped gases within molten metal[13].

Sound castings are dependent upon the considerations such as neglection of steep temperature or stress variations[22], controlled quality of surface [23], appropriate design of air vent, flow stability[24], proper modelling of runner and riser [25] etc. Poor degassing of molten metal leads to residual impurities, mainly entrapment of gaseous form of elements like phosphorous, sulphur, hydrogen, oxygen and nitrogen. These residual elements are chemically unstable by nature and these ultimately form some compounds to gaining stable structure such as nitrides, phosphides, sulphides, carbides etc. The formation of compounds is followed by solidification stage in casting. During solidification, such compounds reside on the grain boundaries due to their limited solubility in matrix[26]. Proper selection and addition of charging materials can resolve such issues [17, 27].

Three dimensional clusters of defects are formed by accumulation and association of tiny individual defects. These defects are created as isolated entities and stay in agglomerated or isolated form based on either operating condition. The presence of non-metallic inclusions, inconsistency in geometry[26], blow holes [28, 29], deviations generated within crystallographic orientations[30], pores [31]etc. greatly cause defects. Failure of casting can be eliminated by taking precautions which minimize such defects. like manipulation of elemental composition, process parameters and geometry of casting. It's difficult to exercise control geometry as is based on its functionality. Thus, the former two parameters need consideration for the generation of sound casting [32].

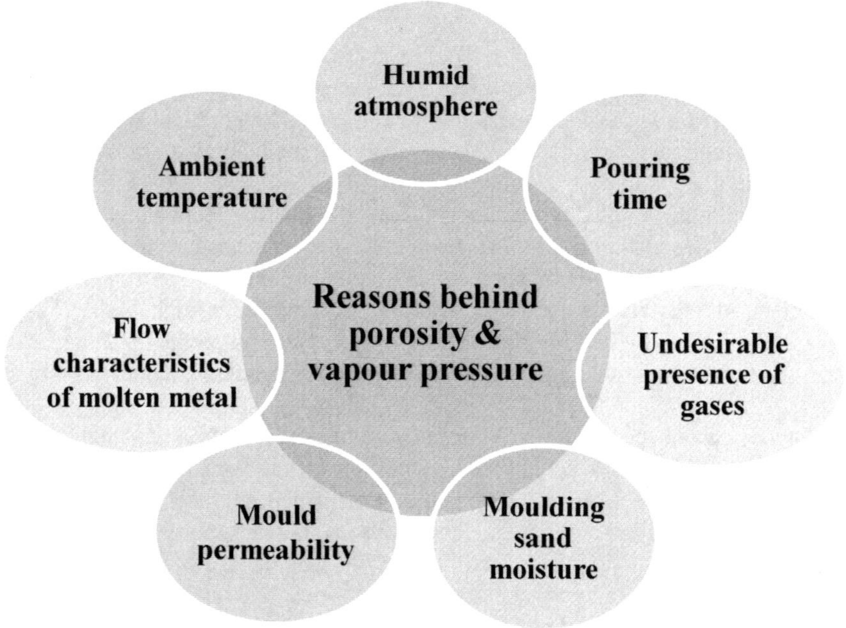

Figure 1. Major causes of porosity and vapour pressure in casting.

3. Degassing

Appropriate melting practices must be followed to eliminate elements like sulphur, phosphorus, silicon, oxygen and nitrogen which cause deleterious impacts[33, 18]. Degassing of molten alloys in is a crucial required for minimizing the generation of defects caused due to the presence of gaseous elements. Killing process is performed in the steel-making process for the removal of H, N, O, P and S from the alloy melt.[34]Need of similar mechanism is necessary for UNS N06003.

Complex role of carbides is observed in superalloys. Carbon is beneficial upto a certain defined range as it provides rupture strength at elevated temperature conditions. If the fraction of carbon elevates beyond a defined limit, it becomes a reason for crack formation due to transformation in nature from ductile to brittle. By removal of such reactive elements, ductility improvement and chemical stability can be achieved. Major types of carbides that are found in superalloys are MC, M_6C and $M_{23}C_6$. Cr_7C_3 is also observed in minor amount. This formation of carbides is beneficial from the perspective of grain boundary precipitation[35, 36, 37].

Mechanical properties such as ductility, toughness, resistance to wear/erosion and friction coefficient are enhanced by optimal addition of phosphorous[38]. Ideally in an alloy, phosphorus content should be below 0.005 wt.% for imparting corrosion/erosion resistance. For many alloys, to maintain their toughness, maximum allowable phosphorous content is 0.020 wt.%. If it exceeds the mentioned amount, intergranular corrosion can be seen which makes it non-usable in applications requiring low temperature or high strength [39, 40].

Considering the metallurgy of alloys, if the content of sulphur does not fall in the range of 0.013-0.078 wt.%, reduction in the solidification temperature can occur[41]. Sulphur causes higher cracking than phosphorus. Increase in brittleness and red-shortness under high temperature conditions is also noticed. Thus, desulphurisation during manufacturing is the need for prolonged life of components[42]. But when the components have high pressure and low temperature application, the content of sulphur should be maintained in the range of 0.0005-0.001 wt.% [43].

Nitrogen is another gaseous element which impacts the properties of alloys. In low temperature conditions, higher content of nitrogen makes the material prone to pitting as well as intergranular corrosion phenomena as the material loses toughness up to some extent. It may also cause health hazards as it causes removal of nickel[44, 45]. During solidification of cast alloys, nitrogen causes nitrogen bubble porosity[46]. Also, dislocated nitrogen atoms interact with the atomic lattices of other major constituting elements creating strain-age embrittlement. This causes diffusion of nitrogen to multiple sites, thereby raising the activation energies and formation of solidification cracks. For materials which are going to be welded, nitrogen content should be less than 0.045wt.% to avoid weld porosities[47].

Researches have shown that better methods of ventilation, control over the curing time or addition of optimum content of resin/clay in sand have been successful methods for degasification [48].Also, for methods such as inert gas fluxing,oxidation-reductionreactions, zinc flaring, vacuum degassing, a solid degassing flux can be used for degassing during casting of alloys. In molten alloys at high temperatures, the addition of solid degassing fluxes is widely exercised in the foundries. It is also a cost-effective way. Limestone ($CaCO_3$), dolomite ((Ca, Mg) CO_3), calcium carbide (CaC_2), fluorspar (CaF_2) and sodium carbonate (Na_2CO_3) are commonly used fluxes for removal of N, P and S during casting of alloys. Limestone liberates CO_2gas when it is added during the casting of nickel-based alloys and elimination of carbon occurs [49, 50, 51].

Thus, by using fluxes based on calcium and/or carbon, such unwanted elements along with slag can be done. The current study investigates the advantages of the same.

4. Materials and Methods

For casting of UNS N06003, sand casting method has been used. A sand standard American Foundry Society has been followed throughout the entire process and sand of size '48' number was used. Mould has been prepared with mixture of silica sand and silica binder. Mixture of fresh and reclaimed sand has been used in the ratio of 40:60. Environment friendly binder was employed for binding the mould, which burns off during pouring stage and does not produce any noxious gases. Carbon dioxide gas was blown over mould cavity for hardening. Proper ramming of sand in mould box was ensured, which avoids misappropriate densification of particles. Probability of happening a damage is high if temperature gradient is introduced to mould in pouring stage. Hence as a precaution, defined range of compaction was applied to the mould sand. Ladle and mould-box were subjected to flame heating before introducing to molten metal to avoid steep temperature difference. Once the temperature of 1550°C was achieved, melt refining and flux treatment was performed. Flux mixture of CaO and CaC_2in equal proportion was added to the molten metal. Properly lined ladle was used to avoid sand erosion and moisture entrapment.

From the large cast component of UNS N06003 material, to be used for marine industry, three small samples of different castings were been taken for this study. During casting of the components, mixture of powders of preheated quicklime (CaO) and calcium carbide (CaC_2) was added to the molten alloy as flux material. Considering the volume of casting, flux mixture was added in ratio of around 10kg/ton. Commonly observed composition of the Ni-Cr alloy (UNS N06003) is as shown in table 1.

Table 1. Percentage of elements in UNS N06003[7].

Element	Ni	Cr	Mn	Si	Cu	Al	P	S	C	Fe
Wt.%	Bal. (~75%)	19.0-21.0	1.0 (max.)	0.5-2.0 (max.)	0.5 (max.)	0.3 (max.)	0.02 (max.)	0.015 (max.)	0.15(max.)	1.0 (max.)

All the samples had cylindrical geometry and dimensions in D × H are approximately 35mm × 25mm. The image of one of the samples is as shown in the Figure 2.

Figure 2. Image of one of the samples in top-view and side-view.

For measuring the compositional details of the samples, optical emission spectroscopy technique was used. The advantage of using this technique is that it requires minimal sample preparation but the results given are accurate with precision up to 0.0001 wt.%. This technique employs a sparking phenomenon on the application of electrical charge on the sample. As a result, small amount of material is vaporized and composition is known[52].An optical emission spectrometer (Make: GNR, Model: S5 SOLARIS CCD Plus) was used for this testing.

5. Results and Discussions

The composition of samples was tested for the samples casted without addition of flux and after the addition of flux. It can be clearly observed that percentage of C, P, S and N has lessened. The results indicate a strong possibility of reduction of porosities and cracks in the produced casting. Also, the effectiveness of produced casting is enhanced by proper mixing of flux with the molten alloy. Degasification is thus achieved within a short span of time with higher reaction rate.

Table 2. Percentage of elements in samples as tested.

Sample	State	Elements (in wt.%)			
		C	P	S	N
S1	Initial	0.230	0.012	0.023	0.034
	Final	0.223	0.010	0.021	0.033
S2	Initial	0.160	0.014	0.025	0.037
	Final	0.150	0.011	0.022	0.034
S3	Initial	0.110	0.013	0.016	0.031
	Final	0.017	0.005	0.010	0.030
	Average% Reduction	31.28	33.21	19.40	4.76

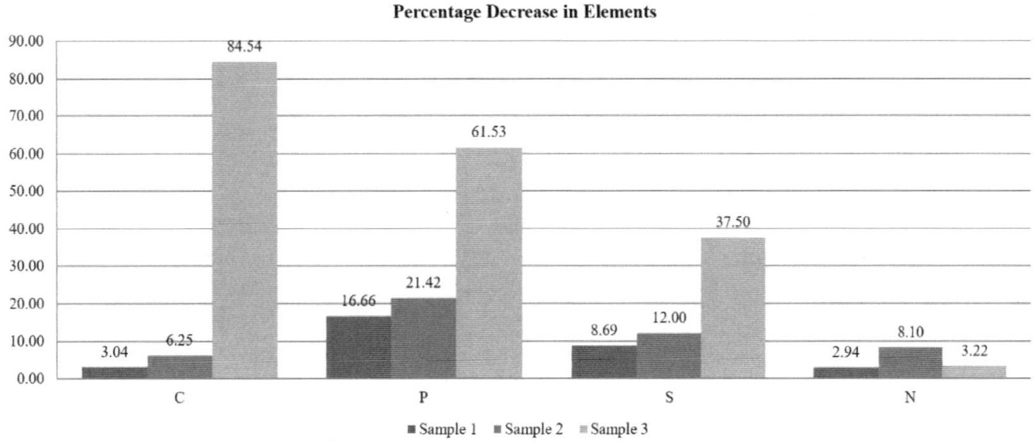

Figure 3. Graph of Percentage Decrease of Elements in UNS N06003.

Table 2 shows the values of elements before and after degassing treatments for samples. On degasification, the average content of C, P, S and N reduces by 31.28%, 33.21%, 19.40% and 4.76% respectively. The individual percentage reduction of samples has been presented in Figure 3. This degassing technique, by decreasing the proportion of C, P, S and N also diminishes the possibility of creating a porosity. The microstructure of the cast alloy shows nickel-based solid solution of Cr and Fe in Ni matrix having austenitic structure.

Carbon, which was initially appearing in higher amount within the alloy appears under the controllable limits by addition of fluxes. This carbon is seen in the form of carbides along the inter-dendritic spacing. The microstructure also shows inter-dendritic porosity.

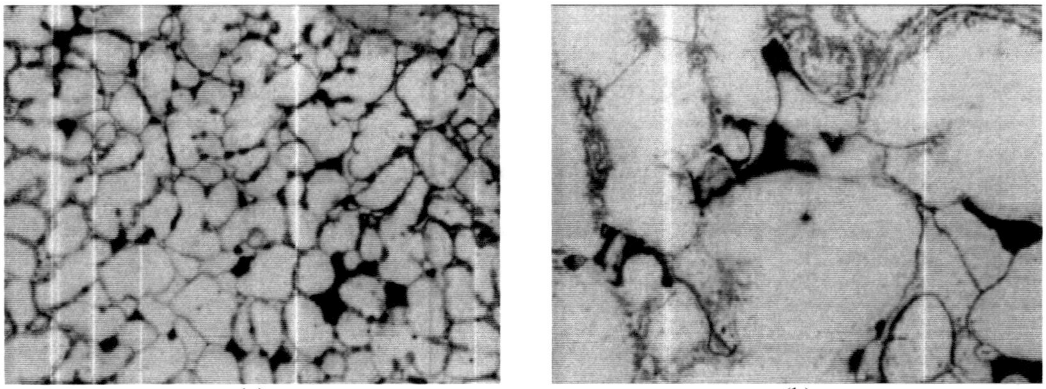

(a) (b)

Figure 4. Optical microstructure of cast UNS N06003, using glyceregia at (a) 100× (b) 400×.

In any casting process, proportion of minor elements is difficult be controlled. Shrinkage-based pores are found near the solidified regions whereas gas-based pores are of distributed nature found throughout the cast material. This may also pose variations of compositions up to 10% within the same casting. Such occurrences cause uncontrolled diffusion and affects the hermiticity of castings[53, 54]. The percentage decrease of gaseous elements indicate that the adopted technique has great influence on the quality of casting.

Holappa et al. [55]and Sanyal et al. [56]had also performed similar study and found that usage of using lime with calcium carbide affected the final composition of cast alloy. Si and Al as minor alloying elements in UNS N06003 and may be externally added as degassing agents[57].By using calcium-based fluxes, reduction metallic-degassing materials is also possible. Minor amount of Al can provide better results if used with calcium-based fluxes[20, 51].

6. Conclusions and Prospects

Analysis of the data obtained from optical emission spectroscopy revealed that the importance of addition of fluxes play a pivital role in the casting of UNS N06003. The percentage reduction of gaseous impurities was found up to 61.53%. Carbon, which is present in higher amount than allowable limit has also reduced up to 84.54% and its content can be observed adhering to ideal chemical composition of alloy. This study asserts that the addition of appropriate fluxes eliminates the impurities/gaseous elements and consequently produces sound castings. This technique attractively upgrades the quality of cast UNS N06003 through induction melting system. In the absence of vacuum technology, better quality castings, devoid of Nitrogen, phosphorous, sulphur and carbon are achieved by degassing. This also reduces the possibility of crack formation, hot shortness, cold shortness and strain-age embrittlement. This method establishes an inexpensive way to fabricate good quality nickel based solid solution strengthened alloys.
This study uses only few samples for research on degasification using CaO with CaC_2casting. It can be further extended for more number of samples varying the proportion of mixture of fluxes and changing the process parameters of casting to understand the respective behaviours. A development of a permanent degassing system can also be explored. Effects of different fluxes on casting of UNS N06003 require further investigations.

Acknowledgments

The authors are greatly thankful to Mr. Pritesh Shah and his team for providing resources and platform for conducting this research at Suryadeep Alloy Steel Castings Pvt. Ltd., V. U. Nagar, Gujarat, India.

References

[1] Elliott P 2011 Overcome The Challenge Of High Temperature Corrosion. *In Corrosion*(Houston, Texas, USA)
[2] Elliott P 2000 Materials Performance in High Temperature Environments - Making the Choice. *In Corrosion*(Orlando, Florida, USA)
[3] John R C and Wright L G 2004 Long-Term Oxidation Commercial Alloys in PO2's of 0.001 - 1 atma, at Temperatures of 200 - 1,150C and After 20,000 Hr *In Corrosion*(New Orleans, Louisiana, USA
[4] Klarstrom D L and Srivastava S 2005 Selection of High Performance Alloys for High Temperature Corrosion Environments *In Corrosion*(Houston, Texas, USA)
[5] Henderson M B, Arrell D, Larsson R, Heobel M and Marchant G2004 Nickel based superalloy welding practices for industrial gas turbine applications *Sci. Technol. Weld. Join.***9**13-21
[6] Nowotnik A, Kubiak K, Sieniawski J, Rokicki P, Pędrak P and Mrówka-Nowotnik G2014 Development of Nickel Based Superalloys for Advanced Turbine Engines *Mater. Sci. Forum.***783–786**,2491–2496
[7] Nichrome 80 2021, Heanjia Super Metals. Available: https://super-metals.com/wp-content/uploads/2015/03/Nichrome-80[Accessed 30 June 2021].
[8] Matthews W J, More K L and Walker L R (2008). Long-Term Microturbine Exposure of an Advanced Alloy for Microturbine Primary Surface Recuperators *In Proc. of ASME Turbo Expo 2008: Power for Land, Sea and Air GT2008, Berlin, Germany*

[9] Rowe M D, Ishwar V R and Klarstrom D L2004 Properties, Weldability, and Applications of Modern, Wrought, Heat-Resistant Alloys for Aerospace and Power Generation Industries.*In ASME Turbo Expo 2004: Power for Land, Sea, and Air*(Vienna, Austria)
[10] Bagga S, Akhtar J and Mishra S2020 Fabrication of coplanar microheater platform for LPG sensing applications. *Microsyst. Technol.***26** 731–38
[11] Haeger A, Schell K G, Nagel A, Bucharsky E C, Oberacker R and Hoffmann M J2012 Fabrication and High Temperature Creep Behaviour of Interpenetrated Nickel–Chromium/Alumina *Compos. Adv. Eng. Mater.***14** 795-801.
[12] Petley V, Sathishkumar S, Thulasi Raman K H, Mohan Rao G and Chandrasekhar U2015Microstructural and mechanical characteristics of Ni–Cr thin films *Mater. Res. Bull.* **66**, 59-64
[13] Pollok T M and Tin S 2006 Nickel-Based Superalloys for Advanced Turbine Engines: Chemistry, Microstructure and Properties *J. Propuls. Power* **22** 361-374.
[14] Fecht H and Furrer D 2000 Processing of Nickel-Base Superalloys for Turbine Engine Disc Applications *Adv. Eng. Mater.* **2**777-87
[15] Zhao X, Qi Y, Wang J, Zhang Z, Zhu J and Quan L2020 Effect of Cast Defects on the Corrosion Behavior and Mechanism of UNS C95810 Alloy in Artificial Seawater *Mater.* **13** 1790
[16] Viswanath A, Dieringa H, Kumar K K A, Pillai U T S and Pai B C 2015 Investigation on mechanical properties and creep behavior of stir cast AZ91-SiCp composites *J. Magnes. Alloy.***3**16-22
[17] Perzyk M and Kochanski A2003 Detection of causes of casting defects assisted by artificial neural networks *Proc Inst Mech Eng B J Eng Manuf***217**1279-83
[18] Long F, Yoo Y S, Jo CY, Seo S M, Jeong H W, Song Y S, Jin T and Hu Z Q2009 Phase transformation of η and σ phases in an experimental nickel-based superalloy *J. Alloys Compd.* **478** 181-187.
[19] Freismuth A1997 The use of calcium carbide in ferrous metallurgy *Metallur.***41** 56-60.
[20] Das N2010 Advances in nickel-based cast superalloys *Trans. Indian Inst. Met.* **63** 265–274
[21] Satyanarayana D V V and Eswara Prasad N2017 Nickel-Based Superalloys. in Aerospace Materials and Material Technologies *Indian Inst. Met. Ser.*199-228.
[22] Mahomed N and Kleynhans H A2019 Reducing Shrinkage Porosity in High-Performance Steel Castings: Case of an ASME B16.34 Gate Valve Body: Part 2—Simulation and Experimental Verification *Int. J. Met.***13**, 463–472
[23] Ramrattan S2019 Evaluating a Ceramic Resin-Coated Sand for Aluminum and Iron Castings. *Int. J. Met.***13**519-27
[24] Wight T A, Grisius R J and Gaugler R W1980. Evaluation of three variables affecting the casting of base metal alloys *J Prosthet Dent* **43** 415-18
[25] Hwang H, Nam C, Choi Y, Hong J and Sun X 2017 A study on effects of ceramic foam filter on flow aspect through water modeling experiment *China Foundry***14** 216–225
[26] Skryabin M L and Likhanov V A 2019 The study of casting defects in steel 35HGSL *J.Phys. Conf. Ser.***1399** 044063
[27] Van Wortel J C 1998 Relaxation cracking in the process industry, an underestimated problem *In VTT Symp.***185**(Espoo, Finland)
[28] Rao B N and Ashokkumar R 2008 Effect of blow-holes on reliability of cast component *Sadhana* **33**,733–751
[29] Latte M R and Chougule P D2017 Blow-hole Defect Analysis of Cylinder Block- A Case Study *Int. J. Eng. Res.***10** 626-631.
[30] Zhang J, Huang T, Liu L and Fu H 2015 Advances In Solidification Characteristics And Typical Casting Defects In Nickel-Based Single Crystal Superalloys *Acta Metall. Sin.* **51** 1163-78
[31] Bocher F, Allison T, Shin B G and Lee S2018 Materials Evaluation and Corrosion Test Needs for a Direct-Fired SCO2 Oxy-Combustion Plant *In The 6th International Supercritical CO2 Power Cycles Symposium*(Pittsburgh, Pennsylvania, USA)

[32] Zang X, Liu L, Yang C, Zhang J, Li Yand Fu H 2012 Advance in Research of Casting Defects of Directionally Solidified Nickel-based Single Superalloys *J. Mater. Eng.* **2** 93-98.
[33] Akca E and Gürsel A2015 A review on superalloys and IN718 nickel-based INCONEL superalloy*Period. Eng. Nat. Sci.* **3**15-27.
[34] Bannenberg N, Bergmann B and Gaye H1992 Combined decrease of sulphur, nitrogen, hydrogen and total oxygen in only one secondary steelmaking operation*Steel Res.* **63** 431-37.
[35] Ezugwu E O, Wang Z M and Machado A R1999The machinability of nickel-based alloys: a review*J. Mater. Process. Technol.* ***86*** 1-16.
[36] Graybill B, Li M, Malawey D, Ma C, Alvarado-Orozco J and Martinez-Franco E2018 Additive Manufacturing of Nickel-Based Superalloys *In Proc. of the ASME 2018 13th International Manufacturing Science and Engineering Conference* (Texas, USA)
[37] Prasad N E,Gokhale AA andWanhill R J H2017*Aluminium–Lithium Alloys, Aerospace Materials and Material Technologies*In: Prasad N., Wanhill R. (eds) Singapore (Springer, Singapore)
[38] Trivedi S and Gupta P 2019 Effect of Phosphorus Addition in Steel: Processed Through Cast Route and Powder Metallurgy Route *J. Mech. Mach. Des. Manuf.* **1** 24-28.
[39] Masumitsu N, Ito K and Fruehan R J1988 Thermodynamics of Ca-CaF2 and Ca-CaCI2 systems for the dephosphorization of steel *Metall. Trans. B* **19** 643–48
[40] Lu L, Devasahayam S and Sahajwalla V2013 Evaluation of coal for metallurgical applications *In The Coal Handbook: Towards Cleaner Production - Coal Utilisation* Woodhead Publishing Limited **2** 352-86
[41] Guo J, Yang W, Shi X, Zheng Z, Liu S, Duan S, Wu J and Guo H2018 Effect of Sulfur Content on the Properties and MnS Morphologies of DH36 *Struct. Steel Met.* **8** 945
[42] Wang W, Luo S, Cai Z and Zhu M2013 The Effect of Phosphorus and Sulfur on the Crack Susceptibility of Continuous Casting Steel *In Mater. Process. Fundamentals* 89-98
[43] Şener B, Hüsken R and Cappel J 2012 Desulphurization Strategies in Oxygen Steelmaking *In AISTech 2012 — The Iron & Steel Technology Conference and Exposition*(Atlanta, Georgia)
[44] Xiong F, Fu R, Li Y, Xu B and Qi X2020 Influences of nitrogen alloying on microstructural evolution and tensile properties of CoCrFeMnNi high-entropy alloy treated by cold-rolling and subsequent annealing *Mater. Sci. Eng. A* **787** 139472
[45] Jiang Z, Li H, Chen Z, Huang Z, Zou D and Liang L 2016 The Nitrogen Solubility in Molten Stainless Steel *Steel Res. Inter.* **76**, 740-45
[46] Svyazhin A G, Kaputkina L M, Bazhenov V E, Skuza Z, Siwka E and Kindop V E2015 Phases and defects upon the solidification of nitrogen-alloyed stainless steels *Phys. Met. Metallogr.***116**552–61
[47] Woo I and Kikuchi Y2002 Weldability of High Nitrogen Stainless Steel *ISIJ Inter.* **42** 1334–43
[48] Desai J, Parikh V and Shah T2013 Reducing Porosity for Alloy Steel Casting *Int. J. Eng. Res.* **2** 871-74.
[49] Srinath Viswanathan S et al. (Eds.) 2008*ASM Handbook* **15** (Casting., Ohio, USA)
[50] Jana B D andChakrabarti A K 2004 A study on the mechanism of desulphurisation and dephosphorisation of liquid steel by Na2CO3 flux treatment *Scand. J. Metall.* **33** 76-79
[51] Freismuth A1997 The use of calcium carbide in ferrous metallurgy *Metallur.* **41** 56-60
[52] Bengtson A2017 Laser Induced Breakdown Spectroscopy compared with conventional plasma optical emission techniques for the analysis of metals – A review of applications and analytical performance*Spectrochim Acta Part B At Spectrosc* **134** 123-132.
[53] Glazoff M V, Khvan A V, Zolotorevsky V S, Belov N A and Dinsdale A T2018 Dependence of Castability and Mechanical Properties on Composition and Microstructure of Aluminum Alloys. *In Casting Aluminum Alloys (Second Edition) Their Physical and Mechanical Metallurgy*313-404 (Butterworth-Heinemann)
[54] Yi C, Zibing H, Wei W and Yun X2016 Longitudinal distribution and mechanism of carbon and manganese elements in high carbon steel billet *Iron and Steel* **51** 43-48.

[55] Holappa L and Wijk O2014 Chapter 1.7 - Inclusion Engineering. in Treatise on Process Metallurgy *Industr. Proc.* 347-72.
[56] Sanyal S, Mishra D, Sambandam M, Panabaka M and Ghorui P K2017 Experiences with Calcium Treatment of Steel *In The 3rd International Conference on Science and Technology of Ironmaking and Steelmaking* (Kanpur, India)
[57] Liu Y and Zhang L2018Relationship Between Dissolved Calcium and Total Calcium in Al-Killed Steels After Calcium Treatment *Metall. Mater. Trans. B* **49** 1624–31

IOP Conf. Series: Materials Science and Engineering 1225 (2022) 012066 doi:10.1088/1757-899X/1225/1/012066

Study of electronic structure and transport coefficients of GaN-MoS_2 heterostructure

Gurpal Singh Khosa[1,*], Ranjan Kumar[1,2,$] and Shuchi Gupta[3,!]

[1]Department of Physics, Panjab University, Chandigarh 160014, India,
[2]Department of Physics, King Abdulaziz University, Jeddah 21589, Kingdom of Saudi Arabia,
[3]University Institute of Engineering and Technology, Panjab University, Chandigarh 160014,
India

*khosagurpal@gmail.com
$ranjan@pu.ac.in
!sgupta@pu.ac.in

Abstract. The electronic structure and transport properties of GaN-MoS_2 heterostructure have been theoretically investigated using first principles calculations. Among the various possibilities of stacking MoS_2 over GaN monolayers to form heterostructure, we studied the case in which the S atom is directly placed over the Ga atom. This type of stacking is energetically favourable. The GaN-MoS_2 heterostructure preserves the direct band gap nature of MoS_2 monolayer i.e. it has direct band gap of 0.97 eV. The valence bands and conduction bands adjoining the fermi level are mostly contributed by N and Mo atoms. To study the performance of both n-type and p-type GaN-MoS_2 heterostructure in thermoelectric applications, we investigate the transport coefficients on which the ZT parameter depends. We found that p-type heterostructure has high electrical conductivity as compared to n-type heterostructure. Due to which it has a high magnitude of power factor which is one of the deciding factors in the evaluation of ZT parameter.

1. Introduction

After the experimental synthesisation of graphene in 2004 [1,2], numerous layered materials have been explored [3,4,5,6,7,8]. The existence of band gap in these layered materials make them potential candidates to be used in the fabrication of electronic devices such as field effect transistors, thermoelectric devices, solar cells, photovoltaic devices etc. [9,10,11]. The several studies proposed that the electronic properties of materials can be modified by alloying [12], by doping [13] and by super lattice formation [14]. In parallel with these investigations of layered materials, the experimental synthesisation and theoretical investigation of vanderwaal (vdw) heterostructures by vertically stacking these layered materials has offered additional possibilities to use these materials in electronic applications. Recently the transition metal dichalcogenide(TMD)/graphene heterostructures have been used in the fabrication of electronic devices such as memory cells, vertical field effect transistors etc. [15,16,17,18]. The existence of vdw heterostructures has attracted much attention because of novel properties as compared to their parent layers. Long *et al.* [19] investigated that the recombination time of photoexcited holes and electrons of bilayer Janus MoSSe heterostructure is large as compared to the single layer. Tong *et al.* [20] studied the electronic properties of MoS_2-WSe_2 heterostructure and found that heterostructure has band gap of 1.48 eV which was smaller as compared to the individuals band gaps of single layer MoS_2 (1.5 eV) and WSe_2 (1.60 eV). Khan *et al.* [21] studied the electronic structure and thermoelectric properties of bilayered MX_2 (M= Zr, Hf and X= S, Se) and found that Seebeck coefficient (electrical conductivity) have enhanced values in n(p)-type doping as compared to monolayers. This enhancement in transport coefficients further boost the performance of layered materials in thermoelectric applications. Zhang *et al.* investigated that MoS_2/GaN heterostructure is energetically stable having enhanced capability in the application of photocatalytic hydrogen generation [22,23]. The GaN-MoS_2 heterostructure is experimentally synthesised using various methods such as enhanced plasma molecular beam technique [24], chemical vapour deposition (CVD) technique [25, 26] and by transferring MoS_2 layer on GaN [27].

Motivated by the above investigations of vanderwaal heterostructures, we study the structural and electronic properties of GaN monolayer, MoS_2 monolayer and GaN-MoS_2 heterostructure. The electronic transport coefficients of GaN-MoS_2 heterostructure have also been explored. The study of electronic transport coefficients such as electrical conductivity, Seebeck coefficients, power factor etc. are very valuable because these coefficients are the deciding factors in the performance of materials in thermoelectric applications.

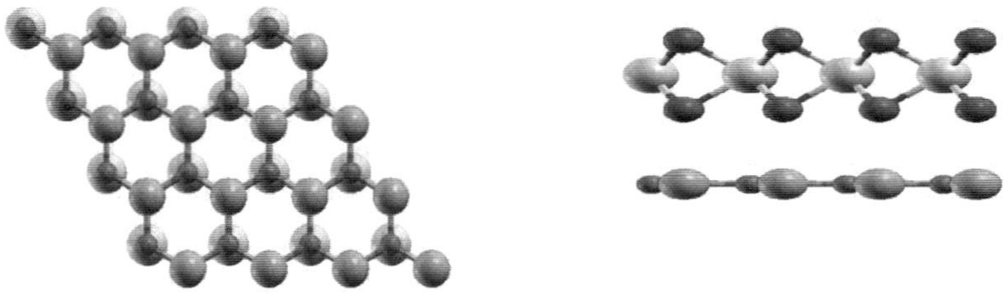

Figure 1. GaN-MoS_2 layered heterostructure viewed from top and side. The green, pink, gold and red balls indicate Ga, N, Mo and S atoms, respectively.

2. Computational Methods

The Density Functional Theory (DFT) based first principles calculations have been done using Quantum Espresso simulation package [28]. The ultrasoft pseudopotential (USPP) of functional type Perdew-Burke-Ernzerhof (PBE) is used for calculations [29]. The kinetic energy cutoff of magnitude 80 Ry and k-mesh of 12×12×1 is used after performing the convergence test of total energy w.r.t to the kinetic energy cutoff and k-mesh. For Density of states (DOS) and projected density of states (PDOS) calculations, a highly densed k-mesh of 28×28×1 is used to achieve smooth curves. The vacuum gap of more than 15 Å is used in the formation of bilayer to avoid interaction effect of adjoining layers. The BoltzTrap code has been used for the determination of electronic transport coefficients [30].

3. Results and Discussion

The GaN-MoS_2 heterostructure is vertically stacked arrangement of two layers i.e.GaN monolayer and MoS_2 monolayer. We calculate the structural parameters and electronic bandstructure of GaN and MoS_2 monolayers. The optimised lattice constants of these monolayers are 3.21 Å and 3.18 Å respectively. The small lattice mismatch between lattice constants depicts that formation of freestanding heterostructure of these monolayers is possible. Figure 1 shows that GaN-MoS_2 heterostructure has a hexagonal lattice geometry similar to their parent monolayers. After optimisation the lattice constant of heterostructure is found to be 3.20 Å. These obtained structural parameters agree with the previous predictions [32,33,23]. Figure 2(a) and (b) show the band structure, density of states (DOS) and projected density of states (PDOS) of GaN monolayer. The GaN monolayer is a semiconductor in which conduction band minima (CBM) and valence band maxima (VBM) reside at Γ and K points respectively. It has indirect energy band gap of 2.17 eV. The PDOS shows that the 'p' orbital of N and Ga atoms mainly contributed in the formation of valence bands and conduction bands respectively. Figure 3(a) and (b) show electronic band structure, DOS and PDOS calculations for MoS_2 monolayer. The CBM and VBM reside at a highly symmetric K point (direct band gap) having a gap of 1.17 eV. The valence bands and conduction bands near the fermi level are mainly contributed by the 'd' and 'p' orbitals of Mo and S atoms, respectively. These results of both monolayers are in good agreement with the previous calculations [8, 34, 35]. Out of different stacking arrangements of GaN-MoS_2 heterostructure, we choose the most stable arrangement as investigated by several groups [31,36,37] in which N, Ga atoms are placed directly below the Mo and S atoms respectively.

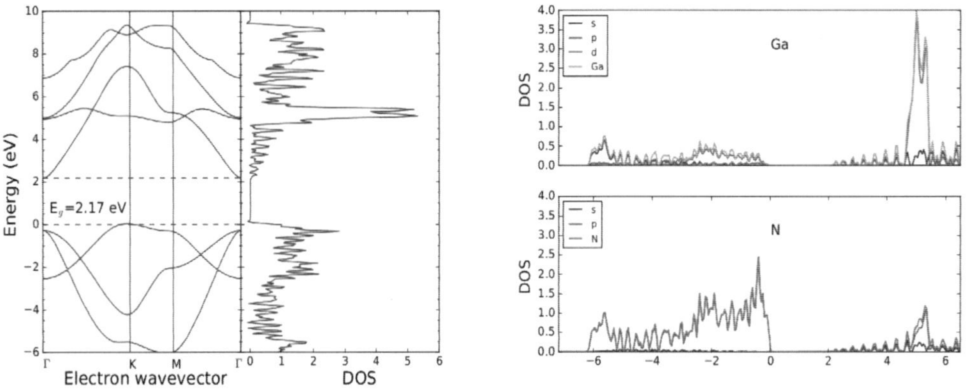

(a) (b)

Figure 2. Electronic band structure, total DOS and projected DOS of GaN monolayer.

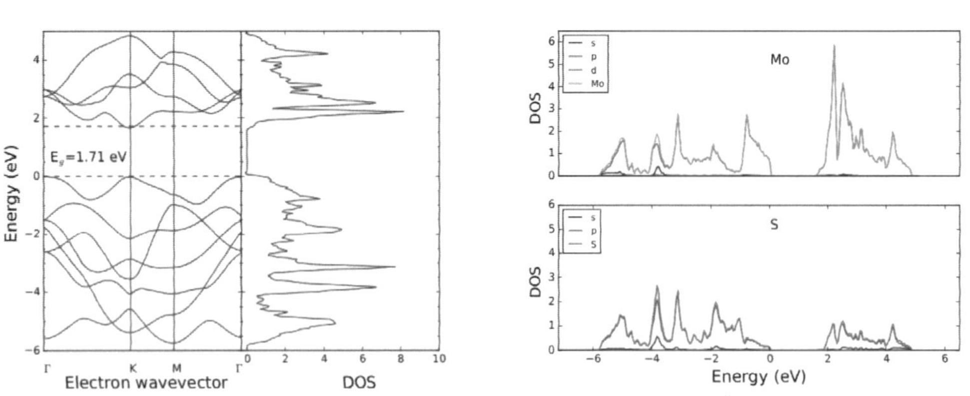

(a) (b)

Figure 3. Electronic band structure, total DOS and projected DOS of MoS_2 monolayer.

IOP Conf. Series: Materials Science and Engineering 1225 (2022) 012066 doi:10.1088/1757-899X/1225/1/012066

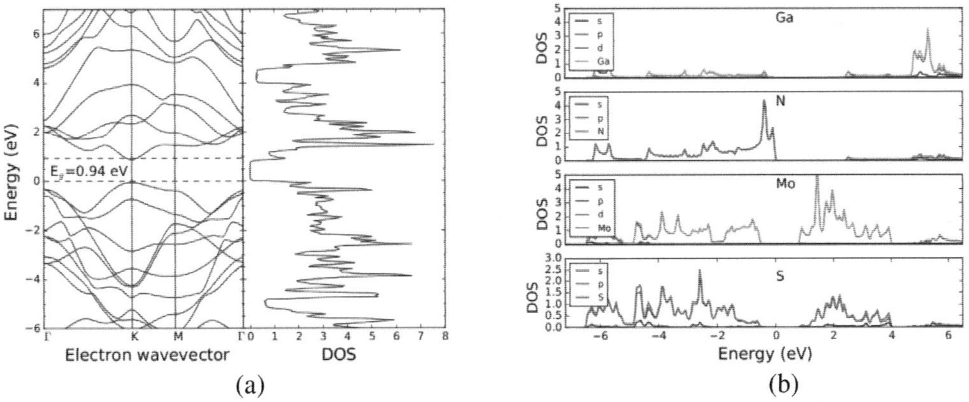

Figure 4. Electronic band structure, total DOS and projected DOS of GaN-MoS_2 heterostructure.

Table 1. The lattice constant (a_0), atomic distance (d_{x-x}), and energy band gap (E_g) for GaN monolayer, MoS_2 monolayer and GaN-MoS_2 heterostructure. **I** stand for indirect band and **D** stands for direct gap. The values in bracket are the results obtained by the work of Wen-Jin and co-authors [31].

	a_0 (Å)	d_{Ga-N} (Å)	d_{Mo-S} (Å)	d_{S-S} (Å)	d_{Ga-S} (Å)	Eg (eV)
GaN	3.21 (3.25)	1.85 (1.87)				2.17 **I** (1.98 **I**)
MoS2	3.18 (3.15)		2.41 (2.40)	3.12		1.71 **D**(1.78 **D**)
GaN-MoS2	3.20 (3.21)	1.84	2.41	3.11	3.37 (2.96)	0.94 **D** (0.77**D**)

Figure 4 shows the band structure, DOS and PDOS calculationsfor GaN-MoS_2 heterostructure. We found that GaN-MoS_2 heterostructure has a direct band of magnitude 0.97 eV at a highly symmetric K point. The band gap of heterostructure is smaller as compared to that of their parent monolayers. It shows that formation of heterostructure leads to the reduction in the band gap. Similar behaviour in the variation of band gap observed in theMoS_2-WS_2 heterostructure [38]. The nature of band gap in heterostructure is similar to that of MoS_2 monolayer which shows that the direct band gap character of MoS_2 monolayer is preserved. From the PDOS profile of GaN-MoS_2 heterostructure, we found that the main contribution in the formation of valence band is of N atom's 'p' orbital whereas conduction bands are formed by the 'd' orbital of Mo atom. So the GaN (MoS_2) monolayer contributes in the formation of valence (conduction) bands of GaN-MoS_2 heterostructure. The structural parameters and energy band gap values for both monolayers and heterostructure are shown in Table 1. After that the electronic transport coefficients of heterostructure has been investigated. Figure 5 shows the behaviour of Seebeck coefficient at three different temperatures under n-type and p-type doping for GaN-MoS_2 heterostructure. The magnitude of Seebeck coefficient varies directly (inversely)w.r.t the temperature (carrier concentration) which strictly follows the equation 1, where T isthe temperature and 'n' is the carrier concentration.

$$S = \frac{8\pi^2 {K_B}^2}{3eh^2} m^* T \left(\frac{\pi}{3n}\right)^{2/3} \quad (1)$$

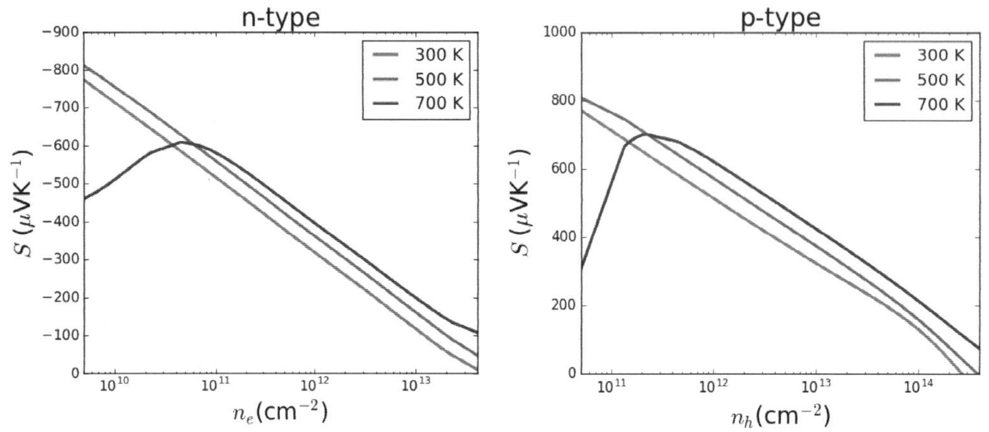

(a) (b)

Figure 5. The Seebeck coefficient as a function of carrier concentration for (a) n-type and (b)p-type GaN-MoS_2 heterostructure at different temperatures.

Figure 6 (a) and (b) show the electrical conductivity (σ) behaviour of GaN-MoS_2 heterostructure under different temperatures w.r.t the carrier concentration. The electrical conductivity obtained from BoltzTrap code depends on the carrier relaxation time. To know the exact behaviour of σ, we calculated the relaxation time (τ) for electrons and holes by solving the equation (2) below

$$\tau = \frac{\mu m^*}{e} \tag{2}$$

For relaxation time calculations, we required carrier effective mass and carrier mobility. We use carrier mobility of electrons and holes reported by Neo and co-authors [31]. The carrier effective mass is calculated by solving equation 3 where the second order derivative of energy w.r.t the k path is obtained by solving the polynomial equation 4.

$$\frac{1}{m^*} = \frac{1}{\hbar^2}\frac{\partial^2 E}{\partial K^2} \tag{3}$$

where the second order derivative of E(k) w.r.t. k is solved by fitting the E(k) to the cubic polynomial

$$E(k) = ak^3+bk^2+ck+d \tag{4}$$

Table 2. Carrier effective mass, mobility and relaxation time of GaN-MoS_2 heterostructure.The values in bracket are the results obtained by the work of Wen-Jin and co-authors [31].

	m^*_e [m_0]	m^*_h [m_0]	μ_e ($cm^2\ V^{-1}\ s^{-1}$)	μ_h ($cm^2\ V^{-1}\ s^{-1}$)	τ_e (fs)	τ_h (fs)
GaN-MoS_2	0.45(0.44)	1.21(1.27)	(266)	(3050.91)	66	2575

The obtained values of carrier's effective mass and relaxation time are shown in table 2. Using the calculated value of relaxation time, we evaluate the behaviour of electrical conductivity (σ) w.r.t increasing carrier concentration under different temperatures as shown in Figure 6. The p-type GaN-MoS_2 heterostructure has high electrical conductivity in contrast to n-type heterostructure because of the high carrier mobility as shown in equation 5 that σ varies directly with the carrier mobility (μ).

$$\sigma = ne\mu = \frac{ne^2\tau}{m^*} \tag{5}$$

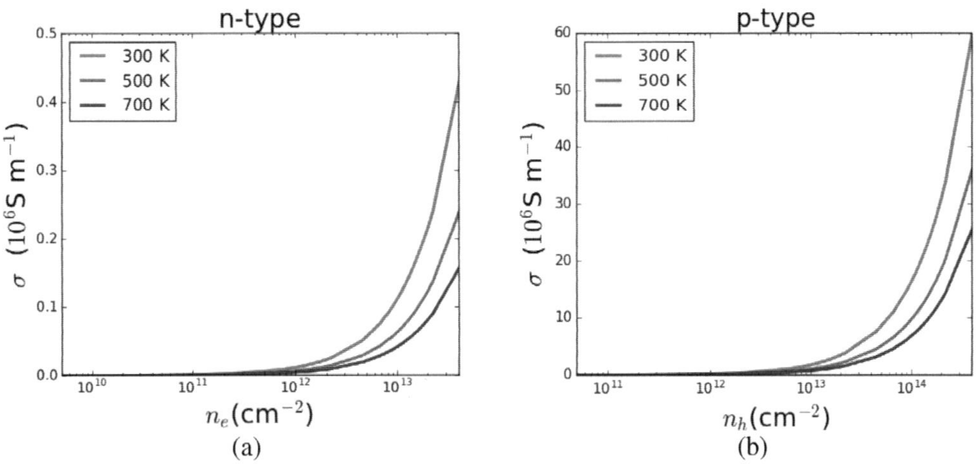

Figure 6. Electrical conductivity as a function of carrier concentration for (a) n-type and (b)p-type GaN-MoS2 heterostructure at different temperatures.

To investigate the performance of this heterostructure in thermoelectric applications, we calculated the power factor (P.F.= $S^2\sigma$) of this heterostructure. The behaviour of power factorfor both n and p-type doped heterostructures are shown in Figure 7. The p-type heterostructure has enhanced power factor in contrast to n-type because of the high values of Seebeck coefficients and electrical conductivity at higher concentration.

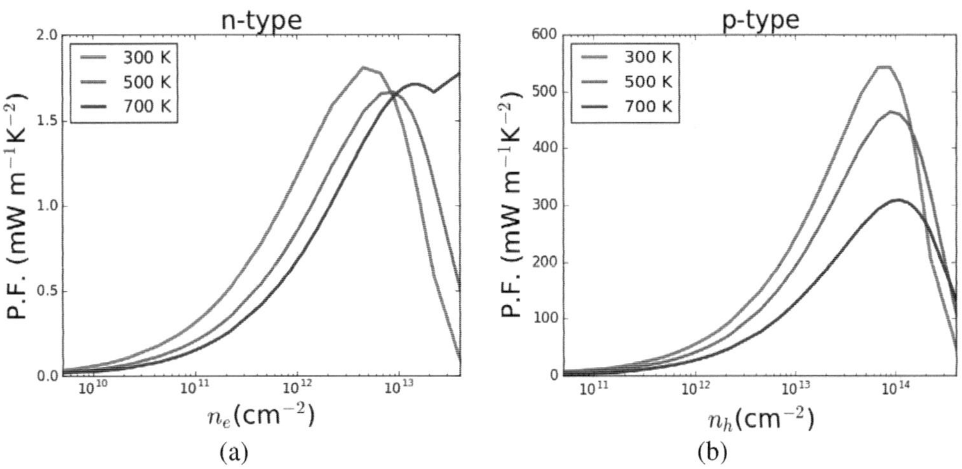

Figure 7. Power factor as a function of carrier concentration for (a) n-type and (b) p-type GaN-MoS2 heterostructure at different temperatures.

As we know the dimensionless figure of merit parameter ZT is calculated by solving equation 6,

$$ZT = \frac{S^2\sigma T}{K} \tag{6}$$

where S is the Seebeck coefficient, σ is the electrical conductivity and K is the total thermal conductivity of the material. As we see that the ZT parameter strongly depends on P.F., from this we conclude that the GaN-MoS_2 heterostructure could be used in the fabrication of thermoelectric devices. The calculations of thermal conductivity of this heterostructure are under investigation, once the calculations complete, we will figure out the ZT parameter of GaN-MoS_2 heterostructure.

4. Conclusion

We systematically examine the structural parameters, electronic band structure and transport coefficients of GaN-MoS_2 heterostructure using the density functional theory based first principles calculations. The GaN-MoS_2 heterostructure is energetically stable, having a directband gap of magnitude 0.94 eV, which is lower than the band gaps of their parent monolayers. The direct band gap character of MoS_2 monolayer is preserved in the GaN-MoS_2 heterostructure.The valence (conduction) bands of heterostructure are formed by the contribution of N(Mo) atoms. The evaluated values of electronic transport coefficients state that GaN-MoS_2 heterostructure is highly favourable material in the thermoelectric applications. From our evaluated results, we conclude that p-type GaN-MoS_2 heterostructure has a high magnitude of power factor as compared to n-type heterostructure and it could be used in the fabrication of thermoelectric devices.

Acknowledgments

This work was funded by the Deanship of Scientific Research (DSR), King Abdulaziz University, Jeddah, under grant No. (KEP-00-130-42). The authors, therefore, acknowledge with thanks DSR technical and financial support. G.S.K. gratefully acknowledges UGC, India for financial support under grant number MANF-2018-19-PUN-98540. We also thank the Department of Physics and University Institute of Engineering and Technology, Panjab University, Chandigarh for providing technical facilities required in this work.

References

[1] Novoselov K S, Geim A K, Morozov S V, Jiang D E, Zhang Y, Dubonos S V, Grigorieva, I V and Firsov A A 2004. Electric field effect in atomically thin carbon films *Sci.* **306** 666-69

[2] Geim A K and Novoselov KS 2010 The rise of graphene. In *Nanoscience and technology: a collection of reviews from nature journals* 11-19

[3] Cahangirov S, Topsakal M, Aktürk E, Şahin H and Ciraci S 2009 Two-and one-dimensional honeycomb structures of silicon and germanium *Phys. Rev. Lett.* **102** 236804.

[4] Bandurin DA, Tyurnina AV, Geliang LY, Mishchenko A, Zólyomi V, Morozov SV, Kumar RK, Gorbachev RV, Kudrynskyi ZR, Pezzini S and Kovalyuk ZD 2017 High electron mobility, quantum Hall effect and anomalous optical response in atomically thin InSe *Nat. Nanotechnol.* **12** 223-27

[5] Wang QH, Kalantar-Zadeh K, Kis A, Coleman JN and Strano MS 2012. Electronics and optoelectronics of two-dimensional transition metal dichalcogenides *Nat. Nanotechnol.* **7** 699-12

[6] Jeong S, Yoo D, Jang JT, Kim M and Cheon J 2012 Well-defined colloidal 2-D layered transition-metal chalcogenide nanocrystals via generalized synthetic protocols *J. Am. Chem. Soc.* **134** 18233-36

[7] Coleman JN, Lotya M, O'Neill A, Bergin SD, King PJ, Khan U, Young K, Gaucher A, De S, Smith R J and Shvets I V 2011 Two-dimensional nanosheets produced by liquid exfoliation of layered materials *Sci.* **331** 568-71

[8] Mak KF, Lee C, Hone J, Shan J and Heinz TF 2010 Atomically thin MoS 2: a new direct-gap

semiconductor *Phys. Rev. Lett.* **105** 136805.

[9] Rehman G, Khan S A, Amin B, Ahmad I, Gan LY and Maqbool M 2018 Intriguing electronic structures and optical properties of two-dimensional van der Waals heterostructures of Zr 2 CT 2 (T= O, F) with MoSe 2 and WSe 2 *J. Mater. Chem. C* **6** 2830-39

[10] Rao C N R, Ramakrishna Matte HSS and Maitra U 2013 Graphene analogues of inorganic layered materials *Angew. Chem. Int. Ed.* **52** 13162-85

[11] Tang Q and Zhou Z 2013 Graphene-analogous low-dimensional materials *Prog. Mater. Sci.* **58** 1244-1315.

[12] Komsa HP and Krasheninnikov AV 2013 Electronic structures and optical properties of realistic transition metal dichalcogenide heterostructures from first principles *Phys. Rev. B* **88** 085318

[13] Komsa HP, Kotakoski J, Kurasch S, Lehtinen O, Kaiser U and Krasheninnikov AV 2012 Two-dimensional transition metal dichalcogenides under electron irradiation: defect production and doping *Phys. Rev. Lett.* **109** 035503

[14] Lu N, Guo H, Wang L, Wu X and Zeng XC 2014. van der Waals trilayers and superlattices: modification of electronic structures of MoS 2 by intercalation *Nanosc.* **6** 4566-71.

[15] Bertolazzi S, Krasnozhon D and Kis A 2013 Nonvolatile memory cells based on MoS2/graphene heterostructures *ACS nano* **7** 3246-52.

[16] Britnell L, Gorbachev RV, Jalil R, Belle BD, Schedin F, Mishchenko A, Georgiou T, Katsnelson M I, Eaves L, Morozov SV and Peres NMR 2012 Field-effect tunneling transistor based on vertical graphene heterostructures *Sci.* **335** 947-950

[17] Georgiou T, Jalil R, Belle BD, Britnell L, Gorbachev RV, Morozov SV, Kim YJ, Gholinia A, Haigh SJ, Makarovsky O and Eaves L 2013 Vertical field-effect transistor based on graphene–WS 2 heterostructures for flexible and transparent electronics *Nat. Nanotechnol.* **8** 100-03

[18] Yu W J, Li Z, Zhou H, Chen Y, Wang Y, Huang Y and Duan X 2013 Vertically stacked multi-heterostructures of layered materials for logic transistors and complementary inverters Nat. Mater. **12** 246-52

[19] Long C, Dai Y, Gong ZR and Jin H 2019. Robust type-II band alignment in Janus-MoSSe bilayer with extremely long carrier lifetime induced by the intrinsic electric field *Phys. Rev. B* **99** 115316

[20] Tong L, Liu T, Liang R, Wang S, Chen J, Dai J and Ye L 2017. Growth of Transition Metal Dichalcogenides and Directly Modulating Their Properties by Chemical Vapor Deposition *Gen. Chem.* **3**

[21] Khan F, Din H U, Khan SA, Rehman G, Bilal M, Nguyen CV, Ahmad I, Gan LY and Amin B 2019 Theoretical investigation of electronic structure and thermoelectric properties of MX2 (M= Zr, Hf; X= S, Se) van der Waals heterostructures *J. Phys. Chem. Solids* **126** 304-09

[22] Zhang Z, Qian Q, Li B and Chen KJ 2018. Interface engineering of monolayer MoS2/GaN hybrid heterostructure: modified band alignment for photocatalytic water splitting application by nitridation treatment *ACS Appl. Mater. Interfaces.* **10** 17419-26.

[23] Liao J, Sa B, Zhou J, Ahuja R and Sun Z 2014 Design of high-efficiency visible-light photocatalysts for water splitting: MoS2/AlN (GaN) heterostructures *J. Phys. Chem. C.* **118** 17594-99.

[24] Yamada A, Ho KP, Maruyama T and Akimoto K 1999 Molecular beam epitaxy of GaN on a substrate of MoS2 layered compound *Appl. Phys. A* **69** 89-92.

[25] Gupta P, Rahman AA, Subramanian S, Gupta S, Thamizhavel A, Orlova T, Rouvimov S, Vishwanath S, Protasenko V, Laskar MR and Xing HG 2016 Layered transition metal dichalcogenides: promising near-lattice-matched substrates for GaN growth *Sci. Rep.* **6** 1-8

[26] Mishra P, Tangi M, Ng TK, Hedhili MN, Anjum DH, Alias MS, Tseng CC, Li LJ and Ooi BS 2017 Impact of N-plasma and Ga-irradiation on MoS2 layer in molecular beam epitaxy *Appl. Phys. Lett.* **110** 012101

[27] Lee EW, Lee CH, Paul PK, Ma L, McCulloch WD, Krishnamoorthy S, Wu Y, Arehart AR and Rajan S 2015 Layer-transferred MoS2/GaN PN diodes *Appl. Phys. Lett.* **107** 103505

[28] Giannozzi P, Baroni S, Bonini N, Calandra M, Car R, Cavazzoni C, Ceresoli D, Chiarotti G L, Cococcioni M, Dabo I and Dal Corso A 2009 QUANTUM ESPRESSO: a modular and open-source software project for quantum simulations of materials J. Condens. Matter Phys. 21 395502.
[29] Perdew JP, Burke K and Ernzerhof M 1996. Generalized gradient approximation made simple *Phys. Rev. Lett.* **77** 3865
[30] Madsen GK and Singh DJ 2006 BoltzTraP. A code for calculating band-structure dependent quantities *Comput. Phys. Commun.* **175** 67-71.
[31] Yin WJ, Zeng XL, Wen B, Ge, Q X, Xu Y, Teobaldi G and Liu LM 2021. The unique carrier mobility of Janus MoSSe/GaN heterostructures Front. Phys. 16 1-9
[32] Zhuang HL, Singh AK and Hennig RG 2013. Computational discovery of single-layer III-V materials Phys. Rev. B 87 165415.
[33] Komsa HP and Krasheninnikov AV 2012 Effects of confinement and environment on the electronic structure and exciton binding energy of MoS 2 from first principles *Phys. Rev. B* **86** 241201.
[34] Splendiani A, Sun L, Zhang Y, Li T, Kim J, Chim CY, Galli G and Wang F 2010 Emerging photoluminescence in monolayer MoS2 *Nano. Lett.* 10 1271-75.
[35] Khosa GS, Kumar R and Gupta S 2020 Effect of biaxial strain on electronic transport coefficients of monolayer MoS2. In AIP Conference Proceedings 2265
[36] Ren D, Tan X, Zhang T and Zhang Y 2019. Electronic and optical properties of GaN–MoS2 heterostructure from first-principles calculations *Chinese Physics B* **28** 086104.
[37] Yelgel C 2017 First-principles modeling of GaN/MoSe $ _ {2} $ van der Waals heterobilayer *Turk. J. Phys.* **41** 463-68.
[38] Kośmider K and Fernández-Rossier J 2013. Electronic properties of the MoS 2-WS 2 heterojunction *Phys. Rev. B* **87** 075451.

Wear and Corrosion Behaviour of Wire Electrical Discharge Machined Ti-6Al-4V alloy

Sanyukt Kumar Manderna, Puneet Katyal, Munish Gupta and Vijender Singh

Department of Mechanical Engineering
Guru Jambheswar University of Science and Technology, Hisar
Haryana, INDIA – 125001

* corresponding author e-mail: katyalgju@gmail.com

Abstract The objective of the current research work is to study the wear and corrosion behavior of Ti-6Al-4V ELI alloy. WEDM is preferred over conventional machining for Ti alloys, as they are hard-to-machine materials. Wear and corrosion tests were performed using wear and friction monitor (Ducom Model TR20-LE) and electrochemical workstation (Autolab PGSTAT204) respectively. SEM and EDS were also carried out on each sample for evaluating the surface morphology and to know the exact composition of the material. WEDM rough and trim-cut modes improve the surface hardness of Ti-6Al-4V ELI alloy and hence affects their c.o.f. and wear rates. The outcomes of the present research work promote WEDM trim-cut mode over rough-cut mode because it results in lower values of c.o.f., wear and corrosion rates.

Keywords: Ti-6Al-4V ELI alloy; WEDM; Coefficient of friction; Wear behaviour, corrosion behaviour

Abbrevations:

SEM- Scanning Electron Microscopy

EDS- Energy Dispersive X-ray Spectroscopy

WEDM- Wire Electrical Discharge Machining

COF- Coefficient of Friction

PDP-Potentiodynamic Polarization Curves

1. Introduction

Titanium based alloys because of their excellent corrosion resistance and top-notch biocompatibility; are potential candidates in biomedical industry as an implant material for orthodontic and orthopaedic applications [1-4]. TiO2 protective layer enhances cell attachment and osteointegration with the Ti based alloy implants [5]. Surface properties for instance chemical composition and surface roughness display a substantial role in enhancing the biocompatibility and cell/tissue adhesion of metallic implants [6,7].

WEDM can machine Ti based alloy implants quite economically and efficiently to produce complex shapes. In WEDM, recast layer phenomenon affects the wear behaviour, fatigue life and other functional properties [8-10]. This surface damage can largely be managed by using trim cutting after rough cutting operations because it produces surface with low crack density and low surface roughness[11-12] and hence improved corrosion resistance property[13-18]. An extensive survey of literature reveals that WEDM modified surface improved the bio-functionality, biocompatibility, osteointegration and cell attachment between implant and human tissues largely. This is because WEDM modified surface consists of nano and micro scale structures which promotes the cell attachment and bone formation leading to short healing time [19-21].

Wear is a crucial factor influencing the presentation and failure of biomedical devices. Surface morphology and lubrication conditions considerably affect their corrosive wear and corrosive rates [22]. It has been observed that wear and c.o.f. values are significantly lower in trim cut mode in comparison to rough-cut mode [23]. Ti alloys because of their good mechanical and chemical properties provides strong adhesion tendency of these alloys.

The main physical properties affecting the long-term durability of the metal-implants are morphology, surface roughness, wettability and surface free energy. Surface roughness of the implant surface serves as a physiochemical aspect that primarily affects the osteointegration of the tissue and the implant. Moreover, surface roughness is a favourable factor that increases cell adhesion and cell growth. A rough surface with a corrosive environment at a larger interfacial area, and consequently, an increase in surface roughness enhanced the corrosion process [24-26].

Above literature-review clearly indicates that only a few numerical studies have found in the literature that considered the effects of WEDM trim-cut mode for Ti-alloys. Therefore, the pretension of the current experimental opus is to analyse wear and corrosion behaviour of Ti-6Al-4V ELI alloy under two modes of WEDM which are trim-cut and rough-cut.

2. Experimental setup

2.1 Material Preparation

In the present experimental work, a cylindrical bar of 11 mm diameter and of 100 cm length Ti6Al4V ELI alloy is used and machined with WEDM Electronica Sprintcut and manufacturing by WEDM then polished by grit 800, 1000,1200, 2000 grit paper and then grade-1 Al_3O_2 paste for getting polished surface. After the polishing of the sample, wear test is done on it and surface is observed using SEM before and after the wear test. Table 1 shows the chemical composition of Ti6Al4V ELI while figure 1 shows the EDS of Ti6Al4V ELI after WED Machining.

Table 1. Chemical configuration of Ti6Al4V ELI

Element	Shell	Weight %	Atomic %
Al	K	6.19	8.03
Ti	K	80.71	58.99
C	K	8.77	29.40
Fe	K	0.05	0.03
V	K	4.22	3.03
Pd	L	0.05	0.02
Sum		100.00	100.00

IOP Conf. Series: Materials Science and Engineering 1225 (2022) 012067 doi:10.1088/1757-899X/1225/1/012067

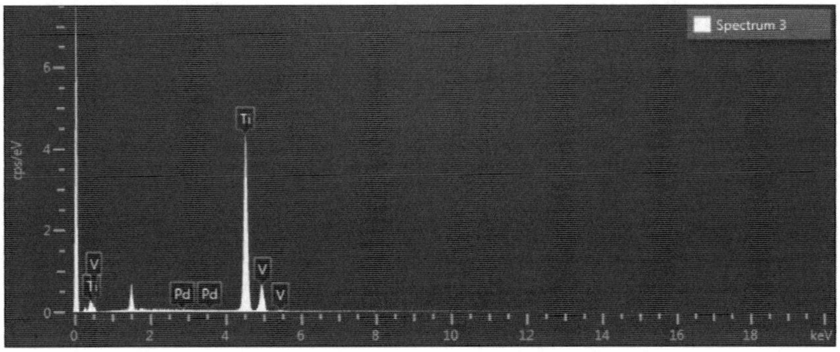

Figure 1. T6Al4V ELI EDS after WEDM

2.2. WEDM Process Parameters and sample preparation

The present experimentation used CNC-WEDM (Sprint Cut) with copper wire electrode (ϕ; 250 μm). Samples were prepared by changing the conditions of discharge under WEDM's rough and trim-cut modes. The objective of first cut i.e., rough-cut is to get rid of the maxima material while Trim cut mode aims to remove a very thick layer on rough cut WEDMed surface, thereby, minimizing the whitish layer or recast layer thickness and surface disfigurement. The process parameters listed in Table 2 were varied under different cutting conditions.

Table 2. WEDM Process Parameters

Sr. No.	WEDM Parameters	Symbol	Units	Rough Cut (RFC)	Trim Cut (TRC) (Number of Trim Pass-2)
1.	Servo voltage	SV	V	20	30
2.	Pulse on time	T_{on}	μs	110	104
3.	Pulse off time	T_{off}	μs	45	40
4.	Peak current	I_p	Ampere	250	120
5.	Wire tension	WT	Grams	500	900
6.	Wire feed	WF	m/min	2	6

3. Results and Discussions

3.1 Wear Analysis

The performance of wear behaviour of Ti based medical implants significantly affects their service life. Therefore, wear tests were performed on Ti-6Al-4V ELI alloy with variable surface integrity using cylindrical pins (ϕ; 3.5mm, length; 6mm). Figure 2(a) and 2(b) showed the SEM before and after the wear test. Futhermore, figure 3 recorded real time variation of c.o.f. for Ti-6Al-4V ELI alloy sample at 10.0 N and 600 seconds. During the sliding period of 30-40 seconds, there is an abrupt rise of c.o.f. and thereafter it decreases with the passage of time and attains the equilibrium condition. Pin prepared under rough cutting mode shows highest value of c.o.f. and reached the condition of equilibrium after the span of 300 seconds. Similar behaviour is observed from pin prepared with trim-cut mode and acquire equilibrium conditions after 100 seconds.

Figure 3 and figure 4 show the c.o.f. and wear rates under different WEDM cutting modes. Trim-cut surfaces show lesser value of wear rates in comparision to rough-cut surfaces due to the presence of higher percentage of rutile-TiO_2 phase and hence higher hardness. Table 4 summarised the c.o.f. and volumetric wear rate under rough and trim cutting modes.
Therefore, the present research work confirms that trim-cut surfaces of WEDM improves wear and fatigue defiance properties and hence promotes non-corrosive and biocompatible forefending layer that enhances their presentation and lifecycle of Ti-6Al-4V ELI Alloy for biomedical building blocks and gadgets.

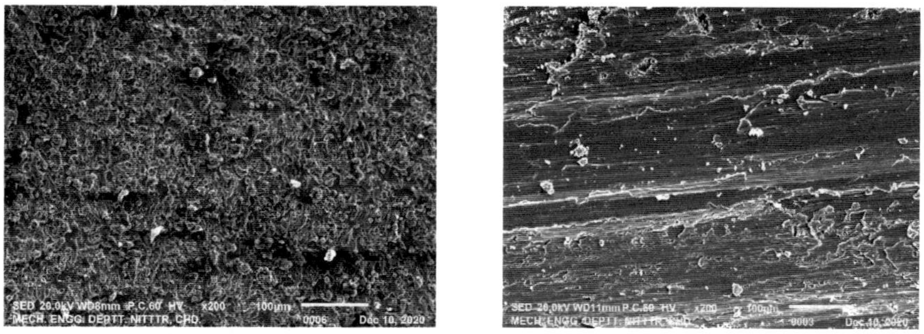

(a) (b)

Figure 2. (a) SEM before wear and (b) SEM after wear

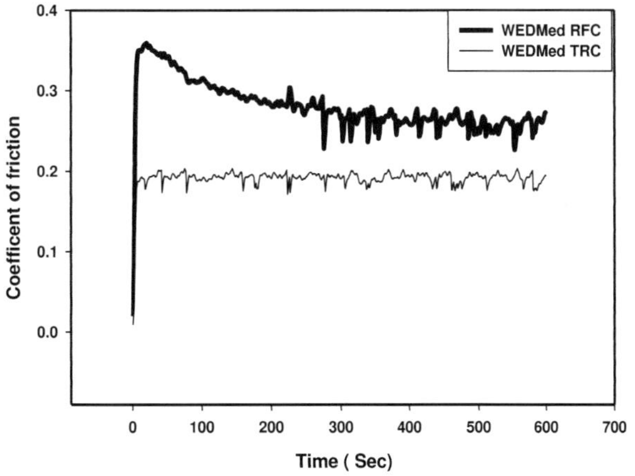

Figure 3. Coefficient of friction versus real time plot at pin-on-disc wear test

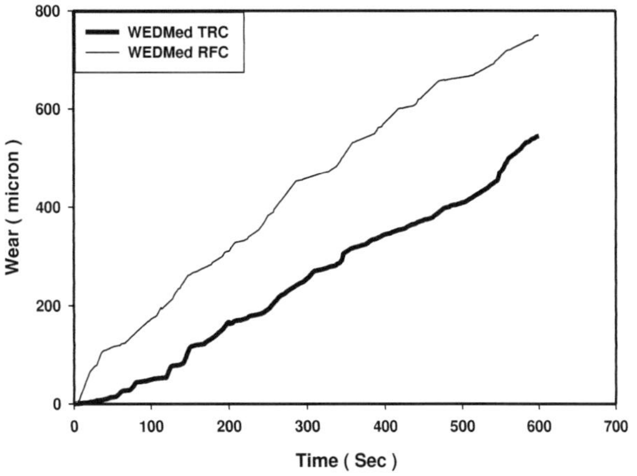

Figure 4. Wear rate analysis of various machined surfaces

Table 3. Surface roughness of various machined surfaces

Sample	Surface Roughness (R_a) Before wear (μm)	Surface Roughness (R_a) After wear (μm)
Rough-Cut (RFC)	2.326	0.54
Trim-Cut (TRC)	2.086	0.56

Table 4. C.O.F and Wear rate of various machined surfaces

Sample	C.O. F	Wear rate (mm^3/min)
Rough-Cut (RFC)	0.385	67.51 x 10^{-2}
Trim-Cut (TRC)	0.35	65.73 x 10^{-2}

3.2. Electrochemical Corrosion Analysis

In the present section, electrochemical measurements were carried out for the duration of two hours. These measurements were obtained from the surface of polished and WEDMed samples and were being compared using potentiodynamic polarization (PDP) curves as shown through figure 5. These plots comparisons for different set of samples clearly showed more negative value of corrosion potential (E_{corr}) for polished surface in comparison to other WEDMed samples and its value changed between the range of -0.4612 V to -0.23692 V (Table 5). Similarly, for the current density (I_{corr}), the minimum and maximum values varies between 7.09x10^{-6} A/cm^2 and 3.14x10^{-5} A/cm^2 for polished and trim-cut surfaces respectively. As far as the corrosion rates were concerned, the minimum and maximum values range between 0.2152 mm/year to 0.6921 mm/year for trim-cut and rough-cut surfaces respectively. Due to the change in chemical composition and morphology, trim-cut surfaces showed lower values of corrosion rates and rough-cut surfaces showed higher values due to increased surface roughness values.

Table 5. Electrochemical parameters for potentiodynamic polarization curves

Sample	E_{corr} (V)	I_{corr} (A/cm^2)	Corrosion Rate (mm/year)
Rough Cut (RFC)	-0.23692	3.14x10^{-5}	0.6921
Trim Cut (TRC)	-0.28132	2.13x10^{-5}	0.2152
Polished Sample (PLC)	-0.4612	7.09x10^{-6}	0.2236

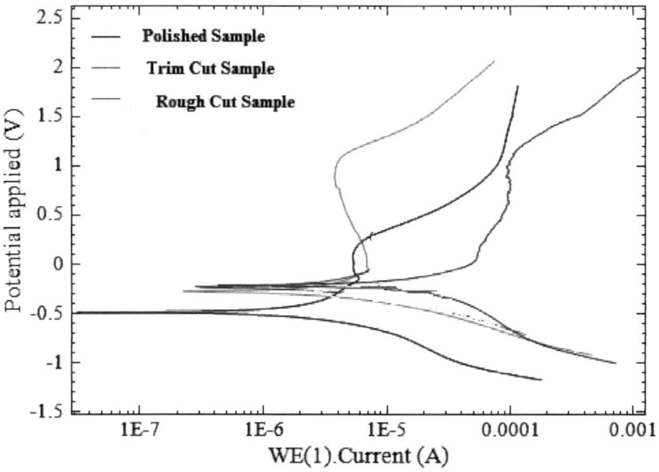

Figure 5. Polarization curve for various machined surfaces

4. Conclusions

The current research work has to do with the wear and corrosion behaviour of Ti-6Al-4V ELI alloy. The conclusions noticed from the existing research work are

- Rough-cut sample due to higher thermal damage on machined surfaces results into higher R_a values, whereas trim-cut modes produce improved surface properties of Ti6Al4V ELI. The R_a values of machining samples under rough-cut and trim-cut samples are 2.326 μm and 2.086 μm respectively.
- Trim-cut mode experience lesser wear rate of 65.73 x 10^{-2} mm^3/min in comparison to rough cut mode as 67.51 x 10^{-2} mm^3/min respectively. This observation is due to higher percentage of rutile-TiO_2 phase and hence higher hardness in trim-cut surfaces in comparison to rough-cut surfaces.
- WEDM trim-cut surfaces improves wear and fatigue resistance properties and hence promotes non-corrosive and biocompatible protective layer that enhances their service life and functioning of Ti-6Al-4V ELI alloy for biomedical practice.
- The corrosion rates for different set of samples vary between minimum and maximum values of 0.2152 mm/year to 0.6921 mm/year for trim-cut and rough-cut surfaces respectively. Trim-cut surfaces shows the least value of corrosion rate due to changes in chemical composition and morphology, while rough-cut surfaces show higher values due to increased surface roughness values.

References

[1]. Xu Y, Liu E, and Wei S 2020 Investigation on tensile fracture properties of TA15 specimen formed by selective electron beam melting *Mater. Sci. Eng. A* **773** 138826

[2]. Wang S, Ma Z, Liao Z, Song J, Yang K and Liu W 2015 Study on improved tribological properties by alloying copper to CP-Ti and Ti–6Al–4V alloy *Mater. Sci. Eng. C* **57** 123-32.

[3]. Li Z, Tang J and Bai J 2020 A novel micro-EDM method to improve microhole machining performances using ultrasonic circular vibration (UCV) electrode, *Int. J. Mech. Sci.***175** 105574.

[4]. Kumar K, Singh V, Katyal P and Sharma N 2019 EDM μ-drilling in Ti-6Al-7Nb: experimental investigation and optimization using NSGA-II *Int. J. Adv. Manuf. Technol.***104** 2727–38.

[5]. Nnamchi P S, Obayi C S, Todd I, and Rainforth M W 2016 Mechanical and electrochemical characterisation of nsew Ti–Mo–Nb–Zr alloys for biomedical applications *J. Mech. Behav. Biomed. Mater.* **60** 68–77.

[6]. Arima Y and Iwata H 2007 Effect of wettability and surface functional groups on protein adsorption and cell adhesion using well-defined mixed self-assembled monolayers *Biomaterials* **28** 3074–82.

[7]. Xu L C and C A Siedlecki 2007 Effects of surface wettability and contact time on protein adhesion to biomaterial surfaces, *Biomaterials* **28** 22 3273–83.

[8]. Renkawitz T and Dendorfer S 2012 Orthopedic navigation technology and biomechanical evaluation for total hip replacement *Proc. Inst. Mech. Eng. [H]* **226** 12 897–98.

[9]. Rahman S S, Ashraf M Z I, Bashar M S, Kamruzzaman M, Amin A K M N and Hossain M M 2017 Crystallinity, surface morphology, and chemical composition of the recast layer and rutile-TiO_2 formation on Ti-6Al-4V ELI by wire-EDM to enhance biocompatibility *Int. J. Adv. Manuf. Technol.* **93** 3285–96.

[10]. Janeček M, Nový F, Stráský J, Harcuba P and Wagner L 2011 Fatigue endurance of Ti-6Al-4V alloy with electro-eroded surface for improved bone in-growth *J. Mech. Behav. Biomed. Mater.* **4** 417–22.

[11]. Kumar V, Kumar V, and Jangra K K 2015 An experimental analysis and optimization of machining rate and surface characteristics in WEDM of Monel-400 using RSM and desirability approach *J. Ind. Eng. Int.* **11** 297–07.

[12]. Jangra K K, Kumar V and Kumar V 2014 An Experimental and Comparative Study on Rough and Trim Cutting Operation in WEDM of Hard to Machine Materials *Procedia Mater. Sci.* **5** 1603–12.

[13]. Prathipati R, Ch R, and Dora S P 2019 Corrosion behavior of surface induced by wire EDM on 316L stainless steel: an experimental investigation *SN Appl. Sci.* **1** 952.

[14]. Bhattacharya S, Abraham G J, Mishra A, Kain V, and Dey G K 2018 Corrosion Behavior of Wire Electrical Discharge Machined Surfaces of P91 Steel *J. Mater. Eng. Perform.* **27** 4561–70.

[15]. Zhang L, Xu J, Chen B, Yu H, and Lian Z 2017 Microhardness and corrosion resistance behaviour of Ti–6Al–4V alloy-coloured surface under WEDM-HS process *Micro Nano Lett.* **12** 618–23.

[16]. Aspinwall D K, Soo S L, Berrisford A E and Walder G 2008 Workpiece surface roughness and integrity after WEDM of Ti–6Al–4V and Inconel 718 using minimum damage generator technology *CIRP Ann.* **57** 187–90.

[17]. Arikatla S P, Tamil Mannan K and Krishnaiah A 2017 Surface Integrity Characteristics in Wire Electrical Discharge Machining of Titanium Alloy during Main cut and Trim cuts *Mater. Today Proc.* **4** 1500–09.

[18]. Sarkar S, Sekh M, Mitra S and Bhattacharyya B 2008 Modeling and optimization of wire electrical discharge machining of γ-TiAl in trim cutting operation *J. Mater. Process. Technol.* **205** 376–87.

[19]. Chen S L, Lin M H, Chen C C and Ou K L 2008 Effect of electro-discharging on formation of biocompatible layer on implant surface *J. Alloys Compd.* **456** 413–18.

[20]. Prakash C, Kansal H K, Pabla B S and Puri S 2016 Multi-objective optimization of powder mixed electric discharge machining parameters for fabrication of biocompatible layer on β-Ti alloy using NSGA-II coupled with Taguchi based response surface methodology *J. Mech. Sci. Technol.* **30** 4195–04.

[21]. Harcuba P, Bačáková L, Stráský J, Bačáková M, Novotná K and Janeček M 2012 Surface treatment by electric discharge machining of Ti–6Al–4V alloy for potential application in orthopaedics *J. Mech. Behav. Biomed. Mater.* **7** 96–105.

[22]. Evgeny B, Hughes T and Eskin D 2016 Effect of surface roughness on corrosion behaviour of low carbon steel in inhibited 4 M hydrochloric acid under laminar and turbulent flow conditions *Corros. Sci.* **103** 196–205

[23]. Bonny K and Baets P 2010 Surface finishing: Impact on tribological characteristics of WC–Co hardmetals *Tribol. Int.* 43, no. 1–2, pp. 40–54.

[24]. Burstein G T and Pistorius P C 1995 Surface Roughness and the Metastable Pitting of Stainless Steel in Chloride Solutions *Corros.* **51** 380–85.

[25]. Alvarez R B 2010 Corrosion relationships as a function of time and surface roughness on a structural AE44 magnesium alloy *Corros. Sci.* **52** 1635–48.

[26]. Sasaki K and Burstein G T 1996 The generation of surface roughness during slurry erosion-corrosion and its effect on the pitting potential *Corros. Sci.* **38** 2111–20.

AUTHOR INDEX